Lecture Notes in Computer Science

Lecture Notes in Artificial Intelligence 13717

Founding Editor

Jörg Siekmann

Series Editors

Randy Goebel, *University of Alberta, Edmonton, Canada*
Wolfgang Wahlster, *DFKI, Berlin, Germany*
Zhi-Hua Zhou, *Nanjing University, Nanjing, China*

The series Lecture Notes in Artificial Intelligence (LNAI) was established in 1988 as a topical subseries of LNCS devoted to artificial intelligence.

The series publishes state-of-the-art research results at a high level. As with the LNCS mother series, the mission of the series is to serve the international R & D community by providing an invaluable service, mainly focused on the publication of conference and workshop proceedings and postproceedings.

Massih-Reza Amini · Stéphane Canu ·
Asja Fischer · Tias Guns · Petra Kralj Novak ·
Grigorios Tsoumakas
Editors

Machine Learning and Knowledge Discovery in Databases

European Conference, ECML PKDD 2022
Grenoble, France, September 19–23, 2022
Proceedings, Part V

Editors
Massih-Reza Amini
Grenoble Alpes University
Saint Martin d'Hères, France

Stéphane Canu
INSA Rouen Normandy
Saint Etienne du Rouvray, France

Asja Fischer
Ruhr-Universität Bochum
Bochum, Germany

Tias Guns
KU Leuven
Leuven, Belgium

Petra Kralj Novak
Central European University
Vienna, Austria

Grigorios Tsoumakas
Aristotle University of Thessaloniki
Thessaloniki, Greece

ISSN 0302-9743 ISSN 1611-3349 (electronic)
Lecture Notes in Artificial Intelligence
ISBN 978-3-031-26418-4 ISBN 978-3-031-26419-1 (eBook)
https://doi.org/10.1007/978-3-031-26419-1

LNCS Sublibrary: SL7 – Artificial Intelligence

© The Editor(s) (if applicable) and The Author(s), under exclusive license
to Springer Nature Switzerland AG 2023
Chapter "Optimization of Annealed Importance Sampling Hyperparameters" is licensed under the terms of the Creative Commons Attribution 4.0 International License (http://creativecommons.org/licenses/by/4.0/). For further details see license information in the chapter.
This work is subject to copyright. All rights are reserved by the Publisher, whether the whole or part of the material is concerned, specifically the rights of translation, reprinting, reuse of illustrations, recitation, broadcasting, reproduction on microfilms or in any other physical way, and transmission or information storage and retrieval, electronic adaptation, computer software, or by similar or dissimilar methodology now known or hereafter developed.
The use of general descriptive names, registered names, trademarks, service marks, etc. in this publication does not imply, even in the absence of a specific statement, that such names are exempt from the relevant protective laws and regulations and therefore free for general use.
The publisher, the authors, and the editors are safe to assume that the advice and information in this book are believed to be true and accurate at the date of publication. Neither the publisher nor the authors or the editors give a warranty, expressed or implied, with respect to the material contained herein or for any errors or omissions that may have been made. The publisher remains neutral with regard to jurisdictional claims in published maps and institutional affiliations.

This Springer imprint is published by the registered company Springer Nature Switzerland AG
The registered company address is: Gewerbestrasse 11, 6330 Cham, Switzerland

Preface

The European Conference on Machine Learning and Principles and Practice of Knowledge Discovery in Databases (ECML–PKDD 2022) in Grenoble, France, was once again a place for in-person gathering and the exchange of ideas after two years of completely virtual conferences due to the SARS-CoV-2 pandemic. This year the conference was hosted for the first time in hybrid format, and we are honored and delighted to offer you these proceedings as a result.

The annual ECML–PKDD conference serves as a global venue for the most recent research in all fields of machine learning and knowledge discovery in databases, including cutting-edge applications. It builds on a highly successful run of ECML–PKDD conferences which has made it the premier European machine learning and data mining conference.

This year, the conference drew over 1080 participants (762 in-person and 318 online) from 37 countries, including 23 European nations. This wealth of interest considerably exceeded our expectations, and we were both excited and under pressure to plan a special event. Overall, the conference attracted a lot of interest from industry thanks to sponsorship, participation, and the conference's industrial day.

The main conference program consisted of presentations of 242 accepted papers and four keynote talks (in order of appearance):

- Francis Bach (Inria), Information Theory with Kernel Methods
- Danai Koutra (University of Michigan), Mining & Learning [Compact] Representations for Structured Data
- Fosca Gianotti (Scuola Normale Superiore di Pisa), Explainable Machine Learning for Trustworthy AI
- Yann Le Cun (Facebook AI Research), From Machine Learning to Autonomous Intelligence

In addition, there were respectively twenty three in-person and three online workshops; five in-person and three online tutorials; two combined in-person and one combined online workshop-tutorials, together with a PhD Forum, a discovery challenge and demonstrations.

Papers presented during the three main conference days were organized in 4 tracks, within 54 sessions:

- Research Track: articles on research or methodology from all branches of machine learning, data mining, and knowledge discovery;
- Applied Data Science Track: articles on cutting-edge uses of machine learning, data mining, and knowledge discovery to resolve practical use cases and close the gap between current theory and practice;
- Journal Track: articles that were published in special issues of the journals *Machine Learning* and *Data Mining and Knowledge Discovery*;

– Demo Track: short articles that propose a novel system that advances the state of the art and include a demonstration video.

We received a record number of 1238 abstract submissions, and for the Research and Applied Data Science Tracks, 932 papers made it through the review process (the remaining papers were withdrawn, with the bulk being desk rejected). We accepted 189 (27.3%) Research papers and 53 (22.2%) Applied Data science articles. 47 papers from the Journal Track and 17 demo papers were also included in the program. We were able to put together an extraordinarily rich and engaging program because of the high quality submissions.

Research articles that were judged to be of exceptional quality and deserving of special distinction were chosen by the awards committee:

– Machine Learning Best Paper Award: "*Bounding the Family-Wise Error Rate in Local Causal Discovery Using Rademacher Averages*", by Dario Simionato (University of Padova) and Fabio Vandin (University of Padova)
– Data-Mining Best Paper Award: "*Transforming PageRank into an Infinite-Depth Graph Neural Network*", by Andreas Roth (TU Dortmund), and Thomas Liebig (TU Dortmund)
– Test of Time Award for highest impact paper from ECML–PKDD 2012: "*Fairness-Aware Classifier with Prejudice Remover Regularizer*", by Toshihiro Kamishima (National Institute of Advanced Industrial Science and Technology AIST), Shotaro Akashi (National Institute of Advanced Industrial Science and Technology AIST), Hideki Asoh (National Institute of Advanced Industrial Science and Technology AIST), and Jun Sakuma (University of Tsukuba)

We sincerely thank the contributions of all participants, authors, PC members, area chairs, session chairs, volunteers, and co-organizers who made ECML–PKDD 2022 a huge success. We would especially like to thank Julie from the Grenoble World Trade Center for all her help and Titouan from Insight-outside, who worked so hard to make the online event possible. We also like to express our gratitude to Thierry for the design of the conference logo representing the three mountain chains surrounding the Grenoble city, as well as the sponsors and the ECML–PKDD Steering Committee.

October 2022

Massih-Reza Amini
Stéphane Canu
Asja Fischer
Petra Kralj Novak
Tias Guns
Grigorios Tsoumakas
Georgios Balikas
Fragkiskos Malliaros

Organization

General Chairs

Massih-Reza Amini	University Grenoble Alpes, France
Stéphane Canu	INSA Rouen, France

Program Chairs

Asja Fischer	Ruhr University Bochum, Germany
Tias Guns	KU Leuven, Belgium
Petra Kralj Novak	Central European University, Austria
Grigorios Tsoumakas	Aristotle University of Thessaloniki, Greece

Journal Track Chairs

Peggy Cellier	INSA Rennes, IRISA, France
Krzysztof Dembczyński	Yahoo Research, USA
Emilie Devijver	CNRS, France
Albrecht Zimmermann	University of Caen Normandie, France

Workshop and Tutorial Chairs

Bruno Crémilleux	University of Caen Normandie, France
Charlotte Laclau	Telecom Paris, France

Local Chairs

Latifa Boudiba	University Grenoble Alpes, France
Franck Iutzeler	University Grenoble Alpes, France

Proceedings Chairs

Wouter Duivesteijn	Technische Universiteit Eindhoven, the Netherlands
Sibylle Hess	Technische Universiteit Eindhoven, the Netherlands

Industry Track Chairs

Rohit Babbar	Aalto University, Finland
Françoise Fogelmann	Hub France IA, France

Discovery Challenge Chairs

Ioannis Katakis	University of Nicosia, Cyprus
Ioannis Partalas	Expedia, Switzerland

Demonstration Chairs

Georgios Balikas	Salesforce, France
Fragkiskos Malliaros	CentraleSupélec, France

PhD Forum Chairs

Esther Galbrun	University of Eastern Finland, Finland
Justine Reynaud	University of Caen Normandie, France

Awards Chairs

Francesca Lisi	Università degli Studi di Bari, Italy
Michalis Vlachos	University of Lausanne, Switzerland

Sponsorship Chairs

Patrice Aknin	IRT SystemX, France
Gilles Gasso	INSA Rouen, France

Web Chairs

Martine Harshé	Laboratoire d'Informatique de Grenoble, France
Marta Soare	University Grenoble Alpes, France

Publicity Chair

Emilie Morvant	Université Jean Monnet, France

ECML PKDD Steering Committee

Annalisa Appice	University of Bari Aldo Moro, Italy
Ira Assent	Aarhus University, Denmark
Albert Bifet	Télécom ParisTech, France
Francesco Bonchi	ISI Foundation, Italy
Tania Cerquitelli	Politecnico di Torino, Italy
Sašo Džeroski	Jožef Stefan Institute, Slovenia
Elisa Fromont	Université de Rennes, France
Andreas Hotho	Julius-Maximilians-Universität Würzburg, Germany
Alípio Jorge	University of Porto, Portugal
Kristian Kersting	TU Darmstadt, Germany
Jefrey Lijffijt	Ghent University, Belgium
Luís Moreira-Matias	University of Porto, Portugal
Katharina Morik	TU Dortmund, Germany
Siegfried Nijssen	Université catholique de Louvain, Belgium
Andrea Passerini	University of Trento, Italy
Fernando Perez-Cruz	ETH Zurich, Switzerland
Alessandra Sala	Shutterstock Ireland Limited, Ireland
Arno Siebes	Utrecht University, the Netherlands
Isabel Valera	Universität des Saarlandes, Germany

Program Committees

Guest Editorial Board, Journal Track

Richard Allmendinger	University of Manchester, UK
Marie Anastacio	Universiteit Leiden, the Netherlands
Ira Assent	Aarhus University, Denmark
Martin Atzmueller	Universität Osnabrück, Germany
Rohit Babbar	Aalto University, Finland

Jaume Bacardit	Newcastle University, UK
Anthony Bagnall	University of East Anglia, UK
Mitra Baratchi	Universiteit Leiden, the Netherlands
Francesco Bariatti	IRISA, France
German Barquero	Universität de Barcelona, Spain
Alessio Benavoli	Trinity College Dublin, Ireland
Viktor Bengs	Ludwig-Maximilians-Universität München, Germany
Massimo Bilancia	Università degli Studi di Bari Aldo Moro, Italy
Ilaria Bordino	Unicredit R&D, Italy
Jakob Bossek	University of Münster, Germany
Ulf Brefeld	Leuphana University of Lüneburg, Germany
Ricardo Campello	University of Newcastle, UK
Michelangelo Ceci	University of Bari, Italy
Loic Cerf	Universidade Federal de Minas Gerais, Brazil
Vitor Cerqueira	Universidade do Porto, Portugal
Laetitia Chapel	IRISA, France
Jinghui Chen	Pennsylvania State University, USA
Silvia Chiusano	Politecnico di Torino, Italy
Roberto Corizzo	Università degli Studi di Bari Aldo Moro, Italy
Bruno Cremilleux	Université de Caen Normandie, France
Marco de Gemmis	University of Bari Aldo Moro, Italy
Sebastien Destercke	Centre National de la Recherche Scientifique, France
Shridhar Devamane	Global Academy of Technology, India
Benjamin Doerr	Ecole Polytechnique, France
Wouter Duivesteijn	Technische Universiteit Eindhoven, the Netherlands
Thomas Dyhre Nielsen	Aalborg University, Denmark
Tapio Elomaa	Tampere University, Finland
Remi Emonet	Université Jean Monnet Saint-Etienne, France
Nicola Fanizzi	Università degli Studi di Bari Aldo Moro, Italy
Pedro Ferreira	University of Lisbon, Portugal
Cesar Ferri	Universität Politecnica de Valencia, Spain
Julia Flores	University of Castilla-La Mancha, Spain
Ionut Florescu	Stevens Institute of Technology, USA
Germain Forestier	Université de Haute-Alsace, France
Joel Frank	Ruhr-Universität Bochum, Germany
Marco Frasca	Università degli Studi di Milano, Italy
Jose A. Gomez	Universidad de Castilla-La Mancha, Spain
Stephan Günnemann	Institute for Advanced Study, Germany
Luis Galarraga	Inria, France

Esther Galbrun	University of Eastern Finland, Finland
Joao Gama	University of Porto, Portugal
Paolo Garza	Politecnico di Torino, Italy
Pascal Germain	Université Laval, Canada
Fabian Gieseke	Westfälische Wilhelms-Universität Münster, Germany
Riccardo Guidotti	Università degli Studi di Pisa, Italy
Francesco Gullo	UniCredit, Italy
Antonella Guzzo	University of Calabria, Italy
Isabel Haasler	KTH Royal Institute of Technology, Sweden
Alexander Hagg	Bonn-Rhein-Sieg University, Germany
Daniel Hernandez-Lobato	Universidad Autónoma de Madrid, Spain
Jose Hernandez-Orallo	Universidad Politecnica de Valencia, Spain
Martin Holena	Neznámá organizace, Czechia
Jaakko Hollmen	Stockholm University, Sweden
Dino Ienco	IRSTEA, France
Georgiana Ifrim	University College Dublin, Ireland
Felix Iglesias	Technische Universität Wien, Austria
Angelo Impedovo	Università degli Studi di Bari Aldo Moro, Italy
Frank Iutzeler	Université Grenoble Alpes, France
Mahdi Jalili	RMIT University, Australia
Szymon Jaroszewicz	Polish Academy of Sciences, Poland
Mehdi Kaytoue	INSA Lyon, France
Raouf Kerkouche	Helmholtz Center for Information Security, Germany
Pascal Kerschke	Westfälische Wilhelms-Universität Münster, Germany
Dragi Kocev	Jožef Stefan Institute, Slovenia
Wojciech Kotlowski	Poznan University of Technology, Poland
Lars Kotthoff	University of Wyoming, USA
Peer Kroger	Ludwig-Maximilians-Universität München, Germany
Tipaluck Krityakierne	Mahidol University, Thailand
Peer Kroger	Christian-Albrechts-University Kiel, Germany
Meelis Kull	Tartu Ulikool, Estonia
Charlotte Laclau	Laboratoire Hubert Curien, France
Mark Last	Ben-Gurion University of the Negev, Israel
Matthijs van Leeuwen	Universiteit Leiden, the Netherlands
Thomas Liebig	TU Dortmund, Germany
Hsuan-Tien Lin	National Taiwan University, Taiwan
Marco Lippi	University of Modena and Reggio Emilia, Italy
Daniel Lobato	Universidad Autonoma de Madrid, Spain

Corrado Loglisci	Università degli Studi di Bari Aldo Moro, Italy
Nuno Lourenço	University of Coimbra, Portugal
Claudio Lucchese	Ca'Foscari University of Venice, Italy
Brian MacNamee	University College Dublin, Ireland
Davide Maiorca	University of Cagliari, Italy
Giuseppe Manco	National Research Council, Italy
Elio Masciari	University of Naples Federico II, Italy
Andres Masegosa	University of Aalborg, Denmark
Ernestina Menasalvas	Universidad Politecnica de Madrid, Spain
Lien Michiels	Universiteit Antwerpen, Belgium
Jan Mielniczuk	Polish Academy of Sciences, Poland
Paolo Mignone	Università degli Studi di Bari Aldo Moro, Italy
Anna Monreale	University of Pisa, Italy
Giovanni Montana	University of Warwick, UK
Gregoire Montavon	Technische Universität Berlin, Germany
Amedeo Napoli	LORIA, France
Frank Neumann	University of Adelaide, Australia
Thomas Nielsen	Aalborg Universitet, Denmark
Bruno Ordozgoiti	Aalto-yliopisto, Finland
Panagiotis Papapetrou	Stockholms Universitet, Sweden
Andrea Passerini	University of Trento, Italy
Mykola Pechenizkiy	Technische Universiteit Eindhoven, the Netherlands
Charlotte Pelletier	IRISA, France
Ruggero Pensa	University of Turin, Italy
Nico Piatkowski	Technische Universität Dortmund, Germany
Gianvito Pio	Università degli Studi di Bari Aldo Moro, Italy
Marc Plantevit	Université Claude Bernard Lyon 1, France
Jose M. Puerta	Universidad de Castilla-La Mancha, Spain
Kai Puolamaki	Helsingin Yliopisto, Finland
Michael Rabbat	Meta Platforms Inc, USA
Jan Ramon	Inria Lille Nord Europe, France
Rita Ribeiro	Universidade do Porto, Portugal
Kaspar Riesen	University of Bern, Switzerland
Matteo Riondato	Amherst College, USA
Celine Robardet	INSA Lyon, France
Pieter Robberechts	KU Leuven, Belgium
Antonio Salmeron	University of Almería, Spain
Jorg Sander	University of Alberta, Canada
Roberto Santana	University of the Basque Country, Spain
Michael Schaub	Rheinisch-Westfälische Technische Hochschule, Germany

Erik Schultheis	Aalto-yliopisto, Finland
Thomas Seidl	Ludwig-Maximilians-Universität München, Germany
Moritz Seiler	University of Münster, Germany
Kijung Shin	KAIST, South Korea
Shinichi Shirakawa	Yokohama National University, Japan
Marek Smieja	Jagiellonian University, Poland
James Edward Smith	University of the West of England, UK
Carlos Soares	Universidade do Porto, Portugal
Arnaud Soulet	Université de Tours, France
Gerasimos Spanakis	Maastricht University, the Netherlands
Giancarlo Sperli	University of Campania Luigi Vanvitelli, Italy
Myra Spiliopoulou	Otto von Guericke Universität Magdeburg, Germany
Jerzy Stefanowski	Poznan University of Technology, Poland
Giovanni Stilo	Università degli Studi dell'Aquila, Italy
Catalin Stoean	University of Craiova, Romania
Mahito Sugiyama	National Institute of Informatics, Japan
Nikolaj Tatti	Helsingin Yliopisto, Finland
Alexandre Termier	Université de Rennes 1, France
Luis Torgo	Dalhousie University, Canada
Leonardo Trujillo	Tecnologico Nacional de Mexico, Mexico
Wei-Wei Tu	4Paradigm Inc., China
Steffen Udluft	Siemens AG Corporate Technology, Germany
Arnaud Vandaele	Université de Mons, Belgium
Celine Vens	KU Leuven, Belgium
Herna Viktor	University of Ottawa, Canada
Marco Virgolin	Centrum Wiskunde en Informatica, the Netherlands
Jordi Vitria	Universität de Barcelona, Spain
Jilles Vreeken	CISPA Helmholtz Center for Information Security, Germany
Willem Waegeman	Universiteit Gent, Belgium
Markus Wagner	University of Adelaide, Australia
Elizabeth Wanner	Centro Federal de Educacao Tecnologica de Minas, Brazil
Marcel Wever	Universität Paderborn, Germany
Ngai Wong	University of Hong Kong, Hong Kong, China
Man Leung Wong	Lingnan University, Hong Kong, China
Marek Wydmuch	Poznan University of Technology, Poland
Guoxian Yu	Shandong University, China
Xiang Zhang	University of Hong Kong, Hong Kong, China

Ye Zhu	Deakin University, USA
Arthur Zimek	Syddansk Universitet, Denmark
Albrecht Zimmermann	Université de Caen Normandie, France

Area Chairs

Fabrizio Angiulli	DIMES, University of Calabria, Italy
Annalisa Appice	University of Bari, Italy
Ira Assent	Aarhus University, Denmark
Martin Atzmueller	Osnabrück University, Germany
Michael Berthold	Universität Konstanz, Germany
Albert Bifet	Université Paris-Saclay, France
Hendrik Blockeel	KU Leuven, Belgium
Christian Böhm	LMU Munich, Germany
Francesco Bonchi	ISI Foundation, Turin, Italy
Ulf Brefeld	Leuphana, Germany
Francesco Calabrese	Richemont, USA
Toon Calders	Universiteit Antwerpen, Belgium
Michelangelo Ceci	University of Bari, Italy
Peggy Cellier	IRISA, France
Duen Horng Chau	Georgia Institute of Technology, USA
Nicolas Courty	IRISA, Université Bretagne-Sud, France
Bruno Cremilleux	Université de Caen Normandie, France
Jesse Davis	KU Leuven, Belgium
Gianmarco De Francisci Morales	CentAI, Italy
Tom Diethe	Amazon, UK
Carlotta Domeniconi	George Mason University, USA
Yuxiao Dong	Tsinghua University, China
Kurt Driessens	Maastricht University, the Netherlands
Tapio Elomaa	Tampere University, Finland
Sergio Escalera	CVC and University of Barcelona, Spain
Faisal Farooq	Qatar Computing Research Institute, Qatar
Asja Fischer	Ruhr University Bochum, Germany
Peter Flach	University of Bristol, UK
Eibe Frank	University of Waikato, New Zealand
Paolo Frasconi	Università degli Studi di Firenze, Italy
Elisa Fromont	Université Rennes 1, IRISA/Inria, France
Johannes Fürnkranz	JKU Linz, Austria
Patrick Gallinari	Sorbonne Université, Criteo AI Lab, France
Joao Gama	INESC TEC - LIAAD, Portugal
Jose Gamez	Universidad de Castilla-La Mancha, Spain
Roman Garnett	Washington University in St. Louis, USA
Thomas Gärtner	TU Wien, Austria

Aristides Gionis	KTH Royal Institute of Technology, Sweden
Francesco Gullo	UniCredit, Italy
Stephan Günnemann	Technical University of Munich, Germany
Xiangnan He	University of Science and Technology of China, China
Daniel Hernandez-Lobato	Universidad Autonoma de Madrid, Spain
José Hernández-Orallo	Universität Politècnica de València, Spain
Jaakko Hollmén	Aalto University, Finland
Andreas Hotho	Universität Würzburg, Germany
Eyke Hüllermeier	University of Munich, Germany
Neil Hurley	University College Dublin, Ireland
Georgiana Ifrim	University College Dublin, Ireland
Alipio Jorge	INESC TEC/University of Porto, Portugal
Ross King	Chalmers University of Technology, Sweden
Arno Knobbe	Leiden University, the Netherlands
Yun Sing Koh	University of Auckland, New Zealand
Parisa Kordjamshidi	Michigan State University, USA
Lars Kotthoff	University of Wyoming, USA
Nicolas Kourtellis	Telefonica Research, Spain
Danai Koutra	University of Michigan, USA
Danica Kragic	KTH Royal Institute of Technology, Sweden
Stefan Kramer	Johannes Gutenberg University Mainz, Germany
Niklas Lavesson	Blekinge Institute of Technology, Sweden
Sébastien Lefèvre	Université de Bretagne Sud/IRISA, France
Jefrey Lijffijt	Ghent University, Belgium
Marius Lindauer	Leibniz University Hannover, Germany
Patrick Loiseau	Inria, France
Jose Lozano	UPV/EHU, Spain
Jörg Lücke	Universität Oldenburg, Germany
Donato Malerba	Università degli Studi di Bari Aldo Moro, Italy
Fragkiskos Malliaros	CentraleSupelec, France
Giuseppe Manco	ICAR-CNR, Italy
Wannes Meert	KU Leuven, Belgium
Pauli Miettinen	University of Eastern Finland, Finland
Dunja Mladenic	Jožef Stefan Institute, Slovenia
Anna Monreale	Università di Pisa, Italy
Luis Moreira-Matias	Finiata, Germany
Emilie Morvant	University Jean Monnet, St-Etienne, France
Sriraam Natarajan	UT Dallas, USA
Nuria Oliver	Vodafone Research, USA
Panagiotis Papapetrou	Stockholm University, Sweden
Laurence Park	WSU, Australia

Andrea Passerini	University of Trento, Italy
Mykola Pechenizkiy	TU Eindhoven, the Netherlands
Dino Pedreschi	University of Pisa, Italy
Robert Peharz	Graz University of Technology, Austria
Julien Perez	Naver Labs Europe, France
Franz Pernkopf	Graz University of Technology, Austria
Bernhard Pfahringer	University of Waikato, New Zealand
Fabio Pinelli	IMT Lucca, Italy
Visvanathan Ramesh	Goethe University Frankfurt, Germany
Jesse Read	Ecole Polytechnique, France
Zhaochun Ren	Shandong University, China
Marian-Andrei Rizoiu	University of Technology Sydney, Australia
Celine Robardet	INSA Lyon, France
Sriparna Saha	IIT Patna, India
Ute Schmid	University of Bamberg, Germany
Lars Schmidt-Thieme	University of Hildesheim, Germany
Michele Sebag	LISN CNRS, France
Thomas Seidl	LMU Munich, Germany
Arno Siebes	Universiteit Utrecht, the Netherlands
Fabrizio Silvestri	Sapienza, University of Rome, Italy
Myra Spiliopoulou	Otto-von-Guericke-University Magdeburg, Germany
Yizhou Sun	UCLA, USA
Jie Tang	Tsinghua University, China
Nikolaj Tatti	Helsinki University, Finland
Evimaria Terzi	Boston University, USA
Marc Tommasi	Lille University, France
Antti Ukkonen	University of Helsinki, Finland
Herke van Hoof	University of Amsterdam, the Netherlands
Matthijs van Leeuwen	Leiden University, the Netherlands
Celine Vens	KU Leuven, Belgium
Christel Vrain	University of Orleans, France
Jilles Vreeken	CISPA Helmholtz Center for Information Security, Germany
Willem Waegeman	Universiteit Gent, Belgium
Stefan Wrobel	Fraunhofer IAIS, Germany
Xing Xie	Microsoft Research Asia, China
Min-Ling Zhang	Southeast University, China
Albrecht Zimmermann	Université de Caen Normandie, France
Indre Zliobaite	University of Helsinki, Finland

Program Committee Members

Amos Abbott	Virginia Tech, USA
Pedro Abreu	CISUC, Portugal
Maribel Acosta	Ruhr University Bochum, Germany
Timilehin Aderinola	Insight Centre, University College Dublin, Ireland
Linara Adilova	Ruhr University Bochum, Fraunhofer IAIS, Germany
Florian Adriaens	KTH, Sweden
Azim Ahmadzadeh	Georgia State University, USA
Nourhan Ahmed	University of Hildesheim, Germany
Deepak Ajwani	University College Dublin, Ireland
Amir Hossein Akhavan Rahnama	KTH Royal Institute of Technology, Sweden
Aymen Al Marjani	ENS Lyon, France
Mehwish Alam	Leibniz Institute for Information Infrastructure, Germany
Francesco Alesiani	NEC Laboratories Europe, Germany
Omar Alfarisi	ADNOC, Canada
Pegah Alizadeh	Ericsson Research, France
Reem Alotaibi	King Abdulaziz University, Saudi Arabia
Jumanah Alshehri	Temple University, USA
Bakhtiar Amen	University of Huddersfield, UK
Evelin Amorim	Inesc tec, Portugal
Shin Ando	Tokyo University of Science, Japan
Thiago Andrade	INESC TEC - LIAAD, Portugal
Jean-Marc Andreoli	Naverlabs Europe, France
Giuseppina Andresini	University of Bari Aldo Moro, Italy
Alessandro Antonucci	IDSIA, Switzerland
Xiang Ao	Institute of Computing Technology, CAS, China
Siddharth Aravindan	National University of Singapore, Singapore
Héber H. Arcolezi	Inria and École Polytechnique, France
Adrián Arnaiz-Rodríguez	ELLIS Unit Alicante, Spain
Yusuf Arslan	University of Luxembourg, Luxembourg
André Artelt	Bielefeld University, Germany
Sunil Aryal	Deakin University, Australia
Charles Assaad	Easyvista, France
Matthias Aßenmacher	Ludwig-Maxmilians-Universität München, Germany
Zeyar Aung	Masdar Institute, UAE
Serge Autexier	DFKI Bremen, Germany
Rohit Babbar	Aalto University, Finland
Housam Babiker	University of Alberta, Canada

Antonio Bahamonde	University of Oviedo, Spain
Maroua Bahri	Inria Paris, France
Georgios Balikas	Salesforce, France
Maria Bampa	Stockholm University, Sweden
Hubert Baniecki	Warsaw University of Technology, Poland
Elena Baralis	Politecnico di Torino, Italy
Mitra Baratchi	LIACS - University of Leiden, the Netherlands
Kalliopi Basioti	Rutgers University, USA
Martin Becker	Stanford University, USA
Diana Benavides Prado	University of Auckland, New Zealand
Anes Bendimerad	LIRIS, France
Idir Benouaret	Université Grenoble Alpes, France
Isacco Beretta	Università di Pisa, Italy
Victor Berger	CEA, France
Christoph Bergmeir	Monash University, Australia
Cuissart Bertrand	University of Caen, France
Antonio Bevilacqua	University College Dublin, Ireland
Yaxin Bi	Ulster University, UK
Ranran Bian	University of Auckland, New Zealand
Adrien Bibal	University of Louvain, Belgium
Subhodip Biswas	Virginia Tech, USA
Patrick Blöbaum	Amazon AWS, USA
Carlos Bobed	University of Zaragoza, Spain
Paul Bogdan	USC, USA
Chiara Boldrini	CNR, Italy
Clément Bonet	Université Bretagne Sud, France
Andrea Bontempelli	University of Trento, Italy
Ludovico Boratto	University of Cagliari, Italy
Stefano Bortoli	Huawei Research Center, Germany
Diana-Laura Borza	Babes Bolyai University, Romania
Ahcene Boubekki	UiT, Norway
Sabri Boughorbel	QCRI, Qatar
Paula Branco	University of Ottawa, Canada
Jure Brence	Jožef Stefan Institute, Slovenia
Martin Breskvar	Jožef Stefan Institute, Slovenia
Marco Bressan	University of Milan, Italy
Dariusz Brzezinski	Poznan University of Technology, Poland
Florian Buettner	German Cancer Research Center, Germany
Julian Busch	Siemens Technology, Germany
Sebastian Buschjäger	TU Dortmund Artificial Intelligence Unit, Germany
Ali Butt	Virginia Tech, USA

Narayanan C. Krishnan	IIT Palakkad, India
Xiangrui Cai	Nankai University, China
Xiongcai Cai	UNSW Sydney, Australia
Zekun Cai	University of Tokyo, Japan
Andrea Campagner	Università degli Studi di Milano-Bicocca, Italy
Seyit Camtepe	CSIRO Data61, Australia
Jiangxia Cao	Chinese Academy of Sciences, China
Pengfei Cao	Chinese Academy of Sciences, China
Yongcan Cao	University of Texas at San Antonio, USA
Cécile Capponi	Aix-Marseille University, France
Axel Carlier	Institut National Polytechnique de Toulouse, France
Paula Carroll	University College Dublin, Ireland
John Cartlidge	University of Bristol, UK
Simon Caton	University College Dublin, Ireland
Bogdan Cautis	University of Paris-Saclay, France
Mustafa Cavus	Warsaw University of Technology, Poland
Remy Cazabet	Université Lyon 1, France
Josu Ceberio	University of the Basque Country, Spain
David Cechák	CEITEC Masaryk University, Czechia
Abdulkadir Celikkanat	Technical University of Denmark, Denmark
Dumitru-Clementin Cercel	University Politehnica of Bucharest, Romania
Christophe Cerisara	CNRS, France
Vítor Cerqueira	Dalhousie University, Canada
Mattia Cerrato	JGU Mainz, Germany
Ricardo Cerri	Federal University of São Carlos, Brazil
Hubert Chan	University of Hong Kong, Hong Kong, China
Vaggos Chatziafratis	Stanford University, USA
Siu Lun Chau	University of Oxford, UK
Chaochao Chen	Zhejiang University, China
Chuan Chen	Sun Yat-sen University, China
Hechang Chen	Jilin University, China
Jia Chen	Beihang University, China
Jiaoyan Chen	University of Oxford, UK
Jiawei Chen	Zhejiang University, China
Jin Chen	University of Electronic Science and Technology, China
Kuan-Hsun Chen	University of Twente, the Netherlands
Lingwei Chen	Wright State University, USA
Tianyi Chen	Boston University, USA
Wang Chen	Google, USA
Xinyuan Chen	Universiti Kuala Lumpur, Malaysia

Yuqiao Chen	UT Dallas, USA
Yuzhou Chen	Princeton University, USA
Zhennan Chen	Xiamen University, China
Zhiyu Chen	UCSB, USA
Zhqian Chen	Mississippi State University, USA
Ziheng Chen	Stony Brook University, USA
Zhiyong Cheng	Shandong Academy of Sciences, China
Noëlie Cherrier	CITiO, France
Anshuman Chhabra	UC Davis, USA
Zhixuan Chu	Ant Group, China
Guillaume Cleuziou	LIFO, France
Ciaran Cooney	AflacNI, UK
Robson Cordeiro	University of São Paulo, Brazil
Roberto Corizzo	American University, USA
Antoine Cornuéjols	AgroParisTech, France
Fabrizio Costa	Exeter University, UK
Gustavo Costa	Instituto Federal de Goiás - Campus Jataí, Brazil
Luís Cruz	Delft University of Technology, the Netherlands
Tianyu Cui	Institute of Information Engineering, China
Wang-Zhou Dai	Imperial College London, UK
Tanmoy Dam	University of New South Wales Canberra, Australia
Thi-Bich-Hanh Dao	University of Orleans, France
Adrian Sergiu Darabant	Babes Bolyai University, Romania
Mrinal Das	IIT Palakaad, India
Sina Däubener	Ruhr University, Bochum, Germany
Padraig Davidson	University of Würzburg, Germany
Paul Davidsson	Malmö University, Sweden
Andre de Carvalho	USP, Brazil
Antoine de Mathelin	ENS Paris-Saclay, France
Tom De Schepper	University of Antwerp, Belgium
Marcilio de Souto	LIFO/Univ. Orleans, France
Gaetan De Waele	Ghent University, Belgium
Pieter Delobelle	KU Leuven, Belgium
Alper Demir	Izmir University of Economics, Turkey
Ambra Demontis	University of Cagliari, Italy
Difan Deng	Leibniz Universität Hannover, Germany
Guillaume Derval	UCLouvain - ICTEAM, Belgium
Maunendra Sankar Desarkar	IIT Hyderabad, India
Chris Develder	University of Ghent - iMec, Belgium
Arnout Devos	Swiss Federal Institute of Technology Lausanne, Switzerland

Laurens Devos	KU Leuven, Belgium
Bhaskar Dhariyal	University College Dublin, Ireland
Nicola Di Mauro	University of Bari, Italy
Aissatou Diallo	University College London, UK
Christos Dimitrakakis	University of Neuchatel, Switzerland
Jiahao Ding	University of Houston, USA
Kaize Ding	Arizona State University, USA
Yao-Xiang Ding	Nanjing University, China
Guilherme Dinis Junior	Stockholm University, Sweden
Nikolaos Dionelis	University of Edinburgh, UK
Christos Diou	Harokopio University of Athens, Greece
Sonia Djebali	Léonard de Vinci Pôle Universitaire, France
Nicolas Dobigeon	University of Toulouse, France
Carola Doerr	Sorbonne University, France
Ruihai Dong	University College Dublin, Ireland
Shuyu Dong	Inria, Université Paris-Saclay, France
Yixiang Dong	Xi'an Jiaotong University, China
Xin Du	University of Edinburgh, UK
Yuntao Du	Nanjing University, China
Stefan Duffner	University of Lyon, France
Rahul Duggal	Georgia Tech, USA
Wouter Duivesteijn	TU Eindhoven, the Netherlands
Sebastijan Dumancic	TU Delft, the Netherlands
Inês Dutra	University of Porto, Portugal
Thomas Dyhre Nielsen	AAU, Denmark
Saso Dzeroski	Jožef Stefan Institute, Ljubljana, Slovenia
Tome Eftimov	Jožef Stefan Institute, Ljubljana, Slovenia
Hamid Eghbal-zadeh	LIT AI Lab, Johannes Kepler University, Austria
Theresa Eimer	Leibniz University Hannover, Germany
Radwa El Shawi	Tartu University, Estonia
Dominik Endres	Philipps-Universität Marburg, Germany
Roberto Esposito	Università di Torino, Italy
Georgios Evangelidis	University of Macedonia, Greece
Samuel Fadel	Leuphana University, Germany
Stephan Fahrenkrog-Petersen	Humboldt-Universität zu Berlin, Germany
Xiaomao Fan	Shenzhen Technology University, China
Zipei Fan	University of Tokyo, Japan
Hadi Fanaee	Halmstad University, Sweden
Meng Fang	TU/e, the Netherlands
Elaine Faria	UFU, Brazil
Ad Feelders	Universiteit Utrecht, the Netherlands
Sophie Fellenz	TU Kaiserslautern, Germany

Stefano Ferilli	University of Bari, Italy
Daniel Fernández-Sánchez	Universidad Autónoma de Madrid, Spain
Pedro Ferreira	Faculty of Sciences University of Porto, Portugal
Cèsar Ferri	Universität Politècnica València, Spain
Flavio Figueiredo	UFMG, Brazil
Soukaina Filali Boubrahimi	Utah State University, USA
Raphael Fischer	TU Dortmund, Germany
Germain Forestier	University of Haute Alsace, France
Edouard Fouché	Karlsruhe Institute of Technology, Germany
Philippe Fournier-Viger	Shenzhen University, China
Kary Framling	Umeå University, Sweden
Jérôme François	Inria Nancy Grand-Est, France
Fabio Fumarola	Prometeia, Italy
Pratik Gajane	Eindhoven University of Technology, the Netherlands
Esther Galbrun	University of Eastern Finland, Finland
Laura Galindez Olascoaga	KU Leuven, Belgium
Sunanda Gamage	University of Western Ontario, Canada
Chen Gao	Tsinghua University, China
Wei Gao	Nanjing University, China
Xiaofeng Gao	Shanghai Jiaotong University, China
Yuan Gao	University of Science and Technology of China, China
Jochen Garcke	University of Bonn, Germany
Clement Gautrais	Brightclue, France
Benoit Gauzere	INSA Rouen, France
Dominique Gay	Université de La Réunion, France
Xiou Ge	University of Southern California, USA
Bernhard Geiger	Know-Center GmbH, Germany
Jiahui Geng	University of Stavanger, Norway
Yangliao Geng	Tsinghua University, China
Konstantin Genin	University of Tübingen, Germany
Firas Gerges	New Jersey Institute of Technology, USA
Pierre Geurts	University of Liège, Belgium
Gizem Gezici	Sabanci University, Turkey
Amirata Ghorbani	Stanford, USA
Biraja Ghoshal	TCS, UK
Anna Giabelli	Università degli studi di Milano Bicocca, Italy
George Giannakopoulos	IIT Demokritos, Greece
Tobias Glasmachers	Ruhr-University Bochum, Germany
Heitor Murilo Gomes	University of Waikato, New Zealand
Anastasios Gounaris	Aristotle University of Thessaloniki, Greece

Antoine Gourru	University of Lyon, France
Michael Granitzer	University of Passau, Germany
Magda Gregorova	Hochschule Würzburg-Schweinfurt, Germany
Moritz Grosse-Wentrup	University of Vienna, Austria
Divya Grover	Chalmers University, Sweden
Bochen Guan	OPPO US Research Center, USA
Xinyu Guan	Xian Jiaotong University, China
Guillaume Guerard	ESILV, France
Daniel Guerreiro e Silva	University of Brasilia, Brazil
Riccardo Guidotti	University of Pisa, Italy
Ekta Gujral	University of California, Riverside, USA
Aditya Gulati	ELLIS Unit Alicante, Spain
Guibing Guo	Northeastern University, China
Jianxiong Guo	Beijing Normal University, China
Yuhui Guo	Renmin University of China, China
Karthik Gurumoorthy	Amazon, India
Thomas Guyet	Inria, Centre de Lyon, France
Guillaume Habault	KDDI Research, Inc., Japan
Amaury Habrard	University of St-Etienne, France
Shahrzad Haddadan	Brown University, USA
Shah Muhammad Hamdi	New Mexico State University, USA
Massinissa Hamidi	PRES Sorbonne Paris Cité, France
Peng Han	KAUST, Saudi Arabia
Tom Hanika	University of Kassel, Germany
Sébastien Harispe	IMT Mines Alès, France
Marwan Hassani	TU Eindhoven, the Netherlands
Kohei Hayashi	Preferred Networks, Inc., Japan
Conor Hayes	National University of Ireland Galway, Ireland
Lingna He	Zhejiang University of Technology, China
Ramya Hebbalaguppe	Indian Institute of Technology, Delhi, India
Jukka Heikkonen	University of Turku, Finland
Fredrik Heintz	Linköping University, Sweden
Patrick Hemmer	Karlsruhe Institute of Technology, Germany
Romain Hérault	INSA de Rouen, France
Jeronimo Hernandez-Gonzalez	University of Barcelona, Spain
Sibylle Hess	TU Eindhoven, the Netherlands
Fabian Hinder	Bielefeld University, Germany
Lars Holdijk	University of Amsterdam, the Netherlands
Martin Holena	Institute of Computer Science, Czechia
Mike Holenderski	Eindhoven University of Technology, the Netherlands
Shenda Hong	Peking University, China

Yupeng Hou	Renmin University of China, China
Binbin Hu	Ant Financial Services Group, China
Jian Hu	Queen Mary University of London, UK
Liang Hu	Tongji University, China
Wen Hu	Ant Group, China
Wenbin Hu	Wuhan University, China
Wenbo Hu	Tsinghua University, China
Yaowei Hu	University of Arkansas, USA
Chao Huang	University of Hong Kong, China
Gang Huang	Zhejiang Lab, China
Guanjie Huang	Penn State University, USA
Hong Huang	HUST, China
Jin Huang	University of Amsterdam, the Netherlands
Junjie Huang	Chinese Academy of Sciences, China
Qiang Huang	Jilin University, China
Shangrong Huang	Hunan University, China
Weitian Huang	South China University of Technology, China
Yan Huang	Huazhong University of Science and Technology, China
Yiran Huang	Karlsruhe Institute of Technology, Germany
Angelo Impedovo	University of Bari, Italy
Roberto Interdonato	CIRAD, France
Iñaki Inza	University of the Basque Country, Spain
Stratis Ioannidis	Northeastern University, USA
Rakib Islam	Facebook, USA
Tobias Jacobs	NEC Laboratories Europe GmbH, Germany
Priyank Jaini	Google, Canada
Johannes Jakubik	Karlsruhe Institute of Technology, Germany
Nathalie Japkowicz	American University, USA
Szymon Jaroszewicz	Polish Academy of Sciences, Poland
Shayan Jawed	University of Hildesheim, Germany
Rathinaraja Jeyaraj	Kyungpook National University, South Korea
Shaoxiong Ji	Aalto University, Finland
Taoran Ji	Virginia Tech, USA
Bin-Bin Jia	Southeast University, China
Yuheng Jia	Southeast University, China
Ziyu Jia	Beijing Jiaotong University, China
Nan Jiang	Purdue University, USA
Renhe Jiang	University of Tokyo, Japan
Siyang Jiang	National Taiwan University, Taiwan
Song Jiang	University of California, Los Angeles, USA
Wenyu Jiang	Nanjing University, China

Zhen Jiang	Jiangsu University, China
Yuncheng Jiang	South China Normal University, China
François-Xavier Jollois	Université de Paris Cité, France
Adan Jose-Garcia	Université de Lille, France
Ferdian Jovan	University of Bristol, UK
Steffen Jung	MPII, Germany
Thorsten Jungeblut	Bielefeld University of Applied Sciences, Germany
Hachem Kadri	Aix-Marseille University, France
Vana Kalogeraki	Athens University of Economics and Business, Greece
Vinayaka Kamath	Microsoft Research India, India
Toshihiro Kamishima	National Institute of Advanced Industrial Science, Japan
Bo Kang	Ghent University, Belgium
Alexandros Karakasidis	University of Macedonia, Greece
Mansooreh Karami	Arizona State University, USA
Panagiotis Karras	Aarhus University, Denmark
Ioannis Katakis	University of Nicosia, Cyprus
Koki Kawabata	Osaka University, Tokyo
Klemen Kenda	Jožef Stefan Institute, Slovenia
Patrik Joslin Kenfack	Innopolis University, Russia
Mahsa Keramati	Simon Fraser University, Canada
Hamidreza Keshavarz	Tarbiat Modares University, Iran
Adil Khan	Innopolis University, Russia
Jihed Khiari	Johannes Kepler University, Austria
Mi-Young Kim	University of Alberta, Canada
Arto Klami	University of Helsinki, Finland
Jiri Klema	Czech Technical University, Czechia
Tomas Kliegr	University of Economics Prague, Czechia
Christian Knoll	Graz, University of Technology, Austria
Dmitry Kobak	University of Tübingen, Germany
Vladimer Kobayashi	University of the Philippines Mindanao, Philippines
Dragi Kocev	Jožef Stefan Institute, Slovenia
Adrian Kochsiek	University of Mannheim, Germany
Masahiro Kohjima	NTT Corporation, Japan
Georgia Koloniari	University of Macedonia, Greece
Nikos Konofaos	Aristotle University of Thessaloniki, Greece
Irena Koprinska	University of Sydney, Australia
Lars Kotthoff	University of Wyoming, USA
Daniel Kottke	University of Kassel, Germany

Anna Krause	University of Würzburg, Germany
Alexander Kravberg	KTH Royal Institute of Technology, Sweden
Anastasia Krithara	NCSR Demokritos, Greece
Meelis Kull	University of Tartu, Estonia
Pawan Kumar	IIIT, Hyderabad, India
Suresh Kirthi Kumaraswamy	InterDigital, France
Gautam Kunapuli	Verisk Inc, USA
Marcin Kurdziel	AGH University of Science and Technology, Poland
Vladimir Kuzmanovski	Aalto University, Finland
Ariel Kwiatkowski	École Polytechnique, France
Firas Laakom	Tampere University, Finland
Harri Lähdesmäki	Aalto University, Finland
Stefanos Laskaridis	Samsung AI, UK
Alberto Lavelli	FBK-ict, Italy
Aonghus Lawlor	University College Dublin, Ireland
Thai Le	University of Mississippi, USA
Hoàng-Ân Lê	IRISA, University of South Brittany, France
Hoel Le Capitaine	University of Nantes, France
Thach Le Nguyen	Insight Centre, Ireland
Tai Le Quy	L3S Research Center - Leibniz University Hannover, Germany
Mustapha Lebbah	Sorbonne Paris Nord University, France
Dongman Lee	KAIST, South Korea
John Lee	Université catholique de Louvain, Belgium
Minwoo Lee	University of North Carolina at Charlotte, USA
Zed Lee	Stockholm University, Sweden
Yunwen Lei	University of Birmingham, UK
Douglas Leith	Trinity College Dublin, Ireland
Florian Lemmerich	RWTH Aachen, Germany
Carson Leung	University of Manitoba, Canada
Chaozhuo Li	Microsoft Research Asia, China
Jian Li	Institute of Information Engineering, China
Lei Li	Peking University, China
Li Li	Southwest University, China
Rui Li	Inspur Group, China
Shiyang Li	UCSB, USA
Shuokai Li	Chinese Academy of Sciences, China
Tianyu Li	Alibaba Group, China
Wenye Li	The Chinese University of Hong Kong, Shenzhen, China
Wenzhong Li	Nanjing University, China

Xiaoting Li	Pennsylvania State University, USA
Yang Li	University of North Carolina at Chapel Hill, USA
Zejian Li	Zhejiang University, China
Zhidong Li	UTS, Australia
Zhixin Li	Guangxi Normal University, China
Defu Lian	University of Science and Technology of China, China
Bin Liang	UTS, Australia
Yuchen Liang	RPI, USA
Yiwen Liao	University of Stuttgart, Germany
Pieter Libin	VUB, Belgium
Thomas Liebig	TU Dortmund, Germany
Seng Pei Liew	LINE Corporation, Japan
Beiyu Lin	University of Nevada - Las Vegas, USA
Chen Lin	Xiamen University, China
Tony Lindgren	Stockholm University, Sweden
Chen Ling	Emory University, USA
Jiajing Ling	Singapore Management University, Singapore
Marco Lippi	University of Modena and Reggio Emilia, Italy
Bin Liu	Chongqing University, China
Bowen Liu	Stanford University, USA
Chang Liu	Institute of Information Engineering, CAS, China
Chien-Liang Liu	National Chiao Tung University, Taiwan
Feng Liu	East China Normal University, China
Jiacheng Liu	Chinese University of Hong Kong, China
Li Liu	Chongqing University, China
Shengcai Liu	Southern University of Science and Technology, China
Shenghua Liu	Institute of Computing Technology, CAS, China
Tingwen Liu	Institute of Information Engineering, CAS, China
Xiangyu Liu	Tencent, China
Yong Liu	Renmin University of China, China
Yuansan Liu	University of Melbourne, Australia
Zhiwei Liu	Salesforce, USA
Tuwe Löfström	Jönköping University, Sweden
Corrado Loglisci	Università degli Studi di Bari Aldo Moro, Italy
Ting Long	Shanghai Jiao Tong University, China
Beatriz López	University of Girona, Spain
Yin Lou	Ant Group, USA
Samir Loudni	TASC (LS2N-CNRS), IMT Atlantique, France
Yang Lu	Xiamen University, China
Yuxun Lu	National Institute of Informatics, Japan

Massimiliano Luca	Bruno Kessler Foundation, Italy
Stefan Lüdtke	University of Mannheim, Germany
Jovita Lukasik	University of Mannheim, Germany
Denis Lukovnikov	University of Bonn, Germany
Pedro Henrique Luz de Araujo	University of Brasília, Brazil
Fenglong Ma	Pennsylvania State University, USA
Jing Ma	University of Virginia, USA
Meng Ma	Peking University, China
Muyang Ma	Shandong University, China
Ruizhe Ma	University of Massachusetts Lowell, USA
Xingkong Ma	National University of Defense Technology, China
Xueqi Ma	Tsinghua University, China
Zichen Ma	The Chinese University of Hong Kong, Shenzhen, China
Luis Macedo	University of Coimbra, Portugal
Harshitha Machiraju	EPFL, Switzerland
Manchit Madan	Delivery Hero, Germany
Seiji Maekawa	Osaka University, Japan
Sindri Magnusson	Stockholm University, Sweden
Pathum Chamikara Mahawaga	CSIRO Data61, Australia
Saket Maheshwary	Amazon, India
Ajay Mahimkar	AT&T, USA
Pierre Maillot	Inria, France
Lorenzo Malandri	Unimib, Italy
Rammohan Mallipeddi	Kyungpook National University, South Korea
Sahil Manchanda	IIT Delhi, India
Domenico Mandaglio	DIMES-UNICAL, Italy
Panagiotis Mandros	Harvard University, USA
Robin Manhaeve	KU Leuven, Belgium
Silviu Maniu	Université Paris-Saclay, France
Cinmayii Manliguez	National Sun Yat-Sen University, Taiwan
Naresh Manwani	International Institute of Information Technology, India
Jiali Mao	East China Normal University, China
Alexandru Mara	Ghent University, Belgium
Radu Marculescu	University of Texas at Austin, USA
Roger Mark	Massachusetts Institute of Technology, USA
Fernando Martínez-Plume	Joint Research Centre - European Commission, Belgium
Koji Maruhashi	Fujitsu Research, Fujitsu Limited, Japan
Simone Marullo	University of Siena, Italy

Elio Masciari	University of Naples, Italy
Florent Masseglia	Inria, France
Michael Mathioudakis	University of Helsinki, Finland
Takashi Matsubara	Osaka University, Japan
Tetsu Matsukawa	Kyushu University, Japan
Santiago Mazuelas	BCAM-Basque Center for Applied Mathematics, Spain
Ryan McConville	University of Bristol, UK
Hardik Meisheri	TCS Research, India
Panagiotis Meletis	Eindhoven University of Technology, the Netherlands
Gabor Melli	Medable, USA
Joao Mendes-Moreira	INESC TEC, Portugal
Chuan Meng	University of Amsterdam, the Netherlands
Cristina Menghini	Brown University, USA
Engelbert Mephu Nguifo	Université Clermont Auvergne, CNRS, LIMOS, France
Fabio Mercorio	University of Milan-Bicocca, Italy
Guillaume Metzler	Laboratoire ERIC, France
Hao Miao	Aalborg University, Denmark
Alessio Micheli	Università di Pisa, Italy
Paolo Mignone	University of Bari Aldo Moro, Italy
Matej Mihelcic	University of Zagreb, Croatia
Ioanna Miliou	Stockholm University, Sweden
Bamdev Mishra	Microsoft, India
Rishabh Misra	Twitter, Inc, USA
Dixant Mittal	National University of Singapore, Singapore
Zhaobin Mo	Columbia University, USA
Daichi Mochihashi	Institute of Statistical Mathematics, Japan
Armin Moharrer	Northeastern University, USA
Ioannis Mollas	Aristotle University of Thessaloniki, Greece
Carlos Monserrat-Aranda	Universität Politècnica de València, Spain
Konda Reddy Mopuri	Indian Institute of Technology Guwahati, India
Raha Moraffah	Arizona State University, USA
Pawel Morawiecki	Polish Academy of Sciences, Poland
Ahmadreza Mosallanezhad	Arizona State University, USA
Davide Mottin	Aarhus University, Denmark
Koyel Mukherjee	Adobe Research, India
Maximilian Münch	University of Applied Sciences Würzburg, Germany
Fabricio Murai	Universidade Federal de Minas Gerais, Brazil
Taichi Murayama	NAIST, Japan

Stéphane Mussard	CHROME, France
Mohamed Nadif	Centre Borelli - Université Paris Cité, France
Cian Naik	University of Oxford, UK
Felipe Kenji Nakano	KU Leuven, Belgium
Mirco Nanni	ISTI-CNR Pisa, Italy
Apurva Narayan	University of Waterloo, Canada
Usman Naseem	University of Sydney, Australia
Gergely Nemeth	ELLIS Unit Alicante, Spain
Stefan Neumann	KTH Royal Institute of Technology, Sweden
Anna Nguyen	Karlsruhe Institute of Technology, Germany
Quan Nguyen	Washington University in St. Louis, USA
Thi Phuong Quyen Nguyen	University of Da Nang, Vietnam
Thu Nguyen	SimulaMet, Norway
Thu Trang Nguyen	University College Dublin, Ireland
Prajakta Nimbhorkar	Chennai Mathematical Institute, Chennai, India
Xuefei Ning	Tsinghua University, China
Ikuko Nishikawa	Ritsumeikan University, Japan
Hao Niu	KDDI Research, Inc., Japan
Paraskevi Nousi	Aristotle University of Thessaloniki, Greece
Erik Novak	Jožef Stefan Institute, Slovenia
Slawomir Nowaczyk	Halmstad University, Sweden
Aleksandra Nowak	Jagiellonian University, Poland
Eirini Ntoutsi	Freie Universität Berlin, Germany
Andreas Nürnberger	Magdeburg University, Germany
James O'Neill	University of Liverpool, UK
Lutz Oettershagen	University of Bonn, Germany
Tsuyoshi Okita	Kyushu Institute of Technology, Japan
Makoto Onizuka	Osaka University, Japan
Subba Reddy Oota	IIIT Hyderabad, India
María Óskarsdóttir	University of Reykjavík, Iceland
Aomar Osmani	PRES Sorbonne Paris Cité, France
Aljaz Osojnik	JSI, Slovenia
Shuichi Otake	National Institute of Informatics, Japan
Greger Ottosson	IBM, France
Zijing Ou	Sun Yat-sen University, China
Abdelkader Ouali	University of Caen Normandy, France
Latifa Oukhellou	IFSTTAR, France
Kai Ouyang	Tsinghua University, France
Andrei Paleyes	University of Cambridge, UK
Pankaj Pandey	Indian Institute of Technology Gandhinagar, India
Guansong Pang	Singapore Management University, Singapore
Pance Panov	Jožef Stefan Institute, Slovenia

Apostolos Papadopoulos	Aristotle University of Thessaloniki, Greece
Evangelos Papalexakis	UC Riverside, USA
Anna Pappa	Université Paris 8, France
Chanyoung Park	UIUC, USA
Haekyu Park	Georgia Institute of Technology, USA
Sanghyun Park	Yonsei University, South Korea
Luca Pasa	University of Padova, Italy
Kevin Pasini	IRT SystemX, France
Vincenzo Pasquadibisceglie	University of Bari Aldo Moro, Italy
Nikolaos Passalis	Aristotle University of Thessaloniki, Greece
Javier Pastorino	University of Colorado, Denver, USA
Kitsuchart Pasupa	King Mongkut's Institute of Technology, Thailand
Andrea Paudice	University of Milan, Italy
Anand Paul	Kyungpook National University, South Korea
Yulong Pei	TU Eindhoven, the Netherlands
Charlotte Pelletier	Université de Bretagne du Sud, France
Jaakko Peltonen	Tampere University, Finland
Ruggero Pensa	University of Torino, Italy
Fabiola Pereira	Federal University of Uberlandia, Brazil
Lucas Pereira	ITI, LARSyS, Técnico Lisboa, Portugal
Aritz Pérez	Basque Center for Applied Mathematics, Spain
Lorenzo Perini	KU Leuven, Belgium
Alan Perotti	CENTAI Institute, Italy
Michaël Perrot	Inria Lille, France
Matej Petkovic	Institute Jožef Stefan, Slovenia
Lukas Pfahler	TU Dortmund University, Germany
Nico Piatkowski	Fraunhofer IAIS, Germany
Francesco Piccialli	University of Naples Federico II, Italy
Gianvito Pio	University of Bari, Italy
Giuseppe Pirrò	Sapienza University of Rome, Italy
Marc Plantevit	EPITA, France
Konstantinos Pliakos	KU Leuven, Belgium
Matthias Pohl	Otto von Guericke University, Germany
Nicolas Posocco	EURA NOVA, Belgium
Cedric Pradalier	GeorgiaTech Lorraine, France
Paul Prasse	University of Potsdam, Germany
Mahardhika Pratama	University of South Australia, Australia
Francesca Pratesi	ISTI - CNR, Italy
Steven Prestwich	University College Cork, Ireland
Giulia Preti	CentAI, Italy
Philippe Preux	Inria, France
Shalini Priya	Oak Ridge National Laboratory, USA

Ricardo Prudencio	Universidade Federal de Pernambuco, Brazil
Luca Putelli	Università degli Studi di Brescia, Italy
Peter van der Putten	Leiden University, the Netherlands
Chuan Qin	Baidu, China
Jixiang Qing	Ghent University, Belgium
Jolin Qu	Western Sydney University, Australia
Nicolas Quesada	Polytechnique Montreal, Canada
Teeradaj Racharak	Japan Advanced Institute of Science and Technology, Japan
Krystian Radlak	Warsaw University of Technology, Poland
Sandro Radovanovic	University of Belgrade, Serbia
Md Masudur Rahman	Purdue University, USA
Ankita Raj	Indian Institute of Technology Delhi, India
Herilalaina Rakotoarison	Inria, France
Alexander Rakowski	Hasso Plattner Institute, Germany
Jan Ramon	Inria, France
Sascha Ranftl	Graz University of Technology, Austria
Aleksandra Rashkovska Koceva	Jožef Stefan Institute, Slovenia
S. Ravi	Biocomplexity Institute, USA
Jesse Read	Ecole Polytechnique, France
David Reich	Universität Potsdam, Germany
Marina Reyboz	CEA, LIST, France
Pedro Ribeiro	University of Porto, Portugal
Rita P. Ribeiro	University of Porto, Portugal
Piera Riccio	ELLIS Unit Alicante Foundation, Spain
Christophe Rigotti	INSA Lyon, France
Matteo Riondato	Amherst College, USA
Mateus Riva	Telecom ParisTech, France
Kit Rodolfa	CMU, USA
Christophe Rodrigues	DVRC Pôle Universitaire Léonard de Vinci, France
Simon Rodríguez-Santana	ICMAT, Spain
Gaetano Rossiello	IBM Research, USA
Mohammad Rostami	University of Southern California, USA
Franz Rothlauf	Mainz Universität, Germany
Celine Rouveirol	Université Paris-Nord, France
Arjun Roy	Freie Universität Berlin, Germany
Joze Rozanec	Josef Stefan International Postgraduate School, Slovenia
Salvatore Ruggieri	University of Pisa, Italy
Marko Ruman	UTIA, AV CR, Czechia
Ellen Rushe	University College Dublin, Ireland

Dawid Rymarczyk	Jagiellonian University, Poland
Amal Saadallah	TU Dortmund, Germany
Khaled Mohammed Saifuddin	Georgia State University, USA
Hajer Salem	AUDENSIEL, France
Francesco Salvetti	Politecnico di Torino, Italy
Roberto Santana	University of the Basque Country (UPV/EHU), Spain
KC Santosh	University of South Dakota, USA
Somdeb Sarkhel	Adobe, USA
Yuya Sasaki	Osaka University, Japan
Yücel Saygın	Sabancı Universitesi, Turkey
Patrick Schäfer	Humboldt-Universität zu Berlin, Germany
Alexander Schiendorfer	Technische Hochschule Ingolstadt, Germany
Peter Schlicht	Volkswagen Group Research, Germany
Daniel Schmidt	Monash University, Australia
Johannes Schneider	University of Liechtenstein, Liechtenstein
Steven Schockaert	Cardiff University, UK
Jens Schreiber	University of Kassel, Germany
Matthias Schubert	Ludwig-Maximilians-Universität München, Germany
Alexander Schulz	CITEC, Bielefeld University, Germany
Jan-Philipp Schulze	Fraunhofer AISEC, Germany
Andreas Schwung	Fachhochschule Südwestfalen, Germany
Vasile-Marian Scuturici	LIRIS, France
Raquel Sebastião	IEETA/DETI-UA, Portugal
Stanislav Selitskiy	University of Bedfordshire, UK
Edoardo Serra	Boise State University, USA
Lorenzo Severini	UniCredit, R&D Dept., Italy
Tapan Shah	GE, USA
Ammar Shaker	NEC Laboratories Europe, Germany
Shiv Shankar	University of Massachusetts, USA
Junming Shao	University of Electronic Science and Technology, China
Kartik Sharma	Georgia Institute of Technology, USA
Manali Sharma	Samsung, USA
Ariona Shashaj	Network Contacts, Italy
Betty Shea	University of British Columbia, Canada
Chengchao Shen	Central South University, China
Hailan Shen	Central South University, China
Jiawei Sheng	Chinese Academy of Sciences, China
Yongpan Sheng	Southwest University, China
Chongyang Shi	Beijing Institute of Technology, China

Zhengxiang Shi	University College London, UK
Naman Shukla	Deepair LLC, USA
Pablo Silva	Dell Technologies, Brazil
Simeon Simoff	Western Sydney University, Australia
Maneesh Singh	Motive Technologies, USA
Nikhil Singh	MIT Media Lab, USA
Sarath Sivaprasad	IIIT Hyderabad, India
Elena Sizikova	NYU, USA
Andrzej Skowron	University of Warsaw, Poland
Blaz Skrlj	Institute Jožef Stefan, Slovenia
Oliver Snow	Simon Fraser University, Canada
Jonas Soenen	KU Leuven, Belgium
Nataliya Sokolovska	Sorbonne University, France
K. M. A. Solaiman	Purdue University, USA
Shuangyong Song	Jing Dong, China
Zixing Song	The Chinese University of Hong Kong, China
Tiberiu Sosea	University of Illinois at Chicago, USA
Arnaud Soulet	University of Tours, France
Lucas Souza	UFRJ, Brazil
Jens Sparsø	Technical University of Denmark, Denmark
Vivek Srivastava	TCS Research, USA
Marija Stanojevic	Temple University, USA
Jerzy Stefanowski	Poznan University of Technology, Poland
Simon Stieber	University of Augsburg, Germany
Jinyan Su	University of Electronic Science and Technology, China
Yongduo Sui	University of Science and Technology of China, China
Huiyan Sun	Jilin University, China
Yuwei Sun	University of Tokyo/RIKEN AIP, Japan
Gokul Swamy	Amazon, USA
Maryam Tabar	Pennsylvania State University, USA
Anika Tabassum	Virginia Tech, USA
Shazia Tabassum	INESCTEC, Portugal
Koji Tabata	Hokkaido University, Japan
Andrea Tagarelli	DIMES, University of Calabria, Italy
Etienne Tajeuna	Université de Laval, Canada
Acar Tamersoy	NortonLifeLock Research Group, USA
Chang Wei Tan	Monash University, Australia
Cheng Tan	Westlake University, China
Feilong Tang	Shanghai Jiao Tong University, China
Feng Tao	Volvo Cars, USA

Youming Tao	Shandong University, China
Martin Tappler	Graz University of Technology, Austria
Garth Tarr	University of Sydney, Australia
Mohammad Tayebi	Simon Fraser University, Canada
Anastasios Tefas	Aristotle University of Thessaloniki, Greece
Maguelonne Teisseire	INRAE - UMR Tetis, France
Stefano Teso	University of Trento, Italy
Olivier Teste	IRIT, University of Toulouse, France
Maximilian Thiessen	TU Wien, Austria
Eleftherios Tiakas	Aristotle University of Thessaloniki, Greece
Hongda Tian	University of Technology Sydney, Australia
Alessandro Tibo	Aalborg University, Denmark
Aditya Srinivas Timmaraju	Facebook, USA
Christos Tjortjis	International Hellenic University, Greece
Ljupco Todorovski	University of Ljubljana, Slovenia
Laszlo Toka	BME, Hungary
Ancy Tom	University of Minnesota, Twin Cities, USA
Panagiotis Traganitis	Michigan State University, USA
Cuong Tran	Syracuse University, USA
Minh-Tuan Tran	KAIST, South Korea
Giovanni Trappolini	Sapienza University of Rome, Italy
Volker Tresp	LMU, Germany
Yu-Chee Tseng	National Yang Ming Chiao Tung University, Taiwan
Maria Tzelepi	Aristotle University of Thessaloniki, Greece
Willy Ugarte	University of Applied Sciences (UPC), Peru
Antti Ukkonen	University of Helsinki, Finland
Abhishek Kumar Umrawal	Purdue University, USA
Athena Vakal	Aristotle University, Greece
Matias Valdenegro Toro	University of Groningen, the Netherlands
Maaike Van Roy	KU Leuven, Belgium
Dinh Van Tran	University of Freiburg, Germany
Fabio Vandin	University of Padova, Italy
Valerie Vaquet	CITEC, Bielefeld University, Germany
Iraklis Varlamis	Harokopio University of Athens, Greece
Santiago Velasco-Forero	MINES ParisTech, France
Bruno Veloso	Porto, Portugal
Dmytro Velychko	Carl von Ossietzky Universität Oldenburg, Germany
Sreekanth Vempati	Myntra, India
Sebastián Ventura Soto	University of Cordoba, Portugal
Rosana Veroneze	LBiC, Brazil

Jan Verwaeren	Ghent University, Belgium
Vassilios Verykios	Hellenic Open University, Greece
Herna Viktor	University of Ottawa, Canada
João Vinagre	LIAAD - INESC TEC, Portugal
Fabio Vitale	Centai Institute, Italy
Vasiliki Voukelatou	ISTI - CNR, Italy
Dong Quan Vu	Safran Tech, France
Maxime Wabartha	McGill University, Canada
Tomasz Walkowiak	Wroclaw University of Science and Technology, Poland
Vijay Walunj	University of Missouri-Kansas City, USA
Michael Wand	University of Mainz, Germany
Beilun Wang	Southeast University, China
Chang-Dong Wang	Sun Yat-sen University, China
Daheng Wang	Amazon, USA
Deng-Bao Wang	Southeast University, China
Di Wang	Nanyang Technological University, Singapore
Di Wang	KAUST, Saudi Arabia
Fu Wang	University of Exeter, UK
Hao Wang	Nanyang Technological University, Singapore
Hao Wang	Louisiana State University, USA
Hao Wang	University of Science and Technology of China, China
Hongwei Wang	University of Illinois Urbana-Champaign, USA
Hui Wang	SKLSDE, China
Hui (Wendy) Wang	Stevens Institute of Technology, USA
Jia Wang	Xi'an Jiaotong-Liverpool University, China
Jing Wang	Beijing Jiaotong University, China
Junxiang Wang	Emory University, USA
Qing Wang	IBM Research, USA
Rongguang Wang	University of Pennsylvania, USA
Ruoyu Wang	Shanghai Jiao Tong University, China
Ruxin Wang	Shenzhen Institutes of Advanced Technology, China
Senzhang Wang	Central South University, China
Shoujin Wang	Macquarie University, Australia
Xi Wang	Chinese Academy of Sciences, China
Yanchen Wang	Georgetown University, USA
Ye Wang	Chongqing University, China
Ye Wang	National University of Singapore, Singapore
Yifei Wang	Peking University, China
Yongqing Wang	Chinese Academy of Sciences, China

Yuandong Wang	Tsinghua University, China
Yue Wang	Microsoft Research, USA
Yun Cheng Wang	University of Southern California, USA
Zhaonan Wang	University of Tokyo, Japan
Zhaoxia Wang	SMU, Singapore
Zhiwei Wang	University of Chinese Academy of Sciences, China
Zihan Wang	Shandong University, China
Zijie J. Wang	Georgia Tech, USA
Dilusha Weeraddana	CSIRO, Australia
Pascal Welke	University of Bonn, Germany
Tobias Weller	University of Mannheim, Germany
Jörg Wicker	University of Auckland, New Zealand
Lena Wiese	Goethe University Frankfurt, Germany
Michael Wilbur	Vanderbilt University, USA
Moritz Wolter	Bonn University, Germany
Bin Wu	Beijing University of Posts and Telecommunications, China
Bo Wu	Renmin University of China, China
Jiancan Wu	University of Science and Technology of China, China
Jiantao Wu	University of Jinan, China
Ou Wu	Tianjin University, China
Yang Wu	Chinese Academy of Sciences, China
Yiqing Wu	University of Chinese Academic of Science, China
Yuejia Wu	Inner Mongolia University, China
Bin Xiao	University of Ottawa, Canada
Zhiwen Xiao	Southwest Jiaotong University, China
Ruobing Xie	WeChat, Tencent, China
Zikang Xiong	Purdue University, USA
Depeng Xu	University of North Carolina at Charlotte, USA
Jian Xu	Citadel, USA
Jiarong Xu	Fudan University, China
Kunpeng Xu	University of Sherbrooke, Canada
Ning Xu	Southeast University, China
Xianghong Xu	Tsinghua University, China
Sangeeta Yadav	Indian Institute of Science, India
Mehrdad Yaghoobi	University of Edinburgh, UK
Makoto Yamada	RIKEN AIP/Kyoto University, Japan
Akihiro Yamaguchi	Toshiba Corporation, Japan
Anil Yaman	Vrije Universiteit Amsterdam, the Netherlands

Hao Yan	Washington University in St Louis, USA
Qiao Yan	Shenzhen University, China
Chuang Yang	University of Tokyo, Japan
Deqing Yang	Fudan University, China
Haitian Yang	Chinese Academy of Sciences, China
Renchi Yang	National University of Singapore, Singapore
Shaofu Yang	Southeast University, China
Yang Yang	Nanjing University of Science and Technology, China
Yang Yang	Northwestern University, USA
Yiyang Yang	Guangdong University of Technology, China
Yu Yang	The Hong Kong Polytechnic University, China
Peng Yao	University of Science and Technology of China, China
Vithya Yogarajan	University of Auckland, New Zealand
Tetsuya Yoshida	Nara Women's University, Japan
Hong Yu	Chongqing Laboratory of Comput. Intelligence, China
Wenjian Yu	Tsinghua University, China
Yanwei Yu	Ocean University of China, China
Ziqiang Yu	Yantai University, China
Sha Yuan	Beijing Academy of Artificial Intelligence, China
Shuhan Yuan	Utah State University, USA
Mingxuan Yue	Google, USA
Aras Yurtman	KU Leuven, Belgium
Nayyar Zaidi	Deakin University, Australia
Zelin Zang	Zhejiang University & Westlake University, China
Masoumeh Zareapoor	Shanghai Jiao Tong University, China
Hanqing Zeng	USC, USA
Tieyong Zeng	The Chinese University of Hong Kong, China
Bin Zhang	South China University of Technology, China
Bob Zhang	University of Macau, Macao, China
Hang Zhang	National University of Defense Technology, China
Huaizheng Zhang	Nanyang Technological University, Singapore
Jiangwei Zhang	Tencent, China
Jinwei Zhang	Cornell University, USA
Jun Zhang	Tsinghua University, China
Lei Zhang	Virginia Tech, USA
Luxin Zhang	Worldline/Inria, France
Mimi Zhang	Trinity College Dublin, Ireland
Qi Zhang	University of Technology Sydney, Australia

Qiyiwen Zhang	University of Pennsylvania, USA
Teng Zhang	Huazhong University of Science and Technology, China
Tianle Zhang	University of Exeter, UK
Xuan Zhang	Renmin University of China, China
Yang Zhang	University of Science and Technology of China, China
Yaqian Zhang	University of Waikato, New Zealand
Yu Zhang	University of Illinois at Urbana-Champaign, USA
Zhengbo Zhang	Beihang University, China
Zhiyuan Zhang	Peking University, China
Heng Zhao	Shenzhen Technology University, China
Mia Zhao	Airbnb, USA
Tong Zhao	Snap Inc., USA
Qinkai Zheng	Tsinghua University, China
Xiangping Zheng	Renmin University of China, China
Bingxin Zhou	University of Sydney, Australia
Bo Zhou	Baidu, Inc., China
Min Zhou	Huawei Technologies, China
Zhipeng Zhou	University of Science and Technology of China, China
Hui Zhu	Chinese Academy of Sciences, China
Kenny Zhu	SJTU, China
Lingwei Zhu	Nara Institute of Science and Technology, Japan
Mengying Zhu	Zhejiang University, China
Renbo Zhu	Peking University, China
Yanmin Zhu	Shanghai Jiao Tong University, China
Yifan Zhu	Tsinghua University, China
Bartosz Zieliński	Jagiellonian University, Poland
Sebastian Ziesche	Bosch Center for Artificial Intelligence, Germany
Indre Zliobaite	University of Helsinki, Finland
Gianlucca Zuin	UFM, Brazil

Program Committee Members, Demo Track

Hesam Amoualian	WholeSoft Market, France
Georgios Balikas	Salesforce, France
Giannis Bekoulis	Vrije Universiteit Brussel, Belgium
Ludovico Boratto	University of Cagliari, Italy
Michelangelo Ceci	University of Bari, Italy
Abdulkadir Celikkanat	Technical University of Denmark, Denmark

Tania Cerquitelli	Informatica Politecnico di Torino, Italy
Mel Chekol	Utrecht University, the Netherlands
Charalampos Chelmis	University at Albany, USA
Yagmur Gizem Cinar	Amazon, France
Eustache Diemert	Criteo AI Lab, France
Sophie Fellenz	TU Kaiserslautern, Germany
James Foulds	University of Maryland, Baltimore County, USA
Jhony H. Giraldo	Télécom Paris, France
Parantapa Goswami	Rakuten Institute of Technology, Rakuten Group, Japan
Derek Greene	University College Dublin, Ireland
Lili Jiang	Umeå University, Sweden
Bikash Joshi	Elsevier, the Netherlands
Alexander Jung	Aalto University, Finland
Zekarias Kefato	KTH Royal Institute of Technology, Sweden
Ilkcan Keles	Aalborg University, Denmark
Sammy Khalife	Johns Hopkins University, USA
Tuan Le	New Mexico State University, USA
Ye Liu	Salesforce, USA
Fragkiskos Malliaros	CentraleSupelec, France
Hamid Mirisaee	AMLRightSource, France
Robert Moro	Kempelen Institute of Intelligent Technologies, Slovakia
Iosif Mporas	University of Hertfordshire, UK
Giannis Nikolentzos	Ecole Polytechnique, France
Eirini Ntoutsi	Freie Universität Berlin, Germany
Frans Oliehoek	Delft University of Technology, the Netherlands
Nora Ouzir	CentraleSupélec, France
Özlem Özgöbek	Norwegian University of Science and Technology, Norway
Manos Papagelis	York University, UK
Shichao Pei	University of Notre Dame, USA
Botao Peng	Chinese Academy of Sciences, China
Antonia Saravanou	National and Kapodistrian University of Athens, Greece
Rik Sarkar	University of Edinburgh, UK
Vera Shalaeva	Inria Lille-Nord, France
Kostas Stefanidis	Tampere University, Finland
Nikolaos Tziortziotis	Jellyfish, France
Davide Vega	Uppsala University, Sweden
Sagar Verma	CentraleSupelec, France
Yanhao Wang	East China Normal University, China

Zhirong Yang — Norwegian University of Science and Technology, Norway
Xiangyu Zhao — City University of Hong Kong, Hong Kong, China

Sponsors

ASML

Google

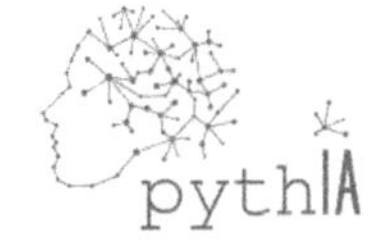

Contents – Part V

Supervised Learning

Probabilistic Inference

Sustainability

Supervised Learning

LCDB 1.0: An Extensive Learning Curves Database for Classification Tasks

Felix Mohr[1], Tom J. Viering[2(✉)], Marco Loog[2,3], and Jan N. van Rijn[4]

[1] Universidad de La Sabana, Chía, Colombia
felix.mohr@unisabana.edu.co
[2] Delft University of Technology, Delft, The Netherlands
{t.j.viering,m.loog}@tudelft.nl
[3] University of Copenhagen, Copenhagen, Denmark
[4] Leiden University, Leiden, The Netherlands
j.n.van.rijn@liacs.leidenuniv.nl

Abstract. The use of learning curves for decision making in supervised machine learning is standard practice, yet understanding of their behavior is rather limited. To facilitate a deepening of our knowledge, we introduce the Learning Curve Database (LCDB), which contains empirical learning curves of 20 classification algorithms on 246 datasets. One of the LCDB's unique strength is that it contains all (probabilistic) predictions, which allows for building learning curves of arbitrary metrics. Moreover, it unifies the properties of similar high quality databases in that it (i) defines clean splits between training, validation, and test data, (ii) provides training times, and (iii) provides an API for convenient access (pip install lcdb). We demonstrate the utility of LCDB by analyzing some learning curve phenomena, such as convexity, monotonicity, peaking, and curve shapes. Improving our understanding of these matters is essential for efficient use of learning curves for model selection, speeding up model training, and to determine the value of more training data.

Keywords: Learning curves · AutoML · Meta-learning

1 Introduction

Learning curves provide the prediction performance of a learning algorithm against the dataset size it has used for training. Such curves provide essential information for decision making [27,35]. For example, they can be extrapolated to determine the value of gathering more data or can be used to speed up training by selecting a smaller dataset size that still reaches sufficient accuracy. In addition, learning curves can provide useful information for model selection [2,26]. Particularly important questions concern the performance in the limit and the training set size at which the learning curves of two algorithms cross, as this can

Supplementary Information The online version contains supplementary material available at https://doi.org/10.1007/978-3-031-26419-1_1.

© The Author(s), under exclusive license to Springer Nature Switzerland AG 2023
M.-R. Amini et al. (Eds.): ECML PKDD 2022, LNAI 13717, pp. 3–19, 2023.
https://doi.org/10.1007/978-3-031-26419-1_1

tell us when one learning algorithm should be preferred over the other. To make use of learning curves for these purposes, it is essential that we know their shape (exponential, power law, etc.) so that we can reliably extrapolate or interpolate them for these tasks.

Unfortunately, there is a gap between common assumptions made by methods used for the above purposes and empirical evidence that would justify those assumptions. For instance, a common assumption is that learning curves behave well, i.e., more data leads to better performance. This assumptions is made for example in the extrapolation of learning curves for decision making [8,17,26]. However, a recent study [35] has collected a variety of results from literature, which illustrate that learning curves can display surprising shapes, such as multiple local minima, peaks, or curves that deteriorate with more data. These examples all illustrate that, in practice, there are clear gaps in our understanding of learning curves and their potential behavior. This limits the practical use of such curves, since correct extrapolations or interpolations depend crucially on the accuracy of learning curve models.

Empirical knowledge on learning curves is surprisingly scarce and often afflicted with severe limitations. Two recent surveys [27,35] give an extensive overview of the learning curve literature including empirical studies. One of the main problems of current studies is that the number of datasets and/or algorithms considered is small: Gu et al. [14] study two classifiers and eight datasets, Perlich et al. [30] examine two classifiers and 36 datasets, Li [23] use eight classifiers on three datasets, and recently Brumen et al. [3] considered four classifiers on 130 datasets. Other issues include that (i) the data acquired is not openly shared, hence not easily accessible, (ii) the focus is on specific performance metrics, or (iii) only a single train/test split is considered, providing a weak estimate for the out-of-sample learning curve and limiting the reliability of analysis.

To overcome these limitations and facilitate research on learning curves and their behavior, this work introduces the Learning Curve Database (LCDB), a high quality and extensive database of classification task learning curves. The current version of LCDB provides already over 150 GB of *ground truth and prediction vectors* of 20 classification algorithms from the scikit-learn library on 246 datasets. These prediction vectors have been recorded for models being trained on an amount of instances (called anchors) that are rounded multiples of $\sqrt{2}$. Instead of training each learner only once at each anchor, we created 25 splits in order to be able to obtain more robust insights into the out-of-sample performance. This makes it, in various respects, the biggest database for learning curves available today.

Note that, in the context of neural networks, a database called LCBench (Learning Curve Benchmark) [37] contains the performance of neural nets versus the number of epochs. Following [35] we call such curves *training curves*, to illustrate the difference with our learning curves which plot performance versus training set size. Because the LCDB investigates learning curves, and LCBench investigates training curves, these databases are incompareable.

LCDB is the first to provide (probabilistic) predictions together with the ground-truth labels. The availability of probabilistic predictions makes it possible to compute one's own choice of metrics, like AUROC or log-loss, rather then have to deal with precomputed ones. Curve data is provided for 25 stratified splits at each anchor for training, validation, and test data, enabling the construction of different curve types. Moreover, runtimes are provided for model training and prediction to study its dependence on sample size.

Other benefits of LCDB are that it enables (i) the a-posteriori analysis of the *shape* of curves for different learners and of function classes describing such shapes, (ii) the study of the relationship between training and test curves, (iii) the simultaneous analysis of learning curves, e.g., whether or not they intersect or if such intersection can be predicted, (iv) research into principled models for the runtime behavior of the algorithms, (v) benchmarking algorithm selection problems, and (vi) quick insights into the "difficulty" of datasets, which can be useful for the design of such benchmarks.

To showcase the utility of LCDB, we provide initial results on three of the above-mentioned analyses: (i) a study of the presence of three shape properties: monotonicity, convexity, and peaking, (ii) the identification of crossing behavior for all pairs of learners on all datasets, and (iii) an analysis of the goodness of fit of a set of model classes, both for the case of capturing the whole curve and predicting higher anchor performance from a set of lower anchor performances. A major insight of the study is that there is great support for the hypothesis that error-rate curves are largely (even though usually not perfectly) monotonic, convex, and mostly free of peaks (double descent). A second major insight is that typical learning curve models considered in literature, such as the 2 or 3 parameter power law, exponential and logarithmic models, may be significantly outperformed by 4 parameter models when used for extrapolation. We find that `mmf4` performs the best overall with `wbl4` a close second, but our results should be interpreted with care, as we ran into various issues with fitting.

2 The Learning Curve Database

To understand and motivate the way how LCDB is designed, Sect. 2.1 briefly recalls the formal definition of the learning curves and terminology. While an intuition on learning curves is common sense, recent surveys [27,35] show that there is a variety of performance curves (with similar terminology), which can quickly lead to confusion on what the exact subject of interest is. Our learning curves plot generalization performance versus training set size.

2.1 Formal Background on Learning Curves

Out-of-Sample Learning Curves. We consider learning curves in the context of supervised machine learning. We follow the definition of Mohr and van Rijn [27], which assumes some *instance space* $\mathcal{X}$ and a *label space* $\mathcal{Y}$. A *dataset* $D \subset \{(x, y) \mid x \in \mathcal{X}, y \in \mathcal{Y}\}$ is a *finite* relation between the instance space and the label space. We denote with $\mathcal{D}$ the set of all possible datasets. A *learning algorithm* is a function $a : \mathcal{D} \times \Omega \to H$, where $H = \{h \mid h : \mathcal{X} \to \mathcal{Y}\}$ is the space of hypotheses and Ω is a source of randomness, such as the random seed.

The performance of a *learning algorithm* is the performance of the hypothesis it produces. For a hypothesis h, this performance is measured as $\mathcal{R}_{out}(h) := \int_{\mathcal{X},\mathcal{Y}} loss(y, h(x))\, d\mathbb{P}_{\mathcal{X}\times\mathcal{Y}}$. Here, $loss(y, h(x)) \in \mathbb{R}$ is the penalty for predicting $h(x)$ for instance $x \in \mathcal{X}$ when the true label is $y \in \mathcal{Y}$, and $\mathbb{P}_{\mathcal{X}\times\mathcal{Y}}$ is the joint probability measure $\mathbb{P}_{\mathcal{X}\times\mathcal{Y}}$ on $\mathcal{X} \times \mathcal{Y}$ underlying the analyzed data. To assess the performance of a learner a, we average the performance once more over the input data and randomness of the learner, which determine the hypothesis:

$$\mathcal{C}(a, s) = \int \mathcal{R}_{out}(a(D_{tr}, \omega))d\mathbb{P}^s_{\mathcal{X}\times\mathcal{Y}}d\mathbb{P}_{\Omega}, \tag{1}$$

where $d\mathbb{P}^s_{\mathcal{X}\times\mathcal{Y}}$ is the distribution over i.i.d. sampled training sets D_{tr} of size s.

Considering Eq. (1) as a function of the number of observations for a fixed learner a yields a *learning curve* of learner a. That is, the observation learning curve is the function $\mathcal{C}(a, \cdot) : \mathbb{N} \to \mathbb{R}$, which is a *sequence* of performances, one for each training size. There are other types of closely related curves, most notably those based on the number of *iterations* or *epochs* used during training (training curves), which we do not consider. See [27,35] for details.

Empirical Learning Curves. While we are generally interested in the *true* learning curve defined in Eq. (1), we cannot determine it in practice. First, the out-of-sample error $\mathcal{R}_{out}$ is unknown because the measure $\mathbb{P}_{\mathcal{X}\times\mathcal{Y}}$ is unknown. Next, the necessity to average over the oftentimes uncountable set of all possible train sets and random seeds adds additional problems.

In practice, we therefore have to rely on *empirical learning curves*. An empirical learning curve is any set of estimates of a true learning curve at different training set sizes, which are called *anchors*. At anchor s, this estimate is obtained by (i) creating one (hold-out) or several (cross-validation) splits (D_{tr}, D_{te}) such that $|D_{tr}| = s$, (ii) obtaining for each such split a hypothesis via $h = a(D_{tr}, \omega)$ using a unique random seed each time, and (iii) then computing the *empirical risk* $\mathcal{R}_{in}(h, \tilde{D}) := \frac{1}{|\tilde{D}|}\sum_{(x,y)\in\tilde{D}} loss(y, h(x))$. Averaging these estimates for $\tilde{D} = D_{te}$ yields an estimate of the curve in Eq. (1) at anchor s:

$$\widehat{\mathcal{C}}(s) := \widehat{\mathcal{C}}(a, D, s) = \frac{1}{k}\sum_{i=1}^{k} \mathcal{R}_{in}(a(D^i_{tr}, \omega_i), D^i_{te}). \tag{2}$$

Since it is usually clear from the context which learner a and dataset D are used, we typically omit them in the notation and only write $\widehat{\mathcal{C}}(s)$. Note that we can also use $\tilde{D} = D_{tr}$ to approximate a train curve, or even use a mixture of both, instead of the test performance curve.

LCDB stores, for all points in D, the ground truth and prediction obtained from a trained learner (hypothesis) $a(D^i_{tr}, \omega_i)$ for various splits. With this, the empirical learning curve in Eq. (2) can be recovered for any concrete measure $\mathcal{R}_{in}$, and replacing D^i_{te} with D^i_{tr} also allows us to compute the curve of training performances. Thereby and in contrast to previous works relying only on one split [3], LCDB can be used to approximate the out-of-sample curve more reliably.

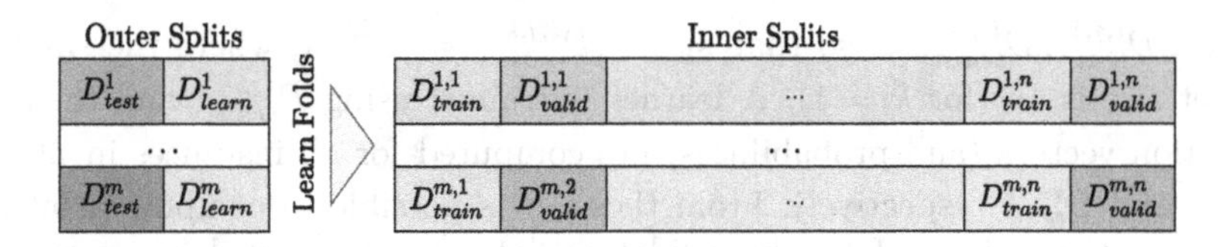

Fig. 1. Procedure to create training data for learners in LCDB: First an 90%/10% outer split is sampled. The 90% learning data fold is further sampled into different independent 90%/10% splits for essential training and validation respectively. Training data at different anchors is sampled independently from $D^{i,j}_{train}$.

2.2 Data Collection Methodology

The goal of LCDB is to provide learning curves for *model selection processes.* In model selection, it is usually not enough to have test data, but at time of selecting a model, one typically evaluates models, which requires *validation data.* If the model selection technique itself is supposed to be cross-validated, one arrives at a concept called *nested cross validation.* To simulate such processes, we need an explicit separation between training, validation and test data.

With this in mind, LCDB does *not* increase the data available for training in a monotonic way, as for example employed in [3], but assumes *independent* training portions at every anchor. That is, when training a classifier on the set S_1 of instances and later on the set S_2 of instances, where $|S_1| < |S_2|$, then we do not want that $S_2 \supset S_1$. Instead, we conceive that the training samples used for S_1 and S_2 are drawn *independently* from a pool of available training instances. While a monotonic approach makes sense in the context of data acquisition, for model selection we are interested in estimates of the out-of-sample performance, which is not compatible with systematic dependencies of the training data used at different points on the learning curve.

To reach this goal and also incorporate the idea of validation data, LCDB is based on nested splits itself. The procedure is illustrated in Fig. 1. On each dataset, we consider m *outer* splits of 90%/10%. These splits help to assess the performance of, say, model selection approaches in a simulated cross-validation. Note that all splits in LCDB are stratified. The test folds do not need to be disjoint and we simply create several independent hold-out splits. Since we need to store all the predictions, the test fold is essentially bounded to 5000 instances in LCDB, which we consider to be large enough for a satisfying approximation. Now let $(D^i_{learn}, D^i_{test})$ the i-th such split. From the learning set D^i_{learn}, n further *inner* splits of 90%/10% are derived. The inner splits can serve, for instance, to simulate cross-validations conducted by the model selection approach itself. Let $D^{i,j} = (D^{i,j}_{train}, D^{i,j}_{valid})$ be the j-th such inner split derived from the i-th outer split. Note that [3] neither compute outer nor several inner splits but basically create *one* 80%/20% split.

For each training set, we compute the prediction vectors and probabilities according to a geometric schedule [31] for all three sets: training, validation, and test. Formally, for each train set $D^{i,j}_{train}$ with $1 \leq i \leq m, 1 \leq j \leq n$, we create a series $D_{train}^{i,j,1}, D_{train}^{i,j,2}, \ldots$ so that $s_k := |D_{train}^{i,j,k}| = \lceil 2^{\frac{7+k}{2}} \rceil$, where the smallest dataset size is 16 (for $k = 1$). A learner is trained using $D_{train}^{i,j,k}$, and then the prediction vectors (and probabilities) are computed for all instances in $D_{train}^{i,j,k}$, $D_{valid}^{i,j}$, and D_{test}^{i}, respectively. From these, it is possible to compute arbitrary metrics on the training data, the validation data, and the test data afterwards. In the initial setup of LCDB, we have $m = n = 5$; further splits can be reliably added in the future thanks to the seeded architecture of the LCDB code.

Note that, at the time of working with LCDB, the distinction between outer and inner splits is optional and can be omitted if one simply wants to get empirical learning curves for learners. To obtain the performance at a particular anchor, one can proceed as follows. For each of the $m \times n$ train sets $D_{train}^{i,j,k}$ for anchor s_k, the prediction vectors of the validation set $D_{valid}^{i,j}$ and the test set D_{test}^{i} after having trained on $D_{train}^{i,j,k}$ are available. These can simply be merged in order to compute some metric, say, the error rate at s_k. This gives $m \cdot n$ performance estimates for each anchor.

2.3 Datasets

In this initial version, LCDB contains learning curves for 246 datasets. The large majority of these datasets was already used in AutoML [9] or is part of published benchmarks [13]. Our main criterion for the source of the data is API-based reproducibility, in order to enable a closed algorithm that is able to fully reproduce results without the need of manually downloading the datasets. To this end, we chose OpenML.org [34] as a source, which, in contrast to other repositories like UCI, offers an official API in both Java and Python. The datasets themselves represent a large range over various properties such as the number of instances, features, and classes. For details, we refer to the supplement.

Many datasets need to be preprocessed before learning can take place. To cope with that, we applied two (admittedly arbitrary) pre-processing steps. First, all missing values were replaced by the column median for numerical and by the column mode for categorical attributes. Then, all categorical attributes were binarized into a Bernoulli encoding. Features were not scaled, as we have not implemented pipelines yet. Classifiers sensitive to scaling of features (such as Nearest Neighbours, SVM's, etc.) may be disadvantaged.

2.4 Classifiers

In this initial work, we considered only classifiers from the scikit-learn library [29]. Using their default hyper-parameters, the 20 considered classifiers are (we here show class names and, in parentheses, abbreviations used in figures and the discussion): SVC (linear, poly, rbf, sigmoid), LinearDiscriminantAnalysis (LDA), QuadraticDiscriminantAnalysis (QDA), ExtraTreesClassifier (Extra Trees), GradientBoostingClassifier (Grad. Boost), RandomForestClassifier (RF), LogisticRegression (LR), PassiveAggressiveClassifier (PA), Perceptron, RidgeClassifier

(Ridge), SGDClassifier (SGD), BernoulliNB, MultinomialNB, KNeighborsClassifier (kNN), MLPClassifier (MLP), DecisionTreeClassifier (DT), ExtraTreeClassifier (Extra Tree). Note the difference between ExtraTreeClassifier (single tree) and ExtraTreesClassifier (ensemble).

2.5 Availability via PyPI (pip)

The learning curves are available[1] in two formats. First, there are the raw sets of ground truth and prediction vectors, which can be downloaded in separate zip files, one for each dataset. These files are space-consuming as for some datasets, the learning curve files accumulate to more than 10GB when unpacked. The second option is to download learning curves that have been pre-compiled for specific metrics such as accuracy, log-loss, and the F1-measure. In the latter format, the data can also be accessed directly through an API in Python via a package that can be installed with `pip install lcdb`.

3 Illustrative Usage of LCDB

To efficiently perform model selection, or to estimate the added value of more data collection, etc., it is essential to develop good and accurate insight into learning curve behaviour. We consider the following three, relevant questions.

1. What is the probability, for each learner, to exhibit a learning curve that is (i) monotone, (ii) convex, or (iii) peaked?
2. What is the probability that learner A starts off worse than learner B, but exhibits better performance on the full data, i.e., the learning curve of A crosses the one of B (from above)?
3. What is the goodness of fit/extrapolation performance of some learning curve models when interpolating or extrapolating the curve?

We focus on error-rate curves, but LCDB can easily facilitate other metrics. Our interest is in answering the above questions for the true, i.e., out-of-sample learning curves, as defined in Eq. (1), and LCDB allows us to estimate these by averaged empirical curves $\widehat{\mathcal{C}}$ as per Eq. (2). The mean is here formed across both the different outer and inner splits of the data. Since the initial version of LCDB comes with $5 \cdot 5 = 25$ training sets, we get a much better estimate of the mean compared to previous works, such as [3] that only consider a single split. In what follows, we formulate the necessary definitions immediately in terms of $\widehat{\mathcal{C}}$ as a proxy of $\mathcal{C}$.

3.1 Curve Properties: Monotonicity, Convexity, and Peaking

Background. Knowledge about properties like monotonicity [24], convexity [26], and peaking is important to justify particular extrapolation techniques

[1] Raw data: https://openml.github.io/lcdb/; Code: https://github.com/fmohr/lcdb.

based on optimistic extrapolation [26] or curve models that assume such shapes, like the inverse power law, exponential, or logarithmic functions [8,17]. We here consider decreasing learning curve models as typically expected for error rates, but the modification of our experiments to alternative models is straightforward.

While all three properties are binary and either are or are not present for a concrete function, it is helpful to consider some *degree* of monotonicity and of convexity. For instance, if the learning curve only has some tiny oscillation somewhere in the curve but is monotone and convex anywhere else, we would still want to consider it largely well-behaved, even though not perfect.

Clearly, there is no unique way in which a degree of monotonicity or convexity can be measured and we merely propose some specific operationalizations. Whether or not these are meaningful eventually depends on the context of application. Now, assuming descending curves (an assumption implicit in its original definition), we take as the degree of violation of monotonicity the highest positive increase observed at any anchor on the empirical curve:

$$\epsilon_{mono} = \max_{s_i, i<T} \left\{ \max\left(0, \widehat{C}(s_{i+1}) - \widehat{C}(s_i)\right) \right\},$$

where T is the highest available anchor index. Similarly, we compute such a violation for convexity:

$$\epsilon_{conv} = \max_{s_i, i<T-1} \left\{ \max\left(0, \widehat{C}(s_{i+1}) - \tfrac{1}{2}\left(\widehat{C}(s_i) + \widehat{C}(s_{i+2})\right)\right) \right\}.$$

We note that, for learning curves over a finite range of anchors, monotonicity and convexity are mutually neither necessary nor sufficient criteria. To illustrate this, the bottom right plot in Fig. 2 shows two examples of learning curves that are monotone but not convex and two examples that are convex but not monotonically decreasing.

Peaking [35], also called sample-wise double descent/ascent, is a classic phenomenon, currently studied in neural networks [25,28]. It refers to the curve showing worsened performance, after initial performance improvements, before eventually starting to consistently improve again at larger anchor sizes. It has been observed for different types of performance curves, including learning curves as functions of the sample size used for training [28]. In a sense, peaking is a special form of non-monotonicity that often occurs at a specific location in the learning curve, typically around the point where model complexity and sample size coincide. Given its prominence, we study it also separately. While one could, in principle, measure the extent of temporary deterioration, we here choose to measure whether such a phenomenon occurs or not in a binary fashion:

$$peak = \left[\!\left[\exists i < T : \widehat{C}(s_i) < \widehat{C}(s_{i+1})) \wedge \left(\exists u > i, \forall v \geq u : \widehat{C}(s_{v-1}) > \widehat{C}(s_v))\right) \right]\!\right].$$

Results. The top row of Fig. 2 summarizes the monotonicty (left) and convexity (right) results over all learners and datasets, effectively giving the general cumulative distribution of the maximum violation. The different colors show

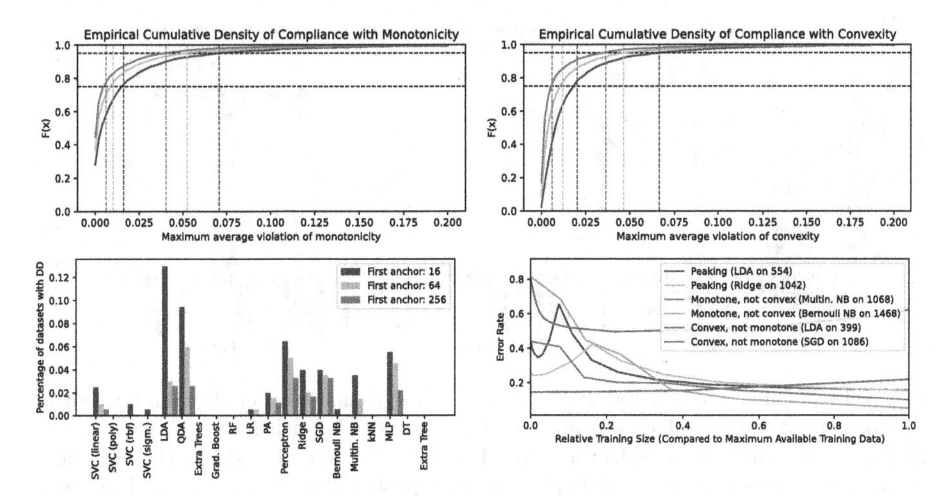

Fig. 2. Insights into shape properties of learning curves. Numbers in legend are dataset IDs at OpenML.org. Top row: cumulative distribution of the degree of violation of monotonicity (left) and concavity (right) for different entry points of analysis (blue, orange green for analysis starting at anchor 16, 64, and 256, respectively). Bottom left: Relative frequency (across datasets) of peaking per learner. Bottom right: Some sample learning curves with non-standard properties. (Color figure online)

different entry points of analysis. Clearly, if we were to consider learning curves starting from a single training instance, we would expect high variation in the observations and possibly non-monotonicities even in the true curve. For this reason, it can be meaningful to start to study the learning curve only past some minimum training size [20,21,32]. As can be seen in the figure, the probability of violations is significantly reduced with increasing entry point from 16 (blue) to 256 (green) via 64 (orange).

The bottom left panel of Fig. 2 provides insights into peaking. It shows, per learner, how often peaking can be observed, i.e., the relative number of datasets on which it is observed. As can be seen, sample-wise peaking in error-rate curves is rather rare and many learners do not exhibit this property at all. The right plot shows some example of curves where peaking was detected. It also shows two examples failing to be monotone and two failing to be convex.

3.2 Learning Curve Crossing

We now consider the probability that some learner A will outperform learner B at the highest available training size (effectively $|D^{i,j}_{train}|$) given that it has a worse initial performance.

The results are summarized in Fig. 3. It shows that there is indeed quite a number of algorithms that start off worse than others, but recover and exceed the performance of the competitor. In particular, for learners commonly considered strong, such as Gradient Boosting or Random Forests, there is a high probability

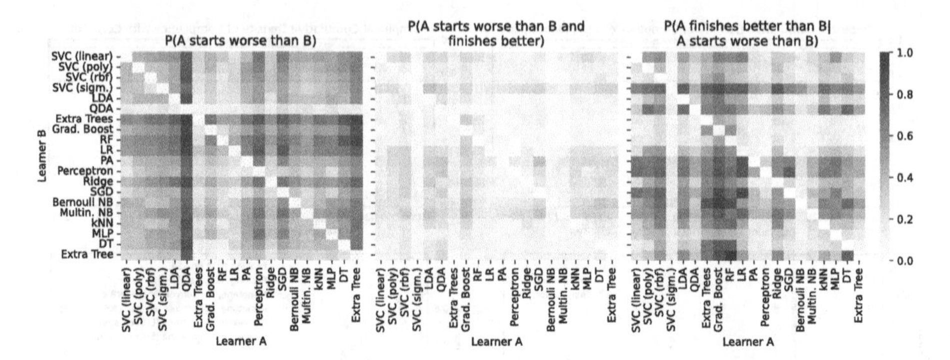

Fig. 3. Probabilities associated with the two events (i) that learner A (columns) starts off worse than learner B (rows) and (ii) that learner A finishes better than learner B. For each combination of learners for A and B, the left plot shows the empirical probability of event (i), the middle plot the joint probability of events (i) and (ii), and the right plot the probability of (ii) given (i). Darker colors show higher probabilities. (Color figure online)

to eventually outperform the other learners (dark green columns). The opposite is hardly ever the case (light green rows). For other algorithms, the opposite is the case, e.g., the Perceptron hardly ever takes the lead if it starts off bad (light green column), but is often outperformed in the long run, even if starts off comparatively well (dark green row).

The above formulation focuses on the value of the learning curve at the maximum training size available, but it could be the case that one or even both of the learning curves have not yet converged at that training size. In another view, one could ask for extrapolation models and then answer the same question as above, based on some potential training size that exceeds the available data.

3.3 Learning Curve Fitting and Extrapolation

Background. Many parametric models have been proposed in literature for modeling learning curves [35]. Typical examples are power laws, exponentials and logarithmic models. Table 1 provides an overview, where `last1` is a baseline that takes the last point on the learning curve that it observes, and always predicts this value. This is a simple baseline that is also used in [22] for example.

Several benchmarks have been performed to determine the best parametric model for empirical learning curves [3,5,10,14,18,19,33], but all have their limitations [35]. First, most studies benchmark only 3 to 6 curve models [3], or leave out models with bias terms, so that non-zero asymptotic cannot be modeled [10]. A comprehensive experiments encompassing all models proposed remains missing, making any conclusive comparison difficult. Additional complications arise from the fact that some studies [6,10,19,33] fit exponentials or power laws by transforming the data and performing the usual least squares procedure in the transformed space [35]. Finally, not all works evaluate the results of curve

Table 1. The 16 learning curve models under consideration. x is the size of the training set and a, b, c, d are parameters to be estimated. Some model performance increase rather than decrease. last1 is a baseline (horizontal line), the number in the abbreviation indicates the number of parameters.

Model	Formula	Used in	Model	Formula	Used in
`last1`	a	[22]	`vap3`	$\exp\left(a + \frac{b}{x} + c\log(x)\right)$	[14]
`pow2`	$-ax^{-b}$	[10,14,19,33]	`expp3`	$c - \exp\left((-b + x)^{a}\right)$	[18]
`log2`	$-a\log(x) + b$	[3,4,10,14,19,33]	`expd3`	$c - (-a + c)\, exp(-bx)$	[18]
`exp2`	$a\exp(-bx)$	[10,19,33]	`logpow3`	$a/\left((x\exp(-b))^{c} + 1\right)$	[8]
`lin2`	$ax + b$	[4,10,19,33]	`pow4`	$a - b\,(d + x)^{-c}$	[18]
`ilog2`	$-a/\log(x) + b$	[18]	`mmf4`	$(ab + cx^{d})/(b + x^{d})$	[14]
`pow3`	$a - bx^{-c}$	[14,18]	`wbl4`	$-b\exp(-ax^{d}) + c$	[14]
`exp3`	$a\exp(-bx) + c$	[3,4,18]	`exp4`	$c - \exp(-ax^{d} + b)$	[18]

fitting with help of statistical tests [14,18,19], which casts some doubt on the significance of their findings. Most studies found evidence that learning curves are modeled best by `pow2`, `pow3`, `log2` or `exp3` [35].

In this work, we therefore aim to compare all parametric models for learning curves so far suggested in the literature, perform model evaluation on unseen curve data (to avoid overfitting concerns [35]), and we aim to use a careful fitting approach, which is the same for all models (except exp4 - see Supplement), and perform statistical tests to determine significance of our findings.

We evaluate curve fitting in terms of the Mean Squared Error (MSE) with (1) the points seen during curve fitting (train anchors, so interpolation), (2) the points that are not seen during fitting (test anchors, extrapolation), (3) extrapolating to the last anchor. This is a common procedure [3,14] and we use averaged curves (25 splits). One thing to note is that curve fitting is more difficult than one may expect, to cope with this we use 5 random restarts and we discard failed fits (MSE>100). In the following, we focus on the main findings; an extensive report with many more details can be found in the Supplement.

Results. Our main findings are summarized in Figs. 4 and 5 and Table 2. Since the semantics differ between the figures, we discuss them one at a time.

Figure 4 shows the performance for the different curve models in the form of (empirical) cumulative densities with respect to the prediction error. Here, we aggregate over models that were fit with all available anchors up to a size of 20% of the maximum available training size for fitting. Both plots show the curves that are at some point of time on top of the others in solid and all the others (somewhat dominated models) in dotted curves. The left plot shows how well the models can accommodate to the anchors they have seen (the training MSE). Unsurprisingly, we find that the more parameters the better the fits, except for `exp3` who performs as well as the 2 parameter models.

The right plot shows how well each model can predict the performance for all upcoming anchors (on average). Here, roles have changed somewhat. It indicates

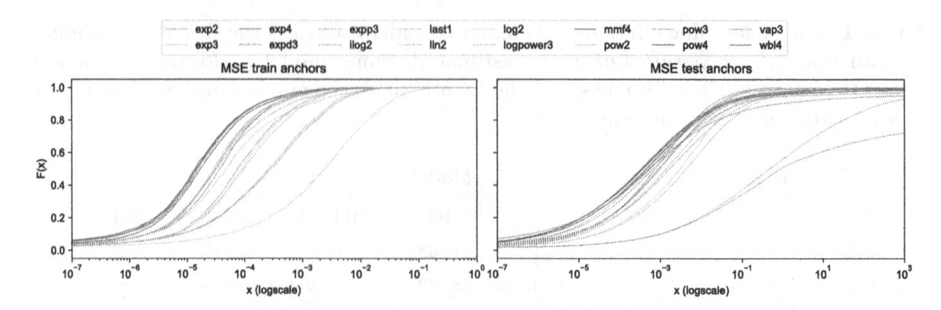

Fig. 4. The Cumulative Density Functions (CDF) for the Mean Squared Error (MSE) for all curve models for interpolation on the train anchors and extrapolation to all test anchors, summarized over all curve fitting experiments.

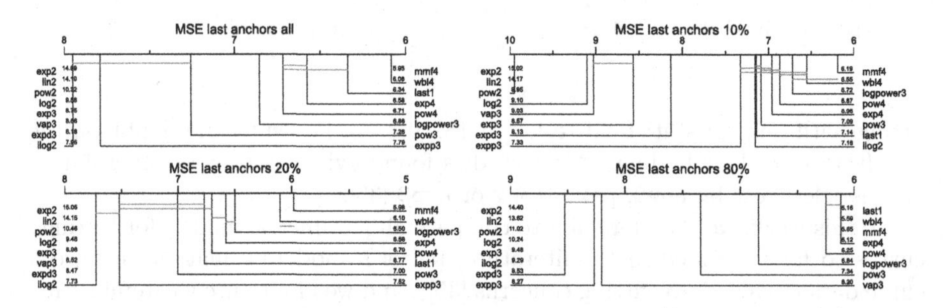

Fig. 5. Critical diagrams for the ranks for the extrapolation to the last anchor. "all" considers all experiments, 10% fits learning curves up to 10% of the total dataset, etc. If two are connected by a red line, the pairwise test did not find significant differences between their performance. If a model doesn't have a line to the axis it's significantly different from all others. Numbers indicate rank (lower is better). (Color figure online)

`lin2`, `exp2` are definitely not suitable for extrapolation, and to a lesser extend we can also already rule out `pow2`.

Since these plots are very aggregated and do not say anything about significance, we visualize ranks in Critical Diagrams (CD) [7] following the approach of [16] in Fig. 5. These figures show the ranking of the different models (lower values are better) while statistically not significantly different ones are tied with red bars. We use the same statistical tests as [3,16], Friedman's test [11] to judge if there are significant differences between curve models, and we use pairwise Wilcoxon signed-rank tests [36] to compare pair of curve models with Holm's alpha correction [15] following [16] with $\alpha = 0.05$.

The obtained ranks from the Friedman tests are given in Table 2. We partition all curve fitting experiments into 6 sets: "all", where all experiments are used, "5%", where all anchors up to exactly 5% of the training set size are used for fitting, "10%", "20%", "40%" and "80%". For all these partitions and performance measures we find significant differences from the Friedman test.

Table 2. Summarized ranks according to the Friedman test for the squared loss to extrapolation to the last anchor. "all" considers all experiments, 10% fits a learning curves up to 10% of the total dataset, etc. Blue/larger numbers means a worse rank, yellow/smaller number indicate a better rank. The last row gives the rank for the MSE on the train anchors over all experiments.

curve	last1	pow2	log2	exp2	lin2	ilog2	pow3	exp3	vap3	expp3	expd3	logp3	pow4	mmf4	wbl4	exp4
all	6.34	10.32	9.58	14.89	14.10	7.95	7.26	8.75	8.66	7.79	8.18	6.86	5.95	6.08	6.58	6.71
5%	7.27	9.28	8.72	14.68	13.93	6.82	7.41	8.23	9.48	7.24	7.88	7.25	6.48	6.91	7.28	7.14
10%	7.14	9.95	9.10	15.02	14.17	7.16	7.09	8.57	9.03	7.33	8.13	6.72	6.19	6.55	6.96	6.87
20%	6.77	10.46	9.48	15.05	14.15	7.73	7.00	8.98	8.52	7.52	8.47	6.50	5.98	6.10	6.58	6.70
40%	6.12	10.74	9.79	14.84	13.92	8.45	7.19	9.30	8.41	7.88	8.54	6.59	5.69	5.77	6.33	6.44
80%	5.16	11.02	10.24	14.40	13.62	9.29	7.34	9.48	8.20	8.27	8.53	6.84	5.65	5.59	6.12	6.25
trn	15.58	11.89	11.51	13.65	12.85	11.25	7.67	8.60	7.04	7.59	7.02	7.60	3.28	2.64	3.76	4.07

In Fig. 5 the critical diagrams are shown (the ranks in this Figure correspond with those in Table 2). However, due to space limitations, the figure only shows the ranks for extrapolations to the *last* anchor; the supplement also contains CD plots for curve fitting and extrapolation performance across the whole upcoming curve (all test anchors) and interpolation on the train anchors. The results for extrapolating to all test anchors is quite similar to extrapolation to the last anchor, but differences between models are larger for the latter.

In accordance with the previous observations, we again see that `exp2` and `lin2` indeed do not perform well, and the performances `log2` and `pow2` are only slightly better. That is unsurprising, since `pow2` and `exp2` cannot converge to a non-zero error rate, and `log2` diverges in the limit. Surprisingly, `exp3`, which can model non-zero error in the limit, obtains a similar ranking and gets progressively worse with more training anchors, indicating it is indeed not suitable. These models are followed by `vap3`, `expp3`, `expd3` that obtain similar ranks.

There is also a group that often tie and attain, generally, much better ranks. `pow3`, `ilog2`, `logpower3` especially works well for small sample sizes and tie often. Nevertheless, for larger sample sizes especially `ilog2` and `logpower3` deteriorate a lot. If little anchors are used it is hard to distinguish performances. However, if more than 20% of the data is used, a new group of overall best model appears, which are `mmf4`, `wbl4`, `pow4`, `exp4`. These models tie often according to the pairwise tests, but `mmf4` and `wbl4` together significantly outperforms the others, especially for 20%-80%. Finally, the baseline `last1` doesn't perform well for less than 20% data, but improves its rank the larger the percentage, and even wins significantly from all others at 80%. This is expected since the curve often plateaus for large sample sizes and no fitting is done. For 'all' `mmf4` performs significantly better than all others with `wbl4` second and `last1` third.

4 Discussion and Conclusion

The learning curve database (LCDB) provides the first extensive collection of learning curves easily accessible and readily deployable for the in-depth study of learning curves behavior. Importantly, the database provides probabilistic predictions that allows the study a wide range of standard and specialized performance metrics. The most important aspects of these learning curves have been utilized in a PyPI package, for easy and fast usage.

Our preliminary study of LCDB already provides some insights: for error-rate learning curves, we found that the large majority of learning curves is, largely, well-behaved, in that they are monotone, convex, and do not show peaking. We also established empirical estimates of the probability that learning curves cross. Furthermore, our curve fitting experiments emphasize the need for more robust fitting. While we have taken some steps to rule out biases from curve fitting issues, our curve fitting results warrant further analysis to rule this out completely. In contrast to other benchmarks that generally extrapolate curves with 2 or 3 parameter power laws, exponentials or logarithmic models, we find that 4 parameter models are quite competitive, with `mmf4` obtaining the best overall results and `wbl4` a close second. Not surprisingly, the amount of anchors used for training seems to influence which curve model performs best. Of course, many research questions remain, which we hope can be successfully addressed with LCDB. For instance, does it make sense to smooth learning curves for model selection or curve extrapolation? Can meta-features [2], like the number of instances, be used to reliably predict (i) whether curves intersect, (ii) if they are monotone or convex, (iii) or which curve model will be accurate? Which non-parametric extrapolation techniques work best and in what case? In how far can training curves support the extrapolation task? Finally, our paper did not touch upon the aspect of training runtimes, which are however also part of LCDB and can be used to develop sophisticated runtime models. These and many additional interesting learning curve problems can now be efficiently investigated.

LCDB is designed to be a continual database that can and will be extended over time and also integrated with other services. Needless to say that we will seek to add new learning curves to LCDB for new datasets. The main challenge here is to assure that the training runtimes will be comparable: while all the data in the initial database has been generated with the same hardware, this is not necessarily the case for upcoming datasets. Addressing this issue, e.g., by normalizing the runtimes based on calibration techniques is an interesting research question in itself. To further improve the exploitation of LCDB, parts of it will be integrated with OpenML.org in the immediate future.

It should be clear that there are also various technical extensions that could be desirable. First and most natural, an extension towards other learning problems like regression would be useful. Next, LCDB should be extended to cover machine learning *pipelines* instead of learners only. This is however a tough undertaking since the space of pipelines is large, and it is not clear how to select a reasonable subset of those; also storing the prediction vectors for an increased number of learners is a logistic issue. Third, an extension of LCDB for monotonically increasing training sets would be great in order to allow analyses for data acquisition situations instead of only model selection situations as covered currently. In all, we expect that LCDB will be of great use to address many further interesting and valuable research questions concerning learning curves.

References

1. Benavoli, A., Corani, G., Mangili, F.: Should we really use post-hoc tests based on mean-ranks? J. Mach. Learn. Res. **17**(1), 152–161 (2016)
2. Brazdil, P., van Rijn, J.N., Soares, C., Vanschoren, J.: Metalearning: Applications to Automated Machine Learning and Data Mining, 2nd edn. Springer, Cham (2022). https://doi.org/10.1007/978-3-030-67024-5
3. Brumen, B., Cernezel, A., Bosnjak, L.: Overview of machine learning process modelling. Entropy **23**(9), 1123 (2021)
4. Brumen, B., Rozman, I., Heričko, M., Černezel, A., Hölbl, M.: Best-fit learning curve model for the C4.5 algorithm. Informatica **25**(3), 385–399 (2014)
5. Cohn, D., Tesauro, G.: Can neural networks do better than the Vapnik-Chervonenkis bounds? In: Advances in Neural Information Processing Systems, vol. 3, pp. 911–917 (1991)
6. Cortes, C., Jackel, L.D., Solla, S.A., Vapnik, V., Denker, J.S.: Learning curves: asymptotic values and rate of convergence. In: Advances in Neural Information Processing Systems, vol. 6, pp. 327–334. Morgan Kaufmann (1993)
7. Demšar, J.: Statistical comparisons of classifiers over multiple data sets. J. Mach. Learn. Res. **7**, 1–30 (2006)
8. Domhan, T., Springenberg, J.T., Hutter, F.: Speeding up automatic hyperparameter optimization of deep neural networks by extrapolation of learning curves. In: Proceedings of the Twenty-Fourth International Joint Conference on Artificial Intelligence, pp. 3460–3468. AAAI Press (2015)
9. Feurer, M., Klein, A., Eggensperger, K., Springenberg, J.T., Blum, M., Hutter, F.: Efficient and robust automated machine learning. In: Advances in Neural Information Processing Systems, vol. 28. pp. 2962–2970 (2015)
10. Frey, L.J., Fisher, D.H.: Modeling decision tree performance with the power law. In: Proceedings of the Seventh International Workshop on Artificial Intelligence and Statistics. Society for Artificial Intelligence and Statistics (1999)
11. Friedman, M.: A comparison of alternative tests of significance for the problem of m rankings. Ann. Math. Stat. **11**(1), 86–92 (1940)
12. Garcia, S., Herrera, F.: An extension on "statistical comparisons of classifiers over multiple data sets" for all pairwise comparisons. J. Mach. Learn. Res. **9**(89), 2677–2694 (2008)
13. Gijsbers, P., LeDell, E., Thomas, J., Poirier, S., Bischl, B., Vanschoren, J.: An open source AutoML benchmark. arXiv preprint arXiv:1907.00909 (2019)
14. Gu, B., Hu, F., Liu, H.: Modelling classification performance for large data sets. In: Wang, X.S., Yu, G., Lu, H. (eds.) WAIM 2001. LNCS, vol. 2118, pp. 317–328. Springer, Heidelberg (2001). https://doi.org/10.1007/3-540-47714-4_29
15. Holm, S.: A simple sequentially rejective multiple test procedure. Scand. J. Stat. **6**, 65–70 (1979)
16. Ismail Fawaz, H., Forestier, G., Weber, J., Idoumghar, L., Muller, P.-A.: Deep learning for time series classification: a review. Data Min. Knowl. Disc. **33**(4), 917–963 (2019). https://doi.org/10.1007/s10618-019-00619-1

17. Klein, A., Falkner, S., Springenberg, J.T., Hutter, F.: Learning curve prediction with Bayesian neural networks. In: 5th International Conference on Learning Representations. OpenReview.net (2017)
18. Kolachina, P., Cancedda, N., Dymetman, M., Venkatapathy, S.: Prediction of learning curves in machine translation. In: The 50th Annual Meeting of the Association for Computational Linguistics, pp. 22–30. The Association for Computer Linguistics (2012)
19. Last, M.: Predicting and optimizing classifier utility with the power law. In: Workshops Proceedings of the 7th IEEE International Conference on Data Mining, pp. 219–224. IEEE Computer Society (2007)
20. Leite, R., Brazdil, P.: Improving progressive sampling via meta-learning on learning curves. In: Boulicaut, J.-F., Esposito, F., Giannotti, F., Pedreschi, D. (eds.) ECML 2004. LNCS (LNAI), vol. 3201, pp. 250–261. Springer, Heidelberg (2004). https://doi.org/10.1007/978-3-540-30115-8_25
21. Leite, R., Brazdil, P.: Selecting classifiers using metalearning with sampling landmarks and data characterization. In: Proceedings of the 2nd Planning to Learn Workshop (PlanLearn) at ICML/COLT/UAI, pp. 35–41 (2008)
22. Li, L., Jamieson, K.G., DeSalvo, G., Rostamizadeh, A., Talwalkar, A.: Hyperband: bandit-based configuration evaluation for hyperparameter optimization. In: 5th International Conference on Learning Representations. OpenReview.net (2017)
23. Li, Y.: An investigation of statistical learning curves: do we always need big data? Master's thesis, University of Canterbury (2017)
24. Loog, M., Viering, T., Mey, A.: Minimizers of the empirical risk and risk monotonicity. In: Advances in Neural Information Processing Systems, vol. 32, pp. 7478–7487 (2019)
25. Loog, M., Viering, T.J., Mey, A., Krijthe, J.H., Tax, D.M.J.: A brief prehistory of double descent. Proc. Natl. Acad. Sci. **117**(20), 10625–10626 (2020)
26. Mohr, F., van Rijn, J.N.: Towards model selection using learning curve cross-validation. In: 8th ICML Workshop on Automated Machine Learning (2021)
27. Mohr, F., van Rijn, J.N.: Learning curves for decision making in supervised machine learning - a survey. CoRR abs/2201.12150 (2022)
28. Nakkiran, P., Kaplun, G., Bansal, Y., Yang, T., Barak, B., Sutskever, I.: Deep double descent: where bigger models and more data hurt. In: 8th International Conference on Learning Representations. OpenReview.net (2020)
29. Pedregosa, F., et al.: Scikit-learn: machine learning in Python. J. Mach. Learn. Res. **12**, 2825–2830 (2011)
30. Perlich, C., Provost, F.J., Simonoff, J.S.: Tree induction vs. logistic regression: a learning-curve analysis. J. Mach. Learn. Res. **4**, 211–255 (2003)
31. Provost, F.J., Jensen, D.D., Oates, T.: Efficient progressive sampling. In: Proceedings of the Fifth ACM SIGKDD International Conference on Knowledge Discovery and Data Mining, pp. 23–32. ACM (1999)
32. van Rijn, J.N., Abdulrahman, S.M., Brazdil, P., Vanschoren, J.: Fast algorithm selection using learning curves. In: Fromont, E., De Bie, T., van Leeuwen, M. (eds.) IDA 2015. LNCS, vol. 9385, pp. 298–309. Springer, Cham (2015). https://doi.org/10.1007/978-3-319-24465-5_26
33. Singh, S.: Modeling performance of different classification methods: deviation from the power law. Project Report, Department of Computer Science, Vanderbilt University, USA (2005)
34. Vanschoren, J., van Rijn, J.N., Bischl, B., Torgo, L.: OpenML: networked science in machine learning. SIGKDD Explor. **15**(2), 49–60 (2014)

35. Viering, T.J., Loog, M.: The shape of learning curves: a review. CoRR abs/2103.10948 (2021)
36. Wilcoxon, F.: Individual comparisons by ranking methods. In: Kotz, S., Johnson, N.L. (eds.) Breakthroughs in Statistics, pp. 196–202. Springer, Cham (1992). https://doi.org/10.1007/978-1-4612-4380-9_16
37. Zimmer, L., Lindauer, M., Hutter, F.: Auto-PyTorch tabular: multi-fidelity MetaLearning for efficient and robust AutoDL. IEEE Trans. Pattern Anal. Mach. Intell. **43**(9), 3079–3090 (2021)

Factorized Structured Regression for Large-Scale Varying Coefficient Models

David Rügamer[1,2](✉), Andreas Bender[1], Simon Wiegrebe[1], Daniel Racek[1], Bernd Bischl[1], Christian L. Müller[1,3,4], and Clemens Stachl[5]

[1] Department of Statistics, LMU Munich, Munich, Germany
david.ruegamer@stat.uni-muenchen.de
[2] Institute of Statistics, RWTH Aachen, Aachen, Germany
[3] ICB, Helmholtz Zentrum Munich, Munich, Germany
[4] CCM, Flatiron Institute, New York, USA
[5] Institute of Behavioral Science and Technology, University of St. Gallen, St. Gallen, Switzerland

Abstract. Recommender Systems (RS) pervade many aspects of our everyday digital life. Proposed to work at scale, state-of-the-art RS allow the modeling of thousands of interactions and facilitate highly individualized recommendations. Conceptually, many RS can be viewed as instances of statistical regression models that incorporate complex feature effects and potentially non-Gaussian outcomes. Such structured regression models, including time-aware varying coefficients models, are, however, limited in their applicability to categorical effects and inclusion of a large number of interactions. Here, we propose Factorized Structured Regression (FaStR) for scalable varying coefficient models. FaStR overcomes limitations of general regression models for large-scale data by combining structured additive regression and factorization approaches in a neural network-based model implementation. This fusion provides a scalable framework for the estimation of statistical models in previously infeasible data settings. Empirical results confirm that the estimation of varying coefficients of our approach is on par with state-of-the-art regression techniques, while scaling notably better and also being competitive with other time-aware RS in terms of prediction performance. We illustrate FaStR's performance and interpretability on a large-scale behavioral study with smartphone user data.

Keywords: Recommender systems · Neural networks · Tensor regression · Time-varying effects · Generalized additive models

1 Introduction

From buying products online to selecting a movie to watch, recommender systems (RS) are part of our everyday life. RS are used to suggest those items that

---

Supplementary Information The online version contains supplementary material available at https://doi.org/10.1007/978-3-031-26419-1_2.

© The Author(s), under exclusive license to Springer Nature Switzerland AG 2023
M.-R. Amini et al. (Eds.): ECML PKDD 2022, LNAI 13717, pp. 20–35, 2023.
https://doi.org/10.1007/978-3-031-26419-1_2

are most appealing to a given user based on the user's past preference data or the similarity of a user to other users. One big advantage of RS is their scalability, as they allow for modeling thousands of interactions, e.g., between users and items, and thereby facilitate individual recommendations in many dimensions (see, e.g., [33] for a recent implementation framework). Many RS can be represented as a regression model with the user and item as covariates. This makes it straightforward to include further features into the model and extend the method by other structural components.

At the same time, the increasing amount of available data and the possibility to model increasingly complex data generating processes calls for efficient methods to fit flexible regression models on large-scale data sets with many observations and features. In the past, several advanced statistical regression models have been proposed to incorporate complex feature effects. One of the most common approaches are generalized additive models (GAMs), widely considered to be state-of-the-art (SotA) for statistical modeling [26]. These models allow the incorporation of time-varying feature effects and spatial effects, among others, and can also deal with non-Gaussian outcomes (see [26] for more details). While well-working adaptions of GAMs for large data scenarios exist [28], both methodology and software reach their limits when modelling categorical effects or categorical interactions of several variables where features comprise hundreds or thousands of categories. An amalgamation of methods from RS and statistical regression can overcome the limitations of statistical regression models on large-scale data sets with many categorical effects and interactions. In this work, we combine smoothing approaches with factorization terms to overcome the limitations of varying coefficient models for categorical features with many factor levels. Our idea arises from the statistical analysis of a large-scale behavioral dataset (Sect. 7). In this dataset, smartphone usage behavior of participants was tracked for several weeks. Domain experts are interested in various structured regression effects, such as the continuous activity levels over time. While standard regression software allows to fit some of these effects, several hundred activities and users make it infeasible to fit a model that learns interaction effects of users and activities or smoothly varying time effects for one or both of these categorical variables.

Our Contribution. We propose Factorized Structured Regression (FaStR) for scalable varying coefficient models. This combined approach has the flexibility that has proven successful in additive regression models, while also being able to deal with high-dimensional categorical effects and interactions. More specifically, we 1) derive a general model formulation in (4) to combine GAMs and factorization approaches, 2) derive a varying factorization interaction in Sect. 4.1 that reduces the number of parameters and therefore computations by a factor of $(I + U)/(IU)$ for given numbers of category levels I and U, and 3) propose an efficient implementation of this fusion approach that a) can reduce the storage cost by a factor of $1/(I + U)$ and b) circumvents computations quadratic in I and U by using stochastic optimization, an array reformulation, as well as dynamic feature encoding. In numerical experiments, we moreover show that our approach 4) leads to an estimation performance comparable with a SotA implementation and 5) has the desired computational complexity. Finally, we 6) demonstrate its interpretability and applicability to large-scale data sets.

2 Related Literature

Multiple different RS have been proposed over the last years, many of them based on matrix factorization (MF) [16] or collaborative filtering [24]. While recent methods increasingly rely on neural network-based factorization or recommendation, e.g., [29], it remains debatable whether they yield superior results, e.g., with respect to performance and efficiency [20]. Factorization Machines (FM) represent another line of research which is closely related to MF. Initially proposed by [19], FM are based on a linear model problem formulation with pairwise (or order $d \geq 3$) interactions between all features. In particular, the formulation as regression model is the basis for extensions to other (prediction) tasks, with many different FM-type models having been developed in recent years (e.g., [4]). An important influence (context variable) in RS is time. Various methods for controlling for short- and long-term temporal dynamics, cyclic patterns, drift or time decay exist [2]. While short-term approaches either divide time into smaller periods or integrate time features into the factorization of the neighbourhood, long-term effects are accounted for by some form of distance calculations between the current and other designated time points. Some approaches also combine the factorization with the time context, and for instance assume smoothness in the factorization, e.g., for video pixel completion [13]. Specific time-aware methods include collaborative filtering with temporal dynamics [15], dynamic MF [7], temporal regularized MF [30], or sequence-aware FM [6]. The common ground of these methods is to account for the time context in the factorization.

Statistical Approaches and Interpretability. Several combinations of statistical approaches and RS have been proposed in past years. Already in 1999, [8] proposed a (Bayesian) generalized mixed-effects model as RS Likelihood approximation approach. [31] proposed the GLMix model that combines the idea of generalized linear models (GLMs; [17]) with RS for large-scale response prediction. Our work is most similar to RS approaches which facilitate interpretability by making connections to generalized additive models (GAMs) [10]. In contrast to our approach, however, past work does not include smoothing splines directly into the models, nor does it address varying coefficient models. An exception is the time-varying tensor decomposition by [32] which is inspired by varying coefficient models. While similar in motivation, their work does not focus on scaling aspects and only considers approximate varying coefficients with separately learned basis coefficients. Our approach implements the full varying coefficient model with exact single-varying coefficients as well as a doubly-varying coefficient with jointly-learned latent basis coefficients.

3 Background

We will first describe the necessary background on factorization approaches in RS, structured additive regression models, and introduce our notation.

3.1 General Notation

In the following, we use $Y \sim \mathcal{F}$ to denote a random outcome value (e.g., a rating) from distribution $\mathcal{F}$ and its observation $y \in \mathcal{Y} \subseteq \mathbb{R}$ for which the model generates a prediction $\hat{y}$. We reserve the indices $i \in \mathcal{I}$ and $u \in \mathcal{U}$ for two categorical features (exemplarily referred to as *item* and *user*) and $t \in \mathcal{T}$ for the context variable *time* on a given time interval $\mathcal{T}$. The features associated with i and u are assumed to be binary indicator variables and are only implicitly referenced using their index. In Sect. 3.3, we however use an integer representation to introduce a memory-efficient storage representation. Other context features are summarized by $\boldsymbol{x} \in \mathbb{R}^p$. We use b, w and v to denote weights in the model that relate to items and users, and make their dependence explicit by indexing these weights correspondingly with i and u. To distinguish between dependencies of categorical features and the (continuous) feature t, we highlight time-dependency by writing objects as functions of the time t. We assume that we are given a dataset $\mathcal{D} = \{(y_{iu}(t), \boldsymbol{x}_{iu}(t))\}_{i \in \mathcal{I}, u \in \mathcal{U}, t \in \mathcal{T}}$ of total size N and allow observations to be sparse, i.e., for $\mathcal{D}$ to be a true subset of $\mathcal{I} \times \mathcal{U} \times \mathcal{T}$. For matrix computations in later sections, let $\boldsymbol{A} \odot \boldsymbol{B} \in \mathbb{R}^{N \times a \cdot b}$ define the row-wise tensor product (RWTP) of matrices $\boldsymbol{A} \in \mathbb{R}^{N \times a}$ and $\boldsymbol{B} \in \mathbb{R}^{N \times b}$, i.e., a Kronecker product applied to every pair of rows of both matrices $[\boldsymbol{A}_{[1,]} \otimes \boldsymbol{B}_{[1,]} \dots \boldsymbol{A}_{[N,]} \otimes \boldsymbol{A}_{[N,]}]$. Further, for $a = b$, let $\boldsymbol{A} \bullet \boldsymbol{B} := (\boldsymbol{A} * \boldsymbol{B})\mathbb{1}_b$, where $*$ is the Hadamard product and $\mathbb{1}_b \in \mathbb{R}^b$ a vector of ones. The operation defined by $\bullet$ can be exploited in models with Kronecker product structures such as array models for fast computation (see Sect. 4.2).

3.2 Model-Based Recommender Systems

The basic MF model generates its predictions as

$$\hat{y}_{iu} = \langle \boldsymbol{v}_{1,i}, \boldsymbol{v}_{2,u} \rangle \tag{1}$$

using a dot product $\langle \boldsymbol{v}_{1,i}, \boldsymbol{v}_{2,u} \rangle = \boldsymbol{v}_{1,i}^\top \boldsymbol{v}_{2,u}$ of two latent factors $\boldsymbol{v}_{1,i}, \boldsymbol{v}_{2,u} \in \mathbb{R}^D$ from a D-dimensional joint latent factor space. After learning the mapping from each item i and user u to the respective latent factor vector, the dot product describes the interplay between user and item and is used to estimate the outcome y (ratings). If the combination $\mathcal{I} \times \mathcal{U}$ is observed completely, common matrix decomposition approaches such as a singular value decomposition can be applied. If the matrix containing the ratings for all user-item combinations is sparse, missing values can be imputed. This, however, can be inaccurate and computationally expensive. The common alternative is to use (1) to only model $(i, u) \in \mathcal{K} \subset (\mathcal{I} \times \mathcal{U})$ in the set $\mathcal{K}$ of observed combinations. The solutions $\hat{y}$ can be found by least squares estimation where an additional L_2-penalty for $\boldsymbol{v}_i$ and $\boldsymbol{v}_u$ is typically added to the objective function [16]. In order to account for systematic user- and item-level trends, biases are further added to (1), yielding

$$\hat{y}_{iu} = \mu + b_i + b_u + \langle \boldsymbol{v}_{1,i}, \boldsymbol{v}_{2,u} \rangle, \tag{2}$$

where μ is a global intercept representing the average rating, and b_i, b_u represent the item and user tendencies. The latter two bias terms are again penalized using a ridge penalty. Together with this penalization, b_i and b_u can also be interpreted as a random effect for the item and user (see, e.g., [31]).

Time-Aware Recommender Systems. Contexts such as the location or time in which data has been observed can make a crucial difference (see, e.g., [3]). RS therefore often include a context dependency. One of the most common context-aware RS are time-aware RS [5]. Time-aware model-based approaches assume the following relationship

$$\hat{y}_{iu}(t) = \mu + b_i(t) + b_u(t) + \boldsymbol{v}_{2,u}^\top(t)\boldsymbol{v}_{1,i}, \tag{3}$$

where both biases and the dot product are time-dependent. The rationale behind a time-varying latent user effect is that users change their behavior over time, whereas influences of items should be time-independent [5,16]. While time is often assumed to be continuous, categorical time-aware models are used if time information is represented as discrete contextual values.

The time-varying latent user effect in (3) has a similar role as varying-coefficients in structured additive regression discussed in the following section.

3.3 Structured Additive Regression and Varying Coefficient Model

In statistical modeling, structured additive regression is a technique for estimating the relationships between a dependent variable (outcome value) and one or more independent variables (predictors, features). While the most common form of regression follows an additive structure as introduced in (2) and (3), in particular including linear effects $\boldsymbol{x}^\top\boldsymbol{\beta}$ of features $\boldsymbol{x}$ or pairwise interactions, factorization terms are usually not present in classical regression models. Instead, to adapt models for complex relationships between features and outcome value, smooth, non-linear, additive terms $f(\cdot)$ of one or more features are included into the model. These terms are represented by a linear combination of L appropriate basis functions. A univariate non-linear effect of feature z is, e.g., approximated by $f(z) \approx \sum_{l=1}^{L} B_l(z)w_l$, where $B_l(z)$ is the l-th basis function (such as regression splines, polynomial bases or B-splines) evaluated at z and w_l is the corresponding basis coefficient. Similarly, tensor product basis representations allow for two- or moderate-dimensional non-linear interactions.

One important part of additive regression is the so-called *varying coefficient* model [12]. The rationale for these models is the same as for time-varying RS: effects of features in the model naturally vary over time. Therefore these models include effects $xf(t)$, such that the effect (coefficient) of x is given by f evaluated at time t, and f is estimated from the data. A special case is a varying coefficient $f_i(t), i = 1, \dots, I$ where a separate function f_i is estimated for all I levels of a categorical feature. Existing software to model varying coefficients with smooth time-effects is, however, not scalable to features with many categories. The bottleneck is an RWTP of the matrix of evaluated basis functions $\boldsymbol{B} := (B_1(t) \dots B_L(t))$ and a (one hot-)encoded matrix for a categorical variable (e.g., item with I levels).

Computational Complexity. Assuming equal number of basis functions L for every smooth term f_i in a varying coefficient model with $i = 1, \dots, I$ levels and N observations, the storage required for the model matrix is $\mathcal{O}(NLI)$ and the computations $\mathcal{O}(N(LI)^2)$ (cf. [27]). Similar, for a model with an interaction effect of, e.g., item and user (with U levels), the storage is $\mathcal{O}(NLIU)$ and $\mathcal{O}(N(LIU)^2)$.

4 Factorized Structured Regression

In order to address the computational limitations of statistical regression techniques, we will first introduce the general idea to obtain predictions from a FaStR model and then go into more specific details and merits. We use the RS notation to define the model by means of a classical recommendation task with items, users, time context and an outcome y such as a rating. As in a typical regression setting, further features $\boldsymbol{x}$ might exist that the modeler is interested in. We assume that the outcome $Y_{iu}(t)$ and all features $\boldsymbol{x}_{iu}(t)$ are observed on a grid for item $i = 1, \dots, I$, user $u = 1, \dots, U$, time $t = 1, \dots, T$. While our approach also works for sparsely and irregularly observed data, we assume a grid of observations to simplify the notation. Conditional on the item, user, time and further features, $Y_{iu}(t)$ is assumed to follow a parametric distribution $\mathcal{F}$. We model the expectation of $Y_{iu}(t)$ as

$$
\begin{aligned}
\mathbb{E}(Y_{iu}(t)|\boldsymbol{x}_{iu}(t)) &= h(\eta_{iu}(t)), \\
\eta_{iu}(t) &= \mu + b_i + b_u + b_{iu} + f^{[0]}(t) + f_i^{[1]}(t) + f_u^{[2]}(t) + f_{iu}^{[3]}(t) \\
&\quad + \textstyle\sum_{o=1}^{O} g_o(\boldsymbol{x}_{iu}(t)).
\end{aligned} \tag{4}
$$

Here, h is an activation or response function mapping the additive predictor $\eta_{iu}(t)$ onto the correct domain (e.g., $\mathbb{R}^+$ for a positive outcome variable). All terms indicated with b are (regularized) bias terms. μ is a global bias term, b_i an item-specific bias, b_u a user-specific bias and b_{iu} an item-user-specific one. Terms denoted by f are smooth non-linear functions of time t represented by (penalized) basis functions. These include a global trend $f^{[0]}$, a subject and activity trend, $f^{[1]}$ and $f^{[2]}$, respectively, and a joint trend $f^{[3]}$. Additional covariates $\boldsymbol{x}$ can be modeled using other (smooth) functions g_o. In the Supplementary Material A we provide further details on smoothness penalties and model optimization.

The model in (4) can be seen as an alternative notation for a varying coefficient model, or also as a time-aware RS with additional exogenous terms. As $\mathcal{F}$ is not required to be Gaussian, it has, however, a more general applicability (e.g., binary, count or interval data). What further distinguishes (4) from existing approaches is the smoothness assumption of terms denoted with f, combined with the efficient implementation of terms $f^{[1]}$, $f^{[2]}$ and a factorization assumption for $f^{[3]}$. These aspects are explained in more detail in the following.

4.1 Varying Factorized Interactions

For high-dimensional data, such as the mobile phone data in our example, estimating the 2- or 3-way interaction terms is computationally not feasible. We thus propose to define $\eta_{iu}(t)$ in (4) using latent factorization representations. We therefore decompose the discrete interaction term(s) into an inner product

$$b_{iu} = \langle \boldsymbol{v}_{1,i}, \boldsymbol{v}_{2,u} \rangle = \sum_{d=1}^{D} v_{1,i,d} \cdot v_{2,u,d} \tag{5}$$

with $D \ll \min(I, U)$ latent factors, resulting in the estimation of $D \cdot (I + U)$ instead of $I \cdot U$ parameters. If $I = U = 1000$ and $D = 5$, for example, this reduces the number of parameters by a factor of 100 from 10^6 to 10^4. While this is the common approach to model interactions in factorization approaches, we here propose to proceed in a similar fashion to model time-dependent interactions and approximate time-varying interactions by a factorization approach:

$$f_{iu}^{[3]}(t) \approx \tilde{f}^{[3]}(t, \boldsymbol{V}_{1,i}, \boldsymbol{V}_{2,u}) = \sum_{l=1}^{L} B_l(t) \sum_{d=1}^{D} v_{1,i,l,d} \cdot v_{2,u,l,d}, \tag{6}$$

where $\boldsymbol{V}_{\cdot,\cdot} \in \mathbb{R}^{L \times D}$ are matrices with rows corresponding to the L basis functions for one categorical effect and columns to the D latent factors. In other words, we approximate the interaction of the smooth effect $f(t)$ of t and categorical variables i, u by a product of the non-linear basis $\boldsymbol{B}$ of dimension L and the two latent matrices $\boldsymbol{B}^\top(\boldsymbol{V}_{1,i} \bullet \boldsymbol{V}_{2,u})$, which can be computed efficiently for all N rows. The representation via latent factors requires the estimation of $L \cdot D \cdot (I + U)$ instead of $L \cdot D \cdot I \cdot U$ parameters (a multiplicative reduction of $(I + U)/(IU)$ in parameters and computations). This principle is general and can be applied to various types of additive effects, also of two or higher dimensions such as tensor-product splines or Markov random field smooths (see [26]). The dimension D is usually tuned to maximize prediction performance or estimation quality. While an increase in D will result in more parameters, the computational cost at this point only plays a minor role, as $D < 20$ often works well in practice.

Penalization. In order to enforce smoothness of the varying coefficients in the time-dimension, a quadratic Kronecker sum penalty J can be added to the loss function [26]. In a similar manner, we can promote smoothness of the latent factors $\boldsymbol{V}_{iu} = (\boldsymbol{V}_{1,i} \bullet \boldsymbol{V}_{2,u}) \in \mathbb{R}^L$ in our adaption using a symmetric difference penalty matrix $\boldsymbol{P}_V \in \mathbb{R}^{L \times L}$. $\boldsymbol{P}_V$ penalizes the time-dimension of the factorized varying-coefficients, where the penalized differences (its entries) depend on the chosen basis B. We further allow for an L_2-regularization of the latent factors in the i- and u-dimension, yielding

$$J = \lambda_t \cdot \boldsymbol{V}_{iu}^\top \boldsymbol{P}_V \boldsymbol{V}_{iu} + \lambda_{iu} \cdot (||\boldsymbol{V}_{1,i}||_F + ||\boldsymbol{V}_{2,u}||_F), \tag{7}$$

where λ_t controls the smoothness of the non-linearity in the direction of the time t, $||\cdot||_F$ is the Frobenius norm, and λ_{iu} the regularization for items and users.

4.2 Efficient Implementation

While the previous section allows to efficiently model (smooth) interactions of two or more categorical features with many categories, the factorization is not a solution for coefficients $f^{[1]}$ and $f^{[2]}$, as these only vary with a single category. Many use cases also require to estimate one effect for each of the levels of a categorical (interaction) effect. One bottleneck if I and/or U is large, is the computation of their dummy-encoded design matrices. For example, for item the matrix $\boldsymbol{X}_I$ of size $N \times I$ contains binary entries indicating which observation (rows) belong to which item category (column), and analogous for a user matrix $\boldsymbol{X}_U$. Second, the execution of operations involving such large matrices is another computational bottleneck. Computations become even more challenging if the model includes interactions, resulting in $\mathcal{O}(NIU)$ storage and $\mathcal{O}(N(IU)^2)$ computations (cf. Sect. 3.3). These interactions are created by calculating the RWTP between both matrices, i.e., $\boldsymbol{X}_I \odot \boldsymbol{X}_U \in \mathbb{R}^{N \times I \cdot U}$. To circumvent creating, storing and processing $\boldsymbol{X}_I$ and $\boldsymbol{X}_U$ as a whole, we propose two simple yet effective implementation tricks explained in the following.

Stochastic Optimization. The first bottleneck in computations of varying coefficient models at scale is the number of observations N. We therefore implement FaStR in a neural network and thereby can make use of stochastic gradient descent optimization routines with mini-batches of size $M \ll N$. This reduces the original cost of computations from $\mathcal{O}(N(IU)^2)$ to $\mathcal{O}(EMIU)$ where E is the number of model updates. It also allows us to leverage high-performance computing platforms such as TensorFlow [1] that support GPU computations.

Array Computations. The second bottleneck is computing the RWTP of two- or higher-dimensional interaction terms. Our proposal is to use an array reformulation that does not require to compute the RWTP design matrix in the first place. More specifically, a two-dimensional interaction effect $(\boldsymbol{X}_I \odot \boldsymbol{X}_U)\boldsymbol{w}$ with weights $\boldsymbol{w}$ can be equally represented by the array reformulation (cf. Sect. 3.1)

$$(\boldsymbol{X}_I \boldsymbol{W}) \bullet \boldsymbol{X}_U, \tag{8}$$

where $\boldsymbol{W} \in \mathbb{R}^{I \times U}$ is a matrix of weights with the ith row and uth column being the weight for the interaction of the ith level in $\boldsymbol{X}_I$ and the uth level in $\boldsymbol{X}_U$. By using (8) instead of a plain linear effect, we circumvent the construction of the large RWTP and the storage cost is reduced from $\mathcal{O}(NIU)$ to $\mathcal{O}(N(I+U))$ without increasing the time complexity (as the $\bullet$ operation for M observations is neglectable with $\mathcal{O}(MU)$ compared to the matrix multiplication with $\mathcal{O}(MIU)$). This array formulation can also be defined for higher-order interactions [9].

Dynamic Feature Encoding. Although array computations can reduce the storage problem notably by not constructing the RWTP in the first place, a third bottleneck is storing the large dummy-encoded matrices $\boldsymbol{X}_I$ and $\boldsymbol{X}_U$ themselves. We circumvent this extra space complexity, by evaluating categorical features dynamically during network training and only constructing the one-hot encoding for categorical features on a given mini-batch m. Thereby, only a matrix $\boldsymbol{X}_I^{(m)}$ of size $M \times I$ needs to be loaded into memory and the full $N \times I$ matrix is never created explicitly. This effectively reduces the storage from $\mathcal{O}(NIU)$ to $\mathcal{O}(N)$ (exactly $2N$ for two integer vectors). While this potentially results in redundant computations as it will create the encoding for a specific observation multiple times (if the number of epochs is greater than 1), deterministic integer encoding is cheap. Hence, the resulting computation overhead is usually neglectable and both forward- and backward pass can make use of the sparse representation.

When evaluating the two matrix operators (matrix product and $\bullet$) in (8) sequentially, encodings can again be created dynamically and the largest matrix involved only contains $\max(I, U)$ instead of $I \cdot U$ columns.

5 Numerical Experiments

Our numerical experiments investigate 1) whether FaStR can estimate (factorized) varying coefficients as proposed in Sect. 4.1 with performance comparable to other SotA methods, and 2) whether the model presented in (4) can be initialized and fitted with constant memory scaling w.r.t. the number of factor levels. In addition, in the Supplementary Material B we investigate whether FaStR can recover GAMs in general using our implementation techniques. Details on the data generating processes for each simulation study, method implementation and the computing environment are given in the Supplementary Material C. In the first experiment, we compare the estimated and true model coefficients using the mean squared error (MSE) and the estimated vs. the true functions f using the squared error, integrated and averaged over the domain of the predictor variables (MISE). We repeat the data generating process 10 times to account for variability in the data and model estimation.

5.1 Estimation of Factorized Smooth Interactions

We first investigate how well our models can recover smooth factorized terms from Sect. 4.1 of two categorical variables j and u, with 4 and 5 levels, respectively, i.e., 20 different smooth effects. We use a relatively small number of levels to be able to fit all possible interactions also in a classic structured regression approach. We define one of the two true factorized smooth effects as a varying coefficient term, i.e., $f_{1,iu}(t)$, and one stemming from an actual factorization, i.e., $f_2(t, \boldsymbol{V}_{1,i}, \boldsymbol{V}_{2,u})$. We use the Bernoulli and Gaussian distributions and investigate factorized terms in the distributions' mean for different data sizes $N \in \{500, 1000, 2000\}$. While the true model is generated using three latent factor dimensions, we additionally investigate a model with six dimensions to see how the misspecification of the latent dimension size influences estimation performance. FaStR is trained using a batch size of 250 with early stopping on 10% of the training data and a patience of 50 epochs. Larger batch sizes do not change estimation performance notably, but slow down the convergence speed.

Results. All results show that FaStR can estimate varying coefficients equally well compared to a classic GAM estimation routine. Figure 1 shows the resulting M(I)SE values for all data settings. Note that the GAM implicitly assumes as many latent dimensions as there are factor levels, but can also shrink single smooth functions to zero. In cases where it is feasible to fit smooth effects for every factor level (as is the case here), GAM can thus be seen as gold standard. For the case of a normal distribution, we observe that GAM yields better results, but also that the performance of FaStR converges to the one from GAM with increasing number of observations. For the Bernoulli case, our approach benefits from the optimization in a neural network and even outperforms the classic GAM estimation routine which requires a multiple of observations compared to a Gaussian setting for good estimation results (as, e.g., also found in [22]).

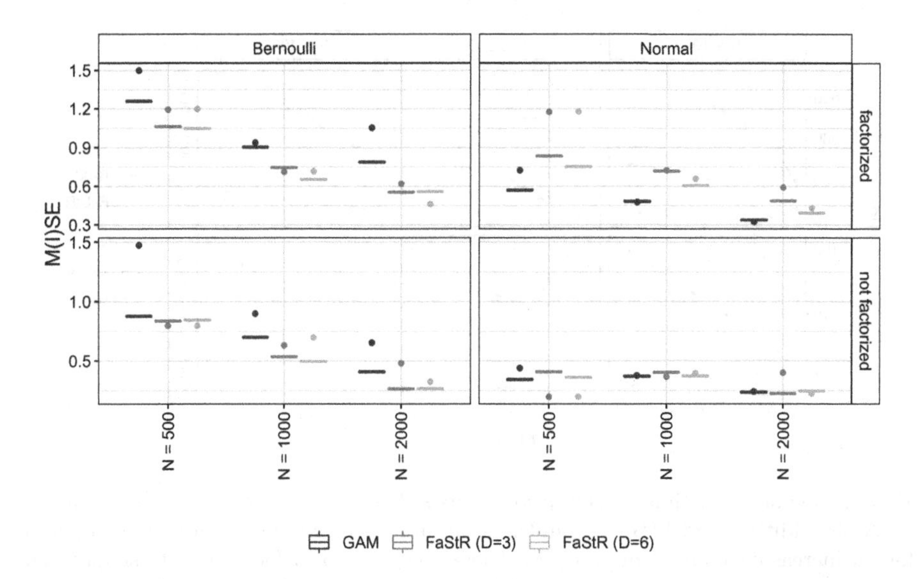

Fig. 1. Comparison of M(I)SE values for the estimated partial effects of different methods (colors) for factorized and non-factorized smooth terms (x-axis), different distributions (columns) and different data sizes (rows). Values > 1.5 are set to the value 1.5 to improve readability.

5.2 Memory Consumption

Finally, we compare the memory consumption of our implementation against the SotA implementation for big additive models (BAM) in the R [18] package `mgcv` [25] for an increasing number of category levels (20, 40, 60, 80) when using a categorical effect or a varying coefficient based on the representation proposed in Sect. 4.2, $N \in \{1000, 2000, 4000\}$ and optimization as in Sect. 5.1. While the improvement in memory consumption is expected to be even larger when using factorized terms instead of interaction terms with weights for each category combination, we do not use factorization in this experiment as there is no equivalent available in software for additive regression. Additionally, we also track the time when running FaStR for 10 epochs to see if there are notable changes in the time consumption for varying data generating settings.

Results. Figure 2 visualizes the results for all different settings and compares run times and memory consumption of the two methods for factor variables (single) and varying coefficient effects (varying). Results show that FaStR has both, almost constant time and memory consumption while the SotA method requires exponentially more memory for growing numbers of factor levels (as the whole encoded matrix must be loaded into memory). These results confirm that our implementation works as intended to allow for the estimation of varying coefficient models in large-scale settings.

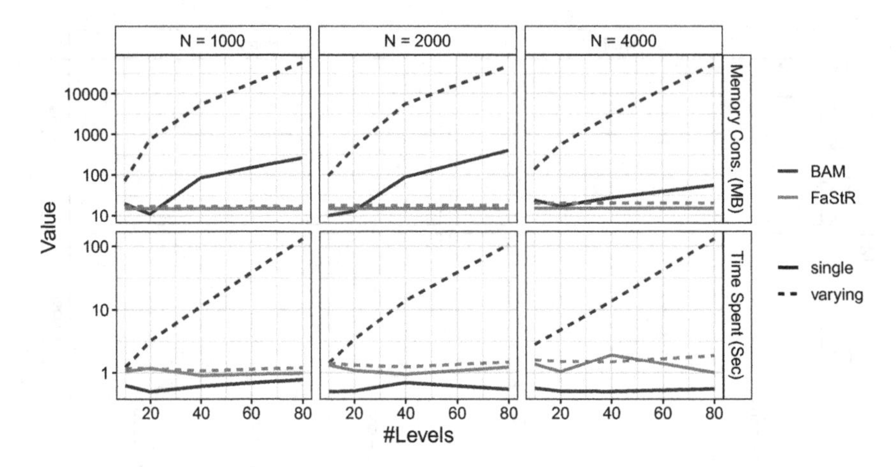

Fig. 2. Memory and time consumption (y-axis; $\log_{10}$-scale) comparison between the SotA big additive model (BAM) implementation and our method (in different colours) for an increasing number of categorical levels (x-axis) of a factor effect (single) and varying coefficient term (varying) for different data sizes (columns).

6 Benchmarks

Although the focus of this work is to provide scalable and interpretable regression models, prediction performance of our models is also of (secondary) interest. We aim for a similar performance compared to SotA time-aware RS techniques, yet without the ambition to outperform these methods. We use the *MovieLens 10M movie ratings* benchmark data set [11], which is sparse in terms of user-item combinations, with *items* corresponding to rated movies. In addition, we benchmark models on a subset of the densely observed *PhoneStudy behavior* data set [23], analyzed in more detail in Sect. 7. In both cases, we use single train-test splits (90%/10% and 70%/30%, respectively) and evaluate the models predictive performance with the root mean squared error (RMSE). The different characteristics of both data sets are given in Table 1.

Table 1. Descriptive statistics of the two benchmark data sets.

	Movies	Phone study
# Observations (N)	$\approx$ 9 m	$\approx$ 8.7 m
# Users (U)	69,878	342
# Movies/Activities (I)	10,677	176
# Unique Time Points (T)	$\approx$ 6.5 m	348

Methods. As comparison we use Bayesian timeSVD and Bayesian timeSVD++ flipped, two variations of the SVD++ method [14], a latent factor model whose key innovative feature is the incorporation of implicit user information. Both Bayesian timeSVD and Bayesian timeSVD++ flipped have been extended to be time-aware [15] and optimized by Gibbs sampling using a Bayesian reformulation [21]. Bayesian timeSVD++ flipped integrates both implicit user and item information and has been reported to be the best-performing model among multiple SotA methods in a recent benchmark study [21]. The second variation, Bayesian timeSVD, is still a time-aware latent factor model, yet it does not incorporate implicit user or item information. As we are mainly interested in the performance of the proposed time-varying coefficient model, the Bayesian timeSVD provides a much fairer comparison with FaStR as it does not include the aforementioned types of implicit information. We use tuning parameter settings as given in [21] for the two benchmark methods (i.e., we use the already tuned models). We compare these two models against our method as proposed in (4) and thereby not only test its predictive performance, but also its capability to scale well to high-dimensional data sets. We do not tune FaStR extensively, but perform a small ablation study by testing different latent dimensions ($D \in 1, 3, 10$) for the factorized varying coefficient term $f^{[3]}$ and by excluding the whole term. All models use early stopping based on a validation data split with the same proportion as the train-test split. Considering further model definitions and tuning the

Table 2. Benchmark results based on the RMSE on the respective test data set for both benchmark data sets (columns) for different methods (rows). Best results are in bold and best results of the respective other method is underlined.

	Movies	PhoneStudy
timeSVD	0.872	0.089
timeSVD++ flipped	**0.856**	<u>0.087</u>
FaStR ($D = 10$)	0.984	**0.076**
FaStR ($D = 3$)	0.975	0.080
FaStR ($D = 1$)	<u>0.890</u>	0.087
FaStR (w/o $f^{[3]}$)	1.027	0.093

smoothing parameters could slightly improve results, but at the cost of having to fit additional models.

Results. Table 2 shows the performance of all methods. Interestingly, FaStR is competitive with the SotA timeSVD approaches, even though the premise of this paper was merely to develop a scalable variant of the varying coefficient model, not to propose a method with SotA performance on RS tasks.

7 User Behavior Phone Study

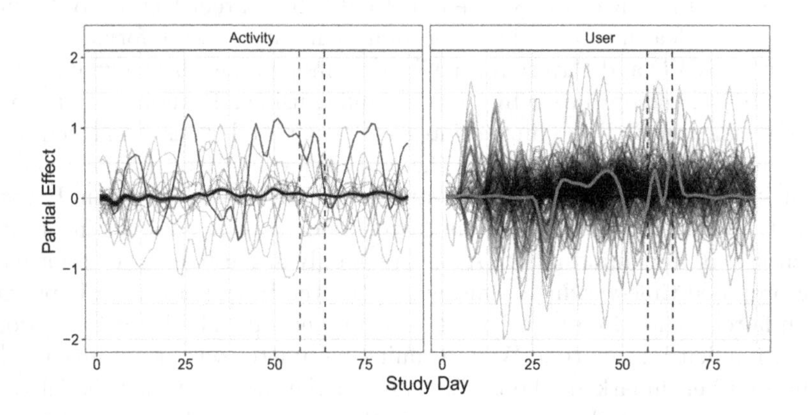

Fig. 3. One dimension of the latent time-varying coefficients for different activities (left plot with "locked screen" in blue) and users (right plot with exemplary user in green) over 87 study days (x-axis). Vertical lines show Christmas and New Year's Eve. (Color figure online)

We finally turn to the motivating case study. A more detailed description of the study and data set can be found in the Supplementary Material D. To analyze the activity levels (a value between 0 and 1, indicating the intensity of the activity in the given aggregation window) of participants in the study, we model the expected activity levels using user and activity effects, their interaction based on a factorization, an effect of the day of the week (Mon–Sun), the time of the day (6-hour windows), a factorized interaction of users and weekday as well as users and daytime, and a smooth time-dependent study day effect varying by user and/or activity. All factorizations use a three-dimensional latent space.

Results. Results are generally plausible and in line with prior expectations. Various model effects are examined in the Supplementary Material. We here briefly analyze the factorized varying coefficient interaction by analyzing its latent factors. One of the dimensions is depicted in Fig. 3. Most activities follow a global pattern (darker concentration of lines), while a few show very specific sequences. The "locked screen" event, e.g., is observed less in the days around New Year's

Eve. At the user level, no general pattern is visible due to the different study starting dates, but we observe still common changes in activity around holidays, e.g., Christmas.

8 Conclusion

In this work we presented an amalgamation of structured statistical regression and RS to allow for large-scale varying coefficient models with many categorical levels. For this, we leveraged factorization approaches combined with an efficient neural network-based implementation. Empirical results confirm the efficacy of our proposal. In order to make the proposed approach as flexible as commonly used statistical regression software, we used a model-based point of view and cast the approach as a generalized additive model.

Acknowledgement. This work has been partially supported by the German Federal Ministry of Education and Research (BMBF) under Grant No. 01IS18036A. We also thank four anonymous reviewers for their helpful suggestions and comments.

References

1. Abadi, M., et al.: TensorFlow: a system for large-scale machine learning. In: 12th USENIX Symposium on Operating Systems Design and Implementation (OSDI 2016), pp. 265–283 (2016)
2. Al-Hadi, I., Sharef, N.M., Sulaiman, M.N., Mustapha, N.: Review of the temporal recommendation system with matrix factorization. Int. J. Innov. Comput. Inf. Control **13**(5), 1579–1594 (2017)
3. Baltrunas, L., Ricci, F.: Experimental evaluation of context-dependent collaborative filtering using item splitting. User Model. User-Adap. Inter. **24**, 7–34 (2013). https://doi.org/10.1007/s11257-012-9137-9
4. Blondel, M., Fujino, A., Ueda, N., Ishihata, M.: Higher-order factorization machines. In: Lee, D., Sugiyama, M., Luxburg, U., Guyon, I., Garnett, R. (eds.) Advances in Neural Information Processing Systems, vol. 29. Curran Associates, Inc. (2016). https://doi.org/10.5555/3157382.3157473
5. Campos, P.G., Díez, F., Cantador, I.: Time-aware recommender systems: a comprehensive survey and analysis of existing evaluation protocols. User Model. User-Adap. Inter. **24**, 67–119 (2013). https://doi.org/10.1007/s11257-012-9136-x
6. Chen, T., Yin, H., Nguyen, Q.V.H., Peng, W.C., Li, X., Zhou, X.: Sequence-aware factorization machines for temporal predictive analytics. In: 2020 IEEE 36th International Conference on Data Engineering (ICDE), pp. 1405–1416. IEEE (2020). https://doi.org/10.1109/ICDE48307.2020.00125
7. Chua, F.C.T., Oentaryo, R.J., Lim, E.P.: Modeling temporal adoptions using dynamic matrix factorization. In: 2013 IEEE 13th International Conference on Data Mining, pp. 91–100 (2013). https://doi.org/10.1109/ICDM.2013.25
8. Condli, M.K., Lewis, D.D., Madigan, D., Posse, C.: Bayesian mixed-effects models for recommender systems. In: ACM SIGIR, vol. 99 (1999)
9. Currie, I.D., Durban, M., Eilers, P.H.: Generalized linear array models with applications to multidimensional smoothing. J. R. Stat. Soc.: Ser. B (Stat. Methodol.) **68**(2), 259–280 (2006). https://doi.org/10.1111/j.1467-9868.2006.00543.x

10. Guo, Y., Su, Y., Yang, Z., Zhang, A.: Explainable recommendation systems by generalized additive models with manifest and latent interactions (2020)
11. Harper, F.M., Konstan, J.A.: The MovieLens datasets: history and context. ACM Trans. Interact. Intell. Syst. (TIIS) **5**(4), 1–19 (2015). https://doi.org/10.1145/2827872
12. Hastie, T., Tibshirani, R.: Varying-coefficient models. J. Roy. Stat. Soc.: Ser. B (Methodol.) **55**(4), 757–779 (1993). https://doi.org/10.1111/j.2517-6161.1993.tb01939.x
13. Imaizumi, M., Hayashi, K.: Tensor decomposition with smoothness. In: International Conference on Machine Learning, pp. 1597–1606. PMLR (2017). https://doi.org/10.5555/3305381.3305546
14. Koren, Y.: Factorization meets the neighborhood: a multifaceted collaborative filtering model. In: Proceedings of the 14th ACM SIGKDD International Conference on Knowledge Discovery and Data Mining, pp. 426–434 (2008). https://doi.org/10.1145/1401890.1401944
15. Koren, Y.: Collaborative filtering with temporal dynamics. In: Proceedings of the 15th ACM SIGKDD International Conference on Knowledge Discovery and Data Mining, pp. 447–456 (2009). https://doi.org/10.1145/1721654.1721677
16. Koren, Y., Bell, R., Volinsky, C.: Matrix factorization techniques for recommender systems. Computer **42**(8), 30–37 (2009). https://doi.org/10.1109/MC.2009.263
17. Nelder, J.A., Wedderburn, R.W.: Generalized linear models. J. R. Stat. Soc. Set. A (Gen.) **135**(3), 370–384 (1972). https://doi.org/10.2307/2344614
18. R Core Team: R: A Language and Environment for Statistical Computing. R Foundation for Statistical Computing, Vienna, Austria (2021)
19. Rendle, S.: Factorization machines. In: 2010 IEEE International Conference on Data Mining, pp. 995–1000. IEEE (2010). https://doi.org/10.1109/ICDM.2010.127
20. Rendle, S., Krichene, W., Zhang, L., Anderson, J.: Neural collaborative filtering vs. matrix factorization revisited. In: Fourteenth ACM Conference on Recommender Systems, pp. 240–248 (2020). https://doi.org/10.1145/3383313.3412488
21. Rendle, S., Zhang, L., Koren, Y.: On the difficulty of evaluating baselines: a study on recommender systems. arXiv preprint arXiv:1905.01395 (2019). https://doi.org/10.48550/ARXIV.1905.01395
22. Rügamer, D., Kolb, C., Klein, N.: Semi-structured deep distributional regression: a combination of additive models and deep learning. arXiv preprint arXiv:2002.05777 (2020). https://doi.org/10.48550/ARXIV.2002.05777
23. Stachl, C., et al.: Predicting personality from patterns of behavior collected with smartphones. Proc. Natl. Acad. Sci. **117**, 17680–17687 (2020). https://doi.org/10.1073/pnas.1920484117
24. Thorat, P.B., Goudar, R., Barve, S.: Survey on collaborative filtering, content-based filtering and hybrid recommendation system. Int. J. Comput. Appl. **110**(4), 31–36 (2015). https://doi.org/10.5120/19308-0760
25. Wood, S.N.: Fast stable restricted maximum likelihood and marginal likelihood estimation of semiparametric generalized linear models. J. R. Stat. Soc. (B) **73**(1), 3–36 (2011). https://doi.org/10.1111/j.1467-9868.2010.00749.x
26. Wood, S.N.: Generalized Additive Models: An Introduction with R. Chapman and Hall/CRC, New York (2017). https://doi.org/10.1201/9781315370279
27. Wood, S.N.: Rejoinder on: Inference and computation with Generalized Additive Models and their extensions. TEST **29**(2), 354–358 (2020). https://doi.org/10.1007/s11749-020-00716-0

28. Wood, S.N., Li, Z., Shaddick, G., Augustin, N.H.: Generalized additive models for GigaData: modeling the u.k. black smoke network daily data. J. Am. Stat. Assoc. **112**(519), 1199–1210 (2017). https://doi.org/10.1080/01621459.2016.1195744
29. Wu, C., Lian, D., Ge, Y., Zhu, Z., Chen, E., Yuan, S.: Fight fire with fire: towards robust recommender systems via adversarial poisoning training. In: Proceedings of the 44th International ACM SIGIR Conference on Research and Development in Information Retrieval, pp. 1074–1083, SIGIR 2021. Association for Computing Machinery, New York, NY, USA (2021). https://doi.org/10.1145/3404835.3462914
30. Yu, H.F., Rao, N., Dhillon, I.S.: Temporal regularized matrix factorization for high-dimensional time series prediction. In: Lee, D., Sugiyama, M., Luxburg, U., Guyon, I., Garnett, R. (eds.) Advances in Neural Information Processing Systems, vol. 29. Curran Associates, Inc. (2016)
31. Zhang, X., Zhou, Y., Ma, Y., Chen, B.C., Zhang, L., Agarwal, D.: GLMix: generalized linear mixed models for large-scale response prediction. In: KDD 2016, pp. 363–372. Association for Computing Machinery, New York, NY, USA (2016). https://doi.org/10.1145/2939672.2939684
32. Zhang, Y., Bi, X., Tang, N., Qu, A.: Dynamic tensor recommender systems. J. Mach. Learn. Res. **22**(65), 1–35 (2021). https://doi.org/10.11159/icsta19.09
33. Zhao, W.X., et al.: RecBole: towards a unified, comprehensive and efficient framework for recommendation algorithms (2020). https://doi.org/10.1145/3459637.3482016

Ordinal Quantification Through Regularization

Mirko Bunse[1(✉)], Alejandro Moreo[2], Fabrizio Sebastiani[2], and Martin Senz[1]

[1] Department of Computer Science, TU Dortmund University, 44227 Dortmund, Germany
{mirko.bunse,martin.senz}@cs.tu-dortmund.de

[2] Istituto di Scienza e Tecnologie dell'Informazione, Consiglio Nazionale delle Ricerche, 56124 Pisa, Italy
{alejandro.moreo,fabrizio.sebastiani}@isti.cnr.it

Abstract. Quantification, i.e., the task of training predictors of the class prevalence values in sets of unlabelled data items, has received increased attention in recent years. However, most quantification research has concentrated on developing algorithms for binary and multiclass problems in which the classes are not ordered. We here study the ordinal case, i.e., the case in which a total order is defined on the set of $n > 2$ classes. We give three main contributions to this field. First, we create and make available two datasets for ordinal quantification (OQ) research that overcome the inadequacies of the previously available ones. Second, we experimentally compare the most important OQ algorithms proposed in the literature so far. To this end, we bring together algorithms that are proposed by authors from very different research fields, who were unaware of each other's developments. Third, we propose three OQ algorithms, based on the idea of preventing ordinally implausible estimates through regularization. Our experiments show that these algorithms outperform the existing ones if the ordinal plausibility assumption holds.

Keywords: Quantification · Ordinal classification · Supervised prevalence estimation

1 Introduction

Quantification is a supervised learning task that consists of training a predictor, on a set of labelled data items, that estimates the relative frequencies $p_\sigma(y_i)$ of the classes of interest $\mathcal{Y} = \{y_1, ..., y_n\}$ in a sample σ of unlabelled data items [16]. In other words, a trained *quantifier* must return a *predicted distribution* $\hat{p}_\sigma$ of the unlabelled data items in σ across the classes in $\mathcal{Y}$, where $\hat{p}_\sigma$ must diverge as little as possible from the true, unknown distribution p_σ. Quantification is also known as "learning to quantify", "supervised class prevalence estimation", and "class prior estimation".

© The Author(s), under exclusive license to Springer Nature Switzerland AG 2023
M.-R. Amini et al. (Eds.): ECML PKDD 2022, LNAI 13717, pp. 36–52, 2023.
https://doi.org/10.1007/978-3-031-26419-1_3

Quantification is important in many disciplines, e.g., market research, political science, the social sciences, and epidemiology. By their own nature, these disciplines are only interested in aggregate, as opposed to individual, data. Hence, classifying individual unlabelled instances is usually not a primary goal, while estimating the prevalence values $p_\sigma(y_i)$ of the classes of interest is. For instance, when classifying the tweets about a certain entity, e.g., about a political candidate, as displaying either a Positive or a Negative stance towards the entity, we are usually not interested in the class of a specific tweet, but in the fraction of these tweets that belong to each class [17].

A predicted distribution $\hat{p}_\sigma$ could, in principle, be obtained via the "classify and count" method (CC), i.e., by training a standard classifier, classifying all the unlabelled data items in σ, counting how many of them have been assigned to each class in $\mathcal{Y}$, and normalizing. However, it has been shown that CC delivers poor prevalence estimates, and especially so when the application scenario suffers from *prior probability shift* [22], the (ubiquitous) phenomenon according to which the distribution $p_U(y_i)$ of the *un-labelled* test documents U across the classes is different from the distribution $p_L(y_i)$ of the *labelled* training documents L. As a result, a plethora of quantification methods has been proposed in the literature [4,14,16,17,33] that aims at accurate class prevalence estimations even in the presence of prior probability shift.

The vast majority of the methods proposed so far deals with quantification tasks in which $\mathcal{Y}$ is a plain, unordered set. Very few methods, instead, deal with *ordinal quantification* (OQ), the task of performing quantification on a set of $n > 2$ classes on which a total order "$\prec$" is defined. Ordinal quantification is important, though, because ordinal scales arise in many applications, especially ones involving human judgements. For instance, in a customer satisfaction endeavour, one may want to estimate how a set of reviews of a certain product distributes across the set of classes $\mathcal{Y}$ ={1Star, 2Stars, 3Stars, 4Stars, 5Stars}, while a social scientist might want to find out how inhabitants of a certain region are distributed in terms of their happiness with health services in the area, $\mathcal{Y}$ ={VeryUnhappy, Unhappy, Happy, VeryHappy}.

In this paper, we contribute to the field of OQ in three ways.

First, we develop and publish two datasets for evaluating OQ algorithms, one consisting of textual product reviews and one consisting of telescope observations. Both datasets stem from scenarios in which OQ arises naturally, and they are generated according to a strong, well-tested protocol for the evaluation of quantifiers. The datasets that were previously used for the evaluation of OQ algorithms were not adequate, for reasons we discuss in Sect. 2.

Second, we perform an extensive experimental comparison among the most important OQ algorithms that have been proposed in the literature. This contribution is important because some algorithms have been evaluated on a testbed that was likely inadequate, while some other algorithms have been developed independently of the previous ones, and have thus never been compared to them.

Third, we propose new OQ algorithms, which introduce regularization into existing quantification methods. We experimentally compare our proposals with the existing state of the art and make the corresponding code publicly available[1].

This paper is organized as follows. In Sect. 2, we review past work on OQ. In Sect. 3, we present quantification algorithms, starting with previously proposed ones and then moving to the ones we propose in this work. Section 4 is devoted to our experimental evaluation and Sect. 5 concludes.

2 Related Work

Quantification, as a task of its own, was first proposed by Forman [16], who observed that some applications of classification only require the estimation of class prevalence values, and that better methods than "classify and count" can be devised for this requirement. Since then, many methods for quantification have been proposed. However, most of these methods tackle the binary and/or multiclass problem with un-ordered classes. While OQ was first discussed in [15] it was not until 2016 that the first true OQ algorithms were developed, the *Ordinal Quantification Tree* (OQT) [10] and *Adjusted Regress and Count* (ARC) [13]. In the same years, the first data challenges that involved OQ were staged [18,25,30]. However, except for OQT and ARC, the participants in these challenges used "classify and count" with highly optimized classifiers, and no true OQ methods; this attitude persisted also in later challenges [39,40], likely due to a general lack of awareness in the scientific community that more accurate methods than "classify and count" exist.

Unfortunately, the data challenges, in which OQT and ARC were evaluated [25,30], tested each quantification method only on a single sample of unlabelled items, which consisted of the entire test set. This evaluation protocol is not adequate for quantification because quantifiers issue predictions for samples of data items, not for individual data items as in classification. Measuring a quantifier's performance on a single sample is thus akin to, and as insufficient as, measuring a classifier's performance on a single data item. As a result, our knowledge of the relative merits of OQT and ARC lacks solidity.

However, even before the previously mentioned developments took place, what we now would call OQ algorithms had been proposed within experimental physics. In this field, we often need to estimate the distribution of a continuous physical quantity. However, a histogram approximation of a continuous distribution is sufficient for many physics-related analyses [6]. This conventional simplification essentially maps the values of a continuous target quantity into a set of classes endowed with a total order, and the problem of estimating the continuous distribution becomes one of OQ. Early on, physicists had termed this problem "unfolding" [5,11], a terminology that prevented quantification researchers from drawing connections between algorithms proposed in the quantification literature and those proposed in the physics literature. In the following, we provide these

[1] Code and supplementary results: https://github.com/mirkobunse/ecml22.

connections, to find that regularization techniques proposed within the physics literature are able to improve well-known quantification methods in ordinal settings. We complete the unification of unfolding and quantification methods in [8].

3 Methods

We use the following notation. By $\mathbf{x} \in \mathcal{X}$ we indicate a data item drawn from a domain $\mathcal{X}$, and by $y \in \mathcal{Y}$ we indicate a class drawn from a set of classes $\mathcal{Y} = \{y_1, ..., y_n\}$, also known as a *codeframe*, on which a total order "$\prec$" is defined. The symbol σ denotes a *sample*, i.e., a non-empty set of unlabelled data items in $\mathcal{X}$, while $L \subset \mathcal{X} \times \mathcal{Y}$ denotes a set of labelled data items $(\mathbf{x}, y)$, which we use to train our quantifiers. By $p_\sigma(y)$, we indicate the true prevalence of class y in sample σ, while by $\hat{p}_\sigma^M(y)$, we indicate an estimate of this prevalence, as obtained by a quantification method M that receives σ as an input, where $0 \leq p_\sigma(y), \hat{p}_\sigma^M(y) \leq 1$ and $\sum_{y \in \mathcal{Y}} p_\sigma(y) = \sum_{y \in \mathcal{Y}} \hat{p}_\sigma^M(y) = 1$.

3.1 Non-ordinal Quantification Methods

We start by introducing some important multiclass quantification methods which do not take ordinality into account. These methods provide the foundation for their ordinal extensions, which we develop in Sect. 3.3.

Classify and Count (CC) [16] is the trivial quantification method, where a "hard" classifier $h : \mathcal{X} \to \mathcal{Y}$ generates predictions for all data items $\mathbf{x} \in \sigma$, and the fraction of predictions is used as a prevalence estimate, i.e.,

$$\hat{p}_\sigma^{\mathrm{CC}}(y_i) = \frac{1}{|\sigma|} \cdot \big|\{\mathbf{x} \in \sigma : h(\mathbf{x}) = y_i\}\big|. \tag{1}$$

In its probabilistic variant, **Probabilistic Classify and Count (PCC)** [4], the hard classifier is replaced by a "soft" classifier $s : \mathcal{X} \to [0,1]^n$ that returns well-calibrated posterior probabilities $[s(\mathbf{x})]_i \equiv \Pr(y_i|\mathbf{x})$, i.e.,

$$\hat{p}_\sigma^{\mathrm{PCC}}(y_i) = \frac{1}{|\sigma|} \cdot \sum_{\mathbf{x} \in \sigma} [s(\mathbf{x})]_i, \tag{2}$$

where $[\mathbf{z}]_i$ denotes the i-th component of some $\mathbf{z}$, and where $\sum_{i=1}^n [s(\mathbf{x})]_i = 1$.

Adjusted Classify and Count (ACC) [16] and **Probabilistic Adjusted Classify and Count (PACC)** [4] are based on the idea of correcting the $\hat{p}_\sigma^{\mathrm{CC}}$ and $\hat{p}_\sigma^{\mathrm{PCC}}$ estimates by using the (hard or soft) misclassification rates estimated on a validation set V, which coincides with L if k-fold cross-validation is used.

In the multiclass setting, we want to estimate a vector of prevalence values $\mathbf{p} \in \mathbb{R}^n$, where $[\mathbf{p}]_i = p_\sigma(y_i)$. In this case, the adjustment of ACC and PACC amounts to solving for $\mathbf{p}$ the system of linear equations

$$\mathbf{q} = \mathbf{M}\mathbf{p}, \tag{3}$$

where $\mathbf{q} \in \mathbb{R}^n$ is a vector of unadjusted prevalence estimates obtained via CC or PCC, i.e., $[\mathbf{q}]_i^{\mathrm{ACC}} = \hat{p}_\sigma^{\mathrm{CC}}(y_i)$ and $[\mathbf{q}]_i^{\mathrm{PACC}} = \hat{p}_\sigma^{\mathrm{PCC}}(y_i)$, and $\mathbf{M} \in \mathbb{R}^{n \times n}$ is a matrix that relates the ground truth labels to the predictions of the employed classifier, i.e.,

$$\mathbf{M}_{ij}^{\mathrm{ACC}} = \frac{|\{(\mathbf{x}, y) \in V : h(\mathbf{x}) = y_i,\ y = y_j\}|}{|\{(\mathbf{x}, y) \in V : y = y_j\}|}, \tag{4}$$

$$\mathbf{M}_{ij}^{\mathrm{PACC}} = \frac{\sum_{(\mathbf{x}, y) \in V : y = y_j} [s(\mathbf{x})]_i}{|\{(\mathbf{x}, y) \in V : y = y_j\}|}. \tag{5}$$

ACC and PACC solve Eq. 3 via the Moore-Penrose pseudo-inverse $\mathbf{M}^\dagger$, i.e.,

$$\hat{\mathbf{p}} = \mathbf{M}^\dagger \mathbf{q}, \tag{6}$$

where $\hat{p}_\sigma^{\mathrm{ACC}}(y_i) \equiv [\hat{\mathbf{p}}]_i$ if Eqs. 1 and 4 are employed, while $\hat{p}_\sigma^{\mathrm{PACC}}(y_i) \equiv [\hat{\mathbf{p}}]_i$ if Eqs. 2 and 5 are employed.

Unlike the true inverse $\mathbf{M}^{-1}$, the pseudo-inverse always exists. If the true inverse exists, the two matrices are identical; if it does not exist, the solution from Eq. 6 amounts to a minimum-norm least-squares estimate of $\mathbf{p}$ [23, Theorem 4.1].

EM-based quantification, also known as the **Saerens-Latinne-Decaestecker (SLD)** method [33], follows an expectation maximization approach, which (i) leverages Bayes' theorem in the E-step, and (ii) updates the prevalence estimates in the M-step. Both steps can be combined in the single update rule

$$\hat{p}_\sigma^{(k)}(y_i) = \frac{1}{|\sigma|} \sum_{\mathbf{x} \in \sigma} \frac{\frac{\hat{p}_\sigma^{(k-1)}(y_i)}{\hat{p}_\sigma^{(0)}(y_i)} \cdot [s(\mathbf{x})]_i}{\sum_{j=1}^{n} \frac{\hat{p}_\sigma^{(k-1)}(y_j)}{\hat{p}_\sigma^{(0)}(y_j)} \cdot [s(\mathbf{x})]_j}, \tag{7}$$

which is applied until the estimates converge. $p_\sigma^{(0)}(y)$ is initialized with the class prevalence values of the training set.

3.2 Existing OQ Methods from the Physics Literature

Similar to the adjustment of ACC, experimental physicists have proposed adjustments that solve for $\mathbf{p}$ the system of linear equations from Eq. 3. However, these "unfolding" quantifiers differ from ACC in two regards.

The first aspect is that the hard classifier h of Eqs. 1 and 4 is often, although not always, replaced by a partition $c : \mathcal{X} \rightarrow \{1, \dots, d\}$ of the feature space, so that

$$\mathbf{q}t \quad _i = \frac{1}{|\sigma|} \cdot |\{\mathbf{x} \in \sigma : c(\mathbf{x}) = i\}|, \tag{8}$$

$$\mathbf{M}_{ij} = \frac{|\{(\mathbf{x}, y) \in V : c(\mathbf{x}) = i,\ y = y_j\}|}{|\{(\mathbf{x}, y) \in V : y = y_j\}|},$$

and $\mathbf{M} \in \mathbb{R}^{d \times n}$. Note that by choosing $c = h$ we obtain exactly Eqs. 1 and 4. Another possible choice for c is to partition the feature space by means of a decision tree; in this case, (i) it typically holds that $d > n$, and (ii) $c(\mathbf{x})$ represents the index of a leaf node [7].

The second difference to ACC is that "unfolding" quantifiers regularize their estimates in order to promote solutions that are the most *plausible* solutions in OQ. Specifically, these methods employ the assumption that neighbouring classes have similar prevalence values; depending on the algorithm, this assumption is encoded in different ways, as we will see in the following paragraphs. This assumption is quite reasonable in OQ because the "smoothness" of the histogram that represents the distribution is arguably *the most important aspect that distinguishes an ordinal distribution from a non-ordinal multiclass distribution.* Without an order of classes, the concept of neighboring classes would even be ill-defined.

Regularized Unfolding (RUN) [5,6] is used by physicists for decades [2,27]. It estimates the vector $\mathbf{p}$ of class prevalence values by minimizing a loss function $\mathcal{L} : \mathbb{R}^n \rightarrow \mathbb{R}$ over the estimate $\hat{\mathbf{p}}$; $\mathcal{L}$ consists of two terms, i.e., a negative log-likelihood term to model the error of $\hat{\mathbf{p}}$, and a regularization term to model the plausibility of $\hat{\mathbf{p}}$.

The negative log-likelihood term in $\mathcal{L}$ builds on a Poisson assumption about the distribution of the data. Namely, this term models the counts $[\bar{\mathbf{q}}]_i = |\sigma| \cdot [\mathbf{q}]_i$, which are observed in the sample σ, as being Poisson-distributed with the rates $\lambda_i = \mathbf{M}_i^\top \bar{\mathbf{p}}$. Here, $\mathbf{M}_i$ is the i-th column vector of $\mathbf{M}$ and $[\bar{\mathbf{p}}]_i = |\sigma| \cdot [\hat{\mathbf{p}}]_i$ are the class counts that would be observed under a prevalence estimate $\hat{\mathbf{p}}$.

The second term of $\mathcal{L}$ is a Tikhonov regularization term $\frac{1}{2}(\mathbf{C}\mathbf{p})^2$. This term introduces an inductive bias towards solutions which are plausible with respect to ordinality. The Tikhonov matrix $\mathbf{C}$ is chosen in such a way that term $\frac{1}{2}(\mathbf{C}\mathbf{p})^2$ measures the smoothness of the histogram that represents the distribution, i.e.,

$$\frac{1}{2}(\mathbf{C}\mathbf{p})^2 = \frac{1}{2}\sum_{i=2}^{n-1} \left(-[\mathbf{p}]_{i-1} + 2[\mathbf{p}]_i - [\mathbf{p}]_{i+1}\right)^2 . \tag{9}$$

Combining the likelihood term and the regularization term, the loss function of RUN is given by

$$\mathcal{L}(\hat{\mathbf{p}};\ \mathbf{M}, \mathbf{q}, \tau, \mathbf{C}) = \sum_{i=1}^{d} \left(\mathbf{M}_i^\top \bar{\mathbf{p}} - [\bar{\mathbf{q}}]_i \cdot \ln(\mathbf{M}_i^\top \bar{\mathbf{p}})\right) + \frac{\tau}{2}(\mathbf{C}\hat{\mathbf{p}})^2 \tag{10}$$

and an estimate $\hat{\mathbf{p}}$ is chosen by minimizing $\mathcal{L}$ numerically over $\hat{\mathbf{p}}$. Here, $\tau \geq 0$ is a hyperparameter which controls the impact of the regularization.

Iterative Bayesian Unfolding (IBU) [11,12] is still popular today [1,24]. This method revolves around an expectation maximization approach with Bayes' theorem, and thus has a common foundation with the SLD method. The E-step and the M-step of IBU can be written as the single, combined update rule

$$\hat{p}_\sigma^{(k)}(y_i) = \sum_{j=1}^{d} \frac{\mathbf{M}_{ij} \cdot \hat{p}_\sigma^{(k-1)}(y_i)}{\sum_{l=1}^{n} \mathbf{M}_{lj} \cdot \hat{p}_\sigma^{(k-1)}(y_l)} [\mathbf{q}]_i. \tag{11}$$

One difference between IBU and SLD is that $\mathbf{q}$ and $\mathbf{M}$ are defined via counts of hard assignments to partitions $c(\mathbf{x})$ (see Eq. 8), while SLD is defined over individual soft predictions $s(\mathbf{x})$ (see Eq. 7).

Another difference between IBU and SLD is regularization. In order to promote solutions which are ordinally plausible, IBU smooths each intermediate estimate $\hat{\mathbf{p}}^{(k)}$ by fitting a low-order polynomial to $\hat{\mathbf{p}}^{(k)}$. A linear interpolation between $\hat{\mathbf{p}}^{(k)}$ and this polynomial is then used as the prior of the next iteration in order to reduce the differences between neighbouring prevalence estimates. The interpolation factor is a hyperparameter of IBU through which the degree of regularization is controlled.

Other methods from the physics literature, which go under the name of "unfolding", are based on similar concepts as RUN and IBU. We focus on these two methods due to their long-standing popularity within physics research. In fact, they are among the first methods that have been proposed in this field, and are still widely adopted today, in astro-particle physics [2,27], high-energy physics [1], and more recently in quantum computing [24]. Moreover, RUN and IBU already cover the most important aspects of unfolding methods with respect to ordinal quantification.

Several other unfolding methods are similar to RUN. The method proposed in [19], for instance, employs the same regularization as RUN, but assumes different Poisson rates, which are simplifications of the rates that RUN uses; in preliminary experiments, here omitted for the sake of conciseness, we have found this simplification to typically deliver less accurate results than RUN. Two other methods [35,36] employ the same simplification as [19] but regularize differently. To this end, [35] regularizes with respect to the deviation from a prior, instead of regularizing with respect to ordinal plausibility; we thus do not perceive this method as a true OQ method. [36] adds to the RUN regularization a second term which enforces prevalence estimates that sum up to one; we use a RUN implementation which already solves this issue through a positivity constraint and normalization. Another line of work evolves around the algorithm of [32] and its extensions [9]. We perceive this algorithm to lie outside the scope of OQ because it does not address the order of classes, like the other "unfolding" methods do. Moreover, the algorithm was shown to exhibit a performance comparable to, but not better than RUN and IBU [9].

3.3 New Ordinal Variants of ACC, PACC, and SLD

In the following, we develop algorithms which extend ACC, PACC, and SLD with the regularizers from RUN and IBU. Through these extensions, we obtain o-ACC, o-PACC, and o-SLD, the OQ counterparts of these well-known non-ordinal quantification algorithms. In doing so, since we employ the regularizers but not any other aspect of RUN and IBU, we preserve the general characteristics of ACC, PACC, and SLD. In particular, our methods continue to work with classifier predictions, i.e., we do not employ the categorical feature representation from Eq. 8, which RUN and IBU employ, and we do not use the Poisson assumption of RUN. Therefore, our extensions are "minimal", in the sense of being directly addressed to ordinality and not introducing any undesired side effects in the original methods.

o-ACC and o-PACC, our ordinal extensions to ACC and PACC, build on the finding reported in [23, Theorem 4.1], which states that the solution from Eq. 6 corresponds to a minimum-norm least-squares solution. Namely, among all least-squares solutions $\hat{\mathbf{p}}^{\mathrm{LSq}} = \arg\min_{\mathbf{p}} \|\mathbf{q} - \mathbf{M}\mathbf{p}\|_2^2$, which by themselves do not need to be unique, the solution to Eq. 6 is the unique one that also minimizes the quadratic norm $\|\mathbf{p}\|_2^2$. Therefore, Eq. 6 is conceptually similar, although not necessarily equal, to a regularized estimate

$$\hat{\mathbf{p}}' = \arg\min_{\mathbf{p}} \|\mathbf{q} - \mathbf{M}\mathbf{p}\|_2^2 + \frac{\tau}{2}\|\mathbf{p}\|_2^2, \tag{12}$$

which employs the quadratic norm for regularization. In particular, both Eqs. 6 and 12 simultaneously minimize a least-squares objective and the norm of their candidate solutions. Note that the regularization function herein is, unlike the regularization from RUN, unrelated to the ordinal nature of the classes.

To obtain the true OQ methods o-ACC and o-PACC, we replace the minimum-norm regularization in Eq. 12 with the regularization term of RUN (see Eq. 9). Through this replacement, we minimize the same objective function as ACC and PACC, i.e., a least-squares objective, but regularize towards solutions that we deem more plausible for OQ. The prevalence estimate is

$$\hat{\mathbf{p}}^{\mathrm{o}} = \arg\min_{\mathbf{p}} \|\mathbf{q} - \mathbf{M}\mathbf{p}\|_2^2 + \frac{\tau}{2}\left(\mathbf{C}\mathbf{p}\right)^2, \tag{13}$$

the minimizer of which can be found through numerical optimization, e.g., through the BFGS optimization technique [26]. The o-ACC variant emerges from plugging in Eqs. 1 and 4 for $\mathbf{q}$ and $\mathbf{M}$, while the o-PACC variant emerges from plugging in Eqs. 2 and 5.

o-SLD leverages the ordinal regularization of IBU in SLD. Namely, our method does not use the latest estimate directly as the prior of the next iteration, but a smoothed version of this estimate. To this end, we fit a low-order polynomial to each intermediate estimate $\hat{\mathbf{p}}^{(k)}$ and use a linear interpolation between this polynomial and $\hat{\mathbf{p}}^{(k)}$ as the prior of the next iteration. Like in IBU, we consider the interpolation factor as a hyperparameter through which the strength of this regularization is controlled.

4 Experiments

The goal of our experiments is to uncover the relative merits of OQ methods that come from different fields. We pursue this goal by carrying out a thorough comparison of these methods on representative OQ data sets.

4.1 Evaluation Measures

The main evaluation measure we use in this paper is the *Normalized Match Distance* (NMD), defined by [34] as

$$\mathrm{NMD}(p, \hat{p}) = \frac{1}{n-1} \mathrm{MD}(p, \hat{p}), \tag{14}$$

where $\frac{1}{n-1}$ is just a normalization factor that allows NMD to range between 0 (best) and 1 (worst). Here MD is the *Match Distance* [38], defined as

$$\mathrm{MD}(p, \hat{p}) = \sum_{i=1}^{n-1} d(y_i, y_{i+1}) \cdot |\hat{P}(y_i) - P(y_i)|, \tag{15}$$

where $d(y_i, y_{i+1})$ is the "distance" between consecutive classes y_i and y_{i+1}, i.e., the cost we incur in assigning to y_i a probability mass that we should instead assign to y_{i+1}, or vice versa. Here, we assume $d(y_i, y_{i+1}) = 1$ for all $i \in \{1, ..., n-1\}$ and $P(y_i) = \sum_{j=1}^{i} p(y_j)$ is the cumulative distribution of p.

MD is a special case of the *Earth Mover's Distance* (EMD) [31], a widely used measure in OQ evaluation [9,10,15,25,30]. Since MD and EMD differ only by a fixed normalization factor, our experiments perfectly follow the tradition in OQ evaluation.

Another proposed measure for evaluating the quality of OQ estimates is the *Root Normalized Order-aware Divergence* (RNOD) [34]. We include a definition of, and an evaluation in terms of, RNOD in the supplementary material (See footnote 1), where we find that RNOD and NMD consistently lead to the same conclusions.

To obtain an overall score for a quantification method on a dataset, we apply this method to each test sample σ. The resulting prevalence estimates are then compared to the ground-truth prevalence values, which yields one NMD (or RNOD) value for each sample. The final score of the method is the average of these values, i.e., the average NMD (or RNOD) across all samples in the test set. We test for statistically significant differences between quantification methods in terms of a paired Wilcoxon signed-rank test.

4.2 Datasets and Preprocessing

We conduct our experiments on two large datasets that we have generated for the purpose of this work, and that we make available to the scientific community (See footnote 1). The first dataset, named AMAZON-OQ-BK, consists of product reviews labelled according to customer's judgments of quality, i.e., 1Star to

5Stars. The second dataset, FACT-OQ, consists of telescope observations labelled by one of 12 totally ordered classes. Hence, these data sets originate in practically relevant and diverse applications of OQ. Each of these data sets consists of a training set, multiple validation samples, and multiple test samples, which are extracted from the original data source according to two extraction protocols that are well suited for OQ.

The Data Sampling Protocol. We start by dividing a set of labelled data items into a training set L, a pool of validation (i.e., development) items, and a pool of test items. These three sets are disjoint from each other, and each of them is obtained through stratified sampling from the original data sources.

From each of the pools, we separately extract samples for quantifier evaluation. This extraction follows the so-called *Artificial Prevalence Protocol* (APP), by now a standard extraction protocol in quantifier evaluation (see, e.g., [16]). This protocol generates each sample in two steps. The first step consists of generating a vector $\mathbf{p}_\sigma$ of class prevalence values. Following [14], we generate this vector by drawing uniformly at random from the set of all legitimate prevalence vectors by using the Kraemer algorithm [37], which (differently from other naive algorithms) ensures that all prevalence values in the unit $(n-1)$ simplex are picked with equal probability. Since each $\mathbf{p}_\sigma$ can be, and typically is, different from the training set prevalences, this approach covers the entire space of prior probability shifts. The second step consists of drawing from the pool of data, be it our validation pool or our test pool, a fixed-size sample σ of data items which obeys the class prevalence values of $\mathbf{p}_\sigma$. The result is a set of samples characterized by uniformly distributed vectors of prevalence values, which give rise to varying levels of prior probability shift. We obtain one such set of samples from the validation pool and another set from the test pool.

For our two datasets, (i) we set the size of the training set to 20,000 data items, (ii) we have each (validation or test) sample consist of 1000 data items, (iii) we have the validation set consist of 1000 such samples, and (iv) we have the test set consist of 5000 such samples. For AMAZON-OQ-BK, a data item corresponds to a single product review, while for FACT-OQ, a data item corresponds to a single telescope recording.

All items in the pool are replaced after the generation of each sample, so that no sample contains duplicate items but samples from the same pool are not necessarily disjoint. Note, however, that our initial split into a training set, a validation pool, and a test pool ensures that each validation sample is disjoint from each test sample, and that the training set is disjoint from all other samples.

Partitioning Samples in Terms of Ordinal Plausibility. In APP, all class prevalence vectors are sampled with the same probability, disregarding of their "plausibility", in the sense of being likely to appear in the practice of OQ. For instance, $\mathbf{p}_1 = (.0, .5, .0, .5, .0)$ and $\mathbf{p}_2 = (.2, .1, .0, .3, .4)$ have the same chances to be generated within APP, despite the fact that $\mathbf{p}_1$ seems much less likely to show up than $\mathbf{p}_2$ in a real OQ application. Indeed, a vector such as $\mathbf{p}_2$ is extremely likely in the realm of product reviews.

We counteract this shortcoming of APP by using APP-OQ, a protocol similar to APP but for the fact that only samples "plausible" in the context of OQ are considered. Namely, in APP-OQ, we retain only the 20% most plausible samples generated by APP. Hence, we perform hyperparameter optimization on the selected 20% validation samples, and perform the evaluation on the selected 20% test samples. We always report the results of both APP and APP-OQ side by side, so as to allow drawing conclusions concerning the OQ-related merits of the different quantification methods.

Motivated by our experience in sentiment quantification and unfolding, we use "smoothness" as an indicator of plausibility. We measure smoothness by applying Eq. 9 to the class prevalence vector $\mathbf{p}$ of each sample, so that the most plausible samples are those with the smallest value of $\frac{1}{2}(\mathbf{Cp})^2$. We recognize that this measure can only be a first step towards assessing the plausibility of prevalence vectors in OQ because plausibility necessarily depends on the use case and on the expected number of data items in each sample.

The AMAZON-OQ-BK dataset is extracted from an existing dataset[2] of 233.1M English-language Amazon product reviews, made available by [21]; here, a data item corresponds to a single product review. As the labels of the reviews, we use their "stars" ratings, and our codeframe is thus $\mathcal{Y}$ ={1Star, 2Stars, 3Stars, 4Stars, 5Stars}, which represents a sentiment quantification task [15].

We restrict our attention to reviews from the Books domain. We then remove (a) all reviews shorter than 200 characters because recognizing sentiment from shorter reviews may be nearly impossible in some cases, and (b) all reviews that have not been recognized as "useful" by any users because many reviews never recognized as "useful" may contain comments, say, on Amazon's speed of delivery, and not on the product itself.

We convert the documents into vectors by using the RoBERTa transformer [20] from the Hugging Face hub[3]. To this aim, we truncate the documents to the first 256 tokens, and fine-tune RoBERTa via prompt learning for a maximum of 5 epochs on our training data, thus taking the model parameters from the epoch which yields the smallest validation loss as monitored on 1000 held-out documents randomly sampled from the training set in a stratified way. For training, we set the learning rate to $2e^{-5}$, the weight decay to 0.01, and the batch size to 16, leaving the other hyperparameters at their default values. For each document, we generate features by first applying a forward pass over the fine-tuned network, and then averaging the embeddings produced for the special token [CLS] across all the 12 layers of RoBERTa. In our initial experiments, this latter approach yielded slightly better results than using the [CLS] embedding of the last layer alone. The embedding size of RoBERTa, and hence the number of dimensions of our vectors, amounts to 768.

[2] http://jmcauley.ucsd.edu/data/amazon/links.html.

[3] https://huggingface.co/docs/transformers/model_doc/roberta.

We make the AMAZON-OQ-BK dataset publicly available (See footnote 1), both in its raw textual form and in its processed vector form.

The FACT-OQ dataset is extracted from the open dataset[4] of the FACT telescope [3]; here, a data item corresponds to a single telescope recording. We represent each data item in terms of the 20 dense features that are extracted by the standard processing pipeline[5] of the telescope. Each of the 1,851,297 recordings is labelled with the energy of the corresponding astro-particle, and our goal is to estimate the distribution of these energy labels via OQ. While the energy labels are originally continuous, astro-particle physicists have established a common practice of dividing the range of energy values into ordinal classes, as argued in Sect. 3.2. Based on discussions with astro-particle physicists, we divide the range of continuous energy values into an ordered set of 12 classes.

4.3 Results with Ordinal Classifiers

In our first experiment, we investigate whether OQ can be solved by non-ordinal quantification methods that embed ordinal classifiers. To this end, we compare the use of a standard multiclass logistic regression (LR) with the use of several ordinal variants of LR. In general, we have found that LR models, trained on the deep RoBERTa embedding of the AMAZON-OQ-BK dataset, are extremely powerful models in terms of quantification performance. Therefore, approaching OQ with ordinal LR variants embedded in non-ordinal quantifiers could be a straightforward solution worth investigating.

The ordinal LR variants we test are the "All Threshold" variant (OLR-AT) and the "Immediate-Threshold variant" (OLR-IT) of [29]. In addition, we try two ordinal classification methods based on discretizing the outputs generated by regression models [28]; the first is based on *Ridge Regression* (ORidge) while the second, called *Least Absolute Deviation* (LAD), is based on linear SVMs.

Table 1 reports the results of this experiment, using the non-ordinal quantifiers of Sect. 3.1. The fact that the best results are almost always obtained by using, as the embedded classifier, non-ordinal LR shows that, in order to deliver accurate estimates of class prevalence values in the ordinal case, it is not sufficient to equip a multiclass quantifier with an ordinal classifier. Moreover, the poor results of PCC, PACC, and SLD, the three methods that make use of posterior probabilities, suggest that the quality of the posterior probabilities returned by the ordinal classifiers may be sub-optimal.

Overall, these results suggest that, in order to tackle OQ, we cannot simply rely on ordinal classifiers embedded in non-ordinal quantification methods. Instead, we need "real" OQ methods.

---

[4] https://factdata.app.tu-dortmund.de/.

[5] https://github.com/fact-project/open_crab_sample_analysis/.

Table 1. Performance of classifiers in terms of average NMD (lower is better) in the AMAZON-OQ-BK dataset. **Boldface** indicates the best classifier for each quantification method, or a classifier not significantly different from the best one in terms of a paired Wilcoxon signed-rank test at a confidence level of $p = 0.01$. For LR we present standard deviations, while for all other classifiers we show the average deterioration in NMD with respect to LR. PCC, PACC, and SLD require a soft classifier, which means that ORidge and LAD cannot be embedded in these methods.

	CC	PCC	ACC	PACC	SLD
LR	**.0526** ±.0190	**.0629** ±.0215	.0247 ±.0096	**.0206** ±.0080	**.0174** ±.0068
OLR-AT	.0527 (+0.2%)	.0657 (+4.4%)	**.0237** (−4.4%)	.0219 (+6.5%)	.0210 (+20.5%)
OLR-IT	**.0526** (+0.0%)	.0695 (+10.4%)	.0256 (+3.6%)	.0215 (+4.5%)	.0648 (+271.8%)
ORidge	.0550 (+4.5%)	–	.0244 (−1.6%)	–	–
LAD	**.0527** (+0.3%)	–	**.0240** (−3.1%)	–	–

4.4 Results of the Quantifier Comparison

In our main experiment, we compare our proposed methods o-ACC, o-PACC, and o-SLD with several baselines, i.e., (i) the existing OQ methods OQT [10] and ARC [13], which we further detail in the supplementary material (See footnote 1), (ii) the "unfolding" OQ methods IBU and RUN (see Sect. 3.2), and (iii) the non-ordinal methods CC, PCC, ACC, PACC, and SLD. We compare these methods on the AMAZON-OQ-BK and FACT-OQ datasets, and under the APP and APP-OQ protocols.

Each method is allowed to tune the hyperparameters of its embedded classifier, using the samples of the validation set. We use logistic regression on AMAZON-OQ-BK and probability-calibrated decision trees on FACT-OQ; this choice of classifiers is motivated by common practice in the fields where these data sets originate, and from our own experience that these classifiers work well on the respective type of data. After the hyperparameters of the classifier are optimized, we apply each method to the samples of the test set.

The results of this experiment are summarized in Table 2. These results show that our proposed methods outperform the competition on both data sets if the ordinal APP-OQ protocol is employed. More specifically, o-SLD is the best method on AMAZON-OQ-BK while o-PACC is the best method on FACT-OQ. Moreover, o-SLD is consistently better or equal to SLD, o-ACC is consistently better or equal to ACC, and o-PACC is consistently better or equal to PACC, also in the standard APP protocol, where smoothness is not imposed.

Using RNOD as an alternative error measure confirms these conclusions, while experiments carried out using additional datasets and using TFIDF as an alternative vectorial representation in AMAZON-OQ-BK, even reinforce these conclusions. We provide these results in the supplementary material (See footnote 1).

Table 2. Average performance in terms of NMD (lower is better). For each data set (Amazon-OQ-BK and Fact-OQ), we present the results of the two protocols APP and APP-OQ. The best performance in each column is highlighted in **boldface**. According to a Wilcoxon signed rank test with $p = 0.01$, all other methods are statistically significantly different from the best method.

Method	Amazon-OQ-BK		Fact-OQ	
	APP	APP-OQ	APP	APP-OQ
CC	$.0526 \pm .019$	$.0344 \pm .013$	$.0534 \pm .012$	$.0494 \pm .011$
PCC	$.0629 \pm .022$	$.0440 \pm .017$	$.0651 \pm .017$	$.0621 \pm .017$
ACC	$.0229 \pm .009$	$.0193 \pm .007$	$.0582 \pm .028$	$.0575 \pm .028$
PACC	$.0209 \pm .008$	$.0176 \pm .007$	$.0791 \pm .048$	$.0816 \pm .049$
SLD	$\mathbf{.0172 \pm .007}$	$.0154 \pm .006$	$.0373 \pm .010$	$.0355 \pm .009$
OQT	$.0775 \pm .026$	$.0587 \pm .027$	$.0746 \pm .019$	$.0731 \pm .020$
ARC	$.0641 \pm .023$	$.0477 \pm .015$	$.0566 \pm .014$	$.0568 \pm .016$
IBU	$.0253 \pm .010$	$.0197 \pm .007$	$\mathbf{.0213 \pm .005}$	$.0187 \pm .004$
RUN	$.0252 \pm .010$	$.0198 \pm .007$	$.0222 \pm .006$	$.0194 \pm .005$
o-ACC	$.0229 \pm .009$	$.0188 \pm .007$	$.0274 \pm .007$	$.0230 \pm .006$
o-PACC	$.0209 \pm .008$	$.0174 \pm .007$	$.0230 \pm .006$	$\mathbf{.0178 \pm .004}$
o-SLD	$.0173 \pm .007$	$\mathbf{.0152 \pm .006}$	$.0327 \pm .008$	$.0289 \pm .007$

5 Conclusion

We have carried out a thorough investigation of ordinal quantification, which includes (i) making available two datasets for OQ, generated according to the strong extraction protocols APP and APP-OQ, which overcome the limitations of existing OQ datasets, (ii) showing that OQ cannot be profitably tackled by simply embedding ordinal classifiers into non-ordinal quantification methods, (iii) proposing three OQ methods (o-ACC, o-PACC, and o-SLD) that combine intuitions from existing, non-ordinal quantification methods and existing, physics-inspired "unfolding" methods, and (iv) experimentally comparing our newly proposed OQ methods with existing non-ordinal quantification methods, ordinal quantification methods, and "unfolding" methods, which we have shown to be OQ methods under a different name. Our newly proposed methods outperform the competition when tested on "ordinally plausible" test data. Our supplementary material (See footnote 1) confirms these results with evaluations under a different error measure and with additional experiments that we have carried out on different datasets and using a different, vectorial representation of the text data.

At the heart of the success of our newly proposed method lies regularization, which is motivated by the assumption that typical OQ class prevalence vectors are smooth. In future work, we plan to attempt using regularization for turning other non-ordinal quantification methods into ordinal ones.

Acknowledgments. The work by M.B., A.M., and F.S. has been supported by the European Union's Horizon 2020 research and innovation programme under grant agreement No. 871042 (SoBigData++). M.B. and M.S. have further been supported by Deutsche Forschungsgemeinschaft (DFG) within the Collaborative Research Center SFB 876 "Providing Information by Resource-Constrained Data Analysis", project C3, https://sfb876.tu-dortmund.de. A.M. and F.S. have further been supported by the AI4Media project, funded by the European Commission (Grant 951911) under the H2020 Programme ICT-48-2020. The authors' opinions do not necessarily reflect those of the European Commission.

References

1. Aad, G., Abbott, B., Abbott, D.C., et al.: Measurements of the inclusive and differential production cross sections of a top-quark-antiquark pair in association with a Z boson at $\sqrt{s} = 13$ TeV with the ATLAS detector. Europ. Phys. J. C **81**(8), 737 (2021)
2. Aartsen, M.G., Ackermann, M., Adams, J., et al.: Measurement of the ν_μ energy spectrum with IceCube-79. Europ. Phys. J. C **77**(10) (2017)
3. Anderhub, H., Backes, M., Biland, A., et al.: Design and operation of FACT, the first G-APD Cherenkov telescope. J. Inst. **8**(06), P06008 (2013)
4. Bella, A., Ferri, C., Hernández-Orallo, J., Ramírez-Quintana, M.J.: Quantification via probability estimators. In: International Conference on Data Mining (2010)
5. Blobel, V.: Unfolding methods in high-energy physics experiments. Technical report, DESY-84-118, CERN, Geneva, CH (1985)
6. Blobel, V.: An unfolding method for high-energy physics experiments. In: Advanced Statistical Techniques in Particle Physics, Durham, UK, pp. 258–267 (2002)
7. Börner, M., Hoinka, T., Meier, M., et al.: Measurement/simulation mismatches and multivariate data discretization in the machine learning era. In: Conference on Astronomical Data Analysis Software and Systems, pp. 431–434 (2017)
8. Bunse, M.: Unification of algorithms for quantification and unfolding. In: Workshop on Machine Learning for Astroparticle Physics and Astronomy. Gesellschaft für Informatik e.V. (2022, to appear)
9. Bunse, M., Piatkowski, N., Morik, K., Ruhe, T., Rhode, W.: Unification of deconvolution algorithms for Cherenkov astronomy. In: Data Science and Advanced Analytics, pp. 21–30 (2018)
10. Da San Martino, G., Gao, W., Sebastiani, F.: Ordinal text quantification. In: International ACM SIGIR Conference on Research and Development in Information Retrieval, pp. 937–940 (2016)
11. D'Agostini, G.: A multidimensional unfolding method based on Bayes' theorem. Nucl. Instr. Meth. Phys. Res.: Sect. A **362**(2–3), 487–498 (1995)
12. D'Agostini, G.: Improved iterative Bayesian unfolding (2010). arXiv:1010.0632
13. Esuli, A.: ISTI-CNR at SemEval-2016 task 4: quantification on an ordinal scale. In: International Workshop on Semantic Evaluation, pp. 92–95 (2016)
14. Esuli, A., Moreo, A., Sebastiani, F.: LeQua@CLEF2022: learning to quantify. In: Hagen, M., et al. (eds.) ECIR 2022. LNCS, vol. 13186, pp. 374–381. Springer, Cham (2022). https://doi.org/10.1007/978-3-030-99739-7_47
15. Esuli, A., Sebastiani, F.: Sentiment quantification. IEEE Intell. Syst. **25**(4), 72–75 (2010)

16. Forman, G.: Counting positives accurately despite inaccurate classification. In: Gama, J., Camacho, R., Brazdil, P.B., Jorge, A.M., Torgo, L. (eds.) ECML 2005. LNCS (LNAI), vol. 3720, pp. 564–575. Springer, Heidelberg (2005). https://doi.org/10.1007/11564096_55
17. Gao, W., Sebastiani, F.: From classification to quantification in tweet sentiment analysis. Soc. Netw. Anal. Min. **6**(1), 1–22 (2016). https://doi.org/10.1007/s13278-016-0327-z
18. Higashinaka, R., Funakoshi, K., Inaba, M., Tsunomori, Y., Takahashi, T., Kaji, N.: Overview of the 3rd dialogue breakdown detection challenge. In: Dialog System Technology Challenge (2017)
19. Hoecker, A., Kartvelishvili, V.: SVD approach to data unfolding. Nucl. Instr. Meth. Phys. Res.: Sect. A **372**(3), 469–481 (1996)
20. Liu, Y., et al.: RoBERTa: a robustly optimized BERT pretraining approach (2019). arXiv:1907.11692
21. McAuley, J.J., Targett, C., Shi, Q., van den Hengel, A.: Image-based recommendations on styles and substitutes. In: International ACM SIGIR Conference on Research and Development in Information Retrieval, pp. 43–52 (2015)
22. Moreno-Torres, J.G., Raeder, T., Alaíz-Rodríguez, R., Chawla, N.V., Herrera, F.: A unifying view on dataset shift in classification. Pattern Recogn. **45**(1), 521–530 (2012)
23. Mueller, J.L., Siltanen, S.: Linear and nonlinear inverse problems with practical applications. SIAM (2012)
24. Nachman, B., Urbanek, M., de Jong, W.A., Bauer, C.W.: Unfolding quantum computer readout noise. NPJ Quant. Inf. **6**(1), 84 (2020)
25. Nakov, P., Ritter, A., Rosenthal, S., Sebastiani, F., Stoyanov, V.: SemEval-2016 task 4: sentiment analysis in Twitter. In: International Workshop on Semantic Evaluation, pp. 1–18 (2016)
26. Nocedal, J., Wright, S.J.: Numerical Optimization. Springer, Cham (2006). https://doi.org/10.1007/978-0-387-40065-5
27. Nöthe, M., Adam, J., Ahnen, M.L., et al.: FACT - performance of the first Cherenkov telescope observing with SiPMs. In: International Cosmic Ray Conference (2018)
28. Pedregosa, F., Bach, F., Gramfort, A.: On the consistency of ordinal regression methods. J. Mach. Learn. Res. **18**, 55:1–55:35 (2017)
29. Rennie, J.D., Srebro, N.: Loss functions for preference levels: regression with discrete ordered labels. In: IJCAI 2005 Workshop on Advances in Preference Handling (2005)
30. Rosenthal, S., Farra, N., Nakov, P.: SemEval-2017 task 4: sentiment analysis in Twitter. In: International Workshop on Semantic Evaluation, pp. 502–518 (2017)
31. Rubner, Y., Tomasi, C., Guibas, L.J.: A metric for distributions with applications to image databases. In: International Conference on Computer Vision, pp. 59–66 (1998)
32. Ruhe, T., Schmitz, M., Voigt, T., Wornowizki, M.: DSEA: a data mining approach to unfolding. In: International Cosmic Ray Conference, pp. 3354–3357 (2013)
33. Saerens, M., Latinne, P., Decaestecker, C.: Adjusting the outputs of a classifier to new a priori probabilities: a simple procedure. Neural Comput. **14**(1), 21–41 (2002)
34. Sakai, T.: Comparing two binned probability distributions for information access evaluation. In: International ACM SIGIR Conference on Research and Development in Information Retrieval, pp. 1073–1076 (2018)

35. Schmelling, M.: The method of reduced cross-entropy: a general approach to unfold probability distributions. Nucl. Instr. Meth. Phys. Res.: Sect. A **340**(2), 400–412 (1994)
36. Schmitt, S.: TUnfold, an algorithm for correcting migration effects in high energy physics. J. Inst. **7**(10), T10003 (2012)
37. Smith, N.A., Tromble, R.W.: Sampling uniformly from the unit simplex. Technical report, Johns Hopkins University (2004)
38. Werman, M., Peleg, S., Rosenfeld, A.: A distance metric for multidimensional histograms. Comput. Vis. Graph. Image Proc. **32**, 328–336 (1985)
39. Zeng, Z., Kato, S., Sakai, T.: Overview of the NTCIR-14 short text conversation task: dialogue quality and nugget detection subtasks. In: NTCIR (2019)
40. Zeng, Z., Kato, S., Sakai, T., Kang, I.: Overview of the NTCIR-15 dialogue evaluation task (DialEval-1). In: NTCIR (2020)

Random Similarity Forests

Maciej Piernik[1,2(✉)], Dariusz Brzezinski[1,2,3], and Pawel Zawadzki[2,4]

[1] Faculty of Computing and Telecommunications, Institute of Computing Science, Poznan University of Technology, Piotrowo 2, 60-965 Poznan, Poland
[2] MNM Bioscience Inc., 1 Broadway, Cambridge, MA 02142, USA
{maciej.piernik,dariusz.brzezinski,pawel.zawadzki}@mnm.bio
[3] Institute of Bioorganic Chemistry of the Polish Academy of Sciences, Zygmunta Noskowskiego 12/14, 61-704 Poznan, Poland
[4] Physics/Biology Department, Faculty of Physics, Adam Mickiewicz University, Uniwersytetu Poznanskiego 2, 61-614 Poznan, Poland

Abstract. The wealth of data being gathered about humans and their surroundings drives new machine learning applications in various fields. Consequently, more and more often, classifiers are trained using not only numerical data but also complex data objects. For example, multi-omics analyses attempt to combine numerical descriptions with distributions, time series data, discrete sequences, and graphs. Such integration of data from different domains requires either omitting some of the data, creating separate models for different formats, or simplifying some of the data to adhere to a shared scale and format, all of which can hinder predictive performance. In this paper, we propose a classification method capable of handling datasets with features of arbitrary data types while retaining each feature's characteristic. The proposed algorithm, called Random Similarity Forest, uses multiple domain-specific distance measures to combine the predictive performance of Random Forests with the flexibility of Similarity Forests. We show that Random Similarity Forests are on par with Random Forests on numerical data and outperform them on datasets from complex or mixed data domains. Our results highlight the applicability of Random Similarity Forests to noisy, multi-source datasets that are becoming ubiquitous in high-impact life science projects.

1 Introduction

Over the last decades, machine learning has matured to the point in which researchers are spoiled for choice with classifiers for *scalar* data. There are hundreds of classifiers to choose from and their performance has been tested on thousands of datasets spanning across various domains [6]. Increasingly, however, researchers are more and more often faced with datasets of *complex* composition, where numerical features are not explicitly available. In such cases, there are two main options to choose from. Either to transform complex objects into many simple features, which can then be used by any traditional feature-based classifier (e.g., Random Forests [3]) or a dedicated one (e.g., CNNs for images [9]), or to rely on a domain-specific distance measure and use distance-based classifiers

© The Author(s), under exclusive license to Springer Nature Switzerland AG 2023
M.-R. Amini et al. (Eds.): ECML PKDD 2022, LNAI 13717, pp. 53–69, 2023.
https://doi.org/10.1007/978-3-031-26419-1_4

(e.g., Similarity Forests [15]). However, there is a third option that relies on the fact that many complex data types can be broken down into both traditional features (e.g., numerical or categorical) and simpler but still complex substructures for which good distance measures exist. For example, sequences can be decomposed into subsequences, sets of elements, distributions of elements, or distributions of lengths of subsequences. These complex substructures can then be combined together with simple, numerical or categorical, features and form a *mixed* type dataset to better capture the detailed characteristics of each example. Unfortunately, the currently available classification methods inherently enable only one of the first two types of analyses, on scalar data or complex objects. This limitation may prohibit us from using the full potential hidden in the data, as it forces us to either reduce all complex structures into simple features (most probably with information loss) or process all features with a single, complex distance measure (which is prone to the curse of dimensionality).

Consider a scenario of building a tumor genome classifier predicting patients responding to a given treatment based on whole-genome sequencing (WGS) [1]. Although, theoretically, a distance-based approach can be used in this scenario, it is unlikely to produce good results because of the curse of dimensionality and computational infeasibility (as the reference human genome is ~3.2 billion base pairs long). Moreover, this approach also neglects decades of research in the field of oncology and genomics, which provide a wide array of potentially useful biomarkers. A more natural approach in this scenario is to use a feature-based classifier that can make use of existing biomarkers as well as explore new ones. Currently however, this approach necessitates a drastic decomposition of the data into a fixed set of scalar features. At the same time, the information in the human genome can be decomposed into many intermediate structures carrying significantly more information than their scalar counterparts. Examples of such structures include variant distributions, sequential patterns, gene sets, interaction graphs. All of these structures have multiple distance measures available and could be used in conjunction with existing numerical biomarkers as a single dataset to produce more informed predictions. However, to the best of our knowledge, there does not exist any classifier capable of handling data with mixed, arbitrarily defined types of features.

In this paper, we propose a new classifier, called Random Similarity Forests (RSF), capable of handling datasets with mixed, arbitrarily defined feature types. Our approach requires a distance measure to be provided for each feature and works as a blend of Random Forests and Similarity Forests. By using a mixture of feature types, Random Similarity Forests inherit the benefits of both feature-based and similarity-based methods. In the paper, we intuitively describe and formally define the proposed classifier. We also perform sensitivity tests with respect to the number of features analyzed in each tree node and the number of trees in the forest. The method is then experimentally evaluated against Random Forests and Similarity Forests on datasets with scalar features, complex objects with a distance measure, and mixed, arbitrarily defined features with distance measures. Our analysis highlights the characteristic properties of each approach and discusses their suitability for different dataset types.

2 Related Work

Random forests [3] are one of the most popular classifiers and have been shown to offer the best quality of predictions across a wide range of datasets when compared to other classifiers [6]. However, like many other approaches, their use is limited to scalar data and, therefore, requires feature extraction when dealing with complex objects. Due to this fact, Similarity Forests [15] have been recently proposed as an alternative to Random Forests, especially when regular features are absent but similarities or distances between examples are attainable.

When dealing with distance-based classification, kernel methods are a popular group to consider, especially since there have been studies exploring the relationship between Random Forests and kernel methods [16]. In general, distance-based classification is often applied when dealing with complex data, e.g., for microbiome, image, or time series classification. A different way of dealing with complex data in classification using distances is through a combination of clustering and distance-based feature extraction. This approach relies on unsupervised clustering of the data and later encoding the clusters as features, either through aggregate inter- and intra-cluster distances [17] or simply by encoding the distances to each cluster center as separate features [13].

Recently, a different combination of feature-based and distance-based methods has been proposed by Liang *et al.* [10], where the authors first use Random Forests to select the most important features according to their impurity-based importance and later use these features in a k-Nearest Neighbors classifier with dynamically selected distance measures.

Although the described existing approaches create forests or calculate distances between objects, they either require scalar features or are based on a single distance calculated for entire examples. Similarly, integrative approaches from biomedical studies that combine data from different domains either require a common feature representation, create separate data type specific models, or perform data re-scaling and simplification [14]. To the best of our knowledge, Random Similarity Forests are the only single-classifier model capable of dealing with datasets with a mixture of arbitrarily defined features without any data transformations.

3 Random Similarity Forest

3.1 Preliminaries

Given a dataset of *training examples* $\mathcal{X} = \{\boldsymbol{x_1}, \boldsymbol{x_2}, \ldots, \boldsymbol{x_n}\}$ and their corresponding *class labels* $\boldsymbol{y} = \{y_1, y_2, \ldots, y_n\}$, the task of a *classifier* is to predict the class label $\hat{y}$ of each *unlabeled example* $\hat{\boldsymbol{x}}$ (i.e., an example for which the class is unknown). Every example $\boldsymbol{x_i}$ is a list $(x_{i1}, \ldots, x_{ij}, \ldots, x_{ip})$ of the same length p, where each position j holds a value a given example has for a particular *feature* $F_1, F_2, \ldots, F_p \in \mathcal{F}$. We also denote all values of $\mathcal{X}$ across a given feature F_j by $\boldsymbol{x_{.j}}$. Each class label y_i falls into one of several categorical class values $\mathcal{C} = \{c_1, c_2, \ldots, c_m\}$. This is the standard definition of a classification problem.

In our case, each feature can be from a different, arbitrary data domain. Additionally, every feature $F_1, F_2, \ldots, F_p \in \mathcal{F}$ has an associated distance measure $\delta_1, \delta_2, \ldots, \delta_p \in \Delta$. Therefore, the distance between any two examples $\boldsymbol{x_i}$, $\boldsymbol{x_j}$ for a given feature F_l is equal to $\delta_l(x_{il}, x_{jl})$, or $\delta_l(\boldsymbol{x_i}, \boldsymbol{x_j})$ for short.

3.2 The Algorithm

Just like Random Forests and Similarity Forests, Random Similarity Forests rely on bagging of a user-defined number (*max_trees*) of single-tree classifiers. A single Random Similarity Tree is constructed in a top-down fashion by recursively splitting the examples in each node into two exclusive subsets. The recursion stops when a given node contains only examples from a single class or when one of the early-stopping conditions is met (discussed in Sect. 3.4). Each split is calculated using a similarity-based 1-dimensional projection of the examples in a given node. First, a subset of *max_features* candidate features is selected, and afterward, *max_pairs* pairs of examples are picked at random for each feature. The examples in a pair are selected so that they come from different classes and have a different value on a given feature.

For a given pair of examples $\boldsymbol{x_p}$, $\boldsymbol{x_q}$, the distance between all other examples $\boldsymbol{x_i} \in \mathcal{X}_{node} \setminus \{\boldsymbol{x_p}, \boldsymbol{x_q}\}$ and the two selected examples is evaluated for each selected feature F_l in order to create a projection $\mathcal{P}(\boldsymbol{x_i})$ of each example $\boldsymbol{x_i}$ into a direction defined by $\boldsymbol{x_p}$ and $\boldsymbol{x_q}$. As shown by Sathe and Aggarwal [15], for the purposes of constructing a split, this projection can be approximated with $\mathcal{P}(\boldsymbol{x_i}) \propto \delta_l(\boldsymbol{x_q}, \boldsymbol{x_i}) - \delta_l(\boldsymbol{x_p}, \boldsymbol{x_i})$, as this approximation preserves the order of the original projection and the order is the only information necessary to construct a split. Therefore, for each pair of examples $\boldsymbol{x_p}$, $\boldsymbol{x_q}$, all remaining examples $\boldsymbol{x_i}$ in a given node are ordered by $\delta_l(\boldsymbol{x_q}, \boldsymbol{x_i}) - \delta_l(\boldsymbol{x_p}, \boldsymbol{x_i})$ along a given feature F_l.

This projection approximation can be viewed as a dynamic feature. After the projection has been computed, it can be used to perform a split just like any typical numerical feature would be used in a regular decision tree. The split point is selected as the point which minimizes the weighted average of the Gini index of the children nodes. For a given node N_i, the Gini index is defined as $G(N_i) = 1 - \sum_{c \in C} p_c^2$, where p_c is the fraction of examples of class c in a given node. Therefore, given two nodes N_i, N_j consisting of n_i, n_j examples, respectively, the weighted Gini index is calculated as:

$$G(N_i, N_j) = \frac{n_i G(N_i) + n_j G(N_j)}{n_i + n_j}. \tag{1}$$

Here, we opted for the Gini-index-based splitting strategy used also in Similarity Forests, however, it is important to note that after the projection is done we are in fact in possession of a dynamically calculated numerical feature. Therefore, other measures, such as entropy, can also be used.

Given the order defined with the projection and an impurity measure, all possible splits along a given feature are evaluated and the split that minimizes the impurity is selected. After performing this procedure for all selected features in a given node, the feature producing the best split is selected. Subsequently, two new tree nodes are created and the examples falling below the split threshold are placed in one node while the remaining examples are placed in the other node, and the whole procedure is repeated until all leaf nodes are pure or an early-stopping criterion is met.

The above-described process is illustrated in Fig. 1. The pseudocode of training a single Random Similarity Tree is presented in Algorithm 1.

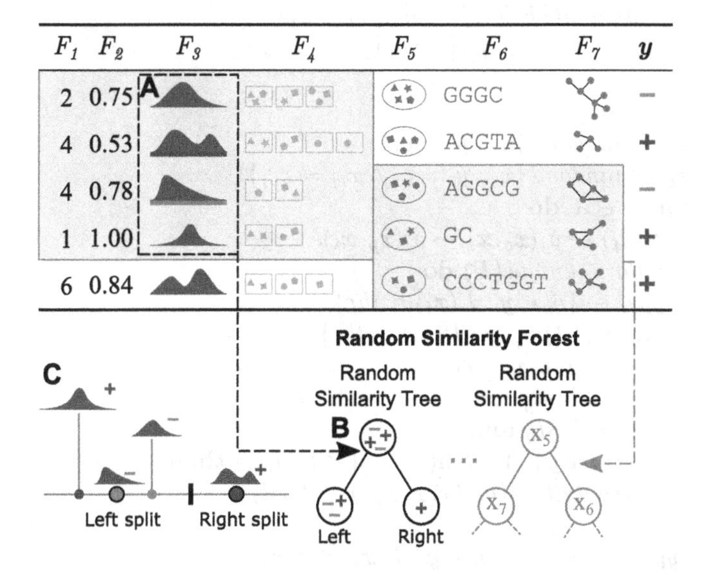

Fig. 1. An example illustrating the training process of Random Similarity Forests. Gray areas symbolize the bootstrap samples and subset of features going into each tree. Feature F_3 (**A**) is selected to perform a split in one of the trees (**B**). The distance-based projection (**C**) for that feature is calculated and the split point yielding the purest split is selected. The reference points (distributions) $\boldsymbol{x_p}, \boldsymbol{x_q}$ used to create the projection are underlined in red and blue in the table. (Color figure online)

After all *max_trees* Random Similarity Trees are computed, the trained Random Similarity Forest is ready to make predictions. For each node in each tree, given an unlabeled example $\hat{\boldsymbol{x}}$, the example is projected along the feature and the two examples stored in that node and assigned to one of the child nodes based on the stored split point. Following this procedure, once the example reaches a leaf node, it returns a weighted prediction of the majority class in that node. After predictions from all trees in the forest are made, the example is assigned to the class with the highest weighted average from all single-tree predictions.

Algorithm 1. Random Similarity Tree: $RST(\cdot)$

Require: $\mathcal{X}$: training examples; $\boldsymbol{y}$: examples' class labels; $\boldsymbol{\delta}$: distance measures for each feature; *max_features*: number of features picked randomly at each node; *max_pairs*: number of example pairs picked at random for each feature at each node
Ensure: a trained Random Similarity Tree model
1: **if** $earlyStopping()$ **or** $len(unique(\boldsymbol{y})) = 1$ **then**
2: $setLeaf()$
3: **return** *self*
4: **else**
5: **for** $1..max\_features$ **do**
6: $F_j \leftarrow random(\{F_j \in \mathcal{F} : var(\boldsymbol{x_{.j}}) > 0\})$
7: $c_1 \leftarrow \text{argmin}_{c \in unique(\boldsymbol{y})}\{var(\{x_{ij} : y_i = c\})\}$
8:
9: **for** $1..max\_pairs$ **do**
10: $\boldsymbol{x_p} \leftarrow random(\{\boldsymbol{x_i} : y_i = c_1\})$
11: $\boldsymbol{x_q} \leftarrow random(\{\boldsymbol{x_i} : y_i \neq c_1 \wedge x_{ij} \neq x_{pj}\})$
12: **for** $\boldsymbol{x_i} \in \mathcal{X}$ **do**
13: $P(\boldsymbol{x_i}) \leftarrow \delta_j(\boldsymbol{x_q}, \boldsymbol{x_i}) - \delta_j(\boldsymbol{x_p}, \boldsymbol{x_i})$
14: **for** $thr \in unique(P)$ **do**
15: $\boldsymbol{y_{left}} \leftarrow \{y_i \in \boldsymbol{y} : P(\boldsymbol{x_i}) \leq thr\}$
16: $\boldsymbol{y_{right}} \leftarrow \{y_i \in \boldsymbol{y} : P(\boldsymbol{x_i}) > thr\}$
17: $g \leftarrow G(\boldsymbol{y_{left}}, \boldsymbol{y_{right}})$
18: **if** $(g < best_g)$ **or** $(g = best_g$ **and** $|len(y_{left}) - len(y_{right})| < best_{bal})$ **then**
19: $update\_best(g, thr, F, \boldsymbol{x_p}, \boldsymbol{x_q}, P, bal)$
20:
21: $\mathcal{X}_{left}, \boldsymbol{y_{left}} \leftarrow \{\boldsymbol{x_i} \in \mathcal{X}, y_i \in \boldsymbol{y} : P(\boldsymbol{x_i}) \leq thr\}$
22: $\mathcal{X}_{right}, \boldsymbol{y_{right}} \leftarrow \{\boldsymbol{x_i} \in \mathcal{X}, y_i \in \boldsymbol{y} : P(\boldsymbol{x_i}) > thr\}$
23: $N_{left} \leftarrow RST(\mathcal{X}_{left}, \boldsymbol{y_{left}}, \boldsymbol{\delta}, max\_features, max\_pairs)$
24: $N_{right} \leftarrow RST(\mathcal{X}_{right}, \boldsymbol{y_{right}}, \boldsymbol{\delta}, max\_features, max\_pairs)$
25: **return** *self*

3.3 Computational Complexity

Let us now discuss the computational complexity of Random Similarity Forests (RSF). Unsurprisingly, it is strongly related to the complexity of Similarity Forests (SF), which is $\mathcal{O}(n \cdot \log n)$. However, RSF additionally adds the cost of checking multiple features at each node, so the complexity is raised by that factor to $\mathcal{O}(p \cdot n \cdot \log n)$, which is equivalent to the complexity of Random Forests (RF). However, it is worth noting that in practice the value of the *max_features* parameter for RSF may need to be higher than for RF as it is inherently more random as it randomly picks both features and pairs of examples. Moreover, even though the number of pairs is a parameter, we recommend picking only a single pair for each feature, as suggested by Sathe and Aggrawal [15].

Another cost hidden in the computational complexity is the cost of calculating the distances between examples. It is an essential factor because, depending on the feature type, it can either be negligible or a very costly operation. For simple numerical features, the distance computation can be usually omitted as for any metric it would produce a projection equivalent to the original feature in terms of the example order. On the other hand, more complex data types may require substantial computation, e.g., edit distance has quadratic complexity for sequential data and is NP-complete for graphs. Therefore, the use of such metrics may be prohibitive for longer sequences or larger graphs.

3.4 Discussion

Since Random Similarity Forests draw from Random Forests and Similarity Forests, they combine their advantages and possess unique properties:

1. By using feature-oriented similarity metrics to split data points, Random Similarity Forests can be used to classify datasets characterized by numerical as well as complex data features. Random Forests cannot classify complex data objects, whereas Similarity Forests treat entire examples as data objects, hiding the characteristics of individual features.
2. By sampling feature subsets for each node split and by analyzing each feature separately, Random Similarity Forests are robust to noisy datasets with a lot of features. Similarity Forests treat entire examples as data objects, which can hinder their performance on datasets with many irrelevant features.
3. Analyzing similarities based on each feature separately allows Random Similarity Forests to use several similarity measures to analyze the same objects and let the forest decide which measures are the most useful for classification.

In addition to combining the advantages from both Random Forests and Similarity Forests, Random Similarity Forests also inherit some of their limitations. The first limitation comes from Similarity Forests and it is the fact that it is designed for binary classification problems. However, one can easily adapt it to multi-class classification or regression problems analogously to the method proposed for Similarity Forests [4], or by using the one-vs-all approach.

Another characteristic worth discussing is the effect of feature variance on node splits. If one would select the pair of examples $(\boldsymbol{x_p}, \boldsymbol{x_q})$ with the same value of feature F_j, then they would produce a completely random ordering of examples for the split (since if $x_{pj} = x_{qj}$ then $\forall \boldsymbol{x_i} : \delta_j(\boldsymbol{x_q}, \boldsymbol{x_i}) - \delta_j(\boldsymbol{x_p}, \boldsymbol{x_i}) = 0$). Similar problems appear in Random Forests when the feature values are the same for examples of different classes, whereas Similarity Forests do not suffer from this particular limitation so often as they always consider each example as a whole. For datasets with many features with low variance, this reduces the effective number of features tested at each node and promotes high-variance features. That is why Random Similarity Forests will perform better on datasets with features with high variance. This limitation can be overcome by tuning the *max_features* hyperparameter. As shown by Geurts *et al.* [7], the best value for this parameter highly depends on the number of correlated and irrelevant features: lower values work better with highly correlated features while higher values with irrelevant features. Indeed, as will be shown in Sect. 4.3, the proposed default value of *max_features* for Random Similarity Forests is $0.5p$, which is higher than $\sqrt{p}$ commonly used for Random Forests [7].

Other issues worth mentioning are the early-stopping conditions (pre-pruning) and pruning (post-pruning). There are many possible options to stop the tree construction process before it produces perfectly pure leaf nodes. These could include, e.g., limiting the maximum depth a tree can reach, specifying the minimum number of elements required to perform a split or the minimum number of elements to form a leaf node. Similarly, there are multiple ways in which a tree can be pruned, with arguably the most popular method being minimal cost complexity pruning [2]. However, pre-and post-pruning methods are not specific to the proposed algorithm or any other forest algorithm and are out of the main scope of this paper.

4 Experimental Evaluation

In this section, we experimentally evaluate the Random Similarity Forests algorithm and illustrate its usefulness in real-world scenarios. Since the proposed classifier is related to both Random Forests and Similarity Forests, the nature of the evaluation is highly comparative. Our goal is to showcase the strengths and limitations of our proposal and indicate when its use would be the most beneficial and when it might be preferable to rely on one of the alternatives. First, we compare the classification performance of the three classifiers on ten publicly available datasets with simple numerical features. Next, we compare the three approaches on complex data, i.e., datasets consisting of arbitrarily defined objects with distance measures. For this purpose, we use sequences of sets, time series, and graph datasets. Finally, we illustrate the usefulness of our proposal by training the classifiers on datasets consisting of mixtures of complex object-like features and simple numerical features.

4.1 Experimental Setup

In all of the experiments, we compare Random Similarity Forests (RSF) against Random Forests (RF) and Similarity Forests (SF) using the area under the ROC curve (AUC). We chose AUC since we focus on binary classification, it is skew invariant, assesses the classifiers' ranking abilities, and it has been shown to be statistically consistent and more discriminating than accuracy [8]. To evaluate each approach on each dataset, we rely on 10 repetitions of stratified 2-fold cross-validation (10×2CV). Afterwards, the results undergo a series of statistical tests [5]. We rely on the Friedman statistic with a post-hoc Nemenyi test to distinguish between the compared approaches across all datasets. Afterwards, we check for differences on each individual dataset. First, we verify the normality assumption using the Shapiro-Wilk test. If the assumption is met, we use the ANOVA statistic with a post-hoc Tukey HSD test. If the assumption is not met, we once again resort to the non-parametric Friedman statistic with a post-hoc Nemenyi test [5]. All tests were performed at significance level $\alpha = 0.05$.

The code used to perform the experiments was written in Python with parts of the implementation of Random Similarity Forests written in Cython to achieve higher efficiency. For Random Forests, we use the scikit-learn implementation [12], whereas for Similarity Forests, we rely on the implementation described in [4]. Unless stated otherwise, we rely on the default hyperparameters for each classifier, as specified in their implementations. The source code, datasets, and reproducible experimental notebook are available upon request.

4.2 Datasets

Since our experimental analysis focuses on data with scalar, complex, and mixed types of features, we use three groups of datasets, each corresponding to a given type of features. Additionally, we have a separate set of datasets dedicated to performing sensitivity analyses. The choice of datasets was made a priori and independently of the results obtained with our methods.

Datasets for Sensitivity Analysis. For sensitivity analysis, we used binarized versions of 12 numerical datasets: ten from the UCI Machine Learning[1] repository and two introduced by Breiman; for a detailed description of these datasets see [7]. We chose these datasets as they were used to discuss the impact of different hyperparameter values for Random Forests and Extra Trees [7]. Importantly, these datasets are independent of those used for assessing the predictive performance of RF, SF, and RSF, to avoid biasing the comparative analysis.

Scalar Data. For the analysis of scalar data, we used ten publicly available datasets with numeric features. The datasets exhibit various conditions in terms of the number of features, learning sample size, and feature variance, and can be accessed through the LIBSVM dataset repository.[2] We used the scaled versions of the datasets, as required by Similarity Forests. Each dataset was additionally preprocessed by removing features containing missing values. Table 1 presents the main characteristics of each dataset after preprocessing.

Complex Data. For the analysis of complex data (i.e., data where each example is an object without any explicitly defined numerical features), we used three groups of datasets consisting of sequences of sets, time series, and graphs, respectively.

[1] https://archive.ics.uci.edu/ml/.
[2] https://www.csie.ntu.edu.tw/~cjlin/libsvmtools/datasets/.

Sequences of sets are a good candidate for complex data because they combine set information, which can be easily encoded with numerical features, and sequential information, which is very difficult to encode as scalars but manifests itself through distance measures on whole objects. Therefore, we used a synthetic data generator[3] to create three datasets of sequences of sets: `items`, `lengths`, and `order`. The sequences in the `items` dataset have the same mean length and mean set size, and are differentiated by the distributions of the selected items between the two classes. Sequences in the `lengths` dataset are generated with identical set size and item distributions but are differentiated by the lengths between the classes. Finally, all the sequences in the `order` dataset are generated from the same length, set size, and item distributions, and are only differentiated by the order of the elements in one of the classes.

Time series datasets were taken from the UCR Time Series Classification Archive[4], with distances between objects calculated with dynamic time warping. The graph datasets were taken from the TU-Dortmund Graph Kernel Benchmarks[5] with distances between objects calculated using the Ipsen-Mikhailov distance. In both cases, we limited our selection to binary classification problems, with varying numbers of examples and features (length, sets, nodes, edges).

Mixed Data. For the analysis of mixed data (i.e., data where each example is described by arbitrarily defined features with a distance measure for each feature), we used four groups of datasets consisting of time series, graphs, multi-omics data, and genome sequencing data.

Time series datasets consisted of three distance measures (euclidean, cosine, dynamic time warping) computed for series from the UCR Repository.[6] Similarly, the graph datasets consisted of four distance measures (portrait divergence, Jaccard distance, degree divergence, Ipsen-Mikhailov distance) computed for graphs from the TU-Dortmund repository[7] using the netrd package [11]. The multi-omics datasets consisted of Alzheimer's Disease and BRCA-mutated breast cancer patients, described by three sets of numeric expression data.[8] Finally, the last group of datasets represented tumor samples of ovarian and breast cancer patients obtained using whole genome sequencing.[9] In this group of datasets, each sample is characterized by 50 gene amplifications represented as simple numerical features and 6 distributions of lengths of various DNA structural variants.

4.3 Sensitivity Analysis

Using the 12 datasets proposed in [7], we have analyzed the effect of *max_features* on Random Similarity Forests. The parameter denotes the number of potential

[3] https://gingerbread.shinyapps.io/SequencesOfSetsGenerator/.
[4] https://www.cs.ucr.edu/~eamonn/time_series_data_2018/.
[5] https://ls11-www.cs.tu-dortmund.de/staff/morris/graphkerneldatasets.
[6] https://www.cs.ucr.edu/~eamonn/time_series_data_2018/.
[7] https://ls11-www.cs.tu-dortmund.de/staff/morris/graphkerneldatasets.
[8] https://github.com/txWang/MOGONET.
[9] https://github.com/MNMdiagnostics/dbfe.

splits screened at each node during the growth of a Random Similarity Tree. It may be chosen in the interval $[1, \ldots, p]$ and the smaller its value the stronger the randomization of the trees. Figure 2 compares the AUC of Random Forests (RF) and Random Similarity Forests (RSF) for increasing values of *max_features.*

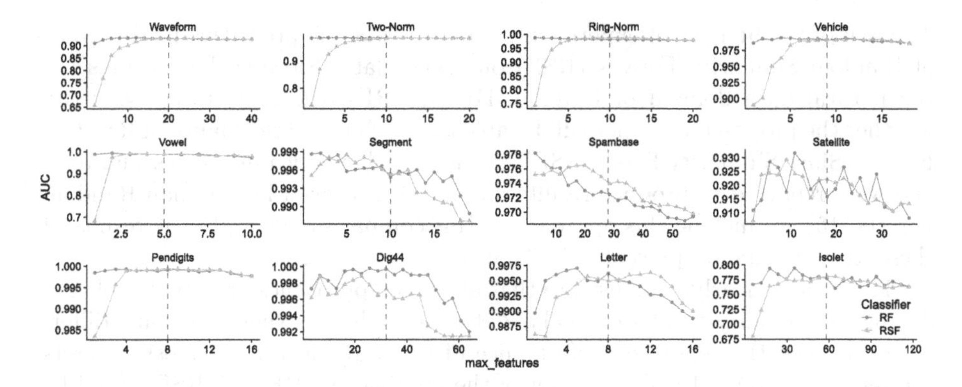

Fig. 2. Evolution of mean AUC of Random Forests (RF) and Random Similarity Forests (RSF) with varying *max_features* on 12 datasets. Dashed line shows the proposed RSF default value of $max\_features = 0.5p$.

One can notice that for all the datasets the performance of RSF first increases and then declines as *max_features* approaches the total number of features in a dataset p. As mentioned in Sect. 3.4, this phenomenon is related to the number of correlated and irrelevant features in a dataset: lower values work better with highly correlated features while higher values of *max_features* work better with irrelevant features. This is due to the fact that a higher value of *max_features* leads to a better chance of filtering out the irrelevant variables. These findings are in accordance with [7], were similar trends were noticed for the Extra Trees classifier. It is worth noting, however, that compared to RF, the increase of predictive performance with *max_features* is slower. This stems from the fact that RSF samples pairs of examples and creates different projections, thus introducing additional randomness. Therefore, we propose to use a default value of $max\_features = 0.5p$, as opposed to $\sqrt{p}$, which is the commonly used default for Random Forests and Extra Trees [7].

We have also analyzed the *max_trees* parameter, responsible for the number of trees in the ensemble (Fig. 3). The results are in accordance with studies on different tree ensembles [7]: the higher the value of *max_trees*, the better the ensemble's accuracy. It can be noticed that all three algorithms (RF, SF, RSF) have the same speed of convergence. In the remaining

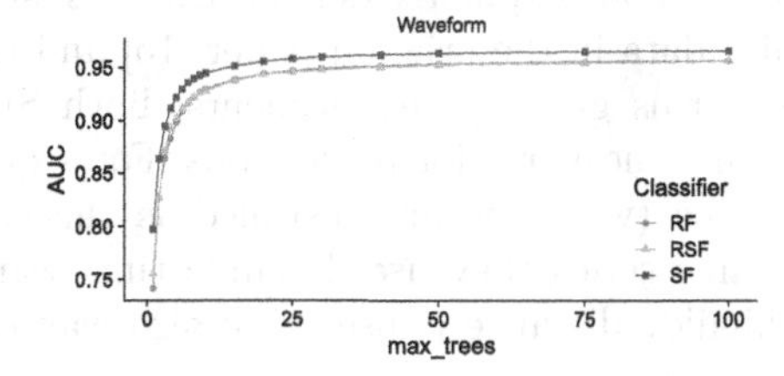

Fig. 3. Mean AUC of Random Forests (RF), Random Similarity Forests (RSF) and Similarity Forests (SF) with varying *max_trees* on the Waveform dataset.

experiments, for all three classifiers, we use *max_trees* = 100, which is large enough to ensure convergence of the ensemble effect on all our datasets.

4.4 Evaluation on Scalar Data

The aim of the first set of experiments was to assess the predictive performance of Random Similarity Forests (RSF) on scalar data. Since scalar features are the natural data domain of Random Forests (RF), the aim is also to verify whether the proposed classifier will be able to match the performance of Random Forests. Since Similarity Forests (SF) are also capable of dealing with scalar data and were reported to produce results on par with or even better than Random Forests [15], we include this approach in this comparison as well. The results of these experiments are presented in Table 1.

The results clearly demonstrate that all three approaches are capable of producing high quality results on scalar data. Nevertheless, there are some differences between the reported performance of the approaches. For most datasets the results are very close, and either of the approaches (RF, SF, RSF) could be used. However, on datasets `splice` and `svmguide3` RSF produces very similar results to RF, but both approaches significantly outperform SF by a large margin. This outcome is probably due to the fact that although both RSF and SF are distance-based, the former still treats each feature separately, while the latter calculates distances over the whole feature space which, in this case, clearly obfuscates the decision boundary. The significantly worse performance of SF on the `splice` and `svmguide3` datasets showcases that it should be used with more caution than RF and RSF, which confirms recent findings of Czekalski and Morzy [4].

4.5 Evaluation on Complex Data

The second experiment was designed to assess the performance of Random Similarity Forests on complex data. This type of data should be a natural environment for Similarity Forests, so analogously to the first experiment, the aim of this comparison is to verify if RSF can match SF on this kind of data. Importantly, this is also the first reported test of Similarity Forests on actual complex data without explicitly defined features, since the algorithm was only tested on scalar data in the original paper [15] and the follow-up study [4].

In this group of experiments, both SF and RSF rely on the same set of distance measures for projections. For sequences of sets the algorithms use edit distance (with Jaccard distance as the cost function for element relabeling), for time series they use dynamic time warping, whereas for graphs the Ipsen-Mikhailov distance is used. We also wanted to include RF in this comparison,

Table 1. Dataset characteristics and mean AUC performance of Random Forests (RF), Similarity Forests (SF), and Random Similarity Forests (RSF) from 10 × 2-fold cross-validation experiments. Features for sequences of sets denoted as [mean set size]/[mean length]. Features for graphs denoted as [mean number of nodes]/[mean number of edges].

Dataset	Type	Features	#Ex.	#Features	Distance	AUC		
						RF	RSF	SF
australian	scalar	numeric	690	8	euclidean	**0.92**[b]	0.91	0.92
breast			683	10		0.99	0.99	**1.00**[b]
diabetes			768	6		**0.80**	0.80	0.80
german			1000	18		**0.73**	0.73	0.73
heart			270	9		0.88	0.87	**0.89**[b]
ionosphere			351	2		0.84	**0.84**	0.84
liver			145	4		**0.77**	0.75	0.76
sonar			208	59		0.89	0.88	**0.91**[b]
splice			1000	60		0.99[b]	**0.99**[b]	0.90
svmguide3			1243	21		0.84[b]	**0.85**[b]	0.77
items	complex	sequences of sets	400	20/20	edit distance	**1.00**[a,c]	0.91[b]	0.85
length			400	20/20		0.86[c]	0.91[b]	**0.92**[b]
order			400	20/20		0.56[c]	**0.79**[a]	0.73[b]
computers		time series	500	720	dynamic time warping	0.68	0.77[b]	**0.78**[b]
housetwenty			159	2000		0.92	0.99[b]	**0.99**[b]
toeseg			268	277		0.85	0.96[b]	**0.97**[b]
cox2		graph	467	41/43	ipsen-mikhailov	0.65[c]	0.70[b]	**0.71**[b]
mutag			188	18/20		0.87[c]	**0.92**[b]	0.92[b]
proteins			1113	39/73		0.77[c]	**0.95**[a]	0.91[b]
ptcfm			349	14/14		0.58[c]	**0.63**[b]	0.61[b]
beetlefly	mixed	time series	40	512	euclidean, cosine, dtw	0.84	**0.85**	0.81[d]
birdchicken			40	512		0.89	**0.90**	0.87[d]
bzr		graphs	405	36/38	portrait, degree, jaccard, ipsen	0.73[c]	**0.90**[a]	0.87[d,b]
dhfr			756	42/45		0.78[c]	**0.85**[a]	0.84[d,b]
rosmap		multiomics (numeric)	351	600	euclidean, cosine	0.78	**0.79**[b]	0.76[d]
brca			875	2503		0.89	0.91[b]	**0.91**[d,b]
wgs_ovarian		distributions, numeric	219	56	euclidean	0.71[e]	**0.76**[a]	0.71[e]
wgs_her2+tnbc			286	56		0.99[e,b]	**0.99**[b]	0.82[e]
wgs_her2+her2-			742	56		**0.99**[e,b]	0.99[b]	0.83[e]
wgs_her2-tnbc			780	56		**0.90**[e,b]	0.90[b]	0.76[e]

[a] result significantly better than both competitors
[b] result significantly better than one of the competitors
[c] RF used a bag of words representation of the data
[d] maximum performance obtained by SF out of all available distance measures
[e] RF and SF used data with distribution features encoded as sums

however, RF is incapable of processing varied sized sequences and graphs in their original form. Therefore, for this classifier, we transform sequences of sets and graphs to a bag-of-words representation, where each sequence of sets/graph is represented as counts of items/nodes. Since time series were of the same length for a given dataset, RF could use individual time points as features, without any preprocessing. The results of this set of experiments are presented in Table 1.

We can observe that for the items dataset a simple bag-of-words representation is able to easily distinguish between the two classes, with RF significantly outperforming SF and RSF. This result clearly demonstrates that for easy problems, relying on approaches based on complex distance measures may actually be detrimental to the quality of predictions. In this case, the sequential information only obfuscates the true source of class information. On the length dataset, where examples in the two classes are differentiated only by the lengths of the sequences, RSF and SF both significantly outperformed RF. This is expected as the information about the length of the sequences is very easy to capture through edit distance. Finally, on the order dataset, where the elements in one of the classes are ordered while in the other they are not, RF was barely able to find any regularities and was significantly outperformed by both SF and RSF.

Results for real-world time series and graphs confirm the above observations. On each time series and graph dataset RSF and SF significantly outperformed RF. The differences between RSF and SF were much smaller and, with the exception of the proteins dataset, were not statistically significant.

4.6 Evaluation on Mixed Data

The final set of experiments was designed to evaluate Random Similarity Forests on datasets with a mix of simple numerical features and complex object-like features. These are the types of datasets that neither Random Forest nor Similarity Forest can process out of the box. Here, RF used only the numerical features, whereas SF treated the examples as complex objects. In cases where multiple distance measures were available for the data (time series, graphs, multiomics), RSF used each measure as an independent feature whereas SF used each of them individually. Therefore, for each dataset with multiple distances, SF was ran multiple times, each time with a different measure. The best performing distance was later taken into account during statistical tests.

The results clearly show that RSF is the best choice for mixed data. For 7 out of 10 datasets RSF achieved the best performance and was never statistically significantly worse than any other approach. It is worth noting that on the bzr and wgs_ovarian datasets, RSF outperforms RF and SF by a large margin. These are two stark examples that on some datasets the combination of numeric and complex data can offer much more than any of the component representations alone, and that multiomics data are potentially a good training ground for algorithms like RSF. Interestingly, there is no clear runner-up in this set of experiments. On the beetlefly, birdchicken, rosmap, and wgs_ datasets SF is the significantly underperforming algorithm, whereas on bzr, dhfr, brca RF takes last spot. This suggests that depending on the application domain, numeric or complex features play a more important role. Algorithms like RSF can take advantage of this property and steer towards the more rewarding feature types for a given dataset.

4.7 Summary of the Results

The main takeaways from the experiments are as follows.

- Due to the increased randomness, Random Similarity Forests usually need to evaluate more features (*max_features*) during splits than Random Forests.
- The proposed algorithm was able to match the performance of Random Forests and match or outperform Similarity Forests on datasets consisting of scalar features.
- Random Similarity Forest was also able to match or outperform Similarity Forests and outperform Random Forests on complex data.
- Random Similarity Forest was the only algorithm capable of using a mixture of scalar and complex features. This property helped Random Similarity Forests outperform Random Forests and Similarity Forests on datasets with both types of features. The experiments have shown that for real-world omics data transforming complex features to scalars or using a single example-wide distance measure may result in information loss.

5 Conclusions

In this paper, we have addressed the problem of classifying data with a mixture of scalar and complex features by proposing a new classifier, called Random Similarity Forest. Like other decision forest algorithms, such as Random Forests and Similarity Forests, the proposed algorithm relies on bagging of decision trees, but in contrast to other approaches splits the tree nodes according to distance-based projections of single feature values. This allows for a very flexible classification approach, in which traditional numerical features can be used alongside complex object-like features with domain-specific distance measures, taking full advantage of each data type. The proposed method was experimentally evaluated against Random Forests and Similarity Forests on datasets with scalar, complex, and mixed data features, showcasing its capabilities across all possible data domain configurations.

Experimental results on 30 datasets show that Random Similarity Forests can be considered a safe alternative to the well-established Random Forests on scalar datasets and are on par or exceed the performance of Similarity Forests on complex data. Most importantly, however, the experiments show that the proposed classifier was the only one inherently capable of dealing with datasets consisting of a mixture of scalar and complex features, and outperformed both of the alternatives on most datasets with mixed feature types. In situations were numeric and distance features offer complementary information, Random Similarity Forests can achieve results that are not attainable by the individual feature types alone. In situations were only numeric or only complex features are informative, the proposed algorithm achieves results that are close to the algorithm using only the informative representation.

The method proposed in this paper opens many new avenues for further research. Regarding the classifier itself, it would be interesting to verify the impact of different proportions of numeric and complex features on classifier performance. To control this effect, one could implement a weighting mechanism that influences how many features of each type are taken into account at a given tree node split. Moreover, the effects of other forms of randomization on the proposed algorithm, such as random splits in the projection space, could also be inspected. Finally, it will be interesting to observe what types of features and distance measures are most useful when applied to different types of multiomics data, which seem to be one of the most promising application domains for Random Similarity Forests.

Acknowledgment. This publication and the underlying study have been made possible by the data that the Hartwig Medical Foundation and the Center of Personalized Cancer Treatment have made available. This work was also supported by data obtained from ICGC Breast Cancer Working Group and The Cancer Genome Atlas. MP and DB acknowledge the support of PUT's Institute of Computing Science Statutory Funds.

References

1. Davies et al., H.: HRDetect is a predictor of BRCA1 and BRCA2 deficiency based on mutational signatures. Nat. Med. **23**(4), 517–525 (2017)
2. Breiman, L., Friedman, J.H., Olshen, R.A., Stone, C.J.: Classification and Regression Trees. Wadsworth International Group, Belmont (1984)
3. Breiman, L.: Random forests. Mach. Learn. **45**(1), 5–32 (2001)
4. Czekalski, S., Morzy, M.: Similarity forests revisited: a Swiss army knife for machine learning. In: Karlapalem, K., et al. (eds.) PAKDD 2021. LNCS (LNAI), vol. 12713, pp. 42–53. Springer, Cham (2021). https://doi.org/10.1007/978-3-030-75765-6_4
5. Demsar, J.: Statistical comparisons of classifiers over multiple data sets. J. Mach. Learn. Res. **7**, 1–30 (2006)
6. Fernández-Delgado, M., Cernadas, E., Barro, S., Amorim, D.: Do we need hundreds of classifiers to solve real world classification problems? J. Mach. Learn. Res. **15**(1), 3133–3181 (2014)
7. Geurts, P., Ernst, D., Wehenkel, L.: Extremely randomized trees. Mach. Learn. **63**(1), 3–42 (2006)
8. Huang, J., Ling, C.X.: Using AUC and accuracy in evaluating learning algorithms. IEEE Trans. Knowl. Data Eng. **17**(3), 299–310 (2005)
9. LeCun, Y., Bengio, Y., Hinton, G.E.: Deep learning. Nature **521**(7553), 436–444 (2015)
10. Liang, J., Liu, Q., Nie, N., Zeng, B., Zhang, Z.: An improved algorithm based on KNN and Random Forest. In: Proceedings of the 3rd International Conference on Computer Science and Application Engineering (2019)
11. McCabe, S., et al.: netrd: a library for network reconstruction and graph distances. J. Open Source Softw. **6**(62), 2990 (2021)
12. Pedregosa, F., et al.: Scikit-learn: machine learning in Python. J. Mach. Learn. Res. **12**, 2825–2830 (2011)
13. Piernik, M., Morzy, T.: A study on using data clustering for feature extraction to improve the quality of classification. Knowl. Inf. Syst. **63**, 1771–1805 (2021)
14. Reel, P.S., Reel, S., Pearson, E., Trucco, E., Jefferson, E.: Using machine learning approaches for multi-omics data analysis: a review. Biotechnol. Adv. **49**, 107739 (2021)

15. Sathe, S., Aggarwal, C.C.: Similarity forests. In: Proceedings of the 23rd ACM SIGKDD International Conference on Knowledge Discovery and Data Mining, pp. 395–403 (2017)
16. Scornet, E.: Random forests and kernel methods. IEEE Trans. Inf. Theory **62**(3), 1485–1500 (2016)
17. Tsai, C.F., Lin, W.Y., Hong, Z.F., Hsieh, C.Y.: Distance-based features in pattern classification. EURASIP J. Adv. Signal Process. **2011**(1), 62 (2011)

Spectral Ranking with Covariates

Siu Lun Chau(✉), Mihai Cucuringu, and Dino Sejdinovic

Department of Statistics, University of Oxford, 29 St Giles', Oxford OX1, UK
{siu.chau,mihai.cucuringu,dino.sejdinovic}@stats.ox.ac.uk

Abstract. We consider spectral approaches to the problem of ranking n players given their incomplete and noisy pairwise comparisons, but revisit this classical problem in light of player covariate information. We propose three spectral ranking methods that incorporate player covariates and are based on *seriation*, *low-rank structure* assumption and *canonical correlation*, respectively. Extensive numerical simulations on both synthetic and real-world data sets demonstrated that our proposed methods compare favourably to existing state-of-the-art covariate-based ranking algorithms.

Keywords: Ranking · Spectral methods · Kernel methods

1 Introduction

We consider the classical problem of ranking n players, given a set of pairwise comparisons, but revisiting this problem in light of available player covariate information. We assume ranking is represented by a vector $\mathbf{r} \in \mathbb{R}^n$, which can be interpreted as a one-dimensional representation of players' underlying skills. In practice, such pairwise comparisons are often incomplete and inconsistent with respect to the underlying skill vector $\mathbf{r}$; therefore, the objective of a ranking problem is often to recover a total ordering of the players that is as consistent as possible with the available noisy and sparse information.

There exists a rich literature on ranking that can be dated back as early as the seminal work of Kendall and Smith [16] in the 1940s. Classical examples such as the Bradley-Terry-Luce model [2], Random Utility model [28], Mallows-Condorcet model [18], and Thurstone-Mosteller model [19] have inspired numerous developments of modern ranking algorithms [4]. These methods take a probabilistic approach, in which they treat ranking as the output generated from a probabilistic model, where maximum likelihood estimators can be built subsequently.

Apart from probabilistic approaches, spectral methods that apply the theory of linear maps (in particular, eigenvalues and eigenvectors) to the pairwise comparison matrix or derivations of it, in order to extract rankings, is also a century-old [29] idea that was made famous by Google's PageRank [23] algorithm. There are several other popular spectral ranking methods proposed in recent years as well. For example, Fogel et al. [11] connected the problem of seriation [1] to ranking and proposed SerialRank algorithm. Cucuringu et al.

© The Author(s), under exclusive license to Springer Nature Switzerland AG 2023
M.-R. Amini et al. (Eds.): ECML PKDD 2022, LNAI 13717, pp. 70–86, 2023.
https://doi.org/10.1007/978-3-031-26419-1_5

[10], on the other hand, proposed SVD-RANK, a simple SVD-based approach to recover rankings when cardinal comparisons are observed. Cucuringu also proposed SYNCRANK [8] based on group synchronization using semidefinite programming and spectral relaxations.

In many instances, it is of interest to investigate whether some player covariates affect the results of the comparisons. It is also natural to believe that incorporating informative covariates will lead to better ranking estimations. Besides improving estimation, we believe there are at least two more reasons why this might be useful in practice:

1. Covariate-free ranking algorithms cannot take into account instances where new players are added to the comparison sets, without first observing new matches with existing players. This can be overcome by building a model of the ranking as a function of player covariates.
2. Traditional ranking methods often break down if the sample complexity of the comparison graph does not scale as $O(n \log n)$, meaning the graph is disconnected with high probability [10]. By utilising the smoothness of rankings across covariate information, one is able to overcome this obstacle and infer meaningful rankings even in the cases of very sparse comparisons that render the graph disconnected.

A number of extensions to incorporate covariates have been proposed for probabilistic approaches [7,22,26]. In contrast, spectral ranking with covariates received much lesser attention. To the best of our knowledge, the recent work of Jain et al. [15] was the first to incorporate covariates into spectral ranking. However, unlike our approaches, their approach cannot be used to predict rankings on new players.

To this end, we introduce a suite of spectral ranking algorithms incorporating covariate information: C-SERIALRANK, SVDCOVRANK and CCRANK. Both C-SERIALRANK and SVDCOVRANK extend on existing spectral ranking techniques, while CCRANK is a new spectral ranking approach motivated by analysing the canonical correlation between covariate information and match outcomes. All proposed methods appeal to a class of expressive non-parametric models based on reproducing kernel Hilbert spaces. Our contributions are summarised as follows:

1. We extend SERIALRANK and SVDRANK to incorporate player covariate information and propose a new spectral ranking algorithm CCRANK. This method extracts rankings based on the canonical correlation between covariate and match information.
2. We provide an extensive set of numerical experiments on simulated and real-world data, showcasing the utility of including covariate information as well as the competitiveness of the proposed methods with state-of-the-art covariate-based ranking algorithms.

Outline. The rest of the paper is organised as follows. Section 2 covers related work in the spectral ranking literature. Our proposed methodologies will be presented in Sect. 3. Sections 4 and 5 cover experiments and concluding remarks.

Notation. Scalars are denoted by lower case letters while vectors and matrices are denoted by bold lower and upper case letters, respectively. In general, there are n players to rank, and we denote the $n \times n$ pairwise comparison matrix by $\mathbf{C}$. We use the notation $\succ$ to denote the direction of preference. For ordinal comparisons, $C_{i,j} = 1$ if $i \succ j$, and -1 otherwise. For simplicity, we assume no ties between players. For *cardinal* comparisons, $C_{i,j} \in \mathbb{R}$ is considered as a proxy or noisy evaluation of the skill offset $r_i - r_j$ as in [8,10]. We denote the player covariate information matrix as $\mathbf{X} \in \mathbb{R}^{n \times d}$ and its subsequent kernel matrix as $\mathbf{K} \in \mathbb{R}^{n \times n}$ with $K_{i,j} = k(\mathbf{x}_i, \mathbf{x}_j)$, where $\mathbf{x}_i$ is the ith row of $\mathbf{X}$ and k the kernel. Given a skill vector $\mathbf{r}$, we define upsets as a pair of items for which the player with lower skill is preferred over the player with higher skill, i.e. there is an upset if the recovered ranking indicates $r_i > r_j$ but the input data has $C_{i,j} < 0$.. Finally, the ranking is inferred by sorting the entries of r, choosing either increasing or decreasing order to *minimise* the number of upsets with respect to the observed $\mathbf{C}$.

2 Background

We give a brief review of both probabilistic and spectral ranking algorithms in this section.

2.1 Probabilistic Ranking

Probabilistic ranking methods typically use a generative model of how the pairwise comparison is conducted: given players' skill vector $\mathbf{r}$, also denoted as *worths* or *weights parameters*, one can model the match outcome probability based on the difference between the skills of the two players, i.e. $\mathbb{P}(i \succ j) = F(r_i - r_j)$, where F is some cumulative distribution function (CDF). The Thurstone [28] model uses a normal CDF for F while the Bradley-Terry-Luce (BT) model uses a logistic CDF. This simple setup has led to various extensions. For example, Huang et al. [14] proposed a generalisation of Bradley-Terry models to tackle paired team comparisons instead of individuals comparisons. Chen et al. [6] studied intransitivity in comparison data by extending ranking from a one-dimensional vector to a multi-dimensional representation. Bayesian approaches taking into account prior beliefs over ranking are developed in [3,7].

Extending BT models to incorporate player covariates has been studied thoroughly in the past decades as well. Springall [26] proposes to describe the skill vector $\mathbf{r}$ as the linear combination of player covariates $\mathbf{X}$, i.e. $\mathbf{r} = \mathbf{X}\boldsymbol{\beta}$ for some regression coefficients $\boldsymbol{\beta} \in \mathbb{R}^d$. Chu et al. [7] later extended this formulation to a more flexible Gaussian Processes [31] regression model as well. We will use these two extensions and the BT model as our baselines in our numerical experiments.

2.2 Spectral Ranking

Spectral ranking is a family of fundamentally different approaches and the core object of interest are the matrices that represent pairwise relationships between the players. It often involves computing the leading eigenvector of the sparse comparison matrix, or similar matrix operators derived from it, which can be computed efficiently in contrast to the optimisation procedures inherent in the MLE [15]. The computational advantage stems from the fact that typical eigen-solvers can compute the leading eigenvectors of a sparse matrix in running time linear in the number of nonzero entries in the matrix (i.e. edges of the comparison graph). We refer the reader to Vigna [29] for a comprehensive review of the early history of spectral ranking. We now present two specific spectral ranking algorithms that our proposed methods are based on, and also discuss Jain et al. [15]'s spectral method that utilises covariates as a baseline in our numerical experiments.

SERIALRANK. Fogel et al. [11] proposed a seriation approach to ranking, building on [1], where they defined similarities between players based on their pairwise comparisons. Given an ordinal pairwise comparison matrix $\mathbf{C}$, we define the similarity $S_{i,j}$ between player i, j as the number of agreeing comparisons between i, j with other players, i.e.

$$S_{i,j} = \sum_{k=1}^{n} \left(\frac{1 + C_{i,k} C_{j,k}}{2} \right), \tag{1}$$

which can be expressed as $\mathbf{S} = \frac{1}{2}(n\mathbf{1}\mathbf{1}^\top + \mathbf{C}\mathbf{C}^\top)$, where $\mathbf{1} \in \mathbb{R}^n$ denotes the all-one vector. The optimal ordering in this seriation process is then recovered by extracting the *Fiedler vector* of $\mathbf{S}$, which is the eigenvector corresponding to the second smallest eigenvalue of the graph Laplacian matrix $\mathcal{L}(\mathbf{S}) = \mathrm{diag}(\mathbf{S}\mathbf{1}) - \mathbf{S}$. This corresponds to the following optimisation

$$\mathbf{r}_* = \arg\min_{\mathbf{r}} \left(\mathbf{r}^\top \mathcal{L}(\mathbf{S}) \mathbf{r} \right) \quad \text{s.t} \quad \mathbf{r}^\top \mathbf{r} = 1, \quad \mathbf{r}^\top \mathbf{1} = 0. \tag{2}$$

$\mathbf{r}_*$ can be interpreted as the smoothest non-trivial graph signal with respect to the similarity graph $\mathbf{S}$ [24].

SVDRANK. Cucuringu et al. [8,10] considered a simple alternative spectral method, arising from models wherein the pairwise comparisons are modelled as noisy entries of a rank 2 skew-symmetric matrix $\mathbf{C} = \mathbf{r}\mathbf{1}^\top - \mathbf{1}\mathbf{r}^\top$, or a proxy of it. This leads to the following optimisation

$$\mathbf{r}_* = \arg\max_{\mathbf{r}} \left(\mathbf{r}^\top \mathbf{C}\mathbf{C}^\top \mathbf{r} \right), \quad \text{s.t.} \quad \mathbf{r}^\top \mathbf{r} = 1, \quad \mathbf{r}^\top \mathbf{1} = 0. \tag{3}$$

One computes the top two singular vectors of $\mathbf{C}$ and extracts the ranking induced by ordering their entries to minimise upsets.

RANK-CENTRALITY. Negahban et al. [21] proposed to estimate player rankings from the stationary distribution of a certain random walk on the graph of players, where edges encode the probabilities of pairwise comparison outcomes. This model was first designed to combine **multiple** *ordinal* pairwise comparisons made on the same pairs of players under the context of *Rank Aggregation* [17]. Cucuringu [25] later combined this approach with SerialRank to handle the single observation setting. Given the similarity matrix $\mathbf{S}$ computed using Eq (1), the key idea behind Rank-Centrality is to construct a transition probability matrix $\mathbf{A}$ where $A_{i,j} = 1 - \frac{S_{i,j}}{2n}$ if $C_{i,j} > 0$, $\frac{S_{i,j}}{2n}$ if $C_{i,j} < 0$, and 0 otherwise. One then extracts the leading top left-eigenvector from $\mathbf{A}$ as the ranking. Jain et al. [15] later extended this to incorporate covariates by setting $\hat{\mathbf{A}} = \mathbf{AV}$ where $\mathbf{V}$ is some row-stochastic player similarity matrix, i.e., a scaled kernel matrix. Jain et al. called this method the Regularised Rank-Centrality algorithm. We note that this method cannot predict the ranking of unseen players because no explicit functional relationships between $\mathbf{r}$ and player covariates $\mathbf{X}$ are modelled.

3 Proposed Methods

In this section, we introduce three lines of spectral ranking algorithms, each having a different flavour in combining the two sources of data: pairwise comparisons and covariate information. C-SerialRank controls the information covariates contribute via a simple hyperparameter (that can be selected based on cross-validation), and therefore can recover a robust ranking even when features are noisy. On the other hand, SVDCovRank connects $\mathbf{r}$ with covariates via a functional model, allowing us to predict the rank of previously unseen players while no comparison with existing players is needed. Finally, CCRank considers the optimal embedding of both match outcomes and covariate information that maximises their canonical correlation and extracts ranking from those embeddings.

3.1 C-SerialRank

Recall in SerialRank, ranking is extracted from the Fiedler vector of the graph Laplacian matrix $\mathcal{L}(\mathbf{S})$, where $\mathbf{S} = \frac{1}{2}(n\mathbf{1}\mathbf{1}^\top + \mathbf{C}\mathbf{C}^\top)$ is the player similarity matrix induced by their match outcomes. However, this is not the only source of player similarity. If we also have access to the covariate matrix $\mathbf{X}$, we can define another source of similarity through the Gram matrix $\mathbf{K}$, with $K_{i,j} = k(\mathbf{x}_i, \mathbf{x}_j)$ for a given kernel k. We propose to merge this extra source of similarity with S via a simple linear combination, i.e. $\hat{\mathbf{S}} = \mathbf{S} + \lambda\mathbf{K}$ with λ controlling the tradeoff between the two sources of data. We call this method C-SerialRank, where C stands for covariates. The ranking can then be recovered by computing the Fiedler vector from the new graph Laplacian matrix $\mathcal{L}(\hat{\mathbf{S}})$

$$\mathbf{r}_* = \arg\min_{\mathbf{r}} \mathbf{r}^T\big(\mathcal{L}(\mathbf{S} + \lambda\mathbf{K})\big)\mathbf{r} \quad \text{s.t} \quad \mathbf{r}^\top\mathbf{r} = 1 \quad \mathbf{r}^\top\mathbf{1} = 0. \tag{4}$$

We summarise the method in Algorithm 1. The tuning parameter λ controls the contribution of the covariates – intuitively, it should be high for covariates that are informative about match outcomes, and low for less informative covariates.

Algorithm 1: C-SerialRank

Require: A set of pairwise comparisons $\mathbf{C}_{i,j} \in \{-1, 0, +1\}$ or $\mathbb{R}$, kernel matrix $\mathbf{K}$, regularisation parameter λ.

1 Compute similarity matrix $\hat{\mathbf{S}} = \frac{1}{2}(n\mathbf{1}\mathbf{1}^T + \mathbf{C}\mathbf{C}^T) + \lambda\mathbf{K}$.

2 Compute the associated graph Laplacian matrix $\mathcal{L}(\hat{\mathbf{S}}) = \text{diag}(\hat{\mathbf{S}}\mathbf{1}) - \hat{\mathbf{S}}$.

3 Compute the Fiedler vector of $\mathbf{S}$ by extracting the eigenvector corresponding to the second smallest eigenvalue of $\mathcal{L}(\hat{\mathbf{S}})$.

4 Output the ranking induced by sorting the Fiedler vector of $\hat{\mathbf{S}}$, with the global ordering chosen to minimise upsets with respect to $\mathbf{C}$.

3.2 SVDCovRank

C-SerialRank provides a simple method to incorporate player covariates into ranking. However, similar to the Regularised Rank-Centrality algorithm, the downside of C-SerialRank is that one cannot perform predictions on unseen players based on available covariate information. To this end, we introduce SVDCovRank, an SVD based ranking algorithm that connects covariates to $\mathbf{r}$ using a functional model.

Recall that when solving for SVDRank, one converts Eq. (3) to a standard eigenvector problem by projecting $\mathbf{r}$ onto $\{\mathbf{1}\}^{\perp}$. This can be done by setting $\mathbf{r} = \mathbf{Zq}$, where $\mathbf{Z} \in \mathbb{R}^{n\times(n-1)}$ is a matrix whose columns form an orthonormal basis of $\{\mathbf{1}\}^{\perp}$. As a result, Eq. (3) can be rewritten as

$$\mathbf{q}_* = \arg\max_{\mathbf{q}} \left(\mathbf{q}^\top \mathbf{Z}^\top \mathbf{C}\mathbf{C}^\top \mathbf{Z}\mathbf{q}\right), \quad \text{s.t. } \mathbf{q}^\top\mathbf{q} = 1. \tag{5}$$

By construction, $\mathbf{r}_* = \mathbf{Z}\mathbf{q}_*$ is orthogonal to $\mathbf{1}$ where $\mathbf{q}_*$ is the top eigenvector of $\mathbf{Z}^\top\mathbf{C}\mathbf{C}^\top\mathbf{Z}$. Now consider the scenario where we also have access to player covariates $\mathbf{X}$, and we assume the final ranking implied by $\mathbf{r}$ varies smoothly as a linear function of these covariates, e.g. $\mathbf{r} = \mathbf{X}\boldsymbol{\beta}$. Following the same line of thought as above, one could set $\mathbf{X}\boldsymbol{\beta} = \mathbf{Zq}$, and substitute $\mathbf{q} = \mathbf{Z}^\top\mathbf{X}\boldsymbol{\beta}$ into Eq. (5) to arrive at

$$\boldsymbol{\beta}_* = \arg\max_{\boldsymbol{\beta}} \left(\boldsymbol{\beta}^\top\mathbf{X}^\top\mathbf{H}\mathbf{C}\mathbf{C}^\top\mathbf{H}\mathbf{X}\boldsymbol{\beta}\right), \quad \text{s.t. } \boldsymbol{\beta}^\top\mathbf{X}\mathbf{H}\mathbf{X}\boldsymbol{\beta} = 1, \tag{6}$$

where $\mathbf{H} = \mathbf{Z}\mathbf{Z}^\top$ is the centering matrix. To solve for $\boldsymbol{\beta}$, we further decompose our feature covariance matrix $\mathbf{X}^\top\mathbf{H}\mathbf{X}$ into $\mathbf{L}\mathbf{L}^\top$ via a Choleskey decomposition or SVD. By setting $\boldsymbol{\beta} = \mathbf{L}^{-\top}\gamma$, we again recover the standard eigenvector problem

$$\gamma_* = \arg\max_{\gamma} \left(\gamma^\top\boldsymbol{\Psi}\gamma\right), \quad \text{s.t. } \gamma^\top\gamma = 1, \tag{7}$$

with $\boldsymbol{\Psi} = \mathbf{L}^{-1}\mathbf{X}^\top\mathbf{H}\mathbf{C}\mathbf{C}^\top\mathbf{H}\mathbf{X}\mathbf{L}^{-\top}$. Finally, the optimal ranking $\mathbf{r}$ is recovered by setting $\mathbf{r}_* = \mathbf{H}\mathbf{X}\mathbf{L}^{-\top}\gamma_*$. We summarise the procedure in Algorithm 2.

Furthermore, if there is reason to believe the skill vector $\mathbf{r}$ relates to the covariates non-linearly, it is straightforward to kernelise the procedure by setting

Algorithm 2: SVDCovRank

Require: A set of pairwise comparisons $\mathbf{C}_{i,j} \in \{-1, 0, 1\}$ or $\mathbb{R}$, covariate matrix $\mathbf{X}$ and kernel function k.

1 **if** *kernelise* **then**
2 Compute the kernel matrix $\mathbf{K}$ with $K_{i,j} = k(\mathbf{x}_i, \mathbf{x}_j)$ and set $\mathbf{B} = \mathbf{K}$
3 **else**
4 Set $\mathbf{B} = \mathbf{X}$.
5 Compute $\boldsymbol{\Psi} = \mathbf{L}^{-1}\mathbf{B}^\top\mathbf{HCC}^\top\mathbf{HBL}^{-\top}$, where $\mathbf{L}$ is the Cholesky decomposition of $\mathbf{B}^\top\mathbf{HB}$.
6 Compute the top eigenvector $\boldsymbol{\gamma}$ of $\boldsymbol{\Psi}$ and set $\mathbf{r} = \mathbf{HBL}^{-\top}\boldsymbol{\gamma}$.
7 Sort $\mathbf{r}$ to minimise the number of upsets with respect to $\mathbf{C}$.
8 **if** *kernelise* **then**
9 Return the skill vector $\mathbf{r}$ and skill function $r(\mathbf{x}) = k(\mathbf{x}, \mathbf{X})\mathbf{L}^\top\boldsymbol{\gamma}$ where $k(\mathbf{x}, \mathbf{X}) = [k(\mathbf{x}, \mathbf{x}_1), ..., k(\mathbf{x}, \mathbf{x}_n)]$.
10 **else**
11 Return the skill vector $\mathbf{r}$ and skill function $r(\mathbf{x}) = \mathbf{x}^\top\mathbf{L}^\top\boldsymbol{\gamma}$.

$\mathbf{r} = \mathbf{K}\boldsymbol{\alpha}$, where $\mathbf{K}$ is the kernel matrix of player covariates and $\boldsymbol{\alpha} \in \mathbb{R}^n$ some dual weights, and obtain a non-parametric variation of SVDCovRank. Similarly by setting $\mathbf{K}\boldsymbol{\alpha} = \mathbf{Z}^\top\mathbf{q}$, we can express Eq. (5) as

$$\boldsymbol{\alpha}_* = \arg\max_{\boldsymbol{\alpha}} \left(\boldsymbol{\alpha}^\top\mathbf{KHCC}^\top\mathbf{HK}\boldsymbol{\alpha}\right), \quad \text{s.t. } \boldsymbol{\alpha}^\top\mathbf{KHK}\boldsymbol{\alpha} = 1. \tag{8}$$

The rest follows in analogy to the linear case and the complete method is given in Algorithm 2.

3.3 CCRank

Besides aggregating match and covariate information via a linear combination or by imposing a regression structure between $\mathbf{r}$ and $\mathbf{X}$, we propose to study the Canonical Correlation between match and covariate information to derive the ranking. We call the method CCRank. Consider the similarity matrix $\mathbf{S} = \frac{1}{2}(n\mathbf{1}\mathbf{1}^\top + \mathbf{CC}^\top)$ from Sect. 2.2 and player covariate matrix $\mathbf{X}$. As each row of $\mathbf{S}$ and $\mathbf{X}$ describe the same player with two distinct sources of information, it is natural to define ranking as the "most agreeing" projections between $\mathbf{S}$ and $\mathbf{X}$. This coincides with the classical Canonical Correlation Analysis (CCA) problem.

CCA is a classical technique developed by H. Hotelling [13] to study the linear relationship between two multidimensional sets of variables. Assume both matrix $\mathbf{S}$ and $\mathbf{X}$ are full rank, CCA can be defined as the problem of finding the optimal projection $\mathbf{z}_S = \mathbf{Sa}$ and $\mathbf{z}_X = \mathbf{Xb}$, with $\mathbf{a} \in \mathbb{R}^n$ and $\mathbf{b} \in \mathbb{R}^d$, such that $\mathbf{z}_S$ and $\mathbf{z}_X$ are maximally correlated, i.e.

$$\hat{\mathbf{a}}, \hat{\mathbf{b}} = \arg\max_{\mathbf{a},\mathbf{b}} \frac{\mathbf{z}_S^\top\mathbf{z}_X}{||\mathbf{z}_S||||\mathbf{z}_X||} = \arg\max_{\mathbf{a},\mathbf{b}} \frac{\mathbf{a}^\top\boldsymbol{\Sigma}_{SX}\mathbf{b}}{\sqrt{\mathbf{a}^\top\boldsymbol{\Sigma}_{SS}\mathbf{a}\mathbf{b}^\top\boldsymbol{\Sigma}_{XX}\mathbf{b}}}, \tag{9}$$

where $\boldsymbol{\Sigma}_{SX} = \mathbf{SX}$ is an estimate of the cross-covariance matrix between $\mathbf{S}$ and $\mathbf{X}$. Equation (9) is equivalent to the following constrained optimisation problem

$$\hat{\mathbf{a}}, \hat{\mathbf{b}} = \arg\max_{\mathbf{a},\mathbf{b}} = \mathbf{a}^\top \boldsymbol{\Sigma}_{SX}\mathbf{b} \quad \text{s.t.} \begin{cases} \mathbf{a}^\top \boldsymbol{\Sigma}_{SS}\mathbf{a} = 1 \\ \mathbf{b}^\top \boldsymbol{\Sigma}_{XX}\mathbf{b} = 1. \end{cases} \tag{10}$$

The solution to this problem can then be obtained by solving a generalised eigenvector problem as detailed in Algorithm 3. We recover ranking by setting $\mathbf{r}$ to be $\mathbf{z}_S$ or $\mathbf{z}_X$ depending on which embedding better minimises the upsets with respect to $\mathbf{C}$. However, note that $\mathbf{z}_S = \mathbf{Sa}$ cannot be used to predict rankings on unseen players by construction.

Similar to SVDCovRank, if we believe the covariates relate to the optimal ranking non-linearly, we could kernelise the procedure and seek projections belonging to RKHSs. Denote ℓ as the kernel on $\mathbf{S}$ and $\mathcal{H}_\ell$ the corresponding RKHS. The objective of Kernel CCA [12] can be expressed using cross-covariance operators, which for simplicity, we use the same notation as the corresponding cross-covariance matrices

$$\hat{f}, \hat{g} = \arg\max_{f\in\mathcal{H}_k, g\in\mathcal{H}_\ell} \langle f, \boldsymbol{\Sigma}_{XS}g\rangle_{\mathcal{H}_\ell} \quad \text{s.t} \begin{cases} \langle f, \boldsymbol{\Sigma}_{XX}f\rangle_{\mathcal{H}_k} = 1 \\ \langle g, \boldsymbol{\Sigma}_{SS}g\rangle_{\mathcal{H}_\ell} = 1. \end{cases} \tag{11}$$

The rest of the algorithm is outlined in Algorithm 3. Different from SVDCovRank which postulates a low-rank structure on $\mathbf{C}$, CCRank does not pose any structure over the match matrix, and thus is more robust to model misspecification. This is illustrated later in our experiments.

4 Experiments

In this section, we compare the performance of C-SerialRank (C-Serial from now on for brevity), SVDCovRank (SVDC), kernelised SVDCovRank (SVDK), CCRank (CC), kernelised CCRank (KCC) with that of a number of other benchmark algorithms from the literature. We compare against the spectral ranking algorithms SerialRank (Serial), SVDRank (SVDN), Rank-Centrality (RC), Regularised Rank-Centrality (RRC) and against the probabilistic ranking algorithms Bradley-Terry-Luce model (BT) as well as its two extension based on logistic regression (BT-LR) and Gaussian processes [7] (BT-GP). For BT-GP we will simply utilise its posterior mean for rank prediction. We note that Bradley-Terry-Luce and Rank Centrality based methods cannot handle *cardinal* comparisons, therefore we first transform each entry of the pairwise comparison matrix to $\text{sign}(C_{i,j})$. See Table 1 for a summary of our methods and benchmarks. All code and implementations are made publicly available.[1]

Throughout our experiments, we will use the Radial Basis Function (RBF) kernel $k(x, x') = \exp\left(\frac{||x-x'||^2}{2\ell_k}\right)$ where the lengthscale ℓ_k is chosen via the median

[1] https://github.com/Chau999/SpectralRankingWithCovariates.

Algorithm 3: CCRank

Require: A set of pairwise comparisons $C_{i,j} \in \{-1, 0, 1\}$ or $\mathbb{R}$, covariate matrix $\mathbf{X}$, kernel functions k, g, centering matrix $\mathbf{H}$, identity matrix $\mathbf{I}_n$ and small perturbation ϵ

1 Compute $\mathbf{S} = \frac{1}{2}(n\mathbf{1}\mathbf{1}^\top + \mathbf{C}\mathbf{C}^\top)$
2 **if** *kernelise* **then**
3 Compute the kernel matrix $\mathbf{K}$ with $K_{i,j} = k(\mathbf{x}_i, \mathbf{x}_j)$ and $\mathbf{G}$ with $G_{i,j} = g(\mathbf{s}_i, \mathbf{s}_j)$ where $\mathbf{s}_i, \mathbf{s}_j$ corresponds to row i, j of $\mathbf{S}$.
4 Compute $\tilde{\mathbf{K}} = \mathbf{HKH}$ and $\tilde{\mathbf{G}} = \mathbf{HGH}$
5 Compute $\boldsymbol{\Sigma}_{XS} = n^{-1}\tilde{\mathbf{K}}\tilde{\mathbf{G}}$, $\boldsymbol{\Sigma}_{XX} = n^{-1}\tilde{\mathbf{K}}(\tilde{\mathbf{K}} + \epsilon\mathbf{I}_n)$ and $\boldsymbol{\Sigma}_{SS} = n^{-1}\tilde{\mathbf{G}}(\tilde{\mathbf{G}} + \epsilon\mathbf{I}_n)$
6 **else**
7 Compute $\boldsymbol{\Sigma}_{SX} = \mathbf{SX}$, $\boldsymbol{\Sigma}_{XX} = \mathbf{X}^\top\mathbf{X}$ and $\boldsymbol{\Sigma}_{SS} = \mathbf{S}^\top\mathbf{S}$.
8 Compute the top generalised eigenvector of the problem

$$\begin{bmatrix} 0 & \boldsymbol{\Sigma}_{XS} \\ \boldsymbol{\Sigma}_{XS}^\top & 0 \end{bmatrix} \begin{bmatrix} \boldsymbol{\alpha} \\ \boldsymbol{\beta} \end{bmatrix} = \rho \begin{bmatrix} \boldsymbol{\Sigma}_{XX} & 0 \\ 0 & \boldsymbol{\Sigma}_{SS} \end{bmatrix} \begin{bmatrix} \boldsymbol{\alpha} \\ \boldsymbol{\beta} \end{bmatrix} \tag{12}$$

if *kernelise* **then**
9 Return the embeddings $\mathbf{z}_S = \mathbf{GH}\boldsymbol{\beta}$ and $\mathbf{z}_X = \mathbf{KH}\boldsymbol{\alpha}$ that minimises upsets with respect to $\mathbf{C}$ and skill function $r(\mathbf{x}) = k(\mathbf{x}, \mathbf{X})\mathbf{H}\boldsymbol{\alpha}$
10 **else**
11 Return the embeddings $\mathbf{z}_S = \mathbf{S}\boldsymbol{\beta}$ and $\mathbf{z}_X = \mathbf{X}\boldsymbol{\alpha}$ that minimises upsets with respect to $\mathbf{C}$ and skill function $r(\mathbf{x}) = \mathbf{x}^\top\boldsymbol{\alpha}$.

heuristic method. Hyperparameters such as λ in C-Serial are tuned using 10-fold cross-validation on withheld match outcomes. In practice, we should select the kernel that reflects the nature of the non-linearity in data. We assess the performance of all algorithms via the following two tasks:

Task 1: Rank Inference. Given n players with a sparse and potentially noisy pairwise comparison matrix $\mathbf{C}$, we pick **70% of the observed matches** as training data $\mathbf{C}^{(1)}_{\text{train}}$, and the remaining observed matches as testing data $\mathbf{C}^{(1)}_{\text{test}}$. Our goal is to train the algorithms on $\mathbf{C}^{(1)}_{\text{train}}$ to recover the rankings, and then measure their performances by reporting the accuracy scores in predicting matches from $\mathbf{C}^{(1)}_{\text{test}}$ based on the estimated rankings.

Task 2: Rank Prediction. Given n players with comparison matrix $\mathbf{C}$, withhold **70% of the players** and observed the induced comparison matrix $\mathbf{C}^{(2)}_{\text{train}} \in \mathbb{R}^{n' \times n'}$ where $n' = \lfloor n \times 0.7 \rfloor$. Set up the testing data $\mathbf{C}^{(2)}_{\text{test}}$ based on the induced comparison matrix of the remaining players. We train on $\mathbf{C}^{(2)}_{\text{train}}$ and report the

Table 1. A summary of our proposed methods (shaded grey) and benchmarks.

Algorithm	Model type	Comparisons	Covariates	Unseen player rank prediction
BT [2]	Probabilistic	Ordinal	✗	✗
BT-LR *n* BT-GP [7]	Probabilistic	Ordinal	✓	✓
RC [20]	Spectral	Ordinal	✗	✗
RRC [15]	Spectral	Ordinal	✓	✗
SERIAL [11]	Spectral	Ordinal + Cardinal	✗	✗
C-SERIAL (Ours)	Spectral	Ordinal + Cardinal	✓	✗
SVD [9]	Spectral	Ordinal + Cardinal	✗	✗
SVDC *n* SVDK (Ours)	Spectral	Ordinal + Cardinal	✓	✓
CC *n* KCC (Ours)	Spectral	Ordinal + Cardinal	✓	✓

accuracy in predicting matches from $\mathbf{C}_{\text{test}}^{(2)}$ based on the predicted rankings on unseen players.

4.1 Simulations on Synthetic Data

We test the performances of the 12 ranking algorithms on both rank inference and prediction problems using synthetic data. We report the Kendall Tau score on test data as our performance metric, as we have access to the ground truth ranking. We simulate $n = 1000$ players with covariates $\mathbf{X}$ drawn from $N(0, 1)$. We set the skill function as $r(x) = \sin(3\pi x) - 1.5x^2 + \epsilon N(0, 1)$ with the noise level $\epsilon \in \{0, 0.05, 0.2\}$. The corresponding comparison matrix $\mathbf{C}$ is given by $C_{i,j} = r(x_i) - r(x_j)$. Besides feature noise, we also consider adding noises to the matches. We follow the *FLIP* noise model deployed in [8] where we flip each match result with probability $p \in \{0, 0.05, 0.1, 0.2, 0.3, 0.4, 0.5\}$. With probability $\phi \in \{0.7, 0.9\}$, we remove an observed match in the comparison matrix $\mathbf{C}$ to yield sparse comparisons matrices as in most practical situations. Experiments are repeated over 10 seeds.

Figure 1a and Fig. 1b demonstrate the performances of the algorithms under different sparsity levels, feature noises and model error rates for the *inference* and *prediction* tasks. For the *inference* tasks, we see a performance drop in the covariate-based ranking algorithms as the feature noise increases. This coincides with our usual "garbage covariate-in, garbage out" intuition in modelling. We also note that as the skill vector non-linearly relates to the covariates by design, all kernelised algorithms outperformed their linear counterparts. We also observe that as the flip error rate p increases, the performance of covariate-free algorithms decreases quickly, while covariate-based methods performed relatively stable until the extreme error rate. For *prediction* tasks, similar trends to the *inference* tasks in terms of algorithmic performances can be seen with respect to model error rates, sparsity and feature noise levels.

Specifically, we observe that the kernelised spectral ranking algorithm SVDK performed competitively to probabilistic ranking BT-GP until a high flip error

Table 2. Accuracy of match outcome predictions based on recovered ranking on each data set. Scores are averaged over 10 seeds, and 1 s.d is reported. † indicates algorithm using covariate information. We shaded our proposed methods in grey. The best performing method is shown in bold red, and second best in bold blue.

Algorithm	Chameleon		FlatLizard		Pokemon		NFL 2000-18	
	Task 1	Task 2	Task 1	Task 2	Task 1	Task 2	Task 1	Task 2
BT	$0.83_{\pm 0.03}$	–	$0.86_{\pm 0.06}$	–	$0.86_{\pm 0.01}$	–	$0.63_{\pm 0.04}$	–
BT-LR†	$0.71_{\pm 0.03}$	$0.82_{\pm 0.14}$	$0.84_{\pm 0.04}$	$0.77_{\pm 0.12}$	$0.88_{\pm 0.01}$	$0.89_{\pm 0.01}$	$0.68_{\pm 0.04}$	$0.62_{\pm 0.09}$
BT-GP†	$0.75_{\pm 0.04}$	$0.74_{\pm 0.13}$	$0.80_{\pm 0.05}$	$0.74_{\pm 0.12}$	$0.72_{\pm 0.01}$	$0.72_{\pm 0.02}$	$0.58_{\pm 0.04}$	$0.61_{\pm 0.08}$
RC	$0.61_{\pm 0.06}$	–	$0.66_{\pm 0.05}$	–	$0.81_{\pm 0.01}$	–	$0.58_{\pm 0.04}$	–
RRC†	$0.61_{\pm 0.03}$	–	$0.66_{\pm 0.01}$	–	$0.85_{\pm 0.01}$	–	$0.62_{\pm 0.04}$	–
SVD	$0.72_{\pm 0.08}$	–	$0.69_{\pm 0.05}$	–	$0.84_{\pm 0.02}$	–	$0.57_{\pm 0.04}$	–
SVDC†	$0.65_{\pm 0.06}$	$0.74_{\pm 0.12}$	$0.81_{\pm 0.05}$	$0.73_{\pm 0.10}$	$0.89_{\pm 0.01}$	$0.90_{\pm 0.01}$	$0.66_{\pm 0.04}$	$0.62_{\pm 0.09}$
SVDK†	$0.76_{\pm 0.06}$	$0.79_{\pm 0.06}$	$0.68_{\pm 0.05}$	$0.65_{\pm 0.09}$	$0.85_{\pm 0.01}$	$0.72_{\pm 0.09}$	$0.59_{\pm 0.04}$	$0.62_{\pm 0.09}$
SERIAL	$0.79_{\pm 0.04}$	–	$0.70_{\pm 0.05}$	–	$0.88_{\pm 0.01}$	–	$0.58_{\pm 0.04}$	–
C-SERIAL†	$0.80_{\pm 0.03}$	–	$0.88_{\pm 0.01}$	–	$0.88_{\pm 0.01}$	–	$0.59_{\pm 0.04}$	–
CC†	$0.66_{\pm 0.10}$	$0.95_{\pm 0.10}$	$0.78_{\pm 0.08}$	$0.92_{\pm 0.15}$	$0.78_{\pm 0.08}$	$0.79_{\pm 0.08}$	$0.61_{\pm 0.05}$	$0.66_{\pm 0.10}$
KCC†	$0.71_{\pm 0.06}$	$0.80_{\pm 0.10}$	$0.78_{\pm 0.03}$	$0.71_{\pm 0.11}$	$0.81_{\pm 0.01}$	$0.72_{\pm 0.09}$	$0.67_{\pm 0.05}$	$0.62_{\pm 0.08}$

rate in both tasks. KCC performed slightly worse than the two, but still much better than the covariate-based spectral method baseline RRC in general. We note that the linear methods do not appear to be sensitive to model error rate because the ranking extracted is dominated by the information in the covariates.

4.2 Experiments on Real-World Data

In this section, we apply all ranking algorithms to a variety of real-world data sets (22 in total) to perform rank *inferences* and *predictions*. We average the accuracy scores over 10 seeds and report the results in Table 2. 5% significant level one-sided paired Wilcoxon tests for all algorithm pairs are conducted for each data set, with results reported in Fig. 2.

Male Cape Dwarf Chameleons Contest. The first data set is used in the study by Stuart-Fox et al. [27]. Physical measurements are made on 35 male Cape dwarf chameleons, and the results of 104 contests are measured and encoded as ordinal comparison data. From Table 2 and Fig. 2 we see that BT outperformed the rest in rank *inferences* (closely followed by C-SERIAL), while CC performed best compared with other covariate based methods in rank *prediction.*

Flatlizard Competition. The data is collected at Augrabies Falls National Park (South Africa) in September–October 2002 [30], on the contest performance and background attributes of 77 male flat lizards (*Platysaurus Broadleyi*). The results of 100 contests were recorded, along with 18 physical measurements made on each lizard, such as *weight* and *head size.* From Fig. 2 we see that C-SERIAL and CC statistically significantly outperformed the rest for inference and prediction, respectively.

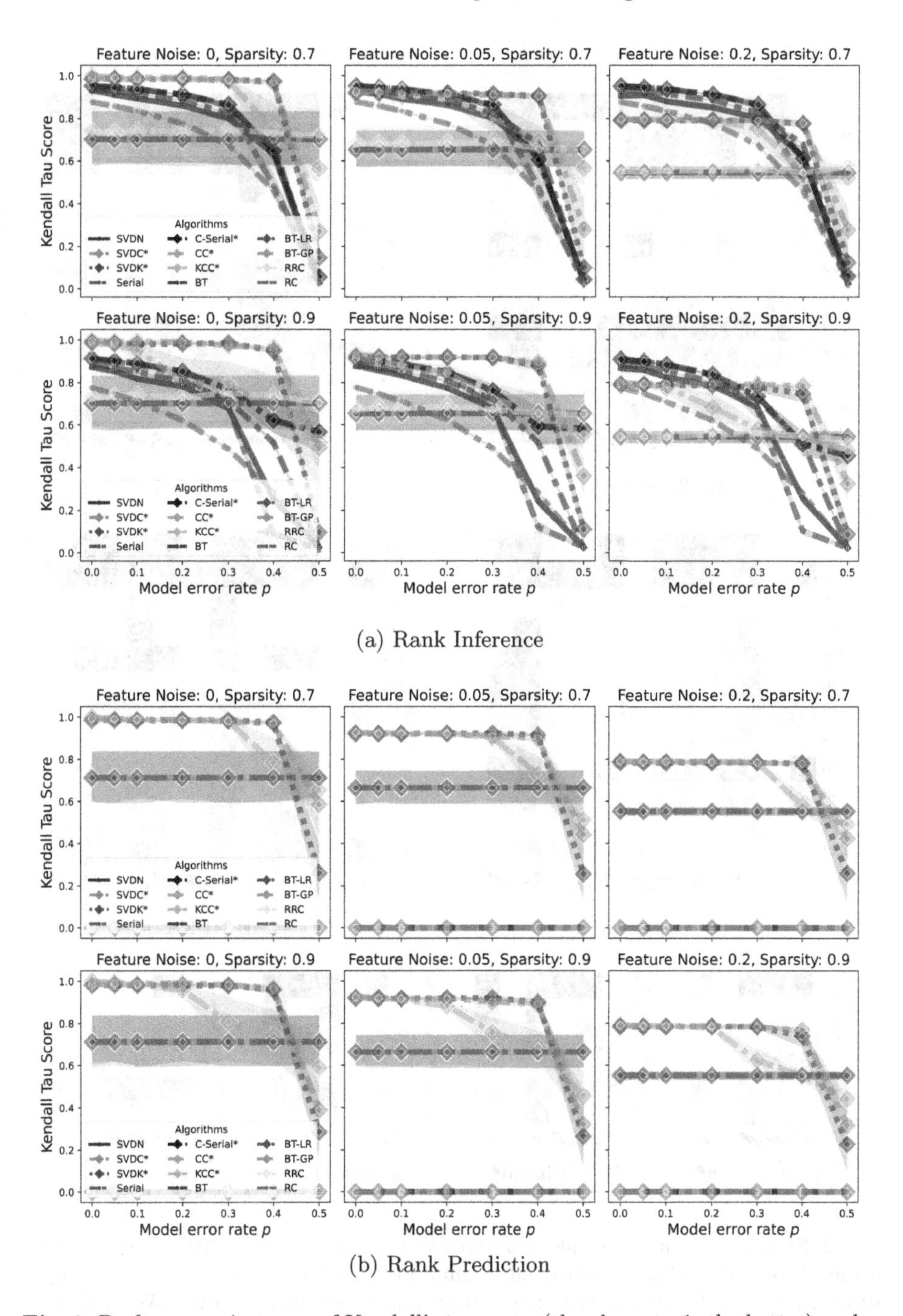

Fig. 1. Performance in terms of Kendall's tau score (the closer to 1, the better) under different synthetic noise models with 1000 players. Results are averaged over 10 iterations and 95% confidence intervals are reported. Methods with diamond markers use covariates, and $*$ indicates our proposed methods.

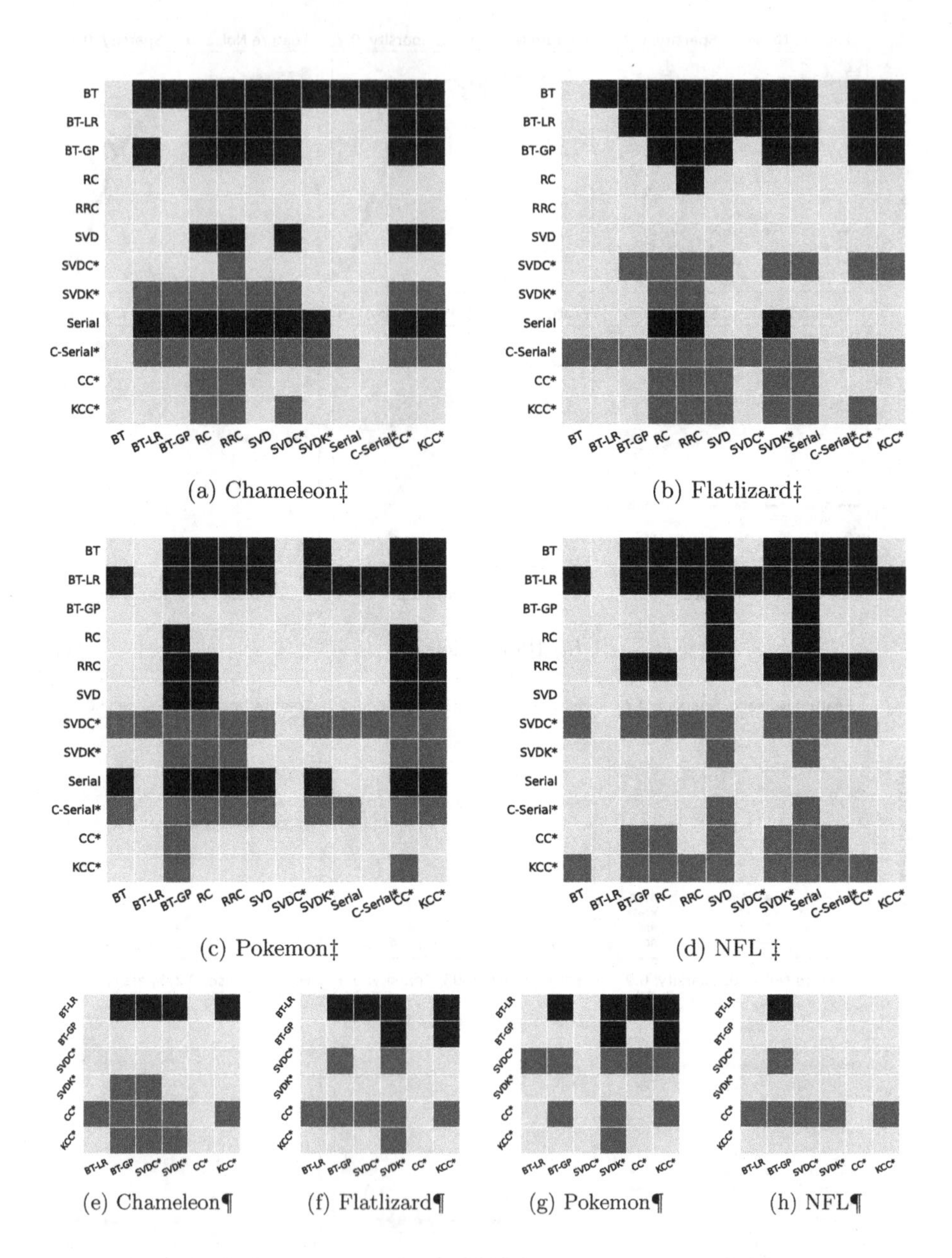

(a) Chameleon‡ (b) Flatlizard‡

(c) Pokemon‡ (d) NFL ‡

(e) Chameleon¶ (f) Flatlizard¶ (g) Pokemon¶ (h) NFL¶

Fig. 2. Statistical Significance plots for our real-world experiments. We shade an entry if the ith row algorithm is statistically significantly better than the jth column algorithm on a one-sided paired Wilcoxon test with a 5% significance level. We shade our algorithms in red and the benchmarks in black. ‡ and ¶ indicate rank inference and prediction tasks respectively.

Pokémon Battle. Our next data set comes from a Kaggle competition[2]. Approximately 45,000 battles among 800 Pokémon are recorded in ordinal format. 25 characteristic of each Pokémon, such as their *type*, *attack* and *defence*, are available. These battles are generated by a custom algorithm written by the Kaggle host that closely reflects the game mechanics. Table 2 shows that linear covariate-based models, in general, outperformed their kernelised counterparts. This is not surprising as the "strength" of a pokemon is linearly related to the covariates provided (e.g. *attack* and *defence*). Given these highly informative features, covariate-based methods, in general, outperformed covariate-free methods as well. We note that SVDC, closely followed by BT-LR, was the top performer for both inference and prediction tasks.

NFL Football 2000–2018. The last data set we use contains the outcome of National Football League (NFL) matches during the regular season, for the years 2000–2018[3]. In addition, 256 matches per year between 32 teams, along with 18 performance metrics, such as *yards per game* and *number of fumbles* are recorded. We used their score differences at each match as the cardinal comparisons. We run our algorithms on each year's comparison graphs separately and average the results. We see for rank *inference*, our covariate-based algorithms outperformed their covariate-free counterparts, with BT-LR the top performer, followed closely by SVDC, KCC and RRC. For rank *prediction*, CC was significantly better than all other methods.

5 Conclusions and Future Directions

We proposed three spectral ranking algorithms to tackle the problem of ranking given noisy and incomplete pairwise comparisons when player covariates are available. We demonstrated the efficacy, strengths and weaknesses of the proposed methods in comparison to baselines through an extensive set of numerical experiments on synthetic and 22 real-world problems.

C-SerialRank extends SerialRank by incorporating covariate-induced player similarities into the seriation process through a linear combination. Although simple, through a straight forward cross-validation scheme on the regularisation parameter, we can ensure the ranking recovered in this way is at least as good as the ranking recovered from SerialRank, as demonstrated in our experiments. The downside of C-SerialRank, similar to Regularised Rank Centrality, is that it cannot be used on rank prediction. However, the results of our experiments show that the former outperformed the latter in the majority of the experiments. Our second contribution SVDCovRank extends SVDRank by relating covariates to the eigenvector of the pairwise comparison matrix $\mathbf{C}$ in a regression fashion. This allows us to predict the rankings of unseen players, which previous spectral ranking with covariates cannot handle. As showed in our

[2] Data collected from https://www.kaggle.com/c/intelygenz-pokemon-challenge.

[3] Data collected from nfl.com.

experiments, SVDCovRank performed comparably to covariate-based probabilistic methods. At last, CCRank approaches the ranking problem using the canonical correlation analysis between match outcomes and player covariates. CCRank places fewer assumptions on the underlying data generating process and is thus more robust in practice. This is demonstrated by its competitive performance against other methods when predicting the rank of unseen players.

Each proposed method will find its application in different practical scenarios. When rank prediction on unseen players is not a concern, we recommend practitioners to use the C-SerialRank algorithm because of its simplicity in balancing covariate and match information. On the other hand, if one believes the pairwise comparisons are caused by a strong underlying skill difference, then SVDCovRank algorithms should be considered. At last, if one seeks to place fewer assumptions on the data, then the CCRank algorithm may give a more robust ranking than the approaches that place specific structures on the match outcome matrix $\mathbf{C}$.

There are several avenues for future work. An interesting direction would be to incorporate more general types of side information into the ranking problem, e.g., prior information on partially observed player relationships. Yet another relevant direction is the extraction of partial rankings using the side covariates. In many real-world scenarios, aiming for a global ordering of the players is unrealistic, and one is often interested in uncovering partial orderings that reflect accurately various subsets/clusters of a less homogeneous population of players. Leveraging features for the discovery of latent structures behind cyclical and inconsistent preferences, as in [5] but using spectral methods, is also an interesting direction to explore.

References

1. Atkins, J.E., Boman, E.G., Hendrickson, B.: A spectral algorithm for seriation and the consecutive ones problem. SIAM J. Comput. **28**(1), 297–310 (1998)
2. Bradley, R.A., Terry, M.E.: Rank analysis of incomplete block designs: I. the method of paired comparisons. Biometrika **39**(3/4), 324–345 (1952)
3. Caron, F., Doucet, A.: Efficient Bayesian inference for generalized Bradley-Terry models. J. Comput. Graph. Stat. **21**(1), 174–196 (2012)
4. Cattelan, M.: Models for paired comparison data: a review with emphasis on dependent data. Stat. Sci. **27**(3), 412–433 (2012)
5. Chau, S., González, J., Sejdinovic, D.: Learning inconsistent preferences with Gaussian processes (2022)
6. Chen, S., Joachims, T.: Modeling intransitivity in matchup and comparison data. In: Proceedings of the Ninth ACM International Conference on Web Search and Data Mining, pp. 227–236 (2016)
7. Chu, W., Ghahramani, Z.: Preference learning with Gaussian processes. In: Proceedings of the 22nd International Conference on Machine Learning, pp. 137–144 (2005)

8. Cucuringu, M.: Sync-Rank: robust ranking, constrained ranking and rank aggregation via eigenvector and semidefinite programming synchronization. IEEE Trans. Netw. Sci. Eng. **3**(1), 58–79 (2016)
9. Cucuringu, M., Koutis, I., Chawla, S., Miller, G., Peng, R.: Simple and scalable constrained clustering: a generalized spectral method. In: Artificial Intelligence and Statistics, pp. 445–454 (2016)
10. d'Aspremont, A., Cucuringu, M., Tyagi, H.: Ranking and synchronization from pairwise measurements via SVD. J. Mach. Learn. Res. **22**(19), 866–928 (2021)
11. Fogel, F., d'Aspremont, A., Vojnovic, M.: SerialRank: spectral ranking using seriation. In: Advances in Neural Information Processing Systems, vol. 27 (2014)
12. Fukumizu, K., Bach, F.R., Gretton, A.: Statistical consistency of Kernel canonical correlation analysis. J. Mach. Learn. Res. **8**(Feb), 361–383 (2007)
13. Hotelling, H.: Relations between two sets of variates. In: Kotz, S., Johnson, N.L. (eds.) Breakthroughs in Statistics, pp. 162–190. Springer, Cham (1992). https://doi.org/10.1007/978-1-4612-4380-9_14
14. Huang, T.K., Weng, R.C., Lin, C.J., Ridgeway, G.: Generalized Bradley-Terry models and multi-class probability estimates. J. Mach. Learn. Res. **7**(1), 85–115 (2006)
15. Jain, L., Gilbert, A., Varma, U.: Spectral methods for ranking with scarce data. In: Conference on Uncertainty in Artificial Intelligence, pp. 609–618. PMLR (2020)
16. Kendall, M.G., Smith, B.B.: On the method of paired comparisons. Biometrika **31**(3–4), 324–345 (1940)
17. Li, X., Wang, X., Xiao, G.: A comparative study of rank aggregation methods for partial and top ranked lists in genomic applications. Brief. Bioinform. **20**(1), 178–189 (2019)
18. Mallows, C.: Non null ranking models I. Biometrika **44**, 114–130 (1957)
19. Mosteller, F., Nogee, P.: An experimental measurement of utility. J. Polit. Econ. **59**(5), 371–404 (1951)
20. Negahban, S., Oh, S., Shah, D.: Iterative ranking from pair-wise comparisons. In: Advances in Neural Information Processing Systems, vol. 25, pp. 2474–2482 (2012)
21. Negahban, S., Oh, S., Shah, D.: Rank centrality: ranking from pairwise comparisons. Oper. Res. **65**(1), 266–287 (2017). https://doi.org/10.1287/opre.2016.1534
22. Niranjan, U., Rajkumar, A.: Inductive pairwise ranking: going beyond the n log (n) barrier. In: Thirty-First AAAI Conference on Artificial Intelligence (2017)
23. Page, L., Brin, S., Motwani, R., Winograd, T.: The PageRank citation ranking: bringing order to the web. In: Proceedings of the 7th International World Wide Web Conference, pp. 161–172 (1998)
24. Shuman, D.I., Narang, S.K., Frossard, P., Ortega, A., Vandergheynst, P.: The emerging field of signal processing on graphs: extending high-dimensional data analysis to networks and other irregular domains. IEEE Signal Process. Mag. **30**(3), 83–98 (2013)
25. Singer, A.: Angular synchronization by eigenvectors and semidefinite programming. Appl. Comput. Harmon. Anal. **30**(1), 20–36 (2011)
26. Springall, A.: Response surface fitting using a generalization of the Bradley-Terry paired comparison model. J. R. Stat. Soc. Ser. C (Appl. Stat.) **22**(1), 59–68 (1973)
27. Stuart-Fox, D.M., Firth, D., Moussalli, A., Whiting, M.J.: Multiple signals in chameleon contests: designing and analysing animal contests as a tournament. Anim. Behav. **71**(6), 1263–1271 (2006)
28. Thurstone, L.: A law of comparative judgement. Psychol. Rev. **34**, 278–286 (1927)
29. Vigna, S.: Spectral ranking. Netw. Sci. **4**(4), 433–445 (2016)

30. Whiting, M.J., Webb, J.K., Keogh, J.S.: Flat lizard female mimics use sexual deception in visual but not chemical signals. Proc. R. Soc. B Biol. Sci. **276**(1662), 1585–1591 (2009)
31. Williams, C.K., Rasmussen, C.E.: Gaussian Processes for Machine Learning. MIT Press, Cambridge (2006)

Truly Unordered Probabilistic Rule Sets for Multi-class Classification

Lincen Yang(✉) and Matthijs van Leeuwen

LIACS, Leiden University, Leiden, The Netherlands
{l.yang,m.van.leeuwen}@liacs.leidenuniv.nl

Abstract. Rule set learning has long been studied and has recently been frequently revisited due to the need for interpretable models. Still, existing methods have several shortcomings: 1) most recent methods require a binary feature matrix as input, while learning rules directly from numeric variables is understudied; 2) existing methods impose orders among rules, either explicitly or implicitly, which harms interpretability; and 3) currently no method exists for learning probabilistic rule sets for multi-class target variables (there is only one for probabilistic rule lists).

We propose TURS, for Truly Unordered Rule Sets, which addresses these shortcomings. We first formalize the problem of learning truly unordered rule sets. To resolve conflicts caused by overlapping rules, i.e., instances covered by multiple rules, we propose a novel approach that exploits the probabilistic properties of our rule sets. We next develop a two-phase heuristic algorithm that learns rule sets by carefully growing rules. An important innovation is that we use a surrogate score to take the global potential of the rule set into account when learning a local rule.

Finally, we empirically demonstrate that, compared to non-probabilistic and (explicitly or implicitly) ordered state-of-the-art methods, our method learns rule sets that not only have better interpretability but also better predictive performance.

1 Introduction

When using predictive models in sensitive real-world scenarios, such as in health care, analysts seek for intelligible and reliable explanations for predictions. Classification rules have considerable advantages here, as they are directly readable by humans. While rules all seem alike, however, some are more interpretable than others. The reason lies in the subtle differences of how rules form a model. Specifically, rules can form an unordered *rule set*, or an explicitly ordered *rule list*; further, they can be categorized as probabilistic or non-probabilistic.

In practice, probabilistic rules should be preferred because they provide information about the uncertainty of the predicted outcomes, and thus are useful when a human is responsible to make the final decision, as the expected "utility" can be calculated. Meanwhile, unordered rule sets should also be preferred, as they have better properties regarding interpretability than ordered rule lists.

Supplementary Information The online version contains supplementary material available at https://doi.org/10.1007/978-3-031-26419-1_6.

© The Author(s), under exclusive license to Springer Nature Switzerland AG 2023
M.-R. Amini et al. (Eds.): ECML PKDD 2022, LNAI 13717, pp. 87–103, 2023.
https://doi.org/10.1007/978-3-031-26419-1_6

While no agreement has been reached on the precise definition of interpretability of machine learning models [14,16], we specifically treat interpretability with domain experts in mind. From this perspective, a model's interpretability intuitively depends on two aspects: the degree of difficulty for a human to comprehend the model itself, and to understand a single prediction. Unordered probabilistic rule sets are favorable with respect to both aspects, for the following reasons. First, comprehending ordered rule lists requires comprehending not only each individual rule, but also the relationship among the rules, while comprehending unordered rule sets requires only the former. Second, the explanation for a single prediction of an ordered rule list must contain the rule that the instance satisfies, together with all of its preceding rules, which becomes incomprehensible when the number of preceding rules is large.

Further, crucially, existing methods for rule set learning claim to learn unordered rule sets, but most of them are not truly unordered. The problem is caused by *overlap*, i.e., a single instance satisfying multiple rules. Ad-hoc schemes are widely used to resolve prediction conflicts caused by overlaps, typically by ranking the involved rules with certain criteria and always selecting the highest ranked rule [12,24] (e.g., the most accurate one). This, however, imposes implicit orders among rules, making them entangled instead of truly unordered.

This can badly harm interpretability: to explain a single prediction for an instance, it is now insufficient to only provide the rules the instance satisfies, because other higher-ranked rules that the instance does *not* satisfy are also part of the explanation. For instance, imagine a patient is predicted to have *Flu* because they have *Fever*. If the model also contains the higher-ranked rule *"Blood in stool → Dysentery"*, the explanation should include the fact that *"Blood in stool"* is not true, because otherwise the prediction would change to *Dysentery*. If the model contains many rules, it becomes impractical to have to go over all higher-ranked rules for each prediction.

Learning truly unordered probabilistic rule sets is a very challenging problem though. Classical rule set learning methods usually adopt a separate-and-conquer strategy, often sequential covering: they iteratively find the next rule and remove instances satisfying this rule. This includes 1) binary classifiers that learn rules only for the "positive" class [8], and 2) its extension to multi-class targets by the one-versus-rest paradigm, i.e., learning rules for each class one by one [2,4]. Importantly, by iteratively removing instances the *probabilistic predictive conflicts* caused by overlaps, i.e., rules having different probability estimates for the target, are ignored. Recently proposed rule learning methods go beyond separate-and-conquer by leveraging discrete optimization techniques [5,12,21,22,24], but this comes at the cost of requiring a binary feature matrix as input. Moreover, these methods are neither probabilistic nor truly unordered, as they still use ad-hoc schemes to resolve predictive conflicts caused by overlaps.

Approach and Contributions. To tackle these challenges and learn truly unordered probabilistic rules, we first formalize rule sets as probabilistic models. We adopt a probabilistic model selection approach for rule set learning, for which we design a criterion based on the minimum description length (MDL) principle [10]. Second, we propose a novel surrogate score based on decision trees that we use to evaluate the potential of incomplete rule sets. Third, we are the first to design a rule learning algorithm that deals with probabilistic conflicts caused by overlaps already during the rule learning process. We point out that rules that have been added to the rule set may become obstacles for new rules, and hence carefully design a two-phase heuristic algorithm, for which we adopt diverse beam search [19]. Last, we benchmark our method, named TURS, for Truly Unordered Rule Sets, against a wide range of methods. We show that the rule sets learned by TURS, apart from being probabilistic and truly unordered, have better predictive performance than existing rule list and rule set methods.

2 Related Work

Rule Lists. Rules in a rule list are connected by IF-THEN-ELSE statements. Existing methods include CBA [13], ordered CN2 [3], PART [6], and the recently proposed CLASSY [17] and Bayesian rule list [23]. We argue that rule lists are more difficult to interpret than rule sets because of their explicit orders.

One-Versus-Rest Learning. This category focuses on only learning rules for a single class label, i.e., the "positive" class, which is already sufficient for binary classification [5,21,22]. For multi-class classification, two approaches exist. The first, taken by RIPPER [4] and C4.5 [18], is to learn each class in a certain order. After all rules for a single class have been learned, all covered instances are removed (or those with this class label). The resulting model is essentially an ordered list of rule sets, and hence is more difficult to interpret than rule set.

The second approach does not impose an order among the classes; instead, it learns a set of rules for each class against all other classes. The most well-known are unordered-CN2 and FURIA [2,11]. FURIA avoids dealing with conflicts of overlaps by using all rules for predicting unseen instances; as a result, it cannot provide a single rule to explain its prediction. Unordered-CN2, on the other hand, handles overlaps by "combining" all overlapping rules into a "hypothetical" rule, which sums up all instances in all overlapping rules and hence ignoring probabilistic conflicts for constructing rules. In Sect. 6, we show that our method learns smaller rule sets with better predictive performance than unordered-CN2.

Multi-class rule sets. Very few methods exist for directly learning rules for multi-class targets, which is algorithmically more challenging than the one-versus-rest paradigm, as the separate-and-conquer strategy is not applicable. To the best of our knowledge, the only existing methods are IDS [12] and DRS [24]. Both are neither probabilistic nor truly unordered. To handle conflicts of overlaps, IDS follows the rule with the highest F1-score, and DRS uses the most accurate rule.

Last, different but related approaches include 1) decision tree based methods such as CART [1], which produce rules that are forced to share many "attributes" and hence are longer than necessary, as we will empirically demonstrate in Sect. 6, and 2) a Bayesian rule mining [9] method, which adopts naive bayes with the mined rules for prediction, and hence does not produce a rule set model in the end. The 'lazy learning' approach for rule-based models can also avoid the conflicts of overlaps [20], but no global rule set model describing the whole dataset is constructed in this case.

3 Rule Sets as Probabilistic Models

We first formalize individual rules as *local* probabilistic models, and then define rule sets as *global* probabilistic models. The key challenge lies in how to define $P(Y = y|X = x)$ for an instance (x, y) that is covered by multiple rules.

3.1 Probabilistic Rules

Denote the input random variables by $X = (X_1, \ldots, X_d)$, where each X_i is a one-dimensional random variable representing one dimension of X, and denote the categorical target variable by $Y \in \mathscr{Y}$. Further, denote the dataset from which the rule set can be induced as $D = \{(x_i, y_i)\}_{i \in [n]}$, or (x^n, y^n) for short. Each (x_i, y_i) is an instance. Then, a probabilistic rule S is written as

$$(X_1 \in R_1 \wedge X_2 \in R_2 \wedge \ldots) \rightarrow P_S(Y), \tag{1}$$

where each $X_i \in R_i$ is called a *literal* of the *condition* of the rule. Specifically, each R_i is an interval (for a quantitative variable) or a set of categorical levels (for a categorical variable).

A probabilistic rule of this form describes a subset S of the full sample space of X, such that for any $x \in S$, the conditional distribution $P(Y|X = x)$ is approximated by the probability distribution of Y conditioned on the event $\{X \in S\}$, denoted as $P(Y|X \in S)$. Since in classification Y is a discrete variable, we can parametrize $P(Y|X \in S)$ by a parameter vector $\boldsymbol{\beta}$, in which the jth element β_j represents $P(Y = j|X \in S)$, for all $j \in \mathscr{Y}$. We therefore denote $P(Y|X \in S)$ as $P_{S,\boldsymbol{\beta}}(Y)$, or $P_S(Y)$ for short. To estimate $\boldsymbol{\beta}$ from data, we adopt the maximum likelihood estimator, denoted as $P_{S,\hat{\boldsymbol{\beta}}}(Y)$, or $\hat{P}_S(Y)$ for short.

Further, if an instance (x, y) satisfies the condition of rule S, we say that (x, y) is *covered* by S. Reversely, the *cover* of S denotes the instances it covers. When clear from the context, we use S to both represent the rule itself and/or its cover, and define the number of covered instances $|S|$ as its *coverage.*

3.2 Truly Unordered Rule Sets as Probabilistic Models

While a rule set is simply a set of rules, the challenge lies in how to define rule sets as probabilistic models while keeping the rules truly unordered. That is, how do we define $P(Y|X = x)$ given a rule set M, i.e., a model, and its parameters? We first explain how to do this for a single instance of the training data, using a simplified setting where at most two rules cover the instance. We then discuss—potentially unseen—test instances and extend to more than two rules covering an instance. Finally, we define a rule set as a probabilistic model.

Class Probabilities for a Single Training Instance. Given a rule set M with K individual rules, denoted $\{S_i\}_{i \in [K]}$, any instance (x, y) falls into one of four cases: 1) exactly one rule covers x; 2) at least two rules cover x and no rule's cover is the subset of another rule's cover (*multiple non-nested*); 3) at least two rules cover x and one rule's cover is the subset of another rule's cover (*multiple nested*); and 4) no rule in M covers x.

To simplify the notation, we here consider at most two rules covering an instance—we later describe how we can trivially extend to more than two rules.
Covered by One Rule. When exactly one rule $S \in M$ covers x, we use $P_S(Y)$ to "approximate" the conditional probability $P(Y|X = x)$. To estimate $P_S(Y)$ from data, we adopt the maximum likelihood (ML) estimator $\hat{P}_S(Y)$, i.e.,

$$\hat{P}_S(Y = j) = \frac{|\{(x, y) : x \in S, y = j\}|}{|S|}, \forall j \in \mathscr{Y}. \tag{2}$$

Note that we do not exclude instances in S that are also covered by other rules (i.e., in overlaps) for estimating $P_S(Y)$. Hence, the probability estimation for each rule is independent of other rules; as a result, each rule is *self-standing*, which forms the foundation of a truly unordered rule set.

Covered by Two Non-nested Rules. Next, we consider the case when x is covered by S_i and S_j, and neither $S_i \subseteq S_j$ nor $S_j \subseteq S_i$, i.e., the rules are non-nested.

When an instance is covered by two non-nested, partially overlapping rules, we interpret this as probabilistic *uncertainty*: we cannot tell whether the instance belongs to one rule or the other, and therefore approximate its conditional probability by the *union* of the two rules. That is, in this case we approximate $P(Y|X = x)$ by $P(Y|X \in S_i \cup S_j)$, and we estimate this with its ML estimator $\hat{P}(Y|X \in S_i \cup S_j)$, using all instances in $S_i \cup S_j$.

This approach is particularly useful when the estimator of $P(Y|X \in S_i \cap S_j)$, i.e., conditioned on the event $\{X \in S_i \cap S_j\}$, is indistinguishable from $\hat{P}(Y|X \in S_i)$ and $\hat{P}(Y|X \in S_j)$. Intuitively, this can be caused by two reasons: 1) $S_i \cap S_j$ consists of very few instances, so the variance of the estimator for $P(Y|X \in S_i \cap S_j)$ is large; 2) $P(Y|X \in S_i \cap S_j)$ is just very similar to $P(Y|X \in S_i)$ and $P(Y|X \in S_i)$, which makes it undesirable to create a separate rule for $S_i \cap S_j$. Our model selection approach, explained in Sect. 4, will ensure that a rule set with non-nested rules has high goodness-of-fit only if this 'uncertainty' is indeed the case.

Covered by Two Nested Rules. When x is covered by both S_i and S_j, and S_i is a subset of S_j, i.e., $x \in S_i \subseteq S_j$, the rules are nested[1]. In this case, we approximate $P(Y|X = x)$ by $P(Y|X \in S_i)$ and interpret S_i as an *exception* of S_j. Having such nested rules to model such exceptions is intuitively desirable, as it allows to have general rules covering large parts of the data while being able to model smaller, deviating parts. In order to preserve the self-standing property of individual rules, for $x \in S_j \setminus S_i$ we still use $P(Y|X \in S_j)$ rather than $P(Y|X \in S_j \setminus S_i)$. Although this might seem counter-intuitive at first glance, using $P(Y|X \in S_j \setminus S_i)$ would implicitly impose an order between S_j and S_i, or—equivalently—implicitly change S_j to another rule that only covers instances in $S_j \wedge \neg S_i$.

Not Covered by Any Rule. When no rule in M covers x, we say that x belongs to the so-called "else rule" that is part of every rule set and equivalent to $x \notin \bigcup_i S_i$.

[1] Note that "nestedness" is based on the rules' covers rather than on their conditions. For instance, if S_i is $X_1 <= 1$ and S_j is $X_2 <= 1$, S_i and S_j could still be nested.

Thus, we approximate $P(Y|X = x)$ by $P(Y|X \notin \bigcup_i S_i)$. We denote the else rule by S_0 and write $S_0 \in M$ for the else rule in M. Observe that the else rule is the only rule in every rule set that depends on the other rules and is therefore not self-standing; however, it will also have no overlap with other rules by definition.

Predicting for a New Instance. When an unseen instance x' comes in, we predict $P(Y|X = x')$ depending on which of the aforementioned four cases it satisfies. An important question is whether we always need access to the training data, i.e., whether the probability estimates we obtain from the training data points are sufficient for predicting $P(Y|X = x')$. Specifically, if x' is covered by non-nested S_i and S_j, $P(Y|X = x')$ is predicted as $\hat{P}(Y|X \in S_i \cup S_j)$. However, if there are no training data points covered both by S_i and S_j, then we would not obtain $\hat{P}(Y|X \in S_i \cup S_j)$ in the training phase. Nevertheless, in this case we have $|S_i \cup S_j| = |S_i| + |S_j|$, and hence

$$\hat{P}(Y|X \in S_i \cup S_j) = \frac{|S_i|\hat{P}(Y|X \in S_i) + |S_j|\hat{P}(Y|X \in S_j)}{|S_i| + |S_j|}. \tag{3}$$

Thus, if x' is covered by one rule, two nested rules, or no rule in M, the corresponding probability estimates are already obtained during training. Thus, we conclude that access to the training data is not necessary for prediction.

Extension to Overlaps of Multiple Rules. Whenever an instance x is covered by multiple rules, denoted $J = \{S_i, S_j, S_k, ...\}$, three cases may happen. The first case is all rules in J are nested. Without loss of generality, assume that $S_i \subseteq S_j \subseteq S_k \subseteq ...$; then, following the rationale for case of two nested rules, $P(Y|X = x)$ should be approximated by $P_{S_i}(Y)$. Therefore, when x is covered by multiple nested rules, only the "smallest" rule matters and we can act as if x is only covered by that single rule.

The second case is that all rules in J are non-nested with each other. Following the solution for modeling two non-nested rules, we use $P(Y|X \in \bigcup_{S \in J} S)$.

The third case is a mix of the previous two cases, i.e., rules in J are partially nested. In this case, we iteratively go over all $S \in J$: if there exists an $S' \in J$ satisfying $S' \subseteq S$ we remove S from J, and continue iterating until no nested overlap in J remains. If one single rule is left, we act as if x is covered by that single rule; otherwise, we follow the paradigm of modeling the non-nested overlaps with the rules left in J.

Probabilistic Rule Sets. We can now build upon the previous to define rule sets as probabilistic models. Formally, the probabilistic model corresponding to a rule set M is a family of probability distributions, denoted $P_{M,\theta}(Y|X)$ and parametrized by θ. Specifically, θ is a parameter vector representing all necessary probabilities of Y conditioned on events $\{X \in G\}$, where G is either a single rule or the union of multiple rules. θ is estimated from data by estimating each $P(Y|X \in G)$ by its maximum likelihood estimator. The resulting estimated vector is denoted as $\hat{\theta}$ and contains $\hat{P}(Y|X \in G)$ for all $G \in \mathscr{G}$, where $\mathscr{G}$ consists of all individual rules and the unions of overlapping rules in M.

Finally, we assume the dataset $D = (x^n, y^n)$ to be i.i.d. Specifically, let us define $(x, y) \vdash G$ for the following two cases: 1) when G is a single rule (including the else rule), then $(x, y) \vdash G \iff x \in G$; and 2) when G is a union of multiple rules, e.g., $G = \bigcup S_i$, then $(x, y) \vdash G \iff x \in \bigcap S_i$. We then have

$$P_{M,\theta}(y^n|x^n) = \prod_{G \in \mathscr{G}} \prod_{(x,y) \vdash G} P(Y = y|X \in G). \tag{4}$$

4 Rule Set Learning as Probabilistic Model Selection

Exploiting the formulation of rule sets as probabilistic models, we define the task of learning a rule set as a probabilistic model selection problem. Specifically, we use the minimum description length (MDL) principle for model selection.

4.1 Normalized Maximum Likelihood Distributions for Rule Sets

The MDL principle is one of the best off-the-shelf model selection methods and has been widely used in machine learning and data mining [10]. Although rooted in information theory, it has been recently shown that MDL-based model selection can be regarded as an extension of Bayesian model selection [10].

The core idea of MDL-based model selection is to assign a single probability distribution to the data given a rule set M, the so-called *universal distribution* denoted by $P_M(Y^n|X^n = x^n)$. Informally, $P_M(Y^n|X^n = x^n)$ should be a representative of the rule set model—as a family of probability distributions—$\{P_{M,\theta}(y^n|x^n)\}_\theta$. The theoretically optimal "representative" is defined to be the one that has minimax regret, i.e.,

$$\arg\min_{P_M} \max_{z^n \in \mathscr{Y}^n} -\log_2 P_M(Y^n = z^n|X^n = x^n) - \Big(-\log_2 P_{\hat{\theta}(x^n,z^n)}(Y^n = z^n|X^n = x^n)\Big). \tag{5}$$

We write the parameter estimator as $\hat{\theta}(x^n, z^n)$ to emphasize that it depends on the values of the target variables Y^n. The unique solution to P_M of Eq. 5 is the so-called normalized maximum likelihood (NML) distribution:

$$P_M^{NML}(Y^n = y^n|X^n = x^n) = \frac{P_{M,\hat{\theta}(x^n,y^n)}(Y^n = y^n|X^n = x^n)}{\sum_{z^n \in \mathscr{Y}^n} P_{M,\hat{\theta}(x^n,z^n)}(Y^n = z^n|X^n = x^n)}. \tag{6}$$

That is, we "normalize" the distribution $P_{M,\hat{\theta}}(.)$ to make it a proper probability distribution, which requires the sum of all possible values of Y^n to be 1. Hence, we have $\sum_{z^n \in \mathscr{Y}^n} P_M^{NML}(Y^n = z^n|X^n = x^n) = 1$ [10].

4.2 Approximating the NML Distribution

A crucial difficulty in using the NML distribution in practice is the computation of the normalizing term $\sum_{z^n} P_{\hat{\theta}(x^n,z^n)}(Y^n = z^n|X^n = x^n)$. Efficient algorithms almost only exist for exponential family models [10], hence we approximate the term by the product of the normalizing terms for the individual rules.

NML distribution for a single rule. For an individual rule $S \in M$, we write all instances covered by S as (x^S, y^S), in which y^S can be regarded as a realization of the random vector $Y^S = (Y, ..., Y)$, and Y^S takes values in $\mathscr{Y}^{|S|}$, the $|S|$-ary Cartesian power of $\mathscr{Y}$. Then, the NML distribution for $P_S(Y)$ equals

$$P_S^{NML}(Y^S = y^S | X^S = x^S) = \frac{\hat{P}_S(Y^S = y^S | X^S = x^S)}{\sum_{z^S \in \mathscr{Y}^S} \hat{P}_S(Y^S = z^S | X^S = x^S)}. \quad (7)$$

Note that $\hat{P}_S$ depends on the values of z^S. As $\hat{P}_S(Y)$ is a categorical distribution, the normalizing term can be written as $\mathcal{R}(|S|, |\mathscr{Y}|)$, a function of $|S|$—the rule's coverage—and $|\mathscr{Y}|$—the number of unique values that Y can take [15]:

$$\mathcal{R}(|S|, |\mathscr{Y}|) = \sum_{z^S \in \mathscr{Y}^S} \hat{P}_S(Y^S = z^S | X^S = x^S), \quad (8)$$

which can be efficiently calculated in sub-linear time [15].

The Approximate NML Distribution. We propose to approximate the normalizing term of P_M^{NML} as the product of the normalizing terms of P_S^{NML} for all $S \in M$, and propose the approximate-NML distribution as our model selection criterion:

$$P_M^{apprNML}(Y^n = y^n | X^n = x^n) = \frac{P_{M, \hat{\theta}(x^n, y^n)}(Y^n = y^n | X^n = x^n)}{\prod_{S \in M} \mathcal{R}(|S|, |\mathscr{Y}|)}. \quad (9)$$

Note that the sum over all $S \in M$ *does* include the "else rule" S_0. Finally, we can formally define the optimal rule set M^* as

$$M^* = \arg\max_M P_M^{apprNML}(Y^n = y^n | X^n = x^n). \quad (10)$$

The rationale of using the approximate-NML distribution is as follows. First, it is equal to the NML distribution for a rule set without any overlap, as follows.

Proposition 1. *Given a rule set M in which for any $S_i, S_j \in M$, $S_i \cap S_j = \emptyset$, then $P_M^{NML}(Y^n = y^n | X^n = x^n) = P_M^{apprNML}(Y^n = y^n | X^n = x^n)$.*

Second, when overlaps exist in M, approximate-NML puts a small extra penalty on overlaps, which is desirable to trade-off overlap with goodness-of-fit: when we sum over all instances in each rule $S \in M$, the instances in overlaps are "repeatedly counted". Third, approximate-NML behaves like the Bayesian information criterion (BIC) asymptotically, which follows from the next proposition.

Proposition 2. *Assume M contains K rules in total, including the else rule, and we have n instances. Then $\log\left(\prod_{S \in M} \mathcal{R}(|S|, |\mathscr{Y}|)\right) = \frac{K(|\mathscr{Y}|-1)}{2} \log n + \mathcal{O}(1)$, where $\mathcal{O}(1)$ is bounded by a constant w.r.t. to n.*

We defer the proofs of the two propositions to the Supplementary Material.

5 Learning Truly Unordered Rule Sets from Data

As our MDL-based model selection criterion unfortunately does not enable efficient search for the optimal model, we resort to heuristics. We first address the challenge of evaluating incomplete rule sets, after which we explain how to grow individual rules in two phases and implement this with beam search. Finally, we show how everything comes together to iteratively learn rule sets from data.

5.1 Evaluating Incomplete Rule Sets with a Surrogate Score

When iteratively searching for the next "best" rule, defining "best" is far from trivial: rule coverage and precision are contradicting factors and typical scores therefore combine those two factors in some—more or less—arbitrary way.

This issue is further aggravated by the iterative rule learning process, in which the intermediate rule set is evaluated as an *incomplete rule set* in each step. Evaluating incomplete rule sets is a challenging task [7], mainly because any good score needs to simultaneously consider two aspects: 1) how well do all the rules currently in the rule set describe the already covered instances; and 2) what is the "potential" for the uncovered instances, in the sense that how well can those uncovered instances be described by rules that might be added later?

Without knowing the rules that will be added later, we cannot compute the NML-based criterion for the complete rule set. Yet, we should take into account the potential of the uncovered instances. We propose to approximate the latter using a *surrogate score*, which we obtain by fitting a decision tree on the uncovered instances and using the leafs of the resulting tree as a surrogate for "future" rules. Formally, we define the tree-based surrogate score as

$$L_{\mathcal{T}}(M) = P_{M \oplus \mathcal{T}}^{apprNML}(Y^n = y^n | X^n = x^n), \tag{11}$$

where $M \oplus \mathcal{T}$ denotes the surrogate rule set obtained by converting the branches of $\mathcal{T}$ to rules and appending those to M (parameters are estimated as usual).

Although the branches of the decision tree learned from the currently uncovered instances may be different from the rules that will later be added to the rule set, using the tree-based surrogate score will make it easier to gradually grow good rule sets. We use decision trees because they are quick to learn and use, and the correspondence of branches to rules makes using them straightforward. We will empirically study the effects of the surrogate score on the predictive performance of rule sets in Sect. 6.

5.2 Two-phase Rule Growth

To avoid having to traverse all possible rules when searching for the rule to add to an incomplete rule set, we resort to a common heuristic: we start with an empty rule and gradually refine it by adding literals—also referred to as *growing* a rule [8]. In contrast to existing methods, we propose a two-phase method.

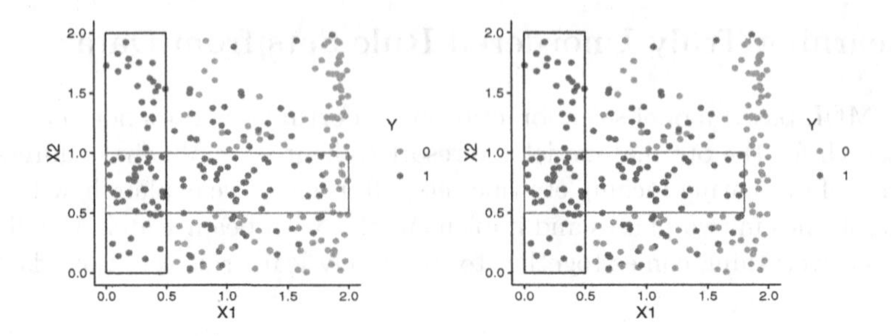

Fig. 1. (Left) Simulated data with two overlapping rules: $S_1 : X_1 < 0.5$ (outlined in black) and $S_2 : 0.5 < X_2 < 1$ (purple). (Right) S_2 has grown to $0.5 < X_2 < 1 \wedge X_1 < 1.8$, which changes $P(Y|X \in S_2)$ and resolves the problematic overlap.

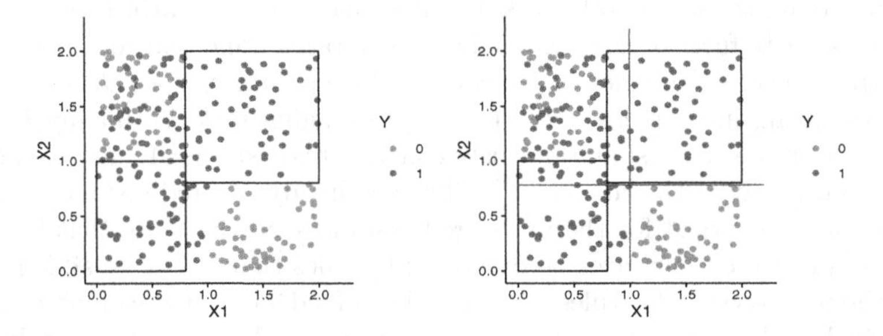

Fig. 2. (Left) Simulated data with a rule set containing two rules (black outlines). (Right) Growing a rule to describe the bottom-right instances will create conflicts with existing rules. I.e., adding either $X_1 > 1$ (vertical purple line) or $X_2 < 0.8$ (horizontal purple line) would create a huge overlap that deteriorates the surrogate score (Eq. 11).

Motivation. A rule can only improve the surrogate score—and thus be added to the rule set—if it achieves two goals: 1) it should improve the likelihood of currently uncovered instances (penalized by the approximate-NML normalizing term); and 2) it should not deteriorate the goodness-of-fit of the rule set by creating "bad" overlaps. These goals can be conflicting though, for two reasons.

First, it is not necessarily bad to have overlaps between a rule being grown and the current rule set, because the rule and its probability estimates for the target variable may still change. For example, consider the left plot of Fig. 1. If the current rule set consists of S_1 (indicated in black), then adding S_2 (in purple) would be problematic: this would strongly deteriorate the likelihood of the instances covered by both rules. However, as we further grow S_2, as shown in the right plot, we get $P(Y|S_1) = P(Y|S_2)$ and the problem is solved.

Second, rules already in the rule set may become obstacles to growing a new rule. For example, consider the data and rule set with two rules (in black) in Fig. 2. If we want to grow a rule that covers the bottom-right instances, the existing rules form a blockade: the right plot shows how adding either $X_1 > 1$ or $X_2 < 0.8$ to the empty rule (in purple) would create a large overlap with the existing rules, with significantly different probability estimates.

Algorithm 1: Find Next Rule Ignoring Overlaps

Input: rule set M, data (x^n, y^n)
Output: A beam that contains the w best rules

```
1  RULE ← ∅; Beam ← [RULE]        // Initialize the empty rule and beam
2  BeamList ← Beam    // Record all the beams in the beam search
3  while length(Beam) ≠ 0 do
4      candidates ← [ ]          // initialized to store all possible refinements
5      for RULE ∈ Beam do
6          Rs ← [Append L to RULE for L ∈ all possible literals]
7          candidates.extend(Rs)
8      Beam ← the w rules in candidates that have 1) the highest positive
         g_unc(), and 2) coverage diversity > α        // w is the beam width
9      if length(Beam) ≠ 0 then
10         BeamList.extend(Beam)  // extend the BeamList as an array
11 for Rule ∈ BeamList do
12     Beam ← w rules in BeamList with best L_T(M ⊕ S_unc)
13 return Beam
```

Therefore, instead of navigating towards the two goals simultaneously, we propose to grow the next rule in two phases: 1) grow the rule as if the instances covered by the (incomplete) rule set are excluded; 2) further grow the rule to eliminate potentially "bad" overlaps, to further optimize the tree-based score.

Method. Given a rule S, define S_{unc} as its uncovered "counterpart", which covers all instances in S not covered by M, i.e., $S_{unc} = S \setminus \cup\{S_i \in M\}$. Then, given M, the search for the next best rule that optimizes the surrogate tree-based score is divided into two phases. First, we aim to find the m rules for which the uncovered counterparts have the highest surrogate scores, defined as

$$L_{\mathcal{T}}(M \oplus S_{unc}) = P^{apprNML}_{M \oplus S_{unc} \oplus \mathcal{T}}(Y^n = y^n | X^n = x^n), \tag{12}$$

where $M \oplus S_{unc} \oplus \mathcal{T}$ denotes M appended with S_{unc} and all branches of $\mathcal{T}$. Here, m is a user-specified hyperparameter that controls the number of candidate rules that are selected for further refinement in the second phase. In the second phase, we further grow each of these m rules to search for the best one rule that optimizes

$$L_{\mathcal{T}}(M \oplus S) = P^{apprNML}_{M \oplus S \oplus \mathcal{T}}(Y^n = y^n | X^n = x^n). \tag{13}$$

Given a rule S and its counterpart S_{unc}, the score of S_{unc} is an upper-bound on the score of S: if S can be further refined to cover exactly what S_{unc} covers, we can obtain $L_{\mathcal{T}}(M \oplus S_{unc}) = L_{\mathcal{T}}(M \oplus S_{unc})$. This is often not possible in practice though, and we therefore generate m candidates in the first phase (instead of 1).

5.3 Beam Search for Two-Phase Rule Growth

In both phases we aim for growing a rule that optimizes the tree-based score (Eq. 11); the difference is that we ignore the already covered instances in the first phase. To avoid growing rules too greedily, i.e., adding literals that quickly reduce the coverage of the rule, we use a heuristic that is based on the NML distribution of a single rule and motivated by Foil's information gain [4].

Phase 1: Rule Growth Ignoring Covered Instances. We propose the NML-gain to optimize $L_T(M \oplus S_{unc})$: given two rules S and Q, where we obtain S by adding one literal to Q, we define the NML-gain as $g_{unc}(S,Q)$:

$$g_{unc}(S,Q) = \left(\frac{P^{NML}_{S_{unc}}(y^{S_{unc}}|x^{S_{unc}})}{|S_{unc}|} - \frac{P^{NML}_{Q_{unc}}(y^{Q_{unc}}|x^{Q_{unc}})}{|Q_{unc}|} \right) |S_{unc}| \tag{14}$$

$$= \left(\frac{\hat{P}_{S_{unc}}(y^{S_{unc}}|x^{S_{unc}})}{\mathcal{R}(|S_{unc}|, |\mathscr{Y}|)\ |S_{unc}|} - \frac{\hat{P}_{Q_{unc}}(y^{Q_{unc}}|x^{Q_{unc}})}{\mathcal{R}(|Q_{unc}|, |\mathscr{Y}|)\ |Q_{unc}|} \right) |S_{unc}|, \tag{15}$$

which we use as the navigation heuristic.

The advantage of having a tree-based score to evaluate rules, besides the navigation heuristic (local score), is that we can adopt beam search, as outlined in Algorithm 1. We start by initializing 1) the rule as an *empty rule* (a rule without any condition), 2) the Beam containing that empty rule, and 3) the BeamRecord to record the rules in the beam search process (Line 1–2). Then, for each rule in the beam, we generate refined candidate rules by adding one literal to it (Ln 5–7). Among all candidates, we select at most w rules with the highest NML-based gain g_{unc}, satisfying two constraints: 1) $g_{unc} > 0$; and 2) for each pair of these (at most) w rules, e.g., S and Q, their "coverage diversity" $\frac{|S_{unc} \cap Q_{unc}|}{|S_{unc} \cup Q_{unc}|} > \alpha$, where α is a user-specified parameter that controls the diversity of the beam search [19]. We update the Beam with these (at most) w rules (Ln 8–10). We repeat the process until we can no longer grow any rule with positive g_{unc} based on all rules in Beam (Ln 3). Last, among the record of all Beams we obtained during the process, we return the best w rules with the highest tree-based score $L(S_{unc} \cup M)$ (Ln 11–13).

Phase 2: Rule Growth Including Covered Instances. We now optimize $L(M \oplus S)$ and select a rule based on the candidates obtained in the previous step. We first define a navigation heuristic: given two rules S and Q, where S is obtained by adding one literal to Q, we define the NML-gain $g(S,Q)$ as

$$g(S,Q) = \left(\frac{\hat{P}_S(y^{S_{unc}}|x^{S_{unc}})}{\mathcal{R}(|S_{unc}|, |\mathscr{Y}|)\ |S_{unc}|} - \frac{\hat{P}_Q(y^{Q_{unc}}|x^{Q_{unc}})}{\mathcal{R}(|Q_{unc}|, |\mathscr{Y}|)\ |Q_{unc}|} \right) |S_{unc}|. \tag{16}$$

Note that the difference between $g(S,Q)$ and $g_{unc}(S,Q)$ is that they use a different maximum likelihood estimator: $\hat{P}_Q$ is the ML estimator based on all instances in Q, while $\hat{P}_{Q_{unc}}$ is based on all instances in Q_{unc}.

Algorithm 2: Find Rule Set

```
  Input: training data (x^n, y^n)
  Output: rule set M
1 M ← ∅; M_record ← [M]
2 scores ← [P_M^{apprNML}(y^n|x^n)]        // Record P_M^{apprNML} while growing
3 while True do
4     S* ← FindNextRule(M, (x^n, y^n))   // find the next best rule S*
5     if S* = ∅ or L_T(M ⊕ S) = P_{M⊕S*}^{apprNML}(y^n|x^n) then
6         Break
7     else
8         M ← M ⊕ S*; M_record.append(M)   // update and record M
9         scores.append(P_M^{apprNML}(y^n|x^n))
10 return the rule set with the maximum score in M_record
```

The algorithm is almost identical to Algorithm 1, with four small modifications: 1) the navigation heuristic is replaced by $g(S, Q)$; 2) $L_{\mathcal{T}}(M \oplus S)$ is used to select the best rule from the BeamRecord instead of $L_{\mathcal{T}}(M \oplus S_{unc})$; and 3) the coverage diversity is based on the rules itself instead of the counterparts; 4) only the best rule is returned.

5.4 Iterative Search for the Rule Set

Algorithm 2 outlines the proposed rule set learner. We start with an empty rule set (Ln 1–2), then iteratively add the next best rule (Ln 3–9) until the stopping criterion is met (Ln 5–6). That is, it stops when 1) the surrogate score equals the 'real' model selection criterion (i.e., the model's NML distribution), or 2) no more rules with positive NML-gain can be found. We record the 'real' criterion when adding each rule to the set, and pick the one maximizing it (Ln 10).

6 Experiments

We demonstrate that TURS learns rule sets with competitive predictive performance, and that using the surrogate score substantially improves the AUC scores. Further, we demonstrate that TURS achieves model complexities comparable to other rule set methods for multi-class targets.

We here discuss the most important parts of the experiment setup; for completeness, additional information can be found in the Supplementary Material[2].

Decision Trees for Surrogate Score. We use CART [1] to learn the trees for the surrogate score. For efficiency and robustness, we do not use any post-pruning for the decision trees but only set a minimum sample size for the leafs.

[2] The source code is available at https://github.com/ylincen/TURS.

Beam Width and Coverage Diversity. We set the coverage diversity $\alpha = 0.05$, and beam width $w = 5$. With the coverage diversity as a constraint, we found that $w \in \{5, 10, 20\}$ all give similar results. Due to the limited space, we leave a formal sensitivity analysis of α as future work.

Benchmark Datasets and Competitor Algorithms. We test on 13 UCI benchmark datasets (shown in Table 1), and compare against the following methods: 1) unordered CN2 [2], the one-versus-rest rule sets method without implicit order among rules; 2) DRS [24], a representative multi-class rule set learning method; 3) BRS [21], the Bayesian rule set method for binary classification; 4) RIPPER [4], the widely used one-versus-rest method with orders among class labels; 5) CLASSY [17], the probabilistic rule list methods using MDL-based model selection; and 6) CART [1], the well-known decision tree method, *with* post-pruning by cross-validation.

Table 1. ROC-AUC scores, averaged over 10 cross-validated folds. The rank (smaller means better) is further averaged over all datasets. Among the four *rule set* methods, TURS is substantially better on 7 out 13 datasets (AUC scores in bold).

Data	TURS	CN2	DRS	BRS	CLASSY	RIPPER	CART	TURS %overlap
Anuran	0.998	1.000	0.858	—	0.983	0.999	0.996	0.395
Avila	0.968	0.978	0.530	—	0.954	0.997	0.988	0.286
Backnote	**0.991**	0.969	0.945	0.957	0.987	0.979	0.984	0.297
Car	**0.978**	0.633	0.924	—	0.945	0.980	0.971	0.063
Chess	**0.995**	0.536	0.823	0.945	0.991	0.995	0.994	0.264
Contracept	**0.667**	0.597	0.544	—	0.630	0.626	0.600	0.074
Diabetes	**0.766**	0.677	0.628	0.683	0.761	0.735	0.661	0.155
Ionosphere	0.914	0.912	0.663	0.837	0.909	0.901	0.845	0.310
Iris	0.964	0.985	0.935	—	0.960	0.973	0.965	0.018
Magic	**0.886**	0.590	0.695	0.794	0.895	0.818	0.800	0.500
tic-tac-toe	0.972	0.826	0.971	0.976	0.983	0.954	0.847	0.231
Waveform	**0.902**	0.775	0.588	—	0.833	0.884	0.803	0.528
Wine	0.954	0.962	0.810	—	0.961	0.945	0.932	0.031
Avg Rank	2.231	4.077	5.846	5.462	3.154	3.000	4.231	/

6.1 Results

Predictive Performance. We report the ROC-AUC scores in Table 1. For multi-class classification, we report the weighted one-versus-rest AUC scores, as was also used for evaluating the recently proposed CLASSY method [17].

Compared to non-probabilistic rule set methods—i.e., CN2, DRS, and BRS (only for binary targets)—TURS is much better in terms of the mean rank of its AUC scores. Specifically, it performs substantially better on about half of the

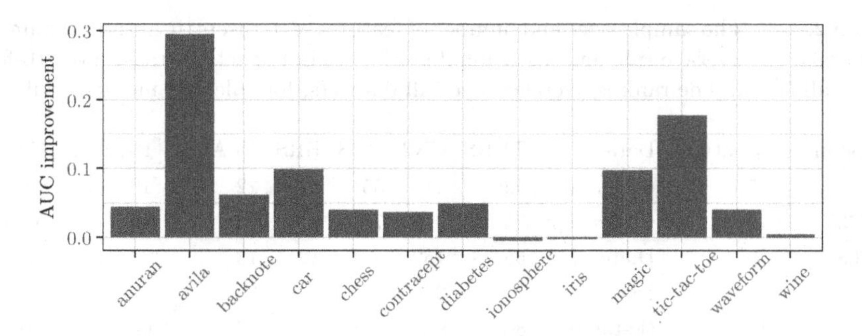

Fig. 3. Improvement in AUC by enabling the surrogate score for TURS.

datasets (shown in bold). Besides, it is ranked better than rule list methods, which produce explicitly ordered rules that may be difficult for domain experts to comprehend and digest in practice. Next, CART attains AUCs generally inferior to TURS, although it helps TURS to get a higher AUC as part of the surrogate score.

Last, we report the percentage of instances covered by more than one rule for TURS in Table 1, and we show that overlaps are common in the rule sets obtained for different datasets. This empirically confirms that our way of formalizing rule sets as probabilistic models, i.e., treating overlaps as uncertainty and exception, can indeed lead to improved predictive performance, as the overlapping rules are a non-negligible part of the model learned from data and hence indeed play a role.

Effects of the Surrogate Score. Fig. 3 shows the difference in AUC obtained by our method with and without using the surrogate score (i.e., without surrogate score means replacing it with the final model selection criterion). We conclude that the surrogate score has a substantial effect on learning better rule sets, except for three "simple" datasets, of which the sample sizes and the number of variables are small, as shown in Table 2 (Left).

Model Complexity. Finally, we compare the 'model complexity' of the rule sets for all methods. As this is hard to quantify in a unified manner, as a proxy we report the *total number of literals in all rules in a rule set*, averaged over 10-fold cross-validation (the same as used for the results reported in Table 1).

We show that among all rule set methods (TURS, CN2, DRS, BRS), TURS has better average ranks than both CN2 and DRS. Although BRS learns very small rule sets, it is only applicable to binary targets and its low model complexity also brings worse AUC scores than TURS. Further, although rule list methods (CLASSY, RIPPER) generally have fewer literals than rule sets methods, this does not make rule lists easy to interpret, as every rule depends on all previous rules. Last, we empirically confirm that tree-based method CART produces much larger rule sets.

Table 2. Left: The sample sizes and number of features of datasets. Right: total number of literals, i.e., average rule lengths × number of rules in the set, averaged over 10-fold cross-validation. The rank is averaged over all datasets, for rule sets methods only.

#instances	#features	Data	TURS	CN2	DRS	BRS	CLASSY	RIPPER	CART
1372	5	backnote	42	41	55	22	22	16	94
1473	10	Contracept	75	275	73	—	14	14	6241
768	9	Diabetes	55	152	131	10	10	6	827
150	5	Iris	7	9	23	—	3	3	9
958	10	tic-tac-toe	86	90	108	24	27	60	816
178	14	Wine	10	6	134	—	6	5	15
1728	7	Car	211	163	325	—	92	111	718
7195	24	Anuran	74	37	407	—	49	7	96
3196	37	Chess	299	316	482	21	37	44	355
351	35	Ionosphere	50	30	261	14	6	5	101
5000	22	Waveform	707	802	60	—	139	115	3928
20867	11	Avila	890	1296	179	—	988	574	8145
19020	11	Magic	1321	2238	48	23	256	69	22566
		Avg Rank	2.15	2.46	2.77	1.00	—	—	—

7 Conclusion

We formalized the problem of learning truly unordered probabilistic rule sets as a model selection task. We also proposed a novel, tree-based surrogate score for evaluating incomplete rule sets. Building upon this, we developed a two-phase heuristic algorithm that learns rule set models that were empirically shown to be accurate in comparison to competing methods.

For future work, we will study the practical use of our method with a case study in the health care domain. This involves investigating how well our method scales to larger datasets. Furthermore, a user study will be performed to investigate whether, and in what degree, the domain experts find the truly unordered property of rule sets obtained by our method helps them comprehend the rules better in practice, in comparison to rule lists/sets with explicit or implicit orders.

Acknowledgements. We are grateful for the very inspiring feedback from the anonymous reviewers. This work is part of the research programme 'Human-Guided Data Science by Interactive Model Selection' with project number 612.001.804, which is (partly) financed by the Dutch Research Council (NWO).

References

1. Breiman, L., Friedman, J., Stone, C.J., Olshen, R.A.: Classification and Regression Trees. CRC Press, Boca Raton (1984)
2. Clark, P., Boswell, R.: Rule induction with CN2: some recent improvements. In: Kodratoff, Y. (ed.) EWSL 1991. LNCS, vol. 482, pp. 151–163. Springer, Heidelberg (1991). https://doi.org/10.1007/BFb0017011

3. Clark, P., Niblett, T.: The CN2 induction algorithm. Mach. Learn. **3**(4), 261–283 (1989)
4. Cohen, W.W.: Fast effective rule induction. In: Machine learning proceedings 1995, pp. 115–123. Elsevier (1995)
5. Dash, S., Gunluk, O., Wei, D.: Boolean decision rules via column generation. Adv. Neural. Inf. Process. Syst. **31**, 4655–4665 (2018)
6. Frank, E., Witten, I.H.: Generating accurate rule sets without global optimization (1998)
7. Fürnkranz, J., Flach, P.A.: Roc 'n'rule learning-towards a better understanding of covering algorithms. Mach. Learn. **58**(1), 39–77 (2005)
8. Fürnkranz, J., Gamberger, D., Lavrač, N.: Foundations of rule learning. Springer Science & Business Media (2012). https://doi.org/10.1007/978-3-540-75197-7
9. Gay, D., Boullé, M.: A Bayesian approach for classification rule mining in quantitative databases. In: Flach, P.A., De Bie, T., Cristianini, N. (eds.) ECML PKDD 2012. LNCS (LNAI), vol. 7524, pp. 243–259. Springer, Heidelberg (2012). https://doi.org/10.1007/978-3-642-33486-3_16
10. Grünwald, P., Roos, T.: Minimum description length revisited. Int. J. Math. Ind. **11**(01), 1930001 (2019)
11. Hühn, J., Hüllermeier, E.: FURIA: an algorithm for unordered fuzzy rule induction. Data Min. Knowl. Disc. **19**(3), 293–319 (2009)
12. Lakkaraju, H., Bach, S.H., Leskovec, J.: Interpretable decision sets: a joint framework for description and prediction. In: Proceedings of the 22nd ACM SIGKDD, pp. 1675–1684 (2016)
13. Liu, B., Hsu, W., Ma, Y., et al.: Integrating classification and association rule mining. In: KDD. vol. 98, pp. 80–86 (1998)
14. Molnar, C.: Interpretable machine learning. https://www.Lulu.com (2020)
15. Mononen, T., Myllymäki, P.: Computing the multinomial stochastic complexity in sub-linear time. In: PGM08, pp. 209–216 (2008)
16. Murdoch, W.J., Singh, C., Kumbier, K., Abbasi-Asl, R., Yu, B.: Interpretable machine learning: definitions, methods, and applications. arXiv preprint arXiv:1901.04592 (2019)
17. Proença, H.M., van Leeuwen, M.: Interpretable multiclass classification by mdl-based rule lists. Inf. Sci. **512**, 1372–1393 (2020)
18. Quinlan, J.R.: C4.5: Programs for machine learning. Elsevier (2014)
19. Van Leeuwen, M., Knobbe, A.: Diverse subgroup set discovery. Data Min. Knowl. Disc. **25**(2), 208–242 (2012)
20. Veloso, A., Meira, W., Zaki, M.J.: Lazy associative classification. In: Sixth International Conference on Data Mining (ICDM'06), pp. 645–654. IEEE (2006)
21. Wang, T., Rudin, C., Doshi-Velez, F., Liu, Y., Klampfl, E., MacNeille, P.: A bayesian framework for learning rule sets for interpretable classification. J. Mach. Learn. Res. **18**(1), 2357–2393 (2017)
22. Yang, F., et al.: Learning interpretable decision rule sets: a submodular optimization approach. In: Advances in Neural Information Processing Systems, vol. 34 (2021)
23. Yang, H., Rudin, C., Seltzer, M.: Scalable Bayesian rule lists. In: International Conference on Machine Learning, pp. 3921–3930. PMLR (2017)
24. Zhang, G., Gionis, A.: Diverse rule sets. In: Proceedings of the 26th ACM SIGKDD, pp. 1532–1541 (2020)

Probabilistic Inference

From Graphs to DAGs: A Low-Complexity Model and a Scalable Algorithm

Shuyu Dong[1(✉)] and Michèle Sebag[2]

[1] TAU, LISN, INRIA, Université Paris-Saclay, 91190 Gif-sur-Yvette, France
shuyu.dong@inria.fr

[2] TAU, LISN, CNRS, INRIA, Université Paris-Saclay, 91190 Gif-sur-Yvette, France
michele.sebag@lri.fr

Abstract. Learning directed acyclic graphs (DAGs) is long known a critical challenge at the core of probabilistic and causal modeling. The NoTears approach of Zheng et al. [23], through a differentiable function involving the matrix exponential trace tr(exp(·)), opens up a way to learning DAGs via continuous optimization, though with a $O(d^3)$ complexity in the number d of nodes. This paper presents a low-complexity model, called LoRAM for Low-Rank Additive Model, which combines low-rank matrix factorization with a sparsification mechanism for the continuous optimization of DAGs. The main contribution of the approach lies in an efficient gradient approximation method leveraging the low-rank property of the model, and its straightforward application to the computation of projections from graph matrices onto the DAG matrix space. The proposed method achieves a reduction from a cubic complexity to quadratic complexity while handling the same DAG characteristic function as NoTears and scales easily up to thousands of nodes for the projection problem. The experiments show that the LoRAM achieves efficiency gains of orders of magnitude compared to the state-of-the-art at the expense of a very moderate accuracy loss in the considered range of sparse matrices, and with a low sensitivity to the rank choice of the model's low-rank component.

Keywords: Bayesian networks · Low-rank matrix factorization · Matrix exponential

1 Introduction

The learning of directed acyclic graphs (DAGs) is an important problem for probabilistic and causal inference [15,17] with important applications in social sciences [11], genome research [21] and machine learning itself [2,16,18]. Through the development of probabilistic graphical models [4,15], DAGs are a most natural mathematical object to describe the causal relations among a number of

Supplementary Information The online version contains supplementary material available at https://doi.org/10.1007/978-3-031-26419-1_7.

© The Author(s), under exclusive license to Springer Nature Switzerland AG 2023
M.-R. Amini et al. (Eds.): ECML PKDD 2022, LNAI 13717, pp. 107–122, 2023.
https://doi.org/10.1007/978-3-031-26419-1_7

variables. In today's many application domains, the estimation of DAGs faces intractability issues as an ever growing number d of variables is considered, due to the fact that estimating DAGs is NP-hard [6]. The difficulty lies in how to enforce the acyclicity of graphs. Shimizu et al. [19] combined independent component analysis with the combinatorial linear assignment method to optimize a linear causal model (LiNGAM) and later proposed a direct and sequential algorithm [20] guaranteeing global optimum of the LiNGAM, for $O(d^4)$ complexities.

Recently, Zheng et al. [23] proposed an optimization approach to learning DAGs. The breakthrough in this work, called NoTears, comes with the characterization of the DAG matrices by the zero set of a real-valued differentiable function on $\mathbb{R}^{d \times d}$, which shows that an $d \times d$ matrix A is the adjacency matrix of a DAG if and only if the *exponential trace* satisfies

$$h(A) := \operatorname{tr}(\exp(A \odot A)) = d, \tag{1}$$

and thus the learning of DAG matrices can be cast as a continuous optimization problem subject to the constraint $h(A) = d$. The NoTears approach broadens the way of learning complex causal relations and provides promising perspectives to tackling large-scale inference problems [10,14,22,24]. However, NoTears is still not suitable for large-scale applications as the complexity of computing the exponential trace and its gradient is $O(d^3)$. More recently, Fang et al. [7] proposed to represent DAGs by low-rank matrices with both theoretical and empirical validation of the low-rank assumption for a range of graph models. However, the adaptation of the NoTears framework [23] to low-rank model still yields a complexity of $O(d^3)$ due to the DAG characteristic function in (1).

The contribution of the paper is to propose a new computational framework and a scalable algorithm to obtain DAGs via low-rank matrix models. We notice that the Hadamard product $\odot$ in characteristic functions as in (1) poses real obstacles to scaling up the optimization of NoTears [23] and NoTears-low-rank [7]. To address these difficulties, we present a low-complexity model, named *Low-Rank Additive Model* (LoRAM), which is a composition of low-rank matrix factorization with sparsification, and then propose a novel approximation method compatible with LoRAM to compute the gradients of the exponential trace in (1). Formally, the gradient approximation—consisting of matrix computation of the form $(A, C, B) \mapsto (\exp(A) \odot C)B$, where $A, C \in \mathbb{R}^{d \times d}$ and B is a thin low-rank matrix—is inspired from the numerical analysis of [1] for the matrix action of $\exp(A)$. We apply the new method to the computation of projections from graphs to DAGs through optimization with the differentiable DAG constraint.

Empirical evidence is presented to identify the cost and benefits of the approximation method combined with Nesterov's accelerated gradient descent [12], depending on the considered range of problem parameters (number of nodes, rank approximation, sparsity of the target graph).

The main contributions are summarized as follows:

- The LoRAM model, combining a low-rank structure with a flexible sparsification mechanism, is proposed to represent DAG matrices, together with a DAG characteristic function generalizing the exponential trace function of NoTears [23].

- An efficient gradient approximation method, exploiting the low-rank and sparse nature of the LoRAM model, is proposed. Under the low-rank assumption ($r \leq C \ll d$), the complexity of the proposed method is quadratic ($O(d^2)$) instead of $O(d^3)$ as shown in Table 1. Large efficiency gains, with insignificant loss of accuracy in some cases, are demonstrated experimentally in the considered range of application.

Table 1. Computational properties of LoRAM and algorithms in related work.

	Search space	Memory req.	Cost for ∇h
NoTears [23]	$\mathbb{R}^{d\times d}$	$O(d^2)$	$O(d^3)$
NoTears-low-rank [7]	$\mathbb{R}^{d\times r} \times \mathbb{R}^{d\times r}$	$O(dr)$	$O(d^3)$
LoRAM (ours)	$\mathbb{R}^{d\times r} \times \mathbb{R}^{d\times r}$	$O(dr)$	$O(d^2 r)$

2 Notation and Formal Background

A graph on d nodes is defined and denoted as a pair $\mathcal{G} = (\mathcal{V}, \mathcal{E})$, where $|\mathcal{V}| = d$ and $\mathcal{E} \subset \mathcal{V} \times \mathcal{V}$. By default, a directed graph is simply referred to as a graph. The adjacency matrix of a graph $\mathcal{G}$, denoted as $\mathbb{A}(\mathcal{G})$, is defined as the matrix such that $[\mathbb{A}(\mathcal{G})]_{ij} = 1$ if $(i, j) \in \mathcal{E}$ and 0 otherwise. The i-th canonical basis vector in $\mathbb{R}^d$ is denoted by e_i. Let $A_{\mathcal{G}} := A \in \mathbb{R}^{d\times d}$ be any *weighted* adjacency matrix of a graph $\mathcal{G}$ on d nodes, then by definition, the adjacency matrix $\mathbb{A}(\mathcal{G})$ indicates the nonzeros of A. The (0–1) adjacency matrix $\mathbb{A}(\mathcal{G})$ corresponds exactly to the *support* of A, denoted as $\mathrm{supp}(A)$, i.e., $(i, j) \in \mathrm{supp}(A)$ if and only if $A_{ij} \neq 0$. The number of nonzeros of A is denoted as $\|A\|_0$. We call a matrix A a *DAG matrix* if $\mathrm{supp}(A)$ is the adjacency matrix of a directed acyclic graph (DAG), and denote the set of DAG matrices as $\mathcal{D}_{d\times d} := \{A \in \mathbb{R}^{d\times d} : \mathrm{supp}(A) \text{ defines a DAG}\}$.

We recall the following theorem that characterizes acyclic graphs using the matrix *exponential trace*—$\mathrm{tr}(\exp(\cdot))$—where $\exp(\cdot)$ denotes the matrix exponential function. The matrix exponential will be denoted as $e^{\cdot}$ and $\exp(\cdot)$ indifferently. The operator $\odot$ denotes the matrix Hadamard product that acts on two matrices of the same size by elementwise multiplications.

Theorem 1 ([23]). *A matrix $A \in \mathbb{R}^{d\times d}$ is a DAG matrix if and only if*

$$\mathrm{tr}(\exp(A \odot A)) = d.$$

The following corollary is a straightforward extension of the theorem above:

Corollary 1 *Let $\sigma : \mathbb{R}^{d\times d} \to \mathbb{R}^{d\times d}$ be an operator such that: (i) $\sigma(A) \geq 0$ and (ii) $\mathrm{supp}(A) = \mathrm{supp}(\sigma(A))$, for any $A \in \mathbb{R}^{d\times d}$. Then, $A \in \mathbb{R}^{d\times d}$ is a DAG matrix if and only if $\mathrm{tr}(\exp(\sigma(A))) = d$.*

In view of the property above, we will refer to the composition of $\mathrm{tr}(\exp(\cdot))$ and the operator σ as a *DAG characteristic function.*

Next, we show some more properties of the exponential trace. All proofs in this paper are given in the supplementary material.

Proposition 1. *The exponential trace* $\tilde{h} : \mathbb{R}^{d\times d} \to \mathbb{R} : A \mapsto \mathrm{tr}(\exp(A))$ *satisfies: (i) For all* $\bar{A} \in \mathbb{R}_+^{d\times d}$, $\mathrm{tr}(\exp(\bar{A})) \geq d$ *and* $\mathrm{tr}(\exp(\bar{A})) = d$ *if and only if* $\bar{A}$ *is a DAG matrix. (ii)* $\tilde{h}$ *is nonconvex on* $\mathbb{R}^{d\times d}$*. (iii) The Fréchet derivative of* $\tilde{h}$ *at* $A \in \mathbb{R}^{d\times d}$ *along any direction* $\xi \in \mathbb{R}^{d\times d}$ *is*

$$\mathrm{D}\tilde{h}(A)[\xi] = \mathrm{tr}(\exp(A)\xi),$$

and the gradient of $\tilde{h}$ *at* A *is* $\nabla\tilde{h}(A) = (\exp(A))^{\mathrm{T}}$.

3 LoRAM: A Low-Complexity Model

In this section, we describe a low-complexity matrix representation of the adjacency matrices of directed graphs, and then a generalized DAG characteristic function for the new matrix model.

In the spirit of searching for best low-rank singular value decomposition and taking inspiration from [7], the search of a full $d \times d$ (DAG) matrix A is replaced by the search of a pair of thin factor matrices (X, Y), in $\mathbb{R}^{d\times r} \times \mathbb{R}^{d\times r}$ for $1 \leq r < d$, and the $d \times d$ candidate graph matrix is represented by the product XY^{T}. This matrix product has a rank bounded by r, with number of parameters $2dr \leq d^2$. However, the low-rank representation $(X, Y) \mapsto XY^{\mathrm{T}}$ generally gives a dense $d \times d$ matrix. Since in many scenarios the sought graph (or Bayesian network) is usually sparse, we apply a *sparsification* operator on XY^{T} in order to trim abundant entries in XY^{T}. Accordingly, we combine the two operations and introduce the following model.

Definition 1 (LoRAM). *Let* $\Omega \subset [d] \times [d]$ *be a given index set. The* low-rank additive model *(LoRAM), noted* A_Ω*, is defined from the matrix product of* $(X, Y) \in \mathbb{R}^{d\times r} \times \mathbb{R}^{d\times r}$ *sparsified according to* Ω*:*

$$A_\Omega(X, Y) = \mathcal{P}_\Omega(XY^{\mathrm{T}}), \tag{2}$$

where $\mathcal{P}_\Omega : \mathbb{R}^{d\times d} \to \mathbb{R}^{d\times d}$ *is a mask operator such that* $[\mathcal{P}_\Omega(A)]_{ij} = A_{ij}$ *if* $(i, j) \in \Omega$ *and* 0 *otherwise. The set* Ω *is referred to as the* candidate set *of* LoRAM.

The candidate set Ω is to be fixed according to the specific problem. In the case of projection from a given graph to the set of DAGs, Ω can be fixed as the index set of the given graph's edges.

The DAG characteristic function on the LoRAM search space is defined as follows:

Definition 2. *Let* $\tilde{h} : \mathbb{R}^{d\times d} \to \mathbb{R} : A \mapsto \mathrm{tr}(\exp(A))$ *denote the exponential trace function. We define* $h : \mathbb{R}^{d\times r} \times \mathbb{R}^{d\times r} \to \mathbb{R}$ *by*

$$h(X, Y) = \mathrm{tr}(\exp(\sigma(A_\Omega(X, Y)))), \tag{3}$$

where $\sigma : \mathbb{R}^{d\times d} \to \mathbb{R}^{d\times d}$ *is one of the following elementwise operators:*

$$\sigma_2(Z) := Z \odot Z \quad \text{and} \quad \sigma_{\mathrm{abs}}(Z) := \sum_{i,j=1}^{d} |Z_{ij}| e_i e_j^{\mathrm{T}}. \tag{4}$$

The operators σ_2 and σ_{abs} (4) are two natural choices for $\text{tr}(\exp(\sigma(\cdot)))$ to be a valid DAG characteristic function (see Corollary 1), since they both produce a nonnegative surrogate matrix of A_Ω while preserving the support of A_Ω.

3.1 Representativity

In the construction of a LoRAM matrix (2), the low-rank component of the model—XY^{T} with $(X,Y) \in \mathbb{R}^{d\times r} \times \mathbb{R}^{d\times r}$—has a rank smaller or equal to r (equality attained when X and Y have full column ranks), and the subsequent sparsification operator $\mathcal{P}_\Omega$ generally induces a change in the rank of the final matrix $A_\Omega = \mathcal{P}_\Omega(XY^{\text{T}})$ (2). Indeed, the rank of $\mathcal{P}_\Omega(XY^{\text{T}})$ depends on an interplay between $(X,Y) \in \mathbb{R}^{d\times r} \times \mathbb{R}^{d\times r}$ and the discrete set $\Omega \subset [d] \times [d]$. The following examples illustrate the two extreme cases of such interplay:

(i) The first extreme case: let Ω be the index set of the edges of a sparse graph $\mathcal{G}_\Omega$, and let $(X,Y) = (\mathbf{1},\mathbf{1}) \in \mathbb{R}^{d\times 1} \times \mathbb{R}^{d\times 1}$ (matrices of ones), for $r = 1$, then the LoRAM matrix $\mathcal{P}_\Omega(XY^{\text{T}}) = \mathbb{A}(\mathcal{G}_\Omega)$, i.e., the adjacency matrix of $\mathcal{G}_\Omega$. Hence $\text{rank}(\mathcal{P}_\Omega(XY^{\text{T}})) = \text{rank}(\mathbb{A}_\Omega)$, which depends solely on Ω and is generally much larger than $r = 1$.
(ii) The second extreme case: let Ω be the full $[d] \times [d]$ index set, then $\mathcal{P}_\Omega$ reduces to identity such that $\mathcal{P}_\Omega(XY^{\text{T}}) = XY^{\text{T}}$ and $\text{rank}(\mathcal{P}_\Omega(XY^{\text{T}})) = \text{rank}(XY^{\text{T}}) \leq r$ for any $(X,Y) \in \mathbb{R}^{d\times r} \times \mathbb{R}^{d\times r}$.

In the first extreme case above, optimizing LoRAM (2) for DAG learning boils down to choosing the most relevant edge set Ω, which is a NP-hard combinatorial problem [6]. In the second extreme case, the optimization of LoRAM (2) reduces to learning the most pertinent low-rank matrices $(X,Y) \in \mathbb{R}^{d\times r} \times \mathbb{R}^{d\times r}$, which coincides with optimizing the NoTears-low-rank [7] model.

In this work, we are interested in settings between the two extreme cases above such that both $(X,Y) \in \mathbb{R}^{d\times r} \times \mathbb{R}^{d\times r}$ and the candidate set Ω have sufficient degrees of freedom. Consequently, the representativity of LoRAM depends on both the rank parameter r and Ω.

Next, we present a way of quantifying the representativity of LoRAM with respect to a subset $\mathcal{D}^\star_{d\times d}$ of DAG matrices. The restriction to a subset $\mathcal{D}^\star_{d\times d}$ is motivated by the revelation that certain types of DAG matrices—such as those with many hubs—can be represented by low-rank matrices [7].

Definition 3. *Let $\mathcal{D}^\star_{d\times d} \subset \mathcal{D}_{d\times d}$ be a given set of nonzero DAG matrices. For $Z_0 \in \mathcal{D}^\star_{d\times d}$, let $A^*_\Omega(Z_0)$ denote any* LoRAM *matrix* (2) *such that $\|A^*_\Omega(Z_0) - Z_0\| = \min_{(X,Y)\in\mathbb{R}^{d\times r}\times\mathbb{R}^{d\times r}} \|A_\Omega(X,Y) - Z_0\|$, then the relative error of* LoRAM *w.r.t. $\mathcal{D}^\star_{d\times d}$ is defined and denoted as $\epsilon^\star_{r,\Omega} = \max_{Z\in\mathcal{D}^\star_{d\times d}} \{\frac{\|A^*_\Omega(Z)-Z\|_{\max}}{\|Z\|_{\max}}\}$, where $\|Z\|_{\max} := \max_{ij} |Z_{ij}|$ denotes the matrix max-norm. For $Z_0 \in \mathcal{D}^\star_{d\times d}$, $A^*_\Omega(Z_0)$ is referred to as an $\epsilon^\star_{r,\Omega}$-quasi DAG matrix.*

Note that the existence of $A^*_\Omega(Z_0)$ for any $Z_0 \in \mathcal{D}^\star_{d\times d}$ is guaranteed by the closeness of the image set of LoRAM (2).

Based on the relative error above, the next proposition provides a quasi DAG characterization of the LoRAM via function h (3).

Proposition 2. *Given a set $\mathcal{D}^{\star}_{d\times d} \subset \mathcal{D}_{d\times d}$ of nonzero DAG matrices. For any $Z_0 \in \mathcal{D}^{\star}_{d\times d}$ such that $\|Z_0\|_{\max} \leq 1$, without loss of generality, the minima of $\min_{(X,Y)\in\mathbb{R}^{d\times r}\times\mathbb{R}^{d\times r}} \|A_\Omega(X,Y) - Z_0\|$ belong to the set*

$$\{(X,Y) \in \mathbb{R}^{d\times r} \times \mathbb{R}^{d\times r} : h(X,Y) - d \leq C_0\epsilon^{\star}_{r,\Omega}\} \tag{5}$$

where $\epsilon^{\star}_{r,\Omega}$ is given in Definition 3 and $C_0 = \left(C_1\|Z_0\|_0 + \sum_{ij}[e^{\sigma(Z_0)}]_{ij}\right)\|Z_0\|_{\max}$ for a constant $C_1 \geq 0$.

Remark 1. The constant C_0 in Proposition 2 can be seen as a measure of total capacity of passing from one node to other nodes, and therefore depends on d, $\|Z_0\|_{\max}$ (bounded by 1), the sparsity and the average degree of the graph of Z_0. For DAG matrices with sparsity $\rho_0 \sim 10^{-3}$ and $d \lesssim 10^3$, one can expect that $C_0 \leq Cd$ for a constant C. □

The result of Proposition 2 establishes that, under the said conditions, a given DAG matrix Z_0 admits low rank approximations $A_\Omega(X,Y)$ satisfying $h(X,Y) - d \leq \delta_\epsilon$ for a small enough parameter δ_ϵ. In other words, the low-rank projection with a relaxed DAG constraint admits solutions.

The general case of projecting a non-acyclic graph matrix Z_0 onto a low-rank matrix under a relaxed DAG constraint is considered in next section.

4 Scalable Projection from Graphs to DAGs

Given a (generally non-acyclic) graph matrix $Z_0 \in \mathbb{R}^{d\times d}$, let us consider the projection of Z_0 onto the feasible set (5):

$$\min_{(X,Y)\in\mathbb{R}^{d\times r}\times\mathbb{R}^{d\times r}} \frac{1}{2}\|A_\Omega(X,Y) - Z_0\|_{\mathrm{F}}^2 \quad \text{subject to } h(X,Y) - d \leq \delta_\epsilon, \tag{6}$$

where $A_\Omega(X,Y)$ is the LoRAM matrix (2), h is given by Definition 2, and $\delta_\epsilon > 0$ is a tolerance parameter. Based on Proposition 2 and given the objective function of (6), the solution to (6) provides a quasi DAG matrix closest to Z_0 and thus enables finding a projection of Z_0 onto the DAG matrix space $\mathcal{D}_{d\times d}$. More precisely, we tackle problem (6) using the penalty method and focus on the primal problem, for a given penalty parameter $\lambda > 0$, followed by elementwise hard thresholding:

$$(X^*, Y^*) = \operatorname*{arg\,min}_{(X,Y)\in\mathbb{R}^{d\times r}\times\mathbb{R}^{d\times r}} h(X,Y) + \frac{1}{2\lambda}\|A_\Omega(X,Y) - Z_0\|_{\mathrm{F}}^2, \tag{7}$$

$$A^* = \mathbb{T}_{\epsilon^{\star}_{r,\Omega}}(A_\Omega(X^*, Y^*)), \tag{8}$$

where $\mathbb{T}_{\epsilon^{\star}_{r,\Omega}}(z) = z\delta_{|z|\geq\epsilon^{\star}_{r,\Omega}}$ is the elementwise hard thresholding operator. The choice of λ and $\epsilon^{\star}_{r,\Omega}$ for problem (6) is discussed in the supplementary.

Remark 2. To obtain any DAG matrix closest to (the non-acyclic) Z_0, it is necessary to break the cycles in $\mathcal{G}(Z_0)$ by suppressing certain edges. Hence, we assume that the edge set of the sought DAG is a strict subset of $\mathrm{supp}(Z_0)$ and thus fix the candidate set Ω to be $\mathrm{supp}(Z_0)$. □

4.1 Gradient of the DAG Characteristic Function and an Efficient Approximation

Problems (6) and (7) are nonconvex due to: (i) the matrix exponential trace $A \mapsto \operatorname{tr}(\exp(A))$ in h (3) is nonconvex (as in [23]), and (ii) the matrix product $(X, Y) \mapsto XY^{\mathrm{T}}$ in $A_\Omega(X, Y)$ (2) is nonconvex.

In view of the DAG characteristic constraint of (6), the augmented Lagrangian algorithms of NoTears [23] and NoTears-low-rank [7] can be applied for the same objective as problem (6); it suffices for NoTears and NoTears-low-rank to replace the LoRAM matrix $A_\Omega(X, Y)$ in (6) by the $d \times d$ matrix variable and the dense matrix product XY^{T} respectively. However, the NoTears-based methods of [7,23] have an $O(d^3)$ complexity due to the composition of the elementwise operations (the Hadamard product $\odot$) with the matrix exponential in function h (1) and (3). We elaborate this argument in the rest of this subsection and then propose a new computational method for computations involving the gradient of function h (3).

Lemma 1. *For any $Z \in \mathbb{R}^{d\times d}$ and $\xi \in \mathbb{R}^{d\times d}$, the differentials of σ_2 and σ_{abs} (4) are*

$$\mathrm{D}\sigma_2(Z)[\xi] = 2Z \odot \xi \quad \textit{and} \quad \hat{\mathrm{D}}\sigma_{\mathrm{abs}}(Z)[\xi] = \operatorname{sign}(Z) \odot \xi, \tag{4b}$$

where $\operatorname{sign}(\cdot)$ *is the element-wise sign function such that* $[\operatorname{sign}(Z)]_{ij} = \frac{Z_{ij}}{|Z_{ij}|}$ *if* $Z_{ij} \neq 0$ *and* 0 *otherwise.*

Theorem 2. *The gradient of h (3) is*

$$\nabla h(X, Y) = (\mathcal{S}Y, \mathcal{S}^{\mathrm{T}}X) \in \mathbb{R}^{d\times r} \times \mathbb{R}^{d\times r}, \tag{9}$$

where $\mathcal{S} \in \mathbb{R}^{d\times d}$ has the following expressions, depending on the choice of σ in (4): with $A_\Omega := A_\Omega(X, Y)$ for brevity,

$$\mathcal{S}_2 = 2(\exp(\sigma_2(A_\Omega))^{\mathrm{T}}) \odot A_\Omega, \tag{10}$$

$$\mathcal{S}_{\mathrm{abs}} = (\exp(\sigma_{\mathrm{abs}}(A_\Omega))^{\mathrm{T}}) \odot \operatorname{sign}(A_\Omega). \tag{11}$$

Proof. From Proposition 1-(iii), the Fréchet derivative of the exponential trace $\tilde{h}$ at $A \in \mathbb{R}^{d\times d}$ is $\mathrm{D}\tilde{h}(A)[\xi] = \operatorname{tr}(\exp(A)\xi)$ for any $\xi \in \mathbb{R}^{d\times d}$. By the chain rule and Lemma 1, the Fréchet derivative of h (3) for $\sigma = \sigma_2$ is as follows: with $A_\Omega = \mathcal{P}_\Omega(XY^{\mathrm{T}})$,

$$\begin{aligned}
\mathrm{D}_X h(X, Y)[\xi] &= \operatorname{tr}\left(\exp(\sigma(A_\Omega))\mathrm{D}\sigma(A_\Omega)\mathrm{D}\mathcal{P}_\Omega(XY^{\mathrm{T}})[\xi Y^{\mathrm{T}}]\right) \\
&= \operatorname{tr}\left(\exp(\sigma(A_\Omega))\mathrm{D}\sigma(A_\Omega)[\mathcal{P}_\Omega(\xi Y^{\mathrm{T}})]\right) && (12) \\
&= 2\operatorname{tr}\left(\exp(\sigma(A_\Omega))(A_\Omega \odot \mathcal{P}_\Omega(\xi Y^{\mathrm{T}}))\right) \\
&= 2\operatorname{tr}\left(\exp(\sigma(A_\Omega))(A_\Omega \odot (\xi Y^{\mathrm{T}}))\right) && (13) \\
&= 2\operatorname{tr}\left((\exp(\sigma(A_\Omega)) \odot {A_\Omega}^{\mathrm{T}})(\xi Y^{\mathrm{T}})\right) && (14) \\
&= 2\operatorname{tr}\left(Y^{\mathrm{T}}(\exp(\sigma(A_\Omega)) \odot {A_\Omega}^{\mathrm{T}})\xi\right),
\end{aligned}$$

where (13) holds, i.e., $\mathcal{P}_\Omega(XY^\mathrm{T}) \odot \mathcal{P}_\Omega(\xi Y^\mathrm{T}) = \mathcal{P}_\Omega(XY^\mathrm{T}) \odot \xi Y^\mathrm{T}$ because $A \odot \mathcal{P}_\Omega(B) = \mathcal{P}_\Omega(A) \odot B$ (for any A and B) and $\mathcal{P}_\Omega^2 = \mathcal{P}_\Omega$, and (14) holds because $\mathrm{tr}(A(B \odot C)) = \mathrm{tr}((A \odot B^\mathrm{T})C)$ for any A, B, C (with compatible sizes). By identifying the above equation with $\mathrm{tr}(\nabla_X h(X,Y)^\mathrm{T} \xi)$, we have $\nabla_X h(X,Y) = \underbrace{2(\exp(\sigma(Z))^\mathrm{T} \odot A_\Omega)}_{\mathcal{S}} Y$, hence the expression (10) for $\mathcal{S}$. The expression of $\mathcal{S}_\mathrm{abs}$ for $\sigma = \sigma_\mathrm{abs}$ can be obtained using the same calculations and (4b). □

The computational bottleneck for the exact gradient (9) lies in the computation of the $d \times d$ matrix $\mathcal{S}$ (10)–(11) and is due to the difference between Hadamard product and matrix multiplication; see the supplementary material for details. Nevertheless, we note that the multiplication $(\mathcal{S}, X) \mapsto \mathcal{S}X$ is similar to the action of a matrix exponential $(A, X) \mapsto \exp(A)X$, which can be computed using only a number of repeated multiplications of a $d \times d$ matrix with the thin matrix $X \in \mathbb{R}^{d \times r}$ based on Al-Mohy and Higham's results [1].

A difficulty in adapting the method of [1] also lies in the presence of Hadamard product in $\mathcal{S}$ (10)–(11). Once the sparse $d \times d$ matrix $A := \sigma(A_\Omega)$ in (10)–(11) is obtained (see the supplementary) the exact computation of $(A, C, B) \mapsto (\exp(A) \odot C)B$, using the Taylor expansion of $\exp(\cdot)$ to a certain order m_*, is to compute $\frac{1}{k!}(A^k \odot C)B$ at each iteration, which inevitably requires the computation of the $d \times d$ matrix product A^k (in the form of $A(A^{k-1})$) before computing the Hadamard product, which requires an $O(d^3)$ cost.

To alleviate the obstacle above, we propose to use inexact[1] incremental multiplications; see Algorithm 1.

Algorithm 1. Approximation of $(A, C, B) \mapsto (\exp(A) \odot C)B$

Input: $d \times d$ matrices A and C, thin matrix $B \in \mathbb{R}^{d \times r}$, tolerance $\mathrm{tol} > 0$
Output: $F \approx (\exp(A) \odot C)B \in \mathbb{R}^{d \times r}$
1: Estimate the Taylor order parameter m_* from A # using [1, Algorithm 3.2]
2: Initialize: let $F = (I \odot C)B$
3: **for** $k = 1, \ldots, m_*$ **do**
4: $\quad B \leftarrow \frac{1}{k+1}(A \odot C)B$
5: $\quad F \leftarrow F + B$
6: $\quad B \leftarrow F$
7: **end for**
8: Return F.

In line 1 of Algorithm 1, the value of m_* is obtained from numerical analysis results of [1, Algorithm 3.2]; often, the value of m_*, depending on the spectrum of A, is a bounded constant (independent of the matrix size d). Therefore, the dominant cost of Algorithm 1 is $2m_*|\Omega|r \lesssim d^2 r$, since each iteration (lines 4–6)

[1] It is inexact because $((A \odot C)^{k+1})B \neq (A^{k+1} \odot C)B$.

costs $(2|\Omega|r+dr) \approx 2|\Omega|r$ flops. Table 1 summarizes this computational property in comparison with existing methods.

Reliability of Algorithm 1. The accuracy of Algorithm 1 with respect to the exact computation of $(A, C, B) \mapsto (\exp(A) \odot C)B$ depends notably on the scale of A, since the differential $\mathrm{D}\exp(A)$ at A has a greater operator norm when the norm of A is greater; see [8, Theorem 2.b] and the supplementary.

To illustrate the remark above, we approximate $\nabla h(X, Y)$ (9) by Algorithm 1 on random points of $\mathbb{R}^{d\times r} \times \mathbb{R}^{d\times r}$ with different scales, with Ω defined from the edges of a random sparse graph; the results are shown in Fig. 1.

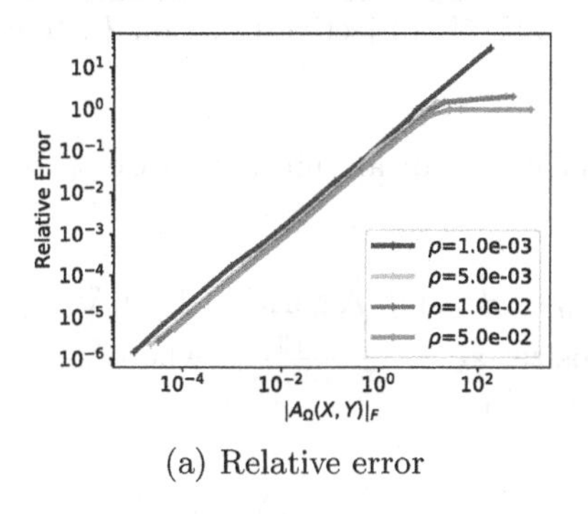

(a) Relative error

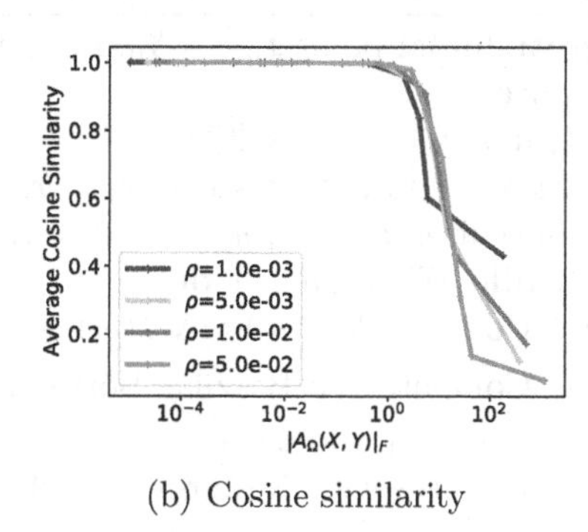

(b) Cosine similarity

Fig. 1. Average accuracy of Algorithm 1 in approximating $\nabla h(X, Y)$ (9) at (X, Y) with different scales and sparsity levels $\rho = \frac{|\Omega|}{d^2}$. Number of nodes $d = 200$, $r = 40$.

We observe from Fig. 1 that Algorithm 1 is reliable, i.e., having gradient approximations with cosine similarity close to 1, when the norm of $A_\Omega(X, Y)$ is sufficiently bounded. More precisely, for $c_0 = 10^{-1}$, Algorithm 1 is reliable in the following set

$$\mathcal{D}(c_0) = \{(X, Y) \in \mathbb{R}^{d\times r} \times \mathbb{R}^{d\times r} : \|A_\Omega(X, Y)\|_\mathrm{F} \le c_0\}. \tag{15}$$

The degrading accuracy of Algorithm 1 outside $\mathcal{D}(c_0)$ (15) can nonetheless be avoided for the projection problem (6), in particular, through *rescaling* of the input matrix Z_0. Note that the edge set of any graph is invariant to the scaling of its weighted adjacency matrix, and that any DAG A^* solution to (6) satisfies $\|A^*\|_\mathrm{F} \lesssim \|Z_0\|_\mathrm{F}$ since $\mathrm{supp}(A^*) \subset \mathrm{supp}(Z_0)$ (see Remark 2). Hence it suffices to rescale Z_0 with a small enough scalar, e.g., replace Z_0 with $Z_0' = \frac{c_0}{10\|Z_0\|_\mathrm{F}} Z_0$, without loss of generality, in (6)–(7). Indeed, this rescaling ensures that both the input matrix Z_0' and matrices like $A^{*\prime} = \frac{c_0}{10\|Z_0\|_\mathrm{F}} A^*$—a DAG matrix equivalent to A^*—stay confined in the image (through LoRAM) of $\mathcal{D}(c_0)$ (15), in which the gradient approximations by Algorithm 1 are reliable.

4.2 Accelerated Gradient Descent

Given the gradient computation method (Algorithm 1) for the h function, we adapt Nesterov's accelerated gradient descent [12,13] to solve (7). The accelerated gradient descent is used for its superior performance than vanilla gradient descent in many convex and nonconvex problems while it also only requires first-order information of the objective function. Details of this algorithm for our LoRAM optimization is given in Algorithm 2.

Algorithm 2. Accelerated Gradient Descent of LoRAM (LoRAM-AGD)

Input: Initial point $x_0 = (X_0, Y_0) \in \mathbb{R}^{d\times r} \times \mathbb{R}^{d\times r}$, objective function F, tolerance ϵ.

Output: $x_t \in \mathbb{R}^{d\times r} \times \mathbb{R}^{d\times r}$

1: Make a gradient descent: $x_1 = x_0 - s_0 \nabla F(x_0)$ for an initial stepsize $s_0 > 0$
2: Initialize: $y_0 = x_0$, $y_1 = x_1$, $t = 1$.
3: **while** $\|\nabla F(x_t)\| > \epsilon$ **do**
4: Compute $\nabla F(y_t) = \nabla f(y_t) + \alpha \nabla h(y_t)$ # using Algorithm 1 for $\nabla h(y_t)$
5: Compute the Barzilai–Borwein stepsize: $s_t = \frac{\|z_{t-1}\|^2}{\langle z_{t-1}, w_{t-1}\rangle}$, where $z_{t-1} = y_t - y_{t-1}$ and $w_{t-1} = \nabla F(y_t) - \nabla F(y_{t-1})$.
6: Updates with Nesterov's acceleration:

$$
\begin{aligned}
x_{t+1} &= y_t - s_t \nabla F(y_t), \\
y_{t+1} &= x_{t+1} + \frac{t}{t+3}(x_{t+1} - x_t).
\end{aligned}
\tag{16}
$$

7: $t = t + 1$
8: **end while**

Specifically, in line 5 of Algorithm 2, the Barzilai–Borwein (BB) stepsize [3] is used for the descent step (16). The computation of the BB stepsize s_t requires evaluating the inner products $\langle z_{t-1}, w_{t-1}\rangle$ and the norm $\|z_{t-1}\|$, where $z_{t-1}, w_{t-1} \in \mathbb{R}^{d\times r} \times \mathbb{R}^{d\times r}$; we choose the Euclidean inner product as the metric on $\mathbb{R}^{d\times r} \times \mathbb{R}^{d\times r}$: for any pair of points $z = (z^{(1)}, z^{(2)})$ and $w = (w^{(1)}, w^{(2)})$ on $\mathbb{R}^{d\times r} \times \mathbb{R}^{d\times r}$, $\langle z, w\rangle = \mathrm{tr}({z^{(1)}}^{\mathrm{T}} w^{(1)}) + \mathrm{tr}({z^{(2)}}^{\mathrm{T}} w^{(2)})$. Note that one can always use backtracking line search based on the stepsize estimation (line 5). We choose to use the BB stepsize directly since it does not require any evaluation of the objective function, and thus avoids non-negligible costs for computing the matrix exponential trace in h (3). We refer to [5] for a comprehensive view on accelerated methods for nonconvex optimization.

Due to the nonconvexity of h (3) (see Proposition 1) and thus (7), we aim at finding stationary points of (7). In particular, empirical results in Sect. 5.2 show that the solutions by the proposed method, with close-to-zero or even zero SHDs to the true DAGs, are close to global optima in practice.

5 Experimental Validation

This section investigates the performance (computational gains and accuracy loss) of the proposed gradient approximation method (Algorithm 1) and thereafter reports on the performance of the LoRAM projection (6), compared to NoTears [23]. Sensitivity to the rank parameter r of the proposed method is also investigated.

The implementation is available at https://github.com/shuyu-d/loram-exp.

5.1 Gradient Computations

We compare the performance of Algorithm 1 in gradient approximations with the exact computation in the following settings: the number of nodes $d \in \{100, 500, 10^3, 2.10^3, 3.10^3, 5.10^3\}$, $r = 40$, and the sparsity $(\frac{|\Omega|}{d^2})$ of the index set Ω tested are $\rho \in \{10^{-3}, 5.10^{-3}, 10^{-2}, 5.10^{-2}\}$. The results shown in Fig. 2 are based on the computation of $\nabla h(X, Y)$ (9) on randomly generated points $(X, Y) \in \mathbb{R}^{d \times r} \times \mathbb{R}^{d \times r}$, where X and Y are Gaussian matrices.

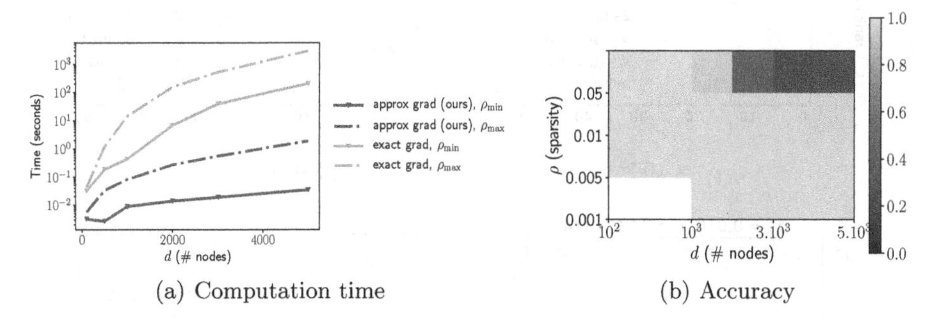

(a) Computation time (b) Accuracy

Fig. 2. (a): Runtime (in log-scale) for computing $\nabla h(X, Y)$. (b): Cosine similarities between the approximate and the exact gradients.

From Fig. 2 (a), Algorithm 1 shows a significant reduction in the runtime for computing the gradient of h (3) at the expense of a very moderate loss of accuracy (Fig. 2 (b)): the approximate gradients are mostly sufficiently aligned with exact gradients in the considered range of graph sizes and sparsity levels.

5.2 Projection from Graphs to DAGs

In the following experiments, we generate the input matrix Z_0 of problem (6) by

$$Z_0 = A^\star + E, \tag{17}$$

where $A^\star$ is a given $d \times d$ DAG matrix and E is a graph containing additive noisy edges that break the acyclicity of the observed graph Z_0.

The ground truth DAG matrix $A^\star$ is generated from the acyclic Erdős-Rényi (ER) subset, in the same way as in [23], with a sparsity rate $\rho \in \{10^{-3}, 5.10^{-3}, 10^{-2}\}$. The noise graph E of (17) is defined as $E = \sigma_E {A^\star}^{\mathrm{T}}$, which consists of edges that create a confusion between causes and effects since these edges are *reversed*, pointing from the (true) effects to their respective causes.

Sensitivity to the Rank Parameter *r*. We evaluate the performance of LoRAM-AGD in the proximal mapping computation (7) for $d = 500$ and different values of the rank parameter r. In all these evaluations, the candidate set Ω is fixed to be $\mathrm{supp}(Z_0)$; see Remark 2. We measure the accuracy of solutions by the false discovery rate (FDR, lower is better), false positive rate (FPR), true positive rate (TPR, higher is better), and the structural Hamming distance (SHD, lower is better) of the solution compared to the DAG $\mathcal{G}(A^\star)$.

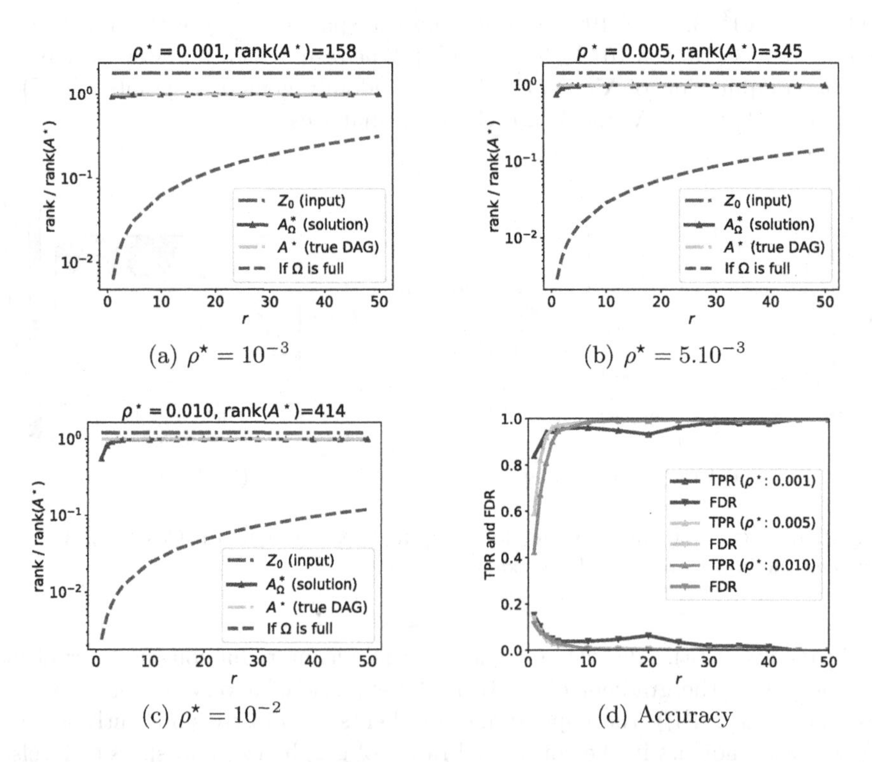

(a) $\rho^\star = 10^{-3}$ (b) $\rho^\star = 5.10^{-3}$

(c) $\rho^\star = 10^{-2}$ (d) Accuracy

Fig. 3. (a)–(c): Rank profiles $\mathrm{rank}(\cdot)/\mathrm{rank}(A^\star)$, and (d) projection accuracy for different values of r (number of columns of X and Y in (2)). The number of nodes $d = 500$. Sparsity of $Z_0 = A^\star + \sigma_E {A^\star}^{\mathrm{T}}$ (17) are $\rho^\star \in \{10^{-3}, 5.10^{-3}, 10^{-2}\}$.

The results in Fig. 3 suggest that:

(i) For each sparsity level, increasing the rank parameter r generally improves the projection accuracy of the LoRAM.

(ii) While the rank parameter r of (2) attains at most around 5, which is only $\frac{1}{100}$-th the rank of the input matrix Z_0 and the ground truth $A^\star$, the rank of the solution $A^*_\Omega = A_\Omega(X^*, Y^*)$ attains the same value as $\text{rank}(A^\star)$. This means that the rank representativity of the LoRAM goes beyond the value of r. This phenomenon is understandable in the present case where the candidate set $\Omega = \text{supp}(Z_0)$ is fairly close to the sparse edge set $\text{supp}(A^\star)$.

(iii) The projection accuracy in TPR and FDR (and also SHD, see the supplementary) of LoRAM-AGD is close to optimum on a wide interval $25 \leq r \leq 50$ of the tested ranks and are fairly stable on this interval.

Scalability. We examine two different types of noisy edges (in E) as follows. Case (a): Bernoulli-Gaussian $E = E(\sigma_E, p)$, where $E_{ij}(\sigma_E, p) \neq 0$ with probability p and all nonzeros of $E(\sigma_E, p)$ are i.i.d. samples of $\mathcal{N}(0, \sigma_E)$. Case (b): cause-effect confusions $E = \sigma_E A^{\star\text{T}}$ as in the beginning of Sect. 5.2.

The initial factor matrices $(X, Y) \in \mathbb{R}^{d\times r} \times \mathbb{R}^{d\times r}$ are random Gaussian matrices. For the LoRAM, we set Ω to be the support of Z_0; see Remark 2. The penalty parameter λ of (7) is varied in $\{2.0, 5.0\}$ with no fine tuning.

In case (a), we test with various noise levels for $d = 500$ nodes. In case (b), we test on various graph dimensions, for $(d, r) \in \{100, 200, \ldots, 2000\} \times \{40, 80\}$. The results are given in Table 2 and Table 3 respectively.

Table 2. Results in case (a): the noise graph is $E(\sigma_E, p)$ for $p = 5.10^{-4}$ and $d = 500$.

σ_E	LoRAM (ours)/NoTears			
	Runtime (sec)	TPR	FDR	SHD
0.1	1.34/5.78	1.0/1.0	9.9e-3/0.0e+0	25.0/0.0
0.2	2.65/11.58	1.0/1.0	9.5e-3/0.0e+0	24.0/0.0
0.3	1.35/28.93	1.0/1.0	9.5e-3/8.0e-4	24.0/2.0
0.4	1.35/18.03	1.0/1.0	9.9e-3/3.2e-3	25.0/9.0
0.5	1.35/12.52	1.0/1.0	9.9e-3/5.2e-3	25.0/13.0
0.6	2.57/16.07	1.0/1.0	9.5e-3/4.4e-3	24.0/11.0
0.7	1.35/18.72	1.0/1.0	9.9e-3/5.2e-3	25.0/13.0
0.8	1.35/32.03	1.0/1.0	9.9e-3/4.8e-3	25.0/15.0

The results in Table 2 and Table 3 show that: (i) in case (a), the solutions of LoRAM-AGD are close to the ground truth despite slightly higher errors than NoTears in terms of FDR and SHD; (ii) in case (b), the solutions of LoRAM-AGD are almost identical to the ground truth $A^\star$ in (17) in all performance indicators (also see the supplementary); (iii) in terms of computation time (see Fig. 4), the proposed LoRAM-AGD achieves significant speedups (around 50 times faster when $d = 2000$) compared to NoTears and also has a smaller growth rate with respect to the problem dimension d, showing good scalability.

Table 3. Results in case (b): the noise graph $E = \sigma_E A^{\star T}$ contains cause-effect confusions for $\sigma_E = 0.4$.

(α, r)	d	LoRAM (ours)/NoTears			
		Runtime (sec)	TPR	FDR	SHD
(5.0, 40)	100	1.82/0.67	1.00/1.00	0.00e+0/0.0	0.0/0.0
(5.0, 40)	200	2.20/3.64	0.98/0.95	2.50e-2/0.0	1.0/2.0
(5.0, 40)	400	2.74/16.96	0.98/0.98	2.50e-2/0.0	4.0/4.0
(5.0, 40)	600	3.40/42.65	0.98/0.96	1.67e-2/0.0	6.0/16.0
(5.0, 40)	800	4.23/83.68	0.99/0.97	7.81e-3/0.0	5.0/22.0
(2.0, 80)	1000	7.63/136.94	1.00/0.96	0.00e+0/0.0	0.0/36.0
(2.0, 80)	1500	13.34/437.35	1.00/0.96	8.88e-4/0.0	2.0/94.0
(2.0, 80)	2000	20.32/906.94	1.00/0.96	7.49e-4/0.0	3.0/148.0

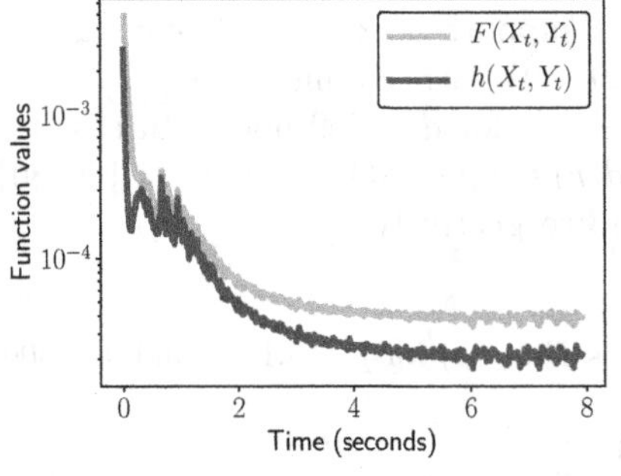

(a) Iterations of LoRAM-AGD (for $d =$ 1000)

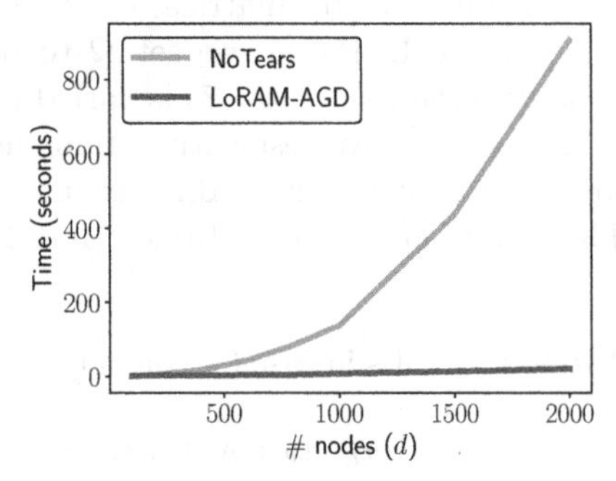

(b) Runtime vs number of nodes

Fig. 4. (a): An iteration history of LoRAM-AGD for (7) with $d = 1000$. (b): Runtime comparisons for different number d of nodes.

6 Discussion and Perspectives

This paper tackles the computation of projection from graphs to DAGs, motivated by problems for causal DAG discovery. The line of research built upon the LiNGAM algorithms [19,20] has recently been revisited through the formal characterization of DAGness in terms of a continuously differentiable constraint by [23]. The NoTears approach of [23] however suffers from an $O(d^3)$ complexity in the number d of variables, precluding its usage for large-scale problems.

Unfortunately, this difficulty is not related to the complexity of the model (number of parameters of the model): the low-rank approach investigated by NoTears-low-rank [7] also suffers from the same $O(d^3)$ complexity, incurred in the gradient-based optimization phase.

The present paper addresses this difficulty by combining a sparsification mechanism with a low-rank matrix model and using a new approximated gradient computation. This approximated gradient takes inspiration from the approach of [1] for computing the action of exponential matrices based on truncated Taylor expansion. This approximation eventually yields a complexity of $O(d^2r)$, where the rank parameter is small ($r \leq C \ll d$). The experimental validation of the approach shows that the approximated gradient entails no significant error with respect to the exact gradient, for LoRAM matrices with a bounded norm, in the considered range of graph sizes (d) and sparsity levels. The proposed algorithm combining the approximated gradient with a Nesterov's acceleration method [12,13] yields gains of orders of magnitude in computation time compared to NoTears on the same artificial benchmark problems. The approximation performance indicators reveal almost no performance loss for the projection problem in the setting of case (b) (where the matrix to be projected is perturbed with anti-causal links), while it incurs minor losses in terms of false discovery rate (FDR) in the setting of case (a) (with random additive spurious links).

The proposed method has the general purpose of computing proximate DAGs from given graphs. Further developments aim to apply this method to the identification of causal DAGs from observational data. A longer term perspective is to extend LoRAM to non-linear causal models, building upon latent causal variables and taking inspiration from non-linear independent component analysis and generalized contrastive losses [9]. Another perspective relies on the use of auto-encoders to yield a compressed representation of high-dimensional data, while constraining the structure of the encoder and decoder modules to enforce the acyclic property.

Acknowledgement. The authors warmly thank Fujitsu Laboratories LTD who funded the first author, and in particular Hiroyuki Higuchi and Maruashi Koji for many discussions.

References

1. Al-Mohy, A.H., Higham, N.J.: Computing the action of the matrix exponential, with an application to exponential integrators. SIAM J. Sci. Comput. **33**(2), 488–511 (2011)
2. Arjovsky, M., Bottou, L., Gulrajani, I., Lopez-Paz, D.: Invariant risk minimization. arXiv preprint arXiv:1907.02893 (2019)
3. Barzilai, J., Borwein, J.M.: Two-point step size gradient methods. IMA J. Numer. Anal. **8**(1), 141–148 (1988)
4. Bühlmann, P., Peters, J., Ernest, J.: CAM: causal additive models, high-dimensional order search and penalized regression. Ann. Stat. **42**(6), 2526–2556 (2014)
5. Carmon, Y., Duchi, J.C., Hinder, O., Sidford, A.: Accelerated methods for non-convex optimization. SIAM J. Optim. **28**(2), 1751–1772 (2018)
6. Chickering, D.M.: Learning bayesian networks is NP-complete. In: Learning from data, pp. 121–130. Springer, New York, NY (1996). https://doi.org/10.1007/978-1-4612-2404-4_12

7. Fang, Z., Zhu, S., Zhang, J., Liu, Y., Chen, Z., He, Y.: Low rank directed acyclic graphs and causal structure learning. arXiv preprint arXiv:2006.05691 (2020)
8. Haber, H.E.: Notes on the Matrix Exponential and Logarithm. Santa Cruz Institute for Particle Physics, University of California, Santa Cruz, CA, USA (2018)
9. Hyvärinen, A., Sasaki, H., Turner, R.: Nonlinear ICA using auxiliary variables and generalized contrastive learning. In: International Conference on Artificial Intelligence and Statistics (AISTATS) (2019)
10. Kalainathan, D., Goudet, O., Guyon, I., Lopez-Paz, D., Sebag, M.: Structural agnostic modeling: adversarial learning of causal graphs. J. Mach. Learn. Res. **23**(219), 1–62 (2022). http://jmlr.org/papers/v23/19-529.html
11. Morgan, S.L., Winship, C.: Counterfactuals and Causal Inference. Cambridge University Press, Cambridge (2015)
12. Nesterov, Y.: A method for solving the convex programming problem with convergence rate $O(1/k^2)$. Soviet Math. Doklady **27**, 372–376 (1983)
13. Nesterov, Y.: Introductory Lectures on Convex Optimization, vol. 87. Springer Publishing Company, Incorporated, 1 edn. (2004). https://doi.org/10.1007/978-1-4419-8853-9
14. Ng, I., Ghassami, A., Zhang, K.: On the role of sparsity and DAG constraints for learning linear DAGs. Adv. Neural. Inf. Process. Syst. **33**, 17943–17954 (2020)
15. Pearl, J.: Causality. Cambridge University Press, Cambridge (2009)
16. Peters, J., Bühlmann, P., Meinshausen, N.: Causal inference by using invariant prediction: identification and confidence intervals. J. Royal Statistical Society. Series B (Statistical Methodology), pp. 947–1012 (2016)
17. Peters, J., Janzing, D., Schölkopf, B.: Elements of Causal Inference: Foundations and Learning Algorithms. The MIT Press, Cambridge (2017)
18. Sauer, A., Geiger, A.: Counterfactual generative networks. In: International Conference on Learning Representations (ICLR) (2021)
19. Shimizu, S., Hoyer, P.O., Hyvärinen, A., Kerminen, A.: A linear non-Gaussian acyclic model for causal discovery. J. Mach. Learn. Res. **7**(10), 2003–2030 (2006)
20. Shimizu, S., et al.: DirectLiNGAM: a direct method for learning a linear non-Gaussian structural equation model. J. Mach. Learn. Res. **12**, 1225–1248 (2011)
21. Stephens, M., Balding, D.J.: Bayesian statistical methods for genetic association studies. Nat. Rev. Genet. **10**(10), 681–690 (2009)
22. Yu, Y., Chen, J., Gao, T., Yu, M.: DAG-GNN: DAG structure learning with graph neural networks. In: International Conference on Machine Learning, pp. 7154–7163. PMLR (2019)
23. Zheng, X., Aragam, B., Ravikumar, P.K., Xing, E.P.: DAGs with NO TEARS: continuous optimization for structure learning. In: Advances in Neural Information Processing Systems, vol. 31 (2018). https://proceedings.neurips.cc/paper/2018/file/e347c51419ffb23ca3fd5050202f9c3d-Paper.pdf
24. Zheng, X., Dan, C., Aragam, B., Ravikumar, P., Xing, E.: Learning sparse nonparametric DAGs. In: International Conference on Artificial Intelligence and Statistics, pp. 3414–3425. PMLR (2020)

Sparse Horseshoe Estimation via Expectation-Maximisation

Shu Yu Tew[1], Daniel F. Schmidt[1](✉), and Enes Makalic[2]

[1] Department of Data Science and AI, Monash University, Clayton, Australia
{shu.tew,daniel.schmidt}@monash.edu
[2] Centre for Epidemiology and Biostatistics, The University of Melbourne, Parkville, Australia
emakalic@unimelb.edu.au

Abstract. The horseshoe prior is known to possess many desirable properties for Bayesian estimation of sparse parameter vectors, yet its density function lacks an analytic form. As such, it is challenging to find a closed-form solution for the posterior mode. Conventional horseshoe estimators use the posterior mean to estimate the parameters, but these estimates are not sparse. We propose a novel expectation-maximisation (EM) procedure for computing the MAP estimates of the parameters in the case of the standard linear model. A particular strength of our approach is that the M-step depends only on the form of the prior and it is independent of the form of the likelihood. We introduce several simple modifications of this EM procedure that allow for straightforward extension to generalised linear models. In experiments performed on simulated and real data, our approach performs comparable, or superior to, state-of-the-art sparse estimation methods in terms of statistical performance and computational cost.

Keywords: Horseshoe regression · Sparse regression · Non-convex penalised regression · Maximum a posteriori estimation · Expectation-maximisation

1 Introduction

Sparse modelling has become increasingly important in statistical learning due to the growing demand for the analysis of high dimensional data with parameters whose dimension exceeds the sample size. Inference under such a setting often involves the assumption that the parameter vector is sparse. This implies that most of the components in the vector are insignificant and should be removed to achieve a good estimate. The lasso (i.e., ℓ_1-penalised regression) [27] is a popular sparsity inducing technique well-known for its ability to simultaneously perform

Supplementary Information The online version contains supplementary material available at https://doi.org/10.1007/978-3-031-26419-1_8.

© The Author(s), under exclusive license to Springer Nature Switzerland AG 2023
M.-R. Amini et al. (Eds.): ECML PKDD 2022, LNAI 13717, pp. 123–139, 2023.
https://doi.org/10.1007/978-3-031-26419-1_8

variable selection and coefficient estimation. The lasso is a convex penalisation technique, and is known to suffer from over-shrinkage and consistency issues [8]; to combat these, authors introduced non-convex penalty to replace the usual ℓ_1-norm used in the lasso. Well-known non-convex penalties include the smoothly clipped absolute deviation (SCAD) [12] and the minimax concave penalty (MCP) [30].

However, these approaches are all frequentist in nature, usually relying on an additional principle such as cross-validation to select the degree of regularisation, and do not have immediate access to reliable measures of uncertainty to complement the point estimates they produce [8]. Bayesian inference, on the other hand, naturally quantifies uncertainty directly through the posterior distribution. Bayesian penalised regression offers readily available uncertainty estimates, automatic estimation of the penalty parameters, and more flexibility in terms of penalties that can be considered [28]. In this paper, we consider the standard Gaussian linear regression model

$$\mathbf{y} = \beta_0 \mathbf{1}_n + \mathbf{X}\boldsymbol{\beta} + \boldsymbol{\epsilon}, \quad \boldsymbol{\epsilon} \sim N(\mathbf{0}, \sigma^2 \mathbf{I}_n) \tag{1}$$

where $\mathbf{y} = (y_1, \cdots, y_n)^T \in \mathbb{R}^n$ is a vector of outcome variable, $\mathbf{X} \in \mathbb{R}^{n \times p}$ is a matrix of predictor variables, $\beta_0 \in \mathbb{R}$ is the intercept, and $\boldsymbol{\epsilon}$ is a vector of i.i.d. normally distributed random error with mean zero and unknown variance σ^2. Given $\mathbf{y}$ and $\mathbf{X}$, our goal is to accurately identify and estimate the non-zero components of the unknown regression coefficients $\boldsymbol{\beta} = (\beta_1, \cdots, \beta_p) \in \mathbb{R}^p$.

1.1 Bayesian Penalised Regression

Most penalised regression methods can be solved in the Bayesian framework by interpreting the estimates of $\boldsymbol{\beta}$ as the posterior point estimate of choice under an appropriate prior distribution [28]. For example, the lasso estimate can be interpreted as posterior mode estimates when the regression parameters $\boldsymbol{\beta}$ follow an independent and identical double exponential (Laplace) prior [27]. Motivated by this relationship, much work has been done in proposing different Bayesian approaches to the sparse estimation problem via a variety of sparsity inducing priors. Some popular examples include the Bayesian Lasso [19], the horseshoe estimator [7] and the normal-gamma estimator [6]. These estimators are frequently implemented using Monte Carlo Markov Chain (MCMC) algorithms to sample from the posterior distribution, and the resulting Bayesian estimates are usually posterior means or medians.

Interestingly, however, while these priors may be sparsity promoting, in the sense that they concentrate probability mass around sparse coefficient vectors, the resulting posterior means or medians will never themselves be sparse. Sparse estimates can be achieved by instead considering the posterior mode, i.e., maximum a posteriori (MAP) estimation [1,4,13]. Unfortunately, most implementations of Bayesian MAP-based sparse estimators are not flexible in the sense that they only work for linear models when paired with a specified prior. These methods do not easily generalise to other regression models and are difficult to apply to priors that lack an analytic form (e.g., horseshoe prior, Strawderman-Berger prior and normal-Gamma prior). Alternatively, simple thresholding ("sparsification") rules for the posterior mean or median of β_js can be used to produce sparse estimates, but these tend to lack theoretical justification, and inference can be highly sensitive to the choice of the threshold [7,16]. To address these limitations:

1. we propose a novel expectation-maximisation (EM) procedure to solve for the *exact* posterior mode of a regression model under the horseshoe prior;
2. we introduce several simple modifications of our base EM procedure that allow for straightforward extension to models beyond the usual Gaussian linear model, e.g., generalised linear models.

Experiments on simulated data and real datasets in Sect. 6 demonstrate that our proposed Bayesian EM sparse horseshoe estimator can give comparable, if not better, performance when compared to state-of-the-art non-convex solvers in terms of statistical performance and computational cost. As far as the authors are aware, this is the first work that explores the posterior mode under the exact representation of the horseshoe prior, instead of using an approximated density function in place of the horseshoe prior.

2 Background and Related Work

The EM algorithm [10] is one of the most widely used methods for MAP estimation of sparse Bayesian linear models, with [13] and [15] being classic papers on this approach. A particular strength, in comparison to approximate methods such as variational Bayes [29], is that it is guaranteed to converge to exact posterior modes (a stationary point in the likelihood) whenever it can be applied [10]. By introducing appropriate latent variables, the hierarchical decomposition of the Laplace prior allows [13] for the derivation of an efficient EM algorithm to recover the lasso estimates. Following this work, many authors have attempted similar procedure to achieve sparse estimation using various priors and likelihood models that admit a scale mixture of normals (SMN) representation. This includes the lasso prior [19], the generalised double Pareto prior [1] and the horseshoe-like prior [4].

In this paper, we focus on the class of global-local shrinkage priors [21], and more specifically the horseshoe prior, which has been recognised as a good default prior choice for Bayesian sparse estimation [3,7]. The horseshoe prior has a pole at $\beta_j = 0$, and heavy, Cauchy-like tails. These properties are desirable in sparse estimation because they allow small coefficients to be heavily shrunk towards zero while ensuring large coefficients are not over-shrunk. This is in contrast to the popular Bayesian lasso and Bayesian ridge hierarchies that apply the same amount of shrinkage to all coefficients and potentially over-shrink coefficients that are far from zero. The Gaussian linear regression model (1) corresponds to the following likelihood

$$\mathbf{y} \mid \mathbf{X}, \boldsymbol{\beta}, \sigma^2 \sim N_n\left(\mathbf{X}\boldsymbol{\beta},\, \sigma^2 \boldsymbol{I}_n\right) \tag{2}$$

where $N_k(\cdot,\cdot)$ is the k-variate Gaussian distribution. In the class of global-local shrinkage priors, each β_j is assigned a continuous shrinkage prior centered at $\beta_j = 0$ that can be represented in an SMN form as

$$\begin{aligned} \beta_j \,|\, \tau^2, \lambda_j^2, \sigma^2 &\sim N\left(0,\, \tau^2 \lambda_j^2 \sigma^2\right) \\ \lambda_j^2 &\sim \pi(\lambda_j^2) d\lambda^2 \\ \tau^2 &\sim \pi(\tau^2) d\tau^2 \end{aligned} \tag{3}$$

where τ is the *global* shrinkage parameter that controls the overall degree of shrinkage, λ_j is the *local* shrinkage parameter associated with the j-th predictor and it controls the shrinkage for individual coefficients, and $\pi(\cdot)$ is an appropriate prior distribution of choice assigned to the shrinkage parameters.

Given the (unormalised) joint posterior distribution of the hierarchy (2)–(3),

$$p(\boldsymbol{\beta}, \tau, \boldsymbol{\lambda}|\mathbf{y}) \propto p(\mathbf{y}|\boldsymbol{\beta}) \cdot \pi(\boldsymbol{\beta}|\boldsymbol{\lambda}, \tau) \cdot \pi(\boldsymbol{\lambda}) \cdot \pi(\tau), \tag{4}$$

the conventional EM approach to find the posterior mode estimates treats the hyperparameters (i.e., the shrinkage parameters) $\boldsymbol{\lambda}$ as "missing data" (i.e., the latent variables). This approach iteratively finds the expected values of the latent variables, and solves for the maximisation problem

$$\underset{\boldsymbol{\beta}}{\operatorname{argmax}} \ \mathrm{E}\left[\log p(\boldsymbol{\beta}, \tau, \boldsymbol{\lambda}|\mathbf{y}) \,|\, \boldsymbol{\beta}\right]$$

to produce a sequence of estimates for the regression coefficients. This approach is effective because the scale-mixture form of the local-global prior means that conditional on $\boldsymbol{\lambda}$, the posterior for $\boldsymbol{\beta}$ is Gaussian and maximisation is (relatively) straightforward. In the case that the likelihood is non-Gaussian, but itself admits a scale-mixture-of-normals representation, such as the Laplace or logistic regression model, this approach can be adapted appropriately.

This approach is not easily applicable if the prior placed on $\boldsymbol{\beta}$ lacks a simple analytic form; in this case, the expected value of the shrinkage parameter will not have a closed form, and may be difficult or impossible to compute. One such prior is the horseshoe discussed previously. To attempt to address this problem [4] introduced a "horseshoe-like" prior that mimics the behavior of the horseshoe density proposed by [7]. This horseshoe-like prior has a closed form density function and a scale-mixture representation that allows for implementation of an EM algorithm to obtain the "horseshoe-like" MAP estimates.

3 Bayesian EM Sparse Linear Regression

In this section, we present a novel, general EM procedure to compute the MAP estimate for the linear model. The key innovation underlying our approach is to treat the coefficients, $\boldsymbol{\beta}$, as latent variables; this is in contrast to the usual application of the EM algorithm [13] that treats the shrinkage hyperparameters, $\boldsymbol{\lambda}$, as missing data.

As is standard in penalised regression, and without any loss of generality, we assume that the predictors are standardised to have mean zero and standard deviation one, and the target has a mean of zero, i.e., the estimate of the intercept is simply $\hat{\beta}_0 = (1/n)\sum y_i$. This means we can simply ignore the intercept when estimating the remaining coefficients $\boldsymbol{\beta}$.

3.1 The Basic EM Algorithm

Here we consider the linear regression model define in hierarchy (2)-(3). The proposed EM algorithm solves for the posterior mode estimates by iterating through the following two steps until convergence is achieved:

E-step. We take the expectation of the complete negative log-posterior (with respect to the missing variable $\boldsymbol{\beta}$), conditional on the current values of $\boldsymbol{\lambda}$, τ^2 and σ^2, and the observed data, $\mathbf{y}$; the resulting quantity is called the "Q-function":

$$\begin{aligned} & Q(\boldsymbol{\lambda}, \tau, \sigma^2 | \hat{\boldsymbol{\lambda}}^{(t)}, \hat{\tau}^{(t)}, \hat{\sigma}^{2^{(t)}}) \\ & = \mathrm{E}\left[-\log p(\boldsymbol{\beta}, \boldsymbol{\lambda}, \tau, \sigma^2 \,|\, \mathbf{y}) \,|\, \hat{\boldsymbol{\lambda}}^{(t)}, \hat{\tau}^{(t)}, \hat{\sigma}^{2^{(t)}}, \mathbf{y} \right] \\ & = \left(\frac{n+p}{2} \right) \log \sigma^2 + \frac{\mathrm{E}\left[||\mathbf{y} - \mathbf{X}\boldsymbol{\beta}||^2 \,|\, \hat{\boldsymbol{\lambda}}^{(t)}, \hat{\tau}^{(t)}, \hat{\sigma}^{2^{(t)}} \right]}{2\sigma^2} + \frac{p}{2} \log \tau^2 \\ & \quad + \frac{1}{2} \sum_{j=1}^{p} \log \lambda_j^2 + \frac{1}{2\sigma^2 \tau^2} \sum_{j=1}^{p} \frac{\mathrm{E}\left[\beta_j^2 \,|\, \hat{\boldsymbol{\lambda}}^{(t)}, \hat{\tau}^{(t)}, \hat{\sigma}^{2^{(t)}} \right]}{\lambda_i^2} - \log \pi(\boldsymbol{\lambda}, \tau) \end{aligned} \tag{5}$$

where $\pi(\boldsymbol{\lambda}, \tau)$ is the joint prior distribution for the hyperparameters. For notational simplicity we use $\mathrm{E}\left[\beta_j^2\right]$ and $\mathrm{E}\left[||\mathbf{y} - \mathbf{X}\boldsymbol{\beta}||^2\right]$ for the conditional expectations of β_j^2 and the sum of squared residuals, respectively, throughout the sequel.

M-step. Update the parameter estimates by minimising the Q-function with respect to the shrinkage hyperparameters and noise variance, i.e.,

$$\{\hat{\boldsymbol{\lambda}}^{(t+1)}, \hat{\tau}^{(t+1)}, \hat{\sigma}^{2^{(t+1)}}\} = \underset{\boldsymbol{\lambda}, \tau, \sigma^2}{\arg\min} \left\{ Q\left(\boldsymbol{\lambda}, \tau, \sigma^2 \,|\, \hat{\boldsymbol{\lambda}}^{(t)}, \hat{\tau}^{(t)}, \hat{\sigma}^{2^{(t)}} \right) \right\} \tag{6}$$

Implementation of this EM algorithm requires only knowledge of the negative log-prior of choice $-\log \pi(\boldsymbol{\lambda}, \tau)$, the conditional expectations $\mathrm{E}\left[\boldsymbol{\beta}^{\mathbf{2}}\right]$ and $\mathrm{E}\left[||\mathbf{y} - \mathbf{X}\boldsymbol{\beta}||^2\right]$. In Sect. 3.2, we discuss several different approaches to compute the conditional expectations for the E-step. This algorithm is quite general, with only the M-step depending on the choice of prior for the coefficients β_j. Application to the specific case of the horseshoe is discussed in Sect. 4.

The overall algorithm iterates the E-step and M-step until a convergence criterion is satisfied. Once convergence is achieved, we use the mode of the posterior distribution of $\boldsymbol{\beta}$, conditional on the final values of $\hat{\boldsymbol{\lambda}}^{(t)}$, $\hat{\tau}^{(t)}$ and $\hat{\sigma}^{2^{(t)}}$ as our point estimate. Given that this conditional distribution is Gaussian, the mode is just the mean of the normal distribution (7) given by (9). Finally we set components of the final conditional posterior mode that are very small (smaller in absolute value than a small fraction of the standard error) to exactly zero. This step is technically not required, as given sufficient iterations the posterior mode will converge to be exactly sparse, but substantially reduces run time. In this paper we used $|\beta_j^{(t)}| < (5\sqrt{n})^{-1}$ as the threshold for all experiments.

3.2 Computing the Conditional Expectations

The conditional expected values $\mathrm{E}\left[\boldsymbol{\beta}^2\right]$ and $\mathrm{E}\left[||y - \mathbf{X}\boldsymbol{\beta}||^2\right]$ depend on the conditional posterior distribution of the regression coefficients $\boldsymbol{\beta} \in \mathbb{R}^p$ [17]:

$$\begin{aligned} \boldsymbol{\beta} \,|\, \boldsymbol{\lambda}, \tau, \sigma, \mathbf{y} &\sim N_p(\mathbf{A}^{-1}\mathbf{X}^T\mathbf{y}, \sigma^2\mathbf{A}^{-1}) \\ \mathbf{A} &= (\mathbf{X}^T\mathbf{X} + \tau^{-2}\boldsymbol{\Lambda}^{-1}) \end{aligned} \tag{7}$$

where $\boldsymbol{\Lambda} = \mathrm{diag}(\lambda_1^2, \cdots, \lambda_\mathrm{p}^2)$. One important point to note here is that the posterior distribution of $\boldsymbol{\beta}$ does not depend on the choice of prior applied to λ_j. This implies that regardless of the (marginal) prior assigned to $\boldsymbol{\beta}$, as long as it has a SMN representation, both $p(\mathbf{y}|\boldsymbol{\beta})$ and $\pi(\boldsymbol{\beta}|\boldsymbol{\lambda}, \tau)$ will always be Gaussian densities, e.g., it does not matter whether the prior assigned to $\boldsymbol{\beta}$ is a Laplace prior (to solve for the Lasso) or a horseshoe prior, the E-step will remain the same. Changing the marginal priors on the regression coefficients only requires straightforward modification to the M-step for updates on the shrinkage parameters (see Sect. 4). This is an interesting advantage of our EM approach as the E-step is frequently very difficult to implement, particularly when dealing with conditional distributions that lack a standard density.

Exact Expectations. The expected value of β_j^2 can be solved for by taking the sum of the variance and the square of the expected value of β_j:

$$\mathrm{E}\left[\beta_j^2 \,|\, \boldsymbol{\lambda}^{(t)}, \tau^{(t)}\right] = \mathrm{Var}[\beta_j] + \mathrm{E}\left[\beta_j\right]^2. \tag{8}$$

Due to the properties of Gaussian distributions, $\mathrm{Var}[\beta_j] = \left(\mathrm{Cov}[\boldsymbol{\beta}]\right)_{j,j}$, and

$$\begin{aligned} \mathrm{Cov}[\boldsymbol{\beta}] &= \sigma^2\mathbf{A}^{-1}, \\ \mathrm{E}\left[\beta_j\right] &= \left(\mathbf{A}^{-1}\mathbf{X}^T\mathbf{y}\right)_j. \end{aligned} \tag{9}$$

Direct computation of this expectation value is potentially computationally expensive and numerically unstable because it involves inverting the $p \times p$ matrix, $\mathbf{A}$. The efficient algorithm proposed in [24] avoids explicitly computing the inverse of $\mathbf{A}$ when sampling from multivariate normal distributions of the form (7), and the approach in [5] is similar but utilises the matrix inversion lemma for improved computational complexity when $p > n$. Both of these algorithms can be trivially modified to compute the exact mean and covariance matrix of the conditional posterior distribution of the hierarchy (7).

Similarly, finding $\mathrm{E}\left[||\mathbf{y} - \mathbf{X}\boldsymbol{\beta}||^2\right]$ involves solving for the expected value of a quadratic form of the regression coefficients, which in turns requires inverting the same matrix, $\mathbf{A}$. From standard results involving expectations of quadratic forms we have:

$$\mathrm{E}\left[||\mathbf{y} - \mathbf{X}\boldsymbol{\beta}||^2\right] = ||\mathbf{y} - \mathbf{X}\mathrm{E}\left[\boldsymbol{\beta}\right]||^2 + \mathrm{tr}(\mathbf{X}^T\mathbf{X} \cdot \mathrm{Cov}[\boldsymbol{\beta}]) \tag{10}$$

where $\mathrm{tr}(\cdot)$ is the usual trace operator.

Approximate Expectations. The two main components required to compute the expectations in the E-step for this linear model are the conditional posterior mean and variance of the regression coefficients $\boldsymbol{\beta}$ under the distribution (7). The conditional posterior mean can be found using the Rue's [24] or Bhattacharya's [5] algorithm without having to explicitly solve for the inverse of $\mathbf{A}$, with a simple extension of the same procedures allowing us to solve for the exact covariance matrix $\sigma^2\mathbf{A}^{-1}$. However, the quantities that we need are the conditional variances (i.e., the diagonal elements of the conditional covariance matrix). Therefore, instead of inverting the full covariance matrix $\mathbf{A}$, the conditional variances may be approximately found by inverting only the diagonal elements of $\mathbf{A}$, i.e.,

$$\begin{aligned}\mathrm{Var}[\beta_j] &\approx \sigma^2 \frac{1}{A_{j,j}} \\ &= \sigma^2 \left(||\mathbf{x}_j||^2 + \frac{1}{\tau^2\lambda_j^2} \right)^{-1}.\end{aligned} \tag{11}$$

where $\mathbf{x}_j$ is the j-th column of $\mathbf{X}$. This approximation takes only $O(p)$ operations. In the specific case that the predictors $\mathbf{X} = (\mathbf{x}_1, \cdots, \mathbf{x}_p)$ are orthogonal, $\mathbf{X}^T\mathbf{X} = c\,\mathbf{I}_p$ and the approximation (11) will recover the exact variance of β_j. This approximation also applies to (10), i.e., instead of computing the matrix multiplication of the components in the trace function, one may instead consider only the diagonal elements of the matrix, yielding the approximation

$$\mathrm{tr}(\mathbf{X}^T\mathbf{X} \cdot \mathrm{Var}[\boldsymbol{\beta}]) \approx \sigma^2 \sum_{j=1}^{p} \left(||\mathbf{x}_j||^2 + \frac{1}{\tau^2\lambda_j^2} \right)^{-1} ||\mathbf{x}_j||^2. \tag{12}$$

In comparison to the $O(p^2)$ operations required to compute the trace of the product of matrices in (10), this approximation requires only $O(p)$ operations.

4 Application to the Horseshoe Estimator

We now demonstrate the application of the EM procedure described in Sect. 3.1 to the horseshoe prior and provide the exact EM updates for each of the shrinkage parameters. Here, we assume no prior knowledge on the sparsity of the regression coefficient and assign the recommended default prior for the global variance parameter [14, 22]:

$$\tau \sim C^+(0,1), \;\; \tau \in (0,1) \tag{13}$$

where $C^+(0,1)$ denotes a standard half-Cauchy distribution with probability density function of

$$p(z) = \frac{2}{\pi(1+z^2)}, \; z > 0. \tag{14}$$

We limit the range of τ to $(0,1)$ following [20]. The degree of sparsity applied to individual regression coefficients now depends on the choice of the prior distribution assigned to the local variance (shrinkage) component. Carvalho et al. [7]

introduced the use of the horseshoe prior in sparse regression and demonstrated its robustness at handling sparsity with large signals. This is achieved by placing a half-Cauchy prior distribution over both the local and global shrinkage parameter. It was subsequently suggested that the horseshoe prior can be generalised by assigning an inverted-beta prior on $\lambda_j^2 \sim \beta^{'}(a, b)$ with probability density function [22]

$$p(\lambda_j^2) = \frac{(\lambda_j^2)^{a-1}(1+\lambda_j^2)^{-a-b}}{B(a,b)} \tag{15}$$

where $B(a, b)$ is the beta function, and $a > 0$, $b > 0$. We refer readers to [26] for further details about the effect of these hyperparameter a and b on the origin and tail properties of the generalised horseshoe prior. Our interest is in the particular case that $a = b = 1/2$ as this corresponds to a usual half Cauchy prior for λ_j. Substituting the negative logarithm of this prior distribution into (5) – (6) and holding τ^2 and σ^2 fixed yields the M-step update for each λ_j^2:

$$\begin{aligned} \hat{\lambda}_j^2 &= \underset{\lambda_j^2}{\operatorname{argmin}} \left\{ \log \lambda_j^2 + \frac{W_j}{\lambda_j^2} + \log(1+\lambda_j^2) \right\} \\ &= \frac{1}{4}\left(\sqrt{1+6W_j+W_j^2} + W_j - 1\right) \qquad (16) \\ (\hat{\tau}^2, \hat{\sigma}^2) &= \underset{\tau}{\overset{2}{\operatorname{argmin}}}, \sigma^2 \; Q(\tau^2, \sigma^2; \hat{\boldsymbol{\lambda}}^2) \qquad (17) \end{aligned}$$

where $W_j = E[\beta_j^2]/(2\sigma^2\tau^2)$. For convenience, the negative log-prior and the M-step update for the horseshoe prior are summarised in Table 1. Given the $\hat{\boldsymbol{\lambda}}^2$ updates (16), the $\hat{\tau}$ and $\hat{\sigma}^2$ estimates can be found using numerical optimisation. This two-dimensional optimisation problem (17) can be further reduced to the following one-dimensional optimisation problem by (approximately) estimating σ^2 using the expected residual sum-of-squares given in Eq. (10) or (11):

$$\begin{aligned} \hat{\sigma}^2 &= \mathrm{E}\left[||\mathbf{y} - \mathbf{X}\boldsymbol{\beta}||^2 \right]/n, \\ \hat{\tau}^2 &= \underset{\tau^2}{\operatorname{argmin}} \; Q(\tau^2; \hat{\boldsymbol{\lambda}}^2, \hat{\sigma}^2). \end{aligned} \tag{18}$$

This approximate update for σ^2 is within $O(n^{-1})$ of the exact solution, and allows us to substantially reduce the complexity of the optimisation problem. Figure 1 compares the shrinkage profiles for the Lasso, horseshoe and horseshoe-like estimator. The plots demonstrate how the three estimators shrink the least-square (unpenalised) estimate $\hat{\beta}_{\mathrm{LS}}$ towards zero. All three procedures shrink small least-squares estimates to zero. The lasso affects all values of $\hat{\beta}_{\mathrm{LS}}$ by translating them towards zero by the same amount, while the horseshoe and horseshoe-like leave large values of $\hat{\beta}_{\mathrm{LS}}$ largely unshrunk. The horseshoe-like prior exerts more of a hard thresholding-like behaviour as $\hat{\beta}_{\mathrm{LS}}$ approaches zero, while the horseshoe prior mimics firm thresholding.

Table 1. Negative log-prior distribution and EM updates for the proposed horseshoe estimator, where $W_j = E[\beta_j^2]/(2\sigma^2\tau^2)$.

$-\log \pi(\boldsymbol{\lambda}, \tau)$	M-step
$\sum_{j=1}^{p} \left[\log(1+\lambda_j^2) + \frac{\log \lambda_j^2}{2} \right] + \log(1+\tau^2)$	$\hat{\lambda}_j^{2\,(t+1)} = \frac{\sqrt{1+6W_j+W_j^2}+W_j-1}{4}$

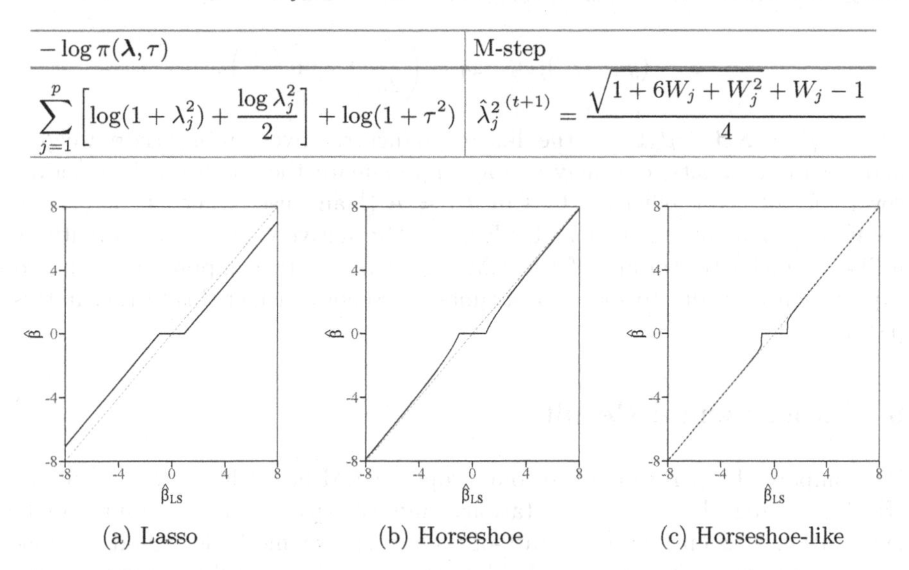

(a) Lasso (b) Horseshoe (c) Horseshoe-like

Fig. 1. The posterior mode estimates $\hat{\beta}$ versus $\hat{\beta}_{\text{LS}}$ for the (a) Lasso, (b) Horseshoe and (c) Horseshoe-like estimator. For illustration purposes, τ is chosen such that all three estimators give nearly identical shrinkage within approximately 1 unit of the origin.

5 Extension to Generalised Linear Models

We now consider an extension of the EM approach proposed in Sect. 3 to non-normal data models through the framework of generalised linear models (GLMs) [18]. A GLM models the conditional mean of the target by an appropriate (potentially) non-linear transformation of the linear predictor; well known examples include binomial logistic regression and Poisson log-linear regression. In general, when the targets are non-Gaussian, the conditional distribution of the coefficients will not have a standard form, and finding the exact expectations $\mathrm{E}\left[\beta_j^2\right]$ is difficult. However, it is frequently the case that the conditional distribution can be approximated by a heteroskedastic Gaussian distribution

$$\begin{aligned} \boldsymbol{\beta} \,|\, \boldsymbol{\lambda}, \tau, \boldsymbol{\omega}, \mathbf{z} &\sim N_p\left(\mathbf{A}_\omega^{-1}\mathbf{X}^T\boldsymbol{\Omega}\mathbf{z},\ \mathbf{A}_\omega^{-1}\right), \\ \mathbf{A}_\omega &= \mathbf{X}^T\boldsymbol{\Omega}\mathbf{X} + \tau^{-2}\boldsymbol{\Lambda}^{-1} \end{aligned} \tag{19}$$

where $\boldsymbol{\Omega} = \mathrm{diag}(\boldsymbol{\omega})$, $\boldsymbol{\omega} = (\omega_1, \ldots, \omega_n)$ is a vector of weights, and $\mathbf{z}$ is an adjusted version of the targets, $\mathbf{y}$. Via standard central-limit theorem arguments, the accuracy of this approximation increases as the sample size n grows. The weights can be obtained via a linearisation argument (i.e., the well known IRLS algorithm) or, preferably, via a scale-mixture of normals representation of the likelihood when available. For example, the logistic regression implementation used in this paper utilises the well-known Polyá-gamma representation of logistic regression [23]; under this scheme, the adjusted targets and weights are

$$z_i = (y - 1/2)/\omega_i, \ \ \omega_i = \left(\frac{1}{2\eta_i}\right) \tanh\left(\frac{\eta_i}{2}\right),$$

where $\boldsymbol{\eta} = \mathbf{X}\boldsymbol{\beta} + \beta_0 \mathbf{1}_n$ is the linear predictor. Given appropriate weights and adjusted targets, one may simply approximate the conditional posterior-covariance of the coefficients by $\text{Cov}[\boldsymbol{\beta}] \approx \mathbf{A}_\omega^{-1}$ and use either (9) or (11) to obtain approximate expressions for $\text{E}\left[\beta_j^2\right]$. Alternatively, one could potentially utilise a stochastic variant of the EM algorithm [9] to compute the required expectations, but due to space constraints we do not consider this further in this paper.

6 Experimental Results

We compare the performance of our proposed EM-based horseshoe estimator (HS-EM, using the exact expectations, and HS-apx, using the approximate expectations) against several state-of-the-art sparse methods including non-convex estimators (SCAD, MCP, HS-like estimator [4]), as well as the well known lasso and ridge estimator. We analyse the performance of the estimators in terms of variables selected and prediction accuracy on both simulated and real data. The experiments in this section are designed for variable selection under high dimensional settings, and we use the following metrics:

- **MSE**. Mean squared prediction error; given by $(\hat{\boldsymbol{\beta}}-\boldsymbol{\beta})'\boldsymbol{\Sigma}(\hat{\boldsymbol{\beta}}-\boldsymbol{\beta})$ for simulated data experiments, and $(1/n)||\hat{\mathbf{y}} - \mathbf{y}||_2^2$ for real data experiments.
- **Time**. Computation time (in seconds).
- **No.V**. Number of variables included in the model.
- **TNZ**. True non-zeros. Number of non-zero coefficients correctly identified.
- **FNZ**. False non-zeros. Number of zero coefficients incorrectly identified as non-zero.

Our proposed estimator terminates when it satisfies the convergence criterion: $\sum_{j=1}^{p}(|\beta_j^{(t)} - \beta_j^{(t+1)}|)/(1+\sum_{j=1}^{p}(|\beta_j^{(t+1)}|)) < \omega$, where $\beta^{(t)}$ denotes the coefficient estimates at iteration t and ω is the tolerance parameter which we set it to 10^{-5}. All experiments are performed in the R statistical platform. The proposed EM horseshoe estimator is implemented in the **bayesreg**[1] Bayesian regression package. We use the `glmnet` package to implement the lasso and ridge, while the `ncvreg` package implements both MCP and SCAD. The hyperparameters of the non-Bayesian techniques are tuned using ten-fold cross-validation. All function arguments are set to the default value unless mentioned otherwise. Datasets and code for the experimental results in this section are publicly available[2]

[1] Available at https://cran.r-project.org/web/packages/bayesreg/index.html.
[2] Available at https://github.com/shuyu-tew/Sparse-Horseshoe-EM.git.

6.1 Simulated Data

We considered $n = 70$, $p = 350$ and simulated 100 different data sets from the linear model, $\mathbf{y} \sim N_n(\mathbf{X}\boldsymbol{\beta}, \sigma^2 \boldsymbol{I}_n)$ with $\sigma^2 \in \{1, 9\}$. The predictor matrix, $\boldsymbol{X} \in \mathbb{R}^{n \times p}$ was generated from a correlated multivariate normal distribution $N_p(0, \boldsymbol{\Sigma})$ with $\Sigma_{ij} = \rho^{|i-j|}$ and $\rho \in \{0, 0.8\}$.

The first ten entries of $\boldsymbol{\beta}$ were set to 3, the next ten were set to -3, and the remaining entries were zero. This setup is similar to the experiments presented in [4]. The results are shown in Table 2. Overall, the HS-like estimator performed the best, obtaining the lowest MSE at the expense of substantial overfitting. The good performance in MSE under this setting is not unexpected, as this problem setting favors estimators that overfit (i.e., are less conservative) to a certain extent. Given a sparse vector with only 20 non-zero coefficients, this setting in which all non-zero coefficients have the same magnitude is one of the hardest settings for a sparse method, as excluding any one of the true coefficients is equally damaging. The prediction risk of this problem is $\mathrm{E}[||\boldsymbol{\beta} - \hat{\boldsymbol{\beta}}||^2]$; if a method incorrectly includes an irrelevant predictor in the model, the risk will approximately be the variance of the estimate associated with this variable, i.e., $\mathrm{E}[\hat{\beta}_j^2] \approx \sigma^2/n$; in contrast, incorrectly excluding non-zero predictors will incur an error of 3^2, which is much larger than $\sigma^2/70$ even for the setting of greater noise, i.e., $\sigma^2 = 9$. In general, approximately half of the coefficients included in the HS-like model are false non-zeros (FNZ); while most of the coefficients included in our HS models are true non-zeros with consistent small difference between TNZ and No.V.

The HS-EM, SCAD and MCP methods all generally select much fewer variables than the HS-like estimator, with the HS-EM method overfitting less than HS-like and underfitting more than MCP or SCAD. In terms of MSE, our HS-EM method is competitive with, or superior to SCAD and MCP for all settings, despite producing substantially sparser estimates. It appears that both SCAD and MCP are sensitive to the correlation structure of the data; their performance is comparable to our HS-EM approach when there is no correlation, but when correlation is introduced to the data, both methods performed substantially worse in terms of MSE. Such behaviour is not observed in our proposed HS estimator. Of the two proposed HS estimators, HS-apx appears to be approximately three times faster than HS-EM. Otherwise, both estimators give roughly similar performance.

6.2 Real Data

We further analyse the performance of our proposed method on six real-world datasets. All datasets are available for download from the UCI machine learning repository [2] unless mentioned otherwise. Each dataset is randomly split such that the training data has a sample size of n with the remaining $N-n$ datapoints used as testing samples. This procedure is repeated 50 times and the averaged summary statistics for the performance measures are presented in Tables 4 and 5 for the linear regression and logistic regression experiments, respectively.

Table 2. Performance of different sparse estimators (with associated standard error in parentheses) for the simulated data experiments.

	HS-EM	HS-apx	HS-like	Lasso	MCP	SCAD	Ridge
	$(\rho = 0, \sigma^2 = 1)$						
MSE	162.7(3.5)	162.8(2.9)	**119.8**(4.7)	139.1(3.4)	159.6(1.9)	149.8(1.8)	176.6(0.5)
Time	5.93(0.19)	1.37(0.04)	0.27(0.02)	0.14(0.01)	0.21(0.01)	0.32(0.01)	0.64(0.01)
No.V	6.11(0.35)	5.83(0.36)	43.4(0.53)	21.4(1.59)	8.61(0.52)	18.7(0.72)	350(0.00)
TNZ	4.37(0.29)	4.27(0.28)	14.2(0.32)	8.77(0.61)	4.89(0.29)	8.31(0.32)	20.0(0.00)
FNZ	1.63(0.16)	1.56(0.16)	29.2(0.72)	12.6(1.06)	3.72(0.28)	10.4(0.51)	330(0.00)
	$(\rho = 0, \sigma^2 = 9)$						
MSE	171.6(3.1)	170.2(2.9)	**141.5**(3.5)	151.9(2.7)	164.2(1.9)	153.6(1.9)	177.1(0.5)
Time	5.89(0.22)	1.37(0.04)	0.27(0.01)	0.15(0.01)	0.21(0.01)	0.33(0.01)	0.64(0.01)
No.V	5.38(0.34)	5.19(0.34)	46.3(0.29)	15.9(1.47)	7.83(0.49)	18.4(0.77)	350(0.00)
TNZ	3.69(0.27)	3.68(0.27)	13.2(0.31)	6.61(0.54)	4.31(0.28)	7.58(0.31)	20.0(0.00)
FNZ	1.69(0.17)	1.51(0.15)	33.1(0.36)	9.35(0.99)	3.52(0.29)	10.8(0.56)	330(0.00)
	$(\rho = 0.7, \sigma^2 = 1)$						
MSE	12.7(1.29)	15.1(1.57)	**3.79**(0.57)	6.48(0.47)	78.1(3.23)	74.3(3.01)	426.6(3.81)
Time	2.83(0.21)	0.97(0.05)	0.22(0.01)	0.11(0.01)	0.14(0.01)	0.16(0.01)	0.65(0.01)
No.V	16.9(0.25)	16.4(0.29)	25.1(0.46)	33.4(0.49)	14.1(0.33)	22.4(0 49)	350(0.00)
TNZ	16.9(0.25)	16.4(0.29)	19.7(0.05)	19.7(0.06)	8.21(0.16)	9.64(0.16)	20.0(0.00)
FNZ	0.04(0.03)	0.04(0.03)	5.27(0.51)	13.7(0.48)	5.89(0.31)	12.8(0.49)	330(0.00)
	$(\rho = 0.7, \sigma^2 = 9)$						
MSE	34.8(1.79)	38.1(1.91)	36.3(1.38)	**27.1**(1.17)	85.9(3.28)	84.8(3.21)	428.7(3.89)
Time	4.41(0.26)	1.17(0.04)	0.24(0.01)	0.12(0.01)	0.14(0.01)	0.16(0.01)	0.64(0.01)
No.V	12.9(0.27)	12.3(0.28)	40.9(0.28)	28.6(0.62)	13.3(0.31)	21.1(0.45)	350(0.00)
TNZ	12.6(0.27)	12.1(0.27)	18.7(0.11)	18.4(0.08)	7.66(0.13)	8.85(0.16)	20.0(0.00)
FNZ	0.24(0.05)	0.24(0.05)	22.2(0.29)	10.3(0.59)	5.61(0.28)	12.2(0.42)	330(0.00))

Linear Regression. Table 3 presents the posterior mean estimates and their 95% credible intervals (CI) for the diabetes data [11] ($N = 442, P = 10$). For comparison, we included the posterior mode estimates computed from the proposed HS-EM estimator. The HS-EM posterior mode estimates are very similar to the corresponding posterior means, and all posterior mode estimates are within the corresponding 95% credible intervals; however, the HS-EM posterior mode provides a sparse point estimate as it excludes all variables (except S3) with 95% CIs that include zero.

Table 3. The posterior mode and mean estimates of the predictors for the diabetes data using the horseshoe estimator. The posterior mean and 95% credible intervals are computed using the `bayesreg` package in R.

	AGE	SEX	BMI	BP	S1	S2	S3	S4	S5	S6
Mean	−0.009	−18.68	5.769	1.034	−0.223	0.013	−0.592	2.419	48.84	0.179
2.50%	−0.341	−30.93	4.371	0.571	−0.937	−0.342	−1.415	−3.462	32.24	−0.225
97.5%	0.326	−5.144	7.109	1.457	0.098	0.656	0.189	11.36	70.14	0.734
Mode	·	−17.54	5.741	1.021	·	·	−0.909	·	43.58	·

In addition to the diabetes data, we also analysed the benchmark Boston housing data ($N = 506, P = 14$), the concrete compressive strength dataset ($N = 1030, P = 9$) and the eye data ($N = 120, P = 200$) [25]. To make the problem more difficult, for each dataset we added a number of noise variables generated using the same procedure described in Sect. 6.1; in all cases we added $p = 15$ additional noise variables, with $\rho = 0.8$. Model fitting is done using all the $P+15$ (original plus noise) predictors as well as all interactions and possible transformations of the variables (logs, squares and cubics).

Overall, our proposed HS estimator performs the best, attaining the lowest MSE on all real data, while generally selecting the simplest models (included the lowest number of variables). Similar to the results observed in Sect. 6.1, the results for HS-apx are virtually identical to the exact HS-EM procedure, while being substantially faster. SCAD and MCP exhibit comparable performance to HS in terms of prediction error for the diabetes data and eye data, but have relatively poor performance for Boston and concrete data. Interestingly, while HS-like dominated the other non-convex estimators in terms of MSE on the simulated experiments, it performs quite poorly on the real data analysis, particularly on the diabetes and Boston housing data. The HS-like estimator tended to include the highest number of variables, suggesting it is potentially prone to overfitting as discussed in Sect. 6.1. Ridge regression generally performs worse than the sparse estimators, as would be expected given the experimental design.

Logistic Regression. We also tested the HS-EM estimator on two binary classification problems: the Pima indians data ($N = 768, P = 8$) and the heart disease data ($N = 302, P = 14$). Similar to the linear regression analysis, we augment the data with 10 noise variables and model the predictors together

Table 4. Performance of different sparse estimators (with associated standard error in parentheses) on the 4 datasets for linear regression model.

	HS-EM	HS-apx	HS-like	Lasso	MCP	SCAD	Ridge
Diabetes ($n = 100, p = 385$)							
MSE	**3383.3**(30.5)	3407.2(32.1)	13068(491.1)	3654.4(47.9)	3624.4(66.2)	3667.6(68.5)	4181.6(42.8)
Time	3.01 (0.02)	1.02 (0.02)	0.32 (0.01)	0.29 (0.01)	1.13 (0.01)	1.47 (0.02)	0.74 (0.01)
No.V	1.62 (0.09)	1.54 (0.09)	95.3 (0.29)	4.14 (0.44)	3.68 (0.41)	6.86 (0.74)	385 (0.00)
Boston Housing ($n = 100, p = 473$)							
MSE	26.76(0.71)	**26.72**(0.71)	58.01(3.05)	31.41(0.89)	293.1(24.3)	323.9(29.7)	49.19(1.19)
Time	4.74(0.34)	1.86(0.03)	0.67(0.03)	0.20(0.01)	1.21(0.01)	1.39(0.01)	1.06(0.01)
No.V	2.82(0.16)	2.84(0.16)	47.9(0.78)	4.60(0.52)	4.52(0.41)	10.9(0.87)	473(0.00)
Concrete ($n = 100, p = 327$)							
MSE	**73.71** (3.67)	73.76 (3.66)	222.4 (8.38)	81.67 (1.98)	113.9 (15.9)	108.3 (16.1)	176.2 (2.49)
Time	2.39 (0.17)	0.86 (0.03)	0.36 (0.02)	0.23 (0.01)	0.98 (0.01)	1.14 (0.01)	0.55 (0.01)
No.V	5.46 (0.13)	5.42 (0.14)	67.8 (0.57)	10.7 (0.61)	6.60 (0.44)	13.8 (0.85)	327 (0.00)
Eye ($n = 100, p = 200$)							
MSE	**0.79** (0.06)	**0.79** (0.06)	1.50 (0.23)	0.84 (0.14)	**0.79** (0.06)	0.86 (0.09)	0.85 (0.09)
Time	0.52 (0.02)	0.24 (0.01)	0.15 (0.01)	0.20 (0.01)	0.13 (0.01)	0.19 (0.01)	0.23 (0.01)
No.V	3.84 (0.11)	3.72 (0.09)	0.00 (0.00)	18.0 (0.55)	5.20 (0.27)	10.5 (0.42)	200 (0.00)

with their interactions and transformations. For the heart disease data, we only included the interactions between the variables and not the transformations, as there were a substantial number of categorical predictors. For this experiment we varied the number of training samples $n \in \{100, 200\}$.

The HS-EM estimator obtains the best classification accuracy and negative log-loss on the Pima data for both training sample sizes. Ridge regression, on the other hand, performs the best in terms of classification accuracy for the heart data. This suggest that many of the predictors and interactions between the variables in the heart data are potentially correlated with the response variable. Therefore, a sparse estimator might not be suitable for modelling this data. Nonetheless, our HS methods gives comparable results to the other sparse methods in terms of log-loss and in general, as the sample size increases, the difference in performance between all estimators becomes less significant.

Table 5. Performance of different sparse estimators on the Pima and heart disease data. CA is the classification accuracy and NLL is the negative log-loss.

Dataset		HS-EM	HS-apx	Lasso	MCP	SCAD	Ridge
		$(n = 100, p = 214)$					
Pima	CA	**0.743**(0.01)	0.741(0.01)	0.704(0.01)	0.741(0.01)	0.735(0.01)	0.674(0.01)
	NLL	**358.1**(2.42)	358.8(2.55)	381.5(3.33)	360.9(4.12)	362.8(3.90)	400.7(2.63)
	Time	0.622(0.02)	0.533(0.01)	0.298(0.01)	1.475(0.07)	2.056(0.07)	0.251(0.01)
	No.V	1.760(0.09)	1.700(0.09)	3.780(0.25)	3.900(0.27)	8.480(0.59)	214.0(0.00)
		$(n = 200, p = 214)$					
	CA	**0.756**(0.01)	**0.756**(0.01)	0.736(0.01)	0.754(0.01)	**0.756**(0.01)	0.706(0.01)
	NLL	**290.8**(1.97)	291.1(1.96)	306.4(1.52)	292.7(2.32)	291.9(1.69)	320.6(1.01)
	Time	0.838(0.03)	0.733(0.02)	0.835(0.02)	3.150(0.16)	4.075(0.19)	0.424(0.01)
	No.V	3.020(0.15)	3.060(0.15)	4.740(0.25)	4.760(0.32)	8.980(0.59)	214.0(0.00)
		$(n = 100, p = 400)$					
Heart	CA	0.758(0.01)	0.751(0.01)	0.792(0.01)	0.786(0.01)	0.797(0.01)	**0.799**(0.01)
	NLL	107.4(1.73)	110.3(1.66)	99.89(0.95)	96.20(1.28)	**94.36**(0.93)	103.2(1.03)
	Time	3.943(0.09)	3.623(0.11)	0.237(0.01)	1.208(0.05)	1.483(0.06)	0.383(0.01)
	No.V	2.820(0.11)	2.800(0.12)	10.10(0.49)	7.080(0.28)	13.76(0.51)	400.0(0.00)
		$(n = 200, p = 400)$					
	CA	0.798(0.01)	0.788(0.01)	0.813(0.01)	0.802(0.01)	0.805(0.01)	**0.828**(0.01)
	NLL	47.86(0.95)	49.53(0.94)	46.31(0.41)	**44.90**(0.71)	45.05(0.73)	48.25(0.32)
	Time	5.629(0.14)	5.022(0.14)	0.570(0.01)	2.755(0.11)	3.517(0.12)	0.621(0.01)
	No.V	5.520(0.15)	5.480(0.14)	14.28(0.67)	9.340(0.37)	17.38(0.58)	400.0(0.00)

7 Discussion

In this paper we introduced a novel EM algorithm that allows us to efficiently find, for the first time, the exact, sparse posterior mode of the horseshoe estimator. The experimental results suggest that in comparison with state-of-the-art non-convex sparse estimators, the HS-EM estimator is quite robust to the underlying structure of the problem. Both the MCP and SCAD algorithm appear sensitive to correlation in the predictors, and the HS-like estimator, based on an approximation to the horseshoe prior, appears sensitive to the signal-to-noise ratio of the problem. In contrast, the HS-EM algorithm, even when not performing the best, appears to always remain competitive with the best performing method while selecting highly sparse models. Comparing the shrinkage profiles of the HS and HS-like estimator (see Fig. 1), the mixed performance of the HS-like procedure, is possibly due to the fact that HS-like approach more aggressively zeros out coefficients, making it more sensitive to misspecification of the global shrinkage parameter. We also conjecture that the conventional EM framework, in which the shrinkage hyperparameters $\boldsymbol{\lambda}$ are treated as missing data, provides a poorer basis for estimation of the global hyperparameter, in comparison to our approach of treating the coefficients as missing data, which potentially integrates the uncertainty of the parameters more effectively into the Q-function. This is a topic for further investigation.

Due to its prior preference for dense models, ridge regression can perform much worse than sparse alternatives if the underlying problem is sparse. However, it remained competitive or superior to other sparse methods on several of the real datasets. This strongly suggests that a method that is adaptive to the unknown degree of sparsity would be beneficial; we anticipate extending our algorithm to the generalised horseshoe discussed in Sect. 4, and automatically tuning our prior to produce a technique that will adapt to the sparsity of the problem.

While our HS-EM algorithm is able to locate, exactly, a sparse posterior mode of the horseshoe estimator, there is an obvious question as to what mode we are finding. Posterior modes (and means) are not invariant under reparameterisation, which suggests that maximising for λ_j (with appropriate transformation of prior distribution) in place of λ_j^2, for example, will produce a new shrinkage procedure with potentially different properties. This is a focus of future research. The EM algorithm in this paper can also be easily extended to other popular priors such as the Bayesian lasso. Finally, given the relative simplicity of the M-step updates in this framework, it is of interest to consider extensions to non-linear models, such as neural networks. In this case, the required conditional expectations are unlikely to be available in closed form, but given access to a Gibbs sampler (or approximate Gibbs sampler) the procedure can easily be extended to utilise a standard stochastic EM implementation. This could potentially provide a simple, and (in expectation) exact, alternative to variational Bayes for these problems.

References

1. Armagan, A., Dunson, D., Lee, J.: Bayesian generalized double Pareto shrinkage. Biometrika (2010)
2. Asuncion, A., Newman, D.: UCI machine learning repository (2007)
3. Bhadra, A., Datta, J., Polson, N.G., Willard, B.: Default Bayesian analysis with global-local shrinkage priors. Biometrika **103**(4), 955–969 (2016)

4. Bhadra, A., Datta, J., Polson, N.G., Willard, B.T.: The horseshoe-like regularization for feature subset selection. Sankhya B **83**(1), 185–214 (2019). https://doi.org/10.1007/s13571-019-00217-7
5. Bhattacharya, A., Chakraborty, A., Mallick, B.K.: Fast sampling with Gaussian scale mixture priors in high-dimensional regression. Biometrika **103**(4), 985–991 (2016)
6. Brown, P.J., Griffin, J.E.: Inference with normal-gamma prior distributions in regression problems. Bayesian Anal. **5**(1), 171–188 (2010)
7. Carvalho, C.M., Polson, N.G., Scott, J.G.: The horseshoe estimator for sparse signals. Biometrika **97**(2), 465–480 (2010)
8. Casella, G., Ghosh, M., Gill, J., Kyung, M.: Penalized regression, standard errors, and Bayesian lassos. Bayesian Anal. **5**(2), 369–411 (2010)
9. Celeux, G.: The SEM algorithm: a probabilistic teacher algorithm derived from the EM algorithm for the mixture problem. Comput. Stat. Quart. **2**, 73–82 (1985)
10. Dempster, A.P., Laird, N.M., Rubin, D.B.: Maximum likelihood from incomplete data via the EM algorithm. J. R. Stat. Soc. Series B Stat. Methodol. **39**(1), 1–22 (1977)
11. Efron, B., Hastie, T., Johnstone, I., Tibshirani, R.: Least angle regression. Ann. Stat. **32**(2), 407–499 (2004)
12. Fan, J., Li, R.: Variable selection via nonconcave penalized likelihood and its oracle properties. J. Am. Stat. Assoc. **96**(456), 1348–1360 (2001)
13. Figueiredo, M.A.: Adaptive sparseness for supervised learning. IEEE Trans. Pattern Anal. Mach. Intell. **25**(9), 1150–1159 (2003)
14. Gelman, A.: Prior distributions for variance parameters in hierarchical models. Bayesian Anal. **1**(3), 515–533 (2006)
15. Kiiveri, H.T.: A Bayesian approach to variable selection when the number of variables is very large. Lecture Notes-Monograph Series, pp. 127–143 (2003)
16. Li, H., Pati, D.: Variable selection using shrinkage priors. Comput. Stat. Data Anal. **107**, 107–119 (2017)
17. Makalic, E., Schmidt, D.F.: A simple sampler for the horseshoe estimator. IEEE Signal Process. Lett. **23**(1), 179–182 (2016)
18. Nelder, J.A., Wedderburn, R.W.: Generalized linear models. J. Royal Stat. Soc.: Series A (General) **135**(3), 370–384 (1972)
19. Park, T., Casella, G.: The Bayesian lasso. J. Am. Stat. Assoc. **103**(482), 681–686 (2008)
20. van der Pas, S., Szabó, B., van der Vaart, A.: Adaptive posterior contraction rates for the horseshoe. Electron. J. Stat. **11**(2), 3196–3225 (2017)
21. Polson, N.G., Scott, J.G.: Shrink globally, act locally: sparse Bayesian regularization and prediction. Bayesian stat. **9**, 501–538 (2010)
22. Polson, N.G., Scott, J.G.: On the half-cauchy prior for a global scale parameter. Bayesian Anal. **7**(4), 887–902 (2012)
23. Polson, N.G., Scott, J.G., Windle, J.: Bayesian inference for logistic models using Pólya-gamma latent variables. J. Am. Stat. Assoc. **108**(504), 1339–1349 (2013)
24. Rue, H.: Fast sampling of Gaussian Markov random fields. J. R. Stat. Soc. Series B Stat. Methodol. **63**(2), 325–338 (2001)
25. Scheetz, T.E., et al.: Regulation of gene expression in the mammalian eye and its relevance to eye disease. Proc. Natl. Acad. Sci. **103**(39), 14429–14434 (2006)
26. Schmidt, D.F., Makalic, E.: Bayesian generalized horseshoe estimation of generalized linear models. In: Brefeld, U., Fromont, E., Hotho, A., Knobbe, A., Maathuis, M., Robardet, C. (eds.) ECML PKDD 2019. LNCS (LNAI), vol. 11907, pp. 598–613. Springer, Cham (2020). https://doi.org/10.1007/978-3-030-46147-8_36

27. Tibshirani, R.: Regression shrinkage and selection via the lasso. J. R. Stat. Soc. Series B Stat. Methodol. **58**(1), 267–288 (1996)
28. Van Erp, S., Oberski, D.L., Mulder, J.: Shrinkage priors for Bayesian penalized regression. J. Math. Psychol. **89**, 31–50 (2019)
29. Wand, M.P., Ormerod, J.T., Padoan, S.A., Frühwirth, R.: Mean field variational Bayes for elaborate distributions. Bayesian Anal. **6**(4), 847–900 (2011)
30. Zhang, C.H., et al.: Nearly unbiased variable selection under minimax concave penalty. Ann. Stat. **38**(2), 894–942 (2010)

Structure-Preserving Gaussian Process Dynamics

Katharina Ensinger[1,2(✉)], Friedrich Solowjow[2], Sebastian Ziesche[1], Michael Tiemann[1], and Sebastian Trimpe[2]

[1] Bosch Center for Artificial Intelligence, Renningen, Germany
katharina.ensinger@bosch.com

[2] Institute for Data Science in Mechanical Engineering, RWTH Aachen University, Aachen, Germany

Abstract. Most physical processes possess structural properties such as constant energies, volumes, and other invariants over time. When learning models of such dynamical systems, it is critical to respect these invariants to ensure accurate predictions and physically meaningful behavior. Strikingly, state-of-the-art methods in Gaussian process (GP) dynamics model learning are not addressing this issue. On the other hand, classical numerical integrators are specifically designed to preserve these crucial properties through time. We propose to combine the advantages of GPs as function approximators with structure-preserving numerical integrators for dynamical systems, such as Runge-Kutta methods. These integrators assume access to the ground truth dynamics and require evaluations of intermediate and future time steps that are unknown in a learning-based scenario. This makes direct inference of the GP dynamics, with embedded numerical scheme, intractable. As our key technical contribution, we enable inference through the implicitly defined Runge-Kutta transition probability. In a nutshell, we introduce an implicit layer for GP regression, which is embedded into a variational inference model learning scheme.

Keywords: Gaussian processes · Bayesian methods · Time-series

1 Introduction

Many physical processes can be described by an autonomous continuous-time dynamical system

$$\dot{x}(t) = f(x(t)) \text{ with } f : \mathbb{R}^d \to \mathbb{R}^d. \tag{1}$$

Dynamics model learning deals with the problem of estimating the function f from sampled data. In practice, it is not possible to observe state trajectories in

Code will be published upon request.

Supplementary Information The online version contains supplementary material available at https://doi.org/10.1007/978-3-031-26419-1_9.

© The Author(s), under exclusive license to Springer Nature Switzerland AG 2023
M.-R. Amini et al. (Eds.): ECML PKDD 2022, LNAI 13717, pp. 140–156, 2023.
https://doi.org/10.1007/978-3-031-26419-1_9

continuous time since data is typically collected on digital sensors and hardware. Thus, we obtain noisy discrete-time observations

$$\{\hat{x}_n\}_{1:N} = \{x_1 + \nu_1, \ldots, x_N + \nu_N\}, \nu_n \sim \mathcal{N}(0, \text{diag}(\sigma_{n,1}^2, \ldots, \sigma_{n,d}^2)). \quad (2)$$

Accordingly, models of dynamical systems are typically learned as of one-step ahead-predictions

$$x_{n+1} = g(x_n). \quad (3)$$

Especially, Gaussian processes (GPs) have been popular for model learning and are predominantly applied to one step ahead predictions (3) [5,7,8]. However, there is a discrepancy between the continuous (1) and discrete-time (3) systems. Importantly, (1) often possesses invariants that represent physical properties. Thus, naively chosen discretizations (such as Eq. (3)) might lead to poor models.

Numerical integrators provide sophisticated tools to efficiently discretize continuous-time dynamics (1). Strikingly, one step ahead predictions (3) correspond to the explicit Euler integrator $x_{n+1} = x_n + hf(x_n)$ with step size h. This follows immediately by identifying $g(x_n)$ with $x_n + hf(x_n)$. It is well-known that the explicit Euler method might lead to problematic behavior and suboptimal performance [13]. Clearly, this raises the immediate question: *can superior numerical integrators be leveraged for dynamics model learning?*

For the numerical integration of dynamical systems, the function f is assumed to be known. The explicit Euler is a popular and straightforward method, which thrives due to its simplicity. No intermediate evaluations of the dynamics are necessary, which makes the integrator also attractive for model learning. While this behavior is tempting when implementing the algorithm, there are theoretical issues [13]. In particular, important physical and geometrical structure is not preserved. In contrast to the explicit Euler, there are also implicit and higher-order methods. However, these generalizations require the evaluation at intermediate and future time steps, which leads to a nonlinear system of equations that needs to be solved. While these schemes become more involved, they yield advantageous theoretical guarantees. In particular, Runge-Kutta (RK) schemes define a rich class of powerful integrators. Despite assuming the dynamics function f to be unknown, we can still benefit from the discretization properties of numerical integrators for model learning. To this end, we propose to combine GP dynamics learning with arbitrary RK integrators, in particular, implicit RK integrators.

Depending on the problem that is addressed, the specific RK method has to be tailored to the system. As an example, we consider structure-preserving integrators, i.e., geometric and symplectic ones. We develop our arguments based on Hamiltonian systems [28,29]. These are an important class of problems that preserve a generalized notion of energy and volume. Symplectic integrators are designed to cope with this type of problems, providing volume-preserving trajectories and accurate approximation of the total energy [12]. In order to demonstrate the flexibility of our method, we also introduce a geometric integrator that is consistent with a mass moving on a surface. For both examples, we show in the experiments section that the predictions with our tailored GP model are indeed preserving the desired structure.

By generalizing to more sophisticated integrators, we have to address the issue of propagating implicitly defined distributions through the dynamics. This is due to the fact that evaluations of the GP at the next time step induce additional implicit evaluations of the dynamics. Depending on the integrator, these might be future or intermediate time steps that are also unknown. On a technical level, sparse GPs provide the necessary flexibility. A decoupled sampling approach allows consistent sampling of functions from the GP posterior [34]. In contrast to previous GP dynamics modeling schemes [7], this yields consistency throughout the entire simulation pipeline. By leveraging these ideas, we derive a recurrent variational inference (VI) model learning scheme.

By addressing integrator-induced implicit transition probabilities, we are essentially proposing implicit layers for probabilistic models. Implicit layers in neural networks (NNs) are becoming increasingly popular and address outputs that can not be calculated explicitly [3,10,23]. Typically, implicit layers in NNs are defined in terms of an optimization problem. However, the idea of implicitly defined layers has (to the best of our knowledge) not yet been generalized to probabilistic models like GPs, introducing the technical challenge of dealing with implicitly defined probability distributions.

In summary, the main contributions of this paper are:

- a general and flexible probabilistic learning framework that combines arbitrary RK integrators with GP dynamics model learning;
- an inference scheme that is able to cope with implicitly defined distributions, thus extending the idea of implicit layers from NNs to GPs; and
- embedding geometrical and symplectic integrators, yielding GP dynamics models that are structure-preserving.

2 Related Work

Dynamics model learning is a very broad field and has been addressed by various communities for decades, e.g., [9,22]. Learning GP dynamics models can be addressed with a parameteric or non-parametric continuous-time GP model. Nonparametric continuous-time GP models were learned by applying sparse GPs [15,16]. A common approach for learning discrete-time models are GP state-space models [32,33]. In this work, we consider fully observable systems in contrast to common state-space models. However, we apply the tools of state-space model literature. In particular, we develop our ideas exemplary for the inference scheme proposed in [7]. At the same time, our contribution is not restricted to that choice of inference scheme and can be combined with other schemes as [19] as well. In constrast to [7], we sample consistent GPs from the posterior (cf. Sect. 4.)

Implicit transitions have become popular for NNs and provide useful tools that we leverage. Implicit layers in neural networks denote layers, in which the output is defined in terms of an optimization problem [10]. On a technical level, we implement related techniques based on the implicit function theorem and backpropagation. A NN with infinitely many layers was trained by implicitly defining the equilibrium of the network [3]. Look et al. [23] propose an efficient backpropagation scheme for implicitly defined neural network layers. We extend these approaches from deterministic NNs to probabilistic GP models.

Including Hamiltonian structure into learning in order to preserve physical properties of the system, is an important problem addressed by many sides. The problem can be tackled by approximating the continuous-time Hamiltonian system from data. This was addressed by applying a NN [11]. Since modern NN approaches provide a challenging benchmark, we compare our method against [11]. In [20], the Hamiltonian structure is learned by stacking multiple symplectic modules. GPs have been combined with symplectic structure as well [4,26]. In contrast to our approach, the focus lies on learning continuous-time dynamics with direct GP inference and afterwards unrolling the dynamics via certain symplectic or structure-preserving integrators. By learning and predicting the identical discrete-time system, we omit an additional numerical error and predictions are computational more efficient.

However, there is literature that addresses discrete-time Hamiltonian systems. The Hamiltonian neural network approach was extended by a recurrent NN coupled with a symplectic integrator [6]. In [27], the symplectic learning approach was extended to variational autoencoders, adding uncertainty to the initial value. Zhong et al. [35] extend previous approaches by adding control input. In contrast to previous approaches, we are able to apply implicit integrators. This allows us to address non-separable Hamiltonians and geometrical invariants. Further, our GP approach allows to sample trajectories from a structure-preserving distribution.

3 Technical Background and Main Idea

Next, we make our problem precise and provide a summary of the preliminaries.

3.1 Gaussian Process Regression

A GP is a distribution over functions [25]. Similar to a normal distribution, a GP is determined by its mean function $m(x)$ and covariance function $k(x, y)$. We assume the prior mean to be zero.

Standard GP Inference: For direct training, the GP predictive distribution is obtained by conditioning on n observed data points. In addition to optimizing the hyperparameters, a system of equations has to be solved, which has a complexity of $\mathcal{O}(n^3)$. Clearly, this is problematic for large datasets.

Variational Sparse GP: The GP can be sparsified by introducing pseudo inputs [31]. Intuitively, we approximate the posterior with a lower number of training points. An elegant approximation strategy is based on casting Bayesian inference as an optimization problem. We consider pseudo inputs $\xi = [\xi_1, \dots, \xi_P]$ and targets $z = [z_1, \dots, z_P]$ as proposed in [17] and applied in [7,19]. Intuitively, the targets can be interpreted as GP observations at ξ. The posterior of pseudo targets is approximated via a variational approximation $q(z) = \mathcal{N}(\mu, \Sigma)$, where μ and Σ are adapted during training. The GP posterior distribution at inputs x^* is conditioned on the pseudo inputs and targets resulting in a normal distribution $f(x^*|z,\xi) \sim \mathcal{N}(\mu(x^*), \Sigma(x^*))$ with

$$\mu(x^*) = k(x^\star, \xi)k(\xi,\xi)^{-1}z, \text{ and } \Sigma(x^\star) = k(x^\star, x^\star) - k(x^\star, \xi)k(\xi,\xi)^{-1}k(\xi, x^\star). \quad (4)$$

Decoupled Sampling: In contrast to standard (sparse) GP conditioning (4) this approach allows to sample functions from the posterior that can be evaluated at arbitrary locations [34]. Thus, iterative sampling at multiple inputs is achieved without conditioning on previous function evaluations. By applying Matheron's rule [18], the GP posterior is decomposed into two parts,

$$\begin{aligned} f(x^\star|z,\xi) &= \underbrace{f(x^\star)}_{\text{prior}} + \underbrace{k(x^\star,\xi)k(\xi,\xi)^{-1}(z - f_z)}_{\text{update}} \\ &\approx \sum_{i=1}^{S} w_i\phi_i(x^\star) + \sum_{j=1}^{P} v_j k(x^\star, \xi_j), \end{aligned} \quad (5)$$

where S Fourier bases ϕ_i and $w_i \sim \mathcal{N}(0,1)$ represent the stationary GP prior [24]. For the update in Eq. (5) it holds that $v = k(\xi,\xi)^{-1}(z - \Phi w)$ with feature matrix $\Phi = \phi(\xi) \in \mathbb{R}^{P \times S}$ and weight vector $w \in \mathbb{R}^S$. The targets z are sampled from the variational distribution $q(z)$ at the inputs ξ. We add technical details and details on the Fourier bases in our setting in the supplementary material.

3.2 Runge-Kutta Integrators

A RK integrator ψ_f for a continuous-time dynamical system f (1) is designed to approximate the solution $x(t_n)$ at discrete time steps t_n via $\bar{x}_n$. Hence,

$$\bar{x}_{n+1} = \psi_f(\bar{x}_n) = \bar{x}_n + h\sum_{j=1}^{s} b_j g_j,\ g_j = f(\bar{x}_n + h\sum_{l=1}^{s} a_{jl}g_l), j = 1, \ldots, s, \quad (6)$$

where g_j are the internal stages and $\bar{x}_0 = x(0)$. We use the notation $\bar{x}$ to indicate numerical error corrupted states and highlight the subtle difference to ground truth data. The parameters $a_{jl}, b_j \in \mathbb{R}$ determine the properties of the method, e.g., the stability radius of the method [14], the geometrical properties, or whether it is symplectic [12].

Implicit Integrators: If $a_{jl} > 0$ for $l \geq j$, Eq. (6) takes evaluations at time steps into account where the state is not yet known. Therefore, the solution of a nonlinear system of equations using a numerical solver is required. A prominent example is the implicit Euler scheme $\bar{x}_{n+1} = \bar{x}_n + hf(\bar{x}_{n+1})$.

3.3 Main Idea

We propose to embed RK methods (6) into GP regression. Since the underlying ground truth dynamics f (1) are given in continuous time, the discretization matters. Naive methods, such as the explicit Euler method, are known to be inconsistent with physical behavior. Therefore, we investigate how to learn more sophisticated models that, by design, are able to preserve physical structure of the original system. Further, we will develop this idea into a tractable inference scheme. In a nutshell, we learn GP dynamics $\hat{f}$ that yield predictions $x_{n+1} = \psi_{\hat{f}}(x_n)$. This enforces the RK (6) instead of explicit Euler (3) structure. Thus, leading to properties like volume preservation. The main technical difficulty lies in making the implicitly defined transition probability $p(\psi_{\hat{f}}(x_n)|x_n)$ tractable.

4 Embbedding Runge-Kutta Integrators in GP Models

Next, we dive into the technical details of merging GPs with RK integrators. We demonstrate how to make the distribution of any (implicit) RK method tractable.

4.1 Efficient Evaluation of the Transition Model

At its core, we consider the problem of evaluating implicitly defined distributions of RK integrators $\psi_{\hat{f}}$. To this end, we derive a sampling-based technique. Leveraging decoupled sampling (5) allows to sample a consistent dynamics function from the GP posterior distribution and thus, proper RK integrator steps. The procedure is illustrated in Fig. 1. We model the dynamics $\hat{f}$ via d variational sparse GPs. Let $z \in \mathrm{R}^{P \times d}$ be a sample from the variational posterior $q(z)$ (cf. Sect. 3.1). The probability of an integrator step $p(\psi_{\hat{f}}(x_n)|z, x_n)$ is formally obtained by integrating over all possible GP dynamics $\hat{f}|z$,

$$p(x_{n+1}|z, x_n) = p(\psi_{\hat{f}}(x_n)|z, x_n) = \int_{\hat{f}} p(\psi_{\hat{f}}(x_n)|\hat{f})p(\hat{f}|z)d\hat{f}. \tag{7}$$

Next, we show how to sample from the distribution $p(\psi_{\hat{f}}(x_n)|z, x_n)$ in Eq. (7). Performing an RK integrator step $\psi_{\hat{f}}(x_n)$ requires the computation of RK stages $g^\star = (g_1^\star, \ldots, g_s^\star)$ (6)

$$g^\star = \arg\min_{g} \|g - \hat{f}(x_n + hAg)\|^2, \tag{8}$$

with $A = (a_{jl})_{j=1,\ldots,s,l=1\ldots j}$ determined by the RK scheme (6). In the explicit case A is a sub-diagonal matrix so g_j can be calculated iteratively. In the implicit case, a minimization problem has to be solved numerically, which requires the iterative evaluation of multiple intermediate function evaluations. In both, the explicit and implicit case, inputs to the dynamics $\hat{f}$ depend on the output of previous dynamics function evaluations. Thus, all evaluations can not directly be drawn from their joint posterior GP distribution. In order to ensure that iterative evaluations of $\hat{f}$ indeed correspond to the identical GP posterior sample, conditioning on prior evaluations is necessary.

In prior work on state-space models, iterative samples at different time-steps were drawn independently, ignoring these correlations [7]. It was shown in [19] that this introduces a non-negligible error in the forward propagation of the

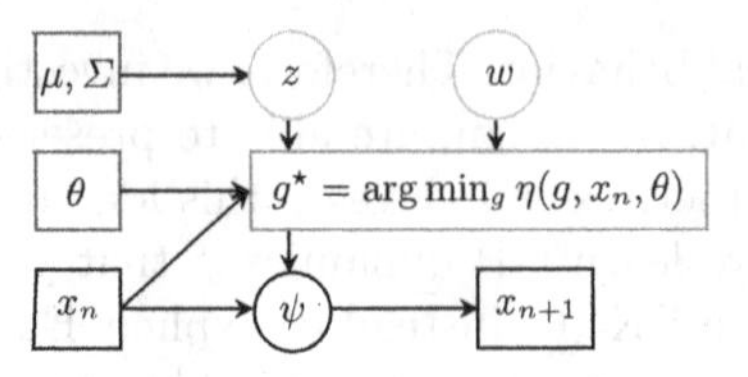

Fig. 1. The evaluation of an integrator step. First, weights $w \sim \mathcal{N}(0, 1)$ and inducing targets $z \sim \mathcal{N}(\mu, \Sigma)$ (green) are sampled. This yields tractable solutions to the minimization problem for the RK stages $g^\star$ (red), which yields RK steps $x_{n+1} = \psi(y_n)$. The trainable parameters are marked in blue.

probability distribution. In our case, the sampling scheme in [7] would lead to an even larger error since consistency would not be ensured along a single integration step. In the implicit case, this would result in a minimization problem that changes while it is solved numerically. However, naive GP conditioning on prior function evaluations is computationally intractable.

We address the problem by leveraging decoupled sampling [34]. This sampling approach allows to compute a consistent GP dynamics function $\hat{f}|z$ via Eq. (5) before applying the RK scheme. Technically, sampling the dynamics function $\hat{f}$ is achieved by sampling inducing targets z from $\mathcal{N}(\mu, \Sigma)$ and weights $w \sim \mathcal{N}(0, 1)$ (cf. Sect. 3.1). We are now able to evaluate the dynamics $\hat{f}$ at arbitrary locations. This allows to define and solve the system of equations (8) with respect to the fixed, but sampled, GP dynamics $\hat{f}$. Combining Eq. (8) and Eq. (5) with $u = u(g) = x_n + hAg$ yields

$$g^\star = \arg\min_g \left\| g - \sum_{i=1}^{S} w_i \phi_i(u) - \sum_{j=1}^{M} v_j k(u, \xi_j) \right\|^2. \tag{9}$$

Next, we give an example. Consider the IA-Radau method [14] $x_{n+1} = x_n + h\left(\frac{1}{4}g_1 + \frac{3}{4}g_2\right)$, with

$$g_1 = \hat{f}\left(x_n + \frac{h}{4}(g_1 - g_2)\right), \; g_2 = \hat{f}\left(x_n + h\left(\frac{1}{4}g_1 + \frac{5}{12}g_2\right)\right). \tag{10}$$

After sampling z and w, the RK scheme (10) is transformed into a minimization problem (9). With $u_1 = x_n + \frac{h}{4}(g_1 - g_2)$, and $u_2 = x_n + h(\frac{1}{4}g_1 + \frac{5}{12}g_2)$ it holds that $\begin{pmatrix} g_1^\star \\ g_2^\star \end{pmatrix} = \arg\min_g F(g)$, with

$$F(g) = \left\| \begin{matrix} g_1 - \sum_{i=1}^{S} w_i \phi_i(u_1) - \sum_{j=1}^{M} v_j k(u_1, \xi_j) \\ g_2 - \sum_{i=1}^{S} w_i \phi_i(u_2) - \sum_{j=1}^{M} v_j k(u_2, \xi_j) \end{matrix} \right\|^2. \tag{11}$$

4.2 Application to Model Learning via Variational Inference

Next, we construct a variational-inference model learning scheme that is based on the previously introduced numerical integrators. Here, we exemplary develop the integrators for an inference scheme similar to [7] and make the method precise. It is also possible to extend the arguments to other inference schemes such as [19]. In contrast to [7] we sample functions $\hat{f}$ instead of independent draws from the GP dynamics. This allows to produce trajectories that are generated by a fixed but probabilistic vector field. Unlike typical state-space models as [7,19] we do not consider transition noise. Thus, the proposed variational posterior is suitable [19]. Structure preservation is in general not possible when adding transition noise to each time step [1, §5].

Factorizing the joint distribution of noisy observations, noise-free states, inducing targets and GP posterior yields

$$p(\hat{x}_{1:N}, x_{1:N}, z, \hat{f}) = \prod_{n=0}^{N-1} p(\hat{x}_{n+1}|x_{n+1}) p(x_{n+1}|x_n, \hat{f}) p(\hat{f}|z) p(z). \tag{12}$$

The posterior distribution $p(x_{1:N}, z, \hat{f}|\hat{x}_{1:N})$ is factorized and approximated by a variational distribution $q(x_{1:N}, z, \hat{f})$. Here, the variational distribution q is chosen

$$q(x_{1:N}, z, \hat{f}) = \prod_{n=0}^{N-1} p(x_{n+1}|x_n, \hat{f}) p(\hat{f}|z) q(z), \tag{13}$$

with the variational distribution $q(z)$ of the inducing targets from Sect. 3.1. The model is adapted by maximizing the Evidence Lower Bound (ELBO)

$$\begin{aligned} \log p(\hat{x}_{1:N}) &\geq \mathbb{E}_{q(x_{1:N}, z, \hat{f})}\left[\log \frac{p(\hat{x}_{1:N}, x_{1:N}, z, \hat{f})}{q(x_{1:N}, z, \hat{f})}\right] \\ &= \sum_{n=1}^{N} \mathbb{E}_{q(x_{1:N}, z, \hat{f})}\left[\log p(\hat{x}_n|x_n)\right] - \mathrm{KL}(p(z)||q(z)) =: \mathcal{L}. \end{aligned} \tag{14}$$

Now, the model can be trained by maximizing the ELBO $\mathcal{L}$ (14) with a sampling-based stochastic gradient descent method that optimizes the sparse inputs and hyperparameters. The expectation $\mathbb{E}_{q(x_{1:N}, z, \hat{f})}\left[\log \frac{p(\hat{x}_{1:N}, x_{1:N}, z, \hat{f})}{q(x_{1:N}, z, \hat{f})}\right]$ is approximated by drawing samples from the variational distribution $q(x_{1:N}, z, \hat{f})$ and evaluating $p(\hat{x}_n|x_n)$ at these samples. Samples from q are drawn by first sampling pseudo targets z and a dynamics function $\hat{f}$ from the GP posterior (5). Trajectories are produced by succesively computing consistent integrator steps $x_{n+1} = \psi_{\hat{f}}(x_n)$ as described in Sect. 4.1. This yields a recurrent learning scheme, by iterating over multiple integration steps. We are able to use our model for predictions by sampling functions from the trained posterior.

4.3 Gradients

The ELBO (14) is minimized by applying stochastic gradient descent to the hyperparameters. When conditioning on the sparse GP (4), the hyperparameters include $\theta = (\mu_{1:d}, \Sigma_{1:d}, \theta_{1:d}^{GP})$ with variational sparse GP parameters $\mu_{1:d}$, $\Sigma_{1:d}$ and GP hyperparameters $\theta_{1:d}^{GP}$. The gradient $\frac{dx_{n+1}}{d\theta}$ depends on $\frac{dg^\star}{d\theta}$ and $\frac{dx_n}{d\theta}$ via the integrator (6). It holds that

$$\frac{dg^\star}{d\theta} = \frac{\partial g^\star}{\partial \theta} + \frac{dg^\star}{dx_n}\frac{dx_n}{d\theta}. \tag{15}$$

By the dependence of x_{n+1} on $g^\star$ and of $g^\star$ on x_n (15), the gradient is back-propagated through time. For an explicit integrator, the gradient $\frac{dg^\star}{d\theta}$ can be computed explicitly, since $g_j^\star$ depends on $g_i^\star$ with $i < j$. For implicit solvers, the implicit functions theorem [21] is applied. It holds that $g^\star = \arg\min_g \eta(g, x_n, \theta)$ with the minimization problem η derived in Eq. (9). For the gradients of $g^\star$ with respect to x_n, respectively θ, it holds with the implicit function theorem [21]

$$\frac{dg^\star}{dx_n} = \left(\frac{\partial^2 \eta}{\partial g^{\star 2}}\right)^{-1}\left(\frac{\partial^2 \eta}{\partial x_n \partial g^\star}\right). \tag{16}$$

5 Application to Symplectic Integrators

In summary, we have first derived how to evaluate the implicitly defined RK distributions. Afterward, we have embedded this technique into a recurrent learning scheme and finally, shown how it is trained. Next, we make the method precise for symplectic integrators and Hamiltonian systems.

5.1 Hamiltonian Systems and Symplectic Integrators

An autonomous Hamiltonian system is given by

$$x(t) = \begin{pmatrix} p(t) \\ q(t) \end{pmatrix} \text{ with } \dot{x}(t) = \begin{pmatrix} \dot{p}(t) \\ \dot{q}(t) \end{pmatrix} = \begin{pmatrix} -H_q(p,q) \\ H_p(p,q) \end{pmatrix} \tag{17}$$

and $p, q \in \mathbb{R}^d$. Here, H_p and H_q denote the partial derivatives of H with respect to p and q. In many applications, q corresponds to the state and p to the velocity. The Hamiltonian H often resembles the total energy and is constant along trajectories. The flow of Hamiltonian systems ψ_t is volume preserving in the sense of $\text{vol}(\psi_t(\Omega)) = \text{vol}(\Omega)$ for each bounded open set Ω. The flow ψ_t describes the solution at time point t for the initial values $x_0 \in \Omega$.

Symplectic integrators are volume preserving for Hamiltonian systems (17) [12]. Thus, $\text{vol}(\Omega) = \text{vol}(\psi_f(\Omega))$ for each bounded Ω. Further, they provide a more accurate approximation of the total energy than standard integrators [12]. When designing the GP, it is critical to respect the Hamiltonian structure (17). Additionally, the symplectic integrator ensures that the volume is preserved.

5.2 Explicit Symplectic Integrators

A broad class of real world systems can be modeled by separable Hamiltonians $H(p,q) = T(p) + V(q)$. For example, ideal pendulums and the two body problem. Then, for the dynamical system it holds that

$$\dot{p}(t) = -V'(q),\, \dot{q}(t) = T'(p), \tag{18}$$

with $V : \mathbb{R}^d \to \mathbb{R}^d$ and $T : \mathbb{R}^d \to \mathbb{R}^d$. For this class of problems, explicit symplectic integrators can be constructed. In order to ensure Hamiltonian structure, $V_1'(q), \dots, V_d'(q)$ and $T_1'(p), \dots, T_d'(p)$ are modeled with independent sparse GPs. Symplecticity is enforced via discretizing with a symplectic integrator.

Consider for example the explicit symplectic Euler method

$$p_{n+1} = p_n - hV'(q_n),\, q_{n+1} = q_n + hT'(p_{n+1}). \tag{19}$$

The symplectic Euler method (19) is a partitioned RK method, meaning that different schemes are applied to different dimensions. Here, the explicit Euler method is applied to p_n and the implicit Euler method to q_n. The integrator (19) is embedded into the sampling scheme (cf. Sect. 4.1) by sampling from V' and T' and the scheme can readily be embedded into the inference scheme (cf. Sect. 4.2).

5.3 General Symplectic Integrators

The general Hamiltonian system (17) requires the application of an implicit symplectic integrator. An example for a symplectic integrator is the midpoint rule applied to (17)

$$x_{n+1} = x_n + hJ^{-1}\nabla H\left(\frac{x_n + x_{n+1}}{2}\right), \text{ with } J^{-1} = \begin{pmatrix} 0 & -1 \\ 1 & 0 \end{pmatrix}. \tag{20}$$

Again, it is critical to embed the Hamiltonian structure into the dynamics model by modeling H with a sparse GP. Sampling from (20) requires evaluating the gradient ∇H, which is again a GP [25].

6 Experiments

In this section, we validate our method numerically. In particular, we i) demonstrate volume-preserving predictions for Hamiltonian systems and the satisfaction of a quadratic geometric constraint for a mechanical system; ii) show that we achieve higher or equal accuracy as state-of-the-art methods; and iii) illustrate that our method can easily deal with different choices of RK integrators.

6.1 Methods

We construct our structure-preserving GP model (SGPD) by tailoring the RK integrator to the underlying problem. We compare with the following methods:

Hamiltonian Neural Network (HNN). [11]: Deep learning approach that is tailored to respect Hamiltonian structure.

Consistent PR-SSM Model (Euler). [7]: Standard variational GP model that corresponds to explicit Euler discretizations. Therefore, we refer to it in the following as Euler. In general all common GP state-space models correspond to the Euler discretization. Here, we use a model similar to [7], but in contrast to [7] we compute consistent predictions via decoupled sampling as discussed in Sect. 4. The general framework is more flexible and can also cope with lower-dimensional state observations. Here, we consider the special case, where we assume noisy state measurements.

6.2 Learning Task and Comparison

In the following, we describe the common setup of all experiments. For each Hamiltonian system, we consider at least one period of a trajectory as training data and all methods are provided with identical training data. For all experiments, we choose the ARD kernel [25]. We apply the training procedure described in Sect. 4.2 on subtrajectories and perform predictions via sampling of trajectories. Since we observed too much influence of the KL-divergence on the ELBO, we include a scaling factor inspired by [2]. In order to draw a fair comparison, we choose similar hyperparameters and number of inducing inputs for our SGPD method and the standard Euler discretization. Details are moved to the appendix. In contrast to our method, the HNN requires additional derivative information, either analytical or as finite differences. Here, we assume that analytical derivative information is not available and thus compute finite differences.

We consider a twofold goal: invariant preservation and accurate predictions in the L^2 sense. Predictions are performed by unrolling the first training point over multiple periods of the system trajectory. The L^2-error is computed via averaging 5 independent samples from the GP posterior $\hat{X}_i = \frac{1}{5}\sum_{j=1}^{5} \hat{X}_i^j$ and computing $\sqrt{\frac{1}{N}\sum_{i=1}^{N} \|X_i - \hat{X}_i\|^2}$ with ground truth $X_{1:N}$. Integrators are volume-preserving if and only if they are divergence free, which requires $\det(\psi') = 1$ [12]. Thus, we evaluate $\det(\psi')$ for the rollouts, which is intractable for the Hamiltonian neural networks. We observed that we could achieve similar results by propagating the GP mean in terms of constraint satisfaction and L^2-error. Here, we focus on trajectory samples in order to highlight that our approach allows to sample structure-preserving rollouts from a structure-preserving GP distribution.

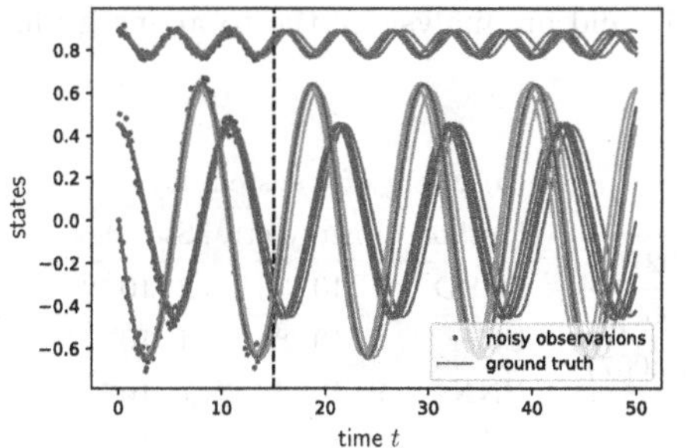

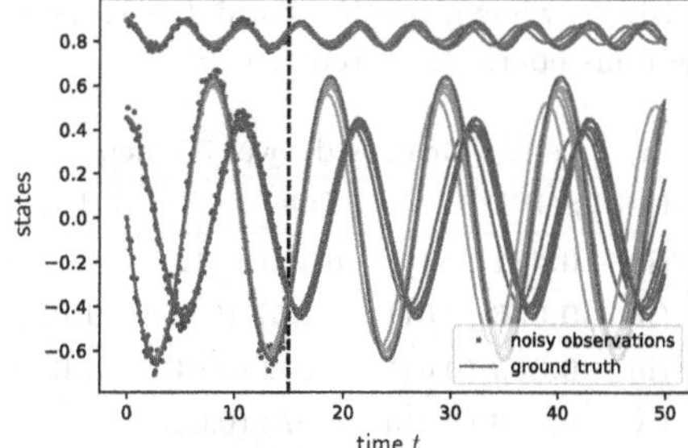

Fig. 2. State trajectory samples for the rigid body dynamics with SGPD (left) and Euler (right). Shown are 5 rollouts from the GP posterior. Both systems are illustrated as a function over time. The training horizon is marked with dotted lines. SGPD visibly enforces structure in contrast to Euler.

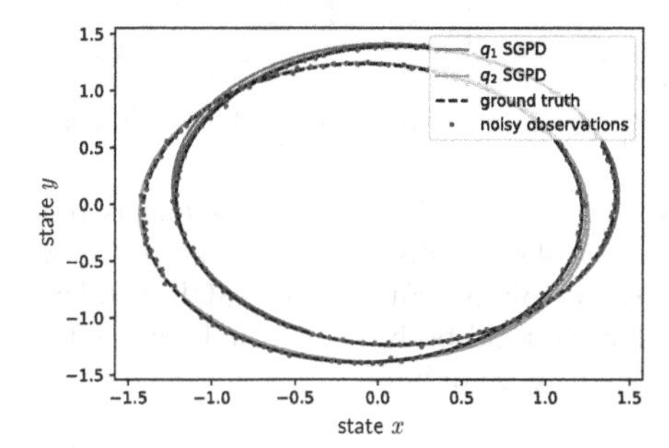

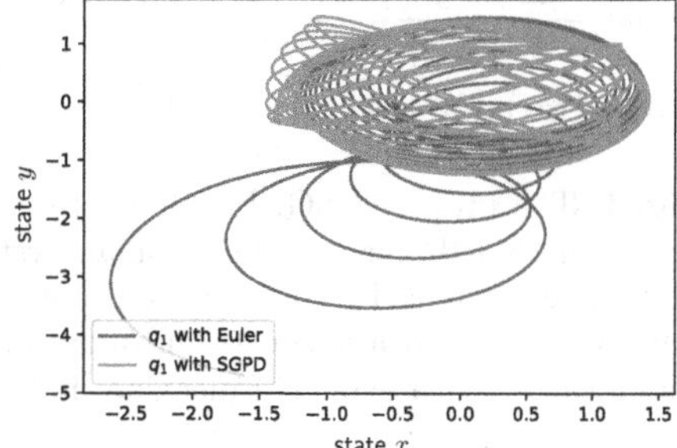

Fig. 3. State trajectory samples for the two-body problem with SGPD (left) and long-term behavior of SGPD and Euler (right). The two-body problem is represented as a phase plot in the two-dimensional space. Shown are single rollouts. The divergent behavior of Euler is clearly visible (right).

6.3 Systems and Integrators

We consider four different systems here i) ideal pendulum; ii) two-body problem; iii) non-separable Hamiltonian; and iv) rigid body dynamics.

Separable Hamiltonians: the systems i) and ii) are both separable Hamiltonians that are also considered as baseline problems in [11]. Due to their structure, we can apply the symplectic Euler method (cf. Sect. 5.2) to both problems.

The Hamiltonian of a pendulum is given by $H(p,q) = (1 - 6\cos(p)) + \frac{p^2}{2}$. Training data is generated from a 10-second ground truth trajectory with discretization $dt = 0.1$ and disturbed with observation noise with variance $\sigma^2 = 0.1$. Predictions are performed on a 40-second interval.

The two body problem models the interaction of two unit-mass particles (p_1, q_1) and (p_2, q_2), where $p_1, p_2, q_1, q_2 \in \mathbb{R}^2$ and $H(p,q) = \frac{1}{2} + \|p_1\|^2 + \|p_2\|^2 + \frac{1}{\|q_1 - q_2\|^2}$. Noisy training data is generated on an interval of 18.75 s, discretization

Table 1. Shown are the total L^2-errors in 1a and an analysis of the total energy for the non-separable system 1b.

(a) L^2-err. (Mean (std) over 5 indep. runs)

task	SGPD	Euler	HNN
(i)	**0.421** (0.1)	0.459 (0.12)	4.69 (0.02)
(ii)	**0.056** (0.01)	**0.057** (0.009)	0.12 (0.009)
(iii)	**0.033** (0.01)	0.062 (0.04)	**0.035** (0.007)
(iv)	**0.046** (0.014)	0.073 (0.02)	–

(b) Energy for system iii)

Method	Energy err.	Std. dev.
SGPD	$9 \cdot 10^{-4}$	$2 \cdot 10^{-3}$
Euler	$2 \cdot 10^{-3}$	$4 \cdot 10^{-3}$
HNN	$9 \cdot 10^{-3}$	$7 \cdot 10^{-5}$

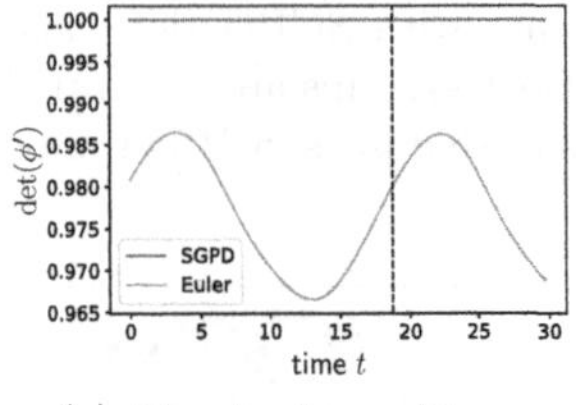

(a) Two-body problem

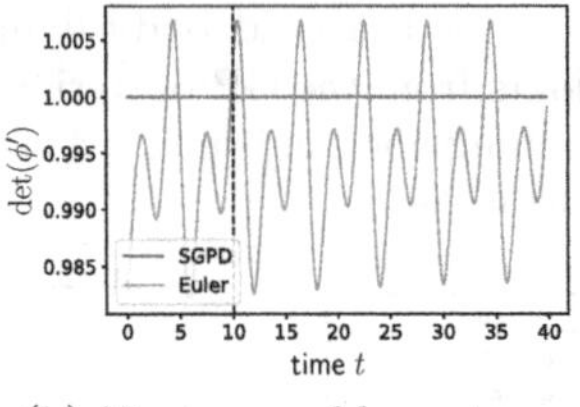

(b) Non-separable system

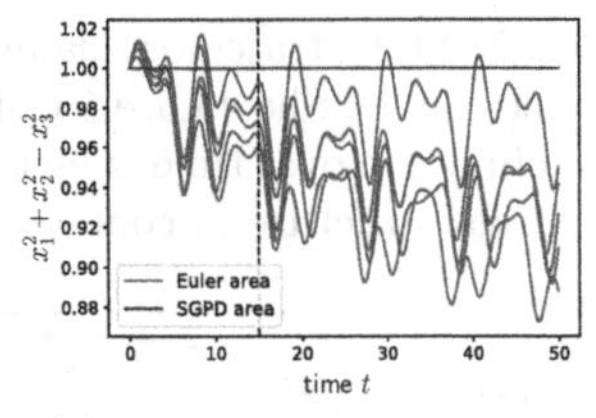

(c) Rigid body dynamics

Fig. 4. The proposed SGPD (blue) preserves structure: The analytic volume is preserved via SGPD for the Hamiltonian systems in 4a and 4b, while simulations show that the explicit Euler (orange) does not preserve volume. Further, the SGPD rollout preserves the quadratic constraint over time for the rigid body system in 4c, while the standard GP with explicit Euler does not. (Color figure online)

level $dt = 0.15$, and variance $\sigma^2 = 1 \cdot 10^{-3}$. Predictions are performed on an interval of 30 s. The orbits of the two bodies q_1 and q_2, predicted by our SGPD method, are shown in Fig. 3 (left) for a single sample from posterior.

Non-Separable Hamiltonian: As an example for a non-separable Hamiltonian system we consider Eq. (17) with $H(p, q) = \frac{1}{2}\left[(q^2 + 1)(p^2 + 1)\right]$ [30]. The implicit midpoint rule (20) is applied as the numerical integrator (cf. Sect. 5.3). The training trajectory is generated on a 10-seconds interval with discretization $dt = 0.1$ and disturbed with noise with variance $\sigma^2 = 5 \cdot 10^{-4}$. Rollouts are performed on an interval of 40 s.

Rigid Body Dynamics: Consider the rigid body dynamics [12]

$$\begin{pmatrix} \dot{x}_1 \\ \dot{x}_2 \\ \dot{x}_3 \end{pmatrix} = \begin{pmatrix} 0 & \frac{3}{2}x_3 & -x_2 \\ -\frac{3}{2}x_3 & 0 & \frac{x_1}{2} \\ x_2 & -\frac{x_1}{2} & 0 \end{pmatrix} \begin{pmatrix} x_1 \\ x_2 \\ x_3 \end{pmatrix} =: f(x) \tag{21}$$

that describe the angular momentum of a body rotating around an axis. The equations of motion can be derived via a constrained Hamiltonian system. We apply the implicit midpoint method. Since the HNN is designed for non-constrained Hamiltonians it requires pairs of p and q and is, thus, not applicable. Training data is generated on a 15-seconds interval with discretization $dt = 0.1$. Due to different scales, x_1 and x_2 are disturbed with noise with variance $\sigma^2 = 1 \cdot 10^{-3}$, and x_3 is disturbed with noise with variance $\sigma^2 = 1 \cdot 10^{-4}$. Predictions are performed on an interval of 50 s (see Fig. 2 (middle) for SGPD and Fig. 2 (right) for Euler). The rigid body dynamics preserve the invariant $x_1^2 + x_2^2 + x_3^2 = 1$, which refers to the ellipsoid determined by the axis of the rotating body. We include this property as prior knowledge in our SGPD model via $x^T \hat{f}(x) = 1$ [12]. The dynamics $\hat{f}$ is again trained with independent sparse GPs, where the third dimension is obtained by solving $\hat{f}(x) = 1 - \frac{\hat{f}_1 x_1 + \hat{f}_2 x_2}{x_3}$.

6.4 Results

For systems (i),(ii), and (iii), we demonstrate volume preservation. Figure 4 depicts that volume is preserved for the symplectic integrator-based SGPD in contrast to the standard explicit Euler method. Shown are the results for samples from the GP posterior in order to highlight the properties of the structure-preserving distribution. However, volume preservation applies to mean predictions as well. For the rigid body dynamics, we consider the invariant $x_1^2 + x_2^2 + x_3^2 = 1$. Figure 4c demonstrates that the implicit midpoint is able to approximately preserve the invariant along 5 samples from the GP posterior. In contrast, the explicit Euler fails on all samples. The results demonstrate that even though the explicit Euler method achieves comparable results in terms of accuracy, it is not able to preserve structure. In summary, our method either shows the smallest L^2-error (see Table 1a) or achieves state-of-the-art accuracy. For the two-body problem, we demonstrate that the Euler long-term predictions are less stable than long-term predictions with SGPD. To this end we compute rollouts with 2500 points in Fig. 3 with SGPD (left) and Euler (right).

The midpoint method-based SGPD furthermore shows accurate approximation of the constant total energy for the systems (iii) and (iv). For system (iii), the total energy corresponds to the Hamiltonian H. Inspired by [11], we average the approximated energy along 5 independent trajectories $H_n = \sum_{i=1}^{5} \frac{H_n^i}{5}$ and compute the average total energy $\hat{H} = \frac{1}{n}\sum_n H_n$. Afterwards, we evaluate the error $\|H - \hat{H}\|$ and the deviation $\sqrt{\sum_n \frac{|H_n - H|^2}{n-1}}$ (see Table 1b). Our SGPD method yields the best approximation to the ground truth energy. In the appendix, we provide additional information for multiple runs of the experiment. For the rigid body dynamics our method yields accurate approximation of the energy as well. Details and an evaluation of higher-order methods is moved to the appendix.

7 Conclusion and Future Work

In this paper we combine numerical integrators with GP regression. Thus, resulting in an inference scheme that preserves physical properties and yields high accuracy. On a technical level, we derive a method that samples from implicitly defined distributions. By the means of empiricial comparison, we show the advantages over Euler-based state-of-the-art methods that are not able to preserve physical structure. An important extension that we want to address in the future are partial observations and control input.

Acknowledgements. The authors thank Barbara Rakitsch, Alexander von Rohr and Mona Buisson-Fenet for helpful discussions.

References

1. Abdulle, A., Garegnani, G.: Random time step probabilistic methods for uncertainty quantification in chaotic and geometric numerical integration. Stat. Comput. **30**(4), 907–932 (2020)
2. Alemi, A., Poole, B., Fischer, I., Dillon, J., Saurous, R.A., Murphy, K.: Fixing a broken ELBO. In: Proceedings of the 35th International Conference on Machine Learning, vol. 80, pp. 159–168, PMLR (2018)
3. Bai, S., Kolter, J.Z., Koltun, V.: Deep equilibrium models. In: Advances in Neural Information Processing Systems vol. 32, pp. 690–701 (2019)
4. Brüdigam, J., Schuck, M., Capone, A., Sosnowski, S., Hirche, S.: Structure-preserving learning using Gaussian processes and variational integrators. In: Proceedings of the 4th Conference on Learning for Dynamics and Control, PMLR (2022)
5. Buisson-Fenet, M., Solowjow, F., Trimpe, S.: Actively learning Gaussian process dynamics. In: Proceedings of the 2nd Conference on Learning for Dynamics and Control, PMLR (2020)
6. Chen, Z., Zhang, J., Arjovsky, M., Bottou, L.: Symplectic recurrent neural networks. In: 8th International Conference on Learning Representations, ICLR 2020 (2020)
7. Doerr, A., Daniel, C., Schiegg, M., Nguyen-Tuong, D., Schaal, S., Toussaint, M., Trimpe, S.: Probabilistic recurrent state-space models. In: Proceedings of the International Conference on Machine Learning (ICML) (2018)
8. Frigola, R., Chen, Y., Rasmussen, C.: Variational Gaussian process state-space models. In: Advances in Neural Information Processing Systems, vol. 27 (NIPS 2014) pp. 3680–3688 (2014)
9. Geist, A., Trimpe, S.: Learning constrained dynamics with Gauss principle adhering Gaussian processes. In: Proceedings of the 2nd Conference on Learning for Dynamics and Control, pp. 225–234, PMLR (2020)
10. Gould, S., Hartley, R., Campbell, D.: Deep declarative networks: A new hope. arXiv:1909.04866 (2019)
11. Greydanus, S., Dzamba, M., Yosinski, J.: Hamiltonian neural networks. In: Advances in Neural Information Processing Systems, vol. 32, pp. 15379–15389 (2019)
12. Hairer, E., Lubich, C., Wanner, G.: Geometric numerical integration: structure-preserving algorithms for ordinary differential equations. Springer. https://doi.org/10.1007/3-540-30666-8 (2006)
13. Hairer, E., Nørsett, S., Wanner, G.: Solving Ordinary Differential Equations I - Nonstiff Problems. Springer (1987). https://doi.org/10.1007/978-3-540-78862-1

14. Hairer, E., Wanner, G.: Solving Ordinary Differential Equations II - Stiff and Differential-Algebraic Problems. Springer (1996). https://doi.org/10.1007/978-3-642-05221-7
15. Hegde, P., Çaatay Yld, Lähdesmäki, H., Kaski, S., Heinonen, M.: Variational multiple shooting for Bayesian ODEs with Gaussian processes. In: Proceedings of the 38th Uncertainty in Artificial Intelligence Conference, PMLR (2022)
16. Heinonen, M., Yildiz, C., Mannerström, H., Intosalmi, J., Lähdesmäki, H.: Learning unknown ODE models with Gaussian processes. In: Proceedings of the 35th International Conference on Machine Learning (2018)
17. Hensman, J., Fusi, N., Lawrence, N.: Gaussian processes for big data. Uncertainty in Artificial Intelligence. In: Proceedings of the 29th Conference, UAI 2013 (2013)
18. Howarth, R.J.: Mining geostatistics. London & New york (academic press), 1978. Mineralogical Mag. **43**, 1–4 (1979)
19. Ialongo, A.D., Van Der Wilk, M., Hensman, J., Rasmussen, C.E.: Overcoming mean-field approximations in recurrent Gaussian process models. In: Proceedings of the 36th International Conference on Machine Learning (ICML) (2019)
20. Jin, P., Zhang, Z., Zhu, A., Tang, Y., Karniadakis, G.E.: Sympnets: Intrinsic structure-preserving symplectic networks for identifying Hamiltonian systems. Neural Netw. **132**(C), 166–179 (2020)
21. Krantz, S., Parks, H.: The implicit function theorem. History, theory, and applications. Reprint of the 2003 hardback edition (2013)
22. Ljung, L.: System identification. Wiley encyclopedia of electrical and electronics engineering, pp. 1–19 (1999)
23. Look, A., Doneva, S., Kandemir, M., Gemulla, R., Peters, J.: Differentiable implicit layers. In: Workshop on machine learning for engineering modeling, simulation and design at NeurIPS 2020 (2020)
24. Rahimi, A., Recht, B.: Random features for large-scale kernel machines. In: Advances in Neural Information Processing Systems, vol. 20 (2008)
25. Rasmussen, C.E., Williams, C.K.I.: Gaussian Processes for Machine Learning (Adaptive Computation and Machine Learning). The MIT Press (2005)
26. Rath, K., Albert, C.G., Bischl, B., von Toussaint, U.: Symplectic Gaussian process regression of maps in Hamiltonian systems. Chaos **31**, 5 (2021)
27. Saemundsson, S., Terenin, A., Hofmann, K., Deisenroth, M.P.: Variational integrator networks for physically structured embeddings. In: Proceedings of the 23rd International Conference on Artificial Intelligence and Statistics (AISTATS), vol. 108 (2020)
28. Sakurai, J.J.: Modern quantum mechanics; rev. ed. Addison-Wesley (1994)
29. Salmon, R.: Hamiltonian fluid mechanics. Annu. Rev. Fluid Mech. **20**, 225–256 (2003)
30. Tao, M.: Explicit symplectic approximation of nonseparable Hamiltonians: Algorithm and long time performance. Phys. Rev. **94**(4), 043303 (2016)
31. Titsias, M.: Variational learning of inducing variables in sparse Gaussian processes. J. Mach. Learn. Res. Proc. Track, pp. 567–574 (2009)
32. Turner, R., Deisenroth, M., Rasmussen, C.: State-space inference and learning with Gaussian processes. In: Proceedings of the Thirteenth International Conference on Artificial Intelligence and Statistics, Proceedings of Machine Learning Research, vol. 9, pp. 868–875, PMLR (2010)
33. Wang, J., Fleet, D., Hertzmann, A.: Gaussian process dynamical models for human motion. IEEE Trans. Pattern Anal. Mach. Intell. **30**, 283–98 (2008)

34. Wilson, J., Borovitskiy, V., Terenin, A., Mostowsky, P., Deisenroth, M.: Efficiently sampling functions from Gaussian process posteriors. In: Proceedings of the 37th International Conference on Machine Learning, vol. 119, pp. 10292–10302 (2020)
35. Zhong, Y.D., Dey, B., Chakraborty, A.: Symplectic ODE-net: Learning Hamiltonian dynamics with control. In: 8th International Conference on Learning Representations, ICLR 2020 (2020)

Summarizing Data Structures with Gaussian Process and Robust Neighborhood Preservation

Koshi Watanabe[1], Keisuke Maeda[2], Takahiro Ogawa[2], and Miki Haseyama[2(✉)]

[1] Graduate School of Information Science and Technology, Hokkaido University, Sapporo, Japan
koshi@lmd.ist.hokudai.ac.jp

[2] Faculty of Information Science and Technology, Hokkaido University, Sapporo, Japan
{maeda,ogawa,mhaseyama}@lmd.ist.hokudai.ac.jp

Abstract. Latent variable models summarize high-dimensional data while preserving its many complex properties. This paper proposes a locality-aware and low-rank approximated Gaussian process latent variable model (LolaGP) that can preserve the global relationship and local geometry in the derivation of the latent variables. We realize the global relationship by imitating the sample similarity non-linearly and the local geometry based on our newly constructed neighborhood graph. Formally, we derive LolaGP from GP-LVM and implement a locality-aware regularization to reflect its adjacency relationship. The neighborhood graph is constructed based on the latent variables, making the local preservation more resistant to noise disruption and the curse of dimensionality than the previous methods that directly construct it from the high-dimensional data. Furthermore, we introduce a new lower bound of a log-posterior distribution based on low-rank matrix approximation, which allows LolaGP to handle larger datasets than the conventional GP-LVM extensions. Our contribution is to preserve both the global and local structures in the derivation of the latent variables using the robust neighborhood graph and introduce the scalable lower bound of the log-posterior distribution. We conducted an experimental analysis using synthetic as well as images with and without highly noise disrupted datasets. From both qualitative and quantitative standpoint, our method produced successful results in all experimental settings.

Keywords: Latent variable model · Gaussian processes · Neighborhood graph · Diffusion map

1 Introduction

Real-world data exist in a high-dimensional data space while including a low-dimensional manifold. This statement is consistent with the manifold hypothesis [5], and determining the meaningful structure is a general task in machine learning. *Dimensionality reduction* [42] is one of the basic approaches to explore the

© The Author(s), under exclusive license to Springer Nature Switzerland AG 2023
M.-R. Amini et al. (Eds.): ECML PKDD 2022, LNAI 13717, pp. 157–173, 2023.
https://doi.org/10.1007/978-3-031-26419-1_10

meaningful structure and estimate the low dimensional manifold, which preserves several properties of the original high dimensional data. The obtained low dimensional representation, particularly two or three-dimensional representation, is frequently useful in interpreting and visualizing complex high dimensional data.

There are numerous approaches to dimensionality reduction. Principal component analysis (PCA) [19,35] is one of the representative approaches to deriving a linear mapping into a low-dimensional subspace. Specifically, PCA selects the dominant components of a sample covariance matrix and treats them as new axes of the low-dimensional subspace. Based on this mapping, we can obtain the low-dimensional representation while preserving the global relationship of the high-dimensional data. However, in general, since this mapping-based approach does not take into account the desired subspace, we cannot obtain an optimal representation for its dimensionality. Probabilistic PCA (PPCA) [37] approaches this problem by explicitly assuming the low-dimensional representation as *latent variables* and learning them linearly by imitating the sample similarity. Gaussian process latent variable model (GP-LVM) [22] incorporates the kernel method into PPCA and non-linearly imitates the similarity. Those conventional latent variable models, on the other hand, still have some limitations, such as the *explainability* [6,18] of the latent variables, the *interpretability* [2,25,33] of the locality within the high-dimensional data, and the *scalability* [27,34] to the sample size.

Previous latent variable models can overcome the limitations of these methods. β-variational autoencoder (β-VAE) [6,18,20] tackles the explainability limitation of the latent variables and extracts them as a factor of variation. Specifically, β-VAE attempts to map one latent feature to one variation factor hidden in the entire dataset (e.g., rotation or shrinking scale). By the *disentangled* representation, β-VAE can generate artificial data where only a single factor has changed [48]. While β-VAE is useful for global analysis of high-dimensional data, it sacrifices *local geometry* within the high-dimensional data. Locally linear embedding (LLE) [7,33] is one of the representative approaches for incorporating local geometry into latent variables. LLE employs a two-step learning process, constructing a neighborhood graph and embedding it into the latent space. LLE can compress high-dimensional data while maintaining high interpretability of their local geometry by embedding local information based on the neighborhood graph. Uniform manifold approximation and projection (UMAP) [27] is a state-of-the-art method in this graph embedding method that addresses the scalability issue. UMAP can reduce the dimensionality of large-scale data reflecting their local manifold (e.g., class separation) and has received a lot of attention for real-data analysis due to its high scalability and visibility, such as genetic analysis [3], human population analysis [12], and social network analysis [31]. These embedding methods, however, have a limitation in their graph construction. They build the neighborhood graph directly from the high-dimensional data, which may result in an undesirable graph construction due to noise disruption [7] or the curse of dimensionality [41].

In this paper, we learn the local geometry under noise disruption and the curse of dimensionality while considering the global relationship via *locality-aware and low-rank approximated GP-LVM (LolaGP)*. LolaGP constructs the neighborhood graph with the low-dimensional latent variables using an iterative learning strategy. Then, LolaGP introduces a new locality-aware regularization into GP-LVM, forcing the latent variables to move closer if they are adjacent on the neighborhood graph. As a result, we can preserve the local geometry more precisely despite noise disruption and the curse of dimensionality. Furthermore, the latent representation in LolaGP holds the property of GP-LVM, which implies LolaGP can compress the high-dimensional data while considering both the global and local structures.

However, Gaussian process-based methods require a matrix inversion for each training step, and their scalability for even thousands of data points is a concern [17,32]. Bayesian GP-LVM [11,39] solves the scalability problem and produces latent variables in a fully Bayesian manner. While Bayesian GP-LVM improves scalability more than previous methods, its scalability and even its closed-form expression collapse with a complex regularization [11]. For this reason, we newly derive a lower bound of a log-posterior distribution based on a combination of the previous research [21,38], and this bound allows for scalable optimization using complex regularization. In summary, LolaGP has the following contributions:

- We construct the neighborhood graph with the low-dimensional latent variables, making the graph construction more robust to the noise disruption and the curse of dimensionality. Furthermore, We can also reflect the global relationship and local geometry into the latent variables by using the robust adjacency relation with GP-LVM.
- We introduce the lower bound of the log-posterior distribution. This bound enables scalable optimization while considering the complex regularization, which is difficult with the previous GP-LVM extensions.

We conducted an experimental analysis in which we visualized and quantified the derived latent variables using one synthetic dataset and two image datasets with and without noise disruptions. In this experiment, we demonstrated that our proposed method could embed high-dimensional data with high interpretability of their local geometry while maintaining the global relationship.

2 Background

In this section, we introduce the previous related methods to clarify the positioning of our proposed method. In Sect. 2.1, we briefly describe the graph embedding methods and discuss their limitations of the learning procedure with a synthetic dataset. In Sect. 2.2, we explain GP-LVM and its extensions and discuss why they cannot handle thousands of data points.

2.1 Graph Embedding Methods

LLE is a representative approach to reducing dimensionality while preserving local geometry, and it is closely related to Manifold Learning or Topological Data Analysis [46]. LLE and its related methods [2,4,15,27] usually construct

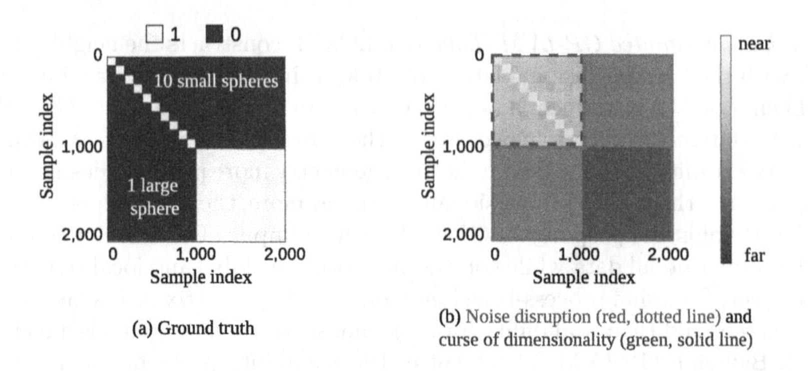

(a) Ground truth

(b) Noise disruption (red, dotted line) and curse of dimensionality (green, solid line)

Fig. 1. An example of a neighborhood graph with a synthetic dataset, Spheres. (a) Ground truth that completely separates each sphere in 101-dimensional space ('1' means an edge exists, and '0' means an edge does not), and (b) an example of a weighted adjacency matrix computed by the Square Exponential kernel.

a neighborhood graph of high-dimensional data to preserve the local geometry while calculating the low dimensional representation based on its adjacency relation. In the language of *topology*, shapes of a point set determined by a graph are *Vietoris-Rips complexes*, and they can recover various manifolds on which the point set is actually distributed [1,14]. The Vietoris-Rips complexes' properties are supported by a solid theoretical foundation, which motivates the effectiveness of the graph-based approaches. Although they produce successful results even in the application settings [47], they have several concerns during the graph construction phase. We demonstrate them using the Spheres dataset, [28] (Fig. 1), which contains ten small spheres and one large sphere surrounding them, and the spheres exist in the 101-dimensional space. Ten small spheres have 100 data points with white Gaussian noise, and one large sphere has as many data points as all small spheres (i.e., 1,000 data points). Under this condition, the most desirable adjacency matrix should separate each sphere, as shown in Fig. 1 (a). However, in Fig. 1 (b), the non-dialog blocks of the small spheres' entries have small values, and the separation of the small spheres is disturbed by the additive Gaussian noise. Furthermore, the distance between the small spheres' points and the large sphere's points is smaller than the distance within the large sphere's points. This implies that we cannot realize the large sphere from the adjacency matrix because of the curse of dimensionality. LolaGP solves these problems by simultaneously learning the low-dimensional latent variables, and the neighborhood graph is constructed based on the low-dimensional representation. This learning strategy can precisely calculate the local geometry-aware representation under the noise disruption and dimensionality problem.

2.2 Gaussian Process Latent Variable Model (GP-LVM)

Let $\mathbf{Y} = [\mathbf{y}_1, \mathbf{y}_2, \ldots, \mathbf{y}_N]^\top \in \mathbb{R}^{N \times D}$ be D dimensional (i.e., high-dimensional) data containing N data points. GP-LVM aims to compress these observed variables into low-dimensional latent variables $\mathbf{X} = [\mathbf{x}_1, \mathbf{x}_2, \ldots, \mathbf{x}_N]^\top \in \mathbb{R}^{N \times Q}$ ($D \gg Q$) and assumes a generative process from the latent space to the observed space as

$$\mathbf{y}_{:,d} = \mathbf{f}(\mathbf{X}) + \boldsymbol{\epsilon},\ \mathbf{f}(\mathbf{X}) \sim \mathcal{N}(\mathbf{0}, \mathbf{K}_{NN}^{(\mathbf{f})}),\ \boldsymbol{\epsilon} \sim \mathcal{N}(\mathbf{0}, \beta^{-1}\mathbf{I}), \tag{1}$$

where $\mathbf{y}_{:,d}$ is $d(= 1, 2, \ldots, D)$-th column vector of $\mathbf{Y}$, $\boldsymbol{\epsilon} \in \mathbb{R}^N$ is a Gaussian noise with a precision β, and $\mathbf{f}(\mathbf{X}) \triangleq \mathbf{f} \in \mathbb{R}^N$ is a Gaussian process prior of the generative process with a covariance matrix $\mathbf{K}_{NN}^{(\mathbf{f})} \in \mathbb{R}^{N \times N}$. Each entry of the matrix $\mathbf{K}_{NN}^{(\mathbf{f})}$ is calculated by a positive definite kernel $k^{(\mathbf{f})}(\mathbf{x}, \mathbf{x}')$. Equation (1) can be rewritten as $p(\mathbf{y}_{:,d}|\mathbf{f}) = \mathcal{N}(\mathbf{y}_{:,d}|\mathbf{f}, \beta^{-1}\mathbf{I})$ and $p(\mathbf{f}|\mathbf{X}) = \mathcal{N}(\mathbf{f}|\mathbf{0}, \mathbf{K}_{NN}^{(\mathbf{f})})$, and we can derive a likelihood function by marginalizing the Gaussian process prior $\mathbf{f}$ as

$$p(\mathbf{Y}|\mathbf{X}) = \prod_{d=1}^{D} \mathcal{N}(\mathbf{y}_{:,d}|\mathbf{0}, \mathbf{K}_{NN}^{(\mathbf{f})} + \beta^{-1}\mathbf{I}). \tag{2}$$

In vanilla GP-LVM, the derivation of the latent variables is performed by maximizing the log-likelihood with respect to $\mathbf{X}$, and this is the same as imitating the sample similarity matrix $\mathbf{Y}\mathbf{Y}^\top$ into the latent precision matrix $(\mathbf{K}_{NN}^{(\mathbf{f})} + \beta^{-1}\mathbf{I})^{-1}$. By this learning strategy, we can preserve the data structure globally in the derivation of the low dimensional latent variables $\mathbf{X}$. The extensions of GP-LVM typically introduce a prior distribution $p(\mathbf{X})$ as a regularization to reflect several properties into the latent variables [13,23,36,40,41,45]. By this expansion, the maximum likelihood estimation of the latent variables $\mathbf{X}$ is replaced by *maximum a posteriori* (MAP) estimation, and a log-posterior distribution is shown as follows:

$$\log p(\mathbf{X}|\mathbf{Y}) = \log p(\mathbf{Y}|\mathbf{X}) + \log p(\mathbf{X}) + C, \tag{3}$$

where $C = -\log p(\mathbf{Y})$ is a log-normalized constant of the distribution $p(\mathbf{X}|\mathbf{Y})$ and usually ignored. Although those extensions perform well when embedding high-dimensional data, they fail to handle thousands of data points because of the matrix inversion of the $N \times N$ matrix $\mathbf{K}^{(\mathbf{f})} + \beta^{-1}\mathbf{I}$. Bayesian GP-LVM [11,39] introduces low-rank matrix approximation to perform fully Bayesian optimization and can handle large datasets as side effects. Bayesian GP-LVM is an efficient method for obtaining the low-dimensional representation. However, with a complex prior distribution, such as the GP-LVM extensions, its scalability and even closed-form expression easily collapse [11]. From the above, we newly derive a scalable lower bound for MAP estimation, and this bound enables us to handle large datasets with a complex prior distribution.

3 Locality-Aware and Low-Rank Approximated GP-LVM

LolaGP employs an iterative learning process to derive the latent variables and construct the neighborhood graph using a low-dimensional representation. In Sect. 3.1, we present a locality-aware regularization based on graph Gaussian process [30] and show how to derive the scalable lower bound in Sect. 3.2. In Sect. 3.3, we describe how to build a neighborhood graph for efficient local preservation using latent variables [25]. Our learning strategy allows for the preservation of both global and local structures and the calculation of latent variables on a scalable basis.

3.1 Locality-aware Regularization

Similar to the previous graph-based methods presented in Sect. 2.1, we preserve the local geometry using the neighborhood graph and its weighted adjacency matrix $\mathbf{W} \in \mathbb{R}^{N \times N}$. For efficient reflection of the local geometry, we assume that the latent variables are weighted averages of their neighbors based on graph Gaussian process [30] by the following equations:

$$\mathbf{x}_{:,q} = \mathbf{g}(\mathbf{X}) + \boldsymbol{\eta}, \tag{4}$$

$$g_n = \frac{\sum_{i=1}^{N} W_{ni} h_i}{D_n} \quad (n = 1, 2, \ldots, N), \tag{5}$$

where

$$\boldsymbol{\eta} \sim \mathcal{N}(\mathbf{0}, \gamma^{-1}\mathbf{I}), \ \mathbf{h}(\mathbf{X}) \sim \mathcal{N}(\mathbf{0}, \mathbf{K}_{NN}^{(\mathbf{h})}),$$

$\mathbf{x}_{:,q}$ is $q\,(= 1, 2, \ldots, Q)$-th column vector of $\mathbf{X}$, W_{ni} is an entry of the weighted adjacency matrix $\mathbf{W}$, and $\mathbf{D} \in \mathbb{R}^{N \times N}$ with a diagonal entry D_n is a degree matrix of $\mathbf{W}$. Furthermore, $\mathbf{g}(\mathbf{X}) = [g_1, g_2, \ldots, g_N]^\top \in \mathbb{R}^N$ is a graph Gaussian process prior, $\mathbf{h}(\mathbf{X}) = [h_1, h_2, \ldots, h_N]^\top \in \mathbb{R}^N$ is a Gaussian process prior with a covariance matrix $\mathbf{K}_{NN}^{(\mathbf{h})}$ with a kernel function $k^{(\mathbf{h})}(\mathbf{x}, \mathbf{x}')$, and $\boldsymbol{\eta}$ is a Gaussian noise with a precision γ. We show how to calculate $\mathbf{W}$ in Sect. 3.3. By the property of Gaussian distribution, we can rewrite Eqs. (4) and (5) as the following probability distributions:

$$p(\mathbf{x}_{:,q}|\mathbf{g}) = \mathcal{N}(\mathbf{x}_{:,q}|\mathbf{g}, \gamma^{-1}\mathbf{I}), \tag{6}$$

$$\begin{aligned} p(\mathbf{g}) &= \mathcal{N}(\mathbf{g}|\mathbf{0}, \mathbf{P}\mathbf{K}_{NN}^{(\mathbf{h})}\mathbf{P}^\top) \\ &\triangleq \mathcal{N}(\mathbf{g}|\mathbf{0}, \mathbf{K}_{NN}^{(\mathbf{g})}), \end{aligned} \tag{7}$$

where $\mathbf{P} = \mathbf{D}^{-1}\mathbf{W}$ is a normalized adjacency matrix of $\mathbf{W}$. Note that we can set up the matrix $\mathbf{P}$ as $\mathbf{D}^{-\frac{1}{2}}\mathbf{W}\mathbf{D}^{-\frac{1}{2}}$ by simple modification of Eq. (5). By marginalizing $\mathbf{g}$, we derive the locality-aware prior distribution $p(\mathbf{X})$ as follows:

$$p(\mathbf{X}) = \prod_{q=1}^{Q} \mathcal{N}(\mathbf{x}_{:,q}|\mathbf{0}, \mathbf{K}_{NN}^{(\mathbf{g})} + \gamma^{-1}\mathbf{I}). \tag{8}$$

By this prior distribution, we can reflect the local geometry represented by the weighted adjacency matrix $\mathbf{W}$ into the latent variables, and this graph Gaussian process-based formulation helps to derive a scalable lower bound. We optimize the latent variables $\mathbf{X}$ by maximizing the posterior distribution $p(\mathbf{X}|\mathbf{Y}) \propto p(\mathbf{Y}|\mathbf{X})p(\mathbf{X})$.

3.2 Lower Bound of Log-posterior Distribution

In this subsection, we demonstrate the objective function of LolaGP. We first describe an **exact** log-posterior distribution as follows:

$$\begin{aligned}
&\log p(\mathbf{X}|\mathbf{Y}) \\
&= -\frac{ND}{2}\log(2\pi) - \frac{D}{2}\log|\mathbf{K}_{NN}^{(\mathbf{f})} + \beta^{-1}\mathbf{I}| - \frac{1}{2}\mathrm{tr}\left[(\mathbf{K}_{NN}^{(\mathbf{f})} + \beta^{-1}\mathbf{I})^{-1}\mathbf{Y}\mathbf{Y}^{\top}\right] \\
&\quad - \frac{NQ}{2}\log(2\pi) - \frac{Q}{2}\log|\mathbf{K}_{NN}^{(\mathbf{g})} + \gamma^{-1}\mathbf{I}| - \frac{1}{2}\mathrm{tr}\left[(\mathbf{K}_{NN}^{(\mathbf{g})} + \gamma^{-1}\mathbf{I})^{-1}\mathbf{X}\mathbf{X}^{\top}\right]. \quad (9)
\end{aligned}$$

This exact log-posterior is not a scalable objective function because its evaluation requires $O(N^3)$ time complexity for the matrix inversion of the $N \times N$ matrices $\mathbf{K}_{NN}^{(\mathbf{f})} + \beta^{-1}\mathbf{I}$ and $\mathbf{K}_{NN}^{(\mathbf{g})} + \gamma^{-1}\mathbf{I}$. From the above, we introduce a newly scalable lower bound based on low-rank approximation of these matrices. Fortunately, both the likelihood in Eq. (2) and the prior distribution in Eq. (8) are Gaussian distributed with the marginalized Gaussian process priors $\mathbf{f} \in \mathbb{R}^N$ and $\mathbf{g} \in \mathbb{R}^N$, and we can select M inducing points $\mathbf{u} \in \mathbb{R}^M$ and $\mathbf{v} \in \mathbb{R}^M$ from the Gaussian process priors with same latent positions $\mathbf{Z} = [\mathbf{z}_1, \mathbf{z}_2, \ldots, \mathbf{z}_M]^{\top} \in \mathbb{R}^{M \times Q}$. The joint probabilities $p(\mathbf{f}, \mathbf{u})$ and $p(\mathbf{g}, \mathbf{v})$ are also Gaussian distributed, and we can write the probability distributions of $\mathbf{f}$ and $\mathbf{g}$ respectively conditioned by $\mathbf{u}$ and $\mathbf{v}$ and the marginal distributions of the inducing points by the following equations:

$$p(\mathbf{f}|\mathbf{u}) = \mathcal{N}(\mathbf{f}|\mathbf{K}_{NM}^{(\mathbf{f})}\mathbf{K}_{MM}^{(\mathbf{f})-1}\mathbf{u}, \mathbf{K}_{NN}^{(\mathbf{f})} - \mathbf{K}_{NM}^{(\mathbf{f})}\mathbf{K}_{MM}^{(\mathbf{f})-1}\mathbf{K}_{MN}^{(\mathbf{f})}), \quad (10)$$

$$p(\mathbf{g}|\mathbf{v}) = \mathcal{N}(\mathbf{g}|\mathbf{K}_{NM}^{(\mathbf{g})}\mathbf{K}_{MM}^{(\mathbf{g})-1}\mathbf{v}, \mathbf{K}_{NN}^{(\mathbf{g})} - \mathbf{K}_{NM}^{(\mathbf{g})}\mathbf{K}_{MM}^{(\mathbf{g})-1}\mathbf{K}_{MN}^{(\mathbf{g})}), \quad (11)$$

$$p(\mathbf{u}|\mathbf{Z}) = \mathcal{N}(\mathbf{u}|\mathbf{0}, \mathbf{K}_{MM}^{(\mathbf{f})}), \quad p(\mathbf{v}|\mathbf{Z}) = \mathcal{N}(\mathbf{v}|\mathbf{0}, \mathbf{K}_{MM}^{(\mathbf{g})}), \quad (12)$$

where $\mathbf{K}_{NM}^{(\cdot)} \in \mathbb{R}^{N \times M}$ and $\mathbf{K}_{MN}^{(\cdot)} \in \mathbb{R}^{M \times N}$ $(\cdot = \mathbf{f}, \mathbf{g})$ are covariance matrices between the latent variables $\mathbf{X}$ and the positions $\mathbf{Z}$, and $\mathbf{K}_{MM}^{(\cdot)} \in \mathbb{R}^{M \times M}$ is a covariance matrix of $\mathbf{Z}$. We define $\mathbf{Q}_{NN}^{(\cdot)} = \mathbf{K}_{NM}^{(\cdot)}\mathbf{K}_{MM}^{(\cdot)-1}\mathbf{K}_{MN}^{(\cdot)}$ to simplify the notation, and they can be regarded as the low-rank (i.e., Nyström) approximation of the full matrix $\mathbf{K}_{NN}^{(\cdot)}$.

By those probability distributions, we introduce a lower bound of the log-posterior distribution based on a combination of the previous research [21,38]. We explicitly marginalize the inducing points $\mathbf{u}$ and $\mathbf{v}$ as

$$\begin{aligned}
&\log p(\mathbf{X}|\mathbf{Y}) \\
&= \sum_{d=1}^{D} \log \int p(\mathbf{y}_{:,d}|\mathbf{u})p(\mathbf{u}|\mathbf{Z})d\mathbf{u} + \sum_{q=1}^{Q} \log \int p(\mathbf{x}_{:,q}|\mathbf{v})p(\mathbf{v}|\mathbf{Z})d\mathbf{v}. \quad (13)
\end{aligned}$$

Next, we evaluate two likelihood functions $p(\mathbf{y}_{:,d}|\mathbf{u})$ and $p(\mathbf{x}_{:,q}|\mathbf{v})$ by the following Jensen's inequality [38]:

$$\log p(\mathbf{y}_{:,d}|\mathbf{u}) \geq \mathbb{E}_{p(\mathbf{f}|\mathbf{u})}\left[\log p(\mathbf{y}_{:,d}|\mathbf{f})\right], \tag{14}$$

$$\log p(\mathbf{x}_{:,q}|\mathbf{v}) \geq \mathbb{E}_{p(\mathbf{g}|\mathbf{v})}\left[\log p(\mathbf{x}_{:,q}|\mathbf{g})\right]. \tag{15}$$

By substituting Eqs. (14) and (15) into Eq. (13), we can derive the following scalable lower bound of the log-posterior distribution:

$$\begin{aligned}
&\log p(\mathbf{X}|\mathbf{Y}) \\
&\geq \sum_{d=1}^{D}\left\{\log \mathcal{N}(\mathbf{y}_{:,d}|\mathbf{0}, \mathbf{Q}_{NN}^{(\mathbf{f})} + \beta^{-1}\mathbf{I}) - \frac{\beta}{2}\mathrm{tr}(\mathbf{K}_{NN}^{(\mathbf{f})} - \mathbf{Q}_{NN}^{(\mathbf{f})})\right\} \\
&\qquad + \sum_{q=1}^{Q}\left\{\log \mathcal{N}(\mathbf{x}_{:,q}|\mathbf{0}, \mathbf{Q}_{NN}^{(\mathbf{g})} + \gamma^{-1}\mathbf{I}) - \frac{\gamma}{2}\mathrm{tr}(\mathbf{K}_{NN}^{(\mathbf{g})} - \mathbf{Q}_{NN}^{(\mathbf{g})})\right\},
\end{aligned} \tag{16}$$

and we can also expand each summation as

$$\begin{aligned}
&\sum_{d=1}^{D}\left\{\log \mathcal{N}(\mathbf{y}_{:,d}|\mathbf{0}, \mathbf{Q}_{NN}^{(\mathbf{f})} + \beta^{-1}\mathbf{I}) - \frac{\beta}{2}\mathrm{tr}(\mathbf{K}_{NN}^{(\mathbf{f})} - \mathbf{Q}_{NN}^{(\mathbf{f})})\right\} \\
&= -\frac{ND}{2}\log(2\pi) - \frac{D}{2}\log|\mathbf{Q}_{NN}^{(\mathbf{f})} + \beta^{-1}\mathbf{I}| \\
&\qquad - \frac{1}{2}\mathrm{tr}\left[(\mathbf{Q}_{NN}^{(\mathbf{f})} + \beta^{-1}\mathbf{I})^{-1}\mathbf{Y}\mathbf{Y}^{\top}\right] - \frac{\beta D}{2}\mathrm{tr}(\mathbf{K}_{NN}^{(\mathbf{f})} - \mathbf{Q}_{NN}^{(\mathbf{f})}), \\
&\sum_{q=1}^{Q}\left\{\log \mathcal{N}(\mathbf{x}_{:,q}|\mathbf{0}, \mathbf{Q}_{NN}^{(\mathbf{g})} + \gamma^{-1}\mathbf{I}) - \frac{\gamma}{2}\mathrm{tr}(\mathbf{K}_{NN}^{(\mathbf{g})} - \mathbf{Q}_{NN}^{(\mathbf{g})})\right\} \\
&= -\frac{NQ}{2}\log(2\pi) - \frac{Q}{2}\log|\mathbf{Q}_{NN}^{(\mathbf{g})} + \gamma^{-1}\mathbf{I}| \\
&\qquad - \frac{1}{2}\mathrm{tr}\left[(\mathbf{Q}_{NN}^{(\mathbf{g})} + \gamma^{-1}\mathbf{I})^{-1}\mathbf{X}\mathbf{X}^{\top}\right] - \frac{\gamma Q}{2}\mathrm{tr}(\mathbf{K}_{NN}^{(\mathbf{g})} - \mathbf{Q}_{NN}^{(\mathbf{g})}).
\end{aligned}$$

Comparing Eqs. (9) and (16), the full covariance matrix $\mathbf{K}_{NN}^{(\cdot)}$ is replaced by the low-rank covariance matrix $\mathbf{Q}_{NN}^{(\cdot)}$ with the additional trace term $\mathrm{tr}(\mathbf{K}_{NN}^{(\cdot)} - \mathbf{Q}_{NN}^{(\cdot)})$. By this replacement, we can avoid computing the $N \times N$ matrix inversion and can reduce the time complexity $O(N^3)$ to $O(NM^2)$ by applying the Woodbury matrix identity to Eq. (16).

3.3 Construction of Neighborhood Graph

We need the neighborhood graph to reflect the local geometry in the latent variables to achieve our objective. Intuitively, The Euclidean distance is appropriate for calculating the adjacency matrix, but this metric cannot precisely realize the local geometry. We visualize this in Fig. 2. From Fig. 2 (a), The distribution of

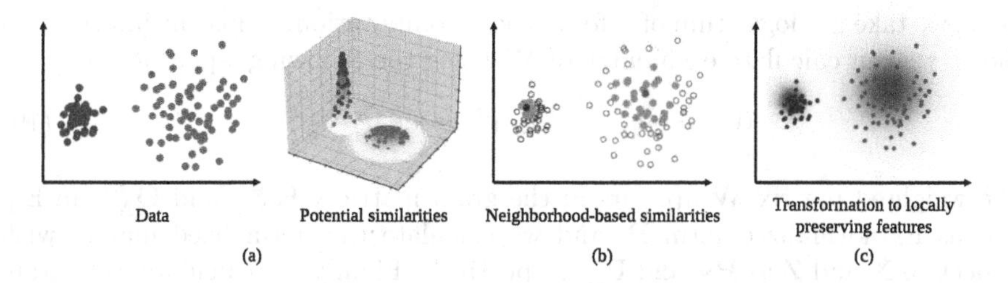

Fig. 2. Transformation of the latent variables into local preserving features. We show (a) data with two different density areas and different potential similarities, (b) neighborhood-based similarities to realize the local geometry and (c) a diffusion process to transform the latent variables into the locally preserving features.

data can be divided into two areas: dense and sparse, and the potential similarities between them are influenced by their local densities. However, we cannot realize the non-linear geometry by the Euclidean distance since it is a simple straight line in the Euclidean space. To overcome this difficulty, we peculiarly focus on *diffusion map* [10,25], which captures the local geometry by propagating neighborhood similarities by the diffusion process, i.e., powering a random walk matrix. In LolaGP, we transform the latent variables into locally preserving features based on the diffusion process and calculate the weighted adjacency matrix using them.

We first calculate a similarity matrix $\mathbf{S} \in \mathbb{R}^{N \times N}$ of the latent variables $\mathbf{X}$ that realizes the local geometry based on the α-decay kernel [25] as

$$k_{\alpha}(\mathbf{x}, \mathbf{x}') = \frac{1}{2} \exp\left[-\left(\frac{||\mathbf{x}-\mathbf{x}'||_2}{\epsilon_k(\mathbf{x})}\right)^{\alpha}\right] + \frac{1}{2} \exp\left[-\left(\frac{||\mathbf{x}-\mathbf{x}'||_2}{\epsilon_k(\mathbf{x}')}\right)^{\alpha}\right], \quad (17)$$

where α is a hyperparameter that controls the decay rate of each exponential value, and $\epsilon_k(\mathbf{x}_*)$ is the Euclidean distance between $\mathbf{x}_*$ and its k-nearest neighbor. The position of $\epsilon_k(\mathbf{x}_*)$ is the same as the *lengthscale* parameter of the Squared Exponential (SE) kernel, and we can reflect the neighborhood geometry of $\mathbf{x}_*$ by setting k to appropriate values (Fig. 2 (b)). We calculate a random walk matrix of $\mathbf{S}$ as $\mathbf{R} = \mathbf{D}_{\mathbf{S}}^{-1}\mathbf{S}$ ($\mathbf{D}_{\mathbf{S}}$ is the degree matrix of $\mathbf{S}$) and derive features after t-step diffusion of the neighborhood-based similarities (Fig. 2 (c)) by powering $\mathbf{R}$ as

$$\mathbf{U}^{(t)} = \mathbf{R}^t, \, \mathbf{U}^{(t)} = [\mathbf{u}_1^{(t)}, \mathbf{u}_2^{(t)}, \ldots, \mathbf{u}_N^{(t)}]^{\top} \in \mathbb{R}^{N \times N}. \quad (18)$$

Each row vector $\mathbf{u}_n^{(t)} \in \mathbb{R}^N$ indicates the strength of the interconnection between sample n and the other samples after the t-step diffusion, and it is reasonable to state that nearby samples on the local manifold have similar vector values. We regard $\mathbf{U}^{(t)}$ as the locally preserving features of $\mathbf{X}$ and calculate the weighted adjacency matrix $\mathbf{W}$ based on them. Since the vector $\mathbf{u}_n^{(t)}$ is the probability value, we take the logarithm of it for efficient computation. Using the information above, we can calculate each entry of $\mathbf{W}$ using the following equation:

$$W_{ij} = \exp\left(-|| \log \mathbf{u}_i^{(t)} - \log \mathbf{u}_j^{(t)} ||_2\right). \tag{19}$$

The weighted matrix $\mathbf{W}$ appears in the gram matrices $\mathbf{K}_{NN}^{(\mathbf{g})}$ and $\mathbf{Q}_{NN}^{(\mathbf{g})}$ in Eq. (16) as its normalized form $\mathbf{P}$, and we calculate the normalized matrix with respect to $\mathbf{X}$ and $\mathbf{Z}$ as $\mathbf{P_X}$ and $\mathbf{P_Z}$, respectively. Finally, we calculate each gram matrix as

$$\mathbf{K}_{NM}^{(\mathbf{g})} = \mathbf{P_X}\mathbf{K}_{NM}^{(\mathbf{h})}\mathbf{P_Z}^\top,$$
$$\mathbf{K}_{MM}^{(\mathbf{g})} = \mathbf{P_Z}\mathbf{K}_{MM}^{(\mathbf{h})}\mathbf{P_Z}^\top,$$
$$\mathbf{K}_{MN}^{(\mathbf{g})} = \mathbf{P_Z}\mathbf{K}_{MN}^{(\mathbf{h})}\mathbf{P_X}^\top.$$

These matrices allow the latent variables to realize the non-linear geometry.

To summarize, we extend GP-LVM by introducing locality-aware regularization via latent variables with the newly constructed neighborhood graph to avoid the influence of noise disruption or the curse of dimensionality. We also use sparse Gaussian process methods to derive the lower bound of the exact log-posterior distribution. Our method can embed high-dimensional data with high scalability and interpretability of both global and local structures, thanks to this innovation.

4 Experiments

In this section, we conduct an experimental analysis to validate the efficacy of our method. We assessed LolaGP and other comparative methods in both qualitative and quantitative ways, viz., by visualizing latent variables and computing quality metrics. We used the Gaussian process open library GPy [16] on an Intel Core i7-10700 CPU to implement our source code.

4.1 Experimental Settings

Datasets. We used two image datasets, COIL20 [29] and MNIST[1] and one synthetic dataset, Spheres [28], described in Sect. 2.1. COIL20 contains 1,440 grayscale images of 20 objects, and each object is captured evenly in a single rotation across 72 images. We selected five objects (indexed as {1, 4, 6, 11, 13}) from COIL20 dataset. Furthermore, in order to get closer to our problem setting, we randomly lost pixels from the selected images with a probability of 35% and regarded them as *Noisy* COIL20 dataset referring to [24] (see Fig. 3). Both COIL20 and *Noisy* COIL20 include the global relationship based on the object separation and the local geometry according to the object rotation. In MNIST, we randomly selected 5,000 images from the training set and also build *Noisy* MNIST by randomly losing their pixels with a probability of 25% which is the limit to remain the characteristics of each digit.

[1] http://yann.lecun.com/exdb/mnist/.

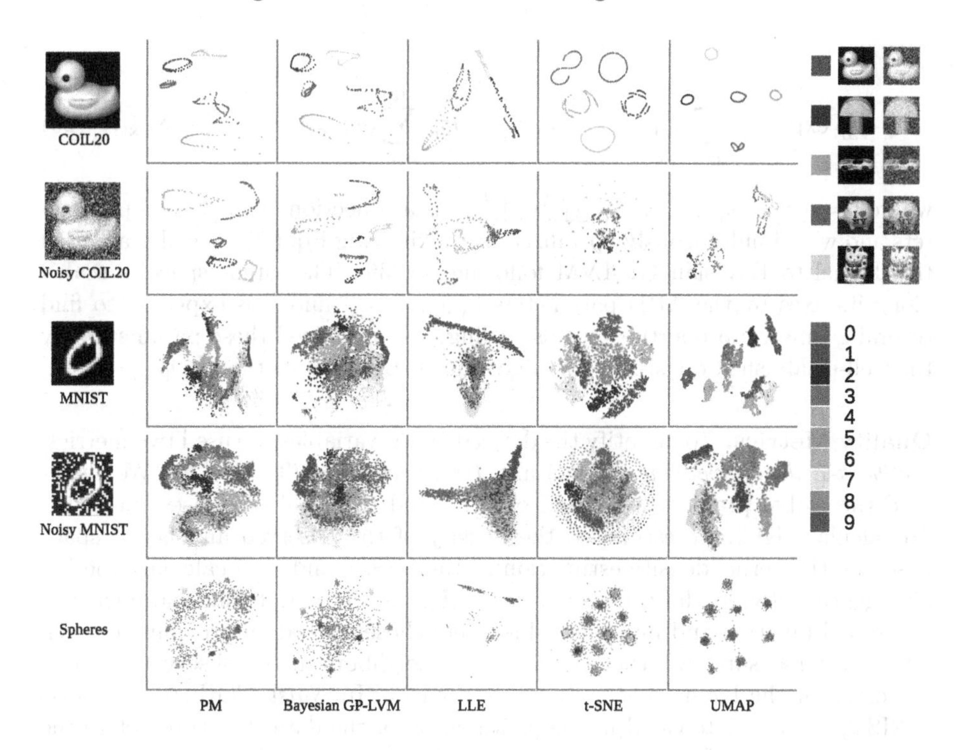

Fig. 3. Visualization results of the five datasets (COIL20, *Noisy* COIL20, MNIST, *Noisy* MNIST, and Spheres). Each color shows the separable class in the dataset, i.e., each object (COIL20 and *Noisy* COIL20), each digit (MNIST and *Noisy* MNIST), and each sphere (Spheres). We ignored some outliers for effficient visualization.

Comparative Methods. We compared our proposed method (PM) with LLE [33] and Bayesian GP-LVM [11] as benchmarks described in Sect. 2. Furthermore, we took the commonly used method, t-SNE [43], and further adopted UMAP [27] as the state-of-the-art method in the same manner as the previous research [28]. All methods were compared from both qualitative and quantitative perspectives after embedding the observed high-dimensional data into two-dimensional latent space.

Training Procedure. The trainable parameters in our proposed method are the latent variables $\mathbf{X}$ and the locations of the inducing points $\mathbf{Z}$. We initialized them by PCA [19] and by randomly picking up from the initialized $\mathbf{X}$, respectively. Furthermore, we need the normalized adjacency matrix $\mathbf{P}$ in 3.3 and initialized it by computing $\mathbf{P}$ from the observed variables $\mathbf{Y}$. Under this initialization, we iterated our method two times and changed $\mathbf{P}$ following the latent variables $\mathbf{X}$ in the middle iteration. We selected the 'SE+linear+whitenoise' kernel with the automatic relevance determination (ARD) [26] as the kernel functions $k^{(\mathbf{f})}(\mathbf{x}, \mathbf{x}')$ and $k^{(\mathbf{h})}(\mathbf{x}, \mathbf{x}')$:

$$k^{(\cdot)}(\mathbf{x}, \mathbf{x}')$$
$$= \sigma_{\text{SE}}^2 \exp\left[-\frac{1}{2}\sum_{q=1}^{Q} a_{\text{SE},q}(x_q - x_q')^2\right] + \sigma_{\text{lin.}}^2 \sum_{q=1}^{Q} a_{\text{lin.},q} x_q x_q' + \sigma_{\text{noise}}^2 \delta_{\mathbf{x},\mathbf{x}'}, \quad (20)$$

where $\boldsymbol{\theta} = \{\sigma_{\{\text{SE,lin.,noise}\}}, \{a_{\{\text{SE,lin.}\},q}\}_{q=1}^{Q}\}$ is a collection of the kernel parameters and was simultaneously optimized by maximizing Eq. (16). We also adopted this kernel to Bayesian GP-LVM following to [39]. The latent space is scaled along its axes by the ARD parameters $a_{\{\text{SE,linear}\},q}$, and it is expected to find several geometric properties by this scaling. We optimized these parameters by the well-established quasi-Newton algorithm L-BFGS-B method [49].

Quality Metrics. To quantify the derived latent variables, we used two metrics, *Kullback-Leibler divergence* (KL_σ) and *trustworthiness* (Trust) [44]. We evaluated the global preservation based on KL_σ and the local one based on Trust. To calculate KL_σ, we estimated the density of the observed and latent space based on the kernel density estimation methods [8,9] and then calculate the KL divergence between those densities. $\sigma \in \mathbb{R}_{>0}$ is the lengthscale parameter of the kernel function and multiple values were chosen based on previous research [28]. Trust measures whether the k-nearest neighbors in the observed space is preserved in the latent space, and we set k to 3. In *Noisy* COIL20 and *Noisy* MNIST, we were interested in the preservation of the data structure before the noise disruption and calculated each metric between the derived latent variables and vanilla COIL20 and MNIST, respectively.

4.2 Results

Visualization. Figure 3 shows the visualization results of all the datasets. In COIL20, all methods can preserve the global relationship and local geometry we expected. However, in *Noisy* COIL20, we confirm that the added missing pixel noise has a significant impact on LLE, t-SNE, and UMAP and that even the global object separation is barely preserved. Since PM and Bayesian GP-LVM learn the global relationship based on the sample similarity, they successfully separated each object under the noise disruption, and PM completely recovered the local rotation geometry. In MNIST and *Noisy* MNIST, we can find a similar tendency to the results of COIL20. Although t-SNE and UMAP behave well in MNIST, we can see the accuracy degradation in the *Noisy* case. PM has the best result under the noise (e.g., the separation of '3', '5', and '8' manifolds) and has better boundary of each digit than Bayesian GP-LVM in MNIST and *Noisy* MNIST. We observe that PM only realizes the enclosing structure hidden in the dataset. Bayesian GP-LVM, t-SNE, and UMAP can preserve the separation of each small sphere. However, they cannot realize the large one due to the curse of dimensionality. LLE cannot separate even small spheres, and we observed they were covered with the point clouds of the large one. LolaGP successfully embeds the five datasets without significant degradation due to noise and dimensionality effects such as t-SNE and UMAP, implying the efficacy of our novelties.

Table 1. Quality evaluation based on the two quality metrics, KL divergence (KL_σ) and Trustworthiness (Trust). The best is boldfaced, and the second best is underlined.

Dataset	Method	$KL_{0.01}$ ↓	$KL_{0.1}$ ↓	KL_1 ↓	Trust↑
COIL20	PM	0.140	**0.0407**	<u>0.00180</u>	0.981
	Bayesian GP-LVM	0.119	<u>0.0527</u>	**0.00161**	0.989
	LLE	0.191	0.0624	0.00336	0.948
	t-SNE	**0.0299**	0.120	0.00677	**0.999**
	UMAP	<u>0.0312</u>	0.113	0.00630	<u>0.998</u>
Noisy COIL20	PM	<u>0.145</u>	**0.0682**	**0.00170**	**0.987**
	Bayesian GP-LVM	**0.104**	<u>0.0751</u>	<u>0.00287</u>	<u>0.979</u>
	LLE	0.246	0.110	0.00419	0.880
	t-SNE	0.265	0.128	0.00392	0.892
	UMAP	0.488	0.0818	0.00456	0.905
MNIST	PM	**0.108**	**0.140**	**0.00155**	0.928
	Bayesian GP-LVM	0.130	<u>0.144</u>	<u>0.00157</u>	0.916
	LLE	0.577	0.309	0.00345	0.825
	t-SNE	**0.108**	0.180	0.00275	**0.995**
	UMAP	<u>0.129</u>	0.178	0.00296	<u>0.977</u>
Noisy MNIST	PM	**0.0765**	**0.171**	0.00254	**0.933**
	Bayesian GP-LVM	0.186	**0.171**	**0.00206**	0.911
	LLE	0.286	0.222	0.00276	0.750
	t-SNE	0.171	0.181	<u>0.00240</u>	<u>0.932</u>
	UMAP	<u>0.101</u>	<u>0.174</u>	0.00253	0.931
Spheres	PM	<u>0.299</u>	0.546	<u>0.0125</u>	0.650
	Bayesian GP-LVM	0.401	0.679	0.0158	0.647
	LLE	0.576	0.696	0.0210	<u>0.659</u>
	t-SNE	**0.294**	**0.516**	**0.0114**	**0.687**
	UMAP	0.339	<u>0.535</u>	0.0131	**0.687**

Quantitative Results. We show the quantitative results in Table 1. In KL_σ, We confirm that our proposed method is the best in eight of fifteen entries and the second-best in four, indicating that PM can preserve the global data structure better than the other methods. In Trust, PM shows successful results on average as opposed to t-SNE and UMAP, which have significant accuracy degradation in *Noisy* COIL20 and *Noisy* MNIST. These results imply the validity of our robust neighborhood preservation via the latent variables. However, in Spheres, PM is slightly inferior to t-SNE and UMAP. Unfortunately, the reliable quality measurement of Spheres is a difficult task because these data contain the curse of dimensionality problem described in 2.1 [28]. Although our method is inferior to t-SNE and UMAP on Trust in the result of Spheres, it is clear that PM outperforms the comparative methods in the other general image datasets and only detects the true neighbors on the manifold in Spheres from the visualization results.

5 Conclusions

We have introduced a novel latent variable model, LolaGP, that can summarize the complex high-dimensional data into the latent variables while preserving global and local structures. We focused on GP-LVM to preserve the global relationship and introduced a novel regularization based on the neighborhood graph to preserve the local geometry. The graph is built with latent variables, which promotes robustness to noise disruption and the curse of dimensionality. Furthermore, we introduced the scalable lower bound of the log-posterior distribution based on the low-rank matrix approximation, which allows us to handle larger datasets than the previous GP-LVM extensions. In the experimental results, we have shown the effectiveness of our proposed method from the qualitative and quantitative perspectives on the natural and even highly disrupted datasets like *Noisy* COIL20 and *Noisy* MNIST.

One drawback of our proposed method is its generativity. Although the latent variables should be visualized discretely for each independent manifold, such as t-SNE and UMAP, the generativity forces the latent variables to change continuously. Our GP-based approach aids in the discovery of the global relationship between the noise effect and the curse of dimensionality, which can be found in various situations. It would be preferable to modify the likelihood function to make it more suitable for visualization in future works.

Acknowledgements. This study was supported in part by JSPS KAKENHI Grant Number JP21H03456 and AMED Grant Number JP21zf0127004.

References

1. Attali, D., Lieutier, A., Salinas, D.: Vietoris-Rips complexes also provide topologically correct reconstructions of sampled shapes. Comput. Geom. **46**(4), 448–465 (2013)
2. Balasubramanian, M., Schwartz, E.L., Tenenbaum, J.B., de Silva, V., Langford, J.C.: The isomap algorithm and topological stability. Science **295**(5552), 7 (2002)
3. Becht, E., et al.: Dimensionality reduction for visualizing single-cell data using UMAP. Nat. Biotechnol. **37**(1), 38–44 (2019)
4. Belkin, M., Niyogi, P.: Laplacian eigenmaps for dimensionality reduction and data representation. Neural Comput. **15**(6), 1373–1396 (2003)
5. Bengio, Y., Courville, A., Vincent, P.: Representation learning: a review and new perspectives. IEEE Trans. Pattern Anal. Mach. Intell. **35**(8), 1798–1828 (2013)
6. Burgess, C.P., et al.: Understanding disentangling in β-VAE. arXiv preprint arXiv:1804.03599 (2018)

7. Chang, H., Yeung, D.Y.: Robust locally linear embedding. Pattern Recognit. **39**(6), 1053–1065 (2006)
8. Chazal, F., Cohen-Steiner, D., Mérigot, Q.: Geometric inference for probability measures. Found. Comput. Math. **11**(6), 733–751 (2011)
9. Chazal, F., et al.: Robust topological inference: distance to a measure and kernel distance. J. Mach. Learn. Res. **18**(1), 5845–5884 (2017)
10. Coifman, R.R., Lafon, S.: Diffusion maps. Appl. Comput. Harmonic Anal. **21**(1), 5–30 (2006)
11. Damianou, A.C., Titsias, M.K., Lawrence, N.D.: Variational inference for latent variables and uncertain inputs in Gaussian processes. J. Mach. Learn. Res. **17**(42), 1–62 (2016)
12. Diaz-Papkovich, A., Anderson-Trocmé, L., Ben-Eghan, C., Gravel, S.: UMAP reveals cryptic population structure and phenotype heterogeneity in large genomic cohorts. PLoS Genet. **15**(11), e1008432 (2019)
13. Ferris, B., Fox, D., Lawrence, N.D.: WiFi-SLAM using Gaussian process latent variable models. In: International Joint Conference on Artificial Intelligence (IJCAI), pp. 2480–2485 (2007)
14. Hausmann, J.C.: On the Vietoris-Rips complexes and a cohomology theory for metric spaces. In: Prospects in Topology: Proceedings of a Conference in Honor of William Browder, pp. 175–188 (1995)
15. He, X., Cai, D., Yan, S., Zhang, H.J.: Neighborhood preserving embedding. In: International Conference on Computer Vision (ICCV), pp. 1208–1213 (2005)
16. Hensman, J., et al.: GPy: A Gaussian process framework in python. https://github.com/SheffieldML/GPy (2012)
17. Hensman, J., Fusi, N., Lawrence, N.D.: Gaussian processes for big data. arXiv preprint arXiv:1309.6835 (2013)
18. Higgins, I., et al.: β-VAE: learning basic visual concepts with a constrained variational framework. In: International Conference on Learning Representations (ICLR), pp. 1–22 (2016)
19. Hotelling, H.: Analysis of a complex of statistical variables into principal components. J. Educ. Psychol. **24**(6), 417–441 (1933)
20. Kingma, D.P., Welling, M.: Auto-encoding variational bayes. arXiv Preprint arXiv:1312.6114 (2013)
21. Lawrence, N.D.: Learning for larger datasets with the Gaussian process latent variable model. In: International Conference on Artificial Intelligence and Statistics (AISTATS), pp. 243–250 (2007)
22. Lawrence, N.D., Hyvärinen, A.: Probabilistic non-linear principal component analysis with Gaussian process latent variable models. J. Mach. Learn. Res. **6**(11), 1783–1816 (2005)
23. Lu, C., Tang, X.: Surpassing human-level face verification performance on LFW with GaussianFace. In: AAAI Conference on Artificial Intelligence (AAAI), pp. 3811–3819 (2015)
24. Lu, Y., Lai, Z., Xu, Y., Li, X., Zhang, D., Yuan, C.: Low-rank preserving projections. IEEE Trans. Cybern. **46**(8), 1900–1913 (2015)
25. Moon, K.R., et al.: Visualizing structure and transitions in high-dimensional biological data. Nat. Biotechnol. **37**(12), 1482–1492 (2019)
26. MacKay, D.J.: Bayesian nonlinear modeling for the prediction competition. ASHRAE Trans. **100**(2), 1053–1062 (1994)
27. McInnes, L., Healy, J., Melville, J.: UMAP: uniform manifold approximation and projection for dimension reduction. arXiv preprint arXiv:1802.03426 (2018)

28. Moor, M., Horn, M., Rieck, B., Borgwardt, K.: Topological autoencoders. In: International Conference on Machine Learning (ICML), pp. 7045–7054 (2020)
29. Nene, S.A., Nayar, S.K., Murase, H.: Columbia object image library (coil-20) (1996). www.cs.columbia.edu/CAVE/software/softlib/coil-20.php
30. Ng, Y.C., Colombo, N., Silva, R.: Bayesian semi-supervised learning with graph Gaussian processes. In: Advances in Neural Information Processing (NeurIPS) (2018)
31. Ordun, C., Purushotham, S., Raff, E.: Exploratory analysis of COVID-19 tweets using topic modeling, UMAP, and DiGraphs. arXiv preprint arXiv:2005.03082 (2020)
32. Quinonero-Candela, J., Rasmussen, C.E.: A unifying view of sparse approximate Gaussian process regression. J. Mach. Learn. Res. **6**(65), 1939–1959 (2005)
33. Roweis, S.T., Saul, L.K.: Nonlinear dimensionality reduction by locally linear embedding. Science **290**(5500), 2323–2326 (2000)
34. Saul, L.K.: A tractable latent variable model for nonlinear dimensionality reduction. Proc. Nat. Acad. Sci. **117**(27), 15403–15408 (2020)
35. Schölkopf, B., Smola, A., Müller, K.R.: Nonlinear component analysis as a kernel eigenvalue problem. Neural Comput. **10**(5), 1299–1319 (1998)
36. Song, G., Wang, S., Huang, Q., Tian, Q.: Harmonized multimodal learning with Gaussian process latent variable models. IEEE Trans. Pattern Anal. Mach. Intell. **43**(3), 858–872 (2021)
37. Tipping, M.E., Bishop, C.M.: Probabilistic principal component analysis. J. Roy. Stat. Soc. Ser. B (Stat. Methodol.) **61**(3), 611–622 (1999)
38. Titsias, M.: Variational learning of inducing variables in sparse Gaussian processes. In: International Conference on Artificial Intelligence and Statistics (AISTATS), pp. 567–574 (2009)
39. Titsias, M., Lawrence, N.D.: Bayesian Gaussian process latent variable model. In: International Conference on Artificial Intelligence and Statistics (AISTATS), pp. 844–851 (2010)
40. Urtasun, R., Darrell, T.: Discriminative Gaussian process latent variable model for classification. In: International Conference on Machine Learning (ICML), 927–934 (2007)
41. Urtasun, R., Fleet, D.J., Geiger, A., Popović, J., Darrell, T.J., Lawrence, N.D.: Topologically-constrained latent variable models. In: International Conference on Machine Learning (ICML), pp. 1080–1087 (2008)
42. Van Der Maaten, L., Postma, E., Van den Herik, J., et al.: Dimensionality reduction: a comparative. Technical Report TiCC-TR 2009–005 (2009)
43. Van der Maaten, L., Hinton, G.: Visualizing data using t-SNE. J. Mach. Learn. Res. **9**(11), 2579–2605 (2008)
44. Venna, J., Kaski, S.: Visualizing gene interaction graphs with local multidimensional scaling. In: European Symposium on Artificial Neural Networks (ESANN), pp. 557–562 (2006)
45. Wang, J.M., Fleet, D.J., Hertzmann, A.: Gaussian process dynamical models for human motion. IEEE Trans. Pattern Anal. Mach. Intell. **30**(2), 283–298 (2007)
46. Wasserman, L.: Topological data analysis. Ann. Rev. Stat. Appl. **5**(1), 501–532 (2018)
47. You, Z.H., Lei, Y.K., Gui, J., Huang, D.S., Zhou, X.: Using manifold embedding for assessing and predicting protein interactions from high-throughput experimental data. Bioinformatics **26**(21), 2744–2751 (2010)

48. Zhang, Y.J., Pan, S., He, L., Ling, Z.H.: Learning latent representations for style control and transfer in end-to-end speech synthesis. In: International Conference on Acoustics, Speech and Signal Processing (ICASSP), pp. 6945–6949 (2019)
49. Zhu, C., Byrd, R.H., Lu, P., Nocedal, J.: Algorithm 778: L-BFGS-B: Fortran subroutines for large-scale bound-constrained optimization. ACM Trans. Math. Softw. **23**(4), 550–560 (1997)

Optimization of Annealed Importance Sampling Hyperparameters

Shirin Goshtasbpour[1,2](✉) and Fernando Perez-Cruz[1,2]

[1] Computer Science Department, ETH Zurich, Ramistrasse 101, 8092 Zurich, Switzerland
shirin.goshtasbpour@inf.ethz.ch
[2] Swiss Data Science Center, Turnerstrasse 1, 8092 Zurich, Switzerland

Abstract. Annealed Importance Sampling (AIS) is a popular algorithm used to estimates the intractable marginal likelihood of deep generative models. Although AIS is guaranteed to provide unbiased estimate for any set of hyperparameters, the common implementations rely on simple heuristics such as the geometric average bridging distributions between initial and the target distribution which affect the estimation performance when the computation budget is limited. In order to reduce the number of sampling iterations, we present a parameteric AIS process with flexible intermediary distributions defined by a residual density with respect to the geometric mean path. Our method allows parameter sharing between annealing distributions, the use of fix linear schedule for discretization and amortization of hyperparameter selection in latent variable models. We assess the performance of Optimized-Path AIS for marginal likelihood estimation of deep generative models and compare it to compare it to more computationally intensive AIS.

Keywords: Annealed importance sampling · Partition function estimation · Generative models

1 Introduction

Deep generative models, developed over a decade, are now capable of simulating complex distributions in high dimensional space and synthesizing high quality samples in various domains such as natural images, text and medical data. Many of these models are built with the assumption that data $x \in \mathcal{X}$ resides close to a low dimension manifold. In this case, data can be represented using latent variable z drawn from prior distribution on $\mathbb{R}^d$ with density $p(z)$ and the observation model $p(x|z) = p(x|f_\theta(z))$ is given by a parametric mapping $f_\theta : \mathbb{R}^d \to \Gamma$ where Γ is the space of parameters of the likelihood $p(x|z)$.

For complex f_θ mappings, evaluation of log marginal likelihood

$$\log p(x) = \log \int p(z)p(x|z)dz$$

© The Author(s) 2023
M.-R. Amini et al. (Eds.): ECML PKDD 2022, LNAI 13717, pp. 174–190, 2023.
https://doi.org/10.1007/978-3-031-26419-1_11

is intractable and the posterior density $p(z|x) \propto p(x|z)p(z)$ is known only up to a constant normalization factor[1]. In this case, latent variable inference, probabilistic model evaluation and model comparison are performed using variational approximation of the posterior [3,24,38] or via sampling methods like Markov Chain Monte Carlo (MCMC) [4], Nested sampling [32,43], Sequential Monte Carlo (SMC) [8] and Annealed Importance Sampling (AIS) [34]. In Variational Inference (VI) the posterior is approximated with the most similar probability density from the family of distributions $\mathcal{Q} = \{q_\phi(z|x) : \phi \in \phi\}$ indexed by parameter ϕ. Success of VI depends on sufficient expressivity of members of $\mathcal{Q}$ and our ability to find the optimal member of this family by, for instance, maximizing the Evidence Lower BOund (ELBO)

$$\mathcal{L}(\phi) = \int \log \left(\frac{p(x,z)}{q_\phi(z|x)} \right) q_\phi(z|x) dz \leq \log p(x) \tag{1}$$

where the equality happens only when $q_\phi(z|x) = p(z|x) \in \mathcal{Q}$.

In MCMC we use Markov kernels with unique stationary distribution $p(z|x)$ and sample a Markov chain $(z_k)_{k\in[M]}$ by iterative application of this kernel on an initial particle $z_0 \sim q_\phi(z|x)$. Contrary to parametric VI, under mild assumptions, it is theoretically guaranteed that distribution of z_M converges to the target distribution $p(z|x)$ as M goes to infinity. However, if the posterior has multiple modes or heavy tails, convergence can require large number of iterations and therefore, be computationally prohibitive. Among the sampling methods, SMC and AIS are of particular interest as they produce unbiased estimation of marginal likelihood regardless of the computational budget by assigning importance weights to samples [31,40,50]. In these algorithms an auxiliary sequence of distributions, $(\pi_k)_{k\in[M]}$, is used to bridge a simple proposal distribution with density $\pi_0(z) = q_\phi(z|x)$ and the target density $\pi_M(z) = p(z|x)$. This sequence is defined via unnormalized densities $(\gamma_k)_{k\in[M]}$ where $\pi_k(z) = \gamma_k(z)/Z_k$ and $\gamma_0(z) = \pi_0(z)$ and $\gamma_M(z) = p(x,z)$. The algorithm produces N Markov chain samples $\left(z_k^j\right)_{k\in[M]}$ for $j \in [N]$ and their corresponding importance weights denoted by w^j as follows: Initially z_0^j is sampled from γ_0. Then z_k^j is approximately sampled from γ_k using a Markov kernel (typically a gaussian distribution around z_{k-1}^j or a transition with invariant distribution γ_k)[2]. The marginal likelihood is approximated with Monte Carlo method

$$p(x) = \mathbb{E}[w] \approx \hat{p}^N(x) := \frac{1}{N} \sum_{j\in[N]} w^j \tag{2}$$

where the expectation is taken over joint distribution of the Markov chains. Variance of $\hat{p}^N(x)$ depends on the selected density sequence and hyperparameters

[1] We use normalization factor and marginal likelihood $p(x)$ interchangeably in this paper.

[2] SMC has an additional resampling step to draw exact samples from γ_k to reduce the variance of importance weight although it sometimes results in insufficient sample diversity.

especially when computation resources are scarce. However, little work is available on optimization of the intermediary distributions. In this paper, we study the impact of optimization of AIS hyperparameters for more accurate estimation of log marginal likelihood with only a few annealing distributions. We optimize the sequence of distributions between proposal and target distributions as well as the hyperparameters of Markov kernels with commonly used tuning and training measures to improve the sampling performance in low budget scenarios. We have made the following contributions:

- We propose the parameterization of a continuous time density path between the proposal and target distributions which can define general density functions. Contrary to [24,25] the densities used do not need to be normalized and we don't require their exact samples since we use sampling algorithms to gradually transition between the intermediary distributions.
- To optimize the bridging distributions we minimize Jefferys and inverse KL divergences between the distributions of AIS process and its auxiliary reverse process (defined in Sect. 2.1). We empirically show that Jefferys divergence captures both bias and variance of the estimation in benchmark data distributions while inverse KL divergence is unable to do so.
- We further implement evaluate our method on deep generative models with different training procedures and achieve comparable results to more computationally expensive version of AIS algorithm.

The rest of this paper is organized as follows: Sect. 2 is dedicated to reviewing vanilla AIS algorithm. In Sect. 3 we present our parameterization of AIS process which results in flexible bridging distributions and motivate out optimization objective. In Sect. 4 we restate a reparameterization method and derivation of the objective gradient estimates previously developed in [45,46]. Finally, we analyze the accuracy of marginal likelihood estimation and its variance on synthetic and image data in Sect. 6.

2 Background

In this Section we give a brief introduction to AIS algorithm and its popular adaptive versions. For the rest of this section, we assume that the observation x is fixed and $\pi(z) = p(z|x)$ is the target density function which we can evaluate up to a normalizing constant $\tilde{\pi}(z) = p(z, x)$ where $\pi(z) = \tilde{\pi}(z)/Z$ and $Z = p(x)$. We also define a proposal distribution with normalized density $\gamma_0(z)$ which is easy to sample and evaluate, such as the variational posterior $\gamma_0(z) = q_\phi(z|x)$.

2.1 Recap of Annealed Importance Sampling

To sample from the target distribution, AIS algorithm requires a sequence of distributions which change from the proposal distribution to the target. To match with our parameterization of the bridging distributions, we consider the generalization of this sequence to a continuously indexed path of density functions. With some abuse of notation, let $\gamma : [0,1] \times \mathbb{R}^d \to \mathbb{R}$ with mapping $(t, z) \mapsto \gamma(t, z)$ denote a path of density functions indexed by $t \in [0,1]$, with fixed endpoints $\gamma(0, \cdot) = q_\phi(\cdot|x)$ and $\gamma(1, \cdot) = \tilde{\pi}(\cdot)$. Assume the density functions have non-increasing support (i.e. $\text{supp}(\gamma(t', \cdot)) \subseteq \text{supp}(\gamma(t, \cdot))$ for $0 \le t \le t' \le 1$). Let $t_k = k/M$ for $k \in \{0, ..., M\}$ be the linear schedule in $[0,1]$ with which the path of density functions is discretized and let $\gamma_k(\cdot) = \gamma(t_k, \cdot)$. We assume that the $\gamma(\cdot, z)$ curves are flexible enough to absorb any increasing mapping $\beta : [0,1] \to [0,1]$ (representing a discretization of $[0,1]$ at $(\beta(t_k))_{k\in\{0,\dots,M\}}$).

AIS algorithm starts by sampling the initial particles $z_0 \sim \gamma_0(z)$ and evaluating the initial weights $w_0 = 1$. Then, particles are gradually moved in iteration k using Markov kernel $\overrightarrow{T}_k(z_k|z_{k-1})$ such that the distribution of particles z_M approaches the target distribution $\pi(z)$. The particle trajectory distribution in AIS has the joint density $\overrightarrow{q}_{\phi,M}(z_{0:M}) = \gamma_0(z_0) \prod_k \overrightarrow{T}_k(z_k|z_{k-1})$. In each iteration a second Markov kernel $\overleftarrow{T}_k(z_{k-1}|z_k)$ is used to approximate the probability density of backward transition $z_k \to z_{k-1}$ such that the ratio $w_k(z_{k-1}, z_k) = \gamma_k(z_k)\overleftarrow{T}_k(z_{k-1}|z_k)/\gamma_{k-1}(z_{k-1})\overrightarrow{T}_k(z_k|z_{k-1})$ is well-defined on $\mathbb{R}^d$ and the importance weights are adjusted by multiplying $w_k(z_{k-1}, z_k)$ with the current weights. The final importance weights can be evaluated as

$$w(z_{0:M}) = \frac{\tilde{\pi}(z_M)\overleftarrow{q}_{\phi,M}(z_{0:M-1}|z_M)}{\overrightarrow{q}_{\phi,M}(z_{0:M})} = \prod_{k=1}^{M} w_k(z_{k-1}, z_k)$$

where $\overleftarrow{q}_{\phi,M}(z_{0:M-1}|z_M) = \prod_k \overleftarrow{T}_k(z_{k-1}|z_k)$.

The optimal backward process $\overleftarrow{q}_{\phi,M}(z_{0:M-1}|z_M)$ with zero variance estimator $\hat{p}^N(x)$ is given by inverse kernels

$$\overleftarrow{T}^*_k(z_{k-1}|z_k) = \frac{\overrightarrow{q}_{\phi,k-1}(z_{k-1})\overrightarrow{T}_k(z_k|z_{k-1})}{\overrightarrow{q}_{\phi,k}(z_k)}$$

where $\overrightarrow{q}_{\phi,k}(z_k)$ is the marginalization of $\overrightarrow{q}_{\phi,M}(z_{0:M})$ over all z_m except $m \neq k$. However it is intractable to use $\overleftarrow{T}^*_k(z_{k-1}|z_k)$ for sampling [8].

Alternatively, we can use a heuristically fixed $(\gamma_k)_k$ to guide the samples towards the target's basins of energy. One may choose $\overrightarrow{T}_k$ from popular MCMC kernels like Random Walk Metropolis-Hastings (RWMH), Hamiltonian Monte Carlo (HMC), Metropolis Adjusted Langevin Algorithm (MALA) or Unadjusted Langevin Algorithm (ULA) with π_k as their stationary distribution [45,46]. MH-corrected kernels typically do not admit a transition probability density. However, due to detailed balance $\gamma_k(z_k)\overrightarrow{T}_k(z_{k-1}|z_k) = \gamma_k(z_{k-1})\overrightarrow{T}_k(z_k|z_{k-1})$ the kernel $\overrightarrow{T}_k$ is reversible with respect to γ_k and with backward kernel set to its reversal $\gamma_k(z_{k-1})\overrightarrow{T}_k(z_k|z_{k-1})/\gamma_k(z_k)$ we can obtain well-defined importance weight updates $w_k(z_{k-1}, z_k) = \gamma_k(z_{k-1})/\gamma_{k-1}(z_{k-1})$ [34]. When $\overrightarrow{T}_k$ is not γ_k-invariant, an approximate reversal kernel with same assumption may be constructed from $\overleftarrow{T}_k(z_{k-1}|z_k) = \overrightarrow{T}_k(z_{k-1}|z_k)$ while the weight updates preserves their original form [46].

In another approach, reparameterizable forward and backward transition kernels can be used where the parameters are optimized to push samples from one predetermined bridging distribution to the next, which limits the complexity of the transition probability density considerably [1,31,49].

3 Optimized Path Annealed Importance Sampling

Typical implementations of AIS use predefined heuristics for the sequence $(\gamma_k)_k$ and the schedule β. Popular choices include the geometric average density path $\gamma(t,z) = \tilde{\pi}(z)^{\beta(t)} q_\phi(z|x)^{1-\beta(t)}$ with linear $\beta(t) = t$ or geometric $\beta(t) = \alpha^{\log t}$ schedule, for some real $\alpha > 1$. Instead of focusing on optimization of the transition kernels for a fix annealing sequence, we propose to optimize the bridging distribution path $\gamma(t,\cdot)$ in a class of positive parametric functions. If the class is sufficiently large the schedule β will be implicit in γ and therefore we can fix the schedule to be linear and only focus on parameterizing the $\gamma(t,\cdot)$. We use Deep Neural Networks (DNN) to define our parametric density function path. We name our procedure Optimized Path Annealed Importance Sampling (OP-AIS).

3.1 Parameterized AIS

We opt to parameterize the time dependent path γ as opposed to parameterizing each intermediary distribution individually. The main reason for this choice is that it results in a density path that is continuous in t and allows parameters to be shared by intermediate distributions. We use a continuous mapping $u_\phi : \mathcal{X} \times \mathcal{Z} \times [0,1] \to \mathbb{R}$ with additional terms that adjust the boundary constraints at $t \in \{0,1\}$. We consider the following boundary adjustment which coincides with geometric average of the proposal and target distributions when $u_\phi \equiv 0$.

$$\begin{aligned}\log \gamma(t,z) =& u_\phi(x,z,t) \\ &+ (1-t)\left[-u_\phi(x,z,0) + \log \gamma_0(z)\right] \\ &+ t\left[-u_\phi(x,z,1) + \log \tilde{\pi}(z)\right]\end{aligned} \tag{3}$$

If $u_\phi(x,z,t) \in (-\infty,\infty)$ is well-defined for all $x \in \mathcal{X}$, $z \in \mathbb{R}^d$ and $t \in (0,1)$ and the proposal is chosen such that $\text{supp}(\gamma(1,\cdot)) \subseteq \text{supp}(\gamma(0,\cdot))$ then $\text{supp}(\gamma(1,\cdot)) = \text{supp}(\gamma(t,\cdot))$ for all $t \in [0,1]$ and the importance weights updates are always defined using the common gaussian or invariant transition kernels. Notably, we can apply this parameterization to any arbitrary path of density functions by twisting the path with multiplication of an arbitrary positive function $\exp(-u_\phi(x,z,t))$ with correct boundary behavior.

3.2 Optimization for the Parametric Distribution

Given an unnormalized target distribution, Effective Sample Size (ESS) and its conditional extension are the de facto measure used to adaptively tune AIS

[8,21,22,34]. ESS of N parallel runs of AIS is given by

$$\mathrm{ESS}(w^1, ..., w^N) = \frac{\left(\sum w^j\right)^2}{N \sum (w^j)^2}$$

and is expected to reflect the estimator variance odds with exact samples from $\pi(z)$ and importance sampling (2). ESS is sensitive to weight degeneracy problem that is prominent in AIS and other sequential importance sampling methods (when only a few particles contribute to the Monte Carlo approximate). However, high ESS does not determine if the particles have sufficient dispersion according to the target distribution. Maximization of this objective corresponds to minimization of a consistent estimate of χ^2-divergence between the sampling and extended target distribution, however typically leads to insufficient dispersion of particles, therefore we omitted the results. See [11] for further discussion on ESS and its defects.

Grosse et. al. in [16] derived a bound on $\log Z$ by running the AIS algorithm in forward and backward directions. Original direction of AIS evaluates $\log \overrightarrow{w} = \log w(z_{0:M})$ which in expectation lower bounds $\log Z$.

$$\mathbb{E}_{\overrightarrow{q}_{\phi,M}}[\log \overrightarrow{w}] \leq \log Z \tag{4}$$

Drawing sample trajectories $z'_{0:M}$ from the reverse direction of AIS algorithm defined by the ancestral sampling in joint distribution $\pi(z'_M)\overleftarrow{q}_{\phi,M}(z'_{0:M-1}|z'_M)$ we can compute random variable $\log \overleftarrow{w} = \log w(z'_{0:M})$ which upper bounds $\log Z$ in expectation. The gap between the two expectations in forward and backward processes is called the BiDirectional Monte Carlo (BDMC) gap and is equivalent to Jefferys divergence between these distributions.

$$\begin{aligned}\mathcal{L}^{\mathrm{BDMC}}(x) &= \mathbb{E}_{\pi\overleftarrow{q}_{\phi,M}}[\log \overleftarrow{w}] - \mathbb{E}_{\overrightarrow{q}_{\phi,M}}[\log \overrightarrow{w}] \\ &= \mathrm{D_{KL}}(\pi\overleftarrow{q}_{\phi,M}||\overrightarrow{q}_{\phi,M}) + \mathrm{D_{KL}}(\overrightarrow{q}_{\phi,M}||\pi\overleftarrow{q}_{\phi,M})\end{aligned} \tag{5}$$

This bound is also frequently used to assess the accuracy of the AIS estimator [15,20,50]. Unbiased estimation of this bound requires exact samples from the target distribution. Therefore, an approximation is made replacing the data samples with the synthesized samples by the generative model, where we have access to an underlying latent variable. This approximation results in more accurate log-likelihood estimation of samples closer to the model which affects the objectivity of the test. Assuming a long chain and close to perfect transitions the final particles from AIS forward process can be used to approximate this bound instead. However, with low computation budget this marginal distribution may be far off from the target especially in the initial optimization iterations. Although, the BDMC gap achieves a natural trade-off between minimization of the bias in ELBO bound and log empirical variance of the importance weights (Fig. 2), the two terms in $\mathcal{L}^{\mathrm{BDMC}}(x)$ have contradicting gradients resulting in unstable optimization of hyperparameters with gradient based methods.

In order to achieve low variance importance weights we need to match the distributions of the forward and backward processes. It is common to minimize the inverse KL-divergence between these distributions similar to maximization of (1) as this results in reducing the bias in (4).

$$\mathcal{L}^{\text{AIS}}(x) = \text{D}_{\text{KL}}(\overrightarrow{q}_{\phi,M}||\tilde{\pi}\overleftarrow{q}_{\phi,M}) = -\mathbb{E}_{\overrightarrow{q}_{\phi,M}}[\log \overrightarrow{w}]$$

We can evaluate an unbiased estimate of inverse KL divergence using samples from $\overrightarrow{q}_{\phi,M}$.

As was shown by authors of [9,28] under some assumptions, the variance of importance weights can be controlled by the inverse KL-divergence between two distributions. In particular, with some adjustment to the AIS setting we have

$$\text{D}_{\text{KL}}(\overrightarrow{q}_{\phi,M}||\tilde{\pi}\overleftarrow{q}_{\phi,M}) \approx \frac{1}{2Z^2}\text{Var}_{\overrightarrow{q}_{\phi,M}}[w(z_{0:M})]$$

Therefore, by minimizing the inverse KL-divergence we minimize the log marginal likelihood bias and the variance of the importance weights, simultaneously. The equality holds when $\pi(z_k)\overleftarrow{q}_{\phi,M}(z_{0:M-1}|z_M) = \overrightarrow{q}_{\phi,M}(z_{0:M})$ which results in zero importance weight variance and $\mathcal{L}^{\text{AIS}}(x) = 0$.

4 Stochastic Optimization for OP-AIS

In order to evaluate the gradients of $\mathcal{L}^{\text{AIS}}(x)$ with respect to the parameters ϕ we rely on the reparameterization of commonly used reversible transition kernels introduced by [45] and their gradient estimation. The MH-corrected Markov kernel $\overrightarrow{T}_k(z_k|z_{k-1})$ is given by

$$\begin{aligned}\overrightarrow{T}_k(z_k|z_{k-1}) &= r(z_k|z_{k-1})\alpha_k(z_{k-1}, z_k)\\ &+ \int (1 - \alpha_k(z|z_{k-1}))\, r(z|z_{k-1})dz\delta_{z_{k-1}}(z_k)\end{aligned}$$

with $r(.|z_{k-1})$ denoting the conditional probability density of proposed particle in the kth iteration, $\alpha_k(z_{k-1}, z) = \left(1 \wedge \frac{\gamma_k(z)r(z_{k-1}|z)}{\gamma_k(z_{k-1})r(z|z_{k-1})}\right)$ the acceptance rate at this iteration and δ_z denoting the Dirac measure at z. We assume that the proposal $r(z|z_{k-1})$ is reparameterizable with a auxiliary random variable $\epsilon_k \sim \eta(\epsilon)$ and transformation $z = T_k(\epsilon_k, z_{k-1})$ which is surjective in second argument. After transformation the proposal state is accepted with probability $\alpha_k(z_{k-1}, T_k(\epsilon_k, z_{k-1}))$ (denoted with α_k as a shorthand) and we set $z_k = T_k(\epsilon_k, z_{k-1})$. Otherwise, we keep the old state and set $z_k = z_{k-1}$. We determine the proper action by sampling a binary random variable $a_k \sim \text{Bern}(\alpha_k(z_{k-1}, T_k(\epsilon_k, z_{k-1})))$.

For instance, for a Random Walk kernel $\overrightarrow{T}_k$, we assume that $\eta(\epsilon)$ is the probability density function of standard normal distribution and $T_k(\epsilon, z) = z + \Sigma^{1/2}\epsilon$ for a positive definite covariance matrix $\Sigma \in \mathbb{R}^{d\times d}$. We refer the readers to the derivation of transformations corresponding to HMC, MALA and ULA by [45] for further details.

Using the above reparameterization we get

$$\mathcal{L}^{\text{AIS}}(x) = -\mathbb{E}_{\gamma_0(z_0)\prod_k \eta(\epsilon_k)\prod_k \zeta_k(a_k|z_0,\epsilon_{1:k},a_{1:k-1})}[\log w]$$

for $\zeta_k(a^k|z_0, \epsilon_{1:k}, a_{1:k-1}) = \alpha_k^{a_k}(1-\alpha_k)^{a_k}$.

The gradient of $\mathcal{L}^{\text{AIS}}(x)$ is given by

$$\begin{aligned}\nabla_\phi \mathcal{L}^{\text{AIS}}(x) = &- \mathbb{E}[\nabla_\phi \log w] \\ &- \sum_k \mathbb{E}[\log w \nabla_\phi \log \zeta_k(a_k|z_0, \epsilon_{1:k}, a_{1:k-1})]\end{aligned}$$

We can derive $\mathcal{L}^{\text{AIS}}(x)$ approximation using the particle and importance weights generated from the AIS.

$$\begin{aligned}\nabla_\phi \hat{\mathcal{L}}^{\text{AIS}}(x) = &- \frac{1}{N}\sum_{i=1}^{N} \nabla_\phi \log w^i \\ &\underbrace{- \frac{1}{N}\sum_{i=1}^{N}\sum_k \log w^i \nabla_\phi \log \zeta(a_k^i|z_0^i, \epsilon_{1:k}^i, a_{1:k-1}^i)}_{\nabla_\phi \hat{\mathcal{L}}^{\text{score}}(x)}\end{aligned}$$

$\nabla_\phi \hat{\mathcal{L}}^{\text{AIS}}(x)$ is a strongly consistent estimator of $\nabla_\phi \mathcal{L}^{\text{AIS}}(x)$ [27].

Score function estimator is notorious for its high variance. It is a standard practice to omit $\nabla_\phi \mathcal{L}^{\text{score}}$ in parameter updates for more stable optimization. However, this results in biased gradient estimate which adversely effects our log marginal likelihood estimation in the latent variable model experiements. To reduce the high variance of $\nabla_\phi \hat{\mathcal{L}}^{\text{score}}(x)$ we can employ the variation reduction techniques [33,36]. For instance, in our experiments we use the common particles and weights to estimate the two expectations in $\mathcal{L}^{\text{AIS}}(x)$ and the one-out control variates proposed by [30] to substitute $\nabla_\phi \hat{\mathcal{L}}^{\text{score}}(x)$ with

$$\sum_{i=1}^{N}\left[\frac{NR^i - \sum_j R^j}{N-1}\right]\sum_k \nabla_\phi \log \zeta_k(a_k^i|z_0^i, \epsilon_{1:k}^i, a_{1:k-1}^i)$$

where $R^i = -\log w^i/N$.

Pseudocode of the optimization algorithm is given in Algorithm (1) for completeness.

5 Literature Review

We can interpret AIS as the discretization of the path sampling algorithm where the thermodynamics integral identity holds

$$\log Z = \int_0^1 \mathbb{E}_{\gamma(t,z)/Z_t}\left[\frac{d}{dt}\log \gamma(t,z)\right] dt$$

Algorithm 1: Optimization of parameterized AIS

input : Target $\tilde{\pi}$, proposal density γ_0, and η
output: Parameters ϕ

1 **for** K *epochs* **do**
2 Set $\Delta \leftarrow 0$
3 **for** $i = 1$ *to* N **do**
4   Set $\log \mathrm{w}^i \leftarrow 0$, $\log \zeta^i \leftarrow 0$
5   Draw $z^i \sim q_0(z)$
6   Draw $\epsilon^i_{1:M} \sim \prod_k \eta(\epsilon_k)$
7 **for** $k = 1$ *to* M **do**
8   **for** $i = 1$ *to* N **do**
9   Set $\log \mathrm{w}^i \leftarrow \log \mathrm{w}^i + \log \frac{\gamma_k(z^i)}{\gamma_{k-1}(z^i)}$
10   Draw $a_k^{(n)} \sim \mathrm{Bern}(\alpha_k)$
11   Set $\log \zeta^i \leftarrow \log \zeta^i + \log \zeta(a^i_k | z^i_0, \epsilon^i_{1:k}, a^i_{1:k-1})$
12   Set $z^i \leftarrow T_k(\epsilon^i_k, z^i)^{a^i_k} (z^i)^{1-a^i_k}$
13 Get R^i for $i \in [N]$
14 Update the annealing path parameters with $\phi \leftarrow \phi - \kappa \nabla_\phi \hat{\mathcal{L}}^{\mathrm{AIS}}(x)$

for $Z_t = \int \gamma(t, z) dz$. Gelman et. al. in [13] derived the optimal path and scaling of the bridging distributions for minimum variance estimation. Evaluation of optimal path is intractable in general. In turn, authors of [2,6] provided closed form optimal path for special cases. Viewing the importance weight updates as the finite difference approximations, the authors in [17] derive the asymptotic bias of $\log w$ estimator and propose moment average path between members of exponential family. A variational representation of the annealing distributions minimizing weighted α-divergences between proposal and target distributions are derived in [17,29]. In [26], the authors minimize the asymptotic variance of estimator $\log w$ for a given parametric path of distributions.

The analysis in these works are derived based on the assumption of reaching equilibrium in each iteration and independence of particles. This is not the case when transition kernels are not invertable. On the other hand, exhaustive heuristic optimization to tune AIS hyperparameters is laborious [40]. Other methods to adapt AIS hyperparameters include monitoring the acceptance rate in Markov transitions to adjust its parameters [21,46] and designing problem specific or heuristic intermediary distribution sequences [17,21,29,40]. On the contrary, while our method requires training before evaluation, we can amortize the overhead by parameter sharing and reduce the computation cost by using only a small subset of validation samples. As a consequence, we find an efficient and shorter annealing process for sampling. Our aim is to enable AIS to achieve high accuracy despite limited computation resources.

Recently, Variational Inference and Monte Carlo methods have been combined to increase the flexibility of parametric variational family $\mathcal{Q}$ and benefit from the convergence properties of sampling methods. Authors of [7,19,42,45,48] combine MCMC and VI by extending the space of latent variables $(z_k)_{k\in[M]}$. A framework with parametric transitions for SMC algorithms and a surrogate lower bound ELBO are provided in [18,31,47]. For example, in [28], the authors derive an approximation of the asymptotic variance of importance weights in Variational Filtering Objectives, and in [1,49], flows are used to overcome energy density barriers. A differentiable alternative to MH-correcting step in HMC transitions are proposed [52], while in [12] the authors derive the importance weights resulting from unadjusted HMC process and its reversal.

Our work in particular bares more similarity to MCVAE proposed by [46] among others as we use the same reparameterization and optimization discribed in Sect. 4. In MCVAE the authors use AIS with heuristic density path for low variance estimation of the log marginal likelihood during training of a latent variable model. Our goal here is to reduce the computation complexity of evaluation of a trained model by optimizing the intermediary distributions in AIS algorithm. We argue that the learnt AIS parameters from concurrent optimization of the model and the sampling algorithm (as performed in MCVAE) are not sufficient since the model is constantly changing during training and the posterior approximator is known to overfit the model [50].

6 Experiments

Here, we evaluate the proposed algorithm on complex synthetic 2d distributions which are often used to benchmark generative models [37]. Target distributions are illustrated in Fig. 1. We compare our algorithm to vanilla AIS [34] with geometric average path and geometric and linear schedule. We use RWMH, HMC and MALA transitions and normal proposal with learnable mean and diagonal covariance matrix. M between 2 and 128 bridging distributions and 256 samples are used for training. u_ϕ is implemented with a DNN with one hidden layer with 4 dimensions and LeakyReLU(0.01) nonlinearty except at the output layer.

The parametric AIS is trained with Adam optimizer [23] with learning rate of 0.03 and betas = (0.5, 0.999) for 100 epochs. The RWMH kernels have local normal proposal steps with learnable diagonal covariance matrix and the step size in all transition kernels are trained along with other parameters. We use cross validation to choose the step size in vanilla AIS from $\{0.01, 0.05, 0.1, 0.5, 1.\}$. The normalization constants are estimated using $N = 4096$ samples. To act as the reference normalization constant we use vanilla AIS with $M = 1024$, 10 MCMC steps per iteration with 3 different step sizes. All the code is written with Pytorch [35] and experiments are run on a GeForce GTX 1080 Ti. Code is available here: https://github.com/shgoshtasb/op_ais.

In Fig. 1 we illustrate particles generated from the samplers trained on each distribution with inverse KL divergence (PKL) and Jefferys divergence (PJ) color coded by their weight for $M = 8$. Warmer colors show particles with higher

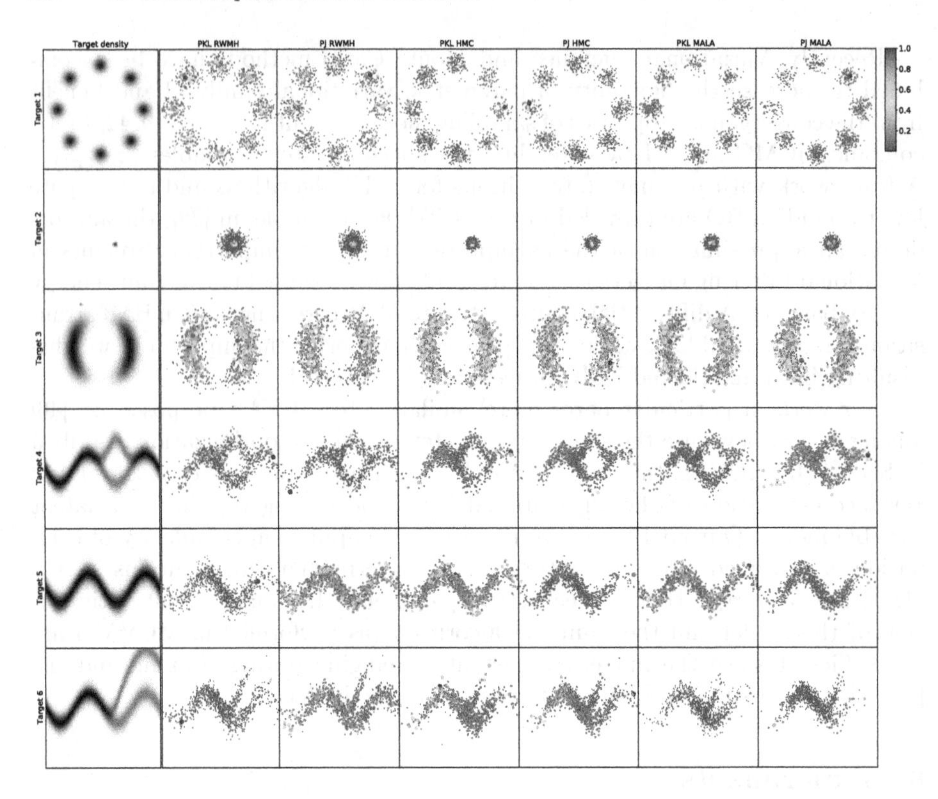

Fig. 1. Target distributions and particles of samplers with optimized path (trained with KL and Jefferys divergence) with RWMH, HMC and MALA transitions and $M = 8$. Particles with larger weights are coded to show comparable scale to maximum weight in each plot. Larger dots with warmer colors show particles with larger weights, with red for maximum weight in each set and dark blue for particles with 0.1 of the maximum weight (Color figure online)

weights which are main contributors to the partition function estimation, while dark blue particles are less effective in the estimation. We observe that training with KL divergence results in smaller bias, while using Jefferys divergence objective results in more effective particles and lower variance estimation in Target 3–6. Figure 2 shows the log partition function estimation and the empirical variance of importance weights for the mentioned samplers. We index the target distributions in the same order they appear in the Fig. 1. In most of the experiments, for small number of bridging distributions ($M \in \{2, 4, 8\}$) our method is able to improve over vanilla AIS algorithm with linear schedule and achieves slightly tighter lower bound in comparison to geometric schedule version. In comparison to training with inverse KL divergence, Jefferys divergence improves the variance of importance weights. We speculate this to be related to higher correlation of Jefferys divergence with weight variances.

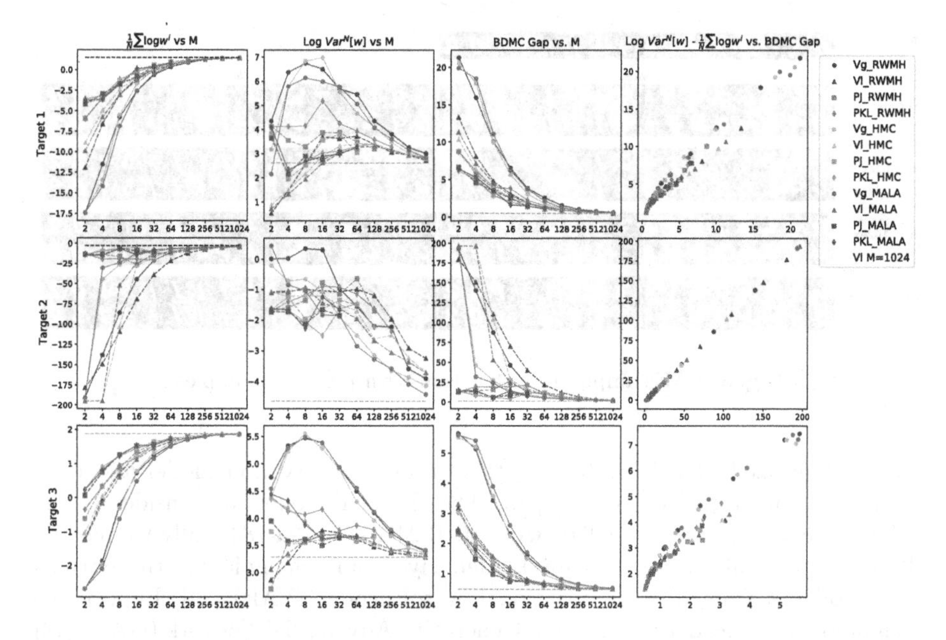

Fig. 2. From left to right: $\log Z$ estimation with Eq. 4, log empirical variance of importance weights and the Jefferys divergence of forward and backward processes vs M . (Vg, Vl for vanilla AIS with geometric and linear schedules, PJ and PKL for OP-AIS with Jefferys and KL-divergence optimization)

Interestingly, in multimodal and heavy tail distributions, empirical variance for small M remains below the variance of exhaustive AIS (Vl M=1024 in dashed Turquoise) and grows with M With only one MCMC step used in the experiments, particles ultimately cover a small area and don't reach the high density regions in the target distribution, resulting in low empirical weight variance despite the fact that the actual variance is much larger than our observation. This is specially problematic with the geometric schedule as the few bridging distributions are placed very close to the proposal and the particles are not encouraged to move far. In the low M setting as it is unlikely to observe high weight particles, it is misleading to use ESS or even empirical χ^2-divergence to tune the hyperparameters, whereas the inverse KL divergence provides better guidance for tuning. Increasing M results in more dispersed particles and higher variance. For large enough M as the particles can reach high density regions in the target, we observe that this variance decreases with M. Variance in parametric samplers is more steady and has a small growth in comparison due to the optimized choice of bridging distributions for each M.

6.1 MNIST Dataset

In this section, we assess the effectiveness of our method on a generative models based on similar DNN decoder architecture trained on MNIST dataset. We

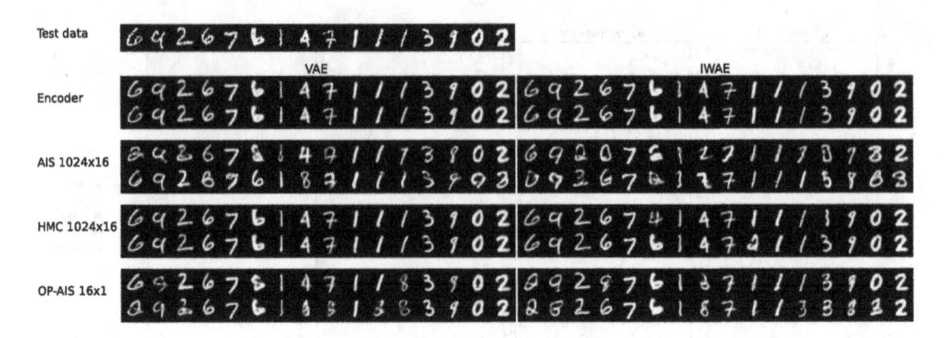

Fig. 3. VAE sample reconstruction with posterior samples

use the binarized MNIST dataset with Bernoulli observation model and data dequantization, rescaled to [0,1] [40]. The decoder model we consider is the "Wide" architecture used in [50] with 50-1024-1024-1024-784 Fully-Connected (FC) layers, Tanh activations for hidden layers and Sigmoid for the output. This model is trained as a Variational AutoEncoder (VAE) [24], Importance Weighted AutoEncoder (IWAE) [5], Generative Adversarial Network (GAN) [14] and Bidirectional GAN (BiGAN) [10]. We compare OP-AIS with $M \in \{8, 16\}$ and $N = 16$ particles per data sample, HMC transition kernels with 1 leapfrog step to MCMC and Vanilla AIS with $M = 1024$ and HMC transitions with 16 leapfrog steps. OP-AIS is trained with Adam optimizer with 0.01 learning rate and betas=(0.5, 0.999) for 100 epochs on 1400 samples and validated on 600 samples for highest log likelihood estimation. Reported log-likelihood values are over remaining 8000 test samples. Here, we use a symmetric encoder with Tanh activation and Dropout(0.2) at each hidden layer to transform x, concatenat with z, and input to u_ϕ with 32-hidden units and LeakyReLU(0.01) activation. The encoder network is similar to the one used in [50] for the "Wide" decoder setup. The training and evaluation time of OP-AIS (2h38 for $M = 16$ on average) is approximately similar to the sampling time required by the MCMC and Vanilla AIS algorithms (2 h 54 on average).

Table 1. Negative Log likelihood estimation of generative models on binarized MNIST

Model	Encoder	OP-AIS		AIS	
$M\times$steps		8×1	16×1	1024×16	$1024 \times 10^{\text{Tuned}}$
VAE	95.7	216.7 ± 29.8	$\mathbf{193 \pm 31.9}$	215.6 ± 35.4	87.8 ± 5.8
IWAE	88.5	$\mathbf{213.9 \pm 27.3}$	221.2 ± 34.1	235.4 ± 35.8	85.8 ± 5.6
GAN		805.9 ± 94.7	$\mathbf{785.4 \pm 67.0}$	902.9 ± 115.8	415.3 ± 48.8
BiGAN		900.5 ± 124.6	885.3 ± 138.3	$\mathbf{864.2 \pm 98.1}$	373.2 ± 44.6

Table 1 shows the negative log marginal likelihood estimation using posterior samples from each sampling algorithm. The ground truth values are approximated using a more exhaustive version of AIS with 10 tuning iterations for each transition kernel on the 1400 separated data samples (AIS $1024 \times 10^{\text{Tuned}}$) (8h12 average tuning + sampling time). Although the values of aquired by ground truth and encoder network (when available) are far from sampler estimations, estimates are more accurate than the sampling counterpart with similar computation time. We also observe that sample reconstruction with OP-AIS achieves similar samples to the test at least in style and are comparable to the other baselines in Fig. 3.

7 Conclusion

AIS yields an unbiased estimation for any annealing path, schedule and transition operator. However, the variance of the Monte Carlo estimator changes considerably with different choices. We propose to optimize the intermediary distribution sequence in AIS algorithm by matching the distribution of the AIS particle trajectory to joint distribution of backward process with the target distribution as its marginal. We compare commonly used measures for tuning and training AIS hyperparameters in their ability to reduce bias and variance of the estimation. Optimization of the annealing sequence improve the tightness of the lower bound and convergence rate on the 2D synthetic benchmarks and competes with more expensive heuristic sampling on latent variable models.

It is important to mention that log-likelihood has to be used with caution for model comparison, since a model with very good or bad generative sample quality or memorization may have high log likelihood [44]. It is recommended to use other sample quality measures (e.g. [41,51]) and model comparison methods (e.g. [20,39]) along with AIS.

Acknowledgements. This work is supported by funding from the European Union's Horizon 2020 research and innovation program under the Marie Sklodowska-Curie grant agreement No 813999 for this project.

References

1. Arbel, M., Matthews, A.G., Doucet, A.: Annealed flow transport Monte Carlo. arXiv preprint arXiv:2102.07501 (2021)
2. Behrens, G., Friel, N., Hurn, M.: Tuning tempered transitions. Stat. Comput. **22**(1), 65–78 (2012)
3. Blei, D.M., Kucukelbir, A., McAuliffe, J.D.: Variational inference: a review for statisticians. J. Am. Stat. Assoc. **112**(518), 859–877 (2017)
4. Brooks, S., Gelman, A., Jones, G., Meng, X.L.: Handbook of Markov Chain Monte Carlo. CRC Press, Boca Raton (2011)
5. Burda, Y., Grosse, R., Salakhutdinov, R.: Importance weighted autoencoders. arXiv preprint arXiv:1509.00519 (2015)

6. Calderhead, B., Girolami, M.: Estimating bayes factors via thermodynamic integration and population MCMC. Comput. Stat. Data Anal. **53**(12), 4028–4045 (2009)
7. Caterini, A.L., Doucet, A., Sejdinovic, D.: Hamiltonian variational auto-encoder. arXiv preprint arXiv:1805.11328 (2018)
8. Del Moral, P., Doucet, A., Jasra, A.: Sequential Monte Carlo samplers. J. Roy. Stat. Soc. Ser. B (Stat. Methodol.) **68**(3), 411–436 (2006)
9. Domke, J., Sheldon, D.: Importance weighting and variational inference. arXiv preprint arXiv:1808.09034 (2018)
10. Donahue, J., Krähenbühl, P., Darrell, T.: Adversarial feature learning. arXiv preprint arXiv:1605.09782 (2016)
11. Elvira, V., Martino, L., Robert, C.P.: Rethinking the effective sample size. arXiv preprint arXiv:1809.04129 (2018)
12. Geffner, T., Domke, J.: MCMC variational inference via uncorrected Hamiltonian annealing. arXiv preprint arXiv:2107.04150 (2021)
13. Gelman, A., Meng, X.L.: Simulating normalizing constants: from importance sampling to bridge sampling to path sampling. Stat. Sci. **13**, 163–185 (1998)
14. Goodfellow, I., et al.: Generative adversarial nets. In: Advances in Neural Information Processing Systems **27** (2014)
15. Grosse, R.B., Ancha, S., Roy, D.M.: Measuring the reliability of MCMC inference with bidirectional Monte Carlo. arXiv preprint arXiv:1606.02275 (2016)
16. Grosse, R.B., Ghahramani, Z., Adams, R.P.: Sandwiching the marginal likelihood using bidirectional Monte Carlo. arXiv preprint arXiv:1511.02543 (2015)
17. Grosse, R.B., Maddison, C.J., Salakhutdinov, R.: Annealing between distributions by averaging moments. In: Advances in Neural Information Processing Systems (NIPS), pp. 2769–2777. Citeseer (2013)
18. Gu, S., Ghahramani, Z., Turner, R.E.: Neural adaptive sequential Monte Carlo. arXiv preprint arXiv:1506.03338 (2015)
19. Hoffman, M.D.: Learning deep latent gaussian models with Markov chain Monte Carlo. In: International Conference on Machine Learning, pp. 1510–1519. PMLR (2017)
20. Huang, S., Makhzani, A., Cao, Y., Grosse, R.: Evaluating lossy compression rates of deep generative models. In: International Conference on Machine Learning, pp. 4444–4454. PMLR (2020)
21. Jasra, A., Stephens, D.A., Doucet, A., Tsagaris, T.: Inference for lévy-driven stochastic volatility models via adaptive sequential Monte Carlo. Scand. J. Stat. **38**(1), 1–22 (2011)
22. Johansen, A.M., Aston, J.A., Zhou, Y.: Towards automatic model comparison: An adaptive sequential Monte Carlo approach (2015)
23. Kingma, D.P., Ba, J.: Adam: a method for stochastic optimization. In: Bengio, Y., LeCun, Y. (eds.) 3rd International Conference on Learning Representations, ICLR 2015, San Diego, CA, USA, 7–9 May 2015, Conference Track Proceedings (2015). http://arxiv.org/abs/1412.6980
24. Kingma, D.P., Welling, M.: Auto-encoding variational bayes. arXiv preprint arXiv:1312.6114 (2013)
25. Kingma, D.P., Salimans, T., Jozefowicz, R., Chen, X., Sutskever, I., Welling, M.: Improved variational inference with inverse autoregressive flow. In: Advances in Neural Information Processing Systems. vol. 29 (2016)
26. Kiwaki, T.: Variational optimization of annealing schedules. arXiv preprint arXiv:1502.05313 (2015)
27. Laubenfels, R.: Feynman-Kac Formulae: Genealogical and Interacting Particle Systems with Applications. Taylor & Francis, Milton Park (2005)

28. Maddison, C.J., et al.: Filtering variational objectives. arXiv preprint arXiv:1705.09279 (2017)
29. Masrani, V., et al.: q-Paths: generalizing the geometric annealing path using power means. arXiv preprint arXiv:2107.00745 (2021)
30. Mnih, A., Rezende, D.: Variational inference for Monte Carlo objectives. In: International Conference on Machine Learning, pp. 2188–2196. PMLR (2016)
31. Naesseth, C., Linderman, S., Ranganath, R., Blei, D.: Variational sequential Monte Carlo. In: International Conference on Artificial Intelligence and Statistics, pp. 968–977. PMLR (2018)
32. Naesseth, C., Lindsten, F., Schon, T.: Nested sequential Monte Carlo methods. In: International Conference on Machine Learning, pp. 1292–1301. PMLR (2015)
33. Naesseth, C., Ruiz, F., Linderman, S., Blei, D.: Reparameterization gradients through acceptance-rejection sampling algorithms. In: Artificial Intelligence and Statistics, pp. 489–498. PMLR (2017)
34. Neal, R.M.: Annealed importance sampling. Stat. Comput. **11**(2), 125–139 (2001)
35. Paszke, A., et al.: Automatic differentiation in pytorch (2017)
36. Ranganath, R., Gerrish, S., Blei, D.: Black box variational inference. In: Artificial intelligence and statistics, pp. 814–822. PMLR (2014)
37. Rezende, D., Mohamed, S.: Variational inference with normalizing flows. In: International Conference on Machine Learning, pp. 1530–1538. PMLR (2015)
38. Rezende, D.J., Mohamed, S., Wierstra, D.: Stochastic backpropagation and approximate inference in deep generative models. In: International conference on machine learning, pp. 1278–1286. PMLR (2014)
39. Sajjadi, M.S., Bachem, O., Lucic, M., Bousquet, O., Gelly, S.: Assessing generative models via precision and recall. In: Advances in Neural Information Processing Systems **31** (2018)
40. Salakhutdinov, R., Murray, I.: On the quantitative analysis of deep belief networks. In: Proceedings of the 25th International Conference on Machine Learning, pp. 872–879 (2008)
41. Salimans, T., Goodfellow, I., Zaremba, W., Cheung, V., Radford, A., Chen, X.: Improved techniques for training GANs. In: Advances in Neural Information Processing Systems **29** (2016)
42. Salimans, T., Kingma, D., Welling, M.: Markov chain Monte Carlo and variational inference: Bridging the gap. In: International Conference on Machine Learning, pp. 1218–1226. PMLR (2015)
43. Skilling, J., et al.: Nested sampling for general bayesian computation. Bayesian Anal. **1**(4), 833–859 (2006)
44. Theis, L., Oord, A.V.D., Bethge, M.: A note on the evaluation of generative models. arXiv preprint arXiv:1511.01844 (2015)
45. Thin, A., et al.: MetFlow: a new efficient method for bridging the gap between Markov chain Monte Carlo and variational inference. arXiv preprint arXiv:2002.12253 (2020)
46. Thin, A., Kotelevskii, N., Doucet, A., Durmus, A., Moulines, E., Panov, M.: Monte Carlo variational auto-encoders. In: International Conference on Machine Learning, pp. 10247–10257. PMLR (2021)
47. Tom, T.A.L.M.I., Wood, J.T.R.F.: Auto-encoding sequential Monte Carlo. stat **1050**, 29 (2017)
48. Wolf, C., Karl, M., van der Smagt, P.: Variational inference with Hamiltonian Monte Carlo. arXiv preprint arXiv:1609.08203 (2016)
49. Wu, H., Köhler, J., Noé, F.: Stochastic normalizing flows. arXiv preprint arXiv:2002.06707 (2020)

50. Wu, Y., Burda, Y., Salakhutdinov, R., Grosse, R.: On the quantitative analysis of decoder-based generative models. CoRR arXiv preprint arXiv:1611.04273 (2016)
51. Yu, Y., Zhang, W., Deng, Y.: Frechet inception distance (fid) for evaluating GANs
52. Zhang, G., Hsu, K., Li, J., Finn, C., Grosse, R.: Differentiable annealed importance sampling and the perils of gradient noise. arXiv preprint arXiv:2107.10211 (2021)

Open Access This chapter is licensed under the terms of the Creative Commons Attribution 4.0 International License (http://creativecommons.org/licenses/by/4.0/), which permits use, sharing, adaptation, distribution and reproduction in any medium or format, as long as you give appropriate credit to the original author(s) and the source, provide a link to the Creative Commons license and indicate if changes were made.

The images or other third party material in this chapter are included in the chapter's Creative Commons license, unless indicated otherwise in a credit line to the material. If material is not included in the chapter's Creative Commons license and your intended use is not permitted by statutory regulation or exceeds the permitted use, you will need to obtain permission directly from the copyright holder.

Bayesian Nonparametrics for Sparse Dynamic Networks

Cian Naik[1(✉)], François Caron[1], Judith Rousseau[1], Yee Whye Teh[1,3], and Konstantina Palla[2]

[1] Department of Statistics, University of Oxford, Oxford, UK
cian.naik@stats.ox.ac.uk
[2] Microsoft Research, Cambridge, UK
[3] Google Deepmind, London, UK

Abstract. In this paper we propose a Bayesian nonparametric approach to modelling sparse time-varying networks. A positive parameter is associated to each node of a network, which models the sociability of that node. Sociabilities are assumed to evolve over time, and are modelled via a dynamic point process model. The model is able to capture long term evolution of the sociabilities. Moreover, it yields sparse graphs, where the number of edges grows subquadratically with the number of nodes. The evolution of the sociabilities is described by a tractable time-varying generalised gamma process. We provide some theoretical insights into the model and apply it to three datasets: a simulated network, a network of hyperlinks between communities on Reddit, and a network of co-occurences of words in Reuters news articles after the September 11th attacks.

Keywords: Bayesian nonparametrics · Poisson random measures · Networks · Random graphs · Sparsity · Point processes

1 Introduction

This article is concerned with the analysis of dynamic networks, where one observes the evolution of links among a set of objects over time. As an example, links may represent interactions between individuals on social media platforms across different days, or the co-occurrence of words across a series of newspaper articles. In each case the pattern of these interactions will generally vary over different time steps. Probabilistic approaches treat the dynamic networks of interest as random graphs, where the vertices (nodes) and edges correspond to objects and links respectively. In the graph setting, sparsity is defined in terms of the rate in which the numbers of edges grows as the number of nodes increases. In a *sparse* graph the number of edges grows sub-quadratically in the number of

Supplementary Information The online version contains supplementary material available at https://doi.org/10.1007/978-3-031-26419-1_12.

© The Author(s), under exclusive license to Springer Nature Switzerland AG 2023
M.-R. Amini et al. (Eds.): ECML PKDD 2022, LNAI 13717, pp. 191–206, 2023.
https://doi.org/10.1007/978-3-031-26419-1_12

nodes. Hence, in a large graphs, two nodes chosen at random are very unlikely to be linked. Sparsity is a property found in many real-world network datasets [34,36], and in our work we are concerned with modelling sparse networks.

Bayesian approaches play an important role in the modelling of random graphs, providing a framework for parameter estimation and uncertainty quantification. However, most of the popular Bayesian random graph models result in dense graphs, i.e. where the number of edges grows quadratically in the number of nodes, see [36] for a review. A recent Bayesian nonparametric approach, proposed by [7] and later developed in a number of articles [4,20,33,39,40], seeks to solve this problem by representing a graph as an infinite point process on $\mathbb{R}_+^2$, giving rise to a class of sparse random graphs. This class of sparse models is projective and admits a representation theorem due to [25].

In this paper, we are interested in the dynamic domain and aim to probabilistically model the evolution of sparse graphs over time, where edges may appear and disappear, and the node popularity may change over time. We build on the sparse graph model of [7] and extend it to deal with time series of network data. We describe a fully generative and projective approach for the construction of sparse dynamic graphs. It is challenging to perform exact inference using the framework we introduce, and thus we consider an approximate inference method, using a finite-dimensional approximation introduced by [30].

The rest of the article is structured as follows. In Sect. 2 we give some background on the sparse network model of [7]. Section 3 describes the novel statistical dynamic network model we introduce in detail, as well as its sparsity properties. The approximate inference method, based on a truncation of the infinite-dimensional model, is described in Sect. 4. In Sect. 5 we present illustrations of our approach to three different dynamic networks with thousands of nodes and edges.

2 Background: Model of Caron and Fox for Sparse Static Networks

Bayesian nonparametrics provides a natural setting for the study of sparse graphs. Parameters can be infinite dimensional, and thus the complexity of models can adapt to data in question. In the context of network modelling, this allows for the consideration of graphs that may have infinitely many nodes, only finitely many of which form connections.

To this end, instead of the standard approach of representing a graph G by a finite dimensional adjacency matrix, [7] instead represent it by a point process. Letting $\alpha > 0$ be a positive parameter tuning the size of the network, a finite multigraph of size $\alpha > 0$ is represented by a point process on $[0, \alpha]^2$

$$N = \sum_{i,j} n_{ij} \delta_{(\theta_i, \theta_j)},$$

where $n_{ij} = n_{ji} \in \{0, 1, 2, \ldots\}$, $i \leq j$, represents the number of interactions between individuals i and j. The $\theta_i \in [0, \alpha]$ can be interpreted as node labels,

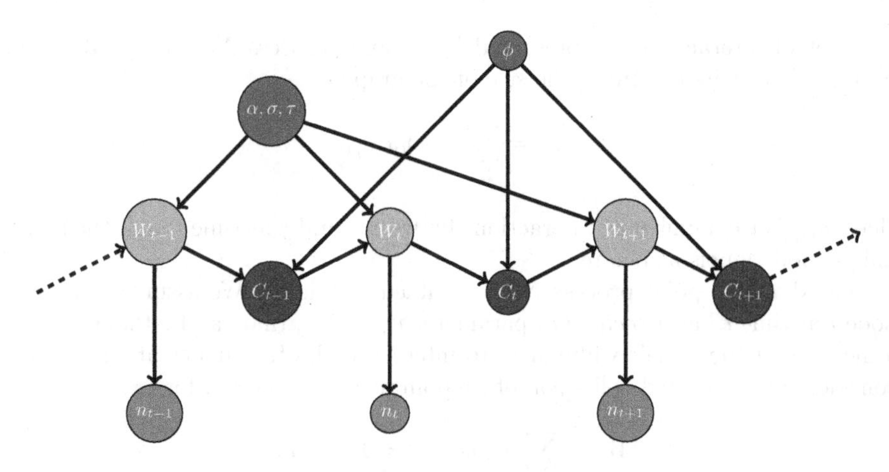

Fig. 1. Graphical representation of the model. The counts n_t are derived from the sociabilities W_t, whose time-evolution depends on the counts C_t and hyperparameters α, σ, τ and ϕ.

and $\delta_{(\theta_i,\theta_j)}$ denotes a point mass at location (θ_i, θ_j). The node labels are introduced for the model's construction, but are not observed nor inferred. Each node i is assigned a sociability parameter $w_i > 0$. Let $W = \sum_i w_i \delta_{\theta_i}$ be the corresponding random measure (CRM) on $[0, \alpha]$. We assume that W is a generalised gamma completely random measure [1,5,27,31]. That is, $\{(w_i, \theta_i)_{i\geq 1}\}$ are the points of a Poisson point process with mean measure $\nu(w)dw\mathbb{1}_{\theta\leq\alpha}d\theta$ where $\mathbb{1}_A = 1$ if the statement A is true and 0 otherwise, and ν is a Lévy intensity on $(0, \infty)$ defined as

$$\nu(w) = \frac{1}{\Gamma(1-\sigma)} w^{-1-\sigma} e^{-\tau w} \tag{1}$$

with hyperparameters $\sigma < 1$ and $\tau > 0$. We write simply $W \sim \text{GG}(\alpha, \sigma, \tau)$ The GGP is a CRM with two interpretable parameters and useful conjugacy properties [10,31]. Importantly, with this GGP construction, [7,8] show that this model yields sparse graphs with a power-law degree distribution when $\sigma > 0$. The advantage of using this construction over a standard gamma process [28] is that the parameter σ allows us to control the sparsity properties of the model, and thus fit to networks with different power-law degree distributions.

To each pair of nodes i, j, we assign a number of latent interactions n_{ij}, where

$$n_{ij} \mid w_i, w_j \sim \begin{cases} \text{Poisson}(2w_iw_j) \; i < j, & n_{ji} = n_{ij} \\ \text{Poisson}(w_iw_j) \;\; i = j \end{cases} \tag{2}$$

Finally, two nodes are said to be connected if they have at least one interaction; let $z_{ij} = \mathbb{1}_{n_{ij}>0}$ be the binary variable indicating if two nodes are connected.

3 Dynamic Statistical Network Model

In order to study dynamically evolving networks, we assume that at each time $t = 1, 2, \ldots, T$, we observe a set of interactions between a number of nodes.

This set of interaction is represented by a point process N_t over $[0, \alpha]^2$ as in Equation (3), where α tunes the size of the graphs.

$$N_t = \sum_{i,j} n_{tij} \delta_{(\theta_i, \theta_j)}. \tag{3}$$

Here, n_{tij} is the number of interactions between i and j at time t, and the θ_i are unique node labels as before.

The dynamic point process N_t is obtained as follows. We assume that each node i at time t has a *sociability* parameter $w_{ti} \in \mathbb{R}_+$, that can be thought of as a measure of the node's willingness to interact with other nodes at time t. We consider the associated collection of random measures on $\mathbb{R}_+$, for $t = 1, \ldots, T$

$$W_t = \sum_i w_{ti} \delta_{\theta_i}, \quad t = 1, \ldots, T.$$

We first describe in Sect. 3.1 the model for the latent interactions. Then we describe in Sect. 3.2 the model for the time-varying sociability parameters $(W_t)_{t \geq 1}$. The overall probabilistic model is summarised in Fig. 1.

3.1 Dynamic Network Model Based on Observed Interactions

In the dynamic setting, what we observe in practice is often counts of interactions between nodes, e.g. hyperlinks, emails or co-occurrences, rather than a binary indicator of whether there is a connection between them. So for each pair of nodes $i \leq j$, we let $(n_{tij})_{t=1,2,\ldots T, j \geq i}$ be the interaction count between them at time t. We assume that n_{tij} can be modelled as

$$n_{tij} \mid w_{ti}, w_{tj} \sim \begin{cases} \text{Poisson}(2 w_{ti} w_{tj}) \; i < j, & n_{tji} = n_{tij} \\ \text{Poisson}(w_{ti} w_{tj}) \;\; i = j \end{cases} \tag{4}$$

This model can be easily adapted to graphs with directed edges, by modifying the Equation (4) to

$$n_{tij} \sim Poisson(w_{ti} w_{tj})$$

for all $i \neq j$, where n_{tij} now represents the number of interactions from i to j. The resulting inference algorithm essentially remains the same, and from now on we assume we are in the undirected edge setting. As in the static case, we can reconstruct the binary graph by letting $z_{tij} = \mathbb{1}_{(n_{tij} > 0)}$ be the binary variable indicating if nodes i and j are connected at time t, i.e. two nodes are connected at time t if and only if $n_{tij} > 0$. To avoid ambiguity, we say that the number of edges in the graph at time t is $\sum_{i > j} z_{tij}$, rather than counting the number of interactions between pairs of nodes. Marginalizing out the interaction counts n_{tij}, we have for $i \neq j$:

$$\Pr(z_{tij} = 1 \mid (w_{t-k,i}, w_{t-k,j})_{k=0,\ldots,t-1}) = 1 - e^{-2 w_{ti} w_{tj}}. \tag{5}$$

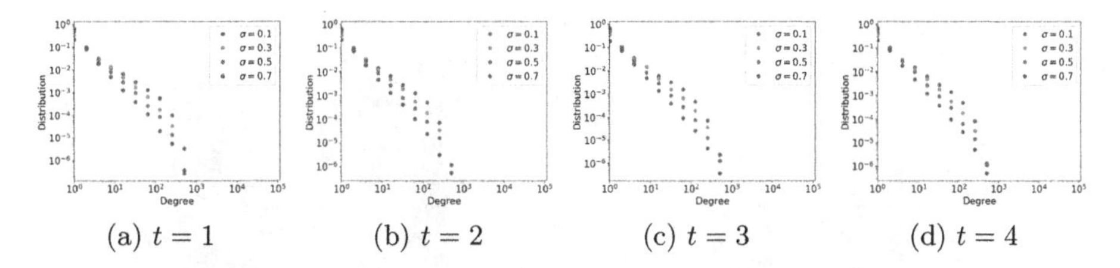

(a) $t = 1$ (b) $t = 2$ (c) $t = 3$ (d) $t = 4$

Fig. 2. Degree distributions over time, for a network simulated from the GG model with $T = 4$, $\alpha = 200$, $\tau = 1$, $\phi = 1$ and varying values of σ.

3.2 A Dependent Generalised Gamma Process for the Sociability Parameters

When modelling the sociability parameters in the dynamic setting, we have two goals. Firstly, we want the resulting graphs at each time to be sparse (ideally fitting within the framework of [7]). Secondly, we want sociability parameters to be dependent over time, so that we may model the smooth evolution of the sociabilities of the nodes. To this end, we consider here that the sequence of random measures $(W_t)_{t=1,2,\ldots}$ follows a Markov model, such that W_t is marginally distributed as $\text{GG}(\alpha, \sigma, \tau)$. In order to do this, we build on the generic construction of [37]. A similar model has been derived by [9] for dependent gamma processes (corresponding to the case $\sigma = 0$ here). As in [7], we use the generalised gamma process here because of the flexibility the sparsity parameter σ gives us. In particular, this setup allows us to capture power-law degree distributions, unlike with the gamma process.

For a sequence of additional latent variables $(C_t)_{t=1,2,\ldots}$, we consider a Markov chain $W_t \to C_t \to W_{t+1}$ starting with $W_1 \sim \text{GG}(\alpha, \sigma, \tau)$ that leaves W_t marginally $\text{GG}(\alpha, \sigma, \tau)$. For $t = 1, \ldots, T-1$, define

$$C_t = \sum_{i=1}^{\infty} c_{ti} \delta_{\theta_i} \quad c_{ti} | W_t \sim \text{Poisson}(\phi w_{ti}) \tag{6}$$

where $\phi > 0$ is a parameter tuning the correlation. Given C_t, the measure W_{t+1} is then constructed as a combination of masses defined by C_t and GG innovation:

$$W_{t+1} = W_{t+1}^* + \sum_{i=1}^{\infty} w_{t+1,i}^* \delta_{\theta_i} \tag{7}$$

with

$$W_{t+1}^* \sim \text{GG}(\alpha, \sigma, \tau + \phi) \tag{8}$$

$$w_{t+1,i}^* | C_t \sim \text{Gamma}(\max(c_{ti} - \sigma, 0), \tau + \phi). \tag{9}$$

By convention, $\text{Gamma}(0, \tau) = \delta_0$ (a point mass at 0). Hence, $w_{t+1,i}^* = 0$ if $c_{ti} = 0$, else $w_{t+1,i}^* > 0$. Because the conditional laws of $W_{t+1} | C_t$ coincide with that of $W_t | C_t$ [23,24,38], the construction guarantees that W_{t+1} has the same marginal

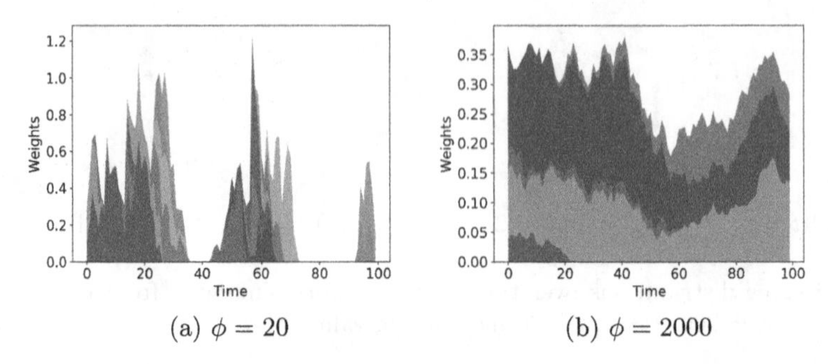

(a) $\phi = 20$ (b) $\phi = 2000$

Fig. 3. Evolution of weights over time, for a network simulated from the GG model with $T = 100$, $\alpha = 1$, $\sigma = 0.01$, $\tau = 1$ and (a) $\phi = 20$ (b) $\phi = 2000$. In each case we plot the largest weights, with the respective nodes represented by different colours.

distribution as W_t, i.e., they are both distributed as $\text{GG}(\alpha, \sigma, \tau)$. Moreover, as proved in Sect. 1.1 of the Supplementary Material, the conditional mean of W_{t+1} given $W_t = \sum_i w_{ti}\delta_{\theta_i}$ has the form

$$E[W_{t+1}|W_t] = \left(\frac{\tau}{\tau+\phi}\right)^{1-\sigma} E[W_t] + \frac{1}{\tau+\phi}\sum_{i=1}^{\infty}[\phi w_{ti} - \sigma(1-e^{-\phi w_{ti}})]\delta_{\theta_i} \quad (10)$$

In the gamma process case ($\sigma = 0$), the above expression reduces to

$$E[W_{t+1}|W_t] = \frac{\tau}{\tau+\phi}E[W_t] + \frac{\phi}{\tau+\phi}W_t.$$

3.3 Summary of the Model's Hyperparameters

The model is parameterised by $(\alpha, \sigma, \tau, \phi)$, where:

- α tunes the overall size of the networks, with a larger value of α corresponding to larger networks.
- σ controls the sparsity and power-law properties of the graph, as will be shown in Sect. 3.4. In Fig. 2 we see that different values of σ give rise to different power-law degree distributions.
- τ induces an exponential tilting of large degrees in the degree distribution.
- ϕ tunes the correlation of the sociabilities of each node over time. As we see in Fig. 3, larger values correspond to higher correlation and smoother evolution of the weights.

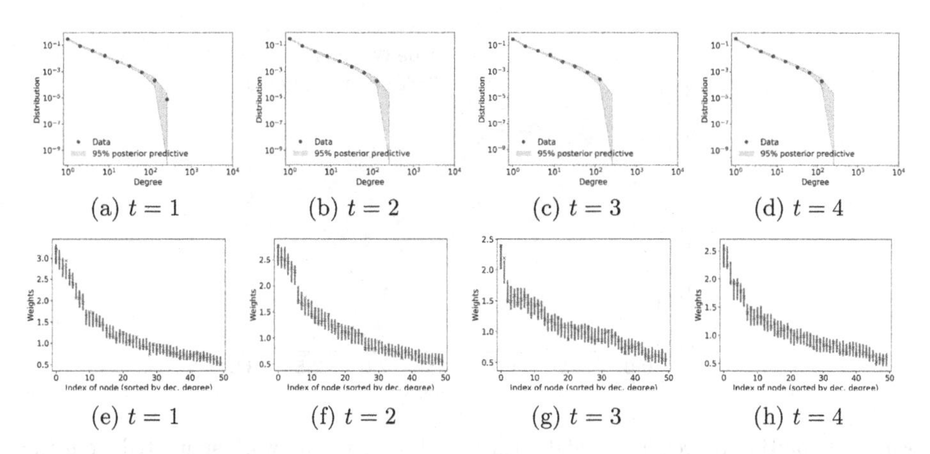

(a) $t = 1$ (b) $t = 2$ (c) $t = 3$ (d) $t = 4$

(e) $t = 1$ (f) $t = 2$ (g) $t = 3$ (h) $t = 4$

Fig. 4. (Top row) Posterior predictive degree distribution and (bottom row) 95% credible intervals for sociabilites of the nodes with the highest degrees, for a network simulated from the GG model. True weights are represented by a green cross. (Color figure online)

3.4 Sparsity and Power-Law Properties of the Model

By construction, the interactions at time t, n_{tij}, are drawn from the same (static) model as in [7], using a generalised gamma process for the Lévy intensity, and so applying Proposition 18 in [8] we obtain the following asymptotic properties

Proposition 1. *Let $N_{t,\alpha}$ be the number of active nodes at time t, $N_{t,\alpha}^{(e)} = \sum_{i \leq j} z_{tij}$ be the number of edges and $N_{t,\alpha,j}$ the number of nodes of degree j in the graph at time t, then as α tends to infinity, almost surely, we have for any t: if $\sigma > 0$, $N_{t,\alpha}^{(e)} \asymp N_{t,\alpha}^{2/(1+\sigma)}$, if $\sigma = 0$, $N_{t,\alpha}^{(e)} \asymp N_{t,\alpha}^2 / \log^2(N_{t,\alpha})$ and if $\sigma < 0$, $N_{t,\alpha}^{(e)} \asymp N_{t,\alpha}^2$. Also, almost surely, for any $t \geq 1$ and $j \geq 1$, if $\sigma \in (0, 1)$,*

$$\frac{N_{t,\alpha,j}}{N_{t,\alpha}} \to p_j, \quad p_j = \frac{\sigma \Gamma(j - \sigma)}{j! \Gamma(1 - \sigma)},$$

while if $\sigma \leq 0$, $N_{t,\alpha,j}/N_{t,\alpha} \to 0$ for all $j \geq 1$.

Hence the graphs are sparse if $\sigma \geq 0$ and dense if $\sigma < 0$. This distinction stems from the properties of the GGP that we use to define the distribution of the sociabilities, as for $\sigma < 0$, the GGP is finite-activity, and for $\sigma \geq 0$ the GGP is infinite activity. For further details see [7,8].

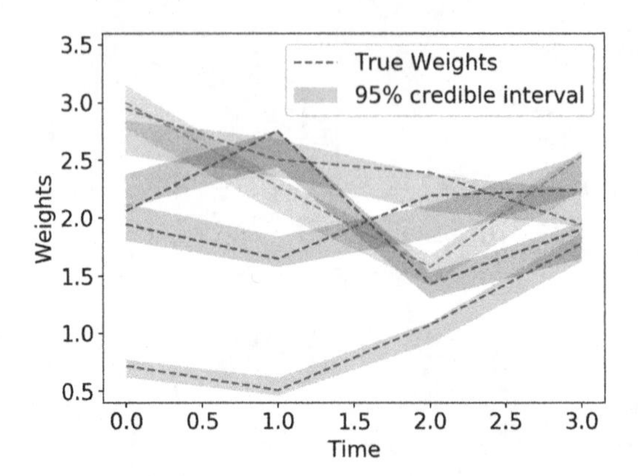

Fig. 5. Evolution of weights of high degree nodes for the network simulated from the GG model. The dotted line shows the true value of the weights in each case.

4 Approximate Inference

4.1 Finite-dimensional Approximation

Performing exact inference using this model is quite challenging, and we consider instead an approximate inference method, using a finite-dimensional approximation to the GG prior, introduced by [30]. This approximation gives rise to a particularly simple conjugate construction, enabling posterior inference to be performed.

Let BFRY (η, τ, σ) denote a (scaled and exponentially tilted) BFRY random variable[1] on $(0, \infty)$ with probability density function

$$g_{\eta,\tau,\sigma}(w) = \frac{\sigma w^{-1-\sigma} e^{-\tau w} \left(1 - e^{-(\sigma/\eta)^{1/\sigma} w}\right)}{\Gamma(1-\sigma)\left\{\left(\tau + (\sigma/\eta)^{1/\sigma}\right)^{\sigma} - \tau^{\sigma}\right\}}$$

with parameters $\sigma \in (0, 1)$, $\tau > 0$ and $\eta > 0$.

At time 1, consider the finite-dimensional measure

$$W_1 = \sum_{i=1}^{K} w_{1i} \delta_{\theta_i}$$

where $w_{1i} \sim$ BFRY $(\alpha/K, \tau, \sigma)$ and $K < \infty$ is the truncation level. As shown by [30], for $\sigma \in (0, 1)$

$$W_1 \xrightarrow{d} \mathrm{GG}(\alpha, \sigma, \tau)$$

as the truncation level K tends to infinity. By this we mean that, if $W \sim GG(\alpha, \sigma, \tau)$, then

$$\lim_{K \to \infty} \mathcal{L}_f(W') = \mathcal{L}_f(W)$$

[1] The name was coined by [13] after [3].

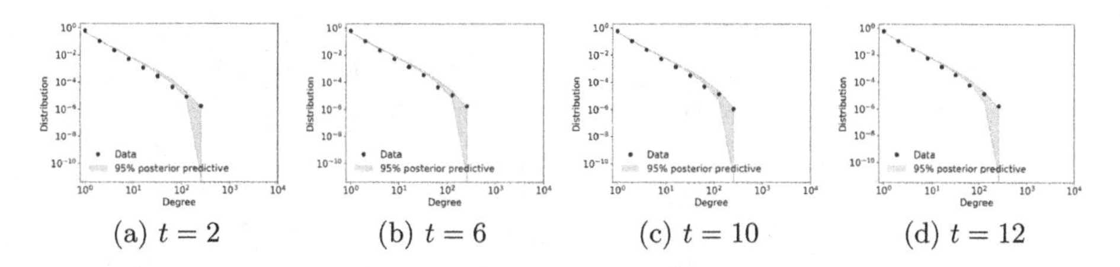

(a) $t = 2$ (b) $t = 6$ (c) $t = 10$ (d) $t = 12$

Fig. 6. Posterior predictive degree distribution over time for the Reddit hyperlink network

for an arbitrary measurable and positive f, where $\mathcal{L}_f(W) := \mathbb{E}\left[e^{-W(f)}\right]$ is the Laplace functional of W as defined by [30].

To use this finite approximation with our dynamic model, we consider Poisson latent variables as in Equation (6). The measure W_{t+1} is then constructed as:

$$W_{t+1} = \sum_{i=1}^{K} w_{t+1,i}\delta_{\theta_i}$$
$$w_{t+1,i}|C_t \sim \mathrm{BFRY}\left(\alpha_t'/K, \tau + \phi, \sigma - c_{ti}\right).$$

where

$$\alpha_t' = K(\sigma - c_{ti})(\sigma K/\alpha)^{\frac{c_{ti}-\sigma}{\sigma}}.$$

This construction mirrors that of Sect. 3.2, with the key difference being that we now obtain a stationary $\mathrm{BFRY}\left(\alpha/K, \tau, \sigma\right)$ distribution for the w_{ti}. We can easily see this by noting that if

$$w_{ti} \sim \mathrm{BFRY}\left(\alpha/K, \tau, \sigma\right)$$
$$c_{ti}|w_{ti} \sim \mathrm{Poisson}(\phi w_{ti})$$

then

$$p(w_{ti}|c_{ti}) \propto w_{ti}^{-1-\sigma+c_{ti}} e^{-(\tau+\phi)w_{ti}} \left(1 - e^{-(\sigma K/\alpha)^{1/\sigma} w_{ti}}\right) \tag{11}$$

which we then recognise as a $\mathrm{BFRY}\left(\alpha_t'/K, \tau + \phi, \sigma - c_{ti}\right)$ with α_t' as above.

Thus, the use of a finite-dimensional approximation BFRY random variables gives us the simple conjugate construction that we desired. The reason that we introduce this non-standard distribution is that, as far as we know, it is not possible to approximate the generalised gamma process using a finite measure with i.i.d. gamma random weights. In the specific case $\sigma = 0$, we could use the simpler finite approximation using $Gamma(\alpha/K, \tau)$ random variables. However, this would preclude us from modelling networks with power-law degree distributions, as discussed previously.

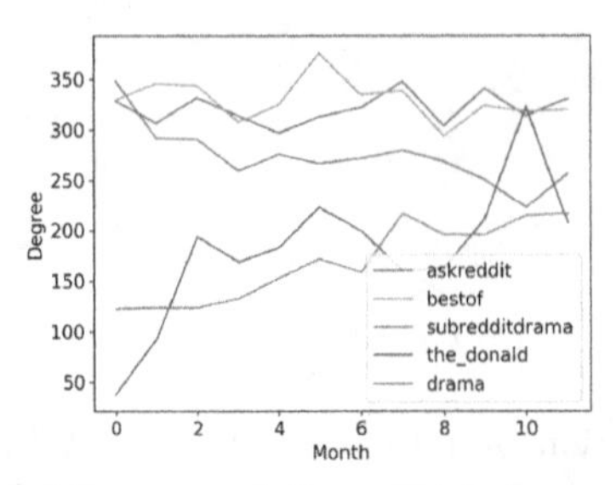

(a) Degree evolution of high degree nodes

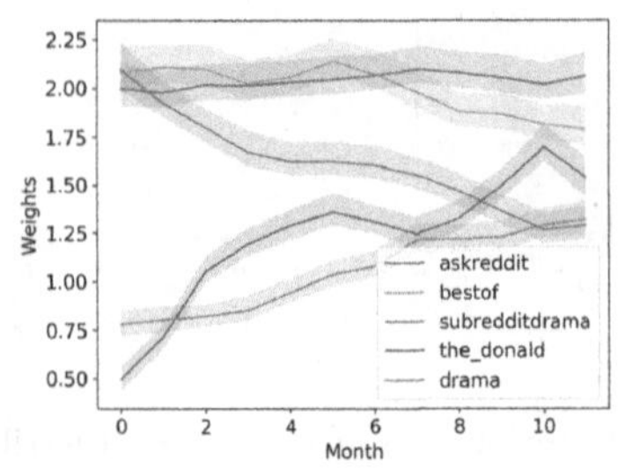

(b) Weight evolution of high degree nodes

Fig. 7. Evolution of (a) degrees and (b) weights of high degree nodes for the Reddit hyperlink network

4.2 Posterior Inference Algorithm

In order to perform posterior inference with the approximate method, we use a Gibbs sampler. We introduce auxiliary variables $\{u_{ti}\}_{i=1}^{K}, t = 1, \ldots, T$ following a truncated exponential distribution. The overall sampler is as follows (see the Supplementary Material for full details):

1. Update the weights w_{ti} given the rest using Hamiltonian Monte Carlo (HMC).
2. Update the latent c_{ti} given the rest using Metropolis Hastings (MH).
3. Update the hyperparameters α, σ, ϕ and τ and the latent variables u_{ti} given the rest. We place gamma priors on α, τ and ϕ, and a beta prior on σ.

5 Experiments

In this section, we use our model to study three dynamic networks. The code used for all experiments is available at https://github.com/ciannaik/SDyNet.

5.1 Simulated Data

In order to assess the approximate inference scheme, we first consider a synthetic dataset. We simulate a network with $T = 4$, $\alpha = 100$, $\sigma = 0.2$, $\tau = 1$, $\phi = 10$ from the exact model (see Sect. 3), and then estimate it using our approximate inference scheme described in Sect. 4. The generated network has 3,096 nodes and 101,571 edges. We run an MCMC chain of length 600,000, with 300,000 samples discarded as burn-in. We set the truncation threshold to $K = 15,000$.

The approximate inference scheme estimates the model's parameters well. This can be seen in terms of the fit of the posterior predictive degree distribution to the empirical as seen in Figs. 4a–4d, and the coverage of the credible intervals for the weights, as we see in Figs. 4e–4h for the 50 nodes with the highest degree. In Fig. 5 we see that the model is also able to capture the weight evolution of the highest degree nodes over time.

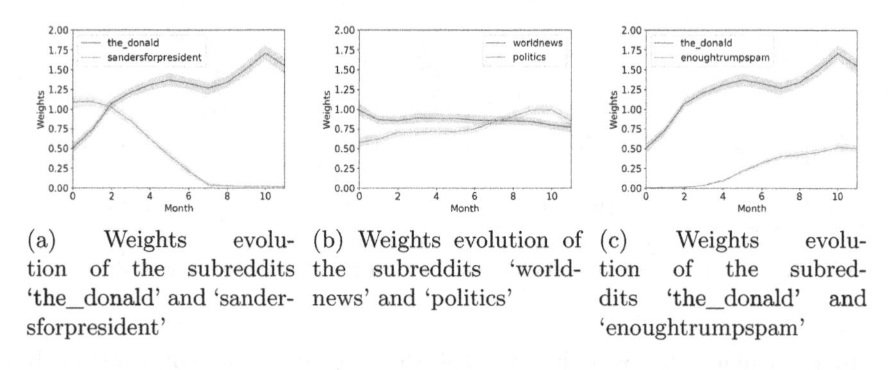

(a) Weights evolution of the subreddits 'the_donald' and 'sandersforpresident'

(b) Weights evolution of the subreddits 'worldnews' and 'politics'

(c) Weights evolution of the subreddits 'the_donald' and 'enoughtrumpspam'

Fig. 8. Evolution of the weights, for the Reddit hyperlink network.

In this simulated data experiment, as well as when looking at the real data experiments, we need to choose the truncation level K. The trade-off here is between a better approximation for larger K, but slower mixing and a longer running time of the MCMC algorithm. We examine the effect that our choice of K has in the Supplementary Material.

5.2 Real Data

We illustrate the use of our model on two more dynamic network datasets: the Reddit Hyperlink Network [29][2] and the Reuters Terror dataset[3].

Reddit Hyperlink Network The Reddit hyperlink network represents hyperlink connections between subreddits (communities on Reddit) over a period of $T = 12$ consecutive months in 2016. Nodes are subreddits (communities) and the symmetric edges represent hyperlinks originating in a post in one community and linking to a post in another community. The network has $N = 28,810$ nodes and $388,574$ interactions. The observations here are hyperlinks between the pair of subreddits i, j at time t. The dataset has been made symmetric by placing an edge between nodes i and j if there is a hyperlink between them in either direction. We also assume that there are no loops in the network, that is $n_{tij} = 0$ for $i = j$. We run the Gibbs sampler $400,000$ samples, with the first $200,000$ discarded as burn in. In this case, we choose a truncation level of $K = 40,000$.

From Fig. 6 we see that our model is capturing the empirical degree distribution well. Furthermore, in Fig. 7 we see that the model is able to capture the evolution of weights associated with each subreddit in a fashion that agrees with the observed frequency of interactions. The high degree nodes here are interpretable as either communities with a very large number of followers - such as "askreddit" or "bestof", or communities which frequently link to others - such as "drama" or "subredditdrama".

[2] https://snap.stanford.edu/data/soc-RedditHyperlinks.html.
[3] http://vlado.fmf.uni-lj.si/pub/networks/data/CRA/terror.htm.

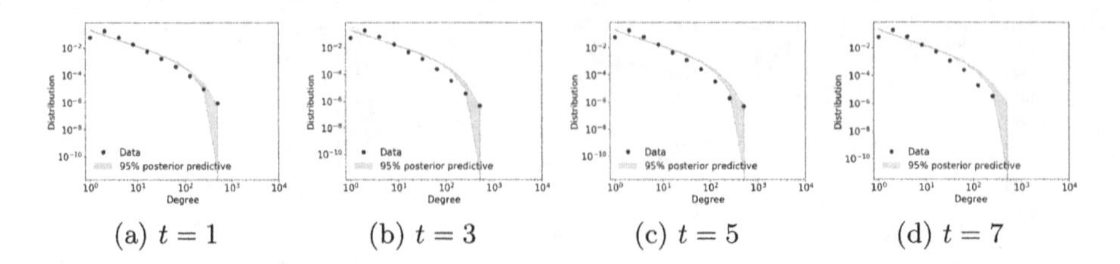

(a) $t = 1$ (b) $t = 3$ (c) $t = 5$ (d) $t = 7$

Fig. 9. Posterior predictive degree distribution for the Reuters terror dataset

In particular, we see in Fig. 8a that the weights of the controversial but popular political subreddit "The Donald" increases to a peak in November, corresponding to the 2016 U.S. presidential election. Conversely, the weights of the subreddit "Sandersforpresident" decrease as the year goes on, corresponding to the end of Senator Bernie Sanders' presidential campaign, a trend that again agrees with the evolution of the corresponding observed degrees.

Reuters Terror Dataset. The final dataset we consider is the Reuters terror news network dataset. It is based on all stories released during $T = 7$ consecutive weeks (the original data was day-by-day, but was shortened and collated for our purposes) by the Reuters news agency concerning the 09/11/01 attack on the U.S.. Nodes are words and edges represent co-occurence of words in a sentence, in news. The network has $N = 13,332$ nodes (different words) and $473,382$ interactions. The observations here are the frequency of co-occurence between the pair of words i, j at time t. We assume that there are no loops in the network, that is $n_{tij} = 0$ for $i = j$. We run the Gibbs sampler with $200,000$ samples, with the first $100,000$ discarded as burn in. In this case, we choose a truncation level of $K = 20,000$.

Figure 9 suggests that the empirical degree distribution does not follow a power law distribution. In particular there are significantly fewer nodes of degree one than we would expect. Our model therefore provides a moderate fit to the empirical degree distribution. The model is however able to capture the evolution of the popularity of the different words, as shown in Fig. 10a. For example, the weights of the words "plane" and "attack" decrease over time after 9/11, while the words "letter" and "anthrax" show a peak a few weeks after the attack. These correspond to the anthrax attacks that occurred over several weeks starting a week after 9/11.

Due to the empirical degree distribution not following a power-law, the estimated value of σ is very close to 0. This causes a slow convergence of the MCMC algorithm, which we see in the Supplementary Material.

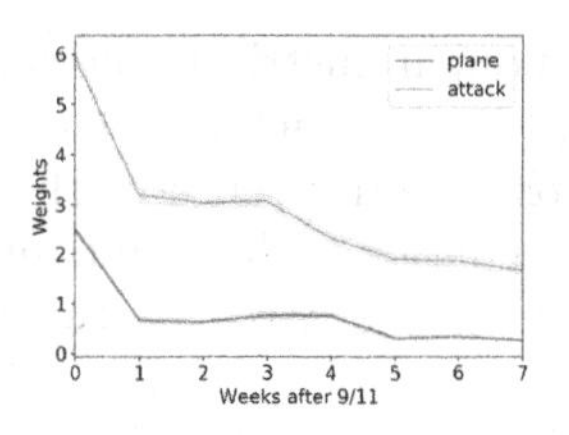

(a) Weights evolution of the words 'plane' and 'attack'

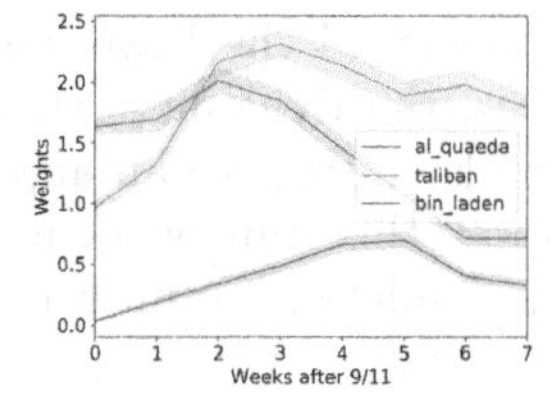

(b) Weights evolution of the words 'al quaeda', 'taliban' and 'bin laden'

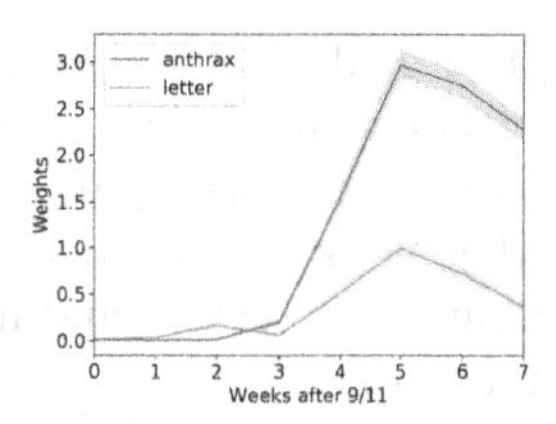

(c) Weights evolution of the words 'anthrax' and 'letter'

Fig. 10. Evolution of the weights, for the Reuters terror dataset

6 Discussion and Extensions

A lot of work has been done on modelling dynamic networks; we restrict ourselves to focussing on Bayesian approaches. Much of this work has centred around extending static models. For example, [14,44] extend the stochastic block model to the dynamic setting by allowing for parameters that evolve over time. There has also been work on extending the mixed membership stochastic blockmodel [16,21,43], the infinite relational model [35] and the latent feature relational model [15,19,26]. The problem inherent in these models is that they lead to networks which are dense almost surely [2,22], a property considered unrealistic for many real-world networks [36] such as social and internet networks.

In order to build models for sparse dynamic networks, [17] build on the framework of edge-exchangeable networks [6,11,42], in which graphs are constructed based on an infinitely exchangeable sequence of edges. As in our case, this framework allows for sparse networks with power-law degree distributions. This work, along with others in this framework [18,35], utilises the mixture of Dirichlet network distributions (MDND) to introduce structure in the networks.

Conversely, our works builds on a different notion of exchangeability [7,25]. Within this framework, [32] use mutually-exciting Hawkes processes to model temporal interaction data. The difference between our work and theirs is that the sociabilities of the nodes are constant throughout time, with the time evolving element driven by previous interactions via the Hawkes process. Their work also builds on that of [39], incorporating community structure to the network. Exploring how communities appear, evolve and merge could have many practical uses. Thus, expanding our model to capture both evolving node popularity and dynamically changing community structure could be a useful extension.

Furthermore, our model assumes that the time between observations of the network is constant. If this is not the case, it may be helpful to use a continuous-time version of our model. This could be done by considering a birth-death process for the interactions between nodes, where each interaction has a certain lifetime distribution. The continuously evolving node sociabilities could then by described by the Dawson-Watanabe superprocess [12,41].

The goal of our work is to extend the Bayesian nonparametric framework for network modelling of [7] to the dynamic setting. Since exact inference is intractable in this regime, we also introduce an approximate inference method. Further work is needed to apply this framework in scenarios such as dynamic community detection and link prediction for unseen time steps. We leave this to future work.

Acknowledgement. We thank the reviewers for their helpful and constructive comments. C. Naik was supported by the Engineering and Physical Sciences Research Council and Medical Research Council [award reference 1930478]. F. Caron was supported by the Engineering and Physical Sciences Council under grant EP/P026753/1. J. Rousseau was supported by the European Research Council (ERC) under the European Union's Horizon 2020 research and innovation programme (grant agreement No 834175). K. Palla and Y.W. Teh were supported by the European Research Council under the European Union's Seventh Framework Programme (FP7/2007–2013) ERC grant agreement no. 617411.

References

1. Aalen, O.: Modelling heterogeneity in survival analysis by the compound Poisson distribution. Annals Appl. Prob. **2**(4), 951–972 (1992)
2. Aldous, D.J.: Representations for partially exchangeable arrays of random variables. J. Multivar. Anal. **11**(4), 581–598 (1981)
3. Bertoin, J., Fujita, T., Roynette, B., Yor, M.: On a particular class of self-decomposable random variables?: the durations of Bessel excursions straddling independent exponential times. Probab. Math. Stat. **26**(2), 315–366 (2006)
4. Borgs, C., Chayes, J., Cohn, H., Holden, N.: Sparse exchangeable graphs and their limits via graphon processes. J. Mach. Learn. Res. **18**(1), 1–71 (2018)
5. Brix, A.: Generalized gamma measures and shot-noise Cox processes. Adv. Appl. Probab. **31**(4), 929–953 (1999)
6. Cai, D., Campbell, T., Broderick, T.: Edge-exchangeable graphs and sparsity. In: Advances in Neural Information Processing Systems, pp. 4249–4257 (2016)
7. Caron, F., Fox, E.: Sparse graphs using exchangeable random measures. J. Royal Stat. Society B **79**, 1–44 (2017)
8. Caron, F., Panero, F., Rousseau, J.: On sparsity, power-law and clustering properties of graphex processes. arXiv pp. arXiv-1708 (2017)
9. Caron, F., Teh, Y.W.: Bayesian nonparametric models for ranked data. In: NIPS (2012)
10. Caron, F., Teh, Y., Murphy, T.: Bayesian nonparametric plackett-luce models for the analysis of preferences for college degree programmes. Annals Appl. Stat. **8**(2), 1145–1181 (2014)
11. Crane, H., Dempsey, W.: Edge exchangeable models for interaction networks. J. Am. Stat. Assoc. **113**(523), 1311–1326 (2018)
12. Dawson, D.A.: Stochastic evolution equations and related measure processes. J. Multivar. Anal. **5**(1), 1–52 (1975)
13. Devroye, L., James, L.: On simulation and properties of the stable law. Stat. Methods Appl. **23**(3), 307–343 (2014). https://doi.org/10.1007/s10260-014-0260-0
14. Durante, D., Dunson, D.: Bayesian logistic gaussian process models for dynamic networks. In: AISTATS, pp. 194–201 (2014)

15. Foulds, J., DuBois, C., Asuncion, A., Butts, C., Smyth, P.: A dynamic relational infinite feature model for longitudinal social networks. In: Proceedings of the fourteenth international conference on artificial intelligence and statistics, pp. 287–295 (2011)
16. Fu, W., Song, L., Xing, E.P.: Dynamic mixed membership blockmodel for evolving networks. In: Proceedings of the 26th Annual International Conference on Machine Learning, pp. 329–336 (2009)
17. Ghalebi, E., Mahyar, H., Grosu, R., Taylor, G.W., Williamson, S.A.: A nonparametric bayesian model for sparse temporal multigraphs. CoRR (2019)
18. Ghalebi, E., Mirzasoleiman, B., Grosu, R., Leskovec, J.: Dynamic network model from partial observations. In: Advances in Neural Information Processing Systems, pp. 9862–9872 (2018)
19. Heaukulani, C., Ghahramani, Z.: Dynamic probabilistic models for latent feature propagation in social networks. In: International Conference on Machine Learning, pp. 275–283 (2013)
20. Herlau, T., Schmidt, M.N., Mørup, M.: Completely random measures for modelling block-structured sparse networks. In: Advances in Neural Information Processing Systems, vol. 29 (2016)
21. Ho, Q., Song, L., Xing, E.: Evolving cluster mixed-membership blockmodel for time-evolving networks. In: Proceedings of the Fourteenth International Conference on Artificial Intelligence and Statistics, pp. 342–350 (2011)
22. Hoover, D.N.: Relations on probability spaces and arrays of random variables. Preprint, Institute for Advanced Study, Princeton, NJ (1979)
23. James, L.F.: Poisson process partition calculus with applications to exchangeable models and bayesian nonparametrics. arXiv preprint math/0205093 (2002)
24. James, L.F., Lijoi, A., Prünster, I.: Posterior analysis for normalized random measures with independent increments. Scand. J. Stat. **36**(1), 76–97 (2009)
25. Kallenberg, O.: Exchangeable random measures in the plane. J. Theor. Probab. **3**(1), 81–136 (1990)
26. Kim, M., Leskovec, J.: Nonparametric multi-group membership model for dynamic networks. In: Advances in Neural Information Processing Systems, pp. 1385–1393 (2013)
27. Kingman, J.: Completely random measures. Pac. J. Math. **21**(1), 59–78 (1967)
28. Klenke, A.: Probability theory: A Comprehensive Course. Springer Science & Business Media (2013)
29. Kumar, S., Hamilton, W.L., Leskovec, J., Jurafsky, D.: Community interaction and conflict on the web. In: Proceedings of the 2018 World Wide Web Conference on World Wide Web, pp. 933–943. International World Wide Web Conferences Steering Committee (2018)
30. Lee, J., James, L.F., Choi, S.: Finite-dimensional bfry priors and variational bayesian inference for power law models. In: Advances in Neural Information Processing Systems, pp. 3162–3170 (2016)
31. Lijoi, A., Mena, R.H., Prünster, I.: Controlling the reinforcement in Bayesian nonparametric mixture models. J. Royal Stat. Society: Series B (Stat. Methodol.) **69**(4), 715–740 (2007)
32. Miscouridou, X., Caron, F., Teh, Y.W.: Modelling sparsity, heterogeneity, reciprocity and community structure in temporal interaction data. In: Advances in Neural Information Processing Systems, pp. 2343–2352 (2018)
33. Naik, C., Caron, F., Rousseau, J.: Sparse networks with core-periphery structure. Electron. J. Stat. **15**(1), 1814–1868 (2021)

34. Newman, M.: Networks: an introduction. OUP Oxford (2009)
35. Ng, Y.C., Silva, R.: A dynamic edge exchangeable model for sparse temporal networks. arXiv preprint arXiv:1710.04008 (2017)
36. Orbanz, P., Roy, D.M.: Bayesian models of graphs, arrays and other exchangeable random structures. IEEE Trans. Pattern Anal. Mach. Intelligence (PAMI) **37**(2), 437–461 (2015)
37. Pitt, M.K., Walker, S.G.: Constructing stationary time series models using auxiliary variables with applications. J. Am. Stat. Assoc. **100**(470), 554–564 (2005)
38. Prünster, I.: Random probability measures derived from increasing additive processes and their application to Bayesian statistics. Ph.D. thesis, University of Pavia (2002)
39. Todeschini, A., Miscouridou, X., Caron, F.: Exchangeable random measures for sparse and modular graphs with overlapping communities. J. Royal Stat. Society: Series B (Stat. Methodol.) **82**(2), 487–520 (2020)
40. Veitch, V., Roy, D.M.: The class of random graphs arising from exchangeable random measures. arXiv preprint arXiv:1512.03099 (2015)
41. Watanabe, S.: A limit theorem of branching processes and continuous state branching processes. J. Math. Kyoto Univ. **8**(1), 141–167 (1968)
42. Williamson, S.A.: Nonparametric network models for link prediction. J. Mach. Learn. Res. **17**(1), 7102–7121 (2016)
43. Xing, E.P., Fu, W., Song, L., et al.: A state-space mixed membership blockmodel for dynamic network tomography. Annals Appl. Stat. **4**(2), 535–566 (2010)
44. Xu, K., Hero, A.O.: Dynamic stochastic blockmodels for time-evolving social networks. IEEE J. Selected Topics Signal Process. **8**(4), 552–562 (2014)

A Pre-screening Approach for Faster Bayesian Network Structure Learning

Thibaud Rahier[1(✉)], Sylvain Marié[2], and Florence Forbes[3]

[1] Criteo AI Lab, 4 Rue des Méridiens, 38130 Echirolles, France
Thibaud.Rahier@criteo.com

[2] Schneider Electric Industries, 160, avenue des Martyrs, 38000 Grenoble, France
Sylvain.Marie@se.com

[3] INRIA, 655 Av. de l'Europe, 38330 Montbonnot-Saint-Martin, France
Florence.Forbes@inria.fr

Abstract. Learning the structure of Bayesian networks from data is a NP-Hard problem that involves optimization over a super-exponential sized space. Still, in many real-life datasets a number of the arcs contained in the final structure correspond to strongly related pairs of variables and can be identified efficiently with information-theoretic metrics. In this work, we propose a meta-algorithm to accelerate any existing Bayesian network structure learning method. It contains an additional arc pre-screening step allowing to narrow the structure learning task down to a subset of the original variables, thus reducing the overall problem size. We conduct extensive experiments on both public benchmarks and private industrial datasets, showing that this approach enables a significant decrease in computational time and graph complexity for little to no decrease in performance score.

Keywords: Bayesian networks · Structure learning · Information theory · Conditional entropy · Determinism · Functional relations · Screening

1 Introduction

Bayesian networks are probabilistic graphical models that present interest both in terms of knowledge discovery and density estimation. Learning Bayesian networks from data has been however proven to be NP-Hard by Chickering (1996). There has been extensive work on tackling the ambitious problem of Bayesian network structure learning (BNSL) from observational data. Algorithms fall under two main categories: *constraint-based* and *score-based*.

Constraint-based structure learning algorithms rely on testing for conditional independence relations that hold in the data in order to reconstruct a Bayesian network encoding these independence relations. The *PC* algorithm by Spirtes

Supplementary Information The online version contains supplementary material available at https://doi.org/10.1007/978-3-031-26419-1_13.

© The Author(s), under exclusive license to Springer Nature Switzerland AG 2023
M.-R. Amini et al. (Eds.): ECML PKDD 2022, LNAI 13717, pp. 207–222, 2023.
https://doi.org/10.1007/978-3-031-26419-1_13

et al. (2000) was the first practical application of this idea, followed by increasingly optimized approaches such as the *fast incremental association* (Fast-IAMB) algorithm.

Score-based structure learning relies on the definition of a network score, then on the search for the best-scoring structure among all possible directed acyclic graphs (DAGs). The number of possible DAG structures with n nodes is of order $2^{\mathcal{O}(n^2)}$, which prevents exhaustive search when n is typically larger than 30. Most of the score-based algorithms used in practice therefore rely on heuristics, as the original approach from Cooper and Herskovits (1992) which assumes a prior ordering of the variables is known, or Bouckaert (1995) who proposed to search through the structure space using greedy hill climbing with random restarts. Since these first algorithms, different approaches have been proposed, increasingly pushing the limits of state-of-the art in BNSL: some based on the search for an optimal ordering (Teyssier and Koller, 2005; Chen et al.,ch13chen2008improving), others on optimizing the search task in accordance to a given score (Scanagatta et al., 2015; Nie et al., 2016), using integer programming (Cussens, 2011) or bio-inspired optimization heuristics (Kareem and Okur, 2019; 2021). See Scutari et al. (2019) for a recent review.

Meanwhile, data itself may contain determinism, for example in the fields of cancer risk identification (de Morais et al.,2010) or nuclear safety (Mabrouk et al., 2014). Data is also increasingly collected and generated by software systems whether in social networks, smart buildings, smart grids, smart cities or the internet of things (IoT) in general (Koo et al., 2016). These systems in their vast majority rely on relational data models or semantic data models (El Kaed et al., 2016) where the same entity may be described with several attributes. It is therefore now common to find deterministically related variables in datasets. Determinism has been shown to interfere with Bayesian network structure learning, notably constraint-based methods as mentioned by Luo (2006).

In this paper, we first remind the background of Bayesian network structure learning (Sect. 2) and bring forward the following contributions:

- we state some theoretical results bridging the gap between the notion of determinism and Bayesian network scoring (Sect. 3),
- we propose and study the complexity of the *quasi-determinism screening BNSL* (`qds-BNSL`) meta-algorithm, whose aim is to accelerate any existing BNSL algorithm by reducing the learning problem to a subset of the original variables via the detection of strong arcs (Sect. 4),
- we conduct experiments on both public benchmarks and private industrial datasets, demonstrating empirically that our meta-algorithm indeed accelerates the overall BNSL procedure with very low performance loss and also leads to sparser and therefore more interpretable graphs (Sect. 5).

2 Bayesian Network Structure Learning

2.1 Bayesian Networks

Let $\mathbf{X} = (X_1, \ldots, X_n)$ be a n-tuple of categorical random variables with respective value sets $Val(X_1), \ldots, Val(X_n)$. The distribution of $\mathbf{X}$ is denoted by, $\forall\, \mathbf{x} = (x_1, \ldots, x_n) \in Val(\mathbf{X})$, $p(\mathbf{x}) = P(X_1 = x_1, \ldots, X_n = x_n)$.

For $I \subset [\![1, n]\!]$, we define $\mathbf{X}_I = \{X_i\}_{i \in I}$, and the notation $p(\cdot)$ and $p(\cdot|\cdot)$ is extended to the marginals and conditionals of any subset of variables: $\forall (\mathbf{x}_I, \mathbf{x}_J) \in Val(\mathbf{X}_{I \cup J})$, $p(\mathbf{x}_I|\mathbf{x}_J) = P(\mathbf{X}_I = \mathbf{x}_I|\mathbf{X}_J = \mathbf{x}_J)$.

Moreover, we suppose that D is a dataset containing M *i.i.d.* instances of $(X_1, \ldots, X_n)$. All quantities empirically computed from D will be written with a $.^D$ exponent (*e.g.* p^D refers to the empirical distribution with respect to D). Finally, D_I refers to the restriction of D to the observations of $\mathbf{X}_I$.

A Bayesian network is an object $\mathcal{B} = (G, \theta)$ where (1) $G = (V, A)$ is a directed acyclic graph (DAG) structure with V the set of nodes and $A \subset V \times V$ the set of arcs. We suppose $V = [\![1, n]\!]$ where each node $i \in V$ is associated with the random variable X_i, and $\pi^G(i) = \{j \in V \text{ s.t. } (j, i) \in A\}$ is the set of i's parents in G[1] and (2) $\theta = \{\theta_i\}_{i \in V}$ is a set of parameters. Each θ_i defines the local conditional distribution of X_i given its parents in the graph, $P(X_i|\mathbf{X}_{\pi(i)})$. More precisely, $\theta_i = \{\theta_{x_i|\mathbf{x}_{\pi(i)}}\}$ where for $i \in V$, $x_i \in Val(X_i)$ and $\mathbf{x}_{\pi(i)} \in Val(\mathbf{X}_{\pi(i)})$, $\theta_{x_i|\mathbf{x}_{\pi(i)}} = p(x_i|\mathbf{x}_{\pi(i)})$.

A Bayesian network $\mathcal{B} = (G, \theta)$ encodes the following factorization of the distribution of $\mathbf{X}$: for $\mathbf{x} = (x_1, \ldots, x_n) \in Val(\mathbf{X})$,

$$p(\mathbf{x}) = \prod_{i=1}^{n} p(x_i|\mathbf{x}_{\pi^G(i)}) = \prod_{i=1}^{n} \theta_{x_i|\mathbf{x}_{\pi^G(i)}}.$$

Such a factorization notably implies that *each variable is independent of its non-descendants given its parents.*

2.2 Score-Based Approach to Bayesian Network Structure Learning

For a given *scoring* function $s : DAG_V \to \mathbb{R}$, where DAG_V is the set of all possible DAG structures with node set V, score-based BNSL aims at solving the following combinatorial optimization problem:

$$G^* \in \underset{G \in DAG_V}{\operatorname{argmax}}\, s(G). \tag{1}$$

It can be shown that $2^{\frac{n(n-1)}{2}} \leq |DAG_V| \leq 2^{n(n-1)}$ where $|V| = n$. There are therefore $2^{\mathcal{O}(n^2)}$ possible DAG structures containing n nodes (Koller and Friedman, 2009): the size of DAG_V is said to be super-exponential in $|V|$. Most scoring functions used in practice are based on the likelihood function. The most straightforward being the *max log-likelihood* sore, that we now present.

[1] The exponent G may be dropped for clarity when the referred graph is obvious from context.

The max log-likelihood (MLL) score Let $l_D(\theta) = \log(p_\theta(D))$ be the log-likelihood of the set of parameters θ given the dataset D. For a given DAG structure $G \in DAG_V$, we define the MLL score of G with respect to D as:

$$s_D^{MLL}(G) = \max_{\theta \in \Theta_G} l_D(\theta).$$

where Θ_G is the set of all θ's such that $\mathcal{B} = (G, \theta)$ is a well defined Bayesian network. The MLL score is very straightforward and intuitive, but it favors denser structures: if $G_1 = (V, A_1)$ and $G_2 = (V, A_2)$ are two graph structures such that $A_1 \subset A_2$, we can show that: $s_D^{MLL}(G_1) \leq s_D^{MLL}(G_2)$. This problem is generally solved by using a score that induces a goodness-of-fit versus complexity tradeoff, such as BIC (Schwarz et al., 1978), which is a penalized version of the MLL score, or BDe (Heckerman et al., 1995), which is derived from the marginalization of the likelihood, which implicitly penalizes the model's complexity through a Dirichlet prior on the parameters. In this paper, we will use the BDe score to evaluate a BN structure's quality, as it is done in several papers as Teyssier and Koller (2005), or Nie et al. (2016). This score is known to be a good indicator of generalization performance, which is what we are aiming to optimize. In the following sections, we propose to look for such a solution by constructing sparse graphs with minimal MLL.

3 Determinism and Bayesian Networks

3.1 Definitions

We propose the following definitions of determinism and deterministic DAGs using the notion of conditional entropy. In this paper, determinism will always be meant empirically, with respect to a dataset D.

Definition 1. *Determinism wrt D*
Given a dataset D containing observations of variables X_i and X_j, the relationship $X_i \rightarrow X_j$ is deterministic with respect to D iff $H^D(X_i|X_j) = 0$,
where

$$H^D(X_i|X_j) = -\sum_{x_i, x_j} p^D(x_i, x_j) \log(p^D(x_i|x_j))$$

is the empirical conditional Shannon entropy.

It is straightforward to prove that Definition 1 relates to a common and intuitive perception of determinism, as the one presented by Luo (2006). Indeed,

$$\begin{aligned} & H^D(X_i|X_j) = 0 \\ & \Leftrightarrow \forall x_j \in Val(X_j), \text{there exists a unique } x_i \in Val(X_i) \ s.t.\ p^D(x_i|x_j) = 1. \end{aligned}$$

This definition is naturally extended to $\mathbf{X}_I$ and $\mathbf{X}_J$ for $I, J \subset V$, *i.e.* $\mathbf{X}_I \rightarrow \mathbf{X}_J$ is deterministic with respect to D iff $H^D(\mathbf{X}_J|\mathbf{X}_I) = 0$.

Definition 2. *Deterministic DAG wrt D*
$G \in DAG_V$ is said to be deterministic with respect to D iff $\forall i \in V$ s.t. $\pi^G(i) \neq \emptyset$, $\mathbf{X}_{\pi^G(i)} \rightarrow X_i$ is deterministic wrt D.

3.2 Deterministic Trees and MLL Score

We first recall a lemma that relates the MLL score presented in Sect. 2 to the notion of empirical conditional entropy. This result is well known and notably stated by Koller and Friedman (2009).

Lemma 1. *For $G \in DAG_V$ associated with variables $X_1, \ldots, X_n$ observed in a dataset D,*

$$s_D^{MLL}(G) = -M \sum_{i=1}^{n} H^D(X_i|\mathbf{X}_{\pi(i)})$$

where by convention $H^D(X_i|\mathbf{X}_\emptyset) = H^D(X_i)$.

The next proposition follows then straightforwardly. We remind that a tree is a DAG in which each node has exactly one parent, except for the root node which has none.

Proposition 1. *If T is a deterministic tree with respect to D, then T is a solution of* (1)*:*

$$s_D^{MLL}(T) = \max_{G \in DAG_V} s_D^{MLL}(G).$$

It is worth noticing that complete DAGs also maximize the MLL score. The main interest of Proposition 1 resides in the fact that, under the (strong) assumption that a deterministic tree exists, we are able to explicit a sparse solution of (1), with $n-1$ arcs instead of $\frac{n(n-1)}{2}$ for a complete DAG.

3.3 Deterministic Forests and the MLL Score

The deterministic tree assumption of Proposition 1 is very restrictive. In this section, it is extended to deterministic forests, defined as follows:

Definition 3. ***Deterministic forest wrt D***
$F \in DAG_V$ is said to be a deterministic forest with respect to D iff $F = \bigcup_{k=1}^{p} T_k$, where $T_1, \ldots, T_p$ are p disjoint deterministic trees wrt $D_{V_{T_1}}, \ldots, D_{V_{T_p}}$ respectively and s.t. $\bigcup_{k=1}^{p} V_{T_k} = V$.

In the expression $\bigcup_{k=1}^{p} T_k$, $\cup$ is the canonical union for graphs: $G \cup G' = (V_G \cup V_{G'}, A_G \cup A_{G'})$. For a given deterministic forest F with respect to D, we define $\mathcal{R}(F) = \{i \in V \mid \pi^F(i) = \emptyset\}$ the set of F's roots (the union of the roots of each of its trees).

Proposition 2. *Suppose F is a deterministic forest wrt D. Let $G^*_{\mathcal{R}(F)}$ be a solution of the BNSL optimization problem* (1) *for $\mathbf{X}_{\mathcal{R}(F)}$ and the MLL score i.e.*

$$s_{D_{\mathcal{R}(F)}}^{MLL}(G^*_{\mathcal{R}(F)}) = \max_{G \in DAG_{\mathcal{R}(F)}} s_{D_{\mathcal{R}(F)}}^{MLL}(G).$$

Then, $G^* = F \cup G^*_{\mathcal{R}(F)}$ *is a solution of* (1) *for* $\mathbf{X}$, *i.e.*

$$s_D^{MLL}(G^*) = \max_{G \in DAG_V} s_D^{MLL}(G).$$

As opposed to Proposition 1, the assumptions of Proposition 2 are always formally verified: if there is no determinism in the dataset D, then $\mathcal{R}(F) = V$, and every tree T_k is formed of a single root node. In that case, solving problem (1) for $G^*_{\mathcal{R}(F)}$ is the same as solving it for G^*. Of course, we are interested in the case where $\mathcal{R}(F) < n$, as this enables us to focus on a smaller structure learning problem while still having the guarantee to learn the optimal Bayesian network with regards to the MLL score.

As seen in Sect. 2, the main issue with the MLL score is that it favors complete graphs. However, a deterministic forest F containing p trees is very sparse ($n-p$ arcs), and even if the graph $G^*_{\mathcal{R}(F)}$ is dense, the graph $G^* = F \cup G^*_{\mathcal{R}(F)}$ may still satisfy sparsity conditions.

4 Structure Learning with Quasi-determinism Screening

4.1 Quasi-determinism

When it comes to BNSL algorithms, even heuristics are computationally intensive. We would like to use the theoretical results presented in Sect. 3 to simplify the structure learning problem.

Our idea is to narrow the structure learning problem down to a subset of the original variables: the roots of a deterministic forest, in order to significantly decrease the overall computation time. This is what we call determinism screening.

However, one does not always observe real empirical determinism, although there are very strong relationships between some of the variables. We therefore propose to *relax* the aforementioned determinism screening to quasi-determinism screening, where *quasi* is meant with respect to a parameter ϵ: we talk about $\epsilon-$*quasi-determinism.*

There are several ways to measure how close a relationship is from deterministic. Huhtala et al. (1999) consider the minimum number of observations that must be dropped from the data for the relationship to be deterministic. Since we are in a score-maximization context, we will rather use ϵ as a threshold on the empirical conditional entropy. The following definition is the natural generalization of Definition 1.

Definition 4. ***$\epsilon-$quasi-determinism ($\epsilon-$qd)***
Given a dataset D containing observations of variables X_i and X_j, the relationship $X_i \rightarrow X_j$ is $\epsilon-$qd wrt D iff $H^D(X_j|X_i) \leq \epsilon$.

It has been seen in Proposition 2 that a deterministic forest is the subgraph of an optimal DAG with respect to the MLL score, while still satisfying sparsity conditions. Such a forest is therefore very promising with regards to the fit-complexity tradeoff (typically evaluated by scores such as BDe or BIC).

Combining this intuition with the $\epsilon-$qd criteria presented in Definition 4, we propose the quasi-determinism screening approach to BNSL, defined in the next subsections. An alternate definition with a relative ϵ is proposed in Sect. 6.

4.2 Quasi-determinism Screening Algorithm

Algorithm 1 details how to find the simplest $\epsilon-$qd forest F_ϵ from a dataset D and a threshold ϵ. Here *simplest* refers to the complexity in terms of number of parameters, which in the case of categorical variables corresponds to the number of states: for each variable X_i that has at least one $\epsilon-$qd parent, we select the one that has the lowest number of states from the set of all potential $\epsilon-$qd parents $\pi_\epsilon(i)$ (line 11).

This algorithm takes for input D (a dataset containing M *i.i.d* realizations of $\mathbf{X}$) and ϵ (a threshold for quasi-determinism). The routine on lines 4–9 makes

---

Algorithm 1. Quasi-determinism screening (qds)

Input: D , ϵ
1: Compute empirical conditional entropy matrix $\mathbb{H}^D = \left(H^D(X_i|X_j)\right)_{1\leq i,j\leq n}$
2: **for** $i = 1$ to n **do**
3: compute $\pi_\epsilon(i) = \{j \in [\![1, n]\!] \setminus \{i\} \mid \mathbb{H}^D_{ij} \leq \epsilon\}$
4: **for** $i = 1$ to n **do**
5: **if** $\exists j \in \pi_\epsilon(i)$ *s.t.* $i \in \pi_\epsilon(j)$ **then**
6: **if** $\mathbb{H}^D_{ij} \leq \mathbb{H}^D_{ji}$ **then**
7: $\pi_\epsilon(j) \leftarrow \pi_\epsilon(j) \setminus \{i\}$
8: **else**
9: $\pi_\epsilon(i) \leftarrow \pi_\epsilon(i) \setminus \{j\}$
10: **for** $i = 1$ to n **do**
11: $\pi^*_\epsilon(i) \leftarrow \underset{j\in\pi_\epsilon(i)}{argmin} \, |Val(X_j)|$ (select one index in case of tie)
12: Compute forest $F_\epsilon = (V_{F_\epsilon}, A_{F_\epsilon})$ where $V_{F_\epsilon} = [\![1, n]\!]$ and $A_{F_\epsilon} = \{(\pi^*_\epsilon(i), i) \mid i \in [\![1, n]\!] \text{ } s.t. \text{ } \pi^*_\epsilon(i) \neq \emptyset\}$
Output: F_ϵ

---

sure no cycle is introduced by the screening phase, and guarantees the next proposition holds (ensuring that Algorithm 1 is indeed well defined):

Proposition 3. *For any rightful input D and ϵ, the output of Algorithm 1 is a forest (i.e. a directed acyclic graph with at most one parent per node).*

4.3 Learning Bayesian Networks Using Quasi-determinism Screening

We now present Algorithm 2, which uses quasi-determinism screening to accelerate Bayesian network structure learning. This algorithm takes the following input: D (a dataset containing M realizations of $\mathbf{X}$), ϵ (a threshold for quasi-determinism) and `ref-BNSL` (a BNSL baseline algorithm, taking for input a dataset, and returning a Bayesian network structure). The extension of the def-

Algorithm 2. Bayesian network structure learning with quasi deterministic screening (qds-BNSL)

Input: D, ϵ, *ref-BNSL*
1: Compute F_ϵ by running **Algorithm 1** with input D and ϵ
2: Identify $R(F_\epsilon) = \{i \in [\![1, n]\!] \mid \pi^{F_\epsilon}(i) = \emptyset\}$, the set of F_ϵ's roots.
3: Compute $G^*_{R(F_\epsilon)}$ by running `ref-BNSL` on $D_{R(F_\epsilon)}$
4: $G^*_\epsilon \leftarrow F_\epsilon \cup G^*_{R(F_\epsilon)}$
Output: G^*_ϵ

inition of determinism to quasi-determinism (Definition 3) prevents us to have 'hard' guarantees as those presented in Proposition 2. However, we are able to explicit bounds for the MLL score of a graph G^*_ϵ returned by Algorithm 2, as stated in the following Proposition.

Proposition 4. *Let ϵ, D and* `ref-BNSL` *be rightful input to Algorithm 2, and G^*_ϵ the associated output.*

Then, if `ref-BNSL` *is exact (i.e. always returns an optimal solution) with respect to the MLL score, we have the following lower bound for $s_D^{MLL}(G^*_\epsilon)$:*

$$s_D^{MLL}(G^*_\epsilon) \geq \left(\max_{G \in DAG_V} s_D^{MLL}(G) \right) - Mn\epsilon.$$

In practice, this bound is not very tight and this result therefore has few applicative potential. However, it shows that:

$$s_D^{MLL}(G^*_\epsilon) \underset{\epsilon \to 0}{\longrightarrow} \max_{G \in DAG_V} s_D^{MLL}(G).$$

In other words, $\epsilon \mapsto s_D^{MLL}(G^*_\epsilon)$ is continuous in 0, and Proposition 4 generalizes Proposition 2.

Algorithm 2 is promising, notably if for small ϵ we have $|R(F_\epsilon)|$ significantly smaller than n. In that case, `ref-BNSL`, that only has to be run on $D_{R(F_\epsilon)}$, can be expected to be much faster and more accurate than if it is run on the entire dataset D.

4.4 Complexity Analysis

Complexity of baseline BNSL learning algorithms The number of possible DAG structures being super exponential in the number of nodes, BNSL algorithms do not entirely explore the structure space but use smart caching and pruning methods to have a good performance & computation time trade-off.

Let `ref-BNSL` be a reference Bayesian network structure learning algorithm and $C_{ref}(M, n)$ be its complexity. $C_{ref}(M, n)$ should typically be thought of as linear in M and exponential, or at least high degree polynomial, in n for the best algorithms.

Complexity of Algorithm 1. We have the following decomposition of the complexity of Algorithm 1:

1. Lines 1–3: $\mathcal{O}(Mn^2)$. Computation of $\mathbb{H}^D$: we need counts for every couple (X_i, X_j) for $i < j$ (each time going through D), which implies $M\frac{n(n-1)}{2}$ operations.
2. lines 4–9: $\mathcal{O}(n^2)$. Going through $\mathbb{H}^D$ once.
3. lines 10–12: $\mathcal{O}(n^2)$. Going through $\mathbb{H}^D$ once.

Overall one has that $C_{Alg1}(M, n) = \mathcal{O}(Mn^2)$.

Complexity of Algorithm 2. For a given dataset D, we define:

$$\forall \epsilon \geq 0,\ n_r(\epsilon) = |R(F_\epsilon)|.$$

The function $n_r(\cdot)$, associates to $\epsilon \geq 0$ the number of roots of the forest F_ϵ returned by Algorithm 1. The complexity of Algorithm 2 then decomposes as:

1. Line 1: $\mathcal{O}(Mn^2)$. Run of Algorithm 1.
2. Lines 2–4: $C_{ref}(M, n_r(\epsilon))$. Run of `ref-BNSL` on reduced dataset $D_{R(F_\epsilon)}$ with $n_r(\epsilon)$ columns.

This yields $C_{Alg2}(M, n) = \mathcal{O}(Mn^2) + C_{ref}(M, n_r(\epsilon))$.

We are interested in how much it differs from $C_{ref}(M, n)$, which depends mainly on:

- how $n_r(\epsilon)$ compares to n,
- how $C_{ref}(M, n)$ varies with respect to n.

$C_{ref}(M, n)$ is known to be typically exponential in n for the best exact structure learning algorithms, as those presented by Silander and Myllymäki (2006) or Cussens (2011), and it is expected to be significantly larger than $\mathcal{O}(Mn^2)$ for high-performing heuristics. We therefore expect an important decrease in computational time when running Algorithm 2 compared to its baseline version, as long as $n_r(\epsilon)$ is sufficiently smaller than n. In the next section, we run a reference structure learning algorithm and Algorithm 1 on benchmark datasets in order to confirm this intuition.

5 Experiments

5.1 Experimental Setup

Data Table 1 summarizes the data used in our experiments. We considered the largest open-source categorical datasets among those presented[2] by Davis and Domingos (2010) and available on the UCI repository (Dheeru and Karra Taniskidou, 2017): `20 newsgroup`, `adult`, `book`, `covertype`, `kddcup 2000`, `msnbc`, `msweb`, `plants`, `reuters-52` and `uscensus`. Moreover, as it was done by Scanagatta et al. (2016), we chose the largest Bayesian networks available in the literature[3], for each of which we simulated 10000 observations: `andes`, `hailfinder`, `hepar 2`, `link`, `munin 1–4`, `pathfinder` and `win95pts`.

We also include two industrial datasets containing descriptive metadata on which we have privileged access, `priv-metadata 1` and `priv-metadata 2`.

Table 1. Datasets presentation

Name	Short name	n	M
20 newsgroups	20ng	930	11293
adult	adult	125	36631
book	book	500	8700
covertype	covertype	84	30000
kddcup 2000	kddcup	64	180092
msnbc	msnbc	17	291326
msweb	msweb	294	29441
plants	plants	69	17412
reuters 52	r52	941	6532
uscensus	uscensus	68	2458285
andes	andes	223	10000
hailfinder	hailfinder	56	10000
hepar 2	hepar2	70	10000
link	link	724	10000
munin 1	munin1	186	10000
munin 2	munin2	1003	10000
munin 3	munin3	1041	10000
munin 4	munin4	1038	10000
pathfinder	pathfinder	109	10000
windows 95 pts	win95pts	76	10000
priv-metadata 1	priv-meta1	43	1000
priv-metadata 2	priv-meta2	41	1000

[2] http://alchemy.cs.washington.edu/papers/davis10a/.

[3] http://www.bnlearn.com/bnrepository/.

Programming details and choice of `ref-BNSL` Most of the code associated with this project was done in `R`, enabling an optimal exploitation of the `bnlearn` package from Scutari (2010), which is a very good reference among open-source packages dealing with Bayesian networks structure learning.

We need a BSNL algorithm to obtain a baseline performance. After carefully evaluating several algorithms implemented in the `bnlearn` package, we chose to use *Greedy Hill Climbing with random restarts and a tabu list*, as it consistently outperformed other built-in algorithms both in time and score, in addition to being also used as a benchmark algorithm in the literature, notably by Teyssier and Koller (2005). In this section, we refer to this algorithm as `ref-BNSL`.

Choice of ϵ for `qds-BNSL`. An approach to choosing ϵ in the case of the `qds-BNSL` algorithm is to pick values for $n_r(\epsilon)$, and manually find the corresponding values for ϵ. For a given dataset and $\rho \in [0, 1]$, we define $\epsilon_\rho = n_r^{-1}(\lfloor \rho n \rfloor)$. In other words, ϵ_ρ is the value of ϵ for which the number of roots of the qd forest F_ϵ represents a proportion ρ of the total number of variables (more details in Rahier (2018)).

The computation of ϵ_ρ is not problematic: once $\mathbb{H}^D$ is computed and stored, evaluating $n_r(\epsilon)$ is done in constant time, and finding one of $n_r(\cdot)$'s quantiles is doable in $\mathcal{O}(\log(n))$ operations (dichotomy), which is negligible compared to the overall complexity of the screening. In the case of the `priv-metadata` datasets, choosing $\epsilon = 0$ leads to a dramatic decrease of the number of variables that are considered by the baseline algorithm, since these datasets contain several truly deterministic relationships by design.

Algorithm evaluation. The algorithms are evaluated using 3 axes of performance:

- BDe score of Sect. 2 with a uniform prior and equivalent sample size (ESS) equal to 5, inspired from Teyssier and Koller (2005) and referred to as BDeu.
- Number of arcs of the learned Bayesian network.
  The BDeu score naturally penalizes overly complex models (in terms of number of parameters), it is however interesting to look at the number of arcs, as it is a straightforward way to evaluate how complex a Bayesian network appears to a human expert (and thus how interpretable this structure is).
- Computing time t_{run} (all algorithms were run on the same machine).
  It is essential to remark that `ref-BNSL` is used both to obtain a baseline performance and inside `qds-BNSL`. In both cases, it is run with the same settings until convergence. The comparison of computing times is thus fair.

We present the obtained results for our selected baseline algorithm `ref-BNSL`, and 3 versions of `qds-BNSL`. For each dataset, we selected $\epsilon \in \{\epsilon_{0.9}, \epsilon_{0.75}, \epsilon_{0.5}\}$), corresponding to a restriction of `ref-BNSL` to 90%, 75% and 50% of the original variables respectively (for the `priv-metadata` datasets, these three choices of ϵ are merged into the single choice $\epsilon = 0$, which results in a decrease of more than 50% of the original variables).

The results are shown in Table 2, one group of columns per evaluation criterion, and each value is the median of 10 runs with different seeds. In each table, the median value of the criterion is displayed for `ref-BNSL` (**ref**), and the relative difference is displayed for the three versions of qds-BNSL we consider ($\mathbf{qds}_{\epsilon_{0.9}}$, $\mathbf{qds}_{\epsilon_{0.75}}$ and $\mathbf{qds}_{\epsilon_{0.5}}$).

5.2 Results

Table 2. For algorithms **ref**, $\mathbf{q}_{0.5}$, $\mathbf{q}_{0.75}$ and $\mathbf{q}_{0.5}$, and benchmark datasets, we display the Bayesian Network's (1) BDeu score averaged by observation, (2) learning time (including prescreening) and (3) number of arcs. Every result that corresponds to a BDeu score less than 5% smaller than ref-BNSL's score is boldfaced.

Dataset	BDeu score				Computation time				Number of arcs			
	ref	$q_{0.9}$ (%)	$q_{0.75}$ (%)	$q_{0.5}$ (%)	ref (s)	$q_{0.9}$ (%)	$q_{0.75}$ (%)	$q_{0.5}$ (%)	ref (nb)	$q_{0.9}$ (%)	$q_{0.75}$ (%)	$q_{0.5}$ (%)
20ng	-143	**−0.7**	**−2.1**	**−4.8**	21495	−1.6	−43	−73	3136	−4.5	−15	−32
adult	−13	**−0.2**	**−0.1**	**−4.0**	102	−6.6	−22	−61	371	+3.2	+7.0	−14
book	−35	**−0.8**	**−1.7**	**−4.6**	7600	−24	−40	−71	2196	−11	−19	−40
covertype	−14	**−0.2**	**−1.2**	−12	565	−6.8	−33	−71	337	−0.9	−11	−38
kddcup	−2.4	**−0.3**	**−1.0**	**−3.8**	2167	−11	−33	−74	285	−5.3	−19	−39
msnbc	−6.2	**−0.1**	**−2.6**	**−4.6**	252	−21	−61	−86	102	−7.8	−33	−64
msweb	−9.8	**+0.0**	**−0.1**	**−1.0**	4701	−6.3	−9.9	−55	1264	−2.5	−3.6	−35
plants	−13	**−2.6**	−7.6	−21	455	−47	−62	−84	320	−6.2	−18	−42
r52	−95	**−0.8**	**−2.0**	−6.1	18630	−14	−38	−77	2713	−3.6	−9.1	−25
uscensus	−23	**−0.3**	**−1.8**	−10	21782	−0.4	−32	−78	220	−10	−20	−38
andes	−93	**−0.5**	−6.2	−17	898	−2.2	−27	−70	336	−0.9	−7.1	−23
hailfinder	−50	**−0.1**	**−2.7**	−10	46	−5.3	−17	−55	64	−1.6	+6.2	−16
hepar2	−33	**−0.3**	**−1.4**	**−3.2**	76	−4.0	−43	−70	92	−3.3	−22	−30
link	−216	**+0.1**	**+1.1**	−17	7240	−12	−11	−61	1146	−1.8	−0.4	−22
munin1	−41	**−0.1**	**−0.2**	−9.9	497	−7.4	−17	−59	208	+0.0	+1.0	−9.6
munin2	−172	**−0.0**	**−0.0**	**−1.8**	7093	−20	−22	−44	879	+0.0	+0.0	−13
munin3	−165	**+0.0**	**+0.0**	**−1.1**	11558	−37	−29	−54	898	+0.0	+0.0	−7.8
munin4	−186	**−0.0**	**−0.0**	**−3.9**	8550	−7.9	−13	−39	903	+0.0	+0.0	−8.5
pathfinder	−27	**−0.7**	**−0.7**	**−4.9**	231	−14	−35	−69	161	−4.3	−8.7	−24
win95pts	−9.2	**+0.1**	−1.1	−9.2	132	−6.0	−31	−69	115	+0.0	−0.9	−12
priv-meta1	−8.72	**+1.1**	**+1.1**	**+1.1**	13794	−99	−99	−99	70	−41	−41	−41
priv-meta2	−8.72	**+12.5**	**+12.5**	**+12.5**	4346	−99	−99	−99	102	−59	−59	−59

Score. It appears in Table 2 that the decrease in BDeu score is smaller than 5% for all the considered datasets when 90% of the variables remain after the pre-screening ($\mathbf{qds}_{\epsilon_{0.9}}$), and for most of them when 75% of the variables remain ($\mathbf{qds}_{\epsilon_{0.75}}$). This is also observed with $\epsilon_{0.5}$ for datasets that contain a lot of very strong pairwise relationships as kddcup, msweb, or munin 2–4. For priv-metadata datasets, our approach increases the score (slightly for priv-metadata1 and of more than 12% for priv-metadata2).

Computing Time. Table 2 shows a significant decrease in computational time for `qds-BNSL`, which is all the more important as ϵ is large. In the best cases, we have both a very small decrease in BDeu score, and an important decrease in computational time. We suspect that this is also due to the presence of many strong pairwise relationships. For example, the algorithm `qds-BNSL` with $\epsilon = \epsilon_{0.5}$ is 55% faster for `msweb`, and 54% for `munin` 3, while implying only around 1% decrease in score compared to `ref-BNSL`. If we allow a 5% score decrease, `qds-BNSL` can be up to 70% faster (`20 newgroups`, `book`, `msnbc`, `kddcup`, `hepar2`, `pathfinder`). On the industrial datasets this computational time decrease is astonishing: it is of more than 2 **orders of magnitude** in the case of `priv-metadata 1` and more than 3 **orders of magnitude** for `priv-metadata 2`[4]. These results confirm the complexity analysis of the previous section, in which we supposed that the screening phase had a very small computational cost compared to the standard structure learning phase.

Complexity. As showed by Table 2, Bayesian networks learned with `qds-BNSL` are consistently less complex than those learned with `ref-BNSL`. Several graphs learned with $\mathbf{qds}_{\epsilon_{0.5}}$ are more than 30% sparser while still scoring less than 5% below the baseline algorithm: `20 newsgroups`, `book`, `kddcup 2000`, `msnbc`, `msweb` and `hepar 2`.

Figure 1 displays two Bayesian networks learned on the `msnbc` dataset.

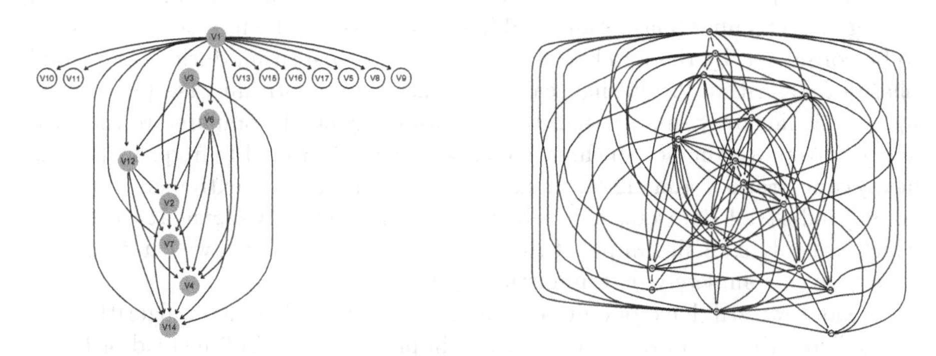

Fig. 1. Bayesian network learned on `msnbc` with `qds-BNSL`, left (resp. `ref-BNSL`, right). BDeu: -6.48 (resp. -6.19), Nb of arcs: 37 (resp. 102), Running time: 36s (resp. 252s)

They provide an interesting example of the sparsity induced by `qds-BNSL`. After the $\mathbf{qd}_{\epsilon_{0.5}}$-screening phase, half of the variables (corresponding to the nodes in white) are considered to be sufficiently explained by $V1$. They are therefore not taken into account by `ref-BNSL`, which is run only on the variables corresponding to the nodes in gray (more details in Rahier et al. (2018)).

[4] These results were so extreme that they could not be fully captured in the 'percentage' format choice of Table 2.

In the msnbc case, this learning problem restriction implies only a small decrease in the final graph's generalization performance (as seen in BDeu scores), while being 7 times faster to compute and enabling a significantly better readability.

In this processed version of the msnbc dataset (Davis and Domingos, 2010), each variable contains a binary information regarding the visit of a given page from the http://msnbc.com/ website[5]. The Bayesian network displayed in Fig. 2 shows in a compact way the influence between the different variables. For instance, we see that visits of the website's pages corresponding to nodes in white (*e.g.* 'weather' ($V8$), 'health' ($V9$) or 'business' ($V11$)) are importantly influenced by whether the user has also visited the frontpage ($V1$). For example, learned parameters show that a user who did not visit the website's frontpage ($V1$) is about 10 times more likely to have visited the website's 'summary' page ($V13$) than a user who did visit the frontpage. Such information is much harder to read from the graph learned with ref-BNSL displayed in Fig. 1 right.

6 Concluding Remarks

We have seen that, both in theory and in practice, the quasi-determinism screening approach enables a decrease in computational time and complexity for a small decrease in graph score. This tradeoff is all the more advantageous as there actually are strong pairwise relationships in the data, that can be detected during the screening phase, thus enabling a decrease in the number of variables to be considered by the baseline structure learning algorithm during the second phase of Algorithm 2. Optimal cases for this meta-algorithm take place when $n_r(\epsilon)$ is significantly smaller than n for ϵ reasonably small compared to the variable's entropies. Among benchmark datasets this is reasonably frequent (*e.g* 20 newsgroup, msnbc, munin2-4, webkb), and we argue it is extremely frequent among industrial datasets, as we have shown in our priv-metadata datasets which are only a small sample of the kind of datasets in which we can find very strong (even completely deterministic) relations.

Besides, we still have potential to improve the qds-BNSL meta-algorithm, by paralellizing the computation of $\mathbb{H}^D$, and implementing it in C instead of R.

Our main research perspectives are (1) to understand how one can anticipate how good the score/computation time/complexity trade-off can be before running any algorithm all the way through, saving us from running qds-BNSL on datasets in which there are no strong pairwise relationships to be detected, (2) find a principled way to choose ϵ and (3) tighten the bound of Proposition 4 and generalize it to the BDeu score.

In another directions, we have some insights on ways to generalize our quasi-determinism screening idea. The proof of Proposition 2 suggests that the result still holds when F is *any kind of deterministic DAG* (and not only a forest). We could therefore use techniques that detect determinism in a broader sense

[5] More details: http://archive.ics.uci.edu/ml/machine-learning-databases/msnbc-mld/msnbc.data.html.

than only pairwise, to make the screening more efficient. For this purpose we could take inspiration from papers of the knowledge discovery in databases (KDD) community, as Huhtala et al. (1999), or more recently Papenbrock et al. (2015) who evaluate functional dependencies discovery methods. We also could broaden our definition of quasi-determinism: instead of considering the information-theoretic quantity $H^D(X|Y)$ to describe the strength of the relationship $Y \rightarrow X$, one could choose $\frac{H^D(X|Y)}{H^D(X)}$, which represents the proportion of X's entropy that is explained by Y. Moreover, $\frac{H^D(X|Y)}{H^D(X)} \leq \epsilon$ can be rewritten as $\frac{MI^D(X,Y)}{H(X)} \geq 1 - \epsilon$, which gives another insight to quasi-determinism screening: for a given variable X, this comes down to finding a variable Y such that $MI^D(X,Y)$ is high. This is connected to the original idea of Chow and Liu (1968), and later Cheng et al. (1997), for whom pairwise empirical mutual information is central. This alternate definition of $\epsilon-$quasi-determinism does not change the algorithms and complexity considerations described in Sect. 4.

References

Bouckaert, R.: Bayesian belief networks: from inference to construction. PhD thesis, Faculteit Wiskunde en Informatica, Utrecht University (1995)

Chen, X.-W., Anantha, G., Lin, X.: Improving Bayesian network structure learning with mutual information-based node ordering in the K2 algorithm. IEEE Trans. Knowl. Data Eng. **20**(5), 628–640 (2008)

Cheng, J., Bell, D.A., Liu, W.: Learning belief networks from data: An information theory based approach. In: Proceedings of the Sixth International Conference on Information and Knowledge Management, pp. 325–331. ACM (1997)

Chickering, D.M.: Learning Bayesian networks is NP-complete. Learning from data: Artif. Intell. Stat. V **112**, 121–130 (1996)

Chow, C., Liu, C.: Approximating discrete probability distributions with dependence trees. IEEE Trans. Inf. Theor. **14**(3), 462–467 (1968)

Cooper, G.F., Herskovits, E.: A Bayesian method for the induction of probabilistic networks from data. Mach. Learn. **9**(4), 309–347 (1992)

Cussens, J.: Bayesian network learning with cutting planes. In: Proceedings of the Twenty-Seventh Conference on Uncertainty in Artificial Intelligence, UAI'11, pp. 153–160, Arlington, Virginia, United States. AUAI Press (2011)

Davis, J., Domingos, P.: Bottom-up learning of Markov network structure. In: Proceedings of the 27th International Conference on Machine Learning (ICML-10), pp. 271–278 (2011)

de Morais, S.R., Aussem, A., Corbex, M.: Handling almost-deterministic relationships in constraint-based Bayesian network discovery: Application to cancer risk factor identification. In: European Symposium on Artificial Neural Networks, ESANN'08 (2010)

Dheeru, D., Karra Taniskidou, E.: UCI machine learning repository (2017)

El Kaed, C., Leida, B., Gray, T.: Building management insights driven by a multi-system semantic representation approach. In: Internet of Things (WF-IoT), 2016 IEEE 3rd World Forum on, pp. 520–525. IEEE (2016)

Heckerman, D., Geiger, D., Chickering, D.M.: Learning Bayesian networks: The combination of knowledge and statistical data. Mach. Learn. **20**(3) (1995). https://doi.org/10.1023/A:1022623210503

Huhtala, Y., Kärkkäinen, J., Porkka, P., Toivonen, H.: Tane: an efficient algorithm for discovering functional and approximate dependencies. Comput. J. **42**(2), 100–111 (1999)

Kareem, S.W., Okur, M.C.: Bayesian network structure learning based on pigeon inspired optimization. Int. J. Adv. Trends Comput. Sci. Eng. **8**, 131–137 (2019)

Kareem, S.W., Okur, M.C.: Falcon optimization algorithm for bayesian networks structure learning. Comput. Sci. **22**(4), 553–569 (2021)

Koller, D., Friedman, N.: Probabilistic graphical models: principles and techniques. MIT press (2009)

Koo, D.D., Lee, J.J., Sebastiani, A., Kim, J.: An internet-of-things (iot) system development and implementation for bathroom safety enhancement. Proc. Eng. **145**, 396–403 (2016)

Luo, W.: Learning Bayesian networks in semi-deterministic systems. In: Lamontagne, L., Marchand, M. (eds.) AI 2006. LNCS (LNAI), vol. 4013, pp. 230–241. Springer, Heidelberg (2006). https://doi.org/10.1007/11766247_20

Mabrouk, A., Gonzales, C., Jabet-Chevalier, K., Chojnacki, E.: An efficient Bayesian network structure learning algorithm in the presence of deterministic relations. In: Proceedings of the Twenty-first European Conference on Artificial Intelligence. IOS Press (2014)

Nie, S., de Campos, C., P., Ji, Q.: Learning Bayesian networks with bounded tree-width via guided search. In: AAAI, pp. 3294–3300 (2016)

Papenbrock, T., et al.: Functional dependency discovery: an experimental evaluation of seven algorithms. Proc. VLDB Endowment **8**(10), 1082–1093 (2015)

Rahier, T., Marie, S., Girard, S., Forbes, F.: Screening strong pairwise relationships for fast Bayesian network structure learning 2nd Italian-French Statistics Seminar-IFSS (2018)

Rahier, T.: Bayesian networks for static and temporal data fusion Université Grenoble Alpes, PhD thesis (2018)

Scanagatta, M., Corani, G., de Campos, C. P., Zaffalon, M.: Learning treewidth-bounded Bayesian networks with thousands of variables. In: Advances in Neural Information Processing Systems, pp. 1462–1470 (2016)

Scanagatta, M., de Campos, C.P., Corani, G., Zaffalon, M.: Learning Bayesian networks with thousands of variables. In: Advances in Neural Information Processing Systems, pp. 1864–1872 (2015)

Schwarz, G., et al.: Estimating the dimension of a model. Ann. Stat. **6**(2), 461–464 (1978)

Scutari, M.: Learning Bayesian networks with the bnlearn R package. J. Stat. Softw. **35**(3), (2010)

Scutari, M., Graafland, C., Gutiérrez, J.: Who learns better bayesian network structures: accuracy and speed of structure learning algorithms. Int. J. Approx. Reason. **115**, 235–253 (2019)

Silander, T., Myllymäki, P.: A simple approach for finding the globally optimal Bayesian network structure. In: Proceedings of the Twenty-Second Conference on Uncertainty in Artificial Intelligence, UAI'06 (2006)

Spirtes, P., Glymour, C. N., Scheines, R. . Causation, prediction, and search. MIT press (2000)

Teyssier, M., Koller, D.: Ordering-based search: a simple and effective algorithm for learning Bayesian networks. In: Proceedings of the Twenty-First Conference on Uncertainty in Artificial Intelligence, pp. 584–590. AUAI Press (2005)

On Projectivity in Markov Logic Networks

Sagar Malhotra[1,2(✉)] and Luciano Serafini[1]

[1] Fondazione Bruno Kessler, Trento, Italy
smalhotra@fbk.eu
[2] University of Trento, Trento, Italy

Abstract. Markov Logic Networks (MLNs) define a probability distribution on relational structures over varying domain sizes. Like most relational models, MLNs do not admit consistent marginal inference over varying domain sizes i.e. marginal probabilities depend on the domain size. Furthermore, MLNs learned on a fixed domain do not generalize to domains of different sizes. In recent works, connections have emerged between domain size dependence, lifted inference, and learning from a sub-sampled domain. The central idea of these works is the notion of *projectivity*. The probability distributions ascribed by projective models render the marginal probabilities of sub-structures independent of the domain cardinality. Hence, projective models admit efficient marginal inference. Furthermore, projective models potentially allow efficient and consistent parameter learning from sub-sampled domains. In this paper, we characterize the necessary and sufficient conditions for a two-variable MLN to be projective. We then isolate a special class of models, namely Relational Block Models (RBMs). In terms of data likelihood, RBMs allow us to learn the best possible projective MLN in the two-variable fragment. Furthermore, RBMs also admit consistent parameter learning over sub-sampled domains.

1 Introduction

Statistical Relational Learning [2,13] (SRL) is concerned with representing and learning probabilistic models over relational structures. Many works have observed that SRL frameworks exhibit unwanted behaviors over varying domain sizes [11,12]. These behaviors make models learned from a fixed or a sub-sampled domain unreliable for inference over larger (or smaller) domains [11]. Drawing on the works of Shalizi and Rinaldo [15] on Exponential Random Graphs (ERGMs), Jaeger and Schulte [4] have recently introduced the notion of *projectivity* as a strong form of guarantee for good scaling behavior in SRL models. A projective model requires that the probability of any given query, over arbitrary m domain objects, is completely independent of the domain size.

Supplementary Information The online version contains supplementary material available at https://doi.org/10.1007/978-3-031-26419-1_14.

© The Author(s), under exclusive license to Springer Nature Switzerland AG 2023
M.-R. Amini et al. (Eds.): ECML PKDD 2022, LNAI 13717, pp. 223–238, 2023.
https://doi.org/10.1007/978-3-031-26419-1_14

Jaeger and Schulte [4] *identify restrictive fragments of SRL models to be projective. But whether these fragments are complete characterization of projectivity, remains an open problem.*

In this paper, our goal is to characterize projectivity for a specific class of SRL models, namely Markov Logic Networks (MLNs) [14]. MLNs are amongst the most prominent template-based SRL models. An MLN is a Markov Random Field with features defined in terms of function-free weighted First Order Logic (FOL) formulae. Jaeger and Schutle [4] show that an MLN is projective if - any pair of atoms in each of its formulae share the same set of variables. We show that this characterization is not complete. Furthermore, we completely characterize projectivity for the class of MLNs with at most 2 variables in their formulae. Our charecterization leads to a parametric restriction that can be easily incorporated into any MLN learning algorithm. We also identify a special class of projective models, namely the Relational Block Models (RBMs). Any projective MLN in the two variable fragment can be expressed as an RBM. We show that the training data likelihood due to the maximum likelihood RBM is greater than or equal to the training data likelihood due to any other projective MLN in the two variable fragment. RBMs also admit consistent maximum likelihood estimation. Hence, RBMs are projective models that admit consistent and efficient learning from sub-sampled domains.

The paper is organized as follows: We first contextualize our work w.r.t the related works in this domain. We then provide some background and notation on FOL and relational structures. We also elaborate on the fragment of FOL with at most two variables i.e. FO^2 and define the notion of FO^2 interpretations as multi-relational graphs. We also overview some results on Weighted First Order Model Counting. In the subsequent section, we provide a parametric representation for any MLN in the two variable fragment. We then dedicate a section to the main result of this paper i.e. the necessary and sufficient conditions for an MLN in the two variable fragment to be projective. Based on the projectivity criterions we identify a special class of models namely Relational Block Models. We dedicate a complete section to RBMs and elaborate on their useful properties. We then move on to a formal comparison between the previous characterizations and the presented characterization of projectivity in MLNs. Finally, we discuss the consistency and efficiency aspects of learning for projective MLNs and RBMs.

2 Related Work

Projectivity has emerged as a formal notion of interest through multiple independent lines of works across ERGM and SRL literature. The key focus of these works have been analyzing [12,17] or mitigating [6,11] the effects of varying domain sizes on relational models. The major step in formalizing the notion of projectivity can be attributed to Shalizi and Rinaldo [15]. The authors both formalize and characterize the sufficient and necessary conditions for ERGMs to be projective. It is interesting to note that their projectivity criterion is strictly structural i.e. they put no restrictions on parameter values but rather inhibit the class of features that can be defined as sufficient statistics in ERGMs. In contrast our results w.r.t MLN are strictly parametric (which may correspond to non-trivial structural restrictions as well). With respect to SRL, the notion of projectivity was first formalized by Jaeger and Schulte [4], they show some restrictive fragments of SRL models to be projective. Jaeger and Schulte [5] significantly extend the scope of projective models by characterizing necessary and sufficient conditions for an arbitrary model on relational structures to be projective. Their characterization is expressed in terms of the so called AHK models. But as they conclude in [5], expressing AHK models in existing SRL frameworks remains an open problem. Hence, a complete characterization of projectivity in most SRL languages is still an open problem. Weitkamper [19] has shown that the characterization of projectivity provided by Jaeger and Schulte [4], for probabilistic logic programs under distribution semantics, is indeed complete. In this work, we will extend this characterization to the two variable fragment of Markov Logic Networks.

Another correlated problem to projectivity is learning from sub-sampled or smaller domains. In the relational setting projectivity is not a sufficient condition for consistent learning from sub-sampled domains [4]. Mittal et. al. have proposed a solution to this problem by introducing domain-size dependent scale-down factors [11] for MLN weights. Although empirically effective, the scale-down factors are not known to be a statistically sound solution. On the other hand, Kuzelka et. al. [9], provide a statistically sound approach to approximately obtain the correct distribution for a larger domain. But their approach requires estimating the relational marginal polytope for the larger domain and hence, offers no computational gains w.r.t learning from a sub-sampled domain. In this work, we will provide a statistically sound approach for efficiently estimating a special class of projective models (namely, RBM) from sub-sampled domains. We also show that our approach provides consistent parameter estimates in an efficient manner and is better than estimating any projective MLN in the two variable fragment (in terms of data likelihood maximisation).

3 Background

Basic Definitions. We use the following basic notation. The set of integers $\{1, ..., n\}$ is denoted by $[n]$. We use $[m : n]$ to denote the set of integers $\{m, ..., n\}$. Wherever the set of integers $[n]$ is obvious from the context we will use $[\overline{m}]$ to represent the set $[m+1 : n]$. We use $\boldsymbol{k} = \langle k_1, ..., k_m \rangle$ to denote an n-partition i.e. $k_i \in \mathbb{Z}^+$ and $\sum_{i\in[m]} k_i = n$. We will also use multinomial coefficients denoted by

$$\binom{n}{k_1, ..., k_m} = \binom{n}{\boldsymbol{k}} = \frac{n!}{\prod_{i\in[m]} k_i!}$$

First Order Logic and Relational Substructures. We assume a function-free First Order Logic (FOL) language $\mathcal{L}$ defined by a set of variables $\mathcal{V}$ and a set of relational symbols $\mathcal{R}$. We use Δ to denote a domain of n constants.

For $a_1, ..., a_k \in \mathcal{V} \cup \Delta$ and $R \in \mathcal{R}$, we call $R(a_1, ...a_k)$ an *atom*. A *literal* is an atom or the negation of an atom. If $a_1, ..., a_k \in \mathcal{V}$, then the atom is called a *first order atom*, whereas if $a_1, ..., a_k \in \Delta$, then it's called a *ground atom*. We use $\mathcal{F}$ to denote the set of first order atoms and $\mathcal{G}$ to denote the set of ground atoms. A *world* or an *interpretation* $\omega : \mathcal{G} \rightarrow \{\mathbf{T}, \mathbf{F}\}$ is a function that maps each ground atom to a boolean. The set of interpretations ω, in the language $\mathcal{L}$ and the domain Δ of size n, is denoted by $\Omega^{(n)}$. We say that $\omega \in \Omega^{(n)}$, has a size n and is also called an n-world. For a subset $\mathrm{I} \subset \Delta$, we use $\omega \downarrow \mathrm{I}$ to denote the partial interpretation induced by I. Hence, $\omega \downarrow \mathrm{I}$ is an interpretation over the ground atoms containing only the domain elements in I.

Example 1. Let us have a language with only one relational symbol R of arity 2 and a domain $\Delta = \{a, b, c\}$. Let us have the following interpretation ω:

$R(a,a)$	$R(a,b)$	$R(a,c)$	$R(b,a)$	$R(b,b)$	$R(b,c)$	$R(c,a)$	$R(c,b)$	$R(c,c)$
T	**T**	**F**	**T**	**T**	**F**	**T**	**T**	**F**

then $\omega \downarrow \{a, b\}$ is given as:

$R(a,a)$	$R(a,b)$	$R(b,a)$	$R(b,b)$
T	**T**	**F**	**T**

For most of our purposes, we will be able to assume w.l.o.g that $\Delta = [n]$.

FO^2, m-Types and m-Tables. FO^2 is the fragment of FOL with two variables. We will use the notion of 1-types, 2-type, and 2-tables as presented in [8]. A 1-type is a conjunction of a maximally consistent set of first order literals containing only one variable. For example, in an FO^2 language on the unary predicate A and binary predicate R, $A(x) \wedge R(x,x)$ and $A(x) \wedge \neg R(x,x)$ are examples of 1-types in variable x. A 2-table is a conjunction of maximally consistent first order literals containing exactly two distinct variables. Extending the previous example, $R(x,y) \wedge \neg R(y,x)$ and $R(x,y) \wedge R(y,x)$ are instances of 2-tables. We assume an arbitrary order on the 1-types and 2-tables, hence, we use $i(x)$ to denote the ith 1-type and $l(x,y)$ to denote the lth 2-table. Finally, a 2-type is a conjunction of the form $i(x) \wedge j(y) \wedge l(x,y) \wedge (x \neq y)$ and we use $ijl(x,y)$ to represent it. In a given interpretation ω, we say a constant c realizes the ith 1-type if $\omega \models i(c)$, we say a pair of constants (c,d) realizes the lth 2-table if $\omega \models l(c,d)$ and (c,d) realizes the 2-type $ijl(x,y)$ if $\omega \models ijl(c,d)$. We call the 2-type $ijl(y,x)$ the dual of $ijl(x,y)$ and denote it by $\underline{ijl}(x,y)$. We will use u to denote the number of 1-types and b to denote the number of 2-tables in a given FO^2 language.

Interpretations as Multi-relational Graphs. Given an FO^2 language $\mathcal{L}$ with interpretations defined over the domain $\Delta = [n]$, we can represent an interpretation $\omega \in \Omega^{(n)}$ as a multi-relational graph $(\boldsymbol{x}, \boldsymbol{y})$. This is achieved by defining $\boldsymbol{x} = (x_1, ..., x_n)$ such that $x_q = i$ if $\omega \models i(q)$ and by defining $\boldsymbol{y} =$

$(y_{12}, y_{13}, ...y_{qr}, ..., y_{n-1,n})$, where $q < r$, such that $y_{qr} = l$ if $\omega \models l(q,r)$. We also define $k_i = k_i(\boldsymbol{x}) = k_i(\omega) := |\{c \in \Delta : c \models i(c)\}|$, $h_l^{ij} = h_l^{ij}(\boldsymbol{y}) = h_l^{ij}(\omega) := |\{(c,d) \in \Delta^2 : \omega \models ijl(c,d)\}|$ and for any $D \subseteq \Delta^2$, $h_l^{ij}(D) = h_l^{ij}(\omega, D) := |\{(c,d) : \omega \models ijl(c,d) \text{ and } (c,d) \in D\}|$. Notice that $\sum_{i \leq j} \sum_{l \in [b]} h_l^{ij} = \binom{n}{2}$ and $\sum_{l \in [b]} h_l^{ij} = \boldsymbol{k}(i,j)$, where $\boldsymbol{k}(i,j)$ is defined in Eq. (3). We use $(\boldsymbol{x}_{\mathrm{I}}, \boldsymbol{y}_{\mathrm{I}})$ to represent the multi-relational graph for $\omega \downarrow \mathrm{I}$. Throughout this paper we will use an interpretation ω and it's multi-relational graph $(\boldsymbol{x}, \boldsymbol{y})$ interchangeably.

Weighted First Order Model Counting in FO^2. We will briefly review Weighted First Order Model Counting (WFOMC) in FO^2 as presented in [10]. WFOMC is formally defined as follows:

$$\text{WFOMC}(\Phi, n) := \sum_{\omega \in \Omega^{(n)} : \omega \models \Phi} w(\omega)$$

where Φ is an FOL formula, n is the size of the domain and w is a weight function that maps each interpretation ω to a positive real. First Order Model Counting (FOMC) is the special case of WFOMC, where for all $\omega \in \Omega^{(n)}$, $w(\omega) = 1$. We assume that w does not depend on individual domain constants, which implies that w assigns same weight to two interpretations which are isomorphic under the permutation of domain elements.

A universally quantified FO^2 formula $\forall xy.\Phi(x,y)$ can be equivalently expressed as $\forall xy.\Phi(\{x,y\})$, where $\Phi(\{x,y\})$ is defined as $\Phi(x,x) \wedge \Phi(x,y) \wedge \Phi(y,x) \wedge \Phi(y,y) \wedge (x \neq y)$. A *lifted interpretation* denoted by $\tau : \mathcal{F} \rightarrow \{\mathbf{T}, \mathbf{F}\}$ assigns boolean values to first order atoms. The truth value of the quantifier free formula $\Phi(x,y)$ under a lifted interpretation τ, denoted by $\tau(\Phi(x,y))$, is computed by applying classical semantics of the propositional connectives to the truth assignments of atoms of $\Phi(\{x,y\})$ under τ. We then define

$$n_{ijl} := |\{\tau \mid \tau \models \Phi(\{x,y\}) \wedge ijl(x,y)\}| \tag{1}$$

and $n_{ij} := \sum_{l \in [b]} n_{ijl}$. First Order Model Counting for a universally quantified formula $\forall xy.\Phi(x,y)$ is then given as:

$$\text{FOMC}(\forall xy.\Phi(x,y), n) = \sum_{\boldsymbol{k}} \binom{n}{\boldsymbol{k}} \prod_{\substack{i \leq j \\ i,j \in [b]}} n_{ij}^{\boldsymbol{k}(i,j)} \tag{2}$$

where $\boldsymbol{k} = \langle k_1, \ldots, k_u \rangle$ is a u-tuple of non-negative integers, $\binom{n}{\boldsymbol{k}}$ is the multinomial coefficient and

$$\boldsymbol{k}(i,j) = \begin{cases} \frac{k_i(k_i - 1)}{2} & \text{if } i = j \\ k_i k_j & \text{otherwise} \end{cases} \tag{3}$$

Intuitively, k_i represents the number of constants c of 1-type i. Also a given constant realizes exactly one 1-type. Hence, for a given $\boldsymbol{k}$, we have $\binom{n}{\boldsymbol{k}}$ possible ways of realizing k_i 1-types. Furthermore, given a pair of constants c and d such that c is of 1-type i and d is of 1-type j, the number of extensions to the binary predicates containing both c and d, such that the extensions are a model of $\forall xy.\Phi(x,y)$, is given by n_{ij} independently of all other constants. Finally, the exponent $\boldsymbol{k}(i,j)$ accounts for all possible pair-wise choices of constants given a $\boldsymbol{k}$ vector. Equation (2) was originally proven in [1], we refer the reader to [10] for the formulation presented here.

Families of Probability Distributions and Projectivity. We will be interested in probability distributions over the set of interpretations or equivalently their multi-relational graphs. A family of probability distributions $\{P^{(n)} : n \in \mathbb{N}\}$ specifies, for each finite domain of size n, a distribution $P^{(n)}$ on the possible n-world set $\Omega^{(n)}$ [5]. We will mostly work with the so-called exchangeable probability distributions [5] i.e. distributions where $P^{(n)}(\omega) = P^{(n)}(\omega')$ if ω and ω' are isomorphic. A distribution $P^{(n)}(\omega)$ over n-worlds induces a marginal probability distribution over m-worlds $\omega' \in \Omega^{(m)}$ as follows:

$$P^{(n)} \downarrow [m](\omega') = \sum_{\omega \in \Omega^{(n)} : \omega \downarrow [m] = \omega'} P^{(n)}(\omega)$$

Notice that due to exchangeability $P^{(n)} \downarrow \mathrm{I}$ is the same for all subsets I of size m, hence we can always assume any induced m-world to be $\omega \downarrow [m]$. We are now able to define projectivity as follows:

Definition 1 ([5])**.** An exchangeable family of probability distributions is called projective if for all $m < n$:

$$P^{(n)} \downarrow [m] = P^{(m)}$$

When dealing with probability distributions over multi-relational representation, we denote by $(\boldsymbol{X}, \boldsymbol{Y})$ the random vector where, $\boldsymbol{X} = (X_1, \ldots, X_n)$ and each X_i takes value in $[u]$; and $\boldsymbol{Y} = (Y_{12}, Y_{13}, \ldots, Y_{qr}, \ldots, Y_{n-1,n})$ where $q < r$ and Y_{qr} takes values in $[b]$.

4 A Parametric Normal Form for MLNs

A Markov Logic Network (MLN) Φ is defined by a set of weighted formulas $\{(\phi_i, a_i)\}_i$, where ϕ_i are quantifier free, function-free FOL formulas with weights $a_i \in \mathbb{R}$. An MLN Φ induces a probability distribution over the set of possible worlds $\omega \in \Omega^{(n)}$:

$$P^{(n)}_{\Phi}(\omega) = \frac{1}{Z(n)} \exp\Big(\sum_{(\phi_i, a_i) \in \Phi} a_i . N(\phi_i, \omega) \Big)$$

where $N(\phi_i, \omega)$ represents the number of true groundings of ϕ_i in ω. The normalization constant $Z(n)$ is called the *partition function* that ensures that $P^{(n)}_{\Phi}$ is a probability distribution.

Theorem 1. *Any Markov Logic Network (MLN) $\Phi = \{(\phi_i, a_i)\}_i$ on a domain of size n, such that ϕ_i contains at-most two variables, can be expressed as follows:*

$$P_{\Phi}^{(n)}(\omega) = \frac{1}{Z(n)} \prod_{i\in[u]} s_i^{k_i} \prod_{\substack{i,j\in[u]\\ i\leq j}} \prod_{l\in[b]} (t_{ijl})^{h_l^{ij}} \tag{4}$$

where s_i and t_{ijl} are positive real numbers and k_i is $k_i(\omega)$ and h_l^{ij} is equal to $h_l^{ij}(\omega)$.

Proof. Let $\Phi = \{(\phi_i, a_i)\}_i$ be an MLN, such that ϕ_i contains at-most two variables. Firstly, every weighted formula $(\phi(x,y), a) \in \Phi$ that contains exactly two variables is replaced by two weighted formulas $(\phi(x,x), a)$ and $(\phi(x,y) \wedge (x \neq y), a)$. The MLN distribution $P_{\Phi}^{(n)}$ is invariant under this transformation. Hence, Φ can be equivalently written as $\{(\alpha_q(x), a_q)\}_q \cup \{(\beta_p(x,y), b_p)\}_p$, where $\{\alpha_q(x)\}_q$ is the set of formulas containing only the variable x and $\{\beta_p(x,y)\}_p$ is the set of formulas containing both the variables x and y. Notice that every $\beta_p(x,y)$ entails $x \neq y$.

Let us have $\omega \in \Omega^{(n)}$, where we have a domain constant c such that $\omega \models i(c)$. Now notice that the truth value of ground formulas $\{\alpha_q(c)\}_q$ in ω is completely determined by $i(c)$ irrespective of all other domain constants. Hence, the (multiplicative) weight contribution of $i(c)$ to the weight of ω can be given as $\exp(\sum_q a_q \mathbb{1}_{i(x)\models\alpha_q(x)})$. We define s_i as follows:

$$s_i = \exp\big(\sum_q a_q \mathbb{1}_{i(x)\models\alpha_q(x)}\big) \tag{5}$$

Clearly, this argument can be repeated for all the domain constants realizing any 1-type in $[u]$. Hence, the (multiplicative) weight contribution due to 1-types of all domain constants and equivalently due to the groundings of all unary formulas, is given as $\prod_{i\in[u]} s_i^{k_i}$.

We are now left with weight contributions due to the binary formulas, given by the set $\{(\beta_p(x,y), b_p)\}_p$. Due to the aforementioned transformation, each binary formula $\beta(x,y)$ contains a conjunct $(x \neq y)$. Hence, all groundings of $\beta(x,y)$ such that both x and y are mapped to the same domain constants evaluate to false. Hence, we can assume that x and y are always mapped to distinct domain constants. Let us have an unordered pair of domain constants $\{c,d\}$ such that $\omega \models ijl(c,d)$. The truth value of any binary ground formula $\beta(c,d)$ and $\beta(d,c)$ is completely determined by $ijl(c,d)$ irrespective of all other domain constants. Hence, the multiplicative weight contribution due to the ground formulas $\{\beta_p(c,d)\}_p \cup \{\beta_p(d,c)\}_p$ is given as t_{ijl}, where t_{ijl} is defined as follows:

$$\exp\left(\sum_p b_p \mathbb{1}_{ijl(x,y)\models\beta_p(x,y)} + \sum_p b_p \mathbb{1}_{ijl(x,y)\models\beta_p(y,x)}\right) \tag{6}$$

Hence, the weight of an interpretation ω under the MLN Φ is given as

$$\prod_{i\in[u]} s_i^{k_i} \prod_{\substack{i,j\in[u]\\ i\leq j}} \prod_{l\in[b]} (t_{ijl})^{h_l^{ij}}$$

Definition 2. Given an MLN in the parametric normal form given by Eq. (4). Then f_{ij} is defined as $\sum_{l\in[b]} t_{ijl}$.

We will now provide the parameterized version of the partition function $Z(n)$ due to Theorem 1.

Proposition 1. *Let Φ be an MLN in the form* (4)*, then the partition function $Z(n)$ is given as:*

$$Z(n) = \sum_{\boldsymbol{k}} \binom{n}{\boldsymbol{k}} \prod_{i\in[u]} s_i^{k_i} \prod_{\substack{i,j\in[u]\\ i\leq j}} (f_{ij})^{\boldsymbol{k}(i,j)} \tag{7}$$

where $\boldsymbol{k}(i,j)$ is defined in Eq. (3).

Proof (Sketch). The proposition is a parameterized version of Eq. (2), where $\prod_{i\in[u]} s_i^{k_i}$ takes into account the weight contributions due to the 1-type realizations and f_{ij} is essentially a weighted version of n_{ij} i.e. given a pair of constants c and d such that they realize the ith and the jth 1-type respectively, then f_{ij} is the sum of the weights due to the 2-types realized by the extensions to the binary predicates containing both c and d.

5 Projectivity in Markov Logic Networks

We present the necessary and sufficient conditions for an MLN to be projective in the two variable fragment. The complete proofs are provided in the appendix.

Lemma 1 (Sufficiency). *A Markov Logic Network in the two variable fragment is projective if all the f_{ij} have the same value i.e. $\forall i,j \in [u] : f_{ij} = F$, for some positive real number F.*

Proof (Sketch). The key idea of the proof is that if $\forall i,j \in [u] : f_{ij} = F$, then the partition function factorizes as $Z(n) = (F)^{\binom{n}{2}}\left(\sum_{i\in[u]} s_i\right)^n$. Now, defining $p_i = \frac{s_i}{\sum_i s_i}$ and $w_{ijl} = \frac{t_{ijl}}{F}$, allows us to re-define the MLN distribution (4) equivalently as follows:

$$P_{\Phi}^{(n)}(\omega) = \prod_{i\in[u]} p_i^{k_i} \prod_{\substack{i,j\in[u]\\ i\leq j}} \prod_{l\in[b]} w_{ijl}^{h_l^{ij}} \tag{8}$$

Here, $\sum_i p_i = 1$ and $\sum_l w_{ijl} = 1$. Hence, $P_{\Phi}^{(n)}(\omega)$ is essentially a (labeled) stochastic block model, which are known to be projective [15].

We will now prove that the aforementioned sufficient conditions are also necessary.

Lemma 2 (Necessary). *If a Markov Logic network in the two variable fragment is projective then, all the* f_{ij} *have the same value i.e.* $\forall i, j \in [u] : f_{ij} = F$, *for some positive real number* F.

Proof (Sketch). We begin by writing the projectivity condition in the multi-relational representation, i.e. $P_{\Phi}^{(n+1)} \downarrow [n](\boldsymbol{X}' = \boldsymbol{x}', \boldsymbol{Y}' = \boldsymbol{y}')$ is equal to:

$$\sum_{\substack{\boldsymbol{x}_{[n]}=\boldsymbol{x}' \\ \boldsymbol{y}_{[n]}=\boldsymbol{y}'}} P_{\Phi}^{(n+1)}(\boldsymbol{X} = \boldsymbol{x}, \boldsymbol{Y} = \boldsymbol{y}) \tag{9}$$

Multiplying and dividing Eq. (9) by $Z(n)$ and using simple algebraic manipulations we get that for all $\boldsymbol{x}'$:

$$\frac{Z(n+1)}{Z(n)} = \sum_{i \in [u]} s_i \prod_{j \in [u]} f_{ij}^{k_j(\boldsymbol{x}')} \tag{10}$$

Now, the LHS of Eq. (10) is completely independent of $\boldsymbol{x}'$, whereas RHS is dependent on $\boldsymbol{x}'$. It can be shown that this is possible iff f_{ij} does not depend on $\boldsymbol{x}'$, which in turn is possible iff f_{ij} does not depend on i and j i.e. $\forall i, j \in [u] :$ $f_{ij} = F$, for some positive real F.

We are finally able to provide the following theorem.

Theorem 2. *A Markov Logic Network (MLN)* $\Phi = \{(\phi_i, a_i)\}_i$, *such that* ϕ_i *contains at-most two variables is projective if and only if all the* f_{ij} *(as given in Definition 2) have the same value i.e.* $\forall i, j \in [u] : f_{ij} = F$, *for some positive real number* F.

In the next section, we will show that the conditions in Theorem 2 correspond to a special type of probability distributions. We will characterize such distributions and then investigate their properties.

6 Relational Block Model

In this section we introduce the Relational Block Model (RBM). We show that any projective MLN in the two variable fragment can be expressed as an RBM. And any RBM can be expressed as a projective MLN. Furthermore, we show that an RBM is a unique characterization of a projective MLN in the two variable fragment.

Definition 3. Let n be a positive integer (the number of domain constants), u be a positive integer (the number of 1-types), b be a positive integer (the number of 2-tables), $p = (p_1, ..., p_u)$ be a probability vector on $[u] = \{1, ..., u\}$ and

$W = (w_{ijl}) \in [0,1]^{u \times u \times b}$, where $w_{ijl} = w_{\underline{ijl}}$ (w_{ijl} is the conditional probability of domain elements (c,d) realizing the lth 2-table, given $i(c)$ and $j(d)$). The multi-relational graph $(\boldsymbol{x}, \boldsymbol{y})$ is drawn under RBM(n, p, W) if $\boldsymbol{x}$ is an n-dimensional vector with $i.i.d$ components distributed under p and $\boldsymbol{y}$ is a random vector with its component $y_{qr} = l$, where $l \in [b]$, with a probability $w_{x_q x_r l}$ independently of all other pair of domain constants.

Thus, the probability distribution of $(\boldsymbol{x}, \boldsymbol{y})$ is defined as follows, where $\boldsymbol{x} \in [u]^n$ and $\boldsymbol{y} \in [b]^{\binom{n}{2}}$

$$
\begin{aligned}
P(\boldsymbol{X} = \boldsymbol{x}) &:= \prod_{q=1}^{n} p_{x_q} = \prod_{i=1}^{u} p_{x_i}^{k_i} \\
P(\boldsymbol{Y} = \boldsymbol{y} | \boldsymbol{X} = \boldsymbol{x}) &:= \prod_{1 \leq q < r \leq n} w_{x_q x_r y_{qr}} \\
&= \prod_{1 \leq i \leq j \leq u} \prod_{1 \leq l \leq b} (w_{ijl})^{h_l^{ij}}
\end{aligned}
$$

In the following example, we show how RBMs can model homophily.

Example 2 (Homophily). Let us have an FO^2 language with a unary predicate C (representing a two colors) and a binary predicate R. We wish to model a distribution on simple undirected graphs i.e. models of the formula $\phi = \forall xy. \neg R(x,x) \wedge (R(x,y) \rightarrow R(y,x))$ such that same color nodes are more likely to have an edge. Due to ϕ the 1-types with $\neg R(x,x)$ as a conjunct have a probability zero. Hence, we can assume we have only two 1-types: $1(x) = C(x) \wedge \neg R(x,x)$ and $2(x) = \neg C(x) \wedge \neg R(x,x)$ (representing two possible colors for a given node). Similarly due to ϕ, we have only two 2-tables $1(x,y) : R(x,y) \wedge R(y,x)$ and $2(x,y) : \neg R(x,y) \wedge \neg R(y,x)$ (representing existence and non existence of edges). We can now easily define homophily by following parameterization of an RBM. $p_1 = p_2 = 0.5$ i.e. any node can have two colors with equal probability. Then we can define $w_{111} = 0.9$, $w_{112} = 0.1$, $w_{221} = 0.9$, $w_{222} = 0.1$, $w_{121} = 0.1$ and $w_{122} = 0.9$.

Theorem 3. *Every projective Markov Logic Network in the two variable fragment can be expressed as an RBM.*

Proof. The proof follows from the sufficiency proof in Lemma 1. Notice that in the proof, we derive Eq. (8) (equivalently, equation (21) in the appendix), which is exactly the expression for RBM. Hence, any projective MLN can be converted to an RBM by defining p_i and w_{ijl} as follows:

$$
p_i = \frac{s_i}{\sum_i s_i} \quad w_{ijl} = \frac{t_{ijl}}{\sum_l t_{ijl}} \tag{11}
$$

Theorem 4. *Every RBM can be expressed as a projective MLN in the two variable fragment.*

Proof. Given an RBM as defined in Definition 3 with parameters $\{p_i, w_{ijl}\}$, let us have a projective MLN Φ such that every 1-type $i(x)$ is a formula in the MLN with a weight $\log p_i$. Φ also has a weighted formula $ijl(x, y)$ for every 2-type, such that $i \leq j$. The weight for $ijl(x, y)$ is $\log(w_{ijl})$ if $ijl(x, y) \neq ijl(y, x)$, and is $0.5 \log(w_{ijl})$ if $ijl(x, y) = ijl(y, x)$. It can be seen from definition of s_i (5) and t_{ijl} (6), that for Φ, $s_i = p_i$ and $t_{ijl} = w_{ijl}$. Hence, due to (4), we have that:

$$P_{\Phi}^{(n)}(\omega) = \frac{1}{Z(n)} \prod_{i \in [u]} p_i^{k_i} \prod_{\substack{i,j \in [u] \\ i \leq j}} \prod_{l \in [b]} (w_{ijl})^{h_l^{ij}} \tag{12}$$

In the MLN Φ, $\sum_i s_i = \sum_i p_i = 1$ and $\sum_l t_{ijl} = \sum_l w_{ijl} = f_{ij} = 1$. Hence, using Proposition 1, we have that $Z(n) = 1$. Hence, completing the proof.

Proposition 2. *Given two RBMs with probability distribution P' and P'' and parameters $\{p'_i, w'_{ijl}\}$ and $\{p''_i, w''_{ijl}\}$. If $P' = P''$, then, $p'_i = p''_i$ and $w'_{ijl} = w''_{ijl}$.*

Proof. The proposition is a consequence of the fact that the parameter p_i is marginal probability of an arbitrary constant c realizing the ith 1-type and w_{ijl} is the conditional probability of an arbitrary pair of constants (c, d) realizing the lth 2-table given $i(c)$ and $j(d)$. Hence, two RBMs that disagree on the p_i and w_{ijl} cannot assign the same probability mass to marginal probability of $i(c)$ and $ijl(c, d)$ and hence, cannot be the same distribution.

Corollary 1 (of Proposition 2). *Given two projective MLNs Φ' and Φ'' such that they have the same probability distributions $P_{\Phi'}$ and $P_{\Phi''}$, with there respective RBMs parameterized by $\{p'_i, w'_{ijl}\}$ and $\{p''_i, w''_{ijl}\}$. Then we must have that $p'_i = p''_i$ and $w'_{ijl} = w''_{ijl}$.*

Hence, RBMs are a unique representation for projective MLNs in the two variable fragment.

7 Previous Characterizations of Projectivity

Jaeger and Schulte [4] show that an MLN is projective if it's formulae ϕ_i satisfy the property that any two atoms appearing in ϕ_i contain exactly the same variables. Such MLNs are also known as *σ-determinate* [16]. We now show that in the two variable fragment, Theorem 2 leads to a strictly more expressive class of MLNs.

Proposition 3. *Given an MLN $\Phi = \{\phi_i, a_i\}_i$ such that any two atoms appearing in ϕ_i contain exactly the same variables or equivalently that the MLN is $\sigma-$determinate. Then:*

$$\forall i, j, i', j' \in [u], \forall l \in [b] : t_{i'j'l} = t_{ijl} \tag{13}$$

Proof. We first write an equivalent MLN $\Phi' = \{\alpha_q(x), a_q\} \cup \{\beta_p(x,y), b_p\}$ as presented in proof of Theorem 1. Due to the conditions provided in the proposition, all the atoms in $\beta_p(x,y)$ contain both the variables x and y. Using the definition of t_{ijl} from (6), and the fact that none of the $\beta_p(x,y)$ have an atom with only one variable, we have that the value of t_{ijl} depends only on the lth 2-table, irrespective of the 1-types i and j. This is because, none of the first order atoms in the ith and the jth 1-type appear in $\beta_p(x,y)$. Hence, t_{ijl} only depends on l, giving us Eq. (13).

Proposition 3 is a stricter condition than Theorem 2. In the following, we prove that σ-determinate MLNs cannot express all the projective MLNs in the two variable fragment.

Theorem 5. *There exists a projective MLN in the two variable fragment which cannot be expressed as a $\sigma-$determinate MLN.*

Proof. Let us have a σ-determinate MLN Φ, since Φ is projective, we can create it's equivalent RBM (due to Theorem 4), say P. Let $\{p_i, w_{ijl}\}$ be the parameters of P. Due to Eq. (11) and Proposition 3, we have that $w_{ijl} = w_{i'j'l}$ for all i, j, i', j'. Due to existence of a projective MLN for every RBM (from Theorem 4), we can always create an MLN Φ' for which the RBM parameters $w_{ijl} \neq w_{i'j'l}$ for some i, j, i', j'. Since, RBMs uniquely characterize the probability distributions due to MLNs (from Corollary 1), Φ' can not be expressed as an MLN such that $w_{ijl} = w_{i'j'l}$. Hence, Φ' can not be expressed as a σ-determinate MLN.

In the following example, we provide an MLN which cannot be written as a $\sigma-$ determinate MLN.

Example 3. Let us have a binary predicate R. We have only two 1-types $R(x,x)$ (say $1(x)$) and $\neg R(x,x)$ (say $2(x)$) and four 2-tables, $R(x,y) \wedge R(y,x)$ (say $1(x,y)$), $R(x,y) \wedge \neg R(y,x)$ (say $2(x,y)$), $\neg R(x,y) \wedge R(y,x)$ (say $3(x,y)$) and $\neg R(x,y) \wedge \neg R(y,x)$ (say $4(x,y)$). An MLN Φ, with the following 2-types as weighted formulas, cannot be expressed as a $\sigma-$determinate MLN:

$$111(x,y) : \log 7 \quad 114(x,y) : \log 4$$
$$124(x,y) : \log 64 \quad 221(x,y) : \log 8$$

In parametric normal form, $t_{111} = \exp(2\log 7)$, $t_{114} = \exp(2\log 4)$, $t_{124} = \exp(\log 64)$ and $t_{221} = \exp(2\log 8)$. All the other t_{ijl}, such that $ijl(x,y)$ is not a dual of $111(x,y)$, $114(x,y)$, $124(x,y)$ or $221(x,y)$, are equal to $\exp(0)$ i.e. 1. It can be verified that $f_{ij} = 67$ for all $i,j \in [2]$, hence, this MLN is projective due to Theorem 2. Using Theorem 3, we can express this distribution as an RBM, such that $w_{111} = \frac{7^2}{67}$ and $w_{114} = \frac{4^2}{67}$. If $w_{111} \neq w_{114}$ then necessarily $t_{111} \neq t_{114}$ (as w_{ijl} is defined as $\frac{t_{ijl}}{f_{ij}}$ and f_{ij} is the same for all i,j in Φ and in any equivalent MLN, due to Theorem 2). Due to uniqueness of RBM parameters for any set of projective MLNs expressing the same distribution (Corollary 1), we have that in all MLNs equivalent to Φ, $t_{111} \neq t_{114}$. Hence, using Proposition 3, we have that any MLN expressing the same distribution as Φ cannot be expressed as a $\sigma-$determinate MLN.

8 Maximum Likelihood Learning

In a learning setting, for an MLN $\{\phi_i, a_i\}$ in the two variable fragment, we are interested in estimating the set of parameters $\boldsymbol{\theta} = \{a_i\}$ that maximize the likelihood of a training example such that the learnt MLN is projective. As analyzed in [9,20], we will focus on the scenario where only a single possible world $\omega \in \Omega^{(n)}$ is observed. We estimate $\boldsymbol{\theta}$ by maximizing the likelihood

$$L^{(n)}(\boldsymbol{\theta}|\omega) = P_{\boldsymbol{\theta}}^{(n)}(\omega) \tag{14}$$

Notice that although every projective MLN can be equivalently defined as an RBM, the maximum likelihood parameter estimate for an RBM is not the same as the parameter estimate for an MLN such that it is projective.

We will now provide, the maximum likelihood estimator for an RBM.

Proposition 4. *Given a training example $\omega \in \Omega^{(n)}$, the maximum likelihood parameter estimate for an RBM is given as, $p_i = \frac{k_i}{n}$ and $w_{ijl} = \frac{h_l^{ij}}{k(i,j)}$.*

Proposition 4 can be derived by maximizing the log likelihood due to the distribution given in Definition 3.

We will now see how maximum likelihood parameter estimate can be obtained for an MLN such that the MLN is projective.

Given an MLN $\{\phi_i, a_i\}_i$ in the two variable fragment, where $\boldsymbol{\theta} = \{a_i\}_i$ are unknown parameters to be estimated, due to Theorem 1, we can define $s_i(\boldsymbol{\theta})$ and $t_{ijl}(\boldsymbol{\theta})$, such that the likelihood is given as:

$$L(\boldsymbol{\theta}|\omega) = \frac{1}{Z(n)} \prod_{i \in [u]} s_i(\boldsymbol{\theta})^{k_i} \prod_{\substack{i,j \in [u] \\ i \leq j}} \prod_{l \in [b]} (t_{ijl}(\boldsymbol{\theta}))^{h_l^{ij}} \tag{15}$$

Defining $F(\boldsymbol{\theta})$ as $\sum_l t_{i'j'l}(\boldsymbol{\theta})$ for some fixed i' and j', the maximum likelihood parameter estimates such that the estimated MLN is projective, can be then obtained by solving the following optimization problem:

$$\begin{aligned} \underset{\boldsymbol{\theta}}{maximize}: \Big[& \sum_{i \in [u]} k_i \log s_i(\boldsymbol{\theta}) + \sum_{\substack{i,j \in [u] \\ i \leq j}} \sum_{l \in [b]} h_l^{ij} \log t_{ijl}(\boldsymbol{\theta}) \\ & - n \log \Big(\sum_{i \in [u]} s_i(\boldsymbol{\theta}) \Big) - \binom{n}{2} \log F(\boldsymbol{\theta}) \Big] \end{aligned} \tag{16}$$

$$subject \quad to: \quad \forall i, j \in [u] : f_{ij}(\boldsymbol{\theta}) = F(\boldsymbol{\theta})$$

Notice that due to factorization of $Z(n)$ under projectivity (see Lemma 1), $-n \log \left(\sum_{i \in [u]} s_i(\boldsymbol{\theta}) \right) - \binom{n}{2} \log F(\boldsymbol{\theta})$ represents $-\log(Z(n))$. The above optimization can be solved through any conventional optimization algorithm. It can be seen that this problem has a much lesser overhead as far as computing $\log(Z(n))$ is concerned. But the additional constraints may counter act this gain. Furthermore, in many cases it may happen that no non-zero weights exist that satisfy the constraints and in that case the problem will return zero weights for the MLN formulas.

Theorem 6. *Given a training example $\omega \in \Omega^{(n)}$, then there is no parameterization for any projective MLN in the two variable fragment that has a higher likelihood for ω than the maximum likelihood RBM for ω.*

Proof. Let L be the likelihood of ω due to the maximum likelihood RBM. Let L' be the likelihood of ω due to a projective MLN Φ, such that $L' > L$. Now, due to Theorem 3, Φ can be expressed as an RBM. Hence, we can have an RBM such that the likelihood of ω is L', but $L' > L$ which is a contradiction. Hence, we cannot have a projective MLN that gives a higher likelihood to ω than the maximum likelihood RBM.

Theorem 6 shows us that if a data source is known to be projective (i.e. we know that marginals in the data will be independent of the domain at large) then in terms of likelihood, specially in the case of large relational datasets, we are better off in using an RBM than an expert defined MLN. This can also be argued from efficiency point of view as RBMs admit much more efficient parameter estimates.

We will now move on to the question: *are parameters learned on a domain of size n, also good for modelling domain of a different size m ?* This question is an abstraction of many real world problems, for example, learning over relational data in presence of incomplete information [7], modelling a social network from only sub-sampled populations [3], modelling progression of a disease in a population by only testing a small set of individuals [18] etc.

Jaeger and Schulte [4] formalized the afore mentioned notions in the following two criterions:

$$E_{\omega}[\operatorname*{argmax}_{\boldsymbol{\theta}} \log L^{(m)}(\boldsymbol{\theta}|\omega')] = \operatorname*{argmax}_{\boldsymbol{\theta}} \log L^{(n)}(\boldsymbol{\theta}|\omega) \tag{17}$$

$$\operatorname*{argmax}_{\boldsymbol{\theta}} E_{\omega}[\log L^{(m)}(\boldsymbol{\theta}|\omega')] = \operatorname*{argmax}_{\boldsymbol{\theta}} \log L^{(n)}(\boldsymbol{\theta}|\omega) \tag{18}$$

It is easy to see, by law of large numbers, that RBMs satisfy both these criterions. On the other hand the same can not be said about the maximum likelihood estimates for projective MLNs as described in (16).

9 Conclusion

In this work, we have characterized the class of projective MLNs in the two-variable fragment. We have also identified a special class of models, namely Relational Block Model. We show that the maximum likelihood RBM maximizes the training data likelihood w.r.t to any projective MLN in the two-variable fragment. Furthermore, RBMs admit consistent parameter learning from sub-sampled domains, potentially allowing them to scale to very large datasets, especially in situations where the test data size is not known or changes over time.

From an applications point of view, the superiority of RBMs in terms of training likelihood maximization and consistent parameter learning can potentially make them a better choice over an expert defined MLN, especially when training set is large and the test domain size is unknown or varies over time. We plan to investigate such capabilities of RBMs and projective MLNs in future work, especially in comparison to models like Adaptive MLNs [6] and Domain Size Aware MLNs [11].

On the theoretical front, the imposed independence structure due to projectivity clearly resembles the AHK models proposed in [5]. In future works, we aim at investigating this resemblance and generalizing our work to capture complete projectivity criterion for all the MLNs.

Acknowledgement. We would like to thank Manfred Jaeger and Felix Weitkämper for their valuable critique and discussion time on the topic.

References

1. Beame, P., den Broeck, G.V., Gribkoff, E., Suciu, D.: Symmetric weighted first-order model counting. In: Milo, T., Calvanese, D. (eds.) Proceedings of the 34th ACM Symposium on Principles of Database Systems, PODS 2015, Melbourne, Victoria, Australia, 31 May–4 June 2015, pp. 313–328. ACM (2015). https://doi.org/10.1145/2745754.2745760
2. Getoor, L., Taskar, B.: Introduction to Statistical Relational Learning (Adaptive Computation and Machine Learning). MIT Press, Cambridge (2007)
3. Handcock, M.S., Gile, K.J.: Modeling social networks from sampled data. Ann. Appl. Stat. **4**(1), 5–25 (2010)
4. Jaeger, M., Schulte, O.: Inference, learning, and population size: projectivity for SRL models. CoRR abs/1807.00564 (2018). https://arxiv.org/abs/1807.00564
5. Jaeger, M., Schulte, O.: A complete characterization of projectivity for statistical relational models. In: Bessiere, C. (ed.) Proceedings of the Twenty-Ninth International Joint Conference on Artificial Intelligence, IJCAI 2020, pp. 4283–4290. ijcai.org (2020). https://doi.org/10.24963/ijcai.2020/591
6. Jain, D., Barthels, A., Beetz, M.: Adaptive Markov logic networks: learning statistical relational models with dynamic parameters. In: Coelho, H., Studer, R., Wooldridge, M.J. (eds.) ECAI 2010–19th European Conference on Artificial Intelligence, Lisbon, Portugal, 16–20 August 2010, Proceedings. Frontiers in Artificial Intelligence and Applications, vol. 215, pp. 937–942. IOS Press (2010). https://doi.org/10.3233/978-1-60750-606-5-937
7. Kossinets, G.: Effects of missing data in social networks. Soc. Netw. **28**(3), 247–268 (2006)
8. Kuusisto, A., Lutz, C.: Weighted model counting beyond two-variable logic. In: Dawar, A., Grädel, E. (eds.) Proceedings of the 33rd Annual ACM/IEEE Symposium on Logic in Computer Science, LICS 2018, Oxford, UK, 09–12 July 2018, pp. 619–628. ACM (2018). https://doi.org/10.1145/3209108.3209168

9. Kuzelka, O., Kungurtsev, V., Wang, Y.: Lifted weight learning of Markov logic networks (revisited one more time). In: Jaeger, M., Nielsen, T.D. (eds.) International Conference on Probabilistic Graphical Models, PGM 2020, 23–25 September 2020, Aalborg, Hotel Comwell Rebild Bakker, Skørping, Denmark. Proceedings of Machine Learning Research, vol. 138, pp. 269–280. PMLR (2020). https://proceedings.mlr.press/v138/kuzelka20a.html
10. Malhotra, S., Serafini, L.: Weighted model counting in FO2 with cardinality constraints and counting quantifiers: a closed form formula. In: Proceedings of AAAI 2022 (2022). https://arxiv.org/abs/2110.05992
11. Mittal, H., Bhardwaj, A., Gogate, V., Singla, P.: Domain-size aware Markov logic networks. In: Chaudhuri, K., Sugiyama, M. (eds.) The 22nd International Conference on Artificial Intelligence and Statistics, AISTATS 2019, 16–18 April 2019, Naha, Okinawa, Japan. Proceedings of Machine Learning Research, vol. 89, pp. 3216–3224. PMLR (2019). https://proceedings.mlr.press/v89/mittal19a.html
12. Poole, D., Buchman, D., Kazemi, S.M., Kersting, K., Natarajan, S.: Population size extrapolation in relational probabilistic modelling. In: Straccia, U., Calì, A. (eds.) SUM 2014. LNCS (LNAI), vol. 8720, pp. 292–305. Springer, Cham (2014). https://doi.org/10.1007/978-3-319-11508-5_25
13. Raedt, L.D., Kersting, K., Natarajan, S., Poole, D.: Statistical Relational Artificial Intelligence: Logic, Probability, and Computation. Synthesis Lectures on Artificial Intelligence and Machine Learning. Morgan & Claypool Publishers, San Rafael (2016). https://doi.org/10.2200/S00692ED1V01Y201601AIM032
14. Richardson, M., Domingos, P.: Markov logic networks. Mach. Learn. **62**(1–2), 107–136 (2006)
15. Shalizi, C.R., Rinaldo, A.: Consistency under sampling of exponential random graph models. Ann. Stat. **41**(2), 508–535 (2013)
16. Singla, P., Domingos, P.M.: Markov logic in infinite domains, pp. 368–375 (2007). https://dslpitt.org/uai/displayArticleDetails.jsp?mmnu=1&smnu=2&article_id=1711&proceeding_id=23
17. Snijders, T.A.B.: Conditional marginalization for exponential random graph models. J. Math. Sociol. **34**(4), 239–252 (2010). https://doi.org/10.1080/0022250X.2010.485707
18. Srinivasavaradhan, S.R., Nikolopoulos, P., Fragouli, C., Diggavi, S.: Dynamic group testing to control and monitor disease progression in a population (2021)
19. Weitkämper, F.Q.: An asymptotic analysis of probabilistic logic programming, with implications for expressing projective families of distributions. Theory Pract. Log. Program. **21**(6), 802–817 (2021)
20. Xiang, R., Neville, J.: Relational learning with one network: an asymptotic analysis. In: Gordon, G.J., Dunson, D.B., Dudík, M. (eds.) Proceedings of the Fourteenth International Conference on Artificial Intelligence and Statistics, AISTATS 2011, Fort Lauderdale, USA, 11–13 April 2011. JMLR Proceedings, vol. 15, pp. 779–788. JMLR.org (2011). https://proceedings.mlr.press/v15/xiang11a/xiang11a.pdf

A Non-parametric Bayesian Approach for Uplift Discretization and Feature Selection

Mina Rafla[1,2](✉), Nicolas Voisine[1], Bruno Crémilleux[2], and Marc Boullé[1]

[1] Orange Labs, 22300 Lannion, France
{mina.rafla,nicolas.voisine,marc.boulle}@orange.com
[2] UNICAEN, ENSICAEN, CNRS - UMR GREYC, Normandie Univ, 14000 Caen, France
bruno.cremilleux@unicaen.fr

Abstract. Uplift modeling aims to estimate the incremental impact of a treatment, such as a marketing campaign or a drug, on an individual's outcome. Bank or Telecom uplift data often have hundreds to thousands of features. In such situations, detection of irrelevant features is an essential step to reduce computational time and increase model performance. We present a parameter-free feature selection method for uplift modeling founded on a Bayesian approach. We design an automatic feature discretization method for uplift based on a space of discretization models and a prior distribution. From this model space, we define a Bayes optimal evaluation criterion of a discretization model for uplift. We then propose an optimization algorithm that finds near-optimal discretization for estimating uplift in $O(n \log n)$ time. Experiments demonstrate the high performances obtained by this new discretization method. Then we describe a parameter-free feature selection method for uplift. Experiments show that the new method both removes irrelevant features and achieves better performances than state of the art methods.

Keywords: Uplift modeling · Feature selection · Discretization · Bayesian methods · Machine learning · Treatment effect estimation

1 Introduction

Uplift modeling aims to estimate the incremental impact of a treatment, such as a marketing campaign or a drug, on an individual's behavior. Uplift models help identify groups of people likely to respond positively to treatment *only because* they received one. This research domain has multiple applications like customer relationship management, personalized medicine, advertising. Uplift estimation is based on groups of people who have received different treatments. A major difficulty is that data are only partially known: it is impossible to know for an individual whether the chosen treatment is optimal because their responses to alternative treatments cannot be observed. Several works address challenges related to the uplift modeling [14,26] or the evaluation of uplift models [21].

© The Author(s), under exclusive license to Springer Nature Switzerland AG 2023
M.-R. Amini et al. (Eds.): ECML PKDD 2022, LNAI 13717, pp. 239–254, 2023.
https://doi.org/10.1007/978-3-031-26419-1_15

Many databases are large and contain hundreds of features [12]. Keeping all the features is costly and inefficient to build uplift models. A feature selection process is then an essential step to remove irrelevant features, improves the estimation accuracy and accelerates the model building. While there are a lot of feature selection methods in machine learning, there are very few propositions for uplift modeling [27]. This observation might be explained since uplift creates new challenges such as the impossibility to observe two treatment outcomes for a same individual. Designing methods for uplift requires overcoming this difficulty. This paper aims to answer the need for feature selection methods for uplift.

We present a parameter-free feature selection method for uplift modeling founded on a Bayesian approach. Following a part of literature on feature selection that performs a discretization of numerical features [16,25], we first describe an automatic feature discretization method for uplift modeling that we call UMODL (for Uplift MODL). UMODL is based on the Bayesian MODL (Minimum Optimized Description Length) criterion [1] that we have extended to the uplift problem. UMODL defines a space of discretization models and a prior distribution on this model space. We construct a Bayes optimal evaluation criterion of a discretization model for uplift modeling. In practice, the best model according to the criterion cannot be computed due to the complexity of the problem and we present a greedy search algorithm in $\mathcal{O}(n \log n)$ to find near-optimal discretizations. Experiments show that the discretization model found by UMODL gives a good estimator of uplift. Then, based on UMODL, we present UMODL feature selection (UMODL-FS in short) a feature selection method for uplift. UMODL-FS computes a score of the features and automatically selects appropriate features for uplift. Experiments demonstrate that UMODL-FS properly removes irrelevant features and clearly outperforms state of the art methods by providing uplift models with the highest and most stable performance. Being a parameter-free method (neither the number of bins in the discretization nor the number of features to keep or remove are given), UMODL-FS can be used without effort.

The remainder of the paper is organized as follows. In the next section, we introduce uplift modeling, feature selection for uplift, MODL and the literature related to our problem setting. Section 3 presents UMODL which is experimentally evaluated in Sect. 4. UMODL-FS and the associated experiments are described in Sect. 5. We conclude in Sect. 6.

2 Background and Literature Review

2.1 Uplift Modeling

Uplift Definition. Uplift is a notion introduced by Radcliffe and Surry [20] and defined in Rubin's causal inference models [23] as the *Individual Treatment Effect.* The uplift modeling literature and a branch of the causal inference literature have recently approached each other [8]. We present the notion of uplift.

Let D be a group of N individuals indexed by $n : 1 \dots N$ where each individual is described by a set of variables $\mathbb{X}$. X_n denotes the set of values of $\mathbb{X}$ for the individual n. Let T be a variable indicating whether or not an individual has received a treatment. Uplift modeling is based on two groups: the individuals having received a treatment (denoted $T = 1$) and those without treatment (denoted $T = 0$). Let Y be the outcome variable (for instance, the purchase or not of a product). We note $Y_n(T = 1)$ the outcome of an individual n when he received a treatment and $Y_n(T = 0)$ his outcome without treatment. The uplift of an individual n, denoted by τ_n, is defined as: $\tau_n = Y_n(T = 1) - Y_n(T = 0)$. The main difficulty is that uplift value is not directly measurable, i.e. for each individual we can either observe $Y_n(T = 1)$ or $Y_n(T = 0)$ but cannot observe simultaneously both outcomes. However, uplift τ_n can be empirically estimated by considering two groups: a treatment group (individuals who received a treatment) and a control group (individuals who did not). The estimated uplift of an individual n denoted by $\hat{\tau}_n$ is then the difference between response rates in both groups and computed by using the CATE[1] (Conditional Average Treatment Effect) [23]: CATE: $\hat{\tau}_n = \mathbb{E}[Y_n(T = 1)|X_n] - \mathbb{E}[Y_n(T = 0)|X_n]$

As the real value of τ_n cannot be observed, it is impossible to directly use machine learning algorithms such as regression to infer a model to predict τ_n. We sketch below how uplift is modeled in the literature. Simple methods such as considering only individuals having received the treatment fail because they do not detect individuals whose response is always positive even without treatment.

Uplift Modeling Approaches. In recent years, several studies on uplift models design have been conducted. One of the most classical and intuitive approach is the *two model approach* [11] (also called *T-learner* in the causal community), which consists of two independent predictive models, one on the treatment group to estimate $P(Y|X, T = 1)$ and another on the control group to estimate $P(Y|X, T = 0)$. The estimated uplift of an individual n is the difference between those values for the given individual, i.e. $\hat{\tau}_n = P(Y = 1|X_n, T = 1) - P(Y = 1|X_n, T = 0)$. *Class transformation approach* [14] is another family of methods that maps the uplift modeling problem to a usual supervised learning problem. With the *Direct-approach*, different machine learning algorithms are modified to suit uplift modeling such as methods based on *decision trees* [24,26], *k nearest neighbors* [7], *logistic regression* [17], etc. The causal inference community defines other methods such as *S-Learner* which includes the outcome variable T in the features with a standard regression, *X-Learner* [13] which performs a two-step regression before the estimation of the CATE, *DR-Learner* [15] which combines a two-model approach and the use of the Inverse Propensity Weighting [18].

Evaluation of Uplift Models. Since real values of uplift cannot be observed, standard performance measures of supervised learning algorithms cannot be used. That is why uplift is evaluated through the ranking of the individuals according to their estimated uplift value. The intuition is that a good uplift model estimates higher uplift values to individuals in the treatment group with positive outcomes than those with negative outcomes and vice versa for the control group. A common approach to evaluate uplift models is the qini measure,

[1] The terms *treatment effect* and *uplift* address the same notion. CATE is an estimation of uplift and we use "CATE" for speaking of the estimated uplift values.

also known as the Area Under Uplift Curve (*AUUC*) [3,19]. It is a variant of the Gini coefficient. Qini values are in $[-1, 1]$, the higher the value, the larger the impact of the predicted optimal treatment.

2.2 Feature Selection for Uplift Models

The accessibility of high dimensional datasets with hundreds of features makes the use of feature selection techniques crucial for machine learning tasks and uplift. The goal of feature selection techniques is to select subset of features that could efficiently describe data while expelling irrelevant features [9]. This can significantly improve models performances and computation time [2]. Regarding uplift modeling, studies addressing feature selection are very limited. To the best of our knowledge, only two research papers deal with this challenge.

Zhao et al. [27] propose filter and embedded feature selection methods for uplift. The principle is to remove features that are not correlated either with the outcome variable or uplift. Filter methods are used in a pre-processing step independently of an uplift model while embedded methods perform feature selection during the training of a model and are specific to an uplift algorithm. In [27], the presented filter methods are *bins methods* (inspired from [24]), *F-filter* and *LR-filter*. Experiments in [27] show that bin-based filter methods have the best performances while *F-filter*, *LR-filter* and embedded methods have poor performances.

We give a few words on the above methods providing the best results as well as *F-filter* and *LR-filter*. The principle of a bins method is to discretize a feature into L bins based on the percentiles of the feature (L is given by the user). The importance of a feature regarding uplift is evaluated by a divergence measure of the treatment effect over the bins. Three divergence measures are used: Kullback-Leibler (KL), squared Euclidean distance (ED), and chi-squared (Chi). The *F-filter* and the *LR-filter* are based respectively on the F-statisic [10] and the likelihood ratio statistic [5] for the coefficients of regression models.

On the other hand, a very recent paper [12] uses some of the filtering methods given in [27] as well as a correlation coefficient to remove redundant features. The paper describes an uplift application on a private database in the bank domain.

2.3 MODL Approach

The MODL (Minimum Optimized Description Length) approach is a non-parametric Bayesian approach for discretization and conditional probability estimation [1]. It is based on the Minimum Description Length (MDL) principle [6,22]. In the MODL approach, a space of discretization models is defined. A discretization model is described by a set of parameters: the number of intervals, the boundaries of the intervals and the frequencies of the classes in each interval. Using these parameters, the MODL approach consists of defining a criterion for a discretization model and with the help of a search algorithm, the MODL approach can score all possible discretization models and selects the one with the best score.

3 UMODL

This section presents UMODL a new criterion for uplift discretization modeling and the search algorithm to find the optimal uplift discretization model.

3.1 UMODL Criterion

While MODL properly exploits discretization for density estimation, it is not suitable for uplift modeling since uplift deals with two treatment groups and the estimation of the conditional probabilities of the outcome variable Y given an attribute X also depends on the treatment variable T.

We now introduce the new criterion that we propose to define the best discretization model for uplift. Let M be an uplift discretization model and D denotes data. From a Bayesian point of view, the best uplift discretization model is found by maximizing the posterior probability of the model given the data $P(M|D)$. Let us consider the Bayes rule:

$$P(M \mid D) = \frac{P(M)P(D \mid M)}{P(D)} \tag{1}$$

Given that $P(D)$ is constant, maximizing $P(M|D)$ is equivalent to maximizing $P(M)P(D|M)$, i.e. the prior probability and the likelihood of the data given the chosen model. Let us first introduce some notations:

- X: explanatory variable to discretize
- Y: binary outcome variable
- N: number of instances in the dataset
- J: number of classes of Y
- I: number of intervals
- N_i: number of instances in the interval i
- $N_{it.}$: number of instances in the interval i of treatment t
- $N_{i.j}$: number of instances in the interval i of class j
- N_{itj}: number of instances in the interval i of class j and the treatment t
- W_i: boolean term indicating if the treatment has an effect in interval i ($W_i = 1$) or not ($W_i = 0$)

We define an uplift discretization model M by the number of intervals, the bounds of the intervals, the presence or absence of a treatment effect, class frequencies per interval or for each treatment per interval. In other words, a model M is defined by the hierarchy of parameters (cf. Fig. 1):

$$\{I, \{N_i\}, \{W_i\}, \{N_{i.j}\}_{W_i=0}, \{N_{itj}\}_{W_i=1}\}$$

The evaluation criterion $C(M)$ which is the cost of an uplift discretization model M is defined then by: $C(M) = -\log\big(P(M) \times P(D|M)\big)$. Taking the negative log turns the maximization problem to a minimization one. M is optimal if $C(M)$ is minimal.

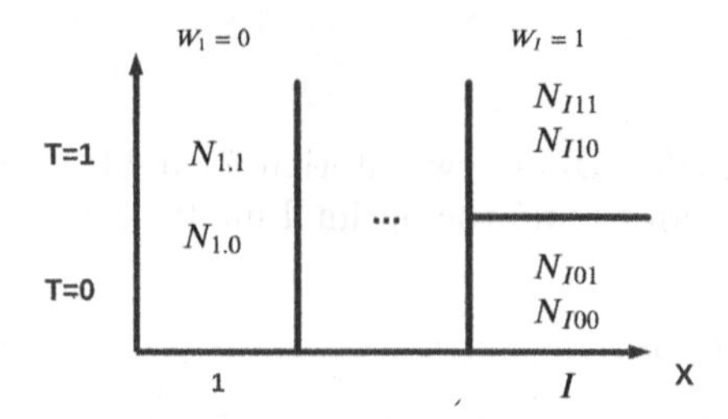

Fig. 1. Parameters of an uplift discretization model. The presence of a treatment effect ($W_i = 1$) in interval i requires describing the distribution of the outcome variable Y separately for each treatment (part right). In contrast, the absence of a treatment effect ($W_i = 0$) indicates to consider the distribution of the outcome variable Y for the interval i independently of the treatment variable (part left).

For the prior distribution of the model parameters, we exploit the hierarchy of the parameters and assume a uniform prior at each stage of the hierarchy with independence across the intervals. Using these assumptions, we express $C(M)$ according to the parameters of an uplift discretization model and we obtain Eq. 2 that we demonstrate below.

$$\begin{aligned} C(M) = {} & \log N + \log \binom{N+I-1}{I-1} + I \times \log 2 \\ & + \sum_{i=1}^{I} (1 - W_i) \log \binom{N_i + J - 1}{J-1} + \underbrace{\sum_{i=1}^{I} (1 - W_i) \log \frac{N_i!}{N_{i.1}!..N_{i.J}!}}_{Likelihood} \\ & + \sum_{i=1}^{I} W_i \sum_t \log \binom{N_{it.} + J - 1}{J-1} + \underbrace{\sum_{i=1}^{I} W_i \sum_t \log \frac{N_{it.}!}{N_{it1}!..N_{itJ}!}}_{Likelihood} \end{aligned} \quad (2)$$

Proof of Eq. 2. We express $P(M)$ and $P(D|M)$ according to the parameters of an uplift discretization model. We introduce a prior distribution by exploiting the hierarchy of the models' parameters. Assuming the independence of the local distributions across the intervals, we obtain:

$$\begin{aligned} P(M) = {} & P(I) \times P(\{N_i\}|I) \times \\ & \prod_i P(W_i|I) \left[(1 - W_i) \times P(\{N_{i.j}\}|I, \{N_i\}) + W_i \times \prod_t P(\{N_{itj}\}|I, \{N_{it.}\}) \right] \end{aligned} \quad (3)$$

We express each of the terms of Eq. 3 according to the parameters of M assuming a uniform distribution for each parameter. Assuming that the number of intervals I is uniformly distributed between 1 and N, the first term in Eq. 3 becomes:

$$P(I) = \frac{1}{N} \tag{4}$$

Given a number of intervals I, all the discretizations into I intervals (i.e. the choices of the bounds) are equiprobable. Computing the probability of an interval set leads to a combinatorial calculation of the number of all possible interval sets or equivalently the number of ways of distributing the N instances in the I intervals, with counts N_i per interval. The second term of Eq. 3 is then:

$$P(\{N_i\}|I) = \frac{1}{\binom{N+I-1}{I-1}} \tag{5}$$

For a given interval i, we assume that a treatment can have an effect or not, with equal probability, i.e. $P(W_i|I) = \frac{1}{2}$. We obtain:

$$\prod_i P(W_i|I) = \left(\frac{1}{2}\right)^I \tag{6}$$

In the case of an interval i where there is not effect of the treatment ($W_i = 0$), UMODL describes one unique distribution of the outcome variable. Given an interval i, its number of examples N_i is known. Assuming that each of the class distributions is equiprobable, we end up also with a combinatorial problem:

$$P(\{N_{i.j}\}|I, N_i) = \frac{1}{\binom{N_i+J-1}{J-1}} \tag{7}$$

In the case of an interval i with an effect of the treatment ($W_i = 1$), UMODL describes two distributions of the outcome variable, with and without the treatment. Given an interval i and a treatment t, we know the number of examples $N_{it.}$. Assuming that each of the distributions of class values is equiprobable, we get:

$$P(\{N_{itj}\}|I, N_{it.}) = \frac{1}{\binom{N_{it.}+J-1}{J-1}} \tag{8}$$

After defining the models' prior, we define the likelihood $P(D|M)$ of the data given the uplift discretization model. For each multinomial distribution of the outcome variable (a single or two distinct distributions per interval depending on whether the treatment has an effect or not), we assume that all possible observed data D_i consistent with the multinomial model are equiprobable. Using multinomial terms, we obtain the following likelihood term:

$$P(D|M) = \prod_i P(D_i|M) \tag{9}$$

$$= \prod_i \left[(1 - W_i) \times \frac{1}{(N_i!/N_{i.1}!..N_{i.J}!)} + W_i \times \prod_t \frac{1}{(N_{it.}!/N_{it1}!..N_{itJ}!)} \right] \tag{10}$$

Combining the prior $P(M)$ (Eq. 4 to 8) with the likelihood $P(D|M)$ (Eq. 10), we obtain $P(M)P(D|M)$. Taking the negative log yields to the UMODL criterion presented in Eq. 2. Coming back to Eq. 2, the prior terms of the first line come from Eq. 4 to 6. In the second line of Eq. 2 (modeling a situation w/o a treatment effect) and the third line (situation with a treatment effect), the first terms are prior terms (Eqs. 7–8) and the second terms are likelihood terms (Eq. 10).

Uplift Estimation. The presented discretization approach is a density estimation approach for uplift modeling. We model the probability of Y conditionally on the explanatory variable X and a binary treatment variable T. The search algorithm we present is looking for the parameters I, $\{W_i\}$, $\{N_i\}$, $\{N_{i.j}\}$, $\{N_{ijt}\}$, and $\{W_i\}$ that minimize the cost of the model. In other words, the search algorithm tries to find the optimal discretization in the Bayes sense that best estimates the real densities of the outcome variable Y conditionally on X and T. Once a discretization and its parameters are defined, the estimation of the CATE for each interval is simple. As shown in Fig. 1, assuming a binary outcome variable Y and given $W_i = 1$, we have $P_i(Y = 1|T = 1) = N_{i11}/(N_{i11} + N_{i01})$ and $P_i(Y = 1|T = 0) = N_{i10}/(N_{i10} + N_{i00})$, therefore $CATE_i = P_i(Y = 1|T = 1) - P_i(Y = 1|T = 0)$. For intervals with $W_i = 0$, $CATE_i$ is considered insignificant.

3.2 Search Algorithm and Post-optimization

We sketch below our search algorithm to find the best model w.r.t. the UMODL criterion. This algorithm finds the optimal values of the parameters that minimize $C(M)$. The principle of this algorithm is inspired by the search algorithm [1] which we adapted to our criterion. As an optimal search algorithm is not practical due to the complexity of the problem, we build a greedy algorithm[2].

Greedy Search Algorithm. The search algorithm is a greedy bottom-up algorithm with the following steps:

- The algorithm starts by making an elementary discretization such that all examples with the same value have their own interval,
- Compute the costs of all possible merges i.e. try to merge adjacent intervals,
- Merge the two adjacent intervals that decrease $C(M)$ the most,
- Recalculate the cost of all possible adjacent merges and select the merge that reduces $C(M)$ the most,
- Repeat until no merge decreases $C(M)$.

While this algorithm is complex, it can be implemented in $O(n \log n)$ time [1].

Post-optimization. This greedy search algorithm can fall into a local minimum, so post-optimization steps are needed to perturb the interval bounds. We used post-optimization steps that consist of recurrent splits, merges, merge splits, and merge splits of adjacent intervals, as described in [1] but designed in this work for uplift.

[2] Our implementation is provided at https://github.com/MinaWagdi/UMODL.

4 UMODL Quality Evaluation Experiments

This section experimentally evaluates whether UMODL is a good estimator of uplift. The principle of the experiments is to generate data with different synthetic uplift patterns in order that results of UMODL can be compared to true uplift. A *synthetic uplift pattern* is a data pattern where $P(Y=1|X,T=1)$ and $P(Y=1|X,T=0)$ are identified for each example. Therefore several indicators can be observed: (1) the number of intervals founded by UMODL w.r.t. the characteristics of the uplift pattern, (2) the RMSE (root mean squared error) between the real uplift and the estimated uplift by UMODL computed for each instance and (3) the number of instances needed by UMODL to find the uplift pattern. We generate synthetic uplift patterns of different characteristics for simulating various situations.

4.1 Description

The experimental protocol is made of the following steps:

1. Define a particular synthetic uplift pattern of one dimension.
2. Generate several train samples according to the defined pattern with 40 different number of instances (also called *data size*) ranging from 10 to 100,000 instances. For each data size, generate ten datasets. All generated data are uniformly distributed on the $[0,10]$ numerical domain for each of the treatment ($T=1$) and control groups ($T=0$).
3. Generate a test set of 10,000 instances based on the defined uplift pattern.
4. For each training sample, apply the UMODL approach to search for the best discretization model.
5. For each experiment, the obtained discretization model is then applied to the test set, and RMSE is computed by comparing for each data point: the CATE estimation in the found interval and the real CATE value.
6. By observing both the number of found intervals for each dataset and the RMSE values, we can determine whether the UMODL approach manages to find the synthetic pattern or not.
7. Repeat these steps with different synthetic uplift patterns.

4.2 Synthetic Uplift Patterns

We generate four bin-based patterns and one continuous pattern. We use patterns of different characteristics[3] to evaluate how UMODL performs both in various situations and different rates of uplift. The patterns are illustrated in Fig. 2 and depicted below.

- *Crenel pattern 1* (cf. Fig. 2a): this crenel pattern is made of 10 intervals containing a repeated sequence of a positive treatment effect followed by a negative one. We generated five versions of this pattern with different treatment effects (uplift).

[3] Other patterns can be found using the github link provided previously.

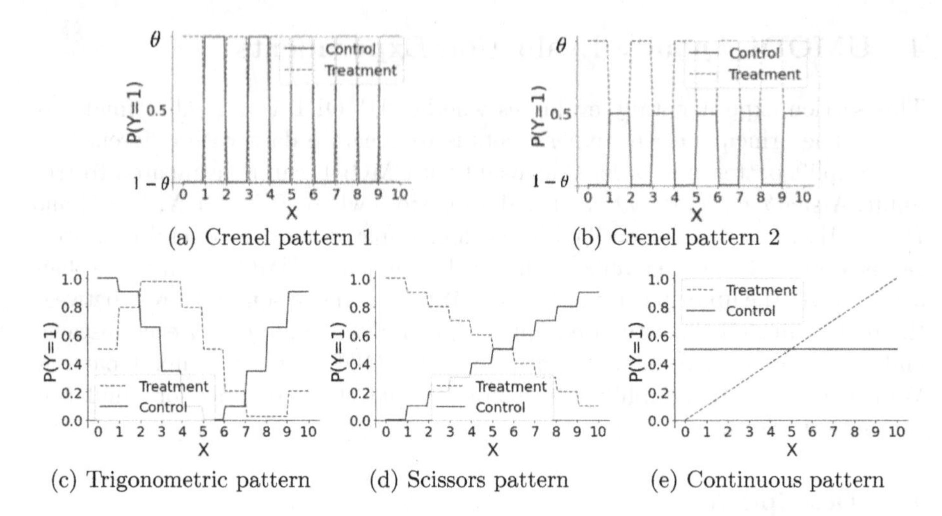

(a) Crenel pattern 1 (b) Crenel pattern 2

(c) Trigonometric pattern (d) Scissors pattern (e) Continuous pattern

Fig. 2. Synthetic uplift patterns. The X-axis represents variable X and the Y-axis represents $P(Y = 1)$. For *Crenel Pattern 1* and *Crenel Pattern 2*, five versions are generated with different values of $\theta \in \{0.6, 0.7, 0.8, 0.9, 1\}$. The difference between $P(Y = 1)$ in the treatment and control groups represents the uplift.

- *Crenel pattern 2* (cf. Fig. 2b): is a slightly different crenel pattern similarly made of 10 intervals containing a repeated sequence of a positive treatment effect followed by no treatment effect. We generated five versions of this pattern with different treatment effects (uplift).
- *Trigonometric pattern* (cf. Fig. 2c) is a particular bin-based pattern with trigonometric shape where: $P(Y = 1|T = 1) = 0.5 + (0.5 \times sin(i \times \frac{2\pi}{10}))$ and $P(Y = 1|T = 0) = 0.5 + (0.5 \times cos(i \times \frac{2\pi}{10}))$
- *Scissors pattern* (cf. Fig. 2d) is a bin-based pattern where $P(Y = 1|T = 1) = \frac{i}{10}$ and $P(Y = 1|T = 0) = 1 - \frac{i}{10}$, where i is the interval number.
- *Continuous pattern* (cf. Fig. 2e) differs from bin-based patterns. Here $P(Y = 1|T = 1) = X/10$ $P(Y = 1|T = 0) = 0.5$.

4.3 Results

Results are given in Figs. 3, 4 and 5. We start by the central question – "Is UMODL a good estimator of uplift?" – and provide complementary observations.

Is UMODL a Good Estimator of Uplift? From Figs. 3 (left) and 4 (left), we clearly see that even when the treatment effect is very small per interval (grey curves), UMODL is able to find the proper number of intervals of the uplift patterns. This is also illustrated by the RMSE curves (Figs. 3 (right) and 4 (right)) showing that RMSE always converges towards 0 for sufficiently large

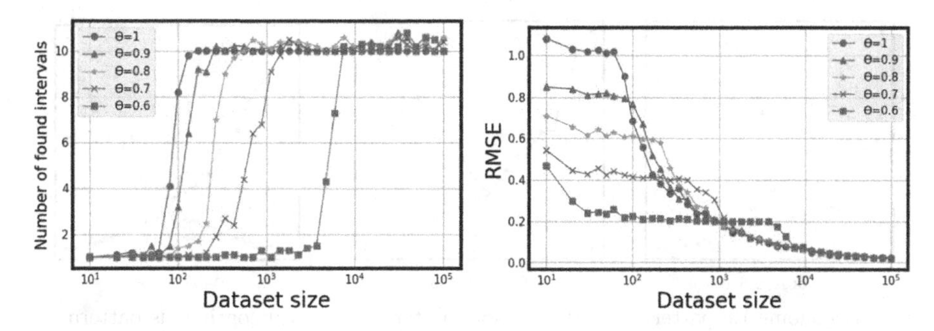

Fig. 3. Results obtained for *Crenel pattern 1.* The left (resp. right) figure shows the mean number of found intervals (resp. the mean value of RMSE) on the test set by UMODL according to the dataset size. Different curve colors correspond to different treatment effects. For example, the blue curve corresponds to the crenel pattern of repeated positive uplift (= 1) followed by negative uplift (= −1).

datasets. Similar performances are reported with the *trigonometric pattern* (cf. Fig. 5a), the *scissors pattern* (cf. Fig. 5b) and the *continuous pattern* (cf. Fig. 5c) except that the number of estimated intervals is not a relevant indicator for the *continuous pattern* because this pattern is continuous.

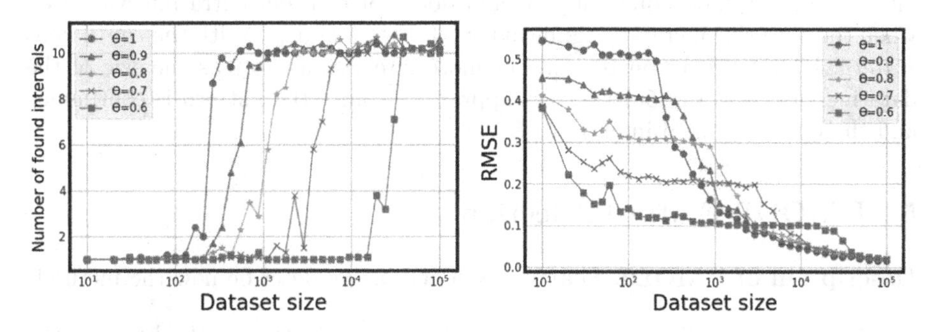

Fig. 4. Results obtained for *Crenel pattern 2.* The left (resp. right) figure shows the mean number of found intervals (resp. the mean value of RMSE) on the test set by UMODL according to the dataset size. Different curve colors correspond to separate treatment effects. For example, the blue curve corresponds to the crenel pattern of repeated positive uplift (=1) followed by zero uplift.

How Many Instances are Needed to Find the Uplift Pattern According to its Characteristics? When the differences of densities between adjacent intervals get smaller, UMODL needs more instances to give prominence to a model with more intervals. This is typically the case with the *scissors pattern* (cf. Fig. 5b). Analogous behaviors are observed in Figs. 3 and 4. For example, in Fig. 3, the

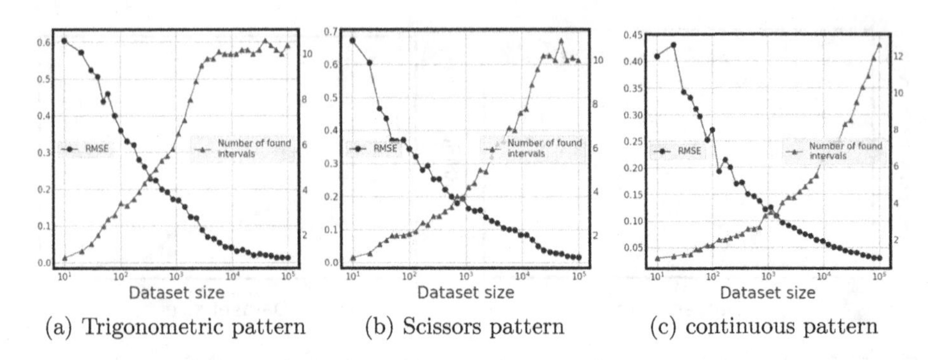

(a) Trigonometric pattern (b) Scissors pattern (c) continuous pattern

Fig. 5. Figures a, b, c present the performances obtained with the *trigonometric pattern, scissors pattern and continuous pattern.* Blue curves depict the mean value of the RMSE per dataset size while the green curves indicate the number of found intervals.

blue curve finds the uplift pattern with less instances than the red curve. Interestingly, UMODL succeeds in finding the proper intervals even when there is no treatment effect (cf. the results with the *crenel pattern 2* in Fig. 4).

Does UMODL Overfit? Another important aspect of the UMODL discretization is that the UMODL method does not overfit, i.e. UMODL always finds the ten intervals of the underlying patterns and does not consider extra intervals even when the data size increases significantly (cf. Figs. 3 and 4). With the *continuous pattern*, UMODL goes on to consider more intervals as long as the size of the data increases (cf. Fig. 5c) which is appropriate since the pattern is continuous and there is no defined intervals.

5 UMODL Feature Selection

Description of UMODL Feature Selection. We describe now the method:

1. Given a feature X, we apply the UMODL discretization method to find the optimal uplift discretization model as presented in Sect. 3.1.
2. Compute for X an importance score (described below), denoted by $imp.s(X)$, which is the divergence measure of the treatment effect over the found intervals.
3. We repeat these steps for each feature of the dataset.
4. All features with $imp.s(X) > 0$ are considered relevant for the uplift estimation, while any feature with $imp.s(X) = 0$ is eliminated.

We define $imp.s(X)$ as follows. Assuming $p_i = P_i(Y = 1|T = 1)$ and $q_i = P_i(Y = 1|T = 0)$. We define:

$$imp.s(X) = \begin{cases} \sum_{i=1}^{I} \frac{N_i}{N} D(p_i : q_i), & \text{if } I > 1 \\ 0, & \text{otherwise} . \end{cases} \tag{11}$$

where the distribution divergence measure D is the squared euclidean distance. We choose the squared euclidean distance for the divergence since it is symmetric and stable [24]. UMODL-FS considers irrelevant for the uplift estimation any feature with $imp.s(X) = 0$ and keeps for the uplift modeling any feature with $imp.s(X) > 0$. When UMODL finds a single interval for a feature, it means there is only one distribution for all instances and thus a non-informative feature (i.e. $imp.s(X) = 0$). Unlike feature selection methods of the literature, our approach does not require parameters to set, and there is no need to give the number of features to keep or delete.

Experimental Protocol. For comparing UMODL-FS to the state-of-art uplift feature selection methods (cf. Sect. 2.2), we design the following experimental protocol:

1. For each dataset, we generate eleven variants of the dataset, each with an incremental total number (from 0 to 100) of noise features. Noise features are sampled from $\mathcal{N}(0, 1)$ for each of the treatment and control groups.
2. For each variant, we apply the following feature selection methods: KL-filter Chi-filter ED-filter LR-filter F-filter UMODL-FS. For KL-filter, Chi-filter and ED-filter, we set the number of bins to 10.
3. To have the same number of features for each feature selection method and perform a fair comparison, we pick the M most important features, where M is the number of all features deemed informative by UMODL-FS.
4. With these sets of features, we build uplift models: a two-model approach with logistic regression [11] and X-Learner with linear regression [13].
5. The learning process is done with stratified ten-fold cross-validation. Test samples are used to evaluate the performance of uplift models based on the selected features.
6. The qini coefficient metric [3] is used to evaluate the performance of the uplift model.

Datasets. Experiments are conducted on two publicly available continuous datasets which are usual on the uplift community:

1. Criteo dataset [4]: a real large scale dataset constructed by assembling data resulting from several incrementality tests in advertising. In the experiments, we use a random sample of 10,000 instances with the 'visit' variable as outcome variable.
2. Zenodo synthetic dataset[4]: this dataset was created for evaluating feature selection methods for uplift modeling. It has three types of features: uplift features influencing the treatment effect on the conversion probability (outcome variable is 'conversion'); classification features influencing the conversion probability independent of the treatment effect; irrelevant features. This dataset consists of 100 trials of different patterns. Each trial has 10,000 instances and 36 features.

[4] https://doi.org/10.5281/zenodo.3653141.

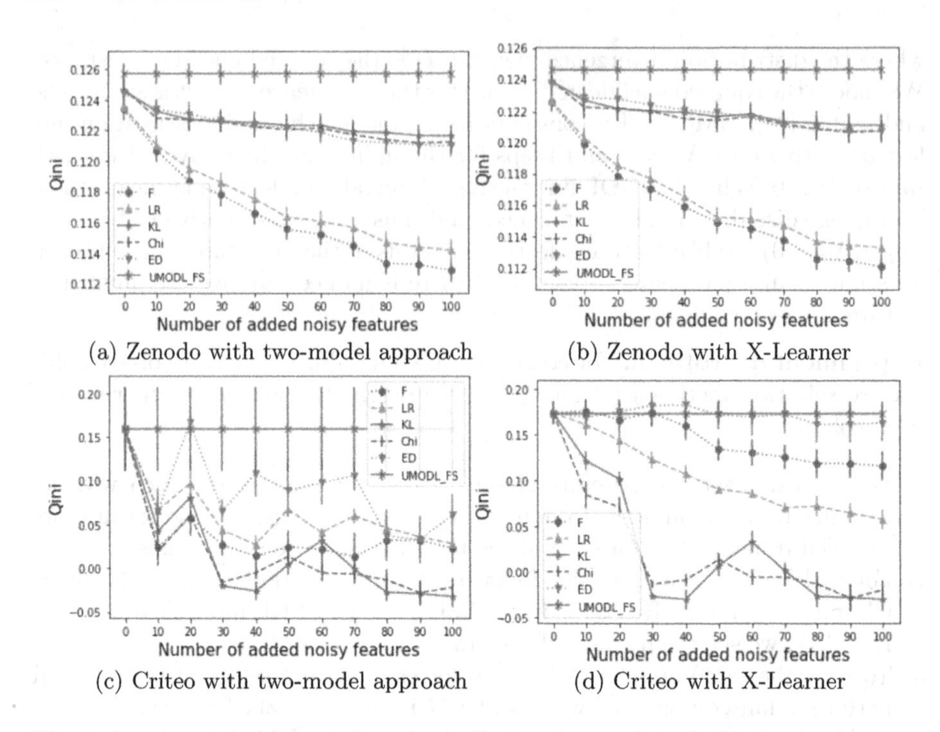

(a) Zenodo with two-model approach (b) Zenodo with X-Learner

(c) Criteo with two-model approach (d) Criteo with X-Learner

Fig. 6. Average qini and its variance according to the number of added noisy features. The X-axis indicates the total number of added noisy features. Y-axis represents the qini values achieved by uplift models.

Results. Figure 6 presents the results on the use of UMODL-FS for uplift modeling. In all experiments, UMODL-FS selects the set of features leading to the uplift model with the best qini (therefore the best uplift model) whatever the used uplift approach. Remarkably, the more noisy features are added, the more the qini difference between UMODL-FS and other feature selection methods increases.

Figure 7 indicates the percentage of added noisy features which are selected by the different feature selection methods according to the number of added noisy features. UMDOL-FS never selects a noisy feature. It illustrates the clear ability of UMODL-FS to remove noisy features. On the contrary, all other methods select noisy features and the percentage of the selected noisy ones increases as the number of added noisy features increases. To sum up, the more the number of added noisy features, the more the feature selection methods of the literature select irrelevant features as informative. In contrast, UMODL-FS always neglects irrelevant features and has the most stable qini. Moreover, UMODL-FS does not require to set a parameter giving the number of features to keep.

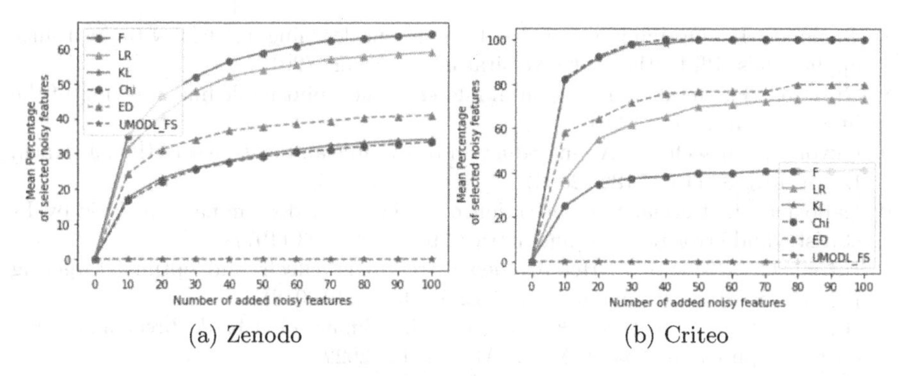

(a) Zenodo (b) Criteo

Fig. 7. Percentage of selected noisy features according to the number of added noisy features.

6 Conclusion and Future Work

In this paper, we have proposed a new non-parametric Bayesian approach for uplift discretization and feature selection. We have defined UMODL, a Bayes optimal evaluation criterion of a discretization model for uplift modeling and a search algorithm to find the best model. We have experimentally shown that UMODL is an efficient and accurate uplift estimator through discretization. Then we have presented UMODL-FS, a feature selection method for uplift. Experiments demonstrate that UMODL-FS properly removes irrelevant features and clearly outperforms state of the art methods by providing uplift models with the highest and most stable qini. The method is parameter free, making it easy to use.

This work opens several perspectives. It is promising to study this approach in the case of multiple treatments and multiple outcomes. On the other hand, as decision trees are based on discretized variables, this approach can be investigated to develop tree-based uplift modeling algorithms.

References

1. Boullé, M.: MODL: a bayes optimal discretization method for continuous attributes. Mach. Learn. **65**(1), 131–165 (2006)
2. Chandrashekar, G., Sahin, F.: A survey on feature selection methods. Comput. Electr. Eng. **40**(1), 16–28 (2014)
3. Devriendt, F., Van Belle, J., Guns, T., Verbeke, W.: Learning to rank for uplift modeling. IEEE Trans. Knowl. Data Eng. **34**(10), 4888–4904 (2020)
4. Diemert, E., Betlei, A., Renaudin, C., Amini, M.R.: A large scale benchmark for uplift modeling. In: KDD, London, United Kingdom (2018)
5. Glover, S., Dixon, P.: Likelihood ratios: a simple and flexible statistic for empirical psychologists. Psychon. Bull. Rev. **11**, 791–806 (2004)
6. Grünwald, P.: The Minimum Description Length Principle. Adaptive Computation and Machine Learning. MIT Press, Cambridge (2007)

7. Guelman, L.: Optimal personalized treatment learning models with insurance applications. Ph.D. thesis, Universitat de Barcelona (2015)
8. Gutierrez, P., Gérardy, J.Y.: Causal inference and uplift modelling: a review of the literature. In: PAPIs (2016)
9. Guyon, I., Elisseeff, A.: An introduction to variable and feature selection. J. Mach. Learn. Res. **3**, 1157–1182 (2003)
10. Habbema, J., Hermans, J.: Selection of variables in discriminant analysis by F-statistic and error rate. Technometrics **19**(4), 487–493 (1977)
11. Hitsch, G.J., Misra, S.: Heterogeneous treatment effects and optimal targeting policy evaluation. Randomized Soc. Exp. eJournal (2018)
12. Hu, J.: Customer feature selection from high-dimensional bank direct marketing data for uplift modeling. J. Mark. Anal. 1–12 (2022)
13. Jacob, D.: Cate meets ML. Digit. Finance **3**(2), 99–148 (2021)
14. Jaskowski, M., Jaroszewicz, S.: Uplift modeling for clinical trial data. In: ICML Workshop on Clinical Data Analysis (2012)
15. Kennedy, E.H.: Towards optimal doubly robust estimation of heterogeneous causal effects (2020). https://arxiv.org/abs/2004.14497
16. Liu, H., Setiono, R.: Feature selection via discretization. IEEE Trans. Knowl. Data Eng. **9**(4), 642–645 (1997)
17. Lo, V.: Pachamanova: from predictive uplift modeling to prescriptive uplift analytics: a practical approach to treatment optimization while accounting for estimation risk. J. Mark. Anal. **3**, 79–95 (2015)
18. Lunceford, J.K., Davidian, M.: Stratification and weighting via the propensity score in estimation of causal treatment effects: a comparative study. Stat. Med. **23**(19), 2937–60 (2004)
19. Radcliffe, N.: Using control groups to target on predicted lift: building and assessing uplift model. Direct Mark. Anal. J. 14–21 (2007)
20. Radcliffe, N., Surry, P.: Differential response analysis: modeling true responses by isolating the effect of a single action. Credit Scoring and Credit Control IV (1999)
21. Radcliffe, N.J., Surry, P.D.: Real-world uplift modelling with significance-based uplift trees. Stochastic Solutions (2011)
22. Rissanen, J.: Modeling by shortest data description. Automatica **14**(5), 465–471 (1978)
23. Rubin, D.B.: Estimating causal effects of treatments in randomized and nonrandomized studies. J. Educ. Psychol. **66**, 688–701 (1974)
24. Rzepakowski, P., Jaroszewicz, S.: Decision trees for uplift modeling with single and multiple treatments. Knowl. Inf. Syst. **32**(2), 303–327 (2012)
25. Sharmin, S., Shoyaib, M., Ali, A.A., Khan, M.A.H., Chae, O.: Simultaneous feature selection and discretization based on mutual information. Pattern Recognit. **91**, 162–174 (2019)
26. Zhao, Y., Fang, X., Simchi-Levi, D.: Uplift modeling with multiple treatments and general response types. In: Chawla, N.V., Wang, W. (eds.) SIAM International Conference on Data Mining, Houston, Texas, USA, 27–29 April 2017, pp. 588–596. SIAM (2017)
27. Zhao, Z., Zhang, Y., Harinen, T., Yung, M.: Feature selection methods for uplift modeling. CoRR abs/2005.03447 (2020). https://arxiv.org/abs/2005.03447

Bounding the Family-Wise Error Rate in Local Causal Discovery Using Rademacher Averages

Dario Simionato and Fabio Vandin(✉)

Department of Information Engineering, University of Padova, Padua, Italy
dario.simionato@phd.unipd.it, fabio.vandin@unipd.it

Abstract. Many algorithms have been proposed to learn local graphical structures around target variables of interest from observational data. The Markov boundary (MB) provides a complete picture of the local causal structure around a variable and is a theoretically optimal solution for the feature selection problem. Available algorithms for MB discovery have focused on various challenges such as scalability and data-efficiency. However, current approaches do not provide guarantees in terms of false discoveries in the MB.

In this paper we introduce a novel algorithm for the MB discovery problem with rigorous guarantees on the Family-Wise Error Rate (FWER), that is, the probability of reporting any false positive. Our algorithm uses Rademacher averages, a key concept from statistical learning theory, to properly account for the multiple-hypothesis testing problem arising in MB discovery. Our evaluation on simulated data shows that our algorithm properly controls for the FWER, while widely used algorithms do not provide guarantees on false discoveries even when correcting for multiple-hypothesis testing. Our experiments also show that our algorithm identifies meaningful relations in real-world data.

Keywords: Local causal discovery · Markov boundary · Rademacher averages · FWER

1 Introduction

One of the most fundamental and challenging problems in science is the discovery of causal relations from observational data [20]. Bayesian networks are a type of graphical models that are widely used to represent causal relations and have been the focus of a large amount of research in data mining and machine learning. Bayesian networks represent random variables or events as vertices of graphical models, and encode conditional-independence relationships according to the (directed) Markov property among the variables or events as directed acyclic graphs (DAGs). They are a fundamental tool to represent causality relations

Supplementary Information The online version contains supplementary material available at https://doi.org/10.1007/978-3-031-26419-1_16.

© The Author(s), under exclusive license to Springer Nature Switzerland AG 2023
M.-R. Amini et al. (Eds.): ECML PKDD 2022, LNAI 13717, pp. 255–271, 2023.
https://doi.org/10.1007/978-3-031-26419-1_16

among variables and events, and have been used to analyze data from several domains, including biology [21,27], medicine [36], and others [13,37].

One of the core tasks in learning Bayesian networks from observational data is the identification of local causal structures around a target variable T. In this work we focus on two related local structures. The first one is the set of parents and children (i.e., the neighbours) of T in the DAG, denoted as the parent-children set $PC(T)$. $PC(T)$ has a natural causal interpretation as the set of *direct* causes and effects of T [30], and the accurate identification of $PC(T)$ is a crucial step for the inference of Bayesian networks. The second structure is the Markov boundary of T, denoted as $MB(T)$. $MB(T)$ is a minimal set of variables that makes T conditionally independent of all the other variables, and comprises the elements of $PC(T)$ and the other parents of the children of T. Thus, $MB(T)$ includes all direct causes, effects, and causes of direct effects of T. Moreover, under certain assumptions, the Markov boundary is the solution of the variable selection problem [32], that is, it is the minimal set of variables with optimal predictive performance for T.

In several real-world applications, such as biology [27] and neuroscience [8], the elements in $PC(T)$ and $MB(T)$ identified from observational data provide *candidate* causal relations explored in follow-up studies and experiments, which often require significant resources (e.g., time or chemical reagents). In other areas, such as algorithmic fairness [13,17], local causal discovery can help in identifying discriminatory relationships in data. In these scenarios, it is crucial to identify *reliable* causal relations between variables, ideally avoiding any false discovery.

While the stochastic nature of random sampling implies that false discoveries cannot be avoided with absolute certainty (when at least a relation is reported), a common approach from statistics to limit false discoveries is to develop methods that rigorously bound the Family-Wise Error Rate (FWER), that is, the probability of reporting one or more false discoveries. However, currently approaches for local causal discovery do not provide guarantees on false discoveries in terms of FWER, and the study of causal discovery with false positive guarantees has received scant attention in general (see Sect. 3).

Our Contributions. In this paper we introduce a novel algorithm, Rademacher Averages for Local structure discovery, or `RAveL-MB`, for the MB discovery problem with rigorous guarantees on the FWER. Our `RAveL-MB` uses a novel algorithm, `RAveL-PC`, that we developed for the identification of the PC of a target variable while bounding the FWER. To the best of our knowledge, our algorithms are the first ones to allow the discovery of the PC set and the MB of a target variable while providing provable guarantees on false discoveries in terms of the FWER. Our algorithms crucially rely on Rademacher averages, a key concept from statistical learning theory [4], to properly account for the multiple-hypothesis testing problem arising in local causal discovery, where a large number of statistical test for conditional independence are performed. To the best of our knowledge, this work is the first one to introduce the use of Rademacher averages in (local) causal discovery. We prove, both analytically and experimentally, that currently used approaches to discover the PC set and the MB of a target variable cannot be adapted to control the FWER simply by correcting for multiple-hypothesis testing. This is due to their additional requirement of conditional dependencies being correctly identified, which is an unreasonable assumption due to the stochastic nature of random sampling and finite sample sizes. Our experimental evaluation shows that our algorithms do control the FWER while allowing for the discovery of elements in the PC set and in the MB of a target variable. On real data, our algorithms return a subset of variables that causally influences the target in agreement with prior knowledge.

The rest of the paper is organized as follows. Section 2 introduces the preliminary concepts used in the rest of the paper. Section 3 describes previous works related to our contribution. Section 4 describes our algorithms and their analysis, and the assumptions required by previously proposed algorithms in order to provide rigorous results in terms of the FWER. For clarity, we describe our algorithms focusing on the case of continuous variables, but our algorithms can be easily adapted to discrete and categorical variables. Section 5 describes our experimental evaluation on synthetic and real data. Finally, Sect. 6 offers some concluding remarks.

2 Preliminaries

In this section, we introduce basic notions and preliminary concepts used in the rest of the paper. More specifically, in Sect. 2.1 we formally define Bayesian networks (BNs) and the sets $PC(T)$ and $MB(T)$ for a target variable T. In Sect. 2.2 we describe the statistical testing procedure commonly used by algorithms for the identification of $PC(T)$ and $MB(T)$. In Sect. 2.3 we introduce the multiple hypotheses testing problem and the family-wise error rate (FWER). Finally, in Sect. 2.4 we introduce the concept of Rademacher averages for supremum deviation estimation.

2.1 Bayesian Networks

Bayesian Networks (BNs) are convenient ways to model the influence among a set of variables $\mathbf{V}$. BNs represent interactions using a *Direct Acyclic Graph (DAG)*, and employ probability distributions to define the strength of the relations. More formally, they are defined as follows.

Definition 1 (Bayesian network [19]**).** *Let p be a joint probability distribution over $\mathbf{V}$. Let $G = (\mathbf{W}, \mathbf{A})$ be a DAG where the vertices $\mathbf{W}$ of G are in a one-to-one correspondence with members of $\mathbf{V}$, and such that $\forall X \in \mathbf{V}$, X is conditionally independent of all non-descendants of X, given the parents of X (i.e., the* Markov condition *holds). A* Bayesian Network (BN) *is defined as a triplet $\langle \mathbf{V}, G, p\rangle$.*

A common assumption for the study of BNs is *faithfulness*, defined as follows.

Definition 2 (Faithfulness [30]**).** *A directed acyclic graph G is* faithful *to a joint probability distribution p over variable set $\mathbf{V}$ if and only if every independence present in p is entailed by G and the Markov Condition. A distribution p is* faithful *if and only if there exists a DAG G such that G is faithful to p.*

The dependencies between variables in a faithful BN can be analyzed through the study of *paths*, which are sequences of consecutive edges of any directionality (i.e. $X \rightarrow Y$ or $X \leftarrow Y$) in G. In particular, the *directional separation*, or *d-separation* [20], criterion can be used to study the dependence between two subsets $\mathbf{X}$ and $\mathbf{Y}$ of variables conditioning on another set $\mathbf{Z}$ of variables, such that $\mathbf{X}, \mathbf{Y}, \mathbf{Z} \subseteq \mathbf{V}$ are disjoint. Informally, the criterion marks a path between any variable in $\mathbf{X}$ and any variable in $\mathbf{Y}$ as *blocked* by $\mathbf{Z}$ if the flow of dependency between the two sets is interrupted and therefore the two sets are *independent* conditioning on $\mathbf{Z}$, written $\mathbf{X} \perp\!\!\!\perp \mathbf{Y}|\mathbf{Z}$. Viceversa, if the two sets $\mathbf{X}$ and $\mathbf{Y}$ are conditionally dependent given $\mathbf{Z}$, denoted with $\mathbf{X} \not\perp\!\!\!\perp \mathbf{Y}|\mathbf{Z}$, the path is marked as *open*. More formally, the definition of d-separated path is the following.

Definition 3 (d-separation [20]). *A path q is* d-separated, *or* blocked, *by a set of nodes $\mathbf{Z}$ if and only if:*

1. *q contains a* chain *$I \rightarrow M \rightarrow J$ or a* fork *$I \leftarrow M \rightarrow J$ such that $M \in \mathbf{Z}$, or*
2. *q contains an* inverted fork *(or* collider*) $I \rightarrow M \leftarrow J$ such that $M \notin \mathbf{Z}$ and no descendant of M is in $\mathbf{Z}$.*

A set $\mathbf{Z}$ is said to d-separate $\mathbf{X}$ from $\mathbf{Y}$ if and only if $\mathbf{Z}$ blocks every path from a node in $\mathbf{X}$ to a node in $\mathbf{Y}$.

A *causal* Bayesian network is a Bayesian network with causally relevant edge semantics [16,20].

Local Causal Discovery. The task of inferring the local region of a causal BN related to a target variable T from data is called *local causal discovery.* Two sets of variables are of major importance in local causal discovery. The first set is the *parent-children set $PC(T)$*.

Definition 4 (Parent-children set of T [16]). *The* parent-children set of T, *or* PC(T), *is the set of all parents and all children of T, i.e., the elements directly connected to T, in the DAG G.*

The elements in $PC(T)$ are the only variables that cannot be d-separated from T, that is, by the Markov property, for each X in $PC(T)$: $X \not\perp\!\!\!\perp T|\mathbf{Z}, \forall \mathbf{Z} \subseteq \mathbf{V} \setminus \{X, T\}$. The second set is the *Markov boundary $MB(T)$* of a target variable T, defined as follows.

Definition 5 (Markov boundary of T [20,33]). *The* Markov boundary of T *or* MB(T) *is the smallest set of variables in $\mathbf{V} \setminus \{T\}$ conditioned on which all other variables are independent of T, that is $\forall Y \in \mathbf{V} \setminus MB(T), Y \neq T, T \perp\!\!\!\perp Y|MB(T)$.*

Given its definition and the d-separation criteria, in a faithful BN $MB(T)$ is composed of all parents, children, and *spouses* (i.e., parents of children) of T [16], that are those variables $X \in \mathbf{V} \setminus \{T\}$ for which $\exists Y \in PC(T)$ such that $X \perp\!\!\!\perp T|\mathbf{Z}$ and $X \not\perp\!\!\!\perp T|\mathbf{Z} \cup \{Y\}$ for all $\mathbf{Z} \subseteq \mathbf{V} \setminus \{X, T\}$. MB is the minimal subset $\mathbf{S} \subseteq \mathbf{V}$ for which $p(T|\mathbf{S})$ is estimated accurately [16,33], therefore is the optimal solution for feature selection tasks.

2.2 Statistical Testing for Independence

The identification of $PC(T)$ and $MB(T)$ is based on the definitions of conditional dependence and independence between two variables X and Y. In practice, given a dataset, the conditional dependencies between variables are assessed using statistical hypothesis testing. Since a universal independence test does not exist [29], a commonly used approach is to compute the *Pearson's linear correlation coefficient* r between two vectors $\mathbf{x}$ and $\mathbf{y}$ of k elements:

$$r_{\mathbf{x},\mathbf{y}} = \frac{\sum_{i=1}^{k} x_i y_i - k\bar{x}\bar{y}}{(k-1)s_{\mathbf{x}}s_{\mathbf{y}}} \tag{1}$$

where $\bar{x}$ and $\bar{y}$ are the sample mean of $\mathbf{x}$ and $\mathbf{y}$, respectively, while $s_{\mathbf{x}}$ and $s_{\mathbf{y}}$ are the sample standard deviations.

The vectors $\mathbf{x}$ and $\mathbf{y}$ correspond to the observations of X and Y in the data, but their definition depends on whether the test is unconditional, or conditional on a set $\mathbf{Z}$ of variables. In the first case, $\mathbf{x}$ and $\mathbf{y}$ are the vectors of observations for variables X and Y, respectively. In the second case, $\mathbf{x}$ and $\mathbf{y}$ represent the residuals of the linear regression of the observations of the variables in $\mathbf{Z}$ on the ones in X (respectively, for $\mathbf{y}$, the ones in Y). For sake of simplicity, in what follows we will use $r_{X,Y,\mathbf{Z}}$ to denote the value of $r_{\mathbf{x},\mathbf{y}}$ when $\mathbf{x}$ and $\mathbf{y}$ are obtained conditioning on the set $\mathbf{Z}$, potentially with $\mathbf{Z} = \emptyset$ (i.e., for unconditional testing), as we just described.

Under the *null hypothesis* of independence between X and Y conditional on $\mathbf{Z}$ (including the case $\mathbf{Z} = \emptyset$), the expected value of $r_{X,Y,\mathbf{Z}}$ is 0, and the statistic $t = \frac{r_{X,Y,\mathbf{Z}}}{\sqrt{(1-r_{X,Y,\mathbf{Z}}^2)/(k-2)}}$ follows a *Student's t* distribution with and $k-2$ degrees of freedom. The dependence between X and Y is then usually assessed by computing (with *Student's t* distribution) the *p-value* for the test statistic t, that is the probability that the statistic is greater or equal than t under the null hypothesis of independence. In practice, algorithms for local causal discovery (e.g., [24,34]) consider X and Y as independent (unconditionally or conditional on $\mathbf{Z}$) if the p-value is greater than a threshold δ (common values for δ are 0.01 or 0.05), while X and Y are considered as dependent otherwise.

2.3 Multiple Hypotheses Testing

As described above, in testing for the independence of two variables X and Y, they are considered dependent if the p-value of the corresponding test is below a threshold δ. It is easy to see that such procedure guarantees that if X and Y are independent, then the probability of a *false discovery*, that is *falsely rejecting* their independence, is at most δ. The situation is drastically different when a large number N of hypotheses are tested, as in the case of local causal discovery. In this case, if the same threshold δ is used for every test, the expected number of false discoveries can be as large as δN. Therefore, it is necessary to correct for multiple hypothesis testing (MHT), with the goal of providing guarantees on false discoveries. A commonly used guarantee is provided by the *Family-Wise Error Rate (FWER)*, which is the probability of having at least one false discovery among all the tests. A common approach to control the FWER is the so called *Bonferroni correction* [9], which performs each test with a corrected threshold $\delta_{test} = \delta/N$ (a simple union bound shows that the resulting FWER is at most δ).

2.4 Supremum Deviation and Rademacher Averages

While Bonferroni correction does control the FWER, it conservatively assumes the worst-case scenario (of independence) between *all* null hypotheses. This often leads to a high number of *false negatives* (i.e. false null hypotheses that are not rejected). We now describe Rademacher averages [4,12], which allow to compute *data-dependent* confidence intervals for *all hypotheses simultaneously*, leading to improved tests for MHT scenarios [22]. Rademacher averages are a concept from statistical learning theory commonly used to measure the complexity of a family of functions and that, in general, also provide a way to probabilistically bound the deviation of the empirical means of the functions in the family from their expected values.

Let $\mathcal{F}$ be a family of functions from a domain $\mathcal{X}$ to $[a,b] \subset \mathbb{R}$ and let S be a sample of m i.i.d. observations from an unknown data generative distribution μ over $\mathcal{X}$. We define the *empirical sample mean* of a function $f \in \mathcal{F}$, $\hat{\mathbb{E}}_S[f]$, and its *expectation* $\mathbb{E}[f]$ as

$$\hat{\mathbb{E}}_S[f] \doteq \frac{1}{m} \sum_{s_i \in S} f(s_i) \text{ and } \mathbb{E}[f] \doteq \mathbb{E}_\mu \left[\frac{1}{m} \sum_{s_i \in S} f(s_i) \right]. \tag{2}$$

Note that $\mathbb{E}[f] = \mathbb{E}_\mu[f]$, that is, the expected value of the empirical mean corresponds to the expectation according to distribution μ. A measure of the maximum deviation of the empirical mean from the (unknown) expectation for every function $f \in \mathcal{F}$ is given by the *supremum deviation* (SD) $D(\mathcal{F}, S)$ that is defined as

$$D(\mathcal{F}, S) = \sup_{f \in \mathcal{F}} \left| \hat{\mathbb{E}}_S[f] - \mathbb{E}[f] \right|. \tag{3}$$

Computing $D(\mathcal{F}, S)$ exactly is not possible given the unknown nature of μ, therefore bounds are commonly used. An important quantity to estimate tight bounds on the SD is the *Empirical Rademacher Average* (ERA) $\hat{R}(\mathcal{F}, \mathcal{S})$ of $\mathcal{F}$ on $\mathcal{S}$, defined as

$$\hat{R}(\mathcal{F}, \mathcal{S}) \doteq \mathbb{E}_\sigma \left[\sup_{f \in \mathcal{F}} \frac{1}{m} \sum_{i=1}^{m} \sigma_i f(s_i) \right] \tag{4}$$

where σ is a vector of m i.i.d. Rademacher random variables, i.e. for which each element σ_i equals 1 or -1 with equal probability. ERA is an alternative of *VC dimension* for computing the expressiveness of a set S over class function $\mathcal{F}$, whose main advantage is that it provides tight *data-dependent* bounds while the

VC dimension provides *distribution-free* bounds that are usually fairly conservative ([18], chap. 14).

Computing the exact value of $\hat{R}(\mathcal{F}, \mathcal{S})$ is often infeasible since the expectation is taken over 2^m elements. A common approach is then to estimate $\hat{R}(\mathcal{F}, \mathcal{S})$ using a Monte-Carlo approach with n samples of σ. The n-samples Monte-Carlo Empirical Rademacher Average (n-MCERA) $\hat{R}^n_m(\mathcal{F}, \mathcal{S}, \sigma)$ is defined as

$$\hat{R}^n_m(\mathcal{F}, \mathcal{S}, \sigma) \doteq \frac{1}{n} \sum_{j=1}^{n} \sup_{f \in \mathcal{F}} \frac{1}{m} \sum_{s_i \in S} \sigma_{j,i} f(s_i) \quad (5)$$

with σ being a $m \times n$ matrix of i.i.d. Rademacher random variables. n-MCERA is useful to derive probabilistic upper bounds to the SD, as the following.

Theorem 1 (Th. 3.1 of [22]). *Let* $\delta \in (0, 1)$. *For ease of notation let*

$$\tilde{R} = \hat{R}^n_m(\mathcal{F}, \mathcal{S}, \sigma) + 2z\sqrt{\frac{\ln \frac{4}{\delta}}{2nm}} \quad (6)$$

With a probability of at least $1 - \delta$ *over the choice of* $\mathcal{S}$ *and* σ, *it holds*

$$D(\mathcal{F}, S) \leq 2\tilde{R} + \frac{\sqrt{c(4m\tilde{R} + c\ln\frac{4}{\delta})\ln\frac{4}{\delta}}}{m} + \frac{c\ln\frac{4}{\delta}}{m} + c\sqrt{\frac{\ln\frac{4}{\delta}}{2m}} \quad (7)$$

where $z = \max\{|a|, |b|\}$ *and* $c = |b - a|$.

Theorem 1 allows us to obtain confidence intervals around the empirical mean containing the expectation with probability at least $1 - \delta$ for all functions in $\mathcal{F}$ simultaneously.

3 Related Work

Given a target variable T, the task of finding $MB(T)$ is strictly related to the discovery of $PC(T)$. A common approach for MB discovery consists of creating a candidate set of elements in $MB(T)$ by running a PC discovery algorithm twice (first on T, and then on all the elements reported as member of $PC(T)$) to find the elements at distance at most 2 from T, and then to eliminate false positives, which are those elements that are not parents, children, or spouses of T. Various algorithms follow this general scheme [1,2,24,33], each one with a different variant that aims at minimizing the number of independence tests *actually* performed and their degrees of freedom to reduce the amount of data required. Note however that, as described in Sect. 4.3, this does not decrease the number of statistical tests to be considered for MHT correction, since *a priori* all tests could *potentially* be performed. Among such algorithms, Pena et al. [24] proposed $PCMB$ and proved its correctness under the assumption of all statistical tests being correct, that is, not returning *any* false positive or false negative. A different approach has been proposed for $IAMB$ [34] that incrementally grows a candidate set of elements in $MB(T)$ without searching for $PC(T)$, and then performs a false positive removal phase. Both $PCMB$ and $IAMB$ do not report false positives only under the assumption of not having any false positive and any false negative. Such assumptions are unrealistic in real-world scenarios due to noise in the data, finite sample sizes, and probabilistic guarantees of statistical tests, especially in multiple hypotheses scenarios. Our algorithms `RAveL-PC` and `RAveL-MB` do not require such assumptions to identify $PC(T)$ and $MB(T)$ with guarantees on the FWER.

To the best of our knowledge, the study of local causal discovery with guarantees on false discoveries has received scant attention. Tsamardinos et al. [35] introduced the problem of MHT in the context of local causal discovery, and proposed to use the Benjamini-Hochberg correction [6] to estimate the False Discovery Rate (FDR) of elements retrieved by $PC(T)$ discovery algorithms. However, such work does not provide an algorithm with guarantees for $MB(T)$. To the best of our knowledge, no method has focused on local causal discovery while bounding the FWER, which is extremely important in domains where false positives are critical or where follow-up studies require significant resources (e.g., biology and medicine).

Additional works focused on the more general task of BN inference. In [3], the authors extended the analysis of [35] from the local discovery task to the BN inference while [14,15,31] re-implemented the PC algorithm for BN structure discovery using the Benjamini-Yekutieli [7] correction for the FDR, the former focusing on the skeleton retrieving and the latter deriving bounds on edge orientation as well. Our work instead focuses on *local* causal discovery tasks.

Rademacher averages have been successfully used to speed-up data mining tasks (e.g., pattern mining [10,22,23,25,26,28]). To the best of our knowledge, ours is the first work to introduce their use in (local) causal discovery.

4 Algorithms for Local Causal Discoveries with FWER Guarantees

In this section we describe algorithms to obtain $PC(T)$ and $MB(T)$ with guarantees on the FWER. First, we discuss in Sect. 4.1 the requirements for previously proposed algorithms $PCMB$ and $IAMB$ to obtain guarantees on the FWER. In particular, we show that they require unrealistic assumptions that are not met in practice, as confirmed by our experimental evaluation (see Sect. 5). We then present in Sect. 4.2 our algorithms `RAveL-PC` and `RAveL-MB` for the computation of $PC(T)$ and $MB(T)$ with guarantees on the FWER. Finally in Sect. 4.3 we describe how Rademacher averages are used by our algorithms for effective independence testing.

4.1 Analysis and Limitations of $PCMB$ and $IAMB$

The algorithms presented in Sect. 3 are correct under the assumption that the independence tests result in no false positive *and* no false negative [24,34]. In this section we determine milder sufficient conditions that allow *GetPC* [24] to control the FWER for the PC discovery task, and *PCMB* [24] and *IAMB* [34] to control the FWER for the MB discovery task. In all cases, a first requirement is that the independence tests performed by the algorithms must be corrected for MHT in order to bound the FWER. However, we also show that an additional requirement on the ability to identify dependent variables (i.e., on the *power* of the tests) is needed. In particular, we refer to the situation where *all tests* on dependent variables correctly reject the null hypothesis of independence as the *infinite power assumption.* In some cases, we consider the infinite power assumption only for independence tests between pairs of variables that are directly connected in the underlying DAG. We refer to such situation as the *local infinite power assumption.*

We start by proving sufficient conditions for bounding the FWER of the elements returned by *GetPC* [24]. (All proofs are in the Appendix.)

Theorem 2. *$GetPC(T)$ outputs a set of elements in $PC(T)$ with $FWER \leq \delta$ if the independence tests performed by GetPC have $FWER \leq \delta$* and *the local infinite power assumption holds.*

The following proves that similar requirements are needed for *PCMB* [24] to have guarantees on the FWER.

Theorem 3. *$PCMB(T)$ outputs a set of elements in $MB(T)$ with $FWER \leq \delta$ if the independence tests performed by PCMB have $FWER \leq \delta$ and the infinite power assumption holds.*

The following result proves analogous requirements for *IAMB*.

Theorem 4. *$IAMB(T)$ outputs a set of elements in $MB(T)$ with $FWER \leq \delta$ if the independence tests performed by IAMB have $FWER \leq \delta$ and the infinite power assumption holds.*

Note that the results above require the (local) infinite power assumption to hold in order to have guarantees on the FWER of the output of previously proposed algorithms. In fact, if the (local) infinite power assumption does not hold, such algorithms may output false positives even when *all* independence tests do not return a single false positive. We present two such examples in the Appendix. Moreover, our experimental evaluation in Sect. 5 shows that this situation does happen in practice.

4.2 Algorithms `RAveL-PC` and `RAveL-MB`

As shown in Sect. 4.1, correcting for the FWER of every independence test is not sufficient for bounding the FWER of the variables returned by current state-of-the-art algorithms for PC and MB discovery. In addition, infinite statistical power is a strong assumption which is impossible to test and ensure in real-world scenarios. Motivated by these observations, we developed `RAveL-PC` and

RAveL-MB, two algorithms for the discovery of elements in PC and MB, respectively, that control the FWER of their outputs without making any assumption on statistical power.

RAveL-MB follows the same overall approach used by previously proposed algorithms (e.g., *PCMB*, see Sect. 3): it first identifies elements in $PC(T)$ and adds them to $MB(T)$, and then tests the spouse condition on elements at distance 2 from T, that are variables $Y \in PC(X)$ with $X \in PC(T)$ and $Y \notin PC(T)$. The pseudocode of RAveL-MB is shown in Algorithm 1. RAveL-MB inizializes MB to the output of the function RAveL-PC(T,$\mathbf{V}$,δ) (line 1), which returns a subset of $PC(T)$. For each element $X \in MB$ (line 2), RAveL-MB computes RAveL-PC(X,$\mathbf{V}$,δ) and, for every returned element Y that is not already in MB (line 3), an independence test of T on Y conditioning on $\mathbf{V} \setminus \{Y, T\}$ using function test_indep(T,Y,$\mathbf{V} \setminus \{Y, T\}$,$\delta$) is performed to test whether Y is a spouse of T with respect to X (line 4). If such test determines the conditional dependence between T and Y, then Y is added to MB (line 5). Finally, after analyzing all variables originally in MB, RAveL-MB outputs the set of elements in the Markov Boundary (line 6).

Note that the spouse condition is tested by conditioning only on the set $\mathbf{V} \setminus \{Y, T\}$. This is sufficient, since it is a set conditioned on which T and Y are d-connected if and only if Y is directly connected or is a spouse of T. In fact, if Y does not belong to any of these elements, then Y is connected to T through paths that contain chains or forks whose middle element is in $\mathbf{V} \setminus \{Y, T\}$. That is, Y is connected to T only through d-blocked paths.

Algorithm 1: RAveL-MB(T,$\mathbf{V}$,δ)

Input: target variable T, set of variables $\mathbf{V}$, threshold $\delta \in (0, 1]$
Output: A subset of $MB(T)$ with FWER lower than δ.

1 $MB \leftarrow$ RAveL-PC(T,$\mathbf{V}$,δ) ;
2 **foreach** $X \in MB$ **do**
3   **foreach** $Y \in$ *RAveL-PC(X,$\mathbf{V}$,δ)* ***and*** $Y \notin MB$ **do**
4   **if** *not test_indep(T,Y,$\mathbf{V} \setminus \{Y, T\}$,$\delta$)* **then**
5             $MB \leftarrow MB \cup \{Y\}$;
6 **return** MB;

RAveL-MB uses algorithm RAveL-PC(X,$\mathbf{V}$,δ) (shown in Algorithm 2) for the discovery of variables of a set $\mathbf{V}$ that are in $PC(X)$. The parameter δ controls the overall FWER of the procedure. RAveL-PC(X,$\mathbf{V}$,δ) identifies $PC(X)$ by using the definition of parent-children set, that is, $Y \in PC(X)$ gets returned if only if all independence tests between X and Y reject the null hypothesis.

Both algorithms RAveL-MB and RAveL-PC employ a function, denoted as test_indep(X,Y,$\mathbf{Z}$,δ), that performs the independence test between $X, Y \in \mathbf{V}$ conditioning on $\mathbf{Z} \subseteq \mathbf{V}$ while controlling the FWER of *all testable hypotheses* with threshold δ, and returns true only if the null hypothesis gets rejected.

Algorithm 2: `RAveL-PC(`T`,`$\mathbf{V}$`,`δ`)`

Input: target variable T, set of variables $\mathbf{V}$, threshold $\delta \in (0, 1]$
Output: A subset of $PC(T)$ with FWER lower than δ.

1 $PC \leftarrow \mathbf{V} \setminus \{T\}$;
2 **foreach** $X \in \mathbf{V} \setminus \{T\}$ **do**
3   **foreach** $\mathbf{Z} \subseteq \mathbf{V} \setminus \{X, T\}$ **do**
4   **if** *test_indep(*T*,*X*,*$\mathbf{Z}$*,*δ*)* **then**
5             $PC \leftarrow PC \setminus \{X\}$;
6 **return** PC;

Practical details on our implementation of `test_indep(`X`,`Y`,`$\mathbf{Z}$`,`δ`)` are provided in Sect. 4.3.

The following results prove that `RAveL-PC` and `RAveL-MB` control the FWER of PC and MB, respectively.

Theorem 5. *`RAveL-PC(`T`,`$\mathbf{V}$`,`δ`)` outputs a set of elements in $PC(T)$ with $FWER \leq \delta$.*

Theorem 6. *`RAveL-MB` outputs a set of elements in $MB(T)$ with $FWER \leq \delta$.*

The choice of $\mathbf{V} \setminus \{Y, T\}$ as conditioning set for testing the spouse condition is a consequence of `RAveL-PC` returning, with probability at least $1 - \delta$, a subset of $PC(T)$, and of any superset of $PC(T)$ allowing the discovery of spouses by `RAveL-MB`. We note that prior knowledge may be incorporated in the algorithm, if available, by conditioning on smaller set of variables, therefore increasing the precision of independence tests.

4.3 Rademacher Averages for Independence Testing

Note that our algorithms `RAveL-PC(`X`,`$\mathbf{V}$`,`δ`)` and `RAveL-MB(`X`,`$\mathbf{V}$`,`δ`)` both rely on the availability of function `test_indep(`X`,`Y`,`$\mathbf{Z}$`,`δ`)`, which assesses the independence between $X, Y \in \mathbf{V}$ conditioning on $\mathbf{Z} \subseteq \mathbf{V}$ and returns `true` only if the null hypothesis gets rejected, while controlling the FWER of *all testable hypotheses* below a threshold δ.

The naïve implementation of `test_indep(`X`,`Y`,`$\mathbf{Z}$`,`δ`)` would be to perform a standard statistical test (see Sect. 2.2) and use Bonferroni correction (see Sect. 2.3) to correct for multiple hypothesis testing. In particular, this requires to use a modified threshold δ/N for every hypothesis, where N is the maximum number of hypotheses that could be tested. Therefore, N is the maximum number of conditional independencies[1] between the variables in $\mathbf{V}$, that is $N = |\mathbf{V}|(|\mathbf{V}| - 1)2^{|\mathbf{V}|-3}$. Note that the value of N grows exponentially with $|\mathbf{V}|$,

[1] N counts, in fact, the total number of possible conditional independencies between any couple of variables by considering the symmetry property of independence tests, that is testing the (conditional) independence of X from Y is equivalent to testing the one of Y from X.

leading to a Bonferroni correction which is very conservative and, therefore, to a high number of false negatives (independence tests between dependent variables for which the null hypothesis does not get rejected).

The high number of tests is not a feature of our algorithms only, but it is, in essence, shared by other widely used algorithms such as IAMB and PCMB (see Sect. 3). In fact, for both algorithms, the potential number of independence tests they perform can be as high as $N = |\mathbf{V}|(|\mathbf{V}| - 1)2^{|\mathbf{V}|-3}$, even if a smaller number of tests may be considered in practice, depending on the output of the tests in previous steps, and a proper MHT correction depends on the maximum number of tests that could be performed.

Our solution to make our algorithms `RAveL-PC(`X`,`$\mathbf{V}$`,`δ`)` and `RAveL-MB(`X`,`$\mathbf{V}$`,`δ`)` practical is to implement `test_indep(`X`,`Y`,`$\mathbf{Z}$`,`δ`)` exploiting Rademacher averages to obtain data-dependent bounds and confidence intervals. The key idea is to estimate confidence intervals around the empirical test statistics $r_{X,Y,\mathbf{Z}}$ so that they contain the true values *simultaneously* with probability $1 - \delta$. In this way, testing for independence corresponds to check whether a confidence interval contains 0, which is the expected value of $r_{X,Y,\mathbf{Z}}$ under the null hypothesis of independence.

To implement the idea described above, we express Eq. 1 as an additive function on the samples as follows. First, we assume that all variables have been centered around 0 and then normalized by dividing for the maximum absolute value (i.e., $\bar{x} = 0$ and $\max(|\mathbf{x}|) = 1$ for all variables). Let $s_1, s_2, \ldots, s_k$ the samples in the dataset $\mathcal{S} = \{s_1, s_2, \ldots, s_k\}$, where each s_i is a collection of observations $s_i = \{v_1^i, v_2^i, \ldots\}$ of variables in $\mathbf{V}$, where v_j^i is the observation of the j-th variable $V_j \in \mathbf{V}$ in sample s_i. Given two variables $X, Y \in \mathbf{V}$, and a set of variables $\mathbf{Z} \subset \mathbf{V}$, we define the following function that, given a sample s_i, provides an estimate of $r_{X,Y,\mathbf{Z}}$ using only s_i

$$r_{X,Y,\mathbf{Z}}(s_i) = k\frac{x_i y_i}{k-1}. \tag{8}$$

Note that the conditioning set $\mathbf{Z}$ does not explicitly appear in the term $k\frac{x_i y_i}{k-1}$, but it is used in the definition of the values in $\mathbf{x}$ and $\mathbf{y}$ (see Sect. 2.2).

Therefore, we have the following modified version of Pearson's coefficient, where $s_{\mathbf{x}}$ is replaced by $\max(|\mathbf{x}|) - \bar{\mathbf{x}}$ (similarly for $s_{\mathbf{y}}$):

$$r_{X,Y,\mathbf{Z}} = \frac{1}{k}\sum_{i=1}^{k} r_{X,Y,\mathbf{Z}}(s_i). \tag{9}$$

By considering the family $\mathcal{F}$ of functions defined by $r_{X,Y,\mathbf{Z}}$ for each pair X, Y of variables and each set $\mathbf{Z} \subseteq \mathbf{V} \setminus \{X, Y\}$, we have that the n-MCERA (Eq. 5) is

$$\hat{R}_k^n(\mathcal{F}, \mathcal{S}, \sigma) \doteq \frac{1}{n}\sum_{j=1}^{n} \sup_{r_{X,Y,\mathbf{Z}} \in \mathcal{F}} \frac{1}{k}\sum_{i=1}^{k} \sigma_{j,i} r_{X,Y,\mathbf{Z}}(s_i). \tag{10}$$

After the n-MCERA has been computed as above, we compute the supremum deviation $D(\mathcal{F}, S)$ according to Theorem 1, which allows us to obtain confidence intervals around the empirical $r_{X,Y,\mathbf{Z}}$ as

$$CI_{X,Y,\mathbf{Z}} = [r_{X,Y,\mathbf{Z}} - D(\mathcal{F}, S), r_{X,Y,\mathbf{Z}} + D(\mathcal{F}, S)] \tag{11}$$

with the guarantee that, *simultaneously* for all $r_{X,Y,\mathbf{Z}} \in \mathcal{F}$, $CI_{X,Y,\mathbf{Z}}$ contains the expected value of $r_{X,Y,\mathbf{Z}}$ with probability at least $1 - \delta$. Then, for a pair X, Y of variables and a set $\mathbf{Z} \subseteq \mathbf{V} \setminus \{X, Y\}$, we reject the null hypothesis of independence between X, Y conditioning on $\mathbf{Z}$ (i.e., `test_indep(`X`,`Y`,`$\mathbf{Z}$`,`δ`)` returns `true`) if $CI_{X,Y,\mathbf{Z}}$ does not contain the value 0.

5 Experimental Evaluation

This section describes the experimental evaluation performed to empirically assess our algorithms. In Sect. 5.1 we compare `RAveL-PC` and `RAveL-MB` performances with other state-of-the-art methods on synthetic data. Section 5.2 presents the analysis on real world data. We implemented[2] `RAveL-PC`, `RAveL-MB`, and the other algorithms considered in this section in Python 3 and R.

5.1 Synthetic Data

We used synthetic data to evaluate `RAveL-PC` and `RAveL-MB` against state-of-the-art algorithms for the task of PC and MB discovery, respectively. In our synthetic data, each variable is a linear combination of its parents values plus a Gaussian noise term. The related structural model includes 13 variables and is shown in the Appendix. We set the rejection threshold $\delta = 0.05$, which is a common value in literature, and we run each algorithm on increasing size datasets. We repeated each trial 100 times and used $n = 1000$ for the n-MCERA. For each dataset, we considered all variables as target variable T in turn and run the algorithms for each choice of T. (Note that the number N of potential hypotheses tested is still the same as defined in Sect. 4.3.)

For the PC discovery task, we compared two versions of *GetPC* [24], the original one (without any correction for MHT) and a modified version with Bonferroni correction, our algorithm `RAveL-PC` (that uses Rademacher averages as described in Sect. 4.3), and a variant of `RAveL-PC` that uses Bonferoni correction instead of Rademacher averages for MHT. Figure 1(a) shows the estimated FWER of each method (that is, the fraction of trials in which at least a false positive is reported). The results confirm our analysis in Sect. 4.2, and we observe that, for the specific BN we consider, algorithm *GetPC* has FWER below the threshold, even if this is not guaranteed from our theoretical analysis. For the MB discovery task, we compared two versions of *PCMB* [24] (which uses *GetPC* as subroutine) and of *IAMB* [34], the original one (without any correction for MHT) and a modified version with Bonferroni correction, our algorithm

[2] Code and appendix available at https://github.com/VandinLab/RAveL.

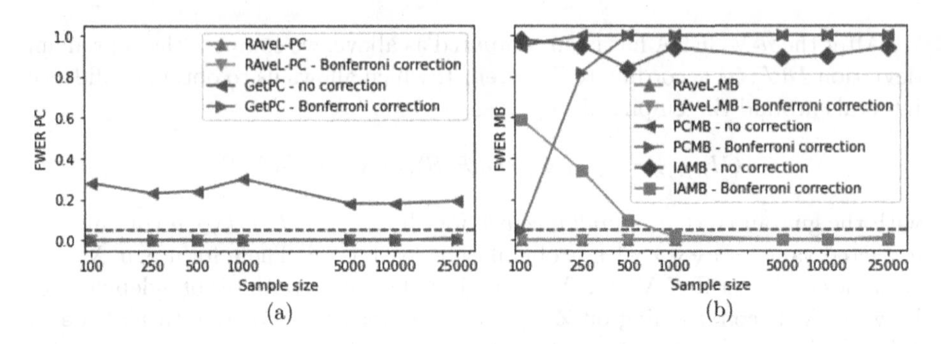

Fig. 1. Empirical FWER of various PC discovery (a) and MB discovery (b) algorithms on synthetic data for different sample sizes. FWER is the fraction of 100 trials in which at least one false positive is reported. The dashed line represents the bound $\delta = 0.05$ to the FWER used in the experiments.

`RAveL-MB`, and a variant of `RAveL-MB` that uses Bonferoni correction instead of Rademacher averages. Figure 1(b) shows the FWER of each method. The results confirms `RAveL-MB` and its variant to be the only algorithms with guarantees on the FWER at any sample size, that is without infinite power assumption. Moreover, note that *PCMB* reports false positives with high probability even if its *GetPC* does not. This is due to elements at distance 2 from T that are correctly identified as candidate spouses, but for which the spouse condition used by *PCMB* results in a false positive due to false negatives in $PC(T)$.

We then assessed the fraction of false negatives for our algorithms, which are the only ones with guarantees on the FWER, on datasets with sample sizes up to 500000 elements by repeating each trial 5 times. We analyzed only the standard version of our algorithms, which is the only practical option in scenarios with a large number of variables (i.e., when the number N of tests is very large). The percentage of false negatives returned by `RAveL-PC` and `RAveL-MB` starts decreasing for datasets with more than 25000 samples, but a simple modification of the test statistic (to be described in the journal version of this paper) greatly improves the performances lowering the data requirement to just 1000 samples. In all such tests our algorithms did not return any false positive.

5.2 Real-World Dataset

We tested our algorithms on the Boston housing dataset [11], which contains data about Boston suburbs, considering the median price of homes in each suburb as target T. Since the number of variables for such dataset is small, we used the Bonferroni variant of our algorithms `RAveL-PC` and `RAveL-MB`, with $\delta = 0.01$. Both algorithms reported in output two variables, one related to the number of rooms per house, and the other to the median income of the suburb residents, that clearly influence the median price of the houses in the neighbourhood. The first variable is a common indicator of the price of a house, while the second confirms the intuition that between two identical houses, the one built in a wealthier neighborhood has a higher price. These results provide empirical evidence that our algorithms identify meaningful causal relations while avoiding false positives.

6 Conclusions

In this paper we presented two algorithms, `RAveL-PC` and `RAveL-MB`, for the task of local causal discovery. In contrast to state-of-the-art approaches, our algorithms provide guarantees on false discoveries in terms of bounding the FWER. Our algorithms use Rademacher averages to properly account for multiple hypothesis testing, and our experimental evaluation shows that our algorithms properly control for false discoveries. Our algorithms can be extended to other (e.g., non-linear) test statistics and to other tests (e.g., based on permutation testing). Interesting research directions include the application of our framework to recently proposed independence tests [5], improving the efficiency of our algorithms, and exploiting them for structure discovery.

Acknowledgements. This work is supported, in part, by the Italian Ministry of Education, University and Research (MIUR), under PRIN Project n. 20174LF3T8 "AHeAD" (efficient Algorithms for HArnessing networked Data) and the initiative "Departments of Excellence" (Law 232/2016), and by University of Padova under project "SID 2020: RATED-X".

References

1. Aliferis, C.F., Statnikov, A., Tsamardinos, I., Mani, S., Koutsoukos, X.D.: Local causal and Markov blanket induction for causal discovery and feature selection for classification Part I: algorithms and empirical evaluation. JMLR **11**, 171–234 (2010)
2. Aliferis, C.F., Tsamardinos, I., Statnikov, A.: Hiton: a novel markov blanket algorithm for optimal variable selection. In: Proceedings of AMIA (2003)
3. Armen, A.P., Tsamardinos, I.: Estimation and control of the false discovery rate of Bayesian network skeleton identification. Technical report, TR-441. University of Crete (2014)
4. Bartlett, P.L., Mendelson, S.: Rademacher and gaussian complexities: risk bounds and structural results. JMLR **3**, 463–482 (2002)
5. Bellot, A., van der Schaar, M.: Conditional independence testing using generative adversarial networks. In: Advances in Neural Information Processing Systems (2019)
6. Benjamini, Y., Hochberg, Y.: Controlling the false discovery rate: a practical and powerful approach to multiple testing. J. Royal Stat. Soc. **57**(1), 289–300 (1995)
7. Benjamini, Y., Yekutieli, D.: The control of the false discovery rate in multiple testing under dependency. Ann. Stat. 1165–1188 (2001)
8. Bielza, C., Larranaga, P.: Bayesian networks in neuroscience: a survey. Front. Comput. Neurosci. **8**, 131 (2014)
9. Bonferroni, C.: Teoria statistica delle classi e calcolo delle probabilita. Istituto Superiore di Scienze Economiche e Commericiali di Firenze (1936)

10. Cousins, C., Wohlgemuth, C., Riondato, M.: BAVARIAN: betweenness centrality approximation with variance-aware rademacher averages. In: ACM SIGKDD (2021)
11. Harrison, D., Jr., Rubinfeld, D.L.: Hedonic housing prices and the demand for clean air. J. Environ. Econ. Manag. **5**(1), 81–102 (1978)
12. Koltchinskii, V., Panchenko, D.: Rademacher processes and bounding the risk of function learning. In: High Dimensional Probability II (2000)
13. Kusner, M.J., Loftus, J.R.: The long road to fairer algorithms. Nature **578**(7793), 34–36 (2020)
14. Li, J., Wang, Z.J.: Controlling the false discovery rate of the association/causality structure learned with the PC algorithm. JMLR (2009)
15. Liu, A., Li, J., Wang, Z.J., McKeown, M.J.: A computationally efficient, exploratory approach to brain connectivity incorporating false discovery rate control, a priori knowledge, and group inference. Comput. Math. Methods Med. (2012)
16. Ma, S., Tourani, R.: Predictive and causal implications of using shapley value for model interpretation. In: KDD Workshop on Causal Discovery. PMLR (2020)
17. Mhasawade, V., Chunara, R.: Causal multi-level fairness. In: Proceedings of AAAI/ACM Conference on AI, Ethics, and Society (2021)
18. Mitzenmacher, M., Upfal, E.: Probability and Computing. Cambridge University Press, Cambridge (2017)
19. Neapolitan, R.E., et al.: Learning Bayesian Networks. Pearson Prentice Hall, Hoboken (2004)
20. Pearl, J.: Causality, 2nd edn. Cambridge University Press, Cambridge (2009)
21. Pe'er, D.: Bayesian network analysis of signaling networks: a primer. Science's STKE (2005)
22. Pellegrina, L., Cousins, C., Vandin, F., Riondato, M.: MCRapper: Monte-Carlo rademacher averages for poset families and approximate pattern mining. In: ACM SIGKDD (2020)
23. Pellegrina, L., Vandin, F.: Silvan: estimating betweenness centralities with progressive sampling and non-uniform rademacher bounds. arXiv:2106.03462 (2021)
24. Pena, J.M., Nilsson, R., Björkegren, J., Tegnér, J.: Towards scalable and data efficient learning of Markov boundaries. Int. J. Approximate Reasoning **45**(2), 211–232 (2007)
25. Riondato, M., Upfal, E.: Mining frequent itemsets through progressive sampling with rademacher averages. In: ACM SIGKDD (2015)
26. Riondato, M., Upfal, E.: ABRA: approximating betweenness centrality in static and dynamic graphs with rademacher averages. ACM TKDD **12**(5), 1–38 (2018)
27. Sachs, K., Perez, O., Pe'er, D., Lauffenburger, D.A., Nolan, G.P.: Causal protein-signaling networks derived from multiparameter single-cell data. Science **308**(5721), 523–529 (2005)
28. Santoro, D., Tonon, A., Vandin, F.: Mining sequential patterns with VC-dimension and rademacher complexity. Algorithms **13**(5), 123 (2020)
29. Shah, R.D. and Peters, J.: The hardness of conditional independence testing and the generalised covariance measure. Ann. Stat. (2020)
30. Spirtes, P., Glymour, C.N., Scheines, R., Heckerman, D.: Causation, Prediction, and Search. MIT Press, Cambridge (2000)
31. Strobl, E.V., Spirtes, P.L., Visweswaran, S.: Estimating and controlling the false discovery rate of the PC algorithm using edge-specific P-values. ACM TIST **10**(5), 1–37 (2019)
32. Tsamardinos, I., Aliferis, C.F.: Towards principled feature selection: relevancy, filters and wrappers. In: International Workshop on AI and Statistics. PMLR (2003)

33. Tsamardinos, I., Aliferis, C.F., Statnikov, A.: Time and sample efficient discovery of Markov blankets and direct causal relations. In: ACM SIGKDD (2003)
34. Tsamardinos, I., Aliferis, C.F., Statnikov, A.R., Statnikov, E.: Algorithms for large scale Markov blanket discovery. In: FLAIRS Conference (2003)
35. Tsamardinos, I., Brown, L.E.: Bounding the false discovery rate in local Bayesian network learning. In: AAAI (2008)
36. Velikova, M., van Scheltinga, J.T., Lucas, P.J., Spaanderman, M.: Exploiting causal functional relationships in Bayesian network modelling for personalised healthcare. Int. J. Approximate Reasoning **55**(1), 59–73 (2014)
37. Yusuf, F., Cheng, S., Ganapati, S., Narasimhan, G.: Causal inference methods and their challenges: the case of 311 data. In: International Conference on on Digital Government Research (2021)

Optimal Transport

Learning Optimal Transport Between Two Empirical Distributions with Normalizing Flows

Florentin Coeurdoux[1(✉)], Nicolas Dobigeon[1], and Pierre Chainais[2]

[1] University of Toulouse, IRIT/INP-ENSEEIHT, 31071 Toulouse, France
{Florentin.Coeurdoux,Nicolas.Dobigeon}@irit.fr

[2] Univ. Lille, CNRS, Centrale Lille, UMR 9189 CRIStAL, 59000 Lille, France
pierre.chainais@centralelille.fr

Abstract. Optimal transport (OT) provides effective tools for comparing and mapping probability measures. We propose to leverage the flexibility of neural networks to learn an approximate optimal transport map. More precisely, we present a new and original method to address the problem of transporting a finite set of samples associated with a first underlying unknown distribution towards another finite set of samples drawn from another unknown distribution. We show that a particular instance of invertible neural networks, namely the normalizing flows, can be used to approximate the solution of this OT problem between a pair of empirical distributions. To this aim, we propose to relax the Monge formulation of OT by replacing the equality constraint on the push-forward measure by the minimization of the corresponding Wasserstein distance. The push-forward operator to be retrieved is then restricted to be a normalizing flow which is trained by optimizing the resulting cost function. This approach allows the transport map to be discretized as a composition of functions. Each of these functions is associated to one sub-flow of the network, whose output provides intermediate steps of the transport between the original and target measures. This discretization yields also a set of intermediate barycenters between the two measures of interest. Experiments conducted on toy examples as well as a challenging task of unsupervised translation demonstrate the interest of the proposed method. Finally, some experiments show that the proposed approach leads to a good approximation of the true OT.

Keywords: Normalizing flows · Optimal transport · Generative model

1 Introduction

The optimal transport (OT) problem was initially formulated by the French mathematician Gaspard Monge. In his seminal paper published in 1781 [18], he raised the following question: how to move a pile of sand to a target location with the least possible effort or cost? The objective was to find the best way to

© The Author(s), under exclusive license to Springer Nature Switzerland AG 2023
M.-R. Amini et al. (Eds.): ECML PKDD 2022, LNAI 13717, pp. 275–290, 2023.
https://doi.org/10.1007/978-3-031-26419-1_17

minimize this cost by a transport plan, without having to list all the possible matches between the starting and ending points. More recently, thanks to recent advances related to computational issues [22], OT has founded notable successes with respect to applications ranging from image processing and computer vision [21] to machine learning [2] and domain adaptation [7].

Normalizing flows (NFs) have also attracted a lot of interest in the machine learning community, motivated in particular by their ability to model high dimensional data [15,19]. These deep generative models are characterized by an invertible operator that associates any input data distribution with a target distribution that is usually chosen to be Gaussian. They have the great advantage of leading to tractable distributions, which eases direct sampling and density estimation. Applications of these generative models include image generation with real-valued non-volume preserving transformations (RealNVP) [10] or generative flows using an invertible 1×1 convolution (GLOW) [14].

Motivated by the similarities between the problem of OT and the training of NF, this paper proposes a neural architecture and a corresponding training strategy that permits to learn an approximate Monge map between any two empirical distributions. The proposed framework is based on a relaxation of the Monge formulation of OT. To adapt the training loss to the flow-based structure of the network, this loss function is supplemented with a Sobolev regularisation to promote minimal efforts achieved by each flow. Numerical simulations show that this regularisation results in a smoother and more efficient trajectory. Interestingly, the discretization inherent to the flow-based structure of the network implicitly provides intermediate transports and, at the same time, Wasserstein barycenters [1]. To the best of our knowledge, this is the first time that NFs are considered to address OT and Wasserstein barycenter computation, up to interesting dimensions.

Contributions. Our contributions are twofold: i) Sect. 2 recalls the Monge formulation of OT and proposes a relaxation in the case of a transport between two empirical distributions. ii) Sect. 3 presents the generic framework based on NFs and describes a particular instance to solve the OT problem. Section 4 presents some experimental results illustrating the performance of the proposed method. Section 6 concludes this paper.

2 Relaxation of the Optimal Transport Problem

Let μ and ν be two probability measures with finite second order moments. More general measures, for example on $\mathcal{X} = \mathbb{R}^d$ (where $d \in \mathbb{N}^*$ is the dimension), can have a density $\mathrm{d}\mu(x) = p_X(x)dx$ with respect to the Lebesgue measure, often noted $p_X = \frac{\mathrm{d}\mu}{\mathrm{d}x}$, which means that

$$\forall h \in \mathcal{C}\left(\mathbb{R}^d\right), \quad \int_{\mathbb{R}^d} h(x)\mathrm{d}\mu(x) = \int_{\mathbb{R}^d} h(x)p_X(x)\mathrm{d}x \tag{1}$$

where $\mathcal{C}(\cdot)$ is the class of continuous functions. In the remainder of this paper, $\mathrm{d}\mu(x)$ and $p_X(x)\mathrm{d}x$ will be used interchangeably.

2.1 Background on Optimal Transport

Let consider $\mathcal{X}$ and $\mathcal{Y}$ two separable metric spaces. Any measurable application $T : \mathcal{X} \to \mathcal{Y}$ can be extended to the so-called push-forward operator $T_\sharp$ which moves a probability measure on $\mathcal{X}$ to a new probability measure on $\mathcal{Y}$. For any measure μ on $\mathcal{X}$, one defines the image measure $\nu = T_\sharp\mu$ on $\mathcal{Y}$ such that

$$\forall h \in \mathcal{C}(\mathcal{Y}), \quad \int_{\mathcal{Y}} h(y)\mathrm{d}\nu(y) = \int_{\mathcal{X}} h(T(x))\mathrm{d}\mu(x). \tag{2}$$

Intuitively, the application $T : \mathcal{X} \to \mathcal{Y}$ can be interpreted as a function moving a single point from one measurable space to another [22]. The operator $T_\sharp$ pushes each elementary mass of a measure μ on $\mathcal{X}$ by applying the function T to obtain an elementary mass in $\mathcal{Y}$. The problem of OT as formulated by Monge is now stated in a general framework. For a given cost function $c : \mathcal{X} \times \mathcal{Y} \to [0, +\infty]$, the measurable application $T : \mathcal{X} \to \mathcal{Y}$ is called the OT map from a measure μ to the image measure $\nu = T_{\#}\mu$ if it reaches the infimum

$$\inf_T \left\{ \int_{\mathcal{X}} c(x, T(x))\mathrm{d}\mu(x) : T_\sharp\mu = \nu \right\}. \tag{3}$$

Alternatively the Kantorovitch formulation of OT results from a convex relaxation of the Monge problem (3). By defining Π as the set of all probabilistic couplings with marginals μ and ν, it yields the optimal π that reaches

$$\min_{\pi \in \Pi} \int_{\mathcal{X} \times \mathcal{Y}} c(\mathbf{x}, \mathbf{y})\, d\pi(\mathbf{x}, \mathbf{y}) \tag{4}$$

Under this formulation, the optimal π, which is a joint probability measure with marginals μ and ν, can be interpreted as the optimal transportation map. It allows the Wasserstein distance of order p between μ and ν to be defined as

$$W_p(\mu, \nu) \stackrel{\text{def}}{=} \inf_{\pi \in \Pi} \left\{ \left(\mathop{\mathbb{E}}_{\substack{\mathbf{x} \sim \mu \\ \mathbf{y} \sim \nu}} d(\mathbf{x}, \mathbf{y})^p \right)^{\frac{1}{p}} \right\} \tag{5}$$

where $d(\cdot, \cdot)$ is a distance defining the cost function $c(\mathbf{x}, \mathbf{y}) = d(\mathbf{x}, \mathbf{y})^p$. The Wasserstein distance is also known as the Earth mover's distance in the computer vision community. It defines a metric over the space of square integrable probability measures.

2.2 Proposed Relaxation of OT

OT boils down to a variational problem, i.e., it requires the minimization of an integral criterion in a class of admissible functions. Given two probability measures μ and ν, the existence and uniqueness of an operator T that belongs to the class of bijective, continuous and differentiable functions such that $T_\sharp\mu = \nu$ is not guaranteed. The difficulty lies in the class defining these admissible functions.

Indeed, even when μ and ν are regular densities on regular subsets of $\mathbb{R}^d$, the search for a transport map such that $T_\sharp\mu = \nu$ makes the problem (3) difficult in a general case. To overcome the difficulty of solving this equation on $T_\sharp$, we propose to reformulate the Monge's OT statement by relaxing the equality on the operator defining the image measure.

More precisely, the equality between the image measure $T_\sharp\mu$ and the target measure ν is replaced by the minimization of their statistical distance $d(T_\sharp\mu, \nu)$. The choice of the distance $d(\cdot, \cdot)$ is crucial because it determines the quality of the approximation of the image measure by the transport map T. In this work, we propose to choose $d(\cdot, \cdot)$ as the Wasserstein distance $W_p(\cdot, \cdot)$. This choice will be motivated by the fact that this distance can be easily approximated numerically without explicit knowledge of the probability distributions μ and ν, in particular when they are empirically described by samples only. The relaxation of the Monge problem (3) can then be written as

$$\inf_T \left\{ W_p(T_\sharp\mu, \nu) + \lambda \int_{\mathcal{X}} c(x, T(x)) \mathrm{d}\mu(x) \right\} \tag{6}$$

where the cost function defined in (3) is interpreted here as a regularisation term adjusted by the hyperparameter λ.

Remark 1. The relaxed formulation (6) relies on the Wasserstein distance between the target measure ν and the image measure $T_\sharp\mu$. This term should not be confused with the Wasserstein distance $W_p(\mu, \nu)$ which is the infimum reached by the solution of the Kantorovitch's formulation of OT (4).

2.3 Discrete Formulation

In a machine learning context, the underlying continuous measures are conventionally approximated by empirical point measures thanks to available data samples. Therefore, in this paper, we are interested in discrete measures and the empirical formulation of the OT problem. Within this framework, we will consider μ and ν two discrete measures described by the respective samples $\mathbf{x} = \{x_n\}_{n=1}^N$ and $\mathbf{y} = \{y_n\}_{n=1}^N$ such that $\mu = \frac{1}{N}\sum_{n=1}^N \delta_{x_n}$ and $\nu = \frac{1}{N}\sum_{n=1}^N \delta_{y_n}$. In the following, an empirical version of the criterion (6) is proposed in the case of discrete measures.

The formulation (6) requires the evaluation of a Wasserstein distance whose computation is not trivial in its original form, especially in high dimension. An alternative consists in considering its rewriting in the form of the *sliced-Wasserstein* (SW) distance. The idea underlying the SW distance is to represent a distribution defined in high dimension thanks to a set of projected one-dimensional distributions for which the computation of the Wasserstein distance is closed-form. Let p_X and p_Y denote the probability distributions of the random variables X and Y. For any vector on the unit sphere $u \in \mathbb{S}^{d-1}$, the projection operator $S_u : \mathbb{R}^d \to \mathbb{R}$ is defined as $S_u(x) \triangleq \langle u, x \rangle$. The SW distance of order $p \in [1, \infty)$ between p_X and p_Y can be written [4]

$$SW_p(p_X, p_Y) = \left(\int_{\mathbb{S}^{d-1}} W_p(S_{u\sharp}p_X, S_{u\sharp}p_Y)^p \, du \right)^{\frac{1}{p}} \tag{7}$$

where the distance $W_p(\cdot,\cdot)$ defining the integrand is now one-dimensional, leading to an explicit computation by inversion of the cumulative distribution functions. In the case where the distributions p_X and p_Y are represented by the respective samples $\mathbf{x}$ and $\mathbf{y}$, a numerical Monte Carlo approximation of the SW distance is

$$\widehat{SW}_p(\mathbf{x}, \mathbf{y}) = \frac{1}{J} \sum_{j=1}^{J} W_p \left(\frac{1}{N} \sum_{n=1}^{N} \delta_{S_{u_j}(x_n)}, \frac{1}{N} \sum_{n=1}^{N} \delta_{S_{u_j}(y_n)} \right) \tag{8}$$

where $u_1, \ldots, u_J$ are drawn uniformly on the sphere $\mathbb{S}^{d-1}$. The empirical form of the relaxation of the Monge problem (6) is then written as

$$\min_T \left\{ \widehat{SW}_p(T(\mathbf{x}), \mathbf{y}) + \lambda \sum_{n=1}^{N} c(x_n, T(x_n)) \right\} \tag{9}$$

where, with a slight abuse of notations, $T(\mathbf{x}) \triangleq \{T(x_n)\}_{n=1}^{N}$.

3 Normalizing Flows to Approximate OT

This section proposes to solve the problem (9) by restricting the class of the operator T to a class of invertible deep networks referred to as normalisation flows. The structure and the main properties of these networks are detailed in Sect. 3.1. The strategy proposed to train these networks to solve the problem (9) is then detailed in Sect. 3.2.

3.1 Normalizing Flows

Normalization flows are a flexible class of deep generative networks that intend to learn a change of variable between two probability distributions p_X and p_Y through an invertible transformation $T_{\boldsymbol{\Theta}} : X \mapsto Y = T_{\boldsymbol{\Theta}}(X)$ parametrized by $\boldsymbol{\Theta}$. In general, the distribution p_X is only known through samples $\mathbf{x} = \{x_n\}_{n=1}^{N}$ and, for tractability purpose, the distribution p_Y is chosen as a centered normal distribution with unit variance. The parameters $\boldsymbol{\Theta}$ defining the operator $T_{\boldsymbol{\Theta}}$ are then adjusted by maximizing the likelihood associated with the observations $\mathbf{x}$ according to the change of variable formula

$$p_X(x) = p_Y(T_\theta(x)) \left| \det J_{T_\theta^{-1}} \right| \tag{10}$$

with $J_{T_\theta^{-1}} = \frac{\partial T_\theta^{-1}}{\partial x}$. NF networks obey a cell-like structure, explicitly defining the operator $T_{\boldsymbol{\Theta}}(\cdot)$ as the composition of M functions $T_{\theta_m}^{(m)}$, usually referred to as *flows*, i.e.,

$$T_{\boldsymbol{\Theta}}(\cdot) = T_{\theta_M}^{(M)} \circ T_{\theta_{M-1}}^{(M-1)} \circ \ldots \circ T_{\theta_1}^{(1)}(\cdot) \tag{11}$$

with $\boldsymbol{\Theta} = \{\boldsymbol{\theta}_1, \ldots, \boldsymbol{\theta}_M\}$. In the following, to lighten notations, each sub-function contributing to the flow will be denoted by $T_m = T^{(m)}_{\boldsymbol{\theta}_m}$. In the present work, these functions are chosen as coupling layers as implemented by flows like RealNVP [10] and nonlinear independent component estimation (NICE) [9]. These coupling layers ensure an invertible transformation and an explicit expression of the Jacobian required in the change of variables (10). The input and output of the mth layer are related as

$$(y_{\text{id}}, y_{\text{ch}}) = T_m(x_{\text{id}}, x_{\text{ch}}) \tag{12}$$

with

$$\begin{cases} y_{\text{id}} &= x_{\text{id}} \\ y_{\text{ch}} &= (x_{\text{ch}} + \mathsf{D}_m(x_{\text{id}})) \odot \exp(\mathsf{E}_m(x_{\text{id}})) \end{cases} \tag{13}$$

where x_{id} and x_{ch} (resp. y_{id} and y_{ch}) are disjoint subsets of components of the input vector x (resp. the output vector y). The splitting of the input x into x_{id} and x_{ch} is achieved by a masking process such that $x_{\text{ch}} = \text{mask}(x)$ is transformed into a function of the unchanged part x_{id}. The scale function $\mathsf{E}_m(\cdot)$ and the offset function $\mathsf{D}_m(\cdot)$ are then described by neural networks whose parameters $\boldsymbol{\theta}_m$ need to be adjusted during the training. It is worth noting that imposing the flow-based architecture detailed in (11) will lead to an explicit discretization scheme of the transport map $T_{\boldsymbol{\Theta}}(\cdot)$ into a sequence of elementary transport functions $T_m(\cdot)$. As it will be shown in Sect. 3.3, this discretization has the great advantage of providing Wasserstein barycenters associated with the two measures μ and ν. Note that the proposed method is not limited to NFs composed of coupling layers such as RealNVP [10], NICE [9] or GLOW [14]. It can be generalized to other types of NFs, including free-form Jacobian of reversible dynamics (FFJORD) [12] and masked autoregressive flows (MAF) [20].

3.2 Loss Function

As mentioned before, the objective of this work is to learn a bijective operator relating any two distributions p_X and p_Y described by samples $\mathbf{x}$ and $\mathbf{y}$. The search for this operator is restricted to the class of invertible deep networks $T_{\boldsymbol{\Theta}}$ described in Sect. 3.1. The conventional strategy to train the network would be to maximize the likelihood defined by (10). However this approach cannot be implemented in the context of interest here since the base distribution p_Y is no longer explicitly given: it is only available through the knowledge of the set of samples $\mathbf{y}$. As a consequence, to adjust the weights of the network, the proposed alternative interprets the underlying learning task as the search for a transport map. Then a first idea would be to adjust these weights by directly solving the problem (9). However, to take advantage of the flow-based architecture of the operator $T_{\boldsymbol{\Theta}}(\cdot)$, it seems legitimate to equally distribute the transport efforts provided by each flow. Thus, the regularization in (9) will be instantiated for each elementary transformation $T_m(\cdot)$ associated to each flow of the network.

Moreover, when fitting deep learning-based models a major challenge arises from the stochastic nature of the optimization procedure, which imposes to use partial information (e.g., as mini-batches) to infer the whole structure of the optimization landscape. On top of that, the cost function to be optimized is not numerically constant since the approximation $\widehat{SW}$ of the SW distance in (9) depends on the precise set of random vectors $\{u_j\}_{j=1}^{J}$ drawn over the unit sphere. To alleviate these optimization difficulties, we propose to further regularize the objective function by penalizing the energy $|J_{T_m}(\cdot)|^2$ of the Jacobians associated with the transformations $T_m(\cdot)$, $m = 1, \ldots, M$. These Sobolev-like penalties promote regular operators $T_m(\cdot)$, promoting an overall operator $T_\theta(\cdot)$ regular itself [13]. In the context of optimal transport, this regularization has already been studied in depth in [17]. In that work, the author focused on the penalization of the Monge's formulation of OT by the ℓ_2-norm of the Jacobian. It stated the existence of an optimal transport map T solving the minimization problem

$$\inf_T \left\{ \int_{\mathcal{X}} \left(|T(x) - x|^2 + \gamma |J_T|^2 \right) T(x) \mathrm{d}x : T_{\#}\mu = \nu \right\} \tag{14}$$

This formulation of OT imposes the transport map T to be regular rather than deducing its regularity from its optimal properties.

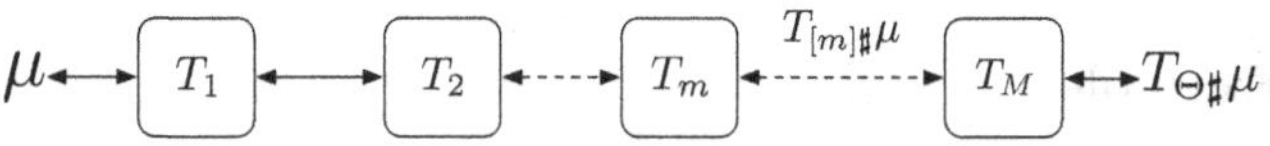

$\mu \leftrightarrow T_1 \leftrightarrow T_2 \leftrightarrow \cdots \leftrightarrow T_m \overset{T_{[m]\sharp}\mu}{\leftrightarrow} \cdots T_M \leftrightarrow T_{\Theta\sharp}\mu$

Fig. 1. Architecture of the proposed SWOT-Flow.

Finally, the training of the NF is carried out by minimizing the loss function

$$\underbrace{\widehat{SW}_p(\mathbf{x}, \mathbf{y})}_{\text{SW}} + \underbrace{\sum_{n=1}^{N} \sum_{m=1}^{M} \left[\lambda c(T_{m-1}(x_n), T_m(x_n)) + \gamma \left| J_{T_m}(x_n) \right|^2 \right]}_{\text{Reg}} \tag{15}$$

with $T_0(x_n) = x_n$. The proposed network, whose general architecture is depicted in Fig. 1, will be referred to as SWOT-Flow in what follows.

3.3 Intermediate Transports and Wasserstein Barycenters

As a consequence of the multiple-flow architecture (11) of the NF, the transport map operated by the proposed SWOT-Flow is a composition of the M individual flows $T_m(\cdot)$ $(m = 1, \ldots, M)$. Thus each flow implements an elementary transport and the composition of the first m flows defined as

$$T_{[m]}(\cdot) \triangleq T_m \circ \ldots \circ T_1(\cdot) \tag{16}$$

can be interpreted as an intermediate step of the transport map from the input measure μ towards the target measure ν, with $T_{[M]}(\cdot) \triangleq T_\Theta(\cdot)$. Interestingly, these intermediate transports can be related to Wasserstein barycenters between μ and ν defined by [1]

$$\inf_{\beta} \left\{ \alpha W_p(\mu, \beta) + (1-\alpha) W_p(\beta, \nu) \right\}. \tag{17}$$

Indeed, the next section dedicated to numerical experiments will empirically show that $T_{[m]\sharp}\mu$ approaches the solution of the problem (17) for the specific choice of the weight $\alpha = \frac{m}{M}$. In other words, the image measures provided by each intermediate transport operated by SWOT-Flow, i.e., as the outputs of each of the M flows, can legitimately be interpreted as Wasserstein barycenters.

4 Numerical Experiments

This section assesses the versatility and the accuracy of SWOT-Flow through two sets of numerical experiments. First, several toy experiments are presented to provide some insights about key ingredients of the proposed approach. Then the performance of SWOT-Flow is illustrated through the more realistic and challenging task of unsupervised alignment of word embeddings in natural language processing. The source code is publicly available on GitHub[1].

4.1 Toy Examples

In these experiments, the proposed framework SWOT-Flow is implemented and tested with synthetic data. In all experiments, the input distributions are described by the respective samples $\mathbf{x} = \{x_n\}_{n=1}^N$ and $\mathbf{y} = \{y_n\}_{n=1}^N$ such that $\mu = \frac{1}{N}\sum_{n=1}^N \delta_{x_n}$ and $\nu = \frac{1}{N}\sum_{n=1}^N \delta_{y_n}$ with $N = 20000$. The cost function $c(\cdot,\cdot)$ is chosen as the squared Euclidean distance, i.e., $c(x, y) = \|x - y\|_2^2$. However, it is worth noting that the proposed method is not limited to this Euclidean distance and can handle other costs defined on $\mathbb{R}^d$ or even on curved domains.

Implementation Details. The stochastic gradient descent used to solve (15) is implemented in Pytorch. We use Adam optimizer with learning rate 10^{-4} and a batch size of 4096 or 8192 samples. The NF implementing $T_\Theta(\cdot)$ is a RealNVP [10] for the example of Fig. 2 and an ActNorm type architecture network [14] for Fig. 3 and Fig. 4. It is composed of $M = 4$ flows, each composed of two four-layer neural networks corresponding to $\mathsf{D}_m(\cdot)$ and $\mathsf{E}_m(\cdot)$ $(d \to 8 \to 8 \to d)$ using hyperbolic tangent activation function. During training, the number J of slices drawn to approximate the SW distance in (8) has been progressively increased, starting from $J = 500$ to $J = 2000$ by step of 50 slices. At each epoch, new slices are uniformly drawn over the unit sphere and 100 epochs are carried out for each number of slices. The training procedure consist in 1) defining the loss function as the sole SW term in (15) from $J = 500$ to 1500 slices and then 2) incorporating the regularization term denoted as Reg in (15) where hyperparameters λ and γ are increased by a factor of 5% every step of 100 slices.

[1] FlorentinCDX/SWOT-Flow.

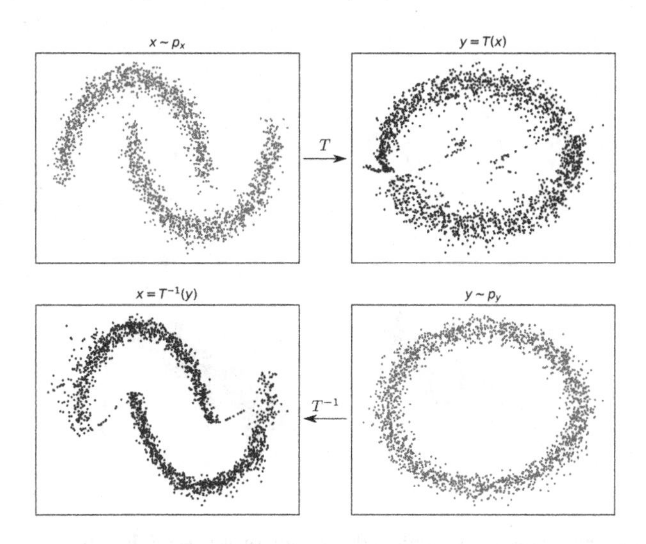

Fig. 2. Operator T learnt by SWOT-Flow when the base distribution p_X is a double-moon (top left) and the target distribution p_Y is a circle (bottom right).

Qualitative Results. As a first illustration of the flexibility of the proposed approach, Fig. 2 shows the results obtained after learning an operator T that transports a double moon-shaped distribution p_X (top left) to a circle-shaped distribution (bottom right). The empirical image measures $T_\sharp p_X$ (top right) and $T_\sharp^{-1} p_Y$ (bottom left) are obtained by applying the estimated $T(\cdot)$ operator or its inverse $T^{-1}(\cdot)$. It is worth noting that the difficulty inherent to this experiment lies in the respective disjoint and non-disjoint supports of the two distributions. Despite the regularity of the trained NF, a very good approximation of the OT is learnt, even in presence of this topological change.

Figure 3 aims at illustrating the relevance of the Sobolev-like regularization (i.e., the ℓ_2-norm of the Jacobian) included into the loss function (15) defined to train the NF. The first simulation protocol considers circle-shaped distributions while the second case considers rectangle-shaped distributions. In what follows, these two cases will be referred to as $\mathcal{P}_1$ and $\mathcal{P}_2$, respectively. In this experiment, the objective is to learn the transport map from an initial distribution p_X (light blue) to a target distribution p_Y (dark blue) which is translated for $\mathcal{P}_1$ and both translated and stretched for $\mathcal{P}_2$. The color gradient shows the outputs of the M successive flows of the network, i.e. the image measures $T_{[m]\sharp} p_X$ for $m = 1, \ldots, M$. In the absence of regularization (left), the successive elementary transports clearly suffer from multiple unexpected deformations (superfluous translations and dilations). In contrast, when the loss is complemented with the proposed Sobolev-type penalty (right), the learnt operator T is decomposed as a sequence of much more regular elementary transports. The resulting transport appears to be very close to optimal. In case $\mathcal{P}_1$, the expected translation is recovered, as well as the combined translation and stretching in case $\mathcal{P}_2$.

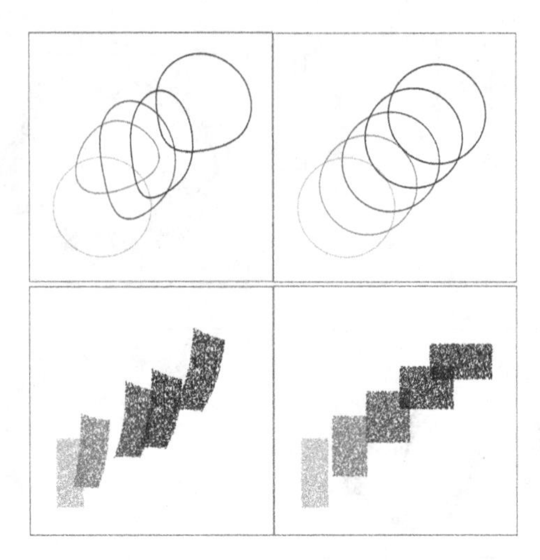

Fig. 3. Elementary transports achieved by the proposed NF when trained without (left) or with (right) the regularization for protocols $\mathcal{P}_1$ (top) and $\mathcal{P}_2$ (bottom). (Color figure online)

Table 1. Overall cost $\bar{C}$ and elementary costs $\bar{c}_m$ required by each flow $T_m(\cdot)$ of the NF trained with or without (w/o) regularization for protocols $\mathcal{P}_1$ (circle-shaped distributions) and $\mathcal{P}_2$ (rectangle-shaped distributions).

		$\bar{c}_1$	$\bar{c}_2$	$\bar{c}_3$	$\bar{c}_4$	$\bar{C}$
$\mathcal{P}_1$	w/o regularization	150.13	110.94	108.41	151.65	521.12
	with regularization	90.20	90.70	90.71	90.22	361.22
$\mathcal{P}_2$	w/o regularization	154.99	98.67	52.49	101.21	407.38
	with regularization	88.77	89.42	89.43	89.38	357.0

To be more precise quantitatively, Table 1 compares some metrics obtained when the NF has been trained using the regularization-free or regularized loss function, as defined in (15). For the two aforementioned simulation protocols, it reports the elementary costs

$$\bar{c}_m = \frac{1}{N}\sum_{n=1}^{N} \|T_{m-1}(x_n) - T_m(x_n)\|_2^2 \tag{18}$$

spent by each of the M flows $T_1(\cdot), \ldots, T_M(\cdot)$ to achieve the transport maps retrieved by SWOT-Flow. This table (last column) also reports the overall cost $\bar{C} = \sum_{m=1}^{M} \bar{c}_m$. For the two simulation protocols $\mathcal{P}_1$ and $\mathcal{P}_2$, these results clearly show cheaper transports when using the proposed regularization. For instance, for the simulation protocol $\mathcal{P}_1$, the overall cost is $\bar{C} = 360$ with the regularization, compared to $\bar{C} = 520$ when it is omitted. Moreover, when using the regularized loss function, this cost is distributed homogeneously over the successive flows, with a variation of at most ±1% from one flow to another, against ±20% otherwise.

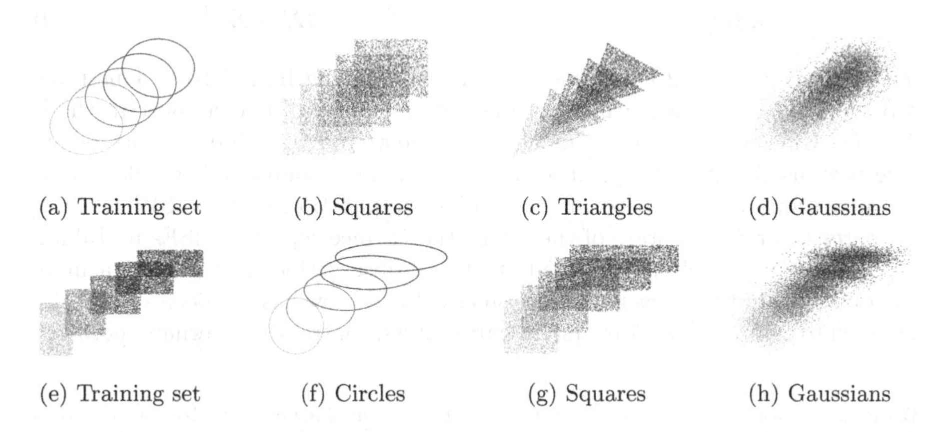

Fig. 4. Examples of transported data sets for protocols $\mathcal{P}_1$ (top) and $\mathcal{P}_1$ (bottom).

Figure 4 aims at illustrating the capacity of generalization of the transport map learnt by SWOT-Flow. In this experiment, SWOT-Flow has been trained following the simulation protocols $\mathcal{P}_1$ (Fig. 4a) or $\mathcal{P}_2$ (Fig. 4e). Once trained on the data set associated with each protocol, the NFs are fed with differently shaped data and the elementary transports are monitored as above. Figure 4b–d and 4g–h show the results when using square-, triangle-, Gaussian-shaped data sets for both protocols, respectively. As expected, all initial distributions are either simply translated in case $\mathcal{P}_1$ or translated and stretched in case $\mathcal{P}_2$. The intermediate distributions correspond to the expected barycenters as well. Figure 4 clearly demonstrates the generalization capacity of the proposed approach.

Multivariate Gaussians with Varying Dimensions. When the source and target distributions μ and ν of a transportation problem are multivariate Gaussians, the Wasserstein barycenters defined by (17) are also multivariate Gaussian distributions. In this case, an efficient fixed-point algorithm can be used to estimate its mean vector $\mathbf{a}$ and covariance matrix $\boldsymbol{\Sigma}$ [11]. This experiment capitalizes on this finding to assess the ability of SWOT-Flow to approximate Wasserstein barycenters, as stated in Sect. 3.3. To this end, the algorithm designed in [11] is implemented to estimate the actual barycenter associated with two prescribed multivariate Gaussian distributions for $\alpha = 1 - \alpha = \frac{1}{2}$. This barycenter is compared to the image measure $T_{[m]\sharp}\mu$ estimated by SWOT-Flow with $m = \frac{M}{2}$. More precisely, the mean vector and the covariance matrix of the barycenter are compared to their maximum likelihood estimates $\hat{\mathbf{a}}$ and $\hat{\boldsymbol{\Sigma}}$ computed from the samples $\{T_{[m]}(x_n)\}_{n=1}^N$ transported by the first m flows. The resulting mean square errors (MSEs)

$$\mathrm{MSE}(\mathbf{a}) = \|\mathbf{a} - \hat{\mathbf{a}}\|_2^2 \quad \text{and} \quad \mathrm{MSE}(\boldsymbol{\Sigma}) = \|\boldsymbol{\Sigma} - \hat{\boldsymbol{\Sigma}}\|_\mathrm{F}^2 \tag{19}$$

are reported in Table 2 for varying dimensions ranging from 2 to 8. This table also reports the MSEs reached by other state-of-the-art free-support methods [5,8,16]. For the methods [8] and [5], $n = 5000$ and $n = 100$ support points have been used, respectively, since these are the maximum numbers allowed for the algorithms to terminate in reasonable computational times. SWOT-Flow compares favorably to state-of-the-art methods since reported MSEs in Table 2 appear to be most often the smallest. These observation may call for a more general study, but remains noticeable since SWOT-Flow has not been specifically designed to compute the Wasserstein barycenters, contrary to alternate methods.

Table 2. Performance of the estimation of the median barycenters. Reported scores result from the average over 5 Monte Carlo runs.

			[8]	[5]	[16]	SWOT-Flow
Dimension	2	MSE($\mathbf{a}$)	$9.99 \cdot 10^{-5}$	$3.14 \cdot 10^{-4}$	$1.17 \cdot 10^{-4}$	$\mathbf{8.09 \cdot 10^{-5}}$
		MSE($\boldsymbol{\Sigma}$)	$7.28 \cdot 10^{-4}$	$2.39 \cdot 10^{-3}$	$1.98 \cdot 10^{-3}$	$\mathbf{1.44 \cdot 10^{-4}}$
	4	MSE($\mathbf{a}$)	$1.73 \cdot 10^{-3}$	$1.68 \cdot 10^{-3}$	$1.44 \cdot 10^{-3}$	$\mathbf{1.44 \cdot 10^{-4}}$
		MSE($\boldsymbol{\Sigma}$)	$1.35 \cdot 10^{-2}$	$2.50 \cdot 10^{-2}$	$1.22 \cdot 10^{-2}$	$\mathbf{3.61 \cdot 10^{-4}}$
	6	MSE($\mathbf{a}$)	$2.04 \cdot 10^{-3}$	$\mathbf{2.58 \cdot 10^{-3}}$	$3.24 \cdot 10^{-3}$	$1.23 \cdot 10^{-2}$
		MSE($\boldsymbol{\Sigma}$)	$4.38 \cdot 10^{-2}$	$8.86 \cdot 10^{-2}$	$2.37 \cdot 10^{-2}$	$\mathbf{5.29 \cdot 10^{-4}}$
	8	MSE($\mathbf{a}$)	$\mathbf{1.23 \cdot 10^{-3}}$	$1.48 \cdot 10^{-3}$	$3.14 \cdot 10^{-3}$	$1.29 \cdot 10^{-2}$
		MSE($\boldsymbol{\Sigma}$)	$8.31 \cdot 10^{-2}$	$1.64 \cdot 10^{-1}$	$4.23 \cdot 10^{-2}$	$\mathbf{2.22 \cdot 10^{-3}}$

4.2 Unsupervised Word Translation

In a second set of experiments, the performance of SWOT-Flow has been assessed on the task of unsupervised word translation. Given word embeddings trained on two monolingual corpora, the goal is to infer a bilingual dictionary by aligning the corresponding word vectors.

Experiment Description. This experiment considers the task of aligning two sets of points in high dimension. More precisely, it aims at inferring a bilingual lexicon, without supervision, by aligning word embeddings trained on monolingual data. FastText [3] has been implemented to learn the word vectors used for representation. It provides monolingual embeddings of dimension 300 trained on Wikipedia corpora. Words are lower-cased, and those that appear less than 5 times are discarded for training. As a post-processing step, only the first 50k most frequent words are selected in the reported experiments.

Architecture. The proposed SWOT-Flow method has been implemented using a RealNVP architecture. The scale function $\mathsf{E}_m(\cdot)$ and the offset function $\mathsf{D}_m(\cdot)$ are multilayer neural networks with two hidden layers of size 512 and hyperbolic tangent activation function. Adam has been used as an optimizer with a learning rate of $1 \cdot 10^{-3}$. The number of slices involved in the Monte Carlo approximation of the SW distance in (8) has been progressively increased from $J = 500$ slices to $J = 3000$ by steps of 50. For each number of slices, 100 epochs have been performed. The hyperparameters λ and γ adjusting the weights of the composite regularization have been increased by a factor of 5% every steps of 500 slices.

Table 3. Comparison of accuracies obtained by SWOT-Flow and adv-net [6] for unsupervised word translation ('en' is English, 'fr' is French, 'de' is German, 'ru' is Russian).

Method		en-es	es-en	en-fr	fr-en	en-de	de-en	en-ru	ru-en
SWOT-Flow	20-NN	37.4	24.2	46.6	34.1	44.4	27.6	14.4	3.8
	10-NN	**33.5**	**22.5**	**42.5**	32.5	39.5	26.8	**10.2**	2.1
adv-net [6]	10-NN	31.4	21.2	39.6	**35.1**	**40.1**	**27.1**	7.1	**2.3**

Main Results. To quantitatively measure the quality of SWOT-Flow, the problem of bilingual lexicon induction is addressed, with the same setting as in [6]. The same evaluation data sets and codes, as well as the same word vectors have been used. Given an input word embedding ($n = 1, \ldots, N$ with $N_{\text{test}} = 1000$) in a given language, the objective is to assess if its counterpart $T(x_n)$ transported by SWOT-Flow belongs to the close neighborhood of the output word embedding y_n in the target language. The neighborhood $\mathcal{V}(y_n)$ is defined as the set of K-nearest neighbors computed in a cosine similarity sense with $K = 10$ or 20 in dimension 300. The overall accuracy is computed as the percentage of correctly transported input samples. Denoting by $\mathbf{1}_{\mathsf{A}}$ the indicator function, i.e., $\mathbf{1}_{\mathsf{A}} = 1$ if the assertion A is true and $\mathbf{1}_{\mathsf{A}} = 0$ otherwise,

$$\text{accuracy} = \frac{1}{N_{\text{test}}} \sum_{n=1}^{N_{\text{test}}} \mathbf{1}_{\{T(x_n) \in \mathcal{V}(y_n)\}} \times 100 \ (\%) \tag{20}$$

Table 3 reports the accuracy scores for several pairs of languages. Although SWOT-Flow has not been specifically designed to perform word translation, these results show that its overall performance is on par with the adversarial network (adv-net) proposed specifically for this task in [6]. In particular, SWOT-Flow seems to perform well for translation between languages with close origins.

Figure 5 qualitatively illustrates this good performance by showing how close a set of translated words $T(x_n)$ are to their true translation y_n. This representation is obtained by a classical projection on the 2 first PCA components of the target embedded space. The translation of 5 specific words from English to French or German fall in the close vicinity of their true counterparts.

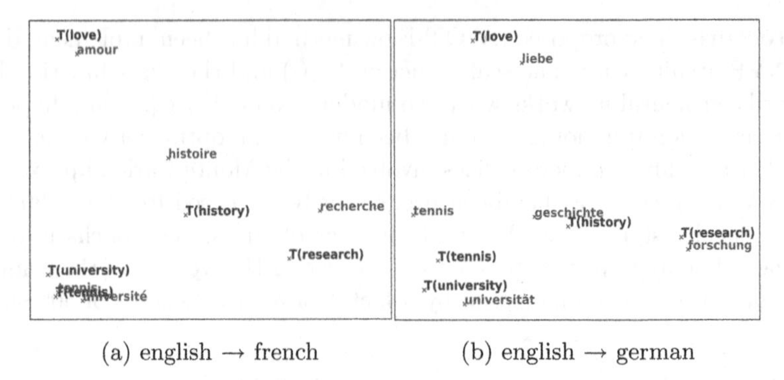

(a) english → french (b) english → german

Fig. 5. 2D PCA representation of the target word embedding space: the targeted translated (in green) and the transported source (in blue) embedded words. (Color figure online)

5 Discussion

Cycle Consistency. Cycle consistency, as proposed in CycleGAN [23], aims at learning meaningful cross-domain mappings such that the data translated from the domain $\mathcal{X}$ to the domain $\mathcal{Y}$ via $T_{\mathcal{X}\to\mathcal{Y}}$ can be mapped back to the original data points in $\mathcal{X}$ via $T_{\mathcal{Y}\to\mathcal{X}}$. That is, $T_{\mathcal{Y}\to\mathcal{X}} \circ T_{\mathcal{X}\to\mathcal{Y}}(x) \approx x$ for all $x \in \mathcal{X}$. For CycleGan, and many other domain transfer models such as [2], this key property should be enforced by including a cycle consistency term into the loss function. Conversely, since NF-based generative models learn bijective mappings, the proposed SWOT-Flow inherits the cycle consistency property by construction.

Semi-discrete Formulation. The proposed SWOT-Flow framework has been explicitly derived to approximate OT between two discrete empirical distributions. It can be instantiated to perform semi-discrete OT, i.e., to handle the case where one of distribution is not described by data points but rather given as an explicit continuous probability measure. Instead of relaxing the Monge formulation (3) as in (6), it would consist in replacing the SW distance with a log-likelihood term $\log f_\nu(\cdot)$ associated with the target continuous measure. The loss function in (15) would be replaced by

$$-\sum_{n=1}^{N} \log f_\nu(T_\Theta(x_n)) + \sum_{n=1}^{N}\sum_{m=1}^{M} \left[\lambda c(T_{m-1}(x_n), T_m(x_n)) + \gamma \left|J_{T_m}(x_n)\right|^2\right] \quad (21)$$

where the log-likelihood term is evaluated at the data points $\{T_\Theta(x_n)\}_{n=1}^{N}$ transported by the NF.

NF to Approximate Barycenters. As discussed in Sect. 3.3 and experimentally illustrated in Sect. 4.1, the flow-based architecture of the SWOT-Flow network leads to intermediate transports, that can be related to Wasserstein barycenters. On the toy Gaussian example considered in Sect. 4.1, SWOT-Flow provides good approximation of the barycenters, although it has not been specifically designed to perform this task. If one is interested in devising a NF approximating these barycenters, the definition (17) would lead to the optimization problem

$$\inf_{T} \left\{ \sum_{m=1}^{M} \alpha_m W_p(\mu, T_{[m]\sharp}\mu) + (1 - \alpha_m) W_p(T_{[m]\sharp}\mu, \nu) \right\} \tag{22}$$

with $\alpha_m = \frac{m}{M}$. When handling empirical measures described by samples, the subsequent discretization would require to replace both terms with Monte Carlo approximations (8) of the SW distances. However, this would lead to a computationally demanding training procedure.

6 Conclusion

We propose a new method to learn the optimal transport map between two empirical distributions from sets of available samples. To this aim, we write a relaxed and penalized formulation of the Monge problem. This formulation is used to build a loss function that balances between the cost of the transport and the proximity in Wasserstein distance between the transported base distribution and the target one. The proposed approach relies on normalizing flows, a family of invertible neural networks. Up to our knowledge, this is the first method that is able to learn such a generalizable transport operator. As a side benefit, the multiple flow architecture of the proposed network interestingly yields intermediate transports and Wasserstein barycenters. The proposed method is illustrated by numerical experiments on toy examples as well as an unsupervised word translation task. Future work will aim at extending these results to high dimensional applications.

Acknowledgments. This work was supported by the Artificial Natural Intelligence Toulouse Institute (ANITI, ANR-19-PI3A-0004), the AI Sherlock Chair (ANR-20-CHIA-0031-01), the ULNE national future investment programme (ANR-16-IDEX-0004) and the Hauts-de-France Region.

References

1. Agueh, M., Carlier, G.: Barycenters in the Wasserstein space. SIAM J. Math. Anal. **43**(2), 904–924 (2011)
2. de Bézenac, E., Ayed, I., Gallinari, P.: Cyclegan through the lens of (dynamical) optimal transport. In: Oliver, N., Pérez-Cruz, F., Kramer, S., Read, J., Lozano, J.A. (eds.) ECML PKDD 2021, pp. 132–147. Springer, Cham (2021). https://doi.org/10.1007/978-3-030-86520-7_9
3. Bojanowski, P., Grave, E., Joulin, A., Mikolov, T.: Enriching word vectors with subword information. Trans. Assoc. Comput. Linguist. **5**, 135–146 (2017)
4. Bonneel, N., Rabin, J., Peyré, G., Pfister, H.: Sliced and radon Wasserstein barycenters of measures. J. Math. Imaging Vis. **51**(1), 22–45 (2015)
5. Claici, S., Chien, E., Solomon, J.M.: Stochastic Wasserstein barycenters. In: Proceedings of International Conference on Machine Learning (ICML) (2018)

6. Conneau, A., Lample, G., Ranzato, M., Denoyer, L., Jégou, H.: Word translation without parallel data. In: Proceedings of IEEE International Conference on Learning Representations (ICLR) (2018)
7. Courty, N., Flamary, R., Tuia, D.: Domain adaptation with regularized optimal transport. In: Calders, T., Esposito, F., Hüllermeier, E., Meo, R. (eds.) ECML PKDD 2014. LNCS (LNAI), vol. 8724, pp. 274–289. Springer, Heidelberg (2014). https://doi.org/10.1007/978-3-662-44848-9_18
8. Cuturi, M., Doucet, A.: Fast computation of Wasserstein barycenters. In: Xing, E.P., Jebara, T. (eds.) Proceedings of International Conference on Machine Learning (ICML) (2014)
9. Dinh, L., Krueger, D., Bengio, Y.: NICE: non-linear independent components estimation. In: Bengio, Y., LeCun, Y. (eds.) Proceedings of IEEE International Conference on Learning Representations (ICLR) (2015)
10. Dinh, L., Sohl-Dickstein, J., Bengio, S.: Density estimation using real NVP. In: Proceedings of IEEE International Conference on Learning Representations (ICLR) (2017)
11. Álvarez Esteban, P.C., del Barrio, E., Cuesta-Albertos, J., Matrán, C.: A fixed-point approach to barycenters in Wasserstein space. J. Math. Anal. Appl. **441**(2), 744–762 (2016)
12. Grathwohl, W., Chen, R.T.Q., Bettencourt, J., Sutskever, I., Duvenaud, D.: FFJORD: free-form continuous dynamics for scalable reversible generative models. In: Proceedings of IEEE International Conference on Learning Representations (ICLR) (2019)
13. Hoffman, J., Roberts, D.A., Yaida, S.: Robust learning with jacobian regularization. arXiv (2020)
14. Kingma, D.P., Dhariwal, P.: Glow: generative flow with invertible 1x1 convolutions. In: Advances in Neural Information Processing Systems (NeurIPS) (2018)
15. Kobyzev, I., Prince, S.J., Brubaker, M.A.: Normalizing flows: an introduction and review of current methods. IEEE Trans. Patt. Anal. Mach. Intell. **43**(11), 3964–3979 (2020)
16. Korotin, A., Li, L., Solomon, J., Burnaev, E.: Continuous Wasserstein-2 barycenter estimation without minimax optimization. In: Proceedings of IEEE International Conference on Learning Representations (ICLR) (2021)
17. Louet, J.: Problèmes de transport optimal avec pénalisation en gradient. Ph.D. thesis, Université Paris-Sud, France (2014)
18. Monge, G.: Mémoire sur la théorie des déblais et des remblais. Histoire de l'Académie Royale des Sciences de Paris (1781)
19. Papamakarios, G., Nalisnick, E., Rezende, D.J., Mohamed, S., Lakshminarayanan, B.: Normalizing flows for probabilistic modeling and inference. J. Mach. Learn. Res. **22**(57), 1–64 (2021)
20. Papamakarios, G., Pavlakou, T., Murray, I.: Masked autoregressive flow for density estimation. In: Advances in Neural Information Processing Systems (NeurIPS) (2017)
21. Paulin, L., et al.: Sliced optimal transport sampling. ACM Trans. Graph. (Proc. SIGGRAPH) (2020)
22. Peyré, G., Cuturi, M.: Computational optimal transport: with applications to data science. Found. Trends® Mach. Learn. **11**(5–6), 355–607 (2019)
23. Zhu, J.Y., Park, T., Isola, P., Efros, A.A.: Unpaired image-to-image translation using cycle-consistent adversarial networks. In: Proceedings of the IEEE International Conference on Computer Vision (ICCV) (2017)

Feature-Robust Optimal Transport for High-Dimensional Data

Mathis Petrovich[1], Chao Liang[2], Ryoma Sato[3], Yanbin Liu[4], Yao-Hung Hubert Tsai[5], Linchao Zhu[4], Yi Yang[2], Ruslan Salakhutdinov[5], and Makoto Yamada[3](✉)

[1] École normale supérieure Paris-Saclay, Cachan, France
[2] Zhejiang University, Hangzhou, China
[3] Kyoto University, RIKEN AIP, Kyoto, Japan
yamada.makoto.8m@kyoto-u.ac.jp
[4] University of Technology Sydney, Ultimo, Australia
[5] Carnegie Mellon University, Pittsburgh, USA

Abstract. Optimal transport is a machine learning problem with applications including distribution comparison, feature selection, and generative adversarial networks. In this paper, we propose feature-robust optimal transport (FROT) for high-dimensional data, which solves high-dimensional OT problems using feature selection to avoid the curse of dimensionality. Specifically, we find a transport plan with discriminative features. To this end, we formulate the FROT problem as a min–max optimization problem. We then propose a convex formulation of the FROT problem and solve it using a Frank–Wolfe-based optimization algorithm, whereby the subproblem can be efficiently solved using the Sinkhorn algorithm. Since FROT finds the transport plan from selected features, it is robust to noise features. To show the effectiveness of FROT, we propose using the FROT algorithm for the layer selection problem in deep neural networks for semantic correspondence. By conducting synthetic and benchmark experiments, we demonstrate that the proposed method can find a strong correspondence by determining important layers. We show that the FROT algorithm achieves state-of-the-art performance in real-world semantic correspondence datasets. Code can be found at https://github.com/Mathux/FROT.

Keywords: Optimal transport · Feature selection

1 Introduction

Optimal transport (OT) is a machine learning problem with several applications in the computer vision and natural language processing communities. The

M. Petrovich and C. Liang—The first two authors contributed equally.

Supplementary Information The online version contains supplementary material available at https://doi.org/10.1007/978-3-031-26419-1_18.

© The Author(s), under exclusive license to Springer Nature Switzerland AG 2023
M.-R. Amini et al. (Eds.): ECML PKDD 2022, LNAI 13717, pp. 291–307, 2023.
https://doi.org/10.1007/978-3-031-26419-1_18

applications include Wasserstein distance estimation (Peyré et al. 2019), domain adaptation (Yan et al. 2018), multitask learning (Janati et al. 2019), barycenter estimation (Cuturi and Doucet 2014), semantic correspondence (Liu et al. 2020), feature matching (Sarlin et al. 2020), and photo album summarization (Liu et al. 2021). The OT problem is extensively studied in the computer vision community as the earth mover's distance (EMD) (Rubner et al. 2000). However, the computational cost of EMD is cubic and highly expensive. Recently, the entropic regularized EMD problem was proposed; this problem can be solved using the Sinkhorn algorithm with a quadratic cost (Cuturi 2013). Owing to the development of the Sinkhorn algorithm, researchers have replaced the EMD computation with its regularized counterparts. However, the optimal transport problem for high-dimensional data has remained unsolved for many years.

Recently, a robust variant of the OT was proposed for high-dimensional OT problems and used for divergence estimation (Paty and Cuturi 2019, 2020). In the robust OT framework, the transport plan is computed with the discriminative subspace of the two data matrices $\boldsymbol{X} \in \mathbb{R}^{d \times n}$ and $\boldsymbol{Y} \in \mathbb{R}^{d \times m}$. The subspace can be obtained using dimensionality reduction. An advantage of the subspace robust approach is that it does not require prior information about the subspace. However, given prior information such as feature groups, we can consider a computationally efficient formulation. The computation of the subspace can be expensive if the dimensionality of data is high (e.g., 10^4).

One of the most common prior information items is a feature group. The use of group features is popular in feature selection problems in the biomedical domain and has been extensively studied in Group Lasso (Yuan and Lin 2006). The key idea of Group Lasso is to prespecify the group variables and select the set of group variables using the group norm (also known as the sum of ℓ_2 norms). For example, if we use a pretrained neural network as a feature extractor and compute OT using the features, then we require careful selection of important layers to compute OT. Specifically, each layer output is regarded as a grouped input. Therefore, using a feature group as prior information is a natural setup and is important for considering OT for deep neural networks (DNNs).

In this paper, we propose a high-dimensional optimal transport method by utilizing prior information in the form of grouped features. Specifically, we propose a feature-robust optimal transport (FROT) problem, for which we select distinct group feature sets to estimate a transport plan instead of determining its distinct subsets, as proposed in (Paty and Cuturi 2019, 2020). To this end, we formulate the FROT problem as a min–max optimization problem and transform it into a convex optimization problem, which can be accurately solved using the Frank–Wolfe algorithm (Frank and Wolfe 1956; Jaggi 2013). The FROT's subproblem can be efficiently solved using the Sinkhorn algorithm (Cuturi 2013). An advantage of FROT is that it can yield a transport plan from high-dimensional data using feature selection, using which the significance of the features is obtained without any additional cost. Therefore, the FROT formulation is highly suited for high-dimensional OT problems. Moreover, we show the connection between FROT and the L1 regularized OT problem; this result supports the ability of FROT to select features and robustness of FROT. Through synthetic

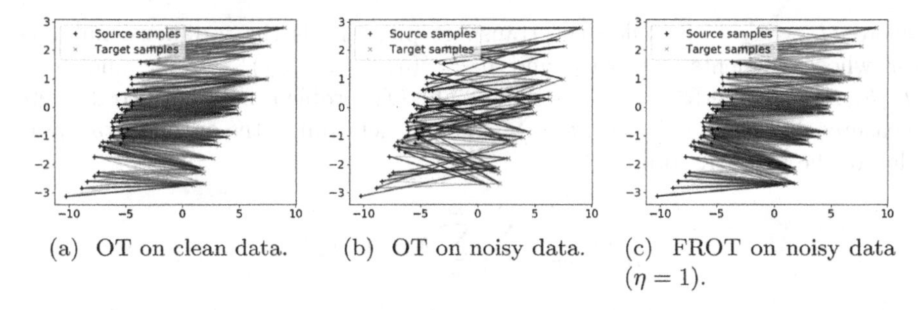

(a) OT on clean data. (b) OT on noisy data. (c) FROT on noisy data ($\eta = 1$).

Fig. 1. Transport plans between two synthetic distributions with 10-dimensional vectors $\widetilde{\boldsymbol{x}} = (\boldsymbol{x}^\top, \boldsymbol{z}_x^\top)^\top$, $\widetilde{\boldsymbol{y}} = (\boldsymbol{y}^\top, \boldsymbol{z}_y^\top)^\top$, where two-dimensional vectors $\boldsymbol{x} \sim N(\boldsymbol{\mu}_x, \boldsymbol{\Sigma}_x)$ and $\boldsymbol{y} \sim N(\boldsymbol{\mu}_y, \boldsymbol{\Sigma}_y)$ are true features; and $\boldsymbol{z}_x \sim N(\boldsymbol{0}_8, \boldsymbol{I}_8)$ and $\boldsymbol{z}_y \sim N(\boldsymbol{0}_8, \boldsymbol{I}_8)$ are noisy features. (a) OT between distribution $\boldsymbol{x}$ and $\boldsymbol{y}$ is a reference. (b) OT between distribution $\widetilde{\boldsymbol{x}}$ and $\widetilde{\boldsymbol{y}}$. (c) FROT transport plan between distribution $\widetilde{\boldsymbol{x}}$ and $\widetilde{\boldsymbol{y}}$ where true features and noisy features are grouped, respectively.

experiments, we initially demonstrate that the proposed FROT is robust to noise dimensions (See Fig. 1). Furthermore, we apply FROT to a semantic correspondence problem (Liu et al. 2020) and show that the proposed algorithm achieves SOTA performance.

Contribution:

- We propose a feature robust optimal transport (FROT) problem and derive a simple and efficient Frank–Wolfe based algorithm. Furthermore, we propose a feature-robust Wasserstein distance (FRWD).
- We show the connection between FROT and the L1 regularized OT problem; this result supports the ability of FROT to select features and robustness of FROT.
- We apply FROT to a high-dimensional feature selection problem and show that FROT is consistent with the Wasserstein distance-based feature selection algorithm with less computational cost than the original algorithm.
- We used FROT for the layer selection problem in a semantic correspondence problem and showed that the proposed algorithm outperforms existing baseline algorithms.

2 Background

In this section, we briefly introduce the OT problem.

Optimal Transport (OT): The following are given: independent and identically distributed (i.i.d.) samples $\boldsymbol{X} = \{\boldsymbol{x}_i\}_{i=1}^n \in \mathbb{R}^{d \times n}$ from a d-dimensional distribution p, and i.i.d. samples $\boldsymbol{Y} = \{\boldsymbol{y}_j\}_{j=1}^m \in \mathbb{R}^{d \times m}$ from the d-dimensional distribution q. In the Kantorovich relaxation of OT, admissible couplings are defined by the set of the transport plan:

$$\boldsymbol{U}(\mu, \nu) = \{\boldsymbol{\Pi} \in \mathbb{R}_+^{n \times m} : \boldsymbol{\Pi} \boldsymbol{1}_m = \boldsymbol{a}, \boldsymbol{\Pi}^\top \boldsymbol{1}_n = \boldsymbol{b}\},$$

where $\boldsymbol{\Pi} \in \mathbb{R}_{+}^{n \times m}$ is called the transport plan, $\mathbf{1}_n$ is the n-dimensional vector whose elements are ones, and $\boldsymbol{a} = (a_1, a_2, \dots, a_n)^\top \in \mathbb{R}_{+}^{n}$ and $\boldsymbol{b} = (b_1, b_2, \dots, b_m)^\top \in \mathbb{R}_{+}^{m}$ are the weights. The OT problem between two discrete measures $\mu = \sum_{i=1}^{n} a_i \delta_{\boldsymbol{x}_i}$ and $\nu = \sum_{j=1}^{m} b_j \delta_{\boldsymbol{y}_j}$ determines the optimal transport plan of the following problem:

$$\min_{\boldsymbol{\Pi} \in U(\mu,\nu)} \sum_{i=1}^{n} \sum_{j=1}^{m} \pi_{ij} c(\boldsymbol{x}_i, \boldsymbol{y}_j), \tag{1}$$

where $c(\boldsymbol{x}, \boldsymbol{y})$ is a cost function. For example, the squared Euclidean distance is used, that is, $c(\boldsymbol{x}, \boldsymbol{y}) = \|\boldsymbol{x} - \boldsymbol{y}\|_2^2$. To solve the OT problem, Eq. (1) (also known as the earth mover's distance) using linear programming requires $O(n^3)$, $(n = m)$ computation, which is computationally expensive. To address this, an entropic-regularized optimal transport is used (Cuturi 2013).

$$\min_{\boldsymbol{\Pi} \in U(\mu,\nu)} \sum_{i=1}^{n} \sum_{j=1}^{m} \pi_{ij} c(\boldsymbol{x}_i, \boldsymbol{y}_j) - \epsilon H(\boldsymbol{\Pi}),$$

where $\epsilon \geq 0$ is the regularization parameter, and $H(\boldsymbol{\Pi}) = -\sum_{i=1}^{n} \sum_{j=1}^{m} \pi_{ij} (\log(\pi_{ij}) - 1)$ is the entropic regularization. If $\epsilon = 0$, then the regularized OT problem reduces to the EMD problem. Owing to entropic regularization, the entropic regularized OT problem can be accurately solved using Sinkhorn iteration (Cuturi 2013) with a $O(nm)$ computational cost (See Algorithm 2 in the supplementary material.).

Wasserstein Distance: If the cost function is defined as $c(\boldsymbol{x}, \boldsymbol{y}) = d(\boldsymbol{x}, \boldsymbol{y})$ with $d(\boldsymbol{x}, \boldsymbol{y})$ as a distance function and $p \geq 1$, then we define the p-Wasserstein distance of two discrete measures $\mu = \sum_{i=1}^{n} a_i \delta_{\boldsymbol{x}_i}$ and $\nu = \sum_{j=1}^{m} b_j \delta_{\boldsymbol{y}_j}$ as

$$W_p(\mu, \nu) = \left(\min_{\boldsymbol{\Pi} \in U(\mu,\nu)} \sum_{i=1}^{n} \sum_{j=1}^{m} \pi_{ij} d(\boldsymbol{x}_i, \boldsymbol{y}_j)^p \right)^{1/p}.$$

Recently, a robust variant of the Wasserstein distance, called the subspace robust Wasserstein distance (SRW), was proposed (Paty and Cuturi 2019). The SRW computes the OT problem in the discriminative subspace. This can be determined by solving dimensionality-reduction problems. Owing to the robustness, it can compute the Wasserstein from noisy data. The SRW is given as

$$\mathrm{SRW}(\mu, \nu) = \left(\min_{\boldsymbol{\Pi} \in U(\mu,\nu)} \max_{\boldsymbol{U} \in \mathbb{R}^{d \times k}} \sum_{i=1}^{n} \sum_{j=1}^{m} \pi_{ij} \|\boldsymbol{U}^\top \boldsymbol{x}_i - \boldsymbol{U}^\top \boldsymbol{y}_j\|_2^2 \right)^{\frac{1}{2}},$$

where $\boldsymbol{U}$ is the orthonormal matrix with $k \leq d$, and $\boldsymbol{I}_k \in \mathbb{R}^{k \times k}$ is the identity matrix. The SRW or its relaxed problem can be efficiently estimated using either eigenvalue decomposition or the Frank–Wolfe algorithm.

3 Proposed Method

This paper proposes FROT. We assume that the vectors are grouped as $\boldsymbol{x} = (\boldsymbol{x}^{(1)^\top}, \ldots, \boldsymbol{x}^{(L)^\top})^\top$ and $\boldsymbol{y} = (\boldsymbol{y}^{(1)^\top}, \ldots, \boldsymbol{y}^{(L)^\top})^\top$. Here, $\boldsymbol{x}^{(\ell)} \in \mathbb{R}^{d_\ell}$ and $\boldsymbol{y}^{(\ell)} \in \mathbb{R}^{d_\ell}$ are the d_ℓ dimensional vectors, where $\sum_{\ell=1}^{L} d_\ell = d$. This setting is useful if we know the explicit group structure for the feature vectors a priori. In an application in L-layer neural networks, we consider $\boldsymbol{x}^{(\ell)}$ and $\boldsymbol{y}^{(\ell)}$ as outputs of the ℓth layer of the network. If we do not have a priori information, we can consider each feature independently (i.e., $d_1 = d_2 = \ldots = d_L = 1$ and $L = d$). All proofs in this section are provided in the supplementary material.

3.1 Feature-Robust Optimal Transport (FROT)

The FROT formulation is given by

$$\mathrm{FROT}(\mu, \nu) = \min_{\boldsymbol{\Pi} \in U(\mu,\nu)} \max_{\boldsymbol{\alpha} \in \boldsymbol{\Sigma}^L} \sum_{i=1}^{n} \sum_{j=1}^{m} \pi_{ij} \sum_{\ell=1}^{L} \alpha_\ell c(\boldsymbol{x}_i^{(\ell)}, \boldsymbol{y}_j^{(\ell)}), \quad (2)$$

where $\boldsymbol{\Sigma}^L = \{\boldsymbol{\alpha} \in \mathbb{R}_+^L : \boldsymbol{\alpha}^\top \mathbf{1}_L = 1\}$ is the probability simplex. The underlying concept of FROT is to estimate the transport plan $\boldsymbol{\Pi}$ using distinct groups with large distances between $\{\boldsymbol{x}_i^{(\ell)}\}_{i=1}^n$ and $\{\boldsymbol{y}_j^{(\ell)}\}_{j=1}^m$. We note that determining the transport plan in nondistinct groups is difficult because the data samples in $\{\boldsymbol{x}_i^{(\ell)}\}_{i=1}^n$ and $\{\boldsymbol{y}_j^{(\ell)}\}_{j=1}^m$ overlap. By contrast, in distinct groups, $\{\boldsymbol{x}_i^{(\ell)}\}_{i=1}^n$ and $\{\boldsymbol{y}_j^{(\ell)}\}_{j=1}^m$ are different, and this aids in determining an optimal transport plan. This is an intrinsically similar idea to the subspace robust Wasserstein distance (Paty and Cuturi 2019), which estimates the transport plan in the discriminative subspace, while our approach selects important groups. Therefore, FROT can be regarded as a feature selection variant of the vanilla OT problem in Eq. (1), whereas the subspace robust version uses dimensionality-reduction counterparts.

Using FROT, we can define a p-feature robust Wasserstein distance (p-FRWD).

Proposition 1. *For the distance function* $d(\boldsymbol{x}, \boldsymbol{y})$,

$$\mathrm{FRWD}_p(\mu, \nu) = \left(\min_{\boldsymbol{\Pi} \in U(\mu,\nu)} \max_{\boldsymbol{\alpha} \in \boldsymbol{\Sigma}^L} \sum_{i=1}^{n} \sum_{j=1}^{m} \pi_{ij} \sum_{\ell=1}^{L} \alpha_\ell d(\boldsymbol{x}_i^{(\ell)}, \boldsymbol{y}_j^{(\ell)})^p \right)^{1/p}, \quad (3)$$

is a distance for $p \geq 1$.

Note that we can show that FRWD_2 is a special case of SRW with $d(\boldsymbol{x}, \boldsymbol{y}) = \|\boldsymbol{x} - \boldsymbol{y}\|_2$ (See the supplementary material). Another difference between SRW and FRWD is that FRWD can use any distance, while SRW can only use $d(\boldsymbol{x}, \boldsymbol{y}) = \|\boldsymbol{x} - \boldsymbol{y}\|_2$. Moreover, FRWD_p can be regarded as a special case of the min-max optimal transport (Dhouib et al. 2020). A contribution of this paper is first to introduce feature selection using min-max optimal transport.

3.2 FROT Optimization

Here, we propose two FROT algorithms based on the Frank–Wolfe algorithm.

Frank–Wolfe: We propose a continuous variant of the FROT algorithm using the Frank–Wolfe algorithm, which is fully differentiable. To this end, we introduce entropic regularization for $\boldsymbol{\alpha}$ and rewrite the FROT as a function of $\boldsymbol{\Pi}$:

$$\min_{\boldsymbol{\Pi} \in U(\mu,\nu)} \max_{\boldsymbol{\alpha} \in \boldsymbol{\Sigma}^L} J_\eta(\boldsymbol{\Pi}, \boldsymbol{\alpha}),$$

$$\text{with } J_\eta(\boldsymbol{\Pi}, \boldsymbol{\alpha}) = \sum_{i=1}^{n} \sum_{j=1}^{m} \pi_{ij} \sum_{\ell=1}^{L} \alpha_\ell c(\boldsymbol{x}_i^{(\ell)}, \boldsymbol{y}_j^{(\ell)}) - \eta H(\boldsymbol{\alpha}),$$

where $\eta \geq 0$ is the regularization parameter, and $H(\boldsymbol{\alpha}) = \sum_{\ell=1}^{L} \alpha_\ell(\log(\alpha_\ell) - 1)$ is the entropic regularization for $\boldsymbol{\alpha}$. An advantage of entropic regularization is that the nonnegative constraint is naturally satisfied, and the entropic regularizer is a strong convex function.

Lemma 2. *The optimal solution of the optimization problem*

$$\boldsymbol{\alpha}^* = \underset{\boldsymbol{\alpha} \in \boldsymbol{\Sigma}^L}{\text{argmax}} \quad J_\eta(\boldsymbol{\Pi}, \boldsymbol{\alpha}), \text{with } J_\eta(\boldsymbol{\Pi}, \boldsymbol{\alpha}) = \sum_{\ell=1}^{L} \alpha_\ell \phi_\ell - \eta H(\boldsymbol{\alpha})$$

with a fixed admissible transport plan $\boldsymbol{\Pi} \in \boldsymbol{U}(\mu, \nu)$, *is given by*

$$\alpha_\ell^* = \frac{\exp\left(\frac{1}{\eta}\phi_\ell\right)}{\sum_{\ell'=1}^{L} \exp\left(\frac{1}{\eta}\phi_{\ell'}\right)}, \text{with } J_\eta(\boldsymbol{\Pi}, \boldsymbol{\alpha}^*) = \eta \log\left(\sum_{\ell=1}^{L} \exp\left(\frac{1}{\eta}\phi_\ell\right)\right) + \eta.$$

Using Lemma 2 (or Lemma 4 in Nesterov (2005)) with the setting $\phi_\ell = \sum_{i=1}^{n} \sum_{j=1}^{m} \pi_{ij} c(\boldsymbol{x}_i^{(\ell)}, \boldsymbol{y}_i^{(\ell)}) = \langle \boldsymbol{\Pi}, \boldsymbol{C}_\ell \rangle$, $[\boldsymbol{C}_\ell]_{ij} = c(\boldsymbol{x}_i^{(\ell)}, \boldsymbol{y}_i^{(\ell)})$, the global problem is equivalent to

$$\min_{\boldsymbol{\Pi} \in U(\mu,\nu)} G_\eta(\boldsymbol{\Pi}) = \eta \log\left(\sum_{\ell=1}^{L} \exp\left(\frac{1}{\eta}\langle \boldsymbol{\Pi}, \boldsymbol{C}_\ell \rangle\right)\right). \tag{4}$$

Note that this is known as a smoothed max-operator (Nesterov 2005; Blondel et al. 2018). The regularization parameter η controls the "smoothness" of the maximum. Moreover, α_ℓ^* becomes an one-hot vector if η is small; we select only one feature if we set $\eta = 0$. In contrast, thanks to the entropic regularization, α_ℓ^* takes non-zero values and we can select multiple features using α_ℓ^*.

Proposition 3. $G_\eta(\boldsymbol{\Pi})$ *is a convex function relative to* $\boldsymbol{\Pi}$.

The derived optimization problem of FROT is convex. Therefore, we can determine globally optimal solutions. Note that the SRW optimization problem is not jointly convex (Paty and Cuturi 2019) for the projection matrix and the

Algorithm 1. FROT with the Frank–Wolfe.

1: **Input:** $\{\boldsymbol{x}_i\}_{i=1}^n$, $\{\boldsymbol{y}_j\}_{j=1}^m$, η, and ϵ.
2: Initialize $\boldsymbol{\Pi}$, compute $\{\boldsymbol{C}_\ell\}_{\ell=1}^L$.
3: **for** $t = 0 \ldots T$ **do**
4: $\quad \widehat{\boldsymbol{\Pi}} = \text{argmin}_{\boldsymbol{\Pi} \in \boldsymbol{U}(\mu,\nu)} \langle \boldsymbol{\Pi}, \boldsymbol{M}_{\boldsymbol{\Pi}^{(t)}} \rangle + \epsilon H(\boldsymbol{\Pi})$
5: $\quad \boldsymbol{\Pi}^{(t+1)} = (1-\gamma)\boldsymbol{\Pi}^{(t)} + \gamma \widehat{\boldsymbol{\Pi}}$
6: $\quad$ with $\gamma = \frac{2}{2+t}$.
7: **end for**
8: **return** $\boldsymbol{\Pi}^{(T)}$

transport plan. In this study, we employ the Frank–Wolfe algorithm (Frank and Wolfe 1956; Jaggi 2013), using which we approximate $G_\eta(\boldsymbol{\Pi})$ with linear functions at $\boldsymbol{\Pi}^{(t)}$ and move $\boldsymbol{\Pi}$ toward the optimal solution in the convex set (See Algorithm 1).

The derivative of $G_\eta(\boldsymbol{\Pi})$ at $\boldsymbol{\Pi}^{(t)}$ is given by

$$\left.\frac{\partial G_\eta(\boldsymbol{\Pi})}{\partial \boldsymbol{\Pi}}\right|_{\boldsymbol{\Pi}=\boldsymbol{\Pi}^{(t)}} = \sum_{\ell=1}^{L} \alpha_\ell^{(t)} \boldsymbol{C}_\ell = \boldsymbol{M}_{\boldsymbol{\Pi}^{(t)}}, \text{with } \alpha_\ell^{(t)} = \frac{\exp\left(\frac{1}{\eta}\langle \boldsymbol{\Pi}^{(t)}, \boldsymbol{C}_\ell\rangle\right)}{\sum_{\ell'=1}^{L} \exp\left(\frac{1}{\eta}\langle \boldsymbol{\Pi}^{(t)}, \boldsymbol{C}_{\ell'}\rangle\right)}.$$

Then, we update the transport plan by solving the EMD problem:

$$\boldsymbol{\Pi}^{(t+1)} = (1-\gamma)\boldsymbol{\Pi}^{(t)} + \gamma\widehat{\boldsymbol{\Pi}}, \text{ with } \widehat{\boldsymbol{\Pi}} = \underset{\boldsymbol{\Pi} \in \boldsymbol{U}(\mu,\nu)}{\text{argmin}} \langle \boldsymbol{\Pi}, \boldsymbol{M}_{\boldsymbol{\Pi}^{(t)}} \rangle,$$

where $\gamma = 2/(2+k)$. Note that $\boldsymbol{M}_{\boldsymbol{\Pi}^{(t)}}$ is given by the weighted sum of the cost matrices. Thus, we can utilize multiple features to estimate the transport plan $\boldsymbol{\Pi}$ for the relaxed problem in Eq. (4).

Using the Frank–Wolfe algorithm, we can obtain the optimal solution. However, solving the EMD problem requires a cubic computational cost that can be expensive if n and m are large. To address this, we can solve the regularized OT problem, which requires $O(nm)$. We denote the Frank–Wolfe algorithm with EMD as FW-EMD and the Frank–Wolfe algorithm with Sinkhorn as FW-Sinkhorn.

Computational Complexity: The proposed method depends on the Sinkhorn algorithm, which requires an $O(nm)$ operation. The computation of the cost matrix in each subproblem needs an $O(Lnm)$ operation, where L is the number of groups. Therefore, the entire complexity is $O(TLnm)$, where T is the number of Frank–Wolfe iterations (in general, $T = 10$ is sufficient).

Proposition 4. *For each $t \geq 1$, the iteration $\boldsymbol{\Pi}^{(t)}$ of Algorithm 1 satisfies*

$$G_\eta(\boldsymbol{\Pi}^{(t)}) - G_\eta(\boldsymbol{\Pi}^*) \leq \frac{4\sigma_{max}(\boldsymbol{\Phi}^\top \boldsymbol{\Phi})}{\eta(t+2)}(1+\delta),$$

where $\sigma_{max}(\boldsymbol{\Phi}^\top\boldsymbol{\Phi})$ *is the largest eigenvalue of the matrix* $\boldsymbol{\Phi}^\top\boldsymbol{\Phi}$ *and* $\boldsymbol{\Phi} = (\mathrm{vec}(\boldsymbol{C}_1), \mathrm{vec}(\boldsymbol{C}_2), \ldots, \mathrm{vec}(\boldsymbol{C}_L))^\top$*; and* $\delta \geq 0$ *is the accuracy to which internal linear subproblems are solved.*

Based on Proposition 4, the number of iterations depends on η, ϵ, and the number of groups. If we set a small η, convergence requires more time. In addition, if we use entropic regularization with a large ϵ, the δ in Proposition 4 can be large. Finally, if we use more groups, the largest eigenvalue of the matrix $\boldsymbol{\Phi}^\top\boldsymbol{\Phi}$ can be larger. Note that the constant term of the upper bound is large; however, the Frank–Wolfe algorithm converges quickly in practice.

3.3 Connection to L1 Regularization

A natural way to select features is to introduce an L1 regularization term for the feature coefficient $\boldsymbol{\alpha}$. We prove the set of features selected by L1-regularized optimal transport is the same as that of FROT. Let the standard optimal transport with feature coefficient $\boldsymbol{\alpha}$ be:

$$\mathrm{OT}(\mu, \nu, \boldsymbol{\alpha}) = \min_{\boldsymbol{\Pi} \in U(\mu,\nu)} \sum_{i=1}^{n}\sum_{j=1}^{m} \pi_{ij} \sum_{\ell=1}^{L} \alpha_\ell c(\boldsymbol{x}_i^{(\ell)}, \boldsymbol{y}_j^{(\ell)}).$$

Then, the L1-regularized optimal transport is defined as follows:

$$\mathrm{L1OT}(\mu, \nu) = \max_{\boldsymbol{\alpha} \in \mathbb{R}^L_{\geq 0}} \mathrm{OT}(\mu, \nu, \boldsymbol{\alpha}) - \lambda \|\boldsymbol{\alpha}\|_1. \tag{5}$$

Note that the regularization is negative because this is a maximization problem. We assume that $\lambda \geq \mathrm{FROT}(\mu, \nu)$ because otherwise L1OT diverges. Let $\mathcal{F}_{\mathrm{L1}}$ be the set of features that the L1 regularization selects. We consider a feature is selected if the corresponding coefficient can take a positive value in the optimal solution. Specifically, $\mathcal{F}_{\mathrm{L1}}$ is the set of indices $f \in \{1, \cdots, L\}$ such that there exists $\boldsymbol{\alpha}$ such that $\alpha_f > 0$ and $\boldsymbol{\alpha}$ takes the optimum value in Eq. (5). Similarly, let $\mathcal{F}_{\mathrm{FROT}}$ be the set of selected features by FROT. To be precise, $\mathcal{F}_{\mathrm{FROT}}$ is the set of indices $f \in \{1, \cdots, L\}$ such that there exists $\boldsymbol{\Pi}$ and α such that $\alpha_f > 0$ and $(\boldsymbol{\Pi}, \alpha)$ takes the optimum value in Eq. (2).

Theorem 5. $\mathcal{F}_{\mathrm{FROT}} = \mathcal{F}_{L1}$ *when* $\lambda = \mathrm{FROT}(\mu, \nu)$.

In other words, FROT and L1 regularization select the same set of features. This result supports the ability of FROT to select features and robustness of FROT.

3.4 Connection to Subspace Robust Wasserstein

Here, we show that 2-FRWD with $d(\boldsymbol{x}, \boldsymbol{y}) = \|\boldsymbol{x} - \boldsymbol{y}\|_2$ is a special case of SRW. Let us define $\boldsymbol{U} = (\sqrt{\alpha_1}\boldsymbol{e}_1, \sqrt{\alpha_2}\boldsymbol{e}_2, \ldots, \sqrt{\alpha_d}\boldsymbol{e}_d)^\top \in \mathbb{R}^{d \times d}$, where $\boldsymbol{e}_\ell \in \mathbb{R}^d$ is the one-hot vector whose ℓth element is 1 and $\boldsymbol{\alpha}^\top \mathbf{1} = 1, \alpha_\ell \geq 0$. Then, the objective function of SRW can be written as

$$\sum_{i=1}^{n}\sum_{j=1}^{m} \pi_{ij} \|\boldsymbol{U}^\top \boldsymbol{x}_i - \boldsymbol{U}^\top \boldsymbol{y}_j\|_2^2 = \sum_{i=1}^{n}\sum_{j=1}^{m} \pi_{ij} (\boldsymbol{x}_i - \boldsymbol{y}_j)^\top \mathrm{diag}(\boldsymbol{\alpha})(\boldsymbol{x}_i - \boldsymbol{y}_j)$$
$$= \sum_{i=1}^{n}\sum_{j=1}^{m} \pi_{ij} \sum_{\ell=1}^{d} \alpha_\ell (x_i^{(\ell)} - y_j^{(\ell)})^2.$$

Therefore, SRW and 2-FRWD are equivalent if we set $\boldsymbol{U} = (\sqrt{\alpha_1}\boldsymbol{e}_1, \sqrt{\alpha_2}\boldsymbol{e}_2, \ldots, \sqrt{\alpha_d}\boldsymbol{e}_d)^\top$ and $d(\boldsymbol{x}, \boldsymbol{y}) = \|\boldsymbol{x} - \boldsymbol{y}\|_2$.

3.5 Application: Semantic Correspondence

We applied our proposed FROT algorithm to semantic correspondence. The semantic correspondence is a problem that determines the matching of objects in two images. That is, given input image pairs (A, B), with common objects, we formulated the semantic correspondence problem to estimate the transport plan from the key points in A to those in B; this framework was proposed in (Liu et al. 2020). In Fig. 2, we show an overview of our proposed framework.

Cost Matrix Computation $\boldsymbol{C}_\ell$: We employed a pretrained convolutional neural network to extract dense feature maps for each convolutional layer. The dense feature map of the ℓth layer output of the sth image is given by

$$\boldsymbol{f}_{s,q+(r-1)h_s}^{(\ell,s)} \in \mathbb{R}^{d_\ell}, \; q \in [\![h_s]\!], r \in [\![w_s]\!], \ell \in [\![L]\!],$$

where $[\![L]\!] = \{1, 2, \ldots, L\}$, w_s and h_s are the width and height of the sth image, respectively, and d_ℓ is the dimension of the ℓth layer's feature map. Note that because the dimension of the dense feature map is different for each layer, we sample feature maps to the size of the 1st layer's feature map size (i.e., $h_s \times w_s$).

The ℓth layer's cost matrix for images s and s' is given by

$$[\boldsymbol{C}_\ell]_{ij} = \|\boldsymbol{f}_i^{(\ell,s)} - \boldsymbol{f}_j^{(\ell,s')}\|_2^2, \quad i \in [\![w_s h_s]\!], j \in [\![w_{s'} h_{s'}]\!].$$

A potential problem with FROT is that the estimation depends significantly on the magnitude of the cost of each layer (also known as a group). Hence, normalizing each cost matrix is important. Therefore, we normalized each feature vector by $\boldsymbol{f}_i^{(\ell,s)} \leftarrow \boldsymbol{f}_i^{(\ell,s)} / \|\boldsymbol{f}_i^{(\ell,s)}\|_2$. Consequently, the cost matrix is given by $[\boldsymbol{C}_\ell]_{ij} = 2 - 2\boldsymbol{f}_i^{(\ell,s)^\top} \boldsymbol{f}_j^{(\ell,s')}$. We can use distances such as the $L1$ distance.

Computation of $\boldsymbol{a}$ and $\boldsymbol{b}$ with Staircase Re-weighting: Setting $\boldsymbol{a} \in \mathbb{R}^{h_s w_s}$ and $\boldsymbol{b} \in \mathbb{R}^{h_{s'} w_{s'}}$ is important because semantic correspondence can be affected by background clutter. Therefore, we generated the class activation maps (Zhou et al. 2016) for the source and target images and used them as $\boldsymbol{a}$ and $\boldsymbol{b}$, respectively. For CAM, we chose the class with the highest classification probability and normalized it to the range $[0, 1]$.

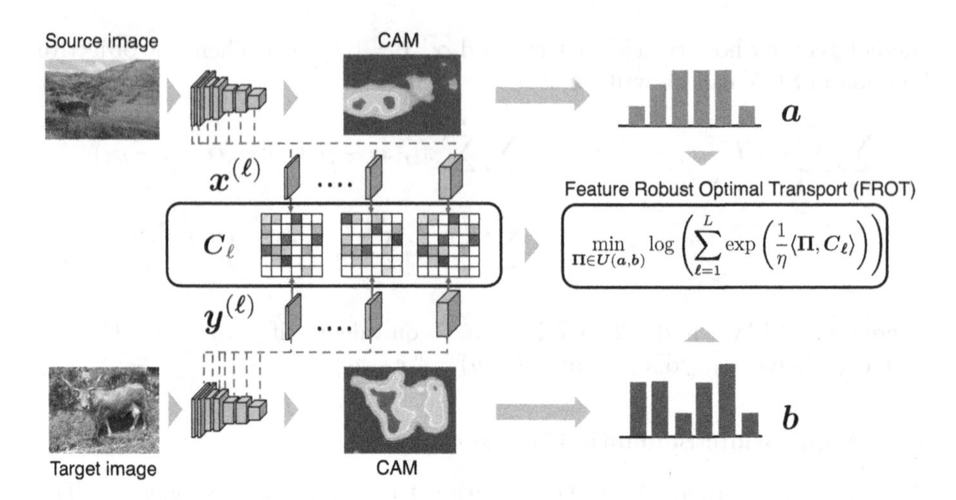

Fig. 2. Semantic correspondence framework based on FROT.

3.6 Application: Feature Selection

Since FROT finds the transport plan and discriminative features between $\boldsymbol{X}$ and $\boldsymbol{Y}$, we can use FROT as a feature-selection method. We considered $\boldsymbol{X} \in \mathbb{R}^{d \times n}$ and $\boldsymbol{Y} \in \mathbb{R}^{d \times m}$ as sets of samples from classes 1 and 2, respectively. The optimal feature importance is given by

$$\widehat{\alpha}_\ell = \frac{\exp\left(\frac{1}{\eta}\langle \widehat{\boldsymbol{\Pi}}, \boldsymbol{C}_\ell \rangle\right)}{\sum_{\ell'=1}^{d} \exp\left(\frac{1}{\eta}\langle \widehat{\boldsymbol{\Pi}}, \boldsymbol{C}_{\ell'} \rangle\right)}, \text{ with } \widehat{\boldsymbol{\Pi}} = \underset{\boldsymbol{\Pi} \in U(\mu,\nu)}{\operatorname{argmin}} \ \eta \log\left(\sum_{\ell=1}^{d} \exp\left(\frac{1}{\eta}\langle \boldsymbol{\Pi}, \boldsymbol{C}_\ell \rangle\right)\right),$$

where $[\boldsymbol{C}_\ell]_{ij} = (x_i^{(\ell)} - y_j^{(\ell)})^2$. Finally, we selected the top K features by the ranking $\widehat{\boldsymbol{\alpha}}$. Hence, $\boldsymbol{\alpha}$ changes to a one-hot vector for a small η and to $\alpha_k \approx \frac{1}{L}$ for a large η.

4 Related Work

The Wasserstein distance can be determined by solving the OT problem, and has many applications in NLP and CV such as measuring document similarity (Kusner et al. 2015; Sato et al. 2022) and finding local feature matching between images (Sarlin et al. 2020; Liu et al. 2020). An advantage of the Wasserstein distance is its robustness to noise; moreover, we can obtain the transport plan, which is useful for many applications. To reduce the computation cost for the Wasserstein distance, the sliced Wasserstein distance is useful (Kolouri et al. 2016. Recently, a tree variant of the Wasserstein distance was proposed (Evans and Matsen 2012; Le et al. 2019; Sato et al. 2020; Takezawa et al. 2021, 2022); the sliced Wasserstein distance is a special case of this algorithm.

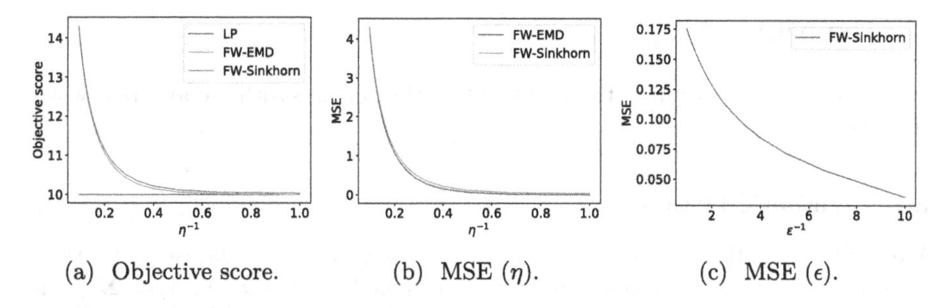

(a) Objective score. (b) MSE (η). (c) MSE (ϵ).

Fig. 3. (a) Objective scores for LP, FW-EMD, and FW-Sinkhorn. (b) MSE between transport plan of LP and FW-EMD and that with LP and FW-Sinkhorn with different η. (c) MSE between transport plan of LP and FW-Sinkhorn with different ϵ.

The approach most closely related to FROT is a robust variant of the Wasserstein distance, called the subspace robust Wasserstein distance (SRW) (Paty and Cuturi 2019). SRW computes the OT problem in a discriminative subspace; this is possible by solving dimensionality-reduction problems. Owing to the robustness, SRW can successfully compute the Wasserstein distance from noisy data. The max–sliced Wasserstein distance (Deshpande et al. 2019) and its generalized counterpart (Kolouri et al. 2019) can also be regarded as subspace-robust Wasserstein methods. Note that SRW (Paty and Cuturi 2019) is a *min–max* based approach, while the max–sliced Wasserstein distances (Deshpande et al. 2019; Kolouri et al. 2019) are *max–min* approaches. The FROT is a feature selection variant of the Wasserstein distance, whereas the subspace approaches are used for dimensionality reduction.

As a parallel work, a general minimax optimal transport problem called the robust Kantorovich problem (RKP) was recently proposed (Dhouib et al. 2020). RKP involves using a cutting-set method for a general minmax optimal transport problem that includes the FROT problem as a special case. The approaches are technically similar. However, we aim to solve a high-dimensional OT problem using feature selection and apply it to semantic correspondence problems, while the RKP approach focuses on providing a general framework and uses it for color transformation problems. As a technical difference, the cutting-set method may not converge to an optimal solution if we use the regularized OT (Dhouib et al. 2020). By contrast, because we use a Frank–Wolfe algorithm, our algorithm converges to a true objective function with regularized OT solvers. The multiobjective optimal transport (MOT) is an approach (Scetbon et al. 2021) parallel to ours. The key difference between FROT and MOT is that MOT tries to use the weighted sum of cost functions, while FROT considers the worst case. Moreover, we focus on the cost matrices computed from subsets of features, while MOT considers cost matrices with different distance functions.

5 Experiments

In this section, we evaluate the FROT algorithm using synthetic and real-world datasets.

5.1 Synthetic Data

We compare FROT with a standard OT using synthetic datasets. In these experiments, we initially generate two-dimensional vectors $\boldsymbol{x} \sim N(\boldsymbol{\mu}_x, \boldsymbol{\Sigma}_x)$ and $\boldsymbol{y} \sim N(\boldsymbol{\mu}_y, \boldsymbol{\Sigma}_y)$. Here, we set $\boldsymbol{\mu}_x = (-5, 0)^\top$, $\boldsymbol{\mu}_y = (5, 0)^\top$, $\boldsymbol{\Sigma}_x = \boldsymbol{\Sigma}_y = ((5, 1)^\top, (4, 1)^\top)$. Then, we concatenate $\boldsymbol{z}_x \sim N(\boldsymbol{0}_8, \boldsymbol{I}_8)$ and $\boldsymbol{z}_y \sim N(\boldsymbol{0}_8, \boldsymbol{I}_8)$ to $\boldsymbol{x}$ and $\boldsymbol{y}$, respectively, to give $\widetilde{\boldsymbol{x}} = (\boldsymbol{x}^\top, \boldsymbol{z}_x^\top)^\top$, $\widetilde{\boldsymbol{y}} = (\boldsymbol{y}^\top, \boldsymbol{z}_y^\top)^\top$.

For FROT, we set $\eta = 1.0$, $T = 10$, and $\epsilon = 0.02$, respectively. To show the proof-of-concept, we set the true features as a group and the remaining noise features as another group.

Figure 1a shows the correspondence from $\boldsymbol{x}$ and $\boldsymbol{y}$ with the vanilla OT algorithm. Figures 1b and c show the correspondence of FROT and OT with $\widetilde{\boldsymbol{x}}$ and $\widetilde{\boldsymbol{y}}$, respectively. Although FROT can identify a suitable matching, the OT fails to obtain a significant correspondence. We observed that the $\boldsymbol{\alpha}$ parameter corresponding to a true group is $\alpha_1 = 0.9999$. Moreover, we compared the objective scores of the FROT with LP, FW-EMD, and FW-Sinkhorn ($\epsilon = 0.1$). Figure 3a shows the objective scores of FROTs with the different solvers, and both FW-EMD and FW-Sinkhorn can achieve almost the same objective score with a relatively small η. Moreover, Fig. 3b shows the mean squared error between the LP method and the FW counterparts. Similar to the objective score cases, it can yield a similar transport plan with a relatively small η. Finally, we evaluated the FW-Sinkhorn by changing the regularization parameter η. In this experiment, we set $\eta = 1$ and varied the ϵ values. The result shows that we can obtain an accurate transport plan with a relatively small ϵ.

5.2 Semantic Correspondence

We evaluated our FROT algorithm for semantic correspondence. In this study, we used the SPair-71k (Min et al. 2019b). The SPair-71k dataset consists of 70, 958 image pairs. For evaluation, we employed a percentage of accurate key points (PCK), which counts the number of accurately predicted key points given a fixed threshold (Min et al. 2019b). All semantic correspondence experiments were run on a Linux server with NVIDIA P100.

For the optimal transport based frameworks, we employed ResNet101 (He et al. 2016) pretrained on ImageNet (Deng et al. 2009) for feature and activation map extraction. The ResNet101 consists of 34 convolutional layers and the entire number of features is $d = 32,576$. Note that we did not fine-tune the network. We compared the proposed method with several baselines (Min et al. 2019b) and the SRW Paty and Cuturi (2019). Owing to the computational cost and the required memory size for SRW, we used the first and the last few convolutional layers of ResNet101 as the input of SRW. In our experiments, we empirically set $T = 3$

Table 1. Per-class PCK ($\alpha_{bbox} = 0.1$) results using SPair-71k. All models use ResNet101. The numbers in the bracket of SRW are the input layer indicies.

Methods		Aero	Bike	Bird	Boat	Bottle	Bus	Car	Cat	Chair	Cow	Dog	Horse	Moto	Person	Plant	Sheep	Train	Tv	All
SPair-71k finetuned models	CNNGeo (Rocco et al. 2017)	23.4	16.7	40.2	14.3	36.4	27.7	26.0	32.7	12.7	27.4	22.8	13.7	20.9	21.0	17.5	10.2	30.8	34.1	20.6
	A2Net (Hongsuck Seo et al. 2018)	22.6	18.5	42.0	16.4	37.9	**30.8**	26.5	35.6	13.3	29.6	24.3	16.0	21.6	22.8	20.5	13.5	31.4	36.5	22.3
	WeakAlign (Rocco et al. 2018a)	22.2	17.6	41.9	15.1	38.1	27.4	**27.2**	31.8	12.8	26.8	22.6	14.2	20.0	22.2	17.9	10.4	32.2	35.1	20.9
	NC-Net (Rocco et al. 2018b)	17.9	12.2	32.1	11.7	29.0	19.9	16.1	39.2	9.9	23.9	18.8	15.7	17.4	15.9	14.8	9.6	24.2	31.1	20.1
SPair-71k validation	HPF (Min et al. 2019a)	25.2	18.9	52.1	15.7	38.0	22.8	19.1	52.9	17.9	33.0	32.8	20.6	24.4	27.9	21.1	15.9	31.5	35.6	28.2
	OT-HPF (Liu et al. 2020)	32.6	18.9	**62.5**	20.7	42.0	26.1	20.4	61.4	**19.7**	**41.3**	**41.7**	**29.8**	**29.6**	**31.8**	25.0	23.5	44.7	37.0	33.9
	FROT($\eta = 0.2, \epsilon = 0.4$)	35.1	20.3	59.8	21.1	**42.9**	27.7	21.2	**63.5**	18.8	39.7	37.9	29.2	28.8	29.9	**28.2**	**24.3**	**52.1**	**39.5**	**34.7**
Without SPair-71k validation	OT	30.1	16.5	50.4	17.3	38.0	22.9	19.7	54.3	17.0	28.4	31.3	22.1	28.0	19.5	21.0	17.8	42.6	28.8	28.3
	FROT ($\eta = 0.3, T = 3$)	35.0	**20.9**	56.3	**23.4**	40.7	27.2	21.9	62.0	17.5	38.8	36.2	27.9	28.0	30.4	26.9	23.1	49.7	38.4	33.7
	FROT ($\eta = 0.3, T = 10$)	34.9	**20.9**	56.4	**23.4**	40.7	27.2	22.0	62.0	17.5	38.8	36.2	27.8	28.2	30.2	26.9	22.9	49.7	38.5	33.7
	FROT ($\eta = 0.5, T = 3$)	34.1	18.8	56.9	19.9	40.0	25.6	19.2	61.9	17.4	38.7	36.5	25.6	26.9	27.2	26.3	22.1	50.3	38.6	32.8
	FROT ($\eta = 0.5, T = 10$)	34.0	18.9	57.0	19.9	40.0	25.6	19.2	61.9	17.3	38.8	36.5	25.6	26.8	27.4	26.4	22.1	50.3	38.8	32.8
	FROT ($\eta = 0.7, T = 3$)	33.4	19.4	56.6	20.0	39.6	26.1	19.1	62.4	17.9	38.0	36.5	26.0	27.5	26.5	25.5	21.6	49.7	38.9	32.7
	FROT ($\eta = 0.7, T = 10$)	33.3	19.5	56.6	19.9	39.5	26.0	19.1	62.4	17.9	38.0	36.5	26.0	27.4	26.5	25.6	21.6	49.6	38.9	32.7
	SRW (layers = {1, 32–34})	29.4	14.0	43.7	15.6	33.8	21.0	17.6	48.0	12.9	23.3	26.5	19.8	25.5	17.6	16.7	15.2	37.1	20.5	24.5
	SRW (layers = {1, 31–34})	29.7	14.3	44.3	15.7	34.2	21.3	17.8	48.5	13.1	23.6	27.1	20.0	25.8	18.1	16.9	15.2	37.3	21.0	24.8
	SRW (layers = {1, 30–34})	29.8	14.7	45.6	15.9	34.8	21.5	18.0	49.3	13.3	24.0	27.7	20.6	25.7	18.7	17.2	15.3	37.7	21.5	25.2
	FROT (layers = {1, 32–34})	32.3	15.7	43.1	18.4	30.7	22.5	20.6	44.5	10.3	23.1	23.9	19.5	23.6	22.0	14.7	15.3	37.4	18.0	24.3
	FROT (layers = {1, 31–34})	35.3	16.5	45.3	20.5	33.0	25.0	21.6	48.1	11.4	25.9	26.9	22.4	25.2	25.0	16.5	17.1	40.5	21.2	26.6
	FROT (layers = {1, 30–34})	**36.7**	18.1	48.8	22.3	34.5	27.5	23.0	51.3	12.9	28.4	30.3	24.2	26.4	27.3	19.5	18.1	43.4	24.9	28.8

and $\epsilon = 0.1$ for FROT and SRW, respectively. For SRW, we set the number of latent dimension as $k = 50$ for all experiments. HPF (Min et al. 2019a) and OT-HPF (Liu et al. 2020) are state-of-the-art methods for semantic correspondence. HPF and OT-HPF required the validation dataset to select important layers, whereas SRW and FROT did not require the validation dataset. OT is a simple Sinkhorn-based method that does not select layers.

Table 1 lists the per-class PCK results obtained using the SPair-71k dataset. FROT ($\eta = 0.3$) outperforms most existing baselines, including HPF and OT. Moreover, FROT ($\eta = 0.3$) is consistent with OT-HPF (Liu et al. 2020), which requires the validation dataset to select important layers. In this experiment, setting $\eta < 1$ results in favorable performance (See Table 2 in the supplementary material). The computational costs of FROT is 0.29, while SRWs are 8.73, 11.73, 15.76, respectively. Surprisingly, FROT outperformed SRWs. However, this is mainly due to the used input layers.

We further evaluated FROT by tuning hyperparameters η and ϵ using validation sets, where the maximum search ranges for η and ϵ are set to 0.2 to 2.0 and 0.1 to 0.6 with intervals of 0.1, respectively. By using hyperparameter search, we selected ($\eta = 0.2, \epsilon = 0.4$) as an optimal parameter. The FROT with optimal parameters outperforms the state-of-the-art method (Liu et al. 2020).

5.3 Feature Selection Experiments

Here, we compared FROT with several baseline algorithms in terms of solving feature-selection problems. In this study, we employed a high-dimensional and a few sample datasets with two class classification tasks (see Table 3 in the supplementary material). All feature selection experiments were run on a Linux server with an Intel Xeon CPU E7-8890 v4 with 2.20 GHz and 2 TB RAM.

In our experiments, we initially randomly split the data into two sets (75% for training and 25% for testing) and used the training set for feature selection and building a classifier. Note that we standardized each feature using the training set. Then, we used the remaining set for the test. The trial was repeated 50 times, and we considered the averaged classification accuracy for all trials. Considered as baseline methods, we computed the Wasserstein distance, maximum mean discrepancy (MMD) (Gretton et al. 2007), and linear correlation[1] for each dimension and sorted them in descending order. Note that the Wasserstein distance is computed via sorting, which is computationally more efficient than the Sinkhorn algorithm when $d = 1$. Then, we selected the top K features as important features. For FROT, we computed the feature importance and selected the features that had significant importance scores. In our experiments, we set $\eta = 1.0$ and $T = 10$. Then, we trained a two-class SVM[2] with the selected features.

Figure 4 shows the average classification accuracy relative to the number of selected features. From Fig. 4, FROT is consistent with the Wasserstein distance-based feature selection and outperforms the linear correlation method and the

[1] https://scikit-learn.org/stable/modules/feature_selection.html.
[2] https://scikit-learn.org/stable/modules/generated/sklearn.svm.SVC.html.

MMD for two datasets. Table 3 in the supplementary file shows the computational time(s) of the methods. FROT is about two orders of magnitude faster than the Wasserstein distance and is also faster than MMD. Note that although MMD is as fast as the proposed method, it cannot determine the correspondence between samples.

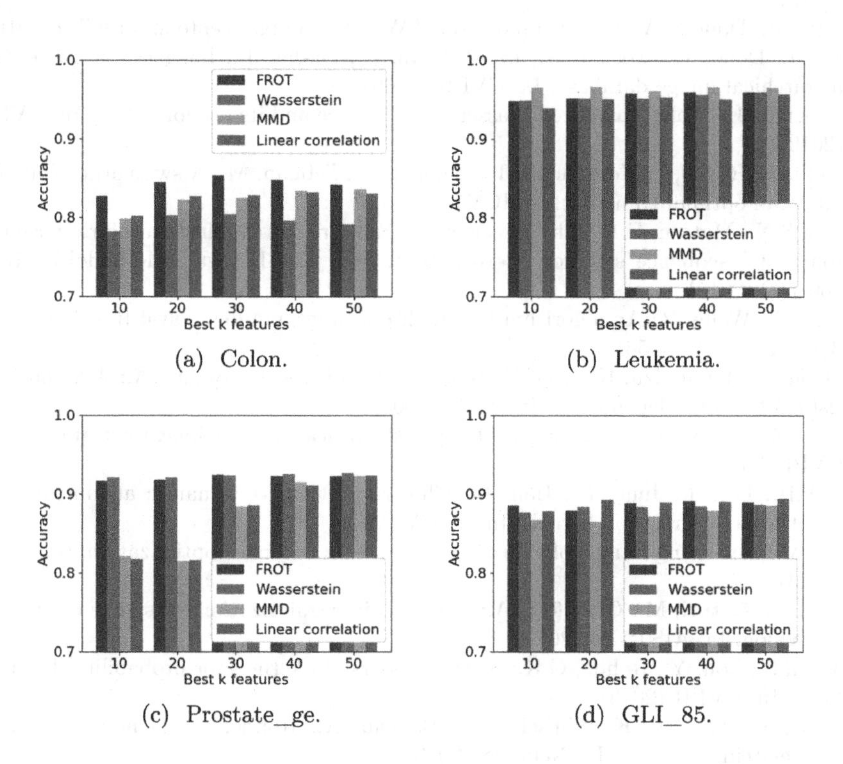

(a) Colon. (b) Leukemia.

(c) Prostate_ge. (d) GLI_85.

Fig. 4. Feature selection results. We average over 50 runs of accuracy (on test set) of SVM trained with top k features selected by several methods.

6 Conclusion

In this paper, we proposed FROT for high-dimensional data. This approach jointly solves feature selection and OT problems. An advantage of FROT is that it is a convex optimization problem and can determine an accurate globally optimal solution using the Frank–Wolfe algorithm. We used FROT for feature selection and semantic correspondence problems. Through experiments, we demonstrated that the proposed algorithm is consistent with state-of-the-art algorithms in both feature selection and semantic correspondence.

Acknowledgement. M.Y. was supported by MEXT KAKENHI Grant Number 20H04243. R.S. was supported by JSPS KAKENHI GrantNumber 21J22490.

References

Blondel, M., Seguy, V., Rolet, A.: Smooth and sparse optimal transport. In: AISTATS (2018)
Cuturi, M.: Sinkhorn distances: lightspeed computation of optimal transport. In: NIPS (2013)
Cuturi, M., Doucet, A.: Fast computation of Wasserstein barycenters. In: ICML (2014)
Deng, J., Dong, W., Socher, R., Li, L. J., Li, K., Fei-Fei, L.: Imagenet: a large-scale hierarchical image database. In: CVPR (2009)
Deshpande, I., et al.: Max-sliced Wasserstein distance and its use for GANs. In: CVPR (2019)
Dhouib, S., Redko, I., Kerdoncuff, T., Emonet, R., Sebban, M.: A swiss army knife for minimax optimal transport. In: ICML (2020)
Evans, S.N., Matsen, F.A.: The phylogenetic Kantorovich-Rubinstein metric for environmental sequence samples. J. Roy. Stat. Soc. Ser. B (Stat. Methodol.) **74**(3), 569–592 (2012)
Frank, M., Wolfe, P.: An algorithm for quadratic programming. Naval Res. Logist. Q. **3**(1–2), 95–110 (1956)
Gretton, A., Fukumizu, K., Teo, C., Song, L., Schölkopf, B., Smola, A.: A kernel statistical test of independence. In: NIPS (2007)
He, K., Zhang, X., Ren, S., Sun, J.: Deep residual learning for image recognition. In: CVPR (2016)
Seo, P.H., Lee, J., Jung, D., Han, B., Cho, M.: Attentive semantic alignment with offset-aware correlation kernels. In: ECCV (2018)
Jaggi, M.: Revisiting frank-wolfe: projection-free sparse convex optimization. In: ICML (2013)
Janati, H., Cuturi, M., Gramfort, A.: Wasserstein regularization for sparse multi-task regression. In: AISTATS (2019)
Kolouri, S., Zou, Y., Rohde, G.K.: Sliced Wasserstein kernels for probability distributions. In: CVPR (2016)
Kolouri, S., Nadjahi, K., Simsekli, U., Badeau, R., Rohde, G.: Generalized sliced Wasserstein distances. In: NeurIPS (2019)
Kusner, M., Sun, Y., Kolkin, N., Weinberger, K.: From word embeddings to document distances. In: ICML (2015)
Le, T., Yamada, M., Fukumizu, K., Cuturi, M.: Tree-sliced approximation of Wasserstein distances. In: NeurIPS (2019)
Liu, Y., Zhu, L., Yamada, M., Yang, Y.: Semantic correspondence as an optimal transport problem. In: CVPR (2020)
Liu, Y., Yamada, M., Tsai, Y.-H.H., Le, T., Salakhutdinov, R., Yang, Y.: Semi-supervised squared-loss mutual information estimation with optimal transport. In: ECML (2021)
Min, J., Lee, J., Ponce, J., Cho, M.: Hyperpixel flow: semantic correspondence with multi-layer neural features. In: ICCV (2019a)
Min, J., Lee, J., Ponce, J., Cho, M.: Spair-71k: a large-scale benchmark for semantic correspondence. arXiv preprint arXiv:1908.10543 (2019b)
Nesterov, Yu.: Smooth minimization of non-smooth functions. Math. Program. **103**(1), 127–152 (2005)
Paty, F.-P., Cuturi, M.: Subspace robust Wasserstein distances. In: ICML (2019)
Paty, F.P., Cuturi, M.: Regularized optimal transport is ground cost adversarial. In: ICML (2020)

Peyré, G., Cuturi, M., et al.: Computational optimal transport. Found. Trends® Mach. Learn. **11**(5–6), 355–607 (2019)
Rocco, I., Arandjelovic, R., Sivic, J.: Convolutional neural network architecture for geometric matching. In: CVPR (2017)
Rocco, I., Arandjelović, R., Sivic, J.: End-to-end weakly-supervised semantic alignment. In: CVPR (2018a)
Rocco, I., Cimpoi, M., Arandjelović, R., Torii, A., Pajdla, T., Sivic, J.: Neighbourhood consensus networks. In: NeurIPS (2018b)
Rubner, Y., Tomasi, C., Guibas, L.J.: The earth mover's distance as a metric for image retrieval. Int. J. Comput. Vis. **40**(2), 99–121 (2000)
Sarlin, P.E., DeTone, D., Malisiewicz, T., Rabinovich, A.: Learning feature matching with graph neural networks. In: CVPR (2020)
Sato, R., Yamada, M., Kashima, H.: Fast unbalanced optimal transport on tree. In: NeurIPS (2020)
Sato, R., Yamada, M., Kashima, H.: Re-evaluating word mover's distance. In: ICML (2022)
Scetbon, M., Meunier, L., Atif, J., Cuturi, M.: Equitable and optimal transport with multiple agents. In: AISTATS (2021)
Takezawa, Y., Sato, R., Yamada, M.: Supervised tree-Wasserstein distance. In: ICML (2021)
Takezawa, Y., Sato, R., Kozareva, Z., Ravi, S., Yamada, M.: Fixed support tree-sliced Wasserstein barycenter. In: AISTATS (2022)
Yan, Y., Li, W., Wu, H., Min, H., Tan, M., Wu, Q.: Semi-supervised optimal transport for heterogeneous domain adaptation. In: IJCAI (2018)
Yuan, M., Lin, Y.: Model selection and estimation in regression with grouped variables. J. Roy. Stat. Soc. Ser. B (Stat. Methodol.) **68**(1), 49–67 (2006)
Zhou, B., Khosla, A., Lapedriza, A., Oliva, A., Torralba, A.: Learning deep features for discriminative localization. In: CVPR (2016)

Optimization

Penalized FTRL with Time-Varying Constraints

Douglas J. Leith[1(✉)] and George Iosifidis[2]

[1] Trinity College Dublin, Dublin, Republic of Ireland
doug.leith@tcd.ie
[2] Delft University of Technology, Delft, The Netherlands

Abstract. In this paper we extend the classical Follow-The-Regularized-Leader (FTRL) algorithm to encompass time-varying constraints, through adaptive penalization. We establish sufficient conditions for the proposed Penalized FTRL algorithm to achieve $\mathcal{O}(\sqrt{t})$ regret and violation with respect to a strong benchmark $\hat{X}_t^{max}$. Lacking prior knowledge of the constraints, this is probably the largest benchmark set that we can reasonably hope for. Our sufficient conditions are necessary in the sense that when they are violated there exist examples where $\mathcal{O}(\sqrt{t})$ regret and violation is not achieved. Compared to the best existing primal-dual algorithms, Penalized FTRL substantially extends the class of problems for which $\mathcal{O}(\sqrt{t})$ regret and violation performance is achievable.

Keywords: FTRL · Online convex optimization · Constrained optimization

1 Introduction

The introduction of online convex optimization (OCO) [15] offered an effective way to tackle online learning and dynamic decision problems, with applications that range from portfolio selection, to routing optimization and ad placement, see [3]. One of the seminal OCO algorithms is the Follow-The-Regularized-Leader (FTRL), which includes online gradient descent and mixture of experts as special cases. Indeed, FTRL is widely used in different contexts, e.g., with linear or non-linear objective functions, composite objectives [8], or budget constraints [1].

The general form of the FTRL update is:

$$x_{\tau+1} \in \arg\min_{x \in X} \left\{ R_\tau(x) + \sum_{i=1}^{\tau} F_i(x) \right\} \tag{1}$$

where action set $X \subset \mathbb{R}^n$ is bounded, function $F_i : X \to \mathbb{R}$ and regularizer $R_\tau : X \to \mathbb{R}$ are strongly convex. When the sum-loss $\sum_{i=1}^{\tau} F_i(x)$ is convex and $F_i(x)$ and $\big(R_i(x) - R_{i-1}(x)\big)$ are uniformly Lipschitz, the FTRL-generated sequence $\{x_\tau\}_{\tau=1}^t$ induces regret $\sum_{i=1}^t \big(F_i(x_i) - F_i(x)\big) \leq \mathcal{O}(\sqrt{t})$, $\forall x \in X$, cf. [8]. Importantly, the set X of admissible actions must be fixed and this is intrinsic to the method of proof, i.e., it is not a minor or incidental assumption.

© The Author(s), under exclusive license to Springer Nature Switzerland AG 2023
M.-R. Amini et al. (Eds.): ECML PKDD 2022, LNAI 13717, pp. 311–326, 2023.
https://doi.org/10.1007/978-3-031-26419-1_19

The focus of this paper is to extend the FTRL algorithm in order to accommodate time-varying action sets, i.e., cases where at each time τ the fixed set action X is replaced by set X_τ which may vary over time. We refer to this extension to FTRL as *Penalized FTRL* or Pen-FTRL.

In general, it is too much to expect to be able to simultaneously achieve $\mathcal{O}(\sqrt{t})$ regret and strict feasibility $x_\tau \in X_\tau$, $\tau = 1, \ldots, t$. We therefore allow limited violation of the action sets $\{X_\tau\}$ and instead aim to simultaneously achieve $\mathcal{O}(\sqrt{t})$ regret and $\mathcal{O}(\sqrt{t})$ constraint violation. That is, defining loss function $f_\tau : D \to \mathbb{R}$ on domain $D \subset \mathbb{R}^n$ and constraint functions $g_\tau^{(j)} : D \to \mathbb{R}$ such that $X_\tau = \{x \in D : g_\tau^{(j)}(x) \leq 0, j = 1, \ldots, m\}$ then we aim to simultaneously achieve regret and violation:

$$\mathcal{R}_t = \sum_{i=1}^{t} \Big(f_i(x_i) - f_i(x)\Big) \leq \mathcal{O}(\sqrt{t}),\ \mathcal{V}_t = \sum_{j=1}^{m} \max\left\{0, \sum_{i=1}^{t} g_i^{(j)}(x_i)\right\} \leq \mathcal{O}(\sqrt{t})$$

for all $x \in X_t^{max} := \Big\{x \in D : \sum_{i=1}^{t} g_i^{(j)}(x) \leq 0, j = 1, \ldots, m\Big\}$.

Importance of Using A Strong Benchmark. We know from [7] that $\mathcal{O}(\sqrt{t})$ regret and violation with respect to benchmark set X_t^{max} is not achievable for all possible sequences of constraints $\{g_i^{(j)}\}$. It is therefore necessary to: *(i)* change the benchmark set X_t^{max} to something more restrictive; or *(ii)* restrict the admissible set of constraint sequences $\{g_i^{(j)}\}$; or *(iii)* both. In the literature, it is common to adopt the weaker benchmark:

$$X_t^{min} := \Big\{x \in D : g_i^{(j)}(x) \leq 0, i = 1, \ldots, t, j = 1, \ldots, m\Big\} \subset X_t^{max}$$

i.e., to focus on actions x which *simultaneously satisfy every constraint at every time.* But this weak benchmark is in fact so restrictive and easy for a learning algorithm to outperform, where the achieved regret $\mathcal{R}_t$ is often negative.

One of our primary interests, therefore, is in retaining a benchmark that is close to X_t^{max}. To this end, we consider the following benchmark:

$$\hat{X}_t^{max} := \left\{x \in D : \sum_{i=1}^{\tau} g_i^{(j)}(x) \leq 0, \forall j = 1, \ldots, m \tau = 1, \ldots, t\right\}.$$

We can see immediately that $X_t^{min} \subset \hat{X}_t^{max}$. The set $\hat{X}_t^{max}$ requires $\sum_{i=1}^{\tau} g_i^{(j)}(x) \leq 0$ to hold at every time $\tau \leq t$ rather than just at the end of the horizon t, and so is still smaller than X_t^{max}. Lacking, however, predictions or prior knowledge of the constraints $g_i^{(j)}$, it is probably the best we can reasonably hope for. To illustrate the difference between $\hat{X}_t^{max}$ and X_t^{min}, suppose the time-varying constraint is $x \leq 1/\sqrt{t}$. Then $X_t^{min} = [0, 1/\sqrt{t}]$ which tends to set $\{0\}$ for t large, $X_t^{max} = D = [0,1]$ for $t \geq 1$, and $\hat{X}_t^{max} = D = [0,1]$.

Role of Pen-FTRL. Almost all of the literature focuses on using primal-dual algorithms to accommodate time-varying constraints (see Sect. 2). In contrast, here we use a direct penalty-based approach [12], which we refer to as *Penalized*

FTRL or `Pen-FTRL`. This has the important advantages of *(i)* conceptually separating the issue of multiplier selection (i.e. λ in the above primal-dual update) from the issue of sum-constraint violation and so facilitating analysis; and *(ii)* maintaining a direct link with the well-established FTRL algorithm.

Contributions. In summary, the main contributions of the present paper are: *(i)* introduction of the `Pen-FTRL` extension to FTRL; and *(ii)* establishing sufficient conditions for `Pen-FTRL` to achieve $\mathcal{O}(\sqrt{t})$ regret and violation with respect to benchmark $\hat{X}_t^{max}$. Lacking prior knowledge of the constraints, $\hat{X}_t^{max}$ is probably the largest benchmark set that we can reasonably hope for. Our sufficient conditions are necessary in the sense that when they are violated there exist examples where $\mathcal{O}(\sqrt{t})$ regret and violation is not achieved. Compared to the best existing primal-dual algorithms, `Pen-FTRL` substantially extends the class of problems for which $\mathcal{O}(\sqrt{t})$ regret and violation performance is achievable.

2 Related Work

The literature on online learning with time-varying constraints focuses on primal-dual algorithms (see update (7) in the sequel), and largely fails to obtain $\mathcal{O}(\sqrt{t})$ regret and violation simultaneously even w.r.t. the weak X_t^{min} benchmark. The standard problem setup consists of a sequence of convex cost functions $f_t : D \to \mathbb{R}$ and constraints $g_t^{(j)} : X \to \mathbb{R}$, $j = 1, \dots, m$, where actions $x \in D \subset \mathbb{R}^n$. The canonical algorithm performs a primal-dual gradient descent iteration, namely:

$$x_{t+1} = \Pi_D\left(x_t - \eta_t(\partial f_t(x_t) + \lambda_t^T \partial g_t(x_t))\right), \quad \lambda_{t+1} = \left[(1-\theta_t)\lambda_t + \mu_t g_t(x_{t+1})\right]^+ \tag{2}$$

with step-size parameters η_t, μ_t and regularization parameter θ_t; while $\Pi_D(\alpha)$ denotes the project of α onto D. Commonly, the parameter $\theta_t \equiv 0$, with exceptions being [4,6], and [10] that employ non-zero θ_t. [14] approximate $g_t(x_{t+1})$ in the λ_{t+1} update by $g_t(x_t) + \partial g_t(x_t)(x_{t+1} - x_t)$.

The $\mathcal{R}_t$ is commonly measured w.r.t. the baseline action set X_t^{min}, with the exception of [11] where a slightly larger set is considered; [14] that considers stochastic constraints and the baseline action set is $\{x \in D : E[g_i^{(j)}(x)] \le 0, j = 1, \dots, m\}$; and [5] which considers a K-slot moving window for the sum-constraint satisfaction. The original work on this topic restricted attention to time-invariant constraints $g_i^{(j)}(x) = g^{(j)}(x)$. With this restriction, the work in [4] achieves $\mathcal{R}_t \le \mathcal{O}(\max\{t^\beta, t^{1-\beta}\})$ and $\mathcal{V}_t \le \mathcal{O}(t^{1-\beta/2})$ constraint violation, which yields $\mathcal{R}_t$, $\mathcal{V}_t \le \mathcal{O}(t^{2/3})$ with $\beta = 2/3$. Similar bounds are derived in [6]. It is worth noting that these results are primarily of interest for their analysis of the primal-dual algorithm rather than the performance bounds per se, since classical algorithms such as FTRL are already known to achieve $\mathcal{O}(\sqrt{t})$ regret and no constraint violation for constant constraints.

For general time-varying cost and constraint functions, [10] achieves $\mathcal{O}(\sqrt{t})$ regret and $\mathcal{O}(t^{3/4})$ constraint violation; [5] achieve $\mathcal{R}_t = \mathcal{O}(\sqrt{t} + Kt/V)$ and $\mathcal{V}_t = \mathcal{O}(\sqrt{Vt})$, with $K = 1$ corresponding to baseline set X_t^{min} and V a design parameter. Selecting $V = \sqrt{t}$ gives $\mathcal{O}(\sqrt{t})$ regret and $\mathcal{O}(t^{3/4})$ constraint violation,

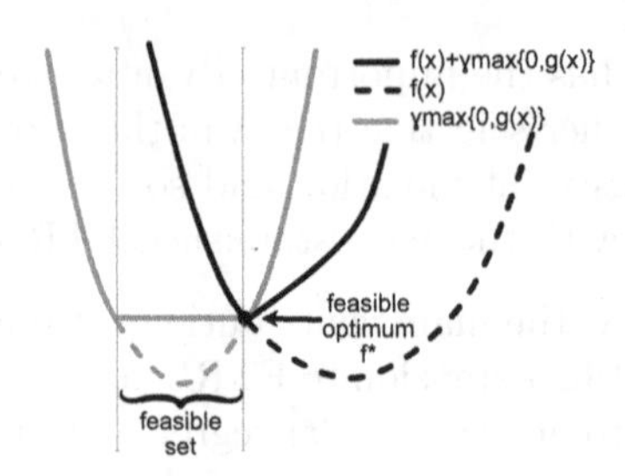

Fig. 1. Illustrating use of a penalty to convert constrained optimization $\min_{x:g(x)\leq 0} f(x)$ into unconstrained optimization $\min_x f(x)+\gamma\max\{0,g(x)\}$, $\gamma>0$. Within the feasible set $g(x)\leq 0$ and $\gamma\max\{0,g(x)\}=0$. Outwith this set $\gamma\max\{0,g(x)\}=\gamma g(x)>0$. The idea is that γ is selected large enough that outwith the feasible set $f(x)+\gamma\max\{0,g(x)\}>f^*$, the min value of f inside the feasible set.

similarly to [10]. By restricting the constraints, [11] improves this to $\mathcal{O}(\sqrt{t})$ regret and constraint violation. As already noted, this requires restricting the constraints to be $g_i^{(j)}(x)=g^{(j)}(x)-b_{i,j}$ with $b_{i,j}\in\mathbb{R}$ i.e. the constraints are $g^{(j)}(x)\leq b_i^{(j)}$ with time-variation confined to threshold $b_i^{(j)}$. Yu et al [14] also achieve $\mathcal{O}(\sqrt{t})$ regret and *expected* constraint violation (i.e. $E[\sum_{i=1}^t g_i^{(j)}(x_t)]\leq\mathcal{O}(\sqrt{t})$), this time by restricting the constraints to be i.i.d. stochastic. Yi et al [13] obtain $\mathcal{R}_t,\mathcal{V}_t=\mathcal{O}(t^{2/3})$ by restricting the cost and constraint functions to be separable, while [2] focuses on a form of dynamic regret that upper bounds the static regret and show $o(t)$ regret and $\mathcal{O}(t^{2/3})$ constraint violation under a slow variation condition on the constraints and dynamic baseline action.

3 Preliminaries

3.1 Exact Penalties

We begin by recalling a classical result of Zangwill [12]. Consider the convex optimization problem P:

$$\min_{x\in D} f(x) \qquad \text{s.t.} \qquad g^{(j)}(x)\leq 0,\ j=1,\cdots,m$$

where $D\subset\mathbb{R}^n$, $f:\mathbb{R}^n\to\mathbb{R}$ and $g^{(j)}:\mathbb{R}^n\to\mathbb{R}$, $j=1,\cdots,m$ are convex. Let $X:=\{x:x\in D, g^{(j)}(x)\leq 0, j=1,\cdots,m\}$ denote the feasible set and $X^*\subset X$ the set of optimal points. Define:

$$F(x):=f(x)+\gamma\sum_{j=1}^m \max\left\{0,g^{(j)}(x)\right\},\quad \gamma\in\mathbb{R}. \tag{3}$$

$F(x)$ is convex since $f(\cdot)$, $g^{(j)}(\cdot)$ are convex and $\max\{\cdot\}$ preserves convexity.

The key idea is that the penalty (second term in (3)) is zero for $x\in X$, but large when $x\notin X$. Provided γ is selected large enough, the penalty forces the minimum of $F(x)$ to *(i)* lie in X and *(ii)* match $\min_{x\in X} f(x)$; see example in Fig. 1. The next lemma, proved in the Appendix, corresponds to [12, Lemma 2].

Lemma 1 (Exact Penalty). *Assume that a Slater point exists i.e. a feasible point $z \in D$ such that $g^{(j)}(z) < 0$, $j = 1, \cdots, m$. Let $f^* := \inf_{x \in X} f(x)$ (the solution to optimization P). Then there exists a finite threshold $\gamma_0 \geq 0$ such that $F(x) \geq f^*$ for all $x \in D$, $\gamma \geq \gamma_0$, with equality only when $x \in X^*$. It is sufficient to choose $\gamma_0 = \frac{f^* - f(z) - 1}{\max_{j \in \{1,\cdots,m\}}\{g^{(j)}(z)\}}$.*

3.2 FTRL Results

We also recall the following standard FTRL results (for proofs see, e.g., [9]).

Lemma 2 (Be-The-Leader). *Let $F_i, i = 1, \ldots, t$ be a sequence of (possibly non-convex) functions $F_i : D \to \mathbb{R}$, $D \subset \mathbb{R}^n$. Assume that $\arg\min_{x \in D} \sum_{i=1}^{\tau} F_i(x)$ is not empty for $\tau = 1, \ldots, t$. Selecting sequence $w_{i+1}, i = 1, \ldots, t$ according to the Follow The Leader (FTL) update $w_{\tau+1} \in \arg\min_{x \in D} \sum_{i=1}^{\tau} F_i(x)$, ensures $\sum_{i=1}^{t} F_i(w_{i+1}) \leq \sum_{i=1}^{t} F_i(y)$ for every $y \in D$.*

Condition 1 (FTRL) *(i) Domain D is bounded (potentially non-convex), (ii) $\sum_{i=1}^{\tau} F_i(x)$ is convex (the individual F_i's need not be convex), (iii) $F_i(x)$ is uniformly L_f-Lipschitz on D i.e. $|F_i(x) - F_i(y)| \leq L_f \|x - y\|$ for all $x, y \in D$ and where L_f does not depend on i, and (iv) $R_\tau(x)$ is $\sqrt{\tau}$-strongly convex and $(R_i(x) - R_{i-1}(x))$ is uniformly Lipschitz, e.g. $\sqrt{\tau}\|x\|_2^2$.*

Lemma 3 (Regret of FTRL). *When Condition 1 holds, the sequence $\{x_\tau\}_{\tau=1}^{t}$ generated by the FTRL update $x_{\tau+1} \in \arg\min_{x \in D} R_\tau(x) + \sum_{i=1}^{\tau} F_i(x)$ has regret $\mathcal{R}_t = \sum_{i=1}^{t} F_i(x_i) - F_i(x) \leq \mathcal{O}(\sqrt{t})$ for all $x \in D$.*

Lemma 4 (σ_τ-Strongly Convex Regularizer). *When $\sum_{i=1}^{\tau} F_i(x)$ is σ_τ-strongly convex, $F_i(x)$ uniformly L_f-Lipschitz over D and $w_{\tau+1} \in \arg\min_{x \in D} \sum_{i=1}^{\tau} F_i(x)$, it holds $\|w_{\tau+1} - w_\tau\| \leq 2L_f/(\sigma_\tau + \sigma_{\tau-1})$*

4 Penalized FTRL

4.1 Exact Penalties for Time-Invariant Constraints

We begin by demonstrating the application of Lemma 1 to FTRL update (1) with time-invariant action set X. Selecting $F_i(x) = f_i(x) + \gamma h(x)$ with $h(x) = \sum_{j=1}^{m} \max\{0, g^{(j)}(x)\}$ and defining the bounded domain D with $X \subset D$, then by standard analysis, cf. [8], the `Pen-FTRL` update[1]:

$$x_{\tau+1} \in \arg\min_{x \in D} \left\{ R_\tau(x) + \sum_{i=1}^{\tau} F_i(x) \right\} \tag{4}$$

ensures regret $\sum_{i=1}^{t}(F_i(x_i) - F_i(x)) \leq \mathcal{O}(\sqrt{t})$ for all $x \in D$, and since $X \subset D$ for all $x \in X$. Of course this says nothing about whether the actions x_i lie in set

[1] Note the subtle yet crucial difference w.r.t. non-`Pen-FTRL` update (1).

X nor anything much about the regret of $f_i(x_i)$, but when set X has a Slater point and γ is selected large enough then by Lemma 1 we have that $x_{\tau+1} \in X$ for all τ. It follows that $F_i(x_i) = f_i(x_i)$ (since $h(x_i) = 0$ when $x_i \in X$) and so regret $\sum_{i=1}^{t}(F_i(x_i) - F_i(x)) = \sum_{i=1}^{t}(f_i(x_i) - f_i(x)) \leq \mathcal{O}(\sqrt{t})$ for all $x \in X$.

4.2 Penalties for Time-Varying Constraints

We now extend consideration to FTRL with time-varying constraints. Our aim is to define a penalty which is zero on a set $\hat{X}_\tau^{max} \approx X_{max}$, and large enough outside this set to force the minimum of $\sum_{i=1}^{\tau} F_i(x)$ to lie in $\hat{X}^{max}$.

Penalties Which Are Zero When $x \in \hat{X}_\tau^{max}$. Consider extending the penalty-based FTRL (4) to time-varying constraints. We might try selecting:

$$F_i(x) = f_i(x) + \gamma h_i(x), \quad \text{with} \quad h_i(x) = \sum_{j=1}^{m} \max\left\{0, g_i^{(j)}(x)\right\},$$

but we immediately run into the following difficulty. We have that $\sum_{i=1}^{\tau} F_i(x) = \sum_{i=1}^{\tau} f_i(x) + \gamma \sum_{i=1}^{\tau} \sum_{j=1}^{m} \max\{0, g_i^{(j)}(x)\}$ and so to make the second term zero requires $g_i^{(j)}(x) \leq 0$ for all $i \leq \tau$ and $j \leq m$, i.e. requires every constraint over all time to simultaneously be satisfied. This penalty choice $h_i(\cdot)$ therefore corresponds to benchmark X_t^{min}, whereas our interest is in set X_t^{max}. It is perhaps worth noting that this corresponds to the penalty used in the primal-dual literature, so it is unsurprising that those results are confined to X^{min}.

With this in mind, consider instead selecting

$$h_\tau(x) = \sum_{j=1}^{m} \max\left\{0, \sum_{i=1}^{\tau} g_i^{(j)}(x)\right\} - \sum_{j=1}^{m} \max\left\{0, \sum_{i=1}^{\tau-1} g_i^{(j)}(x)\right\}$$

with $h_1(x) = \sum_{j=1}^{m} \max\{0, g_i^{(j)}(x)\}$. Then,

$$\sum_{i=1}^{\tau} F_i(x) = \sum_{i=1}^{\tau} f_i(x) + \gamma \sum_{j=1}^{m} \max\left\{0, \sum_{i=1}^{\tau} g_i^{(j)}(x)\right\}.$$

We now have a sum-constraint in the second term, as desired. Unfortunately, this choice of $h_i(\cdot)$ violates the conditions needed for FTRL to achieve $\mathcal{O}(\sqrt{t})$ regret. Namely, it is required that $F_i(\cdot)$ is uniformly Lipschitz but $h_i(\cdot)$ does not satisfy this condition, and so neither does $F_i(\cdot)$. To see this, observe that when $g_i^{(j)}(\cdot)$ is uniformly Lipschitz with constant L_g, then $\sum_{i=1}^{\tau} g_i^{(j)}(x)$ has a Lipschitz constant τL_g that scales with τ, and so there exists no uniform upper bound. The max operator in $h_i(\cdot)$ does not change the Lipschitz constant (see Lemma 5); thus $h_i(\cdot)$ is τL_g Lipschitz, which prevents FTRL achieving $\mathcal{R}_t \leq \mathcal{O}(\sqrt{t})$.

These considerations lead us to the following penalty,

$$h_\tau(x) = \sum_{j=1}^{m} \max\left\{0, \frac{1}{\tau}\sum_{i=1}^{\tau} g_i^{(j)}(x)\right\}. \tag{5}$$

When $g_i^{(j)}(\cdot)$ is uniformly Lipschitz with constant L_g then so is $h_i(\cdot)$ due to the $1/\tau$ prefactor added to the sum and the following Lemma which just states that when a function $h(x)$ is L-Lipschitz then $\max\{0, h(x)\}$ is also L-Lipschitz:

Lemma 5. *When* $|h(x)-h(y)| \leq L\|x-y\|$ *then* $|\max\{0,h(x)\}-\max\{0,h(y)\}| \leq L\|x-y\|$.

Proof. Observe that $2\max\{0,h(x)\} = h(x)+|h(x)|$. Therefore, $2|\max\{0,h(x)\} - \max\{0,h(y)\}| = |h(x)-h(y)+|h(x)|-|h(y)|| \leq |h(x)-h(y)|+||h(x)|-|h(y)|| \leq |h(x)-h(y)| + |h(x)-h(y)| \leq 2L\|x-y\|$.

With this choice, we can write:

$$\sum_{i=1}^{\tau} F_i(x) = \sum_{i=1}^{\tau} f_i(x) + \gamma \sum_{j=1}^{m}\sum_{i=1}^{\tau} \max\left\{0, \frac{1}{i}\sum_{k=1}^{i} g_k^{(j)}(x)\right\}.$$

The second term is zero when

$$x_\tau \in \hat{X}_\tau^{max} := \left\{x \in D : \sum_{i=1}^{\tau} h_i(x) \leq 0\right\} = \left\{x : \sum_{k=1}^{i} g_k^{(j)}(x) \leq 0, j \leq m, i \leq \tau\right\}$$

Penalties Which Are Large When $x \notin \hat{X}_\tau^{max}$. In addition to requiring the penalty for time-varying constraints to be zero for $x \in \hat{X}_\tau^{max}$, we also require the penalty to be large enough when $x \notin \hat{X}_\tau^{max}$ so as to force the minimum of $\sum_{i=1}^{\tau} F_i(x)$ to lie in set $\hat{X}_\tau^{max}$, or at least to only result in $\mathcal{O}(\sqrt{\tau})$ violation.

As already noted, to use FTRL we need $F_i(\cdot)$ to be uniformly Lipschitz, which requires $f_i(\cdot)$ to be uniformly Lipschitz. When $f_i(\cdot)$ is L_f-Lipschitz then $|\sum_{i=1}^{\tau} f_i(x)|$ may grow linearly with τ at rate τL_f. We therefore require the penalty $\sum_{i=1}^{\tau} h_i(x)$ to also grow at least linearly with τ since otherwise for all τ large enough $|\sum_{i=1}^{\tau} f_i(x)| \gg \sum_{i=1}^{\tau} h_i(x)$ and the penalty may become ineffective i.e. we can have $x_\tau \notin \hat{X}_\tau$ for all τ large enough and so end up with $\mathcal{O}(t)$ constraint violation, which is no good.

We formalize the requirement the sum-penalty $\sum_{i=1}^{\tau} h_i(x)$ in (5) needs to grow quickly enough as follows. Let $\partial\hat{X}_\tau^{max}$ denote the boundary of $\hat{X}_\tau^{max}$, and

$$k_\tau := \min_{x\in\partial\hat{X}_\tau^{max}} \left|\left\{(i,j) : \frac{1}{i}\sum_{k=1}^{i} g_k^{(j)}(x) \geq 0, i=1,\ldots,\tau, j=1,\ldots,m\right\}\right|.$$

That is, k_τ is the minimum number of constraints active at the boundary of $\hat{X}_\tau^{max}$. Observe that $1 \leq k_\tau \leq \tau$ with, for example, $k_\tau = \tau$ when $g_i^{(j)}(x) = g^{(j)}(x)$ does not depend on i.

Condition 2 (Penalty Growth). *Let $z \in D$ be a common Slater point such that $\frac{1}{\tau}\sum_{i=1}^{\tau} g_i^{(j)}(z) < -\eta < 0$ for $j = 1, \dots, m$ and $\tau > t_\epsilon$ (the same z must work for all τ and j). We require that $k_\tau \geq \frac{\beta}{\eta}\tau$ for all $\tau > t_\epsilon$, where $\beta > 0$ and the same β must work for all $\tau = 1, \dots, t$.*

Time-Varying Exact Penalties. We are now in a position to extend the penalty approach to time-varying constraints. We begin by applying Lemma 1 to optimization problem P': $\min_{x \in D} f(x)$ *s.t.* $\frac{1}{i}\sum_{k=1}^{i} g_k^{(j)}(x) \leq 0$, $i = 1, \cdots, t$, $j = 1, \cdots, m$ where $f(\cdot)$ and $g_i^{(j)}(\cdot)$, $i = 1, \dots, t$, $j = 1, \cdots, m$ are convex and $D \subset \mathbb{R}^n$ is convex and bounded. Let $C^* = \arg\min_{x \in \hat{X}_t^{max}} f(x)$. Define

$$H(x) := f(x) + \gamma \sum_{i=1}^{t}\sum_{j=1}^{m} \max\left\{0, \frac{1}{i}\sum_{k=1}^{i} g_k^{(j)}(x)\right\}$$

where $\gamma \in \mathbb{R}$. Note that $H(\cdot)$ is convex since $f(\cdot)$, $g_i^{(j)}(\cdot)$ are convex and composition with max preserves convexity.

Lemma 6. *Assume a Slater point exists, i.e. a $z \in D$ such that $\frac{1}{i}\sum_{k=1}^{i} g_k^{(j)}(z) < -\eta < 0$, $i = 1, \dots, t$, $j = 1, \dots, m$. Let $f^* := \min_{x \in \hat{X}_t^{max}} f(x)$. Then there exists a finite threshold $\gamma_0 \geq 0$ such that $H(x) \geq f^*$ for all $x \in D$, $\gamma \geq \gamma_0$, with equality only when $x \in \hat{X}_t^{max}$. It is sufficient to choose $\gamma_0 \geq \frac{f^* - f(z) - 1}{-k_t \eta}$.*

Proof. Setting the expression for γ_0 to one side for now, the result follows from applying Lemma 1 to P'. Turning now to expression $\gamma_0 \geq \frac{f^* - f(z) - 1}{k_t \eta}$, comparing this with the expression in Lemma 1, observe that the only change is in the denominator, which applying Lemma 1 to P' is $\max_{i \leq t, j \leq m}\{\frac{1}{i}\sum_{k=1}^{i} g_k^{(j)}(z)\} = -\eta$. Referring to (8) in the proof of Lemma 1, it is sufficient the denominator G of γ_0 is such that $\frac{\sum_{(i,j)\in A} \frac{1}{i}\sum_{k=1}^{i} g_k^{(j)}(z)}{G} \geq 1$, where $A \subset \{1, \dots, t\} \times \{1, \dots, m\}$. By assumption $g_k^{(j)}(z) \leq -\eta$ and so $\sum_{j \in A} \frac{1}{i}\sum_{k=1}^{i} g_k^{(j)}(z) \leq -|A|\eta$ with $|A| \geq 1$. Now $k_t \in [1, |A|]$, thus suffices to see setting $G = -k_t\eta$ also meets this requirement.

Theorem 1 (Time-Varying Exact Penalty). *The sequence x_τ, $\tau = 1, \dots, t$ generated by the FTRL update (4) with $F_i(x) = f_i(x) + \gamma h_i(x)$ and $h_i(x) := \sum_{j=1}^{m} \max\{0, \frac{1}{i}\sum_{k=1}^{i} g_k^{(j)}(x)\}$ satisfies $x_{\tau+1} \in \hat{X}_\tau^{max}$ for $\tau > t_\epsilon$ when Condition 2 holds and parameter $\gamma > \frac{E+L+1}{\beta}$ where $E \geq \max_{y \in D, i \in \{1,\dots,t\}} (R_i(y) - R_i(z))/i$, $L \geq \max_{y \in D, i \in \{1,\dots,t\}} f_i(y) - f_i(z)$ with $z \in D$ a Slater point.*

Proof. The result follows by application of Lemma 6 at times $\tau > t_\epsilon$ with $h(x) = R_\tau(x) + \sum_{i=1}^{\tau} f_i(x)$. We have that $h(x) - h(z) = R_\tau(x) - R_\tau(z) + \sum_{i=1}^{\tau}(f_i(x) - f_i(z)) \leq E\tau + L\tau$. Hence for $x_{\tau+1} \in \hat{X}_\tau^{max}$ it is sufficient to choose:

$$\gamma > \gamma_0 = \frac{(E+L)\tau - 1}{k_\tau \eta} \leq \frac{E + L + 1/\tau}{\beta} \leq \frac{E+L+1}{\beta}.$$

When Condition 2 holds, $\beta > 0$.

Theorem 1 states a lower bound on γ in terms of constants E, L and β. For a quadratic regularizer $R_\tau(x) = \sqrt{\tau}\|x\|_2^2$ we can choose $E = \max_{y,z\in D}(\|y\|_2^2 - \|z\|_2^2)$. Since functions f_i are uniformly Lipschitz then $|f_i(z) - f_i(y)| \le L_f\|z - y\| \le L_f\|D\|$ and so we can choose $L = L_f\|D\|$. A value for β may be unknown but to apply Theorem 1 in practice we just need to select γ large enough, so a pragmatic approach is simply to make γ grow with time and then freeze it when it is large enough i.e. when the constraint violations are observed to cease.

4.3 Main Result: Penalized FTRL $\mathcal{O}(\sqrt{t})$ Regret & Violation

Our main result extends the standard FTRL analysis to time-varying constraints:

Theorem 2 (Penalized FTRL). *Assume Conditions 1 and 2 hold for $F_i(x) = f_i(x) + \gamma h_i(x)$ with $h_i(x) = \sum_{j=1}^m \max\{0, \frac{1}{i}\sum_{k=1}^i g_k^{(j)}(x)\}$, and the constraints $g_i^{(j)}$ are uniformly Lipschitz. Let the sequence of actions $\{x_\tau\}_{\tau=1}^t$ be generated by the Pen-FTRL update:*

$$x_{\tau+1} \in \arg\min_{x\in D} R_\tau(x) + \sum_{i=1}^{\tau} F_i(x) \tag{6}$$

Then, if γ is sufficiently large, the regret and constraint violation satisfy:

$$\mathcal{R}_t := \sum_{i=1}^t f_i(x_i) - f_i(y) \le \mathcal{O}(\sqrt{t}), \qquad \mathcal{V}_t := \sum_{i=1}^t h_i(x_i) \le \mathcal{O}(\sqrt{t}), \quad \forall y \in \hat{X}_t^{max}$$

$$\hat{X}_t^{max} = \left\{x \in D : \sum_{k=1}^i g_k^{(j)}(x) \le 0, \forall i \le t, j \le m\right\} = \left\{x \in D : \sum_{k=1}^i h_k(x) = 0, \forall i \le t\right\}$$

Proof. Regret: Applying Lemma 3 then $\sum_{i=1}^t F_i(x_i) - F_i(y) \le \mathcal{O}(\sqrt{t})$ for all $y \in D$. This holds in particular for all $y \in \hat{X}_t^{max}$ and for these points $\sum_{i=1}^t F_i(y) = \sum_{i=1}^t f_i(y)$. Therefore, $\sum_{i=1}^t F_i(x_i) - f_i(y) \le \mathcal{O}(\sqrt{t})$ i.e. $\mathcal{R}_t = \sum_{i=1}^t f_i(x_i) - f_i(y) \le \mathcal{O}(\sqrt{t}) - \gamma\sum_{i=1}^t h_i(x_i) \le \mathcal{O}(\sqrt{t})$ since $h_i(x_i) \ge 0$.

Constraint Violation: By Theorem 1, $x_{\tau+1} \in \hat{X}_\tau^{max}$ for $\tau > t_\epsilon$. Our interest is in bounding the violation of $\hat{X}_{\tau+1}^{max}$ by $x_{\tau+1}$. We can ignore the finite interval from 1 to t_ϵ since it will incur at most a finite constraint violation and so not affect an $\mathcal{O}(\sqrt{t})$ bound i.e. when obtaining the $\mathcal{O}(\sqrt{t})$ bound we can take $t_\epsilon = 0$. We follow a "Be-The-Leader" type of approach and apply Lemma 2 with $F_i(x) = h_i(x)$. We have that $h_i(x) \ge 0$ and by Condition 2, there exists a Slater point $z \in D$ such that $h_i(z) = 0$, $i = 1, \dots, t$. Hence, $\min_{x\in D}\sum_{i=1}^\tau F_i(x) = 0$ and $\arg\min_{x\in D}\sum_{i=1}^\tau F_i(x)$ is not empty. Now, $x_{\tau+1} \in \hat{X}_\tau^{max} = \{x \in D : \sum_{i=1}^\tau h_i(x) = 0\} = \arg\min_{x\in D}\sum_{i=1}^\tau h_i(x)$ i.e. $x_{\tau+1}$ is a Follow-The-Leader update with respect to $\sum_{i=1}^\tau h_i(x)$. Hence, by Lemma 2, it is $\sum_{i=1}^t h_i(y) \ge \sum_{i=1}^t h_i(x_{i+1})$, $\forall y \in D$. Multiplying both sides of this inequality by -1 and adding $\sum_{i=1}^t h_i(x_i)$, it follows that:

$$\sum_{i=1}^t \Big(h_i(x_i) - h_i(y)\Big) \le \sum_{i=1}^t \Big(h_i(x_i) - h_i(x_{i+1})\Big) \ \forall y \in D.$$

In particular, for $y \in \hat{X}_t^{max}$ then $\sum_{i=1}^t h_i(y) = 0$ and so

$$\mathcal{V}_t = \sum_{i=1}^t h_i(x_i) \le \sum_{i=1}^t \Big(h_i(x_i) - h_i(x_{i+1})\Big).$$

Since $g_i^{(j)}$ is uniformly Lipschitz then by Lemma 5, we get that h_i is uniformly Lipschitz, i.e. $|h_i(x_i) - h_i(x_{i+1})| \le L_g\|x_i - x_{i+1}\|$ and $\mathcal{V}_t \le L_g \sum_{i=1}^t \|x_i - x_{i+1}\|$, where L_g is the Lipschitz constant. Since the regularizer $R_\tau(x)$ in the `Pen-FTRL` update is $\sqrt{\tau}$-strongly convex, by Lemma 4 we get that $\|x_i - x_{i+1}\|$ is $\mathcal{O}(1/\sqrt{i})$ and so $\sum_{i=1}^t \|x_i - x_{i+1}\|$ is $\mathcal{O}(\sqrt{t})$. Hence, $\mathcal{V}_t \le \mathcal{O}(\sqrt{t})$ as claimed.

We can immediately generalize Theorem 2 by observing that a sequence of constraints $\{g_i^{(j)}\}$ which are active at no more than $\mathcal{O}(\sqrt{t})$ time steps can be violated while still maintaining $\mathcal{O}(\sqrt{t})$ overall sum-violation.

Corollary 1 (Relaxation). *Define the sets*

$$P_- = \{j : \sum_{i=1}^t \max\{0, \frac{1}{i}\sum_{k=1}^i g_k^{(j)}(x)\} \le \mathcal{O}(\sqrt{t})\}, \quad and \quad P_+ = \{1, \dots, m\} \setminus P_-.$$

In Theorem 2 relax Condition 2 so that it only holds for the subset P_+ of constraints. Then the `Pen-FTRL` *update still ensures $\mathcal{O}(\sqrt{t})$ regret and constraint violation with respect to:*

$$\hat{X}_t^{max} = \Big\{x \in D : \sum_{k=1}^i g_k^{(j)}(x) \le 0, i = 1, \dots, t, j \in P_+\Big\}.$$

In effect, Corollary 1 says that we only need Condition 2 to hold for a *subset* of the constraints (i.e. subset P_+). The effect will be to increase the sum-violation, but only by $\mathcal{O}(\sqrt{t})$. This is the key advantage of the penalty-based approach, namely it allows a soft trade-off between sum-constraint satisfaction/violation, Condition 2 and benchmark set $\hat{X}_t^{max}$. Importantly, note that the `Pen-FTRL` update itself remains unchanged and does not require knowledge of the partitioning of constraints into sets P_+ and P_-.

With this in mind, it is worth noting that we also have the flexibility to partition the constraints in other ways. For example:

Corollary 2. *Consider the setup in Theorem 2 but using penalty*

$$h_i(x) = \sum_{j=1}^m \max\left\{0, \frac{1}{i}\sum_{k=1}^i g_k^{(j)}(x)\right\} + \delta_i^{(j)}(x)$$

Then the `Pen-FTRL` *update ensures regret and violation*

$$\mathcal{R}_t := \sum_{i=1}^t \Big(f_i(x_i) - f_i(y)\Big) \le \mathcal{O}(\sqrt{t}) - \sum_{i=1}^t\sum_{j=1}^m \Big(\delta_i^{(j)}(x_i) - \delta_i^{(j)}(y)\Big)$$

$$\mathcal{V}_t := \sum_{i=1}^t h_i(x_i) \le \mathcal{O}(\sqrt{t}) + \sum_{i=1}^t\sum_{j=1}^m \delta_i^{(j)}(x)$$

for all $y \in \hat{X}_t^{max} = \Big\{x \in D : \sum_{k=1}^i g_k^{(j)}(x) \le 0, i = 1, \dots, t, j = 1, \dots, m\Big\}$.

When $\delta_i^{(j)} \leq \mathcal{O}(1/\sqrt{t})$ then Corollary 2 shows that the Pen-FTRL update achieves $\mathcal{O}(\sqrt{t})$ regret and violation, this Corollary will prove useful in the next section. Other variations of this sort are also possible.

4.4 Necessity of Penalty Growth Condition

Condition 2 is necessary for Theorems 1 and 2 to hold in the sense that when the condition is violated then there exist examples where these theorems fail.

Returning again to the example from the Introduction, selecting $h_i(x)$ according to (5) then $h_i(x) = \max\{0, -0.01\} + \max\{0, \frac{n_{2,i}}{i}x\} = \max\{0, \frac{n_{2,i}}{i}x\}$. Hence, the penalty is $\sum_{i=1}^{\tau} h_i(x) \leq \sum_{i=1}^{\tau} \frac{n_{2,i}}{i}x$. When $n_{2,i} < \mathcal{O}(i)$ then $\sum_{i=1}^{\tau} h_i(x) < \mathcal{O}(\tau)$ (since $\sum_{i=1}^{\tau} \frac{1}{i^c} \leq \int_0^{\tau} \frac{1}{i^c} di = \frac{\tau^{1-c}}{1-c}$ for $0 \leq c \leq 1$) and Condition 2 is violated (since $k_\tau \leq n_{2,\tau} < \mathcal{O}(\tau)$ and so there does not exist any $\beta > 0$ such that $k_\tau \geq \frac{\beta}{\eta}\tau$). For τ large enough the penalty $\sum_{i=1}^{\tau} h_i(x)$ therefore inevitably becomes small relative to $\sum_{i=1}^{\tau} f_i(x) = -2\tau x$, which leads to persistent violation of constraint $x \leq 0$ i.e. Theorem 1 fails. This is what we see in Fig. 2(a).

When $n_{2,i} \leq \mathcal{O}(\sqrt{i})$ then $\frac{n_{2,i}}{i} \leq \mathcal{O}(1/\sqrt{i})$ and the constraint sum-violation $\sum_{i=1}^{\tau} h_i(x) \leq \mathcal{O}(\sqrt{i})$. Hence, Corollary 1 still works even though Theorem 1 fails. However, when $n_{2,i}$ greater than $\mathcal{O}(\sqrt{i})$ but less than $\mathcal{O}(i)$ then the constraint violation is greater than $\mathcal{O}(\sqrt{i})$ and so Corollary 1 also fails.

Note that while we might consider gaining penalty growth by scaling γ with t, this is inadmissible because Condition 1 requires $F_t(x) = f_t(x) + \gamma h_t(x)$ to be uniformly Lipschitz, i.e., the same Lipschitz constant to apply at all times t.

4.5 Constraints Satisfying Penalty Growth Condition

A natural question to ask is which classes of time-varying constraints satisfy Condition 2. In this section we present some useful examples. In particular, we consider the classes of constraints considered by [11] and [14], since these are the only previous works for time-varying constraints that report $\mathcal{R}_t, \mathcal{V}_t = \mathcal{O}(\sqrt{t})$.

Perturbed Constraints. In [11] the considered constraints are of the form:

$$g_i^{(j)}(x) = g^{(j)}(x) + b_i^{(j)}$$

with common Slater point and $b_i^{(j)}$ upper bounded by some value, i.e., $b_i^{(j)} \leq \bar{b}^{(j)}, \forall i$. For this class of constraints we have that:

$$h_i(x) = \sum_{j=1}^{m} \max\left\{0, \frac{1}{i}\sum_{k=1}^{i}(g^{(j)}(x) + b_k^{(j)})\right\} = \sum_{j=1}^{m} \max\left\{0, g^{(j)}(x) + \frac{1}{i}\sum_{k=1}^{i} b_k^{(j)}\right\}$$

Defining $\underline{b}_t^{(j)} = \frac{1}{t}\sum_{k=1}^{t} b_k^{(j)}$ and $\Delta_i^{(j)}(x) = \frac{1}{i}\sum_{k=1}^{i}(b_k^{(j)} - \underline{b}_t)$, then we can rewrite the penalty equivalently as

$$h_i(x) = \sum_{j=1}^{m} \max\left\{0, g^{(j)}(x) + \underline{b}_t^{(j)}\right\} + \delta_i^{(j)}(x)$$

with $\delta_i^{(j)}(x) = \max\left\{0, g^{(j)}(x) + \underline{b}_t^{(j)} + \Delta_i^{(j)}(x)\right\} - \max\left\{0, g^{(j)}(x) + \underline{b}_t^{(j)}\right\}$. When $|\Delta_i^{(j)}(x)|$ is $\mathcal{O}(1/\sqrt{i})$ then, by Lemma 5, so is $|\delta_i^{(j)}(x)|$. Hence, when $|\Delta_i^{(j)}(x)|$ is $\mathcal{O}(1/\sqrt{i})$ then we can use the fact that Condition 2 holds for constraints $g^{(j)}(x) + \underline{b}_t^{(j)} \leq 0$ to show, by Corollary 2, that the Pen-FTRL update achieves $\mathcal{O}(\sqrt{t})$ regret and violation with respect to benchmark set $\hat{X}_t^{max} = \{x : g^{(j)}(x) + \underline{b}_t^{(j)} \leq 0\}$. This corresponds to one extreme of [11]'s benchmark but Theorem 1 provides more general conditions under which it is applicable, while [11] only considers constraints that are either time-invariant or i.i.d.

Alternatively, defining $\Delta_i^{(j)}(x) = \frac{1}{i}\sum_{k=1}^{i}(b_k^{(j)} - \bar{b}^{(j)})$, we can rewrite the penalty equivalently as:

$$h_i(x) = \sum_{j=1}^{m} \max\left\{0, g^{(j)}(x) + \bar{b}^{(j)}\right\} + \delta_i^{(j)}(x)$$

with $\delta_i^{(j)}(x) = \max\{0, g^{(j)}(x) + \bar{b}^{(j)} + \Delta_i^{(j)}(x)\} - \max\{0, g^{(j)}(x) + \bar{b}^{(j)}\}$. Observe that $\delta_i^{(j)}(x) \leq 0$ since $\Delta_i^{(j)}(x) \leq 0$. Hence, $\delta_i^{(j)}(x)$ does not add to the upper bound on the sum-constraint violation and so, by Corollary 2, the Pen-FTRL update achieves $\mathcal{O}(\sqrt{t})$ regret and violation with respect to benchmark set $\hat{X}_t^{max} = \{x : g^{(j)}(x) + \bar{b}^{(j)} \leq 0\}$. This corresponds to the other extreme of [11]'s benchmark, and in fact corresponds to the weak benchmark X_t^{min} and so is perhaps less interesting.

Families of Constraints. Suppose the time-varying constraint functions $g_i^{(j)}$ are selected from some family. That is, let $A^{(j)} = \{a_1^{(j)}, \ldots, a_{n_j}^{(j)}\}$ be a family of functions indexed by $k = 1, \ldots, n_j$ with $a_k^{(j)} : D \to \mathbb{R}$ being L_g-Lipschitz and $|a_k^{(j)}(x)| \leq a_{max}$ for all $x \in D$. At time i, constraint $g_i^{(j)} = a_k^{(j)}$ for some $k \in \{1, \ldots, n_j\}$, i.e. at each time step the constraint $g_i^{(j)}$ is selected from family $A^{(j)}$. Let $n_{k,\tau}^{(j)}$ denote the number of times that function $a_k^{(j)}$ is visited up to time τ and $p_{k,\tau}^{(j)} = n_{k,\tau}^{(j)}/\tau$ the fraction of times that $a_k^{(j)}$ is visited. With this setup the penalty is:

$$h_i(x) = \sum_{j=1}^{m} \max\left\{0, \frac{1}{i}\sum_{k=1}^{i} g_i^{(j)}\right\} = \sum_{j=1}^{m} \max\left\{0, \sum_{k=1}^{n_j} p_{k,i}^{(j)} a_k^{(j)}(x)\right\}.$$

We proceed by rewriting the penalty equivalently as

$$h_i(x) = \sum_{j=1}^{m} \max\left\{0, \sum_{k=1}^{n_j} p_k^{(j)} a_k^{(j)}(x)\right\} + \delta_i^{(j)}(x)$$

with $\delta_i^{(j)}(x) = \max\left\{0, \sum_{k=1}^{n_j} p_{k,i}^{(j)} a_k^{(j)}(x)\right\} - \max\left\{0, \sum_{k=1}^{n_j} p_k^{(j)} a_k^{(j)}(x)\right\}$ By Lemma 5, $|\delta_i^{(j)}(x)| \leq |\sum_{k=1}^{n_j}(p_{k,i}^{(j)} - p_k^{(j)})a_k^{(j)}(x)|$. Assume the following condition holds:

Condition 3 ($1/\sqrt{t}$-Convergence). *For $\epsilon > 0$ there exists $t_0 > 0$ and $0 \leq p_k^{(j)} \leq 1$, $\sum_{j=1}^{m}\sum_{k=1}^{n_j} p_k^{(j)} = 1$ such that $|p_{k,\tau}^{(j)} - p_k^{(j)}| \leq \epsilon/\sqrt{\tau}$ for all $\tau > t_0$.*

Then for all $\tau > t_0$, $|\delta_i^{(j)}(x)| \leq n_j \frac{\epsilon}{\sqrt{\tau}} a_{max} \leq \bar{n} \frac{\epsilon}{\sqrt{\tau}} a_{max}$ with $\bar{n} := \max_j n_j$. By Corollary 2 it now follows that Pen-FTRL achieves $\mathcal{O}(\sqrt{t})$ regret and violation with respect to benchmark $\hat{X}_t^{max} = \left\{x : \sum_{k=1}^{n_j} p_k^{(j)} a_k^{(j)}(x) \leq 0\right\}$. Observe that in this case $\hat{X}_t^{max} = X_\infty^{max}$, i.e., we obtain $\mathcal{O}(\sqrt{t})$ regret and violation with respect to the strong benchmark, which is very appealing. Note that we don't need to know the relative frequencies in advance for this analysis to work.

Example. Suppose $D = [-10, 10]$, loss function $f_\tau(x) = -2x$ and constraint $g_\tau(x)$ alternates between $a_1(x) = -0.01$ and $a_2(x) = x$, equaling $a_2(x)$ at time τ with probability[2] $0.1c/\tau^{1-c}$. Figure 2(a) shows the performance vs c of the Pen-FTRL update with quadratic regularizer $R_\tau(x) = \sqrt{\tau}x^2$ and $F_\tau(x) = f_\tau(x) + \gamma \max\{0, p_{1,\tau}a_1(x) + p_{2,\tau}a_2(x)\}$ with parameter $\gamma = 25$. It can be seen that for $c = 1$ and $c = 0.5$ the constraint violation is well-behaved, staying close to zero, but for $c \in (0.5, 1)$ the constraint violation grows with time.

What is happening here is that when $c = 1$ then $p_{1,\tau} \to 0.9$, $p_{2,\tau} \to 0.1$ and the penalty term $\gamma \max\{0, p_{1,\tau}a_1(x) + p_{2,\tau}a_2(x)\}$ in $F_\tau(x)$ ensures the violation $\sum_{i=1}^{t} g_i(x) = t(p_{1,t}a_1(x) + p_{2,t}a_2(x))$ stays small. When $c = 0.5$, then $p_{1,\tau} \to 1$, $p_{2,\tau} \to 0$ and the penalty term ensures $tp_{1,t}a_1(x)$ stays small while $tp_{2,t}a_2(x)$ is $\mathcal{O}(\sqrt{t})$, thus $\sum_{i=1}^{t} g_i(x)$ is $\mathcal{O}(\sqrt{t})$. When $c \in (0.5, 1)$ then again $p_{1,\tau} \to 1$, $p_{2,\tau} \to 0$ and the penalty term ensures $tp_{1,t}a_1(x)$ stays small but now $tp_{2,t}a_2(x)$ is larger than $\mathcal{O}(\sqrt{t})$ and so $\sum_{i=1}^{t} g_i(x)$ is also larger than $\mathcal{O}(\sqrt{t})$.

We claim that $1/\sqrt{t}$-convergence is sufficient for Pen-FTRL to achieve $\mathcal{O}(\sqrt{t})$ regret and violation with respect to X^*, but it remains an open question whether or not it is also a necessary condition. Nevertheless, in simulations we observe that when $1/\sqrt{t}$-convergence does not hold then performance is often poor and that this is not specific to the FTRL algorithm, e.g. Figure 2(b) illustrates the performance of the canonical online primal-dual update (e.g. see [11]),

$$x_{t+1} = \Pi_D\left(x_t - \alpha_t(\partial f_t(x_t) + \lambda_t \partial g_t(x_t))\right), \; \lambda_{t+1} = \left[\lambda_t + \alpha_t g_t(x_{t+1})\right]^+ \quad (7)$$

where Π_D denotes projection onto set D and step size $\alpha_t = 5/\sqrt{t}$.

I.i.d Stochastic Constraints. In [14] i.i.d. constraint functions drawn from a family are considered and a primal-dual algorithm is presented that achieves

[2] Recall that $c\sum_{\tau=0}^{t} \frac{1}{\tau^{1-c}} \approx c\int_0^t \frac{1}{\tau^{1-c}} d\tau = t^c$ for $0 \leq c \leq 1$. Hence, with this choice $E[n_{2,t}] \approx 0.1t^c$ and $E[p_{2,t}] \approx 0.1t^{c-1}$..

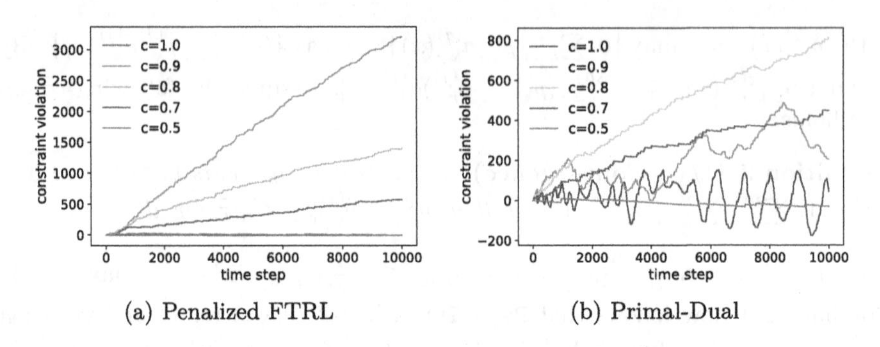

(a) Penalized FTRL (b) Primal-Dual

Fig. 2. Example about the role of $1/\sqrt{t}$-convergence in achieving $\mathcal{V}_t = \mathcal{O}(\sqrt{t})$.

$\mathcal{O}(\sqrt{t})$ regret and expected violation. Since with high probability the empirical mean converges at rate $1/\sqrt{t}$ with high probability, we can immediately apply the foregoing analysis to the sample paths to show that Pen-FTRL achieves $\mathcal{O}(\sqrt{t})$ regret and violation with respect to X_t^{max} with high probability. In more detail, let indicator random variable $I_{k,i}^j = 1$ when constraint function $a_k^{(j)}$ is selected at time i, and otherwise $I_{k,i}^{(j)} = 0$. By the law of large numbers (we can use any convenient concentration inequality, e.g. Chebyshev), with high probability the empirical mean satisfies $|\frac{1}{\tau}\sum_{i=1}^{\tau} I_{k,i}^{(j)} - p_k^j]| \leq 1/\sqrt{\tau}$ with high probability. That is, Condition 3 holds with high probability and we are done.

Periodic Constraints. Let indicator $I_{k,i}^j = 1$ when constraint function $a_k^{(j)}$ is selected at time i, and otherwise $I_{k,i}^{(j)} = 0$. When the constraints are visited in a periodic fashion then

$$I_{k,i}^{(j)} = \begin{cases} 1 & i = nT_k^{(j)}, n = 1, 2, \ldots \\ 0 & \text{otherwise} \end{cases}$$

where $T_k^{(j)}$ is the period of constraint $a_k^{(j)}$. Then $|\frac{1}{\tau}\sum_{i=1}^{\tau} I_{k,i}^{(j)} - \frac{1}{T_k^{(j)}}| = \frac{1}{\tau}|\lfloor\frac{\tau}{T_k^{(j)}}\rfloor - \frac{\tau}{T_k^{(j)}}| \leq \frac{1}{\tau}$. Hence Condition 3 holds and we are done.

5 Summary and Conclusions

In this paper we extend the classical FTRL algorithm to encompass time-varying constraints by leveraging, for the first time in this context, the seminal penalty method of [12]. We establish sufficient conditions for this new Pen-FTRL algorithm to achieve $\mathcal{O}(\sqrt{t})$ regret and violation with respect to a strong benchmark $\hat{X}_t^{max}$ that expands significantly the previously-employed benchmarks in the literature. This result matches the performance of the best existing primal-dual algorithms in terms of regret and constraint violation growth rates, while substantially extending the class of problems covered. The key to this improvement lies in how the time-varying constraints are incorporated into the FTRL algorithm. We conjecture that adopting a similar formulation with a primal-dual algorithm, namely using:

$$x_{t+1} = \Pi_D(x_t - \alpha_t(\partial f_t(x_t) + \lambda_t \partial h_t(x_t))),\ \lambda_{t+1} = [\lambda_t + \alpha_t h_t(x_{t+1})]^+$$

where $h_t(x) = \frac{1}{t}\sum_{i=1}^{t} g_i(x_{t+1})$, would allow similar performance to be achieved by primal-dual algorithms as by FTRL but we leave this to future work.

Acknowledgments. The authors acknowledge support from Science Foundation Ireland (SFI) under grant 16/IA/4610, and from the European Commission through Grant No. 101017109 (DAEMON).

Appendix A: Proofs

A.1 Proof of Lemma 1

Proof. Firstly note that for feasible points $x \in X$ we have that $g^{(j)}(x) \le 0$, $j = 1, \cdots, m$ and so $F(x) = f(x)$. By definition $f(x) \ge f^* = \inf_{x \in X} f(x)$ and so the stated result holds trivially for such points. Now consider an infeasible point $w \notin X$. Let z be an interior point satisfying $g^{(j)}(z) < 0$, $j = 1, \cdots, m$; by assumption such a point exists. Let $\gamma_0 = \frac{f^* - f(z) - 1}{G}$. It is sufficient to show that $F(w) > f^*$ for $\gamma \ge \gamma_0$ and $G = \max_{j \in \{1, \cdots, m\}}\{g^{(j)}(z)\}$.

Let $v = \beta z + (1-\beta)w$ be a point on the chord between points w and z, with $\beta \in (0,1)$ and v on the boundary of X (that is $g^{(j)}(v) \le 0$ for all $j = 1, \cdots, m$ and $g^{(j)}(v) = 0$ for at least one $j \in \{1, \cdots, m\}$). Such a point v exists since z lies in the interior of X and $w \notin X$. Let $A := \{j : j \in \{1, \cdots, m\}, g^{(j)}(v) = 0\}$ and $t(x) := f(x) + \gamma \sum_{j \in A} g^{(j)}(x)$. Then $t(v) = f(v) \ge f^*$. Also, by the convexity of $g^{(j)}(\cdot)$ we have that for $j \in A$ that $g^{(j)}(v) = 0 \le \beta g^{(j)}(z) + (1-\beta)g^{(j)}(w)$. Since $g^{(j)}(z) < 0$, it follows that $g^{(j)}(w) > 0$. Hence, $\sum_{j \in A} g^{(j)}(w) = \sum_{j \in A} \max\{0, g^{(j)}(w)\} \le \sum_{j=1}^{m} \max\{0, g^{(j)}(w)\}$ and so $t(w) \le F(w, \gamma)$. Now, observe that $t(z) = f(z) + \gamma \sum_{j \in A} g^{(j)}(z) \le f(z) + \gamma_0 \sum_{j \in A} g^{(j)}(z)$ since $g^{(j)}(z) < 0$ and $\gamma \ge \gamma_0$. Hence,

$$t(z) \le f(z) + (f^* - f(z) - 1)\frac{\sum_{j \in A} g^{(j)}(z)}{G} \tag{8}$$

Selecting G such that $\frac{\sum_{j \in A} g^{(j)}(z)}{G} \ge 1$ then $t(z) \le f^* - 1 \le t(v) - 1$. So we have established that $f^* \le t(v)$, $t(z) \le t(v) - 1$ and $t(w) \le F(w)$. Finally, by the convexity of $t(\cdot)$, $t(v) \le \beta t(z) + (1-\beta)t(w)$. Since $t(z) \le t(v) - 1$ it follows that $t(v) \le \beta(t(v) - 1) + (1-\beta)t(w)$ i.e. $t(v) \le -\frac{\beta}{1-\beta} + t(w)$. Therefore $f^* \le -\frac{\beta}{1-\beta} + F(w) < F(w)$ as claimed.

References

1. Anderson, D., Iosifidis, G., Leith, D.J.: Lazy lagrangians for optimistic learning With budget constraints. IEEE ACM Trans. Netw. (2023). IEEE
2. Chen, T., Ling, Q., Giannakis, G.B.: An online convex optimization approach to proactive network resource allocation. IEEE Trans. Signal Process. **65**(24), 6350–6364 (2017)
3. Hazan, E.: Introduction to online convex optimization. Found. Trends Optim. **2**, 157–325 (2016)
4. Jenatton, R., Huang, J.C., Archambeau, C.: Adaptive algorithms for online convex optimization with long-term constraints. In: Proceedings of ICML, pp. 402–411 (2016)
5. Liakopoulos, N., Destounis, A., Paschos, G., Spyropoulos, T., Mertikopoulos, P.: Cautious regret minimization: online optimization with long-term budget constraints. In: Proceedings of ICML, pp. 3944–3952 (2019)
6. Mahdavi, M., Jin, R., Yang, T.: Trading regret for efficiency: online convex optimization with long term constraints. J. Mach. Learn. Res. **13**(81), 2503–2528 (2012)
7. Mannor, S., Tsitsiklis, J.N., Yu, J.Y.: Online learning with sample path constraints. J. Mach. Learn. Res. **10**(20), 569–590 (2009)
8. McMahan, H.B.: A survey of algorithms and analysis for adaptive online learning. J. Mach. Learn. Res. **18**, 1–50 (2017)
9. Shalev-Shwartz, S.: Online learning and online convex optimization. Found. Trends Optim. **4**, 107–194 (2011)
10. Sun, W., Dey, D., Kapoor, A.: Safety-aware algorithms for adversarial contextual bandit. In: Proceedings of ICML, pp. 3280–3288 (2017)
11. Valls, V., Iosifidis, G., Leith, D., Tassiulas, L.: Online convex optimization with perturbed constraints: optimal rates against stronger benchmarks. In: Proceedings of AISTATS, pp. 2885–2895 (2020)
12. Zangwill, W.J.: Nonlinear programming via penalty functions. Manag. Sci. **13**(5), 344–358 (1967)
13. Yi, X., Li, X., Xie, L., Johansson, K.H.: Distributed online convex optimization with time-varying coupled inequality constraints. IEEE Trans. Signal Process. **68**, 731–746 (2020)
14. Yu, H., Nelly, M., Wei, X.: Online convex optimization with stochastic constraints. In: Proceedings of NIPS (2017)
15. Zinkevich, M.: Online convex programming and generalized infinitesimal gradient ascent. In: Proceedings of ICML (2003)

Rethinking Exponential Averaging of the Fisher

Constantin Octavian Puiu(✉)

University of Oxford, Mathematical Institute, Oxford, UK
constantin.puiu@maths.ox.ac.uk

Abstract. In optimization for Machine learning (ML), it is typical that curvature-matrix (CM) estimates rely on an exponential average (EA) of local estimates (giving EA-CM algorithms). This approach has little principled justification, but is very often used in practice. In this paper, we draw a connection between EA-CM algorithms and what we call a *"Wake of Quadratic models"*. The outlined connection allows us to understand what EA-CM algorithms are doing from an optimization perspective. Generalizing from the established connection, we propose a new family of algorithms, *KL-Divergence Wake-Regularized Models* (KLD-WRM). We give three different practical instantiations of KLD-WRM, and show numerically that these outperform K-FAC on MNIST.

Keywords: Optimization · Natural gradient · KL Divergence · Overfit

1 Introduction

Recent research in optimization for ML has focused on finding tractable approximations for curvature matrices. In addition to employing ingenious approximating structures [1–3], curvature matrices are typically estimated as an exponential average (EA) of the previously encountered local estimates [1,3–6]. This is in particular true for Natural Gradient (NG) algorithms, which we focus on here. These exponential averages emerge rather heuristically, and have so far only been given incomplete motivations. The main such motivation is *"allowing curvature information to depend on much more data, with exponentially less dependence on older data"* [1]. However, it remains unclear what such an EA algorithm is actually doing from an optimization perspective. In this paper, we show that such EA algorithms can be seen as solving a sequence of local quadratic models whose construction obeys a recursive relationship - which we refer to as a *"Wake of Quadratic Models"*. Inspired by this recursion, we consider a similar, more principled and general recursion for our local models - which we refer to as a *"KL-Divergence Wake-Regularized Models"* (KLD-WRM). We show that under *suitable approximations*, KLD-WRM gives very similar optimization steps to EA-NG. This equivalence raises

Supplementary Information The online version contains supplementary material available at https://doi.org/10.1007/978-3-031-26419-1_20.

© The Author(s), under exclusive license to Springer Nature Switzerland AG 2023
M.-R. Amini et al. (Eds.): ECML PKDD 2022, LNAI 13717, pp. 327–343, 2023.
https://doi.org/10.1007/978-3-031-26419-1_20

the hope that using better approximations in our proposed class of algorithms (KLD-WRM) might lead to algorithms which outperform EA-NG. We propose three different practical instantiations of KLD-WRM (of increasing approximation accuracy) and compare them with K-FAC, the most widely used practical implementation of NG for DNNs. Numerically, KLD-WRM outperforms K-FAC on MNIST: with higher test accuracy, lower test loss, and lower variance.

2 Preliminaries

2.1 Neural Networks, Supervised Learning and Notation

We very briefly look at *fully-connected (FC) nets* (we omit CNN for simplicity). We have $\bar{a}_0 := [x, 1]^T$ (x is the input to the net). The *pre-activation* at layer l: $z_l = W_l \bar{a}_{l-1}$ for $l \in \{1, 2, ..., n_L\}$. The *post-activation* at layer l: $a_l = \phi_l(z_l)$ for $l \in \{1, 2, ..., n_L\}$. The *augmented post-activation* at layer l: $\bar{a}_l = [a_l, 1]^T$ for $l \in \{1, 2, ..., n_L - 1\}$ (we augment the post-activations s.t. we can incorporate the bias into the weight matrix W_{l+1} w.l.o.g.; this is standard practice for K-FAC [1]).

In the above, we consider n_L layers, and $\phi_l(\cdot)$ are nonlinear activation functions. We collect all parameters, namely $\{W_l\}_{l=1}^{n_L}$, in a parameter vector $\theta \in \mathbb{R}^d$. Let us consider $N_o \in \mathbb{Z}^+$ neurons in the output layer. The output of the net is $h_\theta(x) := a_{n_L} \in \mathbb{R}^{N_o}$. The predictive (model) distribution of $y|x$ is $p_\theta(y|x) = p_\theta(y|h_\theta(x))$, that is $p(y|x)$ depends on x only through $h_\theta(x)$.

In supervised learning, where we have labeled pairs of datasets $\mathcal{D} = \{(x_i, y_i)\}_i$. Our objective to minimize is typically some regularized modification of

$$f(\theta) = -\log p(\mathcal{D}|\theta) = \sum_{(x_i, y_i) \in \mathcal{D}} \big(-\log p(y_i|h_\theta(x_i)) \big). \tag{1}$$

Thus, we have our loss function $L(y_i, x_i; \theta) := -\log p(y_i|h_\theta(x_i))$, and $f(\theta) = \sum_{(x_i, y_i) \in \mathcal{D}} L(y_i, x_i; \theta)$. This is what we will focus on here, for simplicity of exposition, but our ideas directly apply to different ML paradigms, such as Variational Inference (VI) in Bayesian Neural Networks (BNNs), and to RL as well.

Let us consider the iterates $\{\theta_k\}_{k=0,1,...}$, with θ_0 initialized in standard manner (perhaps on the edge of Chaos)[1]. Let us denote the optimization steps taken at θ_k as s_k, that is $\theta_{k+1} = \theta_k + s_k$. Let $g(\theta)$ and $H(\theta)$ be the gradient and hessian of our objective $f(\theta)$. We will use the notation $g_k := g(\theta_k)$ and $H_k := H(\theta_k)$.

2.2 KL-Divergence and Fisher Information Matrix

The *symmetric*[2] KL-Divergence is a distance measure between two distributions:

$$\mathbb{D}_{KL}(p, q) := \frac{1}{2}\Big[\mathbb{D}_{KL}(q\,||\,p) + \mathbb{D}_{KL}(p\,||\,q)\Big] = \frac{1}{2}\mathbb{E}_{x \sim p}\Big[\log \frac{p(x)}{q(x)}\Big] + \frac{1}{2}\mathbb{E}_{x \sim q}\Big[\log \frac{q(x)}{p(x)}\Big]. \tag{2}$$

[1] Note that the initialization issue is orthogonal to our purpose.

[2] For convenience, we will refer to the *symmetric KL-Divergence* as *SKL-Divergence*.

In our case, we are interested in the symmetric KL-divergence (SKL) between the data joint distribution with one parameter value and the data joint distribution with a different parameter value. We have

$$\mathbb{D}_{KL}(\theta_1, \theta_2) := \mathbb{D}_{KL}\big(p_{\theta_1}(x, y), p_{\theta_2}(x, y)\big). \tag{3}$$

Since we only model the conditional distribution, we let $p_{\theta_i}(x, y) = \hat{q}(x)p(y|h_{\theta_i}(x))$, where $\hat{q}(x)$ is the marginal (empirical) data distribution of x. This gives

$$\mathbb{D}_{KL}(\theta_1, \theta_2) = \frac{1}{N}\sum_{x_i \in \mathcal{D}} \mathbb{D}_{KL}\big(p_{\theta_1}(y|x_i), p_{\theta_2}(y|x_i)\big). \tag{4}$$

The Fisher has multiple definitions depending on the situation, but here we have

$$F_k := F(\theta_k) := \mathbb{E}_{\substack{x \sim \hat{q}(x) \\ y \sim p_\theta(y|x)}} \left[\nabla_\theta \log p_\theta(y|x) \nabla_\theta \log p_\theta(y|x)^T\right]. \tag{5}$$

This is the Fisher of the joint distribution $p_\theta(x, y) = \hat{q}(x)p(y|h_\theta(x))$. We let $F(\theta_k, x) := \mathbb{E}_{y \sim p_\theta(y|x)}[\nabla_\theta \log p_\theta(y|x) \nabla_\theta \log p_\theta(y|x)^T]$ be the Fisher of the conditional distribution $p_\theta(y|x)$. Then, we have $F_k = \mathbb{E}_{x \sim \hat{q}}[F(\theta_k, x)]$. Using the Fisher, we have the following approximation (see [1,7]) for the SKL-Divergence

$$\mathbb{D}_{KL}\big(p_{\theta_1}(y|x), p_{\theta_2}(y|x)\big) \approx \frac{1}{2}(\theta_2 - \theta_1)^T F(\theta_1, x)(\theta_2 - \theta_1), \tag{6}$$

which is exact as $\theta_2 \to \theta_1$. By plugging (6) into (4) and using linearity we get

$$\mathbb{D}_{KL}(\theta_1, \theta_2) \approx \frac{1}{2}(\theta_2 - \theta_1)^T F_1(\theta_2 - \theta_1). \tag{7}$$

2.3 Natural Gradient and K-FAC

The natural gradient (NG) is defined as [7]

$$\nabla_{NG} f(\theta_k) = F_k^{-1} g_k. \tag{8}$$

The NG descent (NGD) step is then taken to be $s_k^{(NGD)} = -\alpha_k \nabla_{NG} f(\theta_k)$, for some stepsize α_k. NGD has favorable properties, the most notable of which is re-parametrization invariance (when the step-size is infinitesimally small) [8]. The NGD step can be expressed as the solution to the quadratic problem (see [1,8])

$$s_k^{(NGD)} := \arg\min_s s^T g_k + \frac{1}{2\alpha_k} s^T F_k s. \tag{9}$$

K-FAC. Storing and inverting the Fisher is prohibitively expensive. K-FAC is a practical implementation of NG which bypasses this problem by approximating the Fisher as a block-diagonal[3] matrix, with each diagonal block further approximated as the Kronecker product of two smaller matrices [1]. That is, we have

$$F_k^{(KFAC)} := \text{blockdiag}\big(\{A_k^{(l)} \otimes \Gamma_k^{(l)}\}_{l=1,\dots,n_L}\big), \tag{10}$$

where each block corresponds to a layer and $\otimes$ denotes the Kronecker product [1]. For example, for FC nets, the Kronecker factors are given by

$$A_k^{(l)} := \mathbb{E}_{x,y\sim p}[\bar{a}_{l-1}\bar{a}_{l-1}^T], \; \Gamma_k^{(l)} := \mathbb{E}_{x,y\sim p}[\nabla_{z_l} L \nabla_{z_l} L^T]. \tag{11}$$

Note that the Kronecker factors depend on θ_k which influences both the forward and backward pass. Also, note they can be efficiently worked with since $(A_k^{(l)} \otimes \Gamma_k^{(l)})^{-1} = [A_k^{(l)}]^{-1} \otimes [\Gamma_k^{(l)}]^{-1}$, and $([A_k^{(l)}]^{-1} \otimes [\Gamma_k^{(l)}]^{-1})v = \text{vec}\big([\Gamma_k^{(l)}]^{-1} V [A_k^{(l)}]^{-1}\big)$, where v maps to V in the same way $\text{vec}(W_l)$ maps to W_l [1].

2.4 Curvature in Practice: Exponential Averaging

In practice, many algorithms (including ADAM and K-FAC) do not use the curvature matrix estimate as computed. Instead, they maintain an exponential average (EA) of it (eg. [3–6]). In the case of NG, this EA is

$$\bar{F}_k := \rho^k F_0 + (1-\rho)\sum_{i=1}^{k} \rho^{k-i} F_i, \tag{12}$$

where $\rho \in [0, 1)$ is the exponential decay parameter. Let us refer to NG algorithms which replace the Fisher, F_k, with its exponential average, $\bar{F}_k$, as EA-NG.

In a similar spirit, K-FAC maintains an EA of the Kronecker factors

$$\bar{A}_k^{(l)} := \rho^k A_0^{(l)} + (1-\rho)\sum_{i=1}^{k} \rho^{k-i} A_i^{(l)}, \;\; \bar{\Gamma}_k^{(l)} := \rho^k \Gamma_0^{(l)} + (1-\rho)\sum_{i=1}^{k} \rho^{k-i} \Gamma_i^{(l)}, \tag{13}$$

and in practice, $A_k^{(l)}$ and $\Gamma_k^{(l)}$ in (10) are replaced with $\bar{A}_k^{(l)}$ and $\bar{\Gamma}_k^{(l)}$ respectively. We will refer to the practical implementation of K-FAC which uses EA for the K-FAC matrices as EA-KFAC, to emphasize the presence of the EA aspect. However, this is the norm in practice rather than an exception, and virtually any algorithm referred to as K-FAC (or as using K-FAC) is in fact an EA-KFAC algorithm.

3 A Wake of Quadratic Models (WoQM)

The idea behind WoQM is simple. Instead of taking a step s_k, at θ_k which relies only on the local quadratic model, $s_k = \arg\min_s g^T s + (1/2)s^T B_k s$, we take a step which relies on an EA of all previous local models. Formally, let us define

$$M_k^{(Q)}(s) := g_k^T s + \frac{\lambda_k}{2} s^T B_k s, \tag{14}$$

[3] Block tri-diagonal approximation is also possible - but this lies outside our scope.

for an arbitrary *symmetric-positive definite* curvature matrix B_k (typically an approximation of H_k or F_k). Note that $1/\lambda_k$ is a step-size (or learning rate) parameter. Our WoQM step, s_k, is then defined as the solution to

$$\min_s \sum_{i=0}^{k} \rho^{k-i} M_i^{(Q)}\left(s + \sum_{j=i}^{k-1} s_j\right), \quad \text{with} \tag{15}$$

$$\lambda_0 = \lambda, \text{ and } \lambda_k = (1-\rho)\lambda, \ \forall k \in \mathbb{Z}^+, \tag{16}$$

where we set $\sum_{j=k}^{k-1} s_j = 0$ by convention, $\rho \in [0,1)$ is an exponential decay parameter, and $\lambda > 0$ is a hyperparameter. While (15) does not appear to be a proper exponential averaging, missing a $1-\rho$ factor in terms where $i \geq 1$, it can be easily rearranged as such by slightly modifying the definition of $M_0^{(Q)}$ (to disobey (14)). To see this, multiply (15) by $(1-\rho)$ and then absorb the $1-\rho$ factor in the definition of $M_0^{(Q)}$. Our stated definition makes the exposition more compact while preserving intuition.

For our choice of model (14), the WoQM step s_k (at θ_k) is the solution of

$$\min_s s^T\left[\sum_{i=0}^{k} \rho^{k-i}\left(g_i + \lambda\kappa(i)B_i \sum_{j=i}^{k-1} s_j\right)\right] + \frac{\lambda}{2}s^T\left[\sum_{i=0}^{k} \kappa(i)\rho^{k-i}B_i\right]s, \tag{17}$$

where we dropped all the constant terms, and have $\kappa(i) := \exp(\mathbb{I}_{\{i>0\}}\log(1-\rho))$, where $\mathbb{I}_{\mathcal{E}}$ is the indicatior function of event $\mathcal{E}$. We use the $\kappa(i)$ term for notational compactness. Note that definition (15) can be used for general models M_i (rather than quadratic $M_i^{(Q)}$), leading to a larger family of algorithms for which WoQM represents a particular instantiation: the *Wake of Models (WoM)* family.

3.1 Connection with Exponential-Averaging in Curvature Matrices

We now look at how the WoQM step relates to Exponential-Averaging Curvature matrices (EA-CM). EA-CM is standard practice in stochastic optimization, and in particular in training DNNs (see for example [1,3–6]). Formally, using EA-CM boils down to taking a step based on (14), but with B_k replaced by

$$\bar{B}_k := \rho^k B_0 + (1-\rho)\sum_{i=1}^{k} \rho^{k-i}B_i = \sum_{i=0}^{k} \kappa(i)\rho^{k-i}B_i. \tag{18}$$

Note that we used the same exponential decay parameter for convenience, but this is not required.

It is obvious from (17) that we can get an analytic solution for WoQM step s_k, as a function of $\{s_j\}_{j=0}^{k-1}$, $\{(g_j, B_j)\}_{j=0}^{k}$, ρ and λ. Thus, by using the relationship recursively we can get an analytic solution for s_k as a function of $\{(g_j, B_j)\}_{j=0}^{k}$, ρ and λ. When doing this, the connection between WoQM and EA-CM is revealed. The result is presented in *Proposition 3.1.*

Proposition 3.1: Analytic Solution of WoQM step *The* WoQM *step s_k at iterate θ_k can be expressed as*

$$s_k = -\lambda^{-1}\bar{B}_k^{-1}g_k, \quad \forall k \in \mathbb{Z}^+. \tag{19}$$

Proof. Relies on simple inductive argument. See supplementary material. □

Proposition 3.1 tells us that the WoQM step is exactly the step obtained by using an EA curvature matrix $\bar{B}_k$ in a simple quadratic model of the form (14). That is, WoQM (15)-(16) is the principled optimization formulation of a EA-CM step[4], when the EA-CM stepsize is constant and equal to $1/\lambda$. Thus, we see what EA-CM is actually doing from an optimization perspective: instead of perfectly solving for the local quadratic model, it solves for a trade-off between all previously encountered models, where the weights of the trade-off are (almost[5]) given by an exponential average (older models receive exponentially less "attention").

There are two observations to make at this point. First, we began by noting that the justification for EA-CM is largely heuristic, but we ended up explaining EA-CM through some optimization model which involved an EA of local models (the WoQM model). Since the EA was the difficult part to justify in the first place, it might seem that we are sweeping the problem from under one rug to another. However, this is not the case. The WoQM formulation aims to reveal a different perspective on EA-CM algorithms, rather than explain the presence of EA itself. It is indeed true that we did not justify why one should use EA and thus get the WoQM family, but this is not required to enhance our understanding and draw conclusions. We can draw conclusions purely based on the established equivalence. This leads us to the second observation, which is why EA-CM improves stability from a stochastic optimization perspective. Rather than conferring stability because it *"uses more data"* [1,4], EA-CM can alternatively be thought of as conferring stability because it uses the collection of all previous noisy local models to build a better model[6] (in terms of both noise and functional form).

Note that what we have discussed so far applies when using any curvature matrix B_k. In particular, *Proposition 3.1* can be directly applied to establish an equivalence between EA-NG algorithms and FISHER-WoQM algorithms (WoQM algorithms with $B_k = F_k$). In fact, we can replace the Fisher by any approximation and still have the equivalence holding, *if the EA is done in the form of (12).*

---

[4] WoQM with $\rho = 0$ and B_k based on quantities at θ_k only is also the principled optimization formulation of no-EA CM algorithms, which take steps of the form (14).

[5] Can modify the definition of $M_0^{(Q)}$ s.t. WoQM is a proper EA of quadratic models.

[6] Although it is still not clear why putting the previous local models together in an exponentially-averaged fashion is *"right"* for conferring further stability - and this aspect remains heuristic. While this could be informally and partly explained by *"older models should matter less"*, the complete explanation remains an open question.

3.2 Fisher-WoQM and Practical K-FAC Equivalence

We have seen (in *Sect.* 2.4) that K-FAC holds an EA for the *Kronecker factors* (see (13)), rather than for the K-FAC approximation to F_k (as in (12)). Thus, the EA scheme employed by K-FAC is *not* the same as (12) with $F_k \leftarrow F_k^{(KFAC)}$. Therefore, we cannot directly apply *Proposition 3.1* to obtain an equivalence between EA-KFAC and KFAC-WoQM (WoQM with $B_k \leftarrow F_k^{(KFAC)}$). We can loosely establish this equivalence by viewing the *EA over the Kronecker factors* as a convenient (but coarse) approximation to the *EA over* $F_k^{(KFAC)}$ (see *Sect.* 4 in the *supplementary material*). Indeed, carrying an EA for $F_k^{(KFAC)}$ is impractical.

Our equivalence reveals that EA-NG is actually solving an exponentially decaying wake of quadratic models (WoQM) where the curvature matrix (meant to be a Hessian approximation) is taken to be an approximate Fisher. This is in contrast with the typical EA-NG interpretation which says that we take NG steps by solving quadratic models of the form (14) with $B_k = F_k$, but then we further approximate F_k as an EA based on $\{\hat{F}_j\}_{j=1}^{k-1}$. While EA-KFAC does not do the exact same thing, it can be seen as a (very crude) approximation to it.

Dealing with Infrequent Updates. In practice, the curvature matrix is not computed at each location, and the EA update is typically performed every $N_u \approx 100$ steps[7] to save computation cost (at least in supervised learning[8]) [1]. In this circumstance, the final implementation of EA-NG would actually be equivalent to a WoQM algorithm where we set

$$B_k = \begin{cases} \hat{F}_k, & \text{if} \quad \text{mod}\,(k, N_u) = 0 \\ \bar{B}_q, & \text{where}\, q := N_u \lfloor \frac{k}{N_u} \rfloor, \text{ otherwise} \end{cases} . \tag{20}$$

In (20), $\hat{F}_k$ represents an approximation for the Fisher F_k, whose computation has an associated cost. Recall that $\bar{B}_k = \sum_{j=0}^{k} \kappa(j)\rho^{k-j} B_j$. Note that (20) is well defined since $\bar{B}_0 = B_0 = \hat{F}_0$, then $B_1 = B_2 = ... = B_{N_u-1} = B_0$, and hence $\bar{B}_1 = \bar{B}_2 = ... = \bar{B}_{N_u-1} = B_0$, and so on. Further note that by (19), defining our WoQM curvature matirx as in (20) gives us the EA-NG matrix that we want, since we can easily see that: $\bar{B}_{N_u} = \rho \bar{B}_0 + (1-\rho)\hat{F}_{N_u} = \rho \hat{F}_0 + (1-\rho)\hat{F}_{N_u}$. We can then easily extend the argument to show $\bar{B}_{qN_u} = \sum_{j=0}^{q} \kappa(j)\rho^{q-j} \hat{F}_{jN_u}^{(KFAC)}$ which is exactly the form of EA that EA-NG employs when updating statistics every N_u steps. Note that we can apply *Proposition 3.1* irrespectively of our choice of B_k, so in particular, it must hold for B_k as defined in (20).

By extending our reasoning, we see that any heuristic which adapts N_u as a function of observations up until k can be transformed into an equivalent heuristic of picking between $B_k = \bar{B}_q$ (where q is now more generally the previous location where we computed $\hat{F}_k$) and $B_k = \hat{F}_k$. Thus, the equivalence between WoQM and EA-NG holds irrespectively of the heuristic which decides when to update the EA-NG matrix. The exact same reasoning holds for generic EA-CM algorithms.

[7] More complicated heuristics can be designed, see [9].

[8] In RL we may prefer updating the K-FAC EA-matrix at each step [10].

3.3 Fisher-WoQM: A Different Perspective

Since we have the following approximation[9] for $k \geq i$ [8]

$$\mathbb{D}_{KL}(\theta_i, \theta_k + s) \approx \tilde{\mathbb{D}}_{KL}(\theta_i, \theta_k + s) := \frac{1}{2}\Big(s + \sum_{j=i}^{k-1} s_j\Big)^T F_i\Big(s + \sum_{j=i}^{k-1} s_j\Big), \quad (21)$$

one might think that a WoQM model with $B_k = F_k$ and $\rho \in (0,1)$ would in fact be some form of approximate *"KL-Divergence Wake-Regularized*[10] *model"*. This is indeed true, as we can write our FISHER-WoQM as

$$\min_s \Big[\sum_{i=0}^{k} \rho^{k-i} g_i\Big]^T s + \lambda \sum_{i=0}^{k} \kappa(i)\rho^{k-i}\tilde{\mathbb{D}}_{KL}(\theta_i, \theta_k + s). \quad (22)$$

Note that $\tilde{\mathbb{D}}_{KL}(\theta_i, \theta_k + s)$ in (21) is a second-order approximation of the SKL-Divergence between $p_{\theta_i}(x, y)$ and $p_{\theta_k+s}(x, y)$. Thus, we see that FISHER-WoQM (same as EA-NG by *Prop. 3.1*) is in fact solving a *regularized linear model*, where the model gradient is taken to be the *momentum gradient* (with parameter ρ), and the regularization term is an exponentially decaying wake of *(crudely) approximate* SKL-divergences relative to previously encountered distributions.

This equivalence raises scope for a new family of algorithms: perhaps using another model for the objective f, rather than a simple linear model based on momentum-gradient, and/or using a better approximation for the KL divergence could lead to better performance. This is the main topic of this paper, explored formally in *Sect.* 4 and numerically in *Sect.* 5. We now note that WoQM (and thus also EA-NG by *Proposition 3.1*, and approximately so, EA-KFAC) is in fact a particular instantiation of the family proposed in the next section.

4 A KL-Divergence Wake-Regularized Models Approach

We now propose a new family of algorithms, which we call *KL-Divergence Wake-Regularized Models (*KLD-WRM*; reads: "Cold-Worm")*. At each location θ_k, the KLD-WRM step s_k is defined as the solution to the problem

$$\min_s M(s; \mathcal{F}_k) + \lambda \sum_{i=0}^{k} \zeta(i)\rho^{k-i}\mathbb{D}_{KL}(\theta_i, \theta_k + s), \quad (23)$$

where $\rho \in [0,1)$, $M(s; \mathcal{F}_k)$ is a model of the objective which uses *at most* all the information ($\mathcal{F}_k$) encountered up until and including θ_k, and $\zeta(i)$ allows for different 'λ's' at different i's. Simple choices would be $\zeta(i) \equiv 1$, or $\zeta(i) = \kappa(i)$.

[9] Which is exact in the limit as $\theta_k + s \to \theta_i$, but would nevertheless be very crude in practice, particularly for large $k - i$, since the steps taken might be relatively large.

[10] Regularizing w.r.t. SKL divergences relative to all previous distributions, as opposed to just the most recent one - as the quadratic model associated with NG step does.

The motivation behind KLD-WRM is two-fold. First, a wake of SKL regularization allows us to stay close (in a KL sense) to all previously encountered distributions, rather than only to the most recent one. Thus, we might expect KLD-WRM to give more conservative steps in terms of distribution ($p_\theta(y|x)$) change. Second, KLD-WRM can be seen as a generalization[11] of EA-NG (which is also FISHER-WoQM), which also *"undoes"* the approximation[12] of SKL.

Note that (23) is the most general formulation of KLD-WRM, but in order to obtain practical algorithms we have to make further approximations and definitions (of M and ζ). For example, we could set

$$M(s; \mathcal{F}_k) = \sum_{i=0}^{k} \nu^{k-i} M_i\left(s + \sum_{j=i}^{k-1} s_j\right) \tag{24}$$

where the models $M_i(s)$ are general models (not necessary quadratic), constructed only based on local information at θ_i. This would be a *Wake of Models* KLD-WRM (WoM-KLD-WRM). We could make this even more particular, for example by considering instantiations where $\nu = 0$, but $\rho \in [0, 1)$ in (24). This would give a *Local-Model* KLD-WRM (LM-KLD-WRM), whose steps s_k are the solution to

$$\min_s M_k(s) + \lambda \sum_{i=0}^{k} \zeta(i) \rho^{k-i} \mathbb{D}_{KL}(\theta_i, \theta_k + s), \tag{25}$$

In this paper, we focus on LM-KLD-WRM instantiations (a particular sub-family of KLD-WRM), and leave the general case as future work. We give three instantiations of LM-KLD-WRM of increasing complexity, and discuss the links with already existing methods. We investigate their performance in *Sect.* 5.

4.1 Connection Between KLD-WRM and Fisher-WoQM

It is easy to see that setting $\nu = \rho$, $\zeta(i) = \kappa(i)$, $M_i(s) = s^T g_i$ and approximating $\mathbb{D}_{KL}(\theta_i, \theta_k + s) \approx \tilde{\mathbb{D}}_{KL}(\theta_i, \theta_k + s)$ (defined in (21)) in a WoM-KLD-WRM gives the WoQM family. Note that WoQM is *not* an LM-KLD-WRM model as $M_k(s)$ can only include information local to θ_k (eg. cannot include g_{k-1}). However, the simplest instantiation of LM-KLD-WRM takes steps which are formally similar to FISHER-WoQM (and thus EA-NG) steps. We investigate this in *Sect.* 4.2.

4.2 Simplest KLD-WRM Instantiation: Smallest Order KLD-WRM

The simplest practical instantiation of KLD-WRM, *Smallest Order* KLD-WRM (SO-KLD-WRM), uses the most crude approximations to LM-KLD-WRM (25) and sets

[11] The linear model $\left[\sum_{i=0}^{k} \rho^{k-i} g_i\right]^T s$ in (22) becomes arbitrary, and $\kappa(i)$ is replaced by a general $\zeta : \mathbb{Z}^+ \to \mathbb{R}$.

[12] For small $||s||$ we have $\mathbb{D}(\theta, \theta + s) \approx \mathbb{D}(\theta||\theta + s) \approx \mathbb{D}(\theta + s||\theta) \approx (1/2)s^T F(\theta)s$. Thus, the generalization towards our family could use $\xi\mathbb{D}(\theta||\theta + s) + (1 - \xi)\mathbb{D}(\theta + s||\theta)$ $\forall \xi \in \mathbb{R}$, instead of the SKL (i.e. $\xi = 1/2$). We choose SKL for simplicity.

$\zeta(i) = \kappa(i)$. The SO-KLD-WRM step s_k (at θ_k) is given by

$$\min_s g_k^T s + \lambda \sum_{i=0}^{k} \kappa(i)\rho^{k-i}\tilde{\mathbb{D}}_{KL}(\theta_i, \theta_k + s), \tag{26}$$

It is trivial to see that SO-KLD-WRM differs from FISHER-WOQM (22) only through replacing $\left[\sum_{i=0}^{k} \rho^{k-i} g_i\right]$ with g_k. Since FISHER-WOQM is equivalent to EA-NG, one might expect that SO-KLD-WRM steps are *formally* similar to EA-NG steps. *Proposition 4.1* formalizes this result.

Proposition 4.1: Analytic Solution of SO-KLD-WRM step *The* SO-KLD-WRM *step s_k at iterate θ_k can be expressed as*

$$s_k = -\lambda^{-1} \bar{F}_k^{-1}[g_k - \rho g_{k-1}], \tag{27}$$

$\forall k \in \mathbb{Z}^+$, *where $\bar{F}_k := \sum_{i=0}^{k} \kappa(i)\rho^{k-i} F_i$ and we set $g_{-1} := 0$ by convention.*

Proof. By induction. See the *supplementary material.* □

Note that we set $g_{-1} = 0$ to avoid providing two separate cases (for $k = 0$ and for $k \geq 1$). *Proposition 4.1* tells us that the SO-KLD-WRM step is *formally similar* to the FISHER-WOQM step (which we have seen is the EA-NG step). The only (formal) difference between the SO-KLD-WRM step and the EA-NG step is that g_k gets replaced by $g_k - \rho g_{k-1}$ in (19). By *"formally"* here, we mean that the expressions look very similar. However, when considering the two implemented algorithms, the paths taken can be very different.

We have seen that EA-NG (being equivalent to FISHER-WOQM) algorithms are in fact a sub-family of the KLD-WRM family, and obviously SO-KLD-WRM is also a sub-family of KLD-WRM. Thus, the formal difference in the steps between SO-KLD-WRM and EA-NG tells us these two sub-families are distinct. The formal similarity between SO-KLD-WRM and EA-NG, combined with the fact that both are members of the KLD-WRM family raises hopes that more accurate instantiations KLD-WRM might lead to better performance than EA-NG.

Note that basing our SO-KLD-WRM step computation on *Proposition 4.1* gives a tractable algorithm, while solving (26) directly gives an intractable algorithm. To see this, review definition (21), and realize that solving equation (26) *directly* requires storing all previous $\bar{F}_j s_j$ matrix-vector products as a minimum (if the algorithm is written efficiently). Thus, we have an exploding number of vectors that need to be stored when solving (26) *directly*, which eventually will overflow the memory - giving an untractable algorithm. Conversely, using *Proposition 4.1* only requires storing at most 2 matrices ($\bar{F}_k$ and F_k) and 2 gradient-shaped vectors at any one moment in time. Thus, using *Proposition 4.1* with tractable approximations for F_k gives a tractable algorithm.

Reconsidering Gradient Momentum in EA-KFAC. By comparing (22) and (26), we see that FISHER-WOQM (which is also EA-NG) can *almost* be seen as SO-KLD-WRM with added momentum for the gradient. The equivalence would be exact if we would have a $\kappa(i)$ inside the sum of the linear term of (22). In

EA-KFAC, gradient momentum is not added in the standard fashion. A different way to add momentum is proposed[13] and presented as successful, perhaps because trying to add gradient momentum in the standard fashion gives worse performance [1]. From our discussion, this could be because EA-KFAC can be interpreted as an approximate EA-NG, and EA-NG is a SO-KLD-WRM algorithm which already has included gradient momentum. Thus, further adding momentum does not make sense. Note that by adding momentum to gradient in the *standard fashion* we mean replacing g_k by $\sum_{i=0}^{k} \kappa(i)\rho^{k-i} g_k$ (see [3]).

SO-KLD-WRM in Practice. When implementing SO-KLD-WRM in practice, we use the K-FAC approximation of the Fisher. Note that *Proposition 4.1* tells us that we need *not* store all previous K-FAC matrices. Instead, we can only save the EA-KFAC matrix. As we have discussed in *Sect.* 3, we can skip computing the K-FAC matrix at some locations to save on computation. To do that, we just pretend the new K-FAC matrix at θ_k is the EA-KFAC that we currently have stored. For example, if we want to compute K-FAC matrices only once in N_u steps, we use (20), but different heuristics can also be used (for eg. as in [9]). Of course, in practice we store an EA for the Kronecker factors (instead of an EA for the K-FAC matrix), as is standard with practical K-FAC implementation [1].

4.3 Second KLD-WRM Instantiation: Q-KLD-WRM

The *Quadratic* KLD-WRM (Q-KLD-WRM) is an LM-KLD-WRM instantiation which uses a second-order approximation for both the Model and the SKLs, and sets $\zeta(i) = \kappa(i)$. The Q-KLD-WRM step s_k (at θ_k) solves

$$\min_s g_k^T s + \frac{1}{2} s^T B_k s + \lambda \sum_{i=0}^{k} \kappa(i)\rho^{k-i} \tilde{\mathbb{D}}_{KL}(\theta_i, \theta_k + s), \tag{28}$$

where B_k is a curvature matrix which aims to approximate the Hessian H_k. That is, Q-KLD-WRM sets M_k in (25) to be a quadratic model. Since (28) is overall a quadratic, we can obtian an *analytic solution* for the Q-KLD-WRM step.

Proposition 4.2: Analytic Solution of Q-KLD-WRM Step. *The* Q-KLD-WRM *step s_k at location θ_k is given by the solution to the problem*

$$\min_s s^T \hat{g}_k + \frac{\lambda}{2} s^T \left[\bar{F}_k + \frac{1}{\lambda} B_k\right] s \tag{29}$$

where $\hat{g}_k$ is given by the one-step recursion

$$\hat{g}_{k+1} = g_{k+1} + \rho(I - \hat{M}_k)\hat{g}_k - \rho g_k, \quad \forall k \in \mathbb{Z}^+, \tag{30}$$

with $\hat{g}_0 := g_0$, $\bar{F}_k := \sum_{i=0}^{k} \kappa(i)\rho^{k-i} F_i$, and $\hat{M}_k := \left[I + \frac{1}{\lambda} B_k \bar{F}_k^{-1}\right]^{-1}$. That is, the Q-KLD-WRM *step is formally given by $s_k = -\frac{1}{\lambda}\left[\bar{F}_k + \frac{1}{\lambda} B_k\right]^{-1} \hat{g}_k$.*

[13] More akin to a subspace method rather than a standard momentum method (see [1]).

Proof. By induction. See the *supplementary material.* □

By comparing *Propositions 4.1* and *4.2*, we see that, unlike SO-KLD-WRM, the Q-KLD-WRM step deviates significantly from the FISHER-WoQM step (also the EA-NG step). Note that setting $B_k = 0$ in *Proposition 4.2* gives *Proposition 4.1*. That is, SO-KLD-WRM is a particular case of Q-KLD-WRM (with $B_k \equiv 0$).

Note that $\{\hat{g}_k\}$ could explode with k, leading to divergence. A sufficient condition for $\{\hat{g}_k\}$ to stay bounded is $\rho \left\| I - \hat{M}_k \right\|_2 \leq \delta$, $\forall k \in \mathbb{Z}^+$, with $\delta \in (0,1)$ and that $\|\nabla f(\theta)\|_2 \leq K_g$, $\forall \theta$. Under this condition, one can see that $\|\hat{g}_k\|_2 \leq \|g_k - \rho g_{k-1}\|_2 + \delta \|\hat{g}_{k-1}\|_2$. Applying this inequality recursively, and noting that $\|g_k - \rho g_{k-1}\|_2 \leq (1+\rho)K_g$, we get that our sufficient condition yields $\|\hat{g}_k\|_2 \leq \frac{2}{1-\delta}K_g$, $\forall k \in \mathbb{Z}^+$. The condition $\rho \left\| I - \hat{M}_k \right\|_2 = \rho \left\| I - \left[I + \frac{1}{\lambda} B_k \bar{F}_k^{-1}\right]^{-1} \right\|_2 \leq \delta$ can always be achieved in practice, since taking $\lambda \to \infty$ or $\rho \to 0$ gives $\delta = 0$.

In a similar fashion to the role of *Proposition 4.1*, the role of *Proposition 4.2* (besides highlighting any similarity or dissimilarity to EA-NG) is to give a tractable algorithm. Again, as with SO-KLD-WRM, Q-KLD-WRM is not tractable if we implement it by solving (28) directly for the same reasons. On the other hand, implementing *Proposition 4.2* requires simultaneous storage of *at most* 3 matrices and 3 gradient-shaped vectors at any one point in time - that is, the storage cost does not explode with k. However, because we now have two different sets of matrices involved: $\{F_k\}$ and $\{B_k\}$, the situation is more subtle. In particular, if B_k and F_k[14] are block-diagonal, then we can see that all involved matrices are block-diagonal, and thus tractably storable and invertable (eg. choose $B_k = F_k$, and approximate $F_k \approx \hat{F}_k^{(KFAC)}$[15]). On the contrary, B_k and F_k might be tractably storable and invertible in isolation, but if they have different structures, computing s_k from *Proposition 4.2* might be intractable.

Q-KLD-WRM in Practice. In practice, we can in principle employ one of two approximations for B_k. The first option is to use a *BFGS* approximation [11,12]. The second option is to replace B_k by the K-FAC matrix. This latter approximation can be justified through the qualified equivalence between the *Fisher* and the *Generalized Gauss-Newton (GGN)* matrix, the latter of which is an approximation to the Hessian. However, in order for the qualified equivalence to hold, we need our predictive distribution $p(y|h_\theta(x))$ to be in the exponential family with natural parameters $h_\theta(x)$ (thinking of each conditional $y|x$ as a different distribution with its own parameters; see [8]). This qualified equivalence holds for most practically used models (see [8]), so is not of huge practical concern.

As we have discussed, it is not obvious how one could get a tractable algorithm from *Proposition 4.2* if the structures of B_k and F_k are dissimilar. Thus, in this paper we focus on instantiations where $B_k := F_k \approx F_k^{(KFAC)}$. In our experiments, we will choose $p(y|h_\theta(x))$ such that the qualified equivalence between Fisher and GGN matrix holds, and thus use $B_k \approx F_k^{(KFAC)}$, $F_k \approx F_k^{(KFAC)}$. As

[14] Of course, $\bar{F}_k$ will have the same structure as F_k.

[15] Using K-FAC further reduces the storage and computation cost through the K-factors.

is typical with K-FAC, we also choose to store an EA for the Kronecker factors, rather than for the block-diagonal matrix. With these choices, one can efficiently compute the Q-KLD-WRM step form *Proposition 4.2* (details in the *supplementary material*).

4.4 Third KLD-WRM Instantiation: QE-KLD-WRM

The *Quadratic Objective Approximation Exact SKL*[16] KLD-WRM (QE-KLD-WRM) is the final instantiation of LM-KLD-WRM which we propose here. The QE-KLD-WRM step s_k (at θ_k) solves

$$\min_s g_k^T s + \frac{1}{2} s^T B_k s + \lambda \sum_{i=0}^{k} \zeta(i) \rho^{k-i} \mathbb{D}_{KL}(\theta_i, \theta_k + s), \tag{31}$$

where B_k is a curvature matrix at θ_k, treated the same as we did in Q-KLD-WRM.

Practical QE-KLD-WRM for Regression. To be able to work with the exact SKL, we restrict ourselves to a class of $p_\theta(y|x)$ models where the SKL can be expressed in terms of euclidean norms of differences in the network output space. Consider equation (4). For predictive distributions of the form (which are used in regression)

$$p(y|h_{\theta_k}(x_j)) = \mathcal{N}(y|h_{\theta_k}(x_j); I), \tag{32}$$

we have a special form for $D_{KL}\big(p(y|h_{\theta_1}(x_j)), p(y|h_{\theta_2}(x_j))\big)$, namely

$$\mathbb{D}_{KL}\big(p(y|h_{\theta_1}(x_j)), p(y|h_{\theta_2}(x_j))\big) = \frac{1}{2} \left\| h_{\theta_1}(x_j) - h_{\theta_2}(x_j) \right\|_2^2. \tag{33}$$

See the *supplementary material* for derivation details. Note that predictive distributions of the form (32) are the most frequently used in practice with regression. For this choice of predictive distribution, $\mathbb{D}_{KL}(\theta_1, \theta_2)$ becomes

$$\mathbb{D}_{KL}(\theta_1, \theta_2) = \frac{1}{2N} \sum_{j=0}^{N} \left\| h_{\theta_1}(x_j) - h_{\theta_2}(x_j) \right\|_2^2. \tag{34}$$

Thus, the QE-KLD-WRM step solves

$$\min_s g_k^T s + \frac{1}{2} s^T B_k s + \frac{\lambda}{2N} \sum_{i=0}^{k} \zeta(i) \rho^{k-i} \sum_{j=0}^{N} \left\| h_{\theta_i}(x_j) - h_{\theta_k + s}(x_j) \right\|_2^2. \tag{35}$$

Practical QE-KLD-WRM for Classification. A similar practical computation of $\mathbb{D}_{KL}$ is also available for classification (see the *supplementary material*).

[16] Note there is no *tilde* on $\mathbb{D}$ in (31).

QE-KLD-WRM in Practice. While the KL-regularization term in (35) is in principle computable, the amount of associated storage would explode with k (storing old parameters). To bypass this problem, we have two options. We could either choose to discard very old regularization terms, or model them through the approximation (21) (as we did for Q-KLD-WRM, but now only do so for old terms). The latter approach, while convenient, is not well principled - we should really use second-order approximation when we are close (so for recent distributions), not when we are far away. This can be improved, but is left as future work. In this paper, we focus on implementations that use the former approach. Note that we now need to iteratively solve (35) (for eg. with SGD), and get an approximate QE-KLD-WRM step. Further note that the Q-KLD-WRM step given by *Proposition 4.2* is an approximation of the solution to (35), where the SKL-divergence is approximated by its second-order Taylor expansion (21). Thus, the Q-KLD-WRM step is a good initial guess for (35), and we exploit this fact in practice.

Extension to Variable Stepsize. For WoQM, SO-KLD-WRM and Q-KLD-WRM, we have so far considered only cases when the *"step-size"* $1/\lambda$ is fixed across different locations θ_k. Indeed, our established equivalence between FISHER-WoQM and EA-NG holds for fixed λ only. However, we may desire variable $\lambda \leftarrow \lambda^{(k)}$ in our KLD-WRM algorithms. To incorporate variable $\lambda^{(k)}$ in Q-KLD-WRM, all one has to do is to merely replace (30) with $\hat{g}_{k+1} = g_{k+1} + (\lambda^{(k+1)}/\lambda^{(k)})\rho[\hat{g}_k - g_k - \hat{M}_k\hat{g}_k]$ and all λ's in *Proposition 4.2* with $\lambda^{(k)}$. The version of *Proposition 4.2* which includes variable $\lambda^{(k)}$ can be found in the *supplementary material. Proposition 4.1* with variable $\lambda^{(k)}$ then follows trivially as a particular case with $B_k \equiv 0$.

4.5 Connection with Second Order Methods (SOM)

The particular KLD-WRM instantiation in equation (28), most simply illustrates the connection between our proposed class of algorithms and SOM. Setting $\lambda \leftarrow 0$ in (28) reverses Q-KLD-WRM back to a simple second order algorithm. More generally, from (23) we see that relative to SOM, KLD-WRM generalizes the local second order model to an arbitrary model that may also include previous information (in principle), and more importantly, it adds a wake of $\mathbb{D}_{KL}$ regularization. The connection between NG and Generalized Gauss-Newton can be found in [8].

5 Numerical Results

We compare our proposed KLD-WRM instantiations (SO, Q and QE) with K-FAC, on the MNIST classification problem. We investigated 4 different hyper-parameter settings for each solver, but only present the best ones here. The complete results, implementation details, as well as more in depth discussion can be found in *Sect. 8* of the *supplementary material. Figures* 1 *and* 2 show *test loss* and *test accuracy* for our considered solvers. Ten runs were performed for each solver. Important summary statistics of these results are shown in *Table* 1.

Hyper-parameters: The values of ρ and λ are specified in *Figures* 1 and 2. We used $\zeta(i) = \kappa(i)/330$ for QE, and $\zeta(i) = \kappa(i)$ for the other solvers, because the QE estimates of the SKL term were larger. The QE-specific hyper-parameters are: ω/λ – the learning rate of the inner SGD solving (35), N_{IS} – the number of inner SGD steps per iteration, and N_{CAP} – the total number of networks stored. For the results presented here, these were set to $7 \cdot 10^{-4}$, 10 and 4 respectively.

Test Accuracy and Loss: All KLD-WRM variants exceed 97.5% mean test accuracy, outperforming K-FAC by about 1.5%. The SD is 4–5 times lower for KLD-WRM variants, which is desirable. Analogous observations hold for the test loss. Since higher variance may more frequently yield *"favorable"* outliers, we show relevant metrics for this aspect in *Columns 2–6* of *Table* 1. We see that even from the *favorable outliers* point of view, KLD-WRM is mostly preferable.

Robustness and Overall Performance: *If we can only run the training a few times (perhaps once)*, it is preferable (in terms of epochs; both from a *test accuracy* and *test-loss* point of view) to use KLD-WRM rather than K-FAC.

Table 1. MNIST results summary. SO, Q and QE refer to KLD-WRM variants. *"Accuracy"* and *"loss"* are the test-set ones. μ_{acc} and σ_{acc} are the mean and SD of the empirical distribution of *accuracy* at the end of epoch 50. Notation is analogous for μ_{loss} and σ_{loss}, which refer to the *loss*. $\mathcal{N}_C$ is the number of runs for which condition C is satisfied.

	$\mathcal{N}_{acc\geq 98\%}$	$\mathcal{N}_{acc>98\%}$	$\mathcal{N}_{acc\geq 98.5\%}$	$\mathcal{N}_{loss\leq 0.25}$	$\mathcal{N}_{loss\leq 0.2}$	μ_{acc}	σ_{acc}	μ_{loss}	σ_{loss}
K-FAC	3	3	**1**	5	**2**	96.19%	3.2%	0.26	0.10
SO	4	2	0	7	0	97.60%	0.85%	0.25	0.04
Q	5	**4**	0	4	0	97.69%	0.69%	0.26	0.04
QE	**8**	**7**	**2**	**9**	0	**98.01%**	**0.64%**	**0.23**	**0.02**

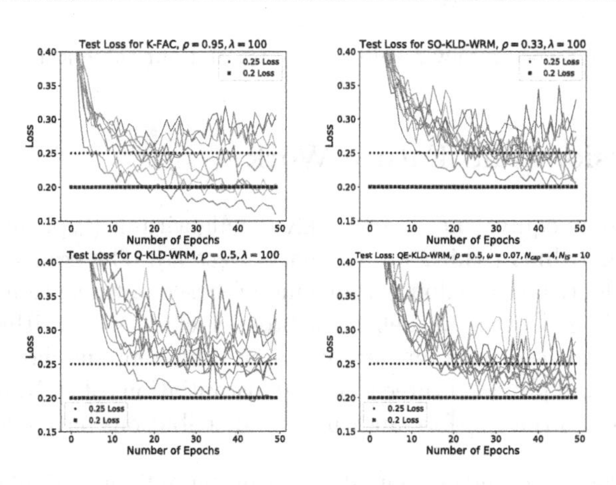

Fig. 1. MNIST test-loss results for K-FAC, and our three KLD-WRM variants.

That is because all our KLD-WRM variants more robustly achieve good test metrics. Conversely, *if we can run the training many times* (and choose the best run) K-FAC's large variance *may* eventually play to our advantage (not guaranteed).

KLD-WRM Variant Selection and Winners: SO-KLD-WRM and Q-KLD-WRM have virtually the same computation cost per epoch as K-FAC. Conversely, QE-KLD-WRM has the same *data acquisition* and *linear algebra* costs as K-FAC, but 3–10 times higher oracle cost (fwd. and bwd. pass cost), owing to approximately solving (35). Thus, when *data cost is relatively low*, SO-KLD-WRM and Q-KLD-WRM will be preferable, as they will have the smallest wall-time per epoch (while having almost the same performance per epoch as QE-KLD-WRM). Conversely, when *data cost dominates*, all 4 solvers will have the same wall-time per epoch. In this case, QE-KLD-WRM is preferable as it gives the best performance per epoch[17]

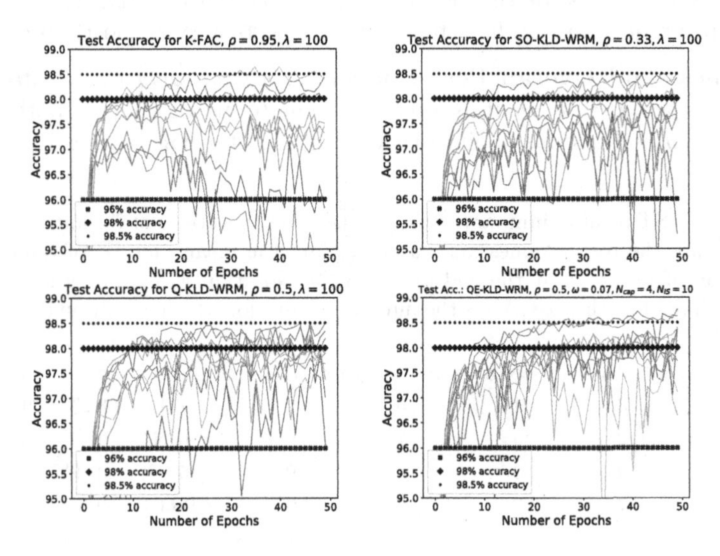

Fig. 2. MNIST test-accuracy results for K-FAC, and our three KLD-WRM variants.

6 Conclusions and Future Work

We established an equivalence between EA-CM algorithms (typically used in ML) and WoQM algorithms (which we defined in *Sect.* 3). The equivalence revealed what EA-CM algorithms are doing from a model-based optimization point of view. Generalizing from WoQM, we defined a broader class of algorithms in *Sect.* 4: KLD-WRM. We then focused our attention on a different subclass of KLD-WRM, LM-KLD-WRM, and provided three practical instantiations of it. Numerical results on MNIST showed that performance-metrics distributions have better mean and

[17] Codes available at: https://github.com/ConstantinPuiu/Rethinking-EA-of-the-Fisher.

lower variance for our KLD-WRM algorithms, indicating they are preferable to K-FAC in practice due to higher robustness.

Future Work: (a) KLD-WRM for VI BNNs and RL; (b) convergence theory; (c) investigate Q and QE variants when B_k and F_k have different structures; (d) consider KLD-WRM algorithms outside the LM-KLD-WRM subfamily (include info. at $\{\theta_j\}_{j<k}$ in $M(s;\mathcal{F}_k)$; see (23)); (e) consider arbitrary $\xi \in \mathbb{R}$ (see footnote 12).

Acknowledgments. Thanks to *Jaroslav Fowkes* for very useful discussions. I am funded by the EPSRC CDT in InFoMM (EP/L015803/1) in collaboration with Numerical Algorithms Group and St. Anne's College (Oxford).

References

1. Martens, J., Grosse, R.: Optimizing neural networks with Kronecker-factored approximate curvature. In: arXiv:1503.05671 (2015)
2. Yang, M., Xu, D., Wen, Z., Chen, M., Xu, P.: Sketchy empirical natural gradient methods for deep learning. In: arXiv:2006.05924 (2021)
3. Ba, J., Kingma, D.: Adam: a method for stochastic optimization. In: ICLR (2015)
4. LeCun, Y., Bottou, L., Orr, G., Muller, K.: Efficient backprop. Neural networks: tricks of the trade, pp. 546–546 (1998)
5. Schaul, T., Zhang, S., LeCun, Y.: No more pesky learning rates. In: ICML (2013)
6. Park, H., Amari, S.-I., Fukumizu, K.: Adaptive natural gradient learning algorithms for various stochastic models. Neural Netw. **13**(7), 755–764 (2000)
7. Amari, S.I.: Natural gradient works efficiently in learning. Neural Comput. **10**(20), 251–276 (1998)
8. Martens, J.: New insights and perspectives on the natural gradient method, arXiv:1412.1193 (2020)
9. Osawa, K., Yuichiro Ueno, T., Naruse, A., Foo, C.-S., Yokota, R.: Scalable and practical natural gradient for large-scale deep learning, arXiv:2002.06015 (2020)
10. Wu, Y., Mansimov, E., Grosse, R.B., Liao, S., Ba, J.: Scalable trust-region method for deep reinforcement learning using kronecker-factored approximation. In: Advances in Neural Information Processing Systems, pp. 5285–5294 (2017)
11. Bottou, L., Curtis, F.E., Nocedal, J.: Optimization methods for large-scale machine learning (2018)
12. Goldfarb, D., Ren, Y., Bahamou, A.: Practical quasi-newton methods for training deep neural networks, arXiv:2006.08877 (2021)

Mixed Integer Linear Programming for Optimizing a Hopfield Network

Bodo Rosenhahn(✉)

Institut für Informationsverarbeitung (tnt/L3S), Leibniz University Hannover, Hanover, Germany
rosenhahn@tnt.uni-hannover.de

Abstract. This work presents an approach to optimize the weights of a discrete Hopfield network as mixed integer linear program (MILP). As the original formulation involves a sign-function, it is not differentiable, but parameter optimization using a (mixed integer) LP is possible. As autoassociative memory, a key question is the amount of patterns which can be stored in such a Hopfield network. In this work it is shown, that the traditional storage description models are far inferior to a globally optimized solution which can be obtained with a MILP. In contrast to a gradient descent based optimization is the proposed approach nearly parameter free and independent from seeding and other factors which are crucial for differentiable programming. Additionally it is possible to enforce sparsity constraints on the weights. Such additional constraints improve the generalization of such a model and make the Hopfield network more stable for the case of outliers or missing values. Several experiments demonstrate the effectiveness of the model.

Keywords: Mixed integer linear program · Hopfield network · Sparse models

1 Introduction

The basic concepts of neural networks go back to the 40th and 50ths, e.g. in 1943, McCulloch and Pitts formulated their idea for logical calculus using concepts from nervous activities [27]. Based on these foundations, binary Hopfield networks were introduced as associative memories in 1982 [18]. It is a one-layer recurrent network that can store and retrieve patterns. In a d-dimensional space, the later explained *Information storage prescription* can store about $0.138d$ patterns [9], whereas the *Pseudoinverse rule* allows to store up to d patterns. The theoretic analysis in [38] estimates $2d$ as the upper limit of patterns which can be stored in a Hopfield model. It will turn out, that our proposed MILP solution can reach this upper limit. To the best of our knowledge, this is the first practical solution for estimating network weights providing maximal storage.

Nearly all existing neural network models are nowadays modified to differentiable neurons and commonly trained using gradient descent based optimization and auto differentiation. Such optimized systems can be summarized with the

© The Author(s), under exclusive license to Springer Nature Switzerland AG 2023
M.-R. Amini et al. (Eds.): ECML PKDD 2022, LNAI 13717, pp. 344–360, 2023.
https://doi.org/10.1007/978-3-031-26419-1_21

term *differentiable programming* [7]. Indeed, it is well known that the gradient descent based optimization is prone to convergence problems, overfitting or getting stuck in local minima, so that many different (differential) approaches exist to overcome these issues, e.g. using special optimizers (adam, sgdm, etc.), drop-out layers, etc. Note, that even though the Hopfield networks are an established topology, they received increasing attention in the past [32]. Please also note, that so-called *modern Hopfield Networks* consist of non-binary output variables which have a significantly higher capacity [2].

In this work, the non-differentiable Rosenblatt perceptron is revisited and used for modeling a Hopfield network. For the optimization of such a network, a mixed integer linear program is proposed. The optimized network weights can then be assembled to a Hopfield network and used for forward inference on unseen data. The formulation as MILP also allows the optimization of integer constraints and therefore the optimization of the network weights in presence of an exact step function. Additionally, integration of additional constraints on the network, e.g. to enforce sparsity, symmetry or binary constraints is possible. Such constraints allow for an optimization of a most efficient memory representation. The commonly used and later described *information storage prescription* or *pseudoinverse rule* to compute the weights of a Hopfield Network as well as differential programming can not fully exploit the storage capacities in contrast to the proposed MILP variants.

To summarize, this work presents the following **contributions**:

1. Formulation and optimization of a non-differentiable Hopfield network as mixed integer LP for given training data.
2. Formulation of additional constraints such as sparseness on weights, symmetry properties or binary network weights.
3. Evaluation of the optimized model with respect to storage capacities, stability with respect to outliers or missing data and experiments on image retrieval or classification.
4. Example code is available online[1].

In the following the general formulation of a Hopfield network as well as the classical storage prescription is described. Then, the basic concepts of mixed integer linear programming (MILP) are introduced and used to formulate such Hopfield networks.

Note, that in this work only discrete Hopfield networks with the output of $\{-1, 1\}$ are investigated. Still, on different datasets a competitive performance of the globally optimized Hopfield network is shown, even though the network is rather simple.

2 Foundations

2.1 The Hopfield Network

Introduced in 1982 [18], a Hopfield network consists of a one-layer recurrent network with the input dimension being equal to the output dimension. Its main

[1] http://www.tnt.uni-hannover.de/staff/rosenhahn/HopfieldNetExamples.zip.

purpose is to act as autoassociative memory, e.g. for a given tiny sample, the memory should retrieve a piece of data. Hopfield networks can be applied in de-noising or removing interference from an input or they can be used to determine whether the given input is *known* or *unknown*.

The units in Hopfield nets are binary threshold units $\in \{-1, 1\}$, the interactions $\omega_{i,j}$ between neurons are defined to be symmetric ($\omega_{i,j} = \omega_{j,i}$) and no unit has a connection with itself ($\omega_{i,i} = 0$). The classical form of a Hopfield network is shown in the left of Fig. 1.

Such a model can be based on the formulations of McCulloch and Pitts [27]. They formulated their idea for logical calculus using concepts from nervous activities. In the original formulation of McCulloch and Pitts, each neuron can emit two states $y_i \in \{0, 1\}$, thus it can *fire* or *not fire*. A neuron consists of n input lines on which the signals $(x_1 \dots x_n)$ are present. Calculation works as follows, the input signals $(x_1 \dots x_n)$ are added to the sum x. The sum x is compared with the threshold (or bias) b. If the sum of the excitations is greater than or equal to b, the neuron returns 1, otherwise it returns 0. In 1958, Frank Rosenblatt published his perceptron model which extends the summation to a scalar product, followed by a step function [34]. This is the basis for neural networks up till now. The perceptron can be summarized as

$$y_i = \begin{cases} 1 : \sum_j \omega_{ij} x_j + b_i > 0 \\ 0 : else \end{cases} \tag{1}$$

The bias value b corresponds to the decision threshold and ω_{ij} are learnable parameters. A combination of such perceptrons in a directed acyclic graph leads to a classic (e.g. fully connected) neural network.

Thus, a Hopfield network can be expressed as a fully connected recurrent layer with the amount of neurons being equal to the input dimension. The matrix of the network weights should be symmetric and form a hollow matrix. Note, that lifting the binary values $y \in \{0, 1\}$ to $y' \in \{-1, 1\}$ can be accomplished by the simple (linear) operation $y' = 2y - 1$.

Let $X = [x_1, \dots, x_n]$ be a matrix of size $m \times n$ containing m-dimensional vectors with binary values $\in \{-1, 1\}$. For the computation of the network weights, the *information storage prescription* proposed in [9] can be applied:

$$\omega = XX^T \tag{2}$$

$$\forall i = 1 \dots n : \ \omega_{i,i} = 0 \tag{3}$$

$$b_i = 0 \tag{4}$$

Note, that ω is a symmetric matrix with zero entries on the diagonal. This formulation is the so-called $Baseline_1$ in the experiments.

An alternative to compute the weight factors is to use the so-called *pseudoinverse* or *projection rule* [31], which is defined as

$$W = XX^+ \tag{5}$$

The matrix X^+ denotes here the pseudoinverse. We call this method $Baseline_2$ in the experiments. Once the weights $\omega_{i,j}, b_j$ are computed, the Hopfield network

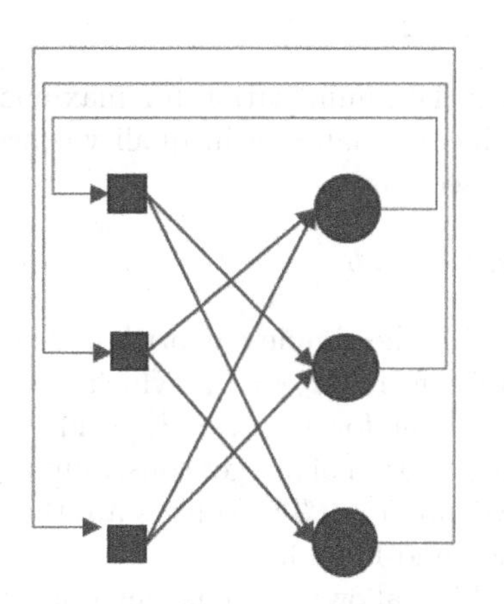

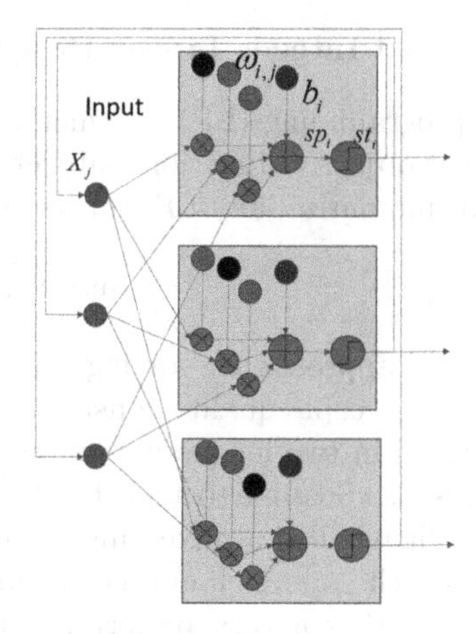

Fig. 1. Left: Visualization of a Hopfield Network. Right: Elements of a Hopfield Network as MILP. The black nodes indicate that these weights are set to zero.

can be evaluated as

$$x'_i = \begin{cases} 1 : \sum_j \omega_{i,j} x_j + b_j > 0 \\ -1 : else \end{cases} \tag{6}$$

Note, that b_j is set to zero for both baselines and later for the neural network and all MILP variants to allow for a fair comparison across the methods. As the output generates a vector with entries $\in \{-1, 1\}$, it can immediately be used again as input, until a convergence (e.g. fix point) is reached.

Hopfield nets have a scalar value associated with each state of the network, referred to as the *energy*, E, of the network, where

$$E = -\frac{1}{2} \sum_{i,j} \omega_{i,j} x_i x_j + \sum_i b_i x_i \tag{7}$$

The optimization of this energy formulation in combination with the involved step-function will be later used in the proposed mixed integer linear program.

The capacity of the Hopfield Network to store patterns depends on the used learning algorithm. The first learning rule applied to store patterns was Hebbian learning [18], which yields $Baseline_1$. Other rules have been formulated and considered in [1] and [12]. The implementation of Hopfield Networks in hardware and learning sparse networks have been considered in [36]. Hopfield neural networks have been applied for image processing [17,30] or solving combinatorial problems [20,33].

2.2 Mixed Integer Linear Programming

Linear programming (LP) is a method for the minimization (or maximization) of a linear objective function, subject to linear equality or inequality constraints [11]. To summarize, any LP can be expressed as

$$\min c^T x \quad s.t. \; Ax \leq b \tag{8}$$

A standard approach for solving such a LP is the simplex algorithm. Note, that solvers also accept equality constraints of the form $A_{eq}x = b_{eq}$ which are directly transformed in two inequality constraints of the form $A_{eq}x \leq b_{eq}$ and $-A_{eq}x \leq -b_{eq}$. It is also possible to allow for the optimization of integer constraints [13,28], which is then called a mixed integer linear program. The solvers are then using concepts such as branch-and-cut or branch-and-bound.

A (MI)LP is a very powerful tool, which allows the efficient optimization of graph problems, e.g. graph cut, graph matching, max-flow or network optimization [4,23], logic inference [25] and even reasonable efficient optimization of NP-hard problems, such as the traveling salesman problem. The work [39] gives a comprehensive introduction to formulate logic calculus (and beyond) using mixed integer linear programs. It is also possible to optimize decision trees [40] or support vector machines [29] using respective LP formulations.

Neural networks and linear programming have been brought together in the context of network fooling and model checking [16], already trained networks can be analyzed using LPs [5]. The work [35] introduces an approach to remove unnecessary units and layers of a neural network while not changing its output using a MILP. The formulation of a linear threshold unit (LTU) as LP, which is similar to the perceptron described above, has been presented in [26]. Based on this formulation, several LTUs are iteratively connected using a so-called multi-surface method tree. The special case of binary neural networks using MILP has been presented in [22]. Interestingly, binarized neural networks (BNNs) which are neural networks with weights and activations in $(-1, 1)$ can gain comparable test performance to standard neural networks but allow for highly efficient implementations on resource limited systems [21]. In [37] a mixed integer linear program to optimize neural network weights is presented. This work also focusses on network compression whereas in Hopfield networks it is not possible to reduce the amount of neurons. Instead, in this work different sparsity constraints are proposed and evaluated.

There are two arguments why a MILP solution may be disadvantageous, as stated by [14] (a) scaling to large datasets is problematic, as the size of the generated equations scales with the size of the training set and (b) solutions with provable optimal training error can overfit, thus training data is perfectly explained, but test performance is much worse. One reason for (a) is, that all data, as well as the network computations, need to be stored simulatenously and expressed as equality and inequality constraints. One option to alleviate this is an iterative training procedure: It is possible to globally optimize for batches and to train the weights iteratively over a couple of epochs [37]. During each iteration it is required to add constraints which enforce a small distance to earlier computed weights. Since it is comparable to a gradient based optimization, this variant is omitted here and only small datasets which can be globally optimized are considered. In this work, the second challenge is adressed by the proposed sparsity constraints.

3 Modeling a Hopfield Network as MILP

The Hopfield network requires the computation of a neuron which yields to the computation of a scalar product, followed by a step function which can also be expressed as a *greater as*, $>$. The formulation for $>$ can be expressed as MILP using a so-called Big-M formulation. Note, that the letter M refers to a (sufficiently) large number associated with the artificial variables (see also [39]). The condition

$$\mathsf{a} > \mathsf{b} \leftrightarrow \delta = 1$$

can be formulated as

$$a \geq b + \epsilon - M(1-\delta) \tag{9}$$

$$a \leq b + M\delta \tag{10}$$

$$\delta \in \{0,1\} \tag{11}$$

The verification of these equations is straight forward:

$$a > b \rightarrow \quad a \geq b + \epsilon - M(1-\delta) \quad \rightarrow \delta = 1 \vee 0$$

$$\text{and} \quad a \leq b + M\delta \quad \rightarrow \delta = 1 \tag{12}$$

$$a \leq b \rightarrow \quad a \geq b + \epsilon - M(1-\delta) \quad \rightarrow \delta = 0$$

$$\text{and} \quad a \leq b + M\delta \quad \rightarrow \delta = 0 \vee 1 \tag{13}$$

Based on Eq. (1), the scalar product and the $>$ formulation it is possible to model the activation of a perceptron with weights ω, bias b and input x. Please note, that the outcome is a binary vector which is later denoted as slack variable *st*. This output can then be uplifted to $\{-1,1\}$ and used for energy minimization.

Let $X = [x_1, \ldots, x_n]$ be a set of m-dimensional vectors with binary values $\in \{-1,1\}$. The parameter I indicates the example of the dataset and i the index of a neuron. Note, that the amount of neurons corresponds to the input dimension. Then the MILP looks as follows:

$$\min f^T x \quad s.t. \tag{14}$$

$$\forall I = 1 \ldots n, i = 1 \ldots m$$

$$\sum_{j=1}^{m} \omega_{ij} x_{j,I} + b_i - sp_{I,i} = 0 \tag{15}$$

$$\epsilon - M(1 - st_{I,i}) - sp_{I,i} \leq 0 \tag{16}$$

$$sp_{I,i} - M st_{I,i} \leq 0 \tag{17}$$

$$2st_{I,i} - \delta_{I,i} = 1 \tag{18}$$

$$\sum_{j=1}^{m} \frac{1}{2} \delta_{I,j} x_{j,I} - E_I = 0 \tag{19}$$

$$\forall j = 1 \ldots m : \omega_{j,j} = 0 \tag{20}$$

$$\omega_{i,j} - \omega_{j,i} = 0 \tag{21}$$

$$st_{I,i} \in \{0, 1\} \tag{22}$$

$$f(ind(E_I)) = -1 \tag{23}$$

The slack variables *sp* encode results of scalar products Eq. (15), *st* encode the evaluation of the step function Eqs. (16–17), δ the lifting from $\{0, 1\}$ to $\{-1, 1\}$ Eq. (18) and E Eq. (19) the energy to optimize. Equations (20–21) enforce the symmetry properties and the 0-entries on the diagonal matrix. The vector f contains only zero entries and only a factor -1 at the positions of the slack variables containing the loss E_I, Eq. (23). All used variables can be ordered as a large vector and accessed via an index-operation $ind(.)$. Thus, minimizing the objective function $f^T x$ means to minimize the loss. Note, that the variables $\omega_{i,i}$ can also be removed from the system but they are kept for readability.

The equations lead to sparse vector expressions and describe a mixed integer linear program which can be optimized with standard methods [15].

3.1 Sparsity Constraints

As mentioned in [14], the MILP solution can be problematic for generalization as no margin or other factor is optimized jointly with the optimization function. Motivated from other works, such as [3,6,8,10] sparsity can be beneficial for generalization and fortunately it is easy to enforce during the MILP optimization. In the following, two variants are proposed, one is enforcing the sparsity integer weights e.g. $\in [-1, 1]$, whereas the other variant can be applied to images. It is explicitly exploiting the grid-structure and enforcing only weights in the neighborhood. For the first variant, the MILP-constraints from the earlier section need to be extended with

$$\forall i, j = 1 \ldots m : \quad \omega_{i,j} - \omega^B_{i,j} \leq 0 \tag{24}$$

$$-\omega_{i,j} - \omega^B_{i,j} \leq 0 \tag{25}$$

$$f(\omega^B_{i,j}) = 0.001 \tag{26}$$

The additional slack variable $\omega^B_{i,j}$ contains the absolute value of the weight $\omega_{i,j}$. The vector f is modified with a small penalizer (with a scale factor of 0.001) to enforce the sparsity. Minimizing $\omega^B_{i,j}$ means to maximize the amount of zero-entries in $\omega_{i,j}$.

In the case of images or data with a clear topological structure it is also possible to enforce only connectivity of weights in a close neighborhood. This is known as local receptive field in image processing. The constraints simply enforce all weights outside a specific local distance d to be zero:

$$\forall i, j = 1 \ldots m: \quad \omega_{i,j} = 0 \leftrightarrow dist(i, j) > d \tag{27}$$

Note, that these variables can also be completely removed from the MILP. For the experiments on image data (e.g. mnist examples) the pixel distance d has been set to the value $d = 3$. The function $dist(i, j)$ is defined as the Euclidean distance of the point positions.

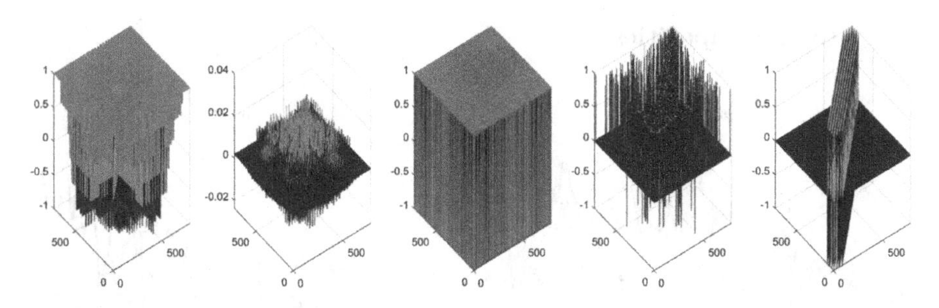

Fig. 2. Obtained weight-matrices ω for the different proposed variants. (From left to right): (a) Baseline ω_{B1}, (b) Baseline ω_{B2}, (c) MILP-optimized weights ω, (d) MILP-optimized weights with Sparsity constraints ω_{SP} (e) MILP-optimized weights as local receptive field ω_{LRF}

Figure 2 visualizes the outcome of different weight matrices obtained from the five presented variants, from left to right: (a) Baseline1 ω_{B1} (Eq. (2)), (b) Baseline ω_{B2} (Eq. (5)), (c) MILP-optimized weights ω (Eqs. (14–23)), (d) MILP-optimized weights with Sparsity constraints ω_{SP} (Eqs. (14–23 + 24–26)) (e) MILP-optimized weights as local receptive field (LRF) ω_{LRF} (Eqs. (14–23 + 27)). As dataset examples from mnist have been used. The dataset contains images of size 28 × 28 which are reordered as a 784 dimensional vector. Whereas the Baseline and the MILP optimized weight matrices are fully occupied, the sparse and the LRF variant have a completely different shape on locality and sparseness. The visualization shows that the constraints work exactly as intended.

4 Experiments

Two main research questions are analyzed in the following experiments: **(Q1)** What is the maximum storage capacity which can be achieved with a MILP optimized Hopfield network? (**Q2**) How does the MILP optimized Hopfield network perform for applications such as classification under missing data? In the following respective experiments on the MILP optimized Hopfield networks are presented to answer the raised research questions, to analyze the storage capacities and the stability with respect to noisy data. Afterwards we use the Hopfield networks for classification and compare the generalization and the compensation of missing values on different datasets. Here, the focus is specifically on learning with only a few examples.

For the experiments, matlab2021(b) has been used to generate the (in)equality constraints, boundary conditions and the objective function. For optimization itself, gurobi v9.1.2 has been used. The timelimit has been set to three hours and the standard optimization parameters (MIPFocus, Presolve, etc.) have not been changed. The computation has been performed on 10 physical cores on a local linux computer.

4.1 Storage Capacities

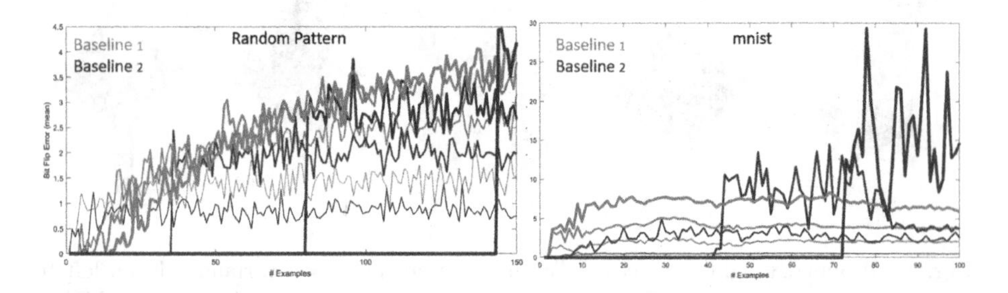

Fig. 3. Capacities for the two baseline methods with increasing amount of neurons (indicated by the line thickness). The x-axis shows the amount of training data to store. The y-axis shows the average amount of bitflips as error measure ($\frac{1}{N}\sum_{i=1}^{N} sum(x_i \neq x_i'))$. The left image shows the capacities for storing random patterns and the right image the capacities for the same dimensions on rescaled mnist data. As the mnist data is highly correlated, the storage capacities decrease significantly. The red line shows the memory capacities of $Baseline_1$ and the blue line the capacities of $Baseline_2$. For random patterns, $Baseline_2$ can store $N-1$ patterns successfully. The largest model consists of 144 neurons and manages to store exactly 143 samples when using $Baseline_2$. Afterwards the model completely fails. For mnist data, for 144 given neurons the $Baseline_2$ only manages to store approx. 73 data samples.

In the first experiment, the general capacities of the baseline methods for an increasing amount of neurons are analyzed. As stated in the introduction, the Hopfield Network can store up to N uncorrelated patterns. To verify this, we generate random binary patches of sizes $3\times 3, 6\times 6, 9\times 9, 12\times 12$ and apply both baselines on a different amount of training images. Figure 3 shows in the left the outcome for storing random (uncorrelated) patterns, whereas the right

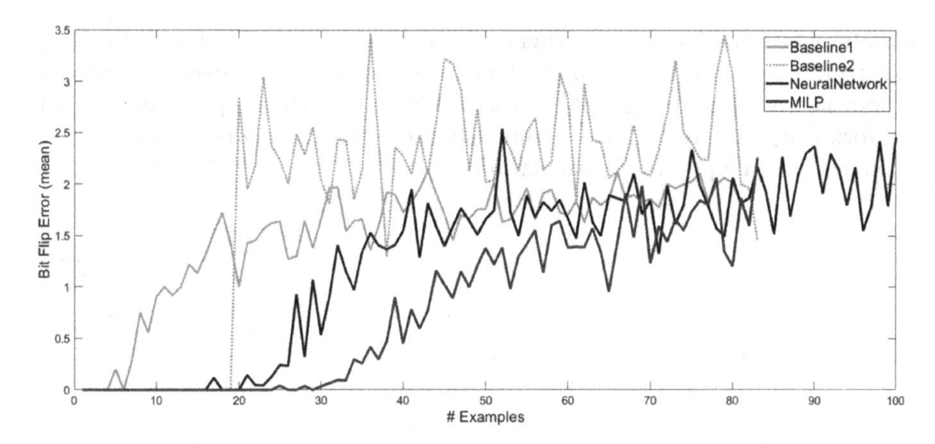

Fig. 4. Capacities for the two baseline methods, a shallow neural network and the MILP solution. The amount of neurons is set to 20. As observed before, $Baseline_1$ (red) can only store a few random patterns. $Baseline_2$ (red dashed) exactly 19 (and then it fails), the neural network (black) slightly more ($\sim$ 25), whereas the MILP version can go up to around 35 elements. (Color figure online)

diagram shows the outcome on mnist data [24] of similar size. The x-axis denotes the amount of example images ranging from 1 to 150 which are independently used for training. Figure 3 (left) shows that the model can capture exactly $N-1$ (random, uncorrelated) patterns with $Baseline_2$. The storage capacities decrease significantly for mnist data (right). For example on the image size $12 \times 12 = 144$ only around 72 images can be memorized effectively. This effect is one of the reasons, why Hopfield and colleagues also discussed the concept of unlearning in Hopfield networks [19].

The work [38] estimates an upper bound of $2N$ as storage capacity of a shallow neural network with N neurons. Furthermore, the author notes that the result is not tied to any particular algorithmic formulation of neural networks. Figure 4 shows the storage capacities for 20 memory elements on both baselines, a neural network implementation and the MILP optimized model for random patterns. The outcome for the baselines is in line with the outcome of Fig. 3 (left), for 20 neurons, exactly 19 patterns can be stored with $Baseline_2$. The neural network is a shallow network with 20 neurons and a tanh activation function. It is optimized with classical back propagation using the sgdm optimizer (100 epochs, InitialLearnRate 0.001, MiniBatchSize 5). Here, the symmetry of the weight matrix, as well as the zero entries on the diagonal matrix have been enforced during optimization. This has been achieved by using weight sharing and by setting the diagonal entries and its learning rate to zero. The neural network performs better than the baselines. In contrast to these approaches, the MILP-Model performs superior. One reason is, that the MILP can take all information simultaneously into account for an optimal arrangement of the weights. As the patterns are based on random samples, locality and sparsity have been omitted for this experiment. Note, that the MILP-optimized network is getting very close to the upper bound of $2N$ which has been theoretically proven nearly 40 years ago [38]. Please note, that a MILP-optimized Hopfield network can generate a global optimal solution, thus the storage capacities are maximized in the proposed formulation.

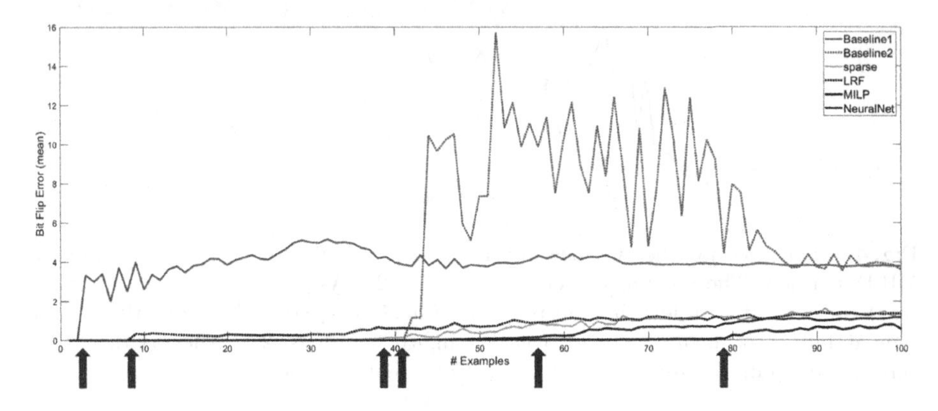

Fig. 5. Capacities of the Hopfield Networks on mnist examples for the two Baselines, a neural network, as well as the standard MILP, the MILP with Sparsity constraints and the MILP with LRF.

In the following we perform a similar analysis on the mnist dataset and summarize the outcome in Fig. 5: For the first experiment, mnist images have been downscaled to the size 9×9 pixels resulting in 81 neurons for memorization. Then the Hopfield network weights for the two baselines, an iteratively trained neural network (as before) as well as the standard MILP, the MILP with Sparsity constraints and the MILP with LRF have been trained for an increasing amount of data points. Afterwards, the memorization for each training sample is recalled and the average amount of bit flips is used as error metric, similar as before. Whereas the $Baseline_1$ can only memorize a few images, does $Baseline_2$ perform much better for a small amount of data. $Baseline_2$ manages to store around 40 images. Once a certain limit has been reached, $Baseline_2$ starts to fail completely, as described before. The neural network performs better and can store approx. 60 patterns before errors are introduced. The MILP can store the most amount of patterns, around 80. Additionally, the storage capacities of the sparse variants are shown. The overall capacities are (slightly) smaller, as many entries are enforced to be zero. Especially the LRF variant has significantly less parameters available for optimization which causes a slight error reasonable early. For an increasing amount of examples, this error is getting similar to the other variants (including the neural network).

Figure 6 shows the amount of zero-entries of the computed weight matrices in percent. As can be seen, the Neural network, the standard MILP and the baselines make use of the full matrix and $Baseline_2$ degenerates after 40 samples.

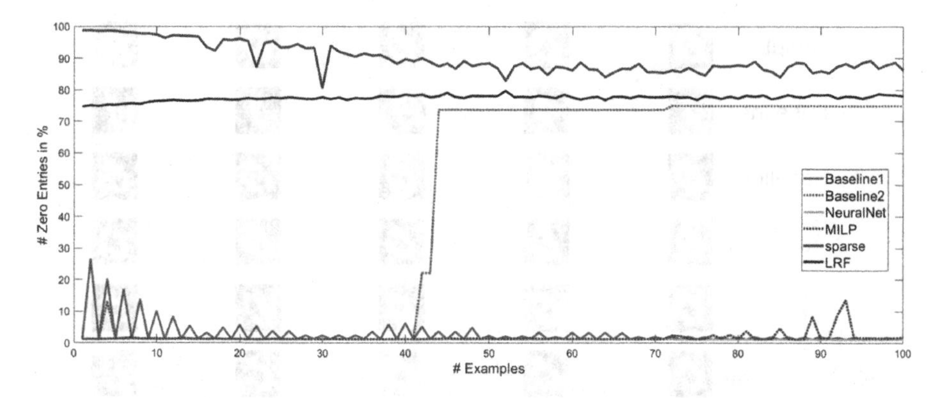

Fig. 6. Sparsity (y-axis) of the Hopfield Networks for different amounts of mnist examples (x-axis) for the two Baselines, a neural network, as well as the standard MILP, the MILP with Sparsity constraints and the MILP with LRF.

Only the sparse and the LRF variant heavily reduce the amount of entries while maintaining a competitive performance.

To summarize, all MILP solutions perform competitive, especially for a larger amount of data to memorize and also the error increase behaves more linear and incremental. Note, that the MILP provides a global optimal solution for the network weights, thus a better performance in memorization is not possible. Compared to $Baseline_2$, the amount of stored data (Fig. 5) is nearly doubled which is consistent with Fig. 4.

Figure 7 shows memorization results in presence of noise. The first row contains the input image with increasing noise (starting form 0). The other rows show the results using (row 2) the Baseline 1 ω_B (Eq. (2)), (row 3) the Baseline 2 (Eq. (5)), (row 4) a shallow neural network, trained with backpropagation, (row 5) the MILP-optimized weights ω (Eqs. (12–21)), (row 6) the MILP-optimized weights with Sparsity constraints ω_{SP} (Eqs. (12–21 + 22–24)) and (row 7) the MILP-optimized weights as local receptive field ω_{LRF} (Eqs. (12–21 + 25)). The amount of data to memorize has been selected appropriately so that (except Basline 1) all approaches can successfully memorize the input data (left column). Even though the memorization capacities of the MILP versions (red) are much higher compared to the neural network, especially the sparse variant generalizes in a similar quality of the neural network in the presence of noise. Note, that once the memory limits of the baselines have been reached, they produce useless results. Only the MILP and the neural network show an incremental error behavior.

In this section the storage capacities of MILP optimized Hopfield networks are analyzed. It is shown that the proposed three variants (MILP, MILP with sparsity and MILP with LRF) can store more data compared to the baseline implementations or differentiable programming.

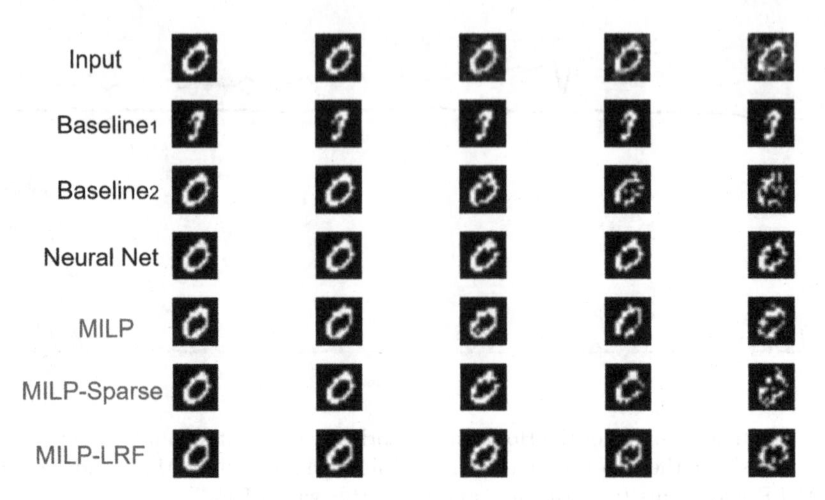

Fig. 7. Memorization results in presence of noise: First row: Input image with increasing noise (starting form 0). Reconstruction using: (2nd row) the Baseline 1, (3rd row) the Baseline 2, (4th row) a shallow neural network, (5th row) the MILP-optimized weights ω, (6th row) the MILP-optimized weights with Sparsity constraints ω_{SP}, (7th row) the MILP-optimized weights as local receptive field ω_{LRF}. The amount of training data was selected to keep (except Baslinel) competitive in the original memorization (left column).

4.2 Classification Using Hopfield Networks

For the next series of experiments, a learned Hopfield network is applied to a classification task. The classical *wine*, *zoo* and *breastEW* dataset have been used. The first two datasets are multicriterial classification tasks, with 3 categories for the wine dataset and 7 categories for the zoo dataset, whereas the breastEW dataset is a binary task.

To transform a numerical feature in one dimension into a binary one, k-means-clustering on this dimension (e.g. $k = 3$) is applied and used for a one-hot encoding. If e.g. the value 0.78 falls into cluster 2 (out of 3), this value is encoded as $(-1, 1, -1)$. Thus, e.g. the 30-dimensional features of the breastEW dataset are converted into a 90-dimensional binary dataset. From this dataset examples are used to train the Hopfield network and an unseen example is classified by feeding it into the Network and by performing a matching of the obtained values with the memorized dataset.

In the following, two factors are analyzed, (a) the performance behavior when using an increasing amount of samples for training, and (b) the stability with respect to missing data (from 0 to 50%). For these experiments we use a decision tree, a random forest and a neural network for comparison (on exactly the same train and test data). All experiments have been repeated 10 times and the diagram in Fig. 8 shows the obtained mean performance values. The neural network is a shallow network with the same amount of neurons as the Hopfield network and a softmax classification layer. As can be seen, already with two or

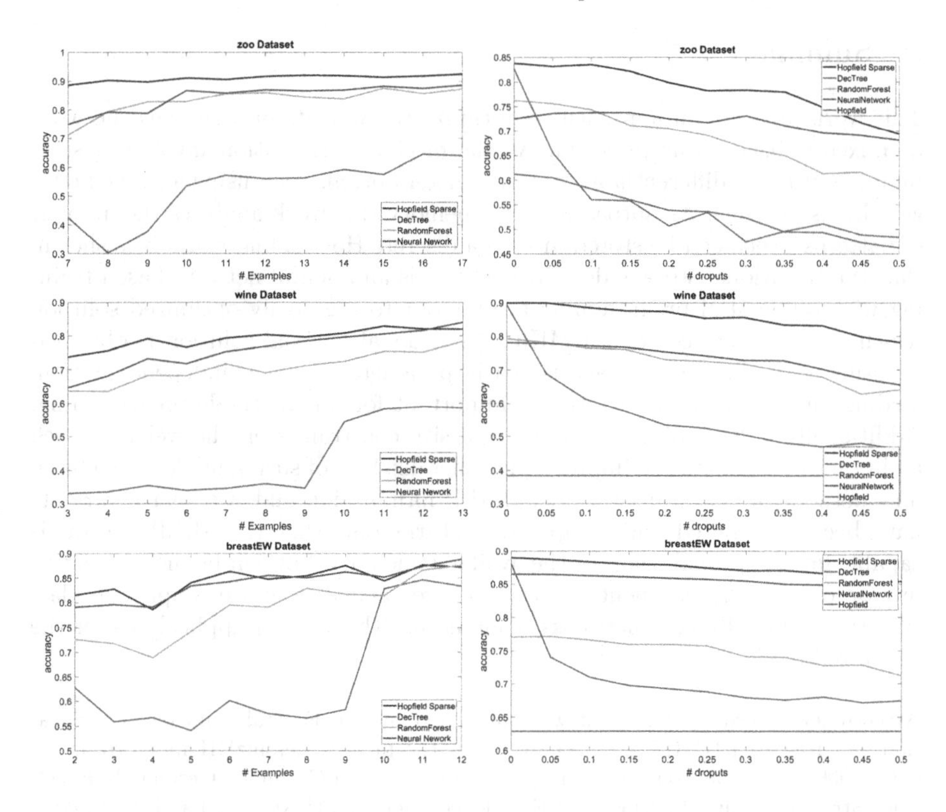

Fig. 8. Left: Classification performance for learning from small data. The x-axis shows the number of examples which have been used for training and the y-axis the classification performance on the unseen test data. Right: Classification performance for missing data. The x-axis shows the amount of missing data (up to 50 %). The Hopfield network can memorize from the missing data and therefore classify the test data better then the decision tree, random forest or a neural network. The sparse MILP solution is superior to the non-sparse MILP solution which is highly sensitive to missing data.

three examples (for each class), a competitive classification performance can be obtained which is pretty constant for the Hopfield network and increasing for a decision tree (which is known to be sensitive to overfitting) and a random forest. For the experiment in Fig. 8 (right), the number of examples has been kept fixed to the number 4. For inference, the samples have been disturbed with a random amount of zero-entries, ranging from 0 to 50%. Here the outcome is even more clear: The (sparse) Hopfield networks can memorize from missing data and yield a much better classification accuracy on the test sets. The sparse MILP solution is superior to the non-sparse MILP solution which is in line with the discussion in [14].

5 Summary

This work presents an approach to optimize the weights of a Hopfield network as mixed-integer linear program. As the original formulation involves a sign-function, it is not differentiable, but parameter optimization using a (mixed integer) LP is possible. As autoassociative memory this work analyzes the amount of patterns which can be stored in such a discrete Hopfield network. It is shown, that the traditional storage description models and neural network based training with differential programming are inferior to a globally optimized solution which can be obtained with a MILP. As a global optimal solution is obtained by using a MILP, the approach is nearly parameter free and independent from seeding and other factors which are important for differentiable programming. Additionally it is possible to enforce sparsity constraints on the weights. Such additional constraints can improve the generalization of such a model. Two baseline methods, a neural network, the MILP formulation and two sparse variants have been formulated and compared. It turns out, that the MILP variant is capable for an optimal usage of the available memory, but can be more sensitive in case of noise. Experiments on the storage capacities and examples on classification using Hopfield networks demonstrate the general applicability of the optimized model.

Acknowledgments. This work was supported by the Federal Ministry of Education and Research (BMBF), Germany under the project LeibnizKILabor (grant no. 01DD20003), the Deutsche Forschungsgemeinschaft (DFG) under Germany's Excellence Strategy within the Cluster of Excellence PhoenixD (EXC 2122) and the Erskine Programme at the University of Canterbury, New Zealand. The author also thanks Dr. Roberto Henschel for fruitful discussions and hints.

References

1. Abbott, L.F.: Learning in neural network memories. Netw. Comput. Neural Syst. **1**, 105–122 (1990)
2. Abu-Mostafa, Y., St. Jacques, J.: Information capacity of the hopfield model. IEEE Trans. Inf. Theory **31**(4), 461–464 (1985)
3. Aharon, M., Elad, M., Bruckstein, A.: K-SVD: an algorithm for designing overcomplete dictionaries for sparse representation. IEEE Trans. Signal Process. **54**(11), 4311–4322 (2006)
4. Almohamad, H., Duffuaa, S.: A linear programming approach for the weighted graph matching problem. IEEE Trans. Pattern Anal. Mach. Intell. **15**(5), 522–525 (1993)
5. Anderson, R., Huchette, J., Ma, W.: Strong mixed-integer programming formulations for trained neural networks. Math. Program. **183**, 3–39 (2020)
6. Bae, W., Lee, S., Lee, Y., Park, B., Chung, M., Jung, K.-H.: Resource optimized neural architecture search for 3D medical image segmentation. In: Shen, D., et al. (eds.) MICCAI 2019. LNCS, vol. 11765, pp. 228–236. Springer, Cham (2019). https://doi.org/10.1007/978-3-030-32245-8_26

7. Baydin, A.G., Pearlmutter, B.A., Radul, A.A., Siskind, J.M.: Automatic differentiation in machine learning: a survey. J. Mach. Learn. Res. **18**(1), 5595–5637 (2017)
8. Bellec, G., Kappel, D., Maass, W., Legenstein, R.: Deep rewiring: training very sparse deep networks. arXiv abs/1711.05136 (2018)
9. Cooper, L., Liberman, F., Oja, E.: A theory for the acquisition and loss of neuron specificity in visual cortex. Biol. Cybern. **33**, 9–28 (1979)
10. Cun, Y.L., Denker, J.S., Solla, S.A.: Optimal brain damage. In: Advances in Neural Information Processing Systems, pp. 598–605. Morgan Kaufmann (1990)
11. Dantzig, G.B.: Maximization of a linear function of variables subject to linear inequalities. Act. Anal. Prod. Allocation **13**, 339–347 (1951)
12. Davey, N., Adams, R.: High capacity associative memories and connection constraints. Connect. Sci. **16**, 47–65 (2004)
13. Dennis, J.E., Schnabel, R.B.: Numerical Methods for Unconstrained Optimization and Nonlinear Equations. Society for Industrial and Applied Mathematics (1996)
14. Gambella, C., Ghaddar, B., Naoum-Sawaya, J.: Optimization problems for machine learning: a survey. Eur. J. Oper. Res. **290**(3), 807–828 (2021)
15. Gurobi Optimization: Gurobi optimizer reference manual (2021). http://www.gurobi.com
16. Heo, J., Joo, S., Moon, T.: Fooling neural network interpretations via adversarial model manipulation. In: Wallach, H., Larochelle, H., Beygelzimer, A., Alche-Buc, F., Fox, E., Garnett, R. (eds.) Advances in Neural Information Processing Systems, vol. 32. Curran Associates, Inc. (2019)
17. Hillar, C., Mehta, R., Koepsell, K.: A hopfield recurrent neural network trained on natural images performs state-of-the-art image compression. In: 2014 IEEE International Conference on Image Processing (ICIP), pp. 4092–4096 (2014)
18. Hopfield, J.J.: Neural networks and physical systems with emergent collective computational abilities. Proc. Natl. Acad. Sci. **79**(8), 2554–2558 (1982)
19. Hopfield, J., Feinstein, D., Palmer, R.: 'Unlearning' has a stabilizing effect in collective memories. Nature **304**(5922), 158–159 (1983)
20. Hopfield, J., Tank, D.: Neural computation of decisions in optimization problems. Biol. Cybern. **52**, 141–52 (1985). https://doi.org/10.1007/BF00339943
21. Hubara, I., Courbariaux, M., Soudry, D., El-Yaniv, R., Bengio, Y.: Binarized neural networks. In: Lee, D., Sugiyama, M., Luxburg, U., Guyon, I., Garnett, R. (eds.) Advances in Neural Information Processing Systems, vol. 29. Curran Associates, Inc. (2016)
22. Toro Icarte, R., Illanes, L., Castro, M.P., Cire, A.A., McIlraith, S.A., Beck, J.C.: Training binarized neural networks using MIP and CP. In: Schiex, T., de Givry, S. (eds.) CP 2019. LNCS, vol. 11802, pp. 401–417. Springer, Cham (2019). https://doi.org/10.1007/978-3-030-30048-7_24
23. Komodakis, N., Tziritas, G.: Approximate labeling via graph cuts based on linear programming. IEEE Trans. Pattern Anal. Mach. Intell. **29**(8), 1436–1453 (2007)
24. LeCun, Y., Cortes, C., Burges, C.: MNIST handwritten digit database. ATT Labs (2010). http://yann.lecun.com/exdb/mnist
25. Makhortov, S., Ivanov, I.: Equivalent transformation of the reasoning model in production zeroth-order logic. In: 2020 International Conference on Information Technology and Nanotechnology (ITNT), pp. 1–4 (2020)
26. Mangasarian, O.L.: Mathematical programming in neural networks. ORSA J. Comput. **5**, 349–360 (1993)
27. McCulloch, W., Pitts, W.: A logical calculus of ideas immanent in nervous activity. Bull. Math. Biophys. **5**, 127–147 (1943)

28. Murty, K.: Linear Programming. Wiley, Hoboken (1983)
29. Nguyen, H.T., Franke, K.: A general Lp-norm support vector machine via mixed 0-1 programming. In: Perner, P. (ed.) MLDM 2012. LNCS (LNAI), vol. 7376, pp. 40–49. Springer, Heidelberg (2012). https://doi.org/10.1007/978-3-642-31537-4_4
30. Pajares, G., Guijarro, M., Ribeiro, A.: A hopfield neural network for combining classifiers applied to textured images. Neural Netw. **23**(1), 144–153 (2010)
31. Personnaz, L., Guyon, I., Dreyfus, G.: Information storage and retrieval in spin-glass like neural networks. J. Phys. Lett. **46**, 359–365 (1985)
32. Ramsauer, H., et al.: Hopfield networks is all you need. arXiv 2008.02217 (2021)
33. Recanatesi, S., Katkov, M., Romani, S., Tsodyks, M.: Neural network model of memory retrieval. Front. Comput. Neurosci. **9**, 149 (2015)
34. Rosenblatt, F.: The perceptron: a probabilistic model for information storage and organization in the brain. Psychol. Rev. **65**(6), 386–408 (1958)
35. Serra, T., Kumar, A., Ramalingam, S.: Lossless compression of deep neural networks. In: Hebrard, E., Musliu, N. (eds.) CPAIOR 2020. LNCS, vol. 12296, pp. 417–430. Springer, Cham (2020). https://doi.org/10.1007/978-3-030-58942-4_27
36. Tanaka, G., et al.: Spatially arranged sparse recurrent neural networks for energy efficient associative memory. IEEE Trans. Neural Netw. Learn. Syst. **31**(1), 24–38 (2020)
37. Thorbjarnarson, T., Yorke-Smith, N.: On training neural networks with mixed integer programming. arXiv 2009.03825 (2020). https://www.cmu.edu/epp/patents/events/aaai21/aaaicontent/papers/training-integer-valued-neural-networks-with-mixed-integer-programming.pdf
38. Venkatesh, S.S.: Epsilon capacity of neural networks. In: AIP Conference Proceedings 1986, vol. 151, pp. 440–445 (2014)
39. Williams, H.P.: Logic and Integer Programming, 1st edn. Springer, New York (2009). https://doi.org/10.1007/978-0-387-92280-5
40. Zhu, H., Murali, P., Phan, D., Nguyen, L., Kalagnanam, J.: A scalable MIP-based method for learning optimal multivariate decision trees. In: Larochelle, H., Ranzato, M., Hadsell, R., Balcan, M.F., Lin, H. (eds.) Advances in Neural Information Processing Systems, vol. 33, pp. 1771–1781. Curran Associates, Inc. (2020)

Learning to Control Local Search for Combinatorial Optimization

Jonas K. Falkner[(✉)], Daniela Thyssens, Ahmad Bdeir, and Lars Schmidt-Thieme

Institute for Computer Science, University of Hildesheim, Hildesheim, Germany
{falkner,thyssens,bdeir,schmidt-thieme}@ismll.uni-hildesheim.de

Abstract. Combinatorial optimization problems are encountered in many practical contexts such as logistics and production, but exact solutions are particularly difficult to find and usually NP-hard for considerable problem sizes. To compute approximate solutions, a zoo of generic as well as problem-specific variants of local search is commonly used. However, which variant to apply to which particular problem is difficult to decide even for experts.

In this paper we identify three independent algorithmic aspects of such local search algorithms and formalize their sequential selection over an optimization process as Markov Decision Process (MDP). We design a deep graph neural network as policy model for this MDP, yielding a learned controller for local search called NeuroLS. Ample experimental evidence shows that NeuroLS is able to outperform both, well-known general purpose local search controllers from the field of Operations Research as well as latest machine learning-based approaches.

Keywords: Combinatorial optimization · Local search · Neural networks

1 Introduction

Combinatorial optimization problems (COPs) arise in many research areas and applications. They appear in many forms and variants like vehicle routing [43], scheduling [41] and constraint satisfaction [44] problems but share some general properties. One of these properties is that most COPs are proven to be NP-hard which makes their solution very complex and time consuming. Over the years many different solution approaches were proposed. *Exact methods* like *branch-and-bound* [26] attempt to find the global optimum of a COP based on smart and efficient ways of searching the solution space. While they are often able to find the optimal solution to small scale COPs in reasonable time, they require a significant amount of time to tackle larger instances of sizes relevant for practical applications. For that reason, *heuristic methods* were proposed which usually cannot guarantee to find the global optimum but sometimes can define a lower bound on the performance,

---

Supplementary Information The online version contains supplementary material available at https://doi.org/10.1007/978-3-031-26419-1_22.

© The Author(s), under exclusive license to Springer Nature Switzerland AG 2023
M.-R. Amini et al. (Eds.): ECML PKDD 2022, LNAI 13717, pp. 361–376, 2023.
https://doi.org/10.1007/978-3-031-26419-1_22

which can be achieved at a minimum, and have shown good empirical performance. One common and well-known heuristic method is *local search* (LS) [1]. The main concept of LS is to iteratively explore the search space of candidate solutions in the close neighborhood of the current solution by applying small (local) changes. Simple LS procedures, like hill climbing, easily get stuck in bad local optima from which it cannot escape anymore. Therefore, LS is commonly used in combination with *meta-heuristics* which enable the procedure to escape from local optima and achieve better final performance. The meta-heuristics introduced in Sect. 2.2 are well established in the optimization community and have demonstrated very good performance on a plenitude of different COPs.

Since a few years however, there is an increasing interest in leveraging methods from the field of machine learning (ML) to solve COPs, as recent developments in neural network architectures, involving Transformers [47] and Graph Neural Networks (GNNs) [52], have led to significant progress in this domain. Most work based on ML is concerned with auto-regressive approaches to construct feasible solutions [17,24,49,53], but there is also some work which focuses on iterative improvement [6,28,51] or exact solutions [10,33] for COPs. While the existing improvement approaches share some similarities with meta-heuristics and local search, the exact formulation of the methods is often problem specific and misses the generality of the LS framework as well as a clear definition of the intervention points which can be used to control the meta-heuristic procedure.

In this work we present a consistent formulation of learned meta-heuristics in local search procedures and show on two representative problems how it can be successfully applied. In our experiments we compare our method to well-known meta-heuristics commonly used with LS at the example of capacitated vehicle routing (CVRP) and job shop scheduling (JSSP). The results show that our GNN-based learned meta-heuristic consistently outperforms these methods in terms of speed and final performance. In further experiments we also establish our method in the context of existing ML solution approaches.

Contributions

1. We identify and describe three independent algorithmic aspects of local search for COPs, each with several alternatives, and formalize their sequential selection during an iterative search run as Markov Decision Process.
2. We design a deep graph neural network as policy model for this MDP, yielding a learned controller for local search called *NeuroLS*.
3. We provide ample experimental evidence that NeuroLS outperforms both, well-known general purpose local search controllers from the Operations Research literature (so called meta-heuristics) as well as earlier machine learning-based approaches.

2 Related Work

There is a plenitude of approaches and algorithms to tackle combinatorial optimization problems. One common heuristic method is *Local Search* (LS) [1]. In the classical discrete optimization literature it is often embedded into a meta-heuristic procedure to escape local optima.

2.1 Construction Heuristics

Construction algorithms are concerned with finding a first feasible solution for a given COP. They usually start with an empty solution and consecutively assign values to the respective decision variables to construct a full solution. Well-known methods are e.g. the *Savings* heuristic [7] for vehicle routing problems or *priority dispatching rules* (PDR) [5] for scheduling. Most improvement and meta-heuristic methods require a feasible initial solution from which they can start to improve.

2.2 Meta-Heuristics

Meta-heuristics are the go-to method for complex discrete optimization problems. They effectively balance the exploration of the solution space and the exploitation of promising solutions. There are two major types of methods:

Trajectory-Based Methods. Trajectory-based methods include many well-known approaches which are used in combination with LS. During the search they only maintain a single solution at a time which is iteratively changed and adapted. *Simulated Annealing* (SA) is a probabilistic acceptance strategy based on the notion of controlled cooling of materials first proposed in [22]. The idea is to also accept solutions which are worse than the best solution found so far to enable exploration of the search space but with a decreasing probability to increasingly focus on exploitation the further the search advances. *Iterated Local Search* (ILS) [27] alternates between a diversification and an intensification phase. In the diversification step the current solution is perturbed while the alternating intensification step executes a greedy local search with a particular neighborhood. *Variable Neighborhood Search* (VNS) [31] employs a similar diversification step but changes the type of the applied LS move after each perturbation (in a predefined order) to systematically control the LS neighborhood in the intensification phase. *Tabu Search* (TS) was proposed by Glover [12] and is based on the possible acceptance of non-improving moves and a so called tabu list, a kind of filter preventing moves which would keep the search stuck in local optima. This list acts as a memory which in the simplest case stores the solutions of the last k iterations and prevents changes which would move the current solution back to solutions encountered in recent steps. Instead of using a tabu list, *Guided Local Search* (GLS) [50] relies on penalties for different moves to guide the search. These penalties are often based on problem specific features in the solution, e.g. the edges between nodes in a routing problem, and are added to the original objective function of the problem when the LS gets stuck.

Population-Based Methods. In comparison to trajectory-based approaches population-based methods maintain a whole pool of different solutions throughout the search. Methods include different evolutionary algorithms, particle swarm optimization and other bio-inspired algorithms such as ant colony optimization [11]. Their main idea is based on different adaption, selection and recombination schemes to refine the solution pool during search in order to find better solutions. While a learned population based meta-heuristic is interesting and potentially promising, in this paper we focus on the impact and effectiveness of a learned trajectory-based approach. More information on advanced meta-heuristics can be found in [11].

2.3 Machine Learning Based Methods

In recent years an increasing number of ML-based methods has been proposed. While some work relies on supervised [19,42,49] or unsupervised [20] learning, most current state-of-the-art methods use reinforcement learning (RL). A large fraction of the work focuses on auto-regressive models which learn to sequentially construct feasible solutions from scratch. Such methods have been proposed for many common COPs including the TSP [3,21], CVRP [24,25,34], CVRP-TW [9] and JSSP [15,36,37,53]. The second type of methods is concerned with improvement approaches. Hudson, Malencia and Prorok [18] propose an LS approach guided by an underlying GNN for the TSP. Chen and Tian [6] design a model to rewrite sub-sequences of the problem solution for the CVRP and JSSP based on a component which selects a specific element of the solution and a second component parameterizing a heuristic move to change that part of the solution. However, their model is limited to specific problem settings with a fixed number of jobs and machines or customers while our approach works seamlessly for different problem sizes, as we show in the experiments in Sect. 5.3. In contrast, the authors in [28] learn a policy that selects a specific LS move at each iteration. However, their method incurs prohibitively large computation times and for this reason is not competitive with any of the recent related work [23,25,29]. The authors in [16] learn a repair operator to re-construct heuristically destroyed solutions in a Large Neighborhood Search. Finally, da Costa et al. [35] learn a model to select node pairs for 2-opt moves in the TSP while Wu et al. [51] learn a similar pair-wise selection scheme for 2-opt, node swap and relocation moves in TSP and CVRP. The authors in [29] further improve on the method in [51] by introducing the Dual-aspect collaborative Transformer (DACT) model. Although there exist advanced inference approaches [15] to further improve the performance of auto-regressive ML methods on COPs, here we focus on the vanilla inference via greedy decoding or sampling.

While these methods share some similarities with our approach, they ignore the importance of being able to reject unpromising moves to escape local optima, whereas our approach specifically focuses on this important decision point to effectively control the search. Moreover, our approach is also able to learn when to apply a particular perturbation if the rejection of a sequence of moves is not sufficient for exploration. A detailed overview of the current machine learning approaches to COPs is given in [4,30].

3 Preliminaries

3.1 Problem Formulation

A Combinatorial Optimization Problem Ω is defined on its domain D_Ω which is the set of its instances $x \in D_\Omega$. A COP instance x is usually given by a pair

(S_Ω, f_Ω) of the solution space S consisting of all feasible solutions to Ω and a corresponding cost function $f : S \rightarrow \mathbb{R}$. Combinatorial optimization problems normally are either to be minimized or maximized. In this paper we consider all COPs to be problems for which the cost of a corresponding objective function has to be minimized. The main concern is to find a solution $s^* \in S$ representing a *global optimum*, i.e. $f(s^*) \leq f(s) \; \forall s \in S$.

3.2 Local Search

LS is a heuristic search method which is based on the concept of neighborhoods. A neighborhood $\mathcal{N}(s) \subseteq S$ of solution $s \in S$ represents a set of solutions which are somehow close to s. This "closeness" is defined by the neighborhood function $\mathcal{N}^\varphi$ w.r.t. some problem specific operator $\varphi \in \Phi_\Omega$ (e.g. all solutions which can be reached from the current solution by an exchange of nodes). Moreover, we always consider s to be part of its own neighborhood, i.e. $s \in \mathcal{N}(s)$. Then the *local optimum* $\hat{s}$ in the neighborhood $\mathcal{N}$ satisfies $f(\hat{s}) \leq f(s) \; \forall s \in \mathcal{N}(\hat{s})$. A general LS procedure (see Algorithm 1) iterates through the neighborhood $\mathcal{N}(s)$ of the current solution s until it finds the local optimum $\hat{s}$.

Algorithm 1: Local Search

```
  input: cost function f, solution s, neighborhood function N,
         acceptance rule accept, stopping rule stop,
1 while not stop(s) do
2 |   find s' ∈ N(s) for which accept(s, s')
3 |   s ← s'
4 return s
```

3.3 Meta-Heuristics

Meta-heuristics wrap an LS procedure to enable it to escape from local optima and to explore the solution space more efficiently. Some of the most common meta-heuristic strategies were described in Sect. 2.2. Algorithm 2 describes a general formulation of a trajectory-based meta-heuristic procedure. Each particular strategy takes different decisions about restarting or perturbing the current solution, configuring the local search and accepting intermediate candidate solutions s' (see Algorithm 1). Some decisions can be fixed and are treated as hyper-parameters for some methods. For example, an SA procedure normally does not select a specific neighborhood but just decides about acceptance during the LS. Other approaches like VNS greedily accept all improving moves but apply a perturbation and select a new operator neighborhood every time a local optimum has been reached.

Algorithm 2: Meta-Heuristic (trajectory-based)

input: Solution space S, cost function f, stopping criterion
1 $s \leftarrow$ `construct`(S) `// Construct initial solution`
2 **while** *not stopping criterion* **do**
3 $\quad s \leftarrow$ `perturb`(S, s) `// Decide if to perturb/restart`
4 $\quad \mathcal{N} \leftarrow$ `GetNeighborhood`(S, s) `// Define search neighborhood`
5 $\quad s \leftarrow$ `LocalSearch`$(f, s, \mathcal{N},$ `accept`, `stop`$)$ `// Execute local search`
6 **return** s

4 Proposed Method

4.1 Intervention Points of Meta-Heuristics for Local Search

The application of meta-heuristic strategies to an underlying LS involves several points of intervention, at which decisions can be made to help the search escape local optima in order to guide it towards good solutions. In the following, we define three such intervention points that have a significant impact on the search:

1. *Acceptance*: The first intervention point is the acceptance of candidate solutions s' in an LS step (Algorithm 1, line 2). A simple hill climbing heuristic is completely greedy and only accepts improving moves, which often leaves the search stuck in local optima very quickly. In contrast, other approaches like SA will also accept non-improving moves with some probability.
2. *Neighborhood*: The second possible decision a meta-heuristic can make is the selection of a particular operator φ that defines the neighborhood function $\mathcal{N}^{\varphi}$ for the LS (Algorithm 2, line 4). Possible operators for scheduling or routing problems could for example be a node exchange. While many standard approaches like SA and ILS only use one particular neighborhood which they treat as a hyper-parameter, VNS is an example for a method that selects a different operator defining a particular neighborhood at each step.
3. *Perturbation*: Many meta-heuristics like ILS and VNS employ perturbations $\psi \in \Psi_{\Omega}$ to the current solution to move to different regions of the search space and escape particularly persistent local optima. Such a perturbation can simply be a restart from a new stochastically constructed initial solution, a random permutation of (a part of) the current solution or a random sequence of operations. The decision *when* to employ a perturbation (Algorithm 2, line 5) is commonly done w.r.t. a specific pre-defined number of steps without improvement.

4.2 Meta-Heuristics as Markov Decision Process

In this section we formulate meta-heuristics in terms of an MDP [40] to enable the use of RL approaches to learn a parameterized policy to replace them. In general an MDP is given by a tuple $(\mathcal{S}, \mathcal{A}, \mathcal{P}(s_t, a_t), \mathcal{R}(s_t, a_t))$ representing the set of states $\mathcal{S}$, the set of actions $\mathcal{A}$, transition probability function $\mathcal{P}(s_t, a_t)$ and reward function $\mathcal{R}(s_t, a_t)$. For our method we define these entities in terms of meta-heuristic decisions as follows:

States. We define the state s_t of the problem at time step t with slight abuse of notation as the solution s at time step t, combined with 1) its cost $f(s)$, 2) the cost $f(\hat{s}_t)$ of the best solution found so far, 3) the last acceptance decision, 4) the last operator used, 5) current time step t, 6) number of LS steps without improvement and 7) the number of perturbations or restarts.

Actions. Depending on the policy we want to train, we define the action set as the combinatorial space of

1. *Acceptance* decisions: a boolean decision variable of either accepting or rejecting the last LS step
$$\mathcal{A}_{\mathrm{A}} := \{0, 1\}, \tag{1}$$
2. *Acceptance-Neighborhood* decisions: the joint space of the acceptance of the last move and the set of possible operators $\varphi \in \Phi$ which define the search neighborhood(s) $\mathcal{N}^{\varphi}$ for the next step
$$\mathcal{A}_{\mathrm{AN}} := \{0, 1\} \times \Phi, \tag{2}$$
3. *Acceptance-Neighborhood-Perturbation* decisions: the joint space of acceptance and the combined sets of operators $\varphi \in \Phi$ and perturbations $\psi \in \Psi$
$$\mathcal{A}_{\mathrm{ANP}} := \{0, 1\} \times \{\Phi \cup \Psi\}. \tag{3}$$

Transitions. The transition probability function $\mathcal{P}(s_t, a_t)$ models the state transition from state s_t to the next state s_{t+1} depending on action a_t representing the acceptance decision, the next operator φ and a possible perturbation or restart, which is part of the problem state (or action in case of $\mathcal{A}_{\mathrm{ANP}}$).

Rewards. The reward function $\mathcal{R}(s_t, a_t)$ gives the reward for a transition from state s_t to the next state s_{t+1}. Here we define the reward r_t as the relative improvement of the last LS step (defined by action a_t) w.r.t. the cost of the best solution found until t and clamped at 0 to avoid negative rewards:

$$r_t := \max(f(\hat{s}_t) - f(s_{t+1}), 0) \tag{4}$$

Policy. We employ Deep Q-Learning [46] and parameterize the learned policy π_θ via a softmax over the corresponding Q-function $Q_\theta(s_t, a_t)$ which is represented in turn by our GNN-based encoder-decoder model with trainable parameters θ:

$$\pi_\theta(a_t \mid s_t) = \frac{\exp(Q_\theta(s_t, a_t))}{\sum_{\mathcal{A}} \exp(Q_\theta(s_t, a_t))}. \tag{5}$$

4.3 Model Architecture

In this section we describe our encoder and decoder models to parameterize $Q_\theta(s_t, a_t)$. Many COPs like routing and scheduling problems have an underlying graph structure which can be used when encoding these problems, providing an effective inductive bias for the corresponding learning methods. In general we assume a graph $\mathcal{G} = (\mathcal{V}, \mathcal{E})$ with the set of nodes $\mathcal{V}$, $N = |\mathcal{V}|$ and the set of directed edges $\mathcal{E} \subseteq \{(i,j) \subseteq \mathcal{V}\}$. Moreover, we assume an original node feature matrix $X \in \mathbb{R}^{N \times d_{\text{in}}}$ and edge features $e_{i,j} \in \mathbb{R}$ for each edge (i,j). To leverage this structural information many authors have used Recurrent Neural Networks [3,6,49], Transformers [24,25,29,47] or Graph Neural Networks (GNN) [19,35,53].

Encoder

Since edge weights are very important in COP graphs, we employ a simple version of the GNN operator proposed in [32] with GELU [14] activations which can directly work with edge weights and outperformed GAT [48] and ReLU in preliminary experiments. The resulting GNN layer is defined as:

$$\begin{aligned} h_i^{(l)} &= \text{GNN}^{(l)}(h_i^{(l-1)}) \\ &= \text{GELU}\left(\text{MLP}_1^{(l)}(h_i^{(l-1)}) + \text{MLP}_2^{(l)}\big(\sum_{j \in \mathcal{H}(i)} e_{j,i} \cdot h_j^{(l-1)}\big)\right), \end{aligned} \tag{6}$$

where $h_i^{(l-1)} \in \mathbb{R}^{1 \times d_{\text{emb}}}$ is the latent feature embedding of node i at the previous layer $l-1$, $\mathcal{H}(i)$ is the 1-hop graph neighborhood of node i, $\text{MLP}_1^{(l)}$ and $\text{MLP}_2^{(l)}$ are Multi-Layer Perceptrons $\text{MLP} : \mathbb{R}^{d_{\text{emb}}} \to \mathbb{R}^{d_{\text{emb}}}$. Furthermore, we add residual connections and layer normalization [2] to each layer.

In the first layer the latent feature vector $h_i^{(0)}$ is created by feeding the original node features x_i into an $\text{MLP} : \mathbb{R}^{d_{\text{in}}} \to \mathbb{R}^{d_{\text{emb}}}$:

$$h_i^{(0)} = \text{MLP}^{(0)}(x_i). \tag{7}$$

and another $\text{MLP}^{(L)} : \mathbb{R}^{d_{\text{emb}}} \to \mathbb{R}^{d_{\text{emb}}}$ is placed at the end of the GNN stack.

In order to further leverage structural information, we introduce 3 stages to compute the latent embeddings. The first stage uses the edge set $\mathcal{E}^{\text{stat}}$ of the *static* problem graph. For the CVRP we use the graph induced by the K nearest neighbors of each node, for scheduling problems the Directed Acyclic Graph (DAG) representing the predefined order of operations for each job. This edge set is fixed and does not change throughout the search. In contrast, the second stage utilizes the edge set $\mathcal{E}^{\text{dyna}}$ representing the *dynamic* problem graph which usually changes in every LS step, e.g. the edges constituting the different tours in routing problems or the machine graph which represents the sequence of jobs on each machine for scheduling. Our proposed network architecture consists of L^{stat} GNN layers for the static graph, followed by L^{dyna} layers which propagate over the dynamic graph. Finally, we add another layer, again using the static edges, to consolidate the dynamic information over the static graph, leading to a total of $L = L^{\text{stat}} + L^{\text{dyna}} + 1$ GNN layers.

The final stage serves to refine the embedding via aggregation based on the dynamic information of group membership which is present in the solution. Each node normally belongs to one of K (not necessarily disjoint) groups $\mathcal{M}_k$ of the solution, e.g. to a particular tour or machine. Following this idea we pool the

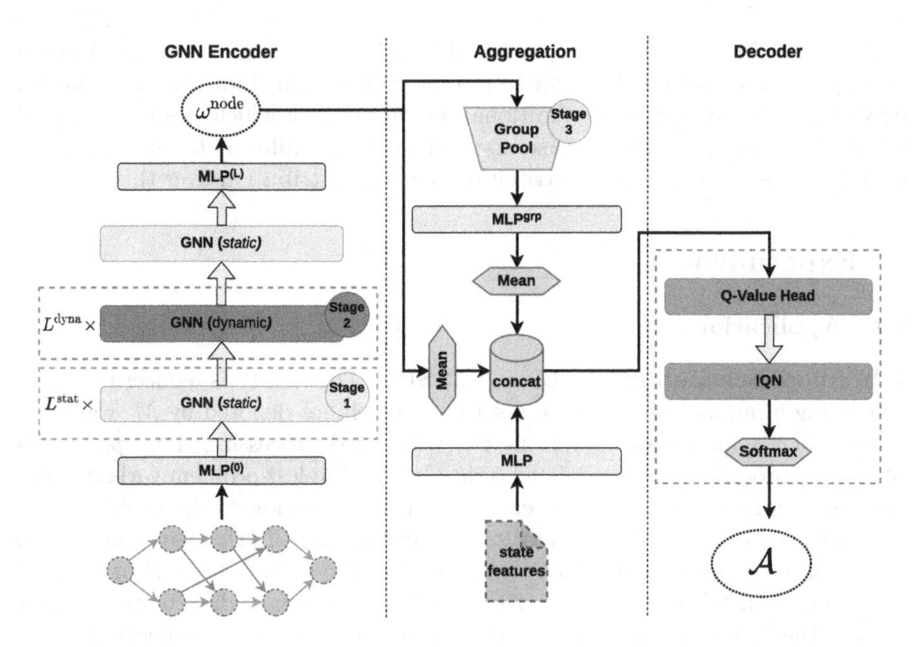

Fig. 1. Visualization of the NeuroLS model architecture.

final embeddings ω_i^{node} from the GNN stack w.r.t. their membership and feed them through another MLP:

$$\omega_k^{\text{grp}} = \text{MLP}^{\text{grp}}\big([\text{MAX}(\omega_i^{\text{node}} \mid i \in \mathcal{M}_k); \text{MEAN}(\omega_i^{\text{node}} \mid i \in \mathcal{M}_k)]\big), \tag{8}$$

with max and mean pooling $\mathbb{R}^{N\times d_{\text{emb}}} \rightarrow \mathbb{R}^{K\times d_{\text{emb}}}, K << N$ over the node dimension N, $\mathcal{M}_k$ as membership to the k-th group (k-th tour, k-th machine, etc.) and [;] representing concatenation in the embedding dimension d_{emb}.

Finally, the additional features of the state representation (current cost, best cost, last acceptance, etc.) described in the last section, are concatenated and projected by a simple linear layer to create an additional latent feature vector $\omega^{\text{feat}} \in \mathbb{R}^{d_{\text{emb}}}$. We provide a runtime and memory analysis in Appendix B.

Decoder

Our decoder takes the different embeddings created in the encoder, aggregates the node embeddings $\omega^{\text{node}} \in \mathbb{R}^{N\times d_{\text{emb}}}$ and group embeddings $\omega^{\text{grp}} \in \mathbb{R}^{K\times d_{\text{emb}}}$ via a simple mean over the node and group dimension and concatenates them with the feature embedding $\omega^{\text{feat}} \in \mathbb{R}^{d_{\text{emb}}}$. This representation is the input to a final 2-layer MLP regression head $\mathbb{R}^{3*d_{\text{emb}}} \rightarrow \mathbb{R}^{|\mathcal{A}|}$ which outputs the value predictions of the Q-function. The full architecture is shown in Fig. 1.

4.4 Reinforcement Learning Algorithm

To train our policy model we employ Double Deep Q-Learning [46] with n-step returns [40] and Implicit Quantile Networks (IQN) [8]. IQNs enable a distributional formulation of Q-Learning where the Deep Q-Network is trained w.r.t. an underlying value distribution represented by a learned quantile function instead of single point estimates. In order to represent this learned quantile function an IQN is introduced as a small additional neural network which is jointly trained to transform samples from a base distribution (e.g. uniform) to the respective quantile values of the target distribution, i.e. the distribution over the returns.

5 Experiments

5.1 Applications

Job Shop Scheduling Problem (JSSP). The JSSP is concerned with scheduling a number of jobs J on a set of K machines denoted by M. Each job consists of a sequence of operations O_{ij} with fixed processing times p_{ij} which need to be processed in a predefined order. In the simplest problem variant every job has exactly one operation on each machine. A solution to the problem consists of the exact order of the operations on all machines. In this paper we choose to minimize the *makespan*, which is the longest time span from start of the first operation until the end of the last one to finish, corresponding to the longest path in the respective DAG representation of the problem. We denote the size of a JSSP instances as $|J| \times |M|$ and follow [53] in creating instances for training and validation by sampling processing times from a uniform distribution.

Capacitated Vehicle Routing Problem (CVRP). The CVRP consists of a set of N customer nodes and a depot node. It is concerned with serving the demands q_i of the customer nodes in tours starting and ending at the depot node by employing K homogeneous vehicles with a fixed capacity $Q > 0$. The tour of vehicle $k \in K$ is a sequence of indices w.r.t. a subset of all customers nodes representing the order in which vehicle k visits the respective nodes. A set of feasible tours serving all customer nodes represents a solution to the problem, whereas the objective is to minimize the total length of all tours. We follow [24] in creating the training and validation sets by generating instances with coordinates uniformly sampled in the unit square.

5.2 Setup

For the JSSP we implement a custom LS solver in python. It implements four different node moves and a perturbation operator based on a sequence of such moves. As construction heuristic and for restarts we implement several stochastic variants of common PDRs (see appendix D for more details). The LS for the CVRP uses the C++ based open source solver VRPH [13] for which we implement a custom wrapper and interface to expose and support the necessary intervention points. VRPH includes several different LS moves and perturbation operators including 2-opt, 3-opt and different exchange and point moves.

Preliminary experiments showed that the CET and 2-opt moves performed best for the JSSP and CVRP respectively, when used for meta-heuristics which do not select the operator. Thus, we employ these operators in our main experiments. We train all our models for 80 epochs with 19200 transitions each and pick the model checkpoint with the best validation performance.

Hyperparameters for NeuroLS and all meta-heuristics are tuned on a validation set consisting of 512 generated random instances. This is in contrast to most classical approaches which fine tune their hyper-parameters directly on the benchmark dataset. We argue that our approach facilitates a better and more objective comparison between these methods. Further details on hyper-parameters can be found in Appendix C and we provide our code on github[1].

We train 3 different types of policies for NeuroLS, one for each of the action spaces described in Sect. 4.2. In the experiments we denote these policies as NLS_{A}, NLS_{AN} and NLS_{ANP}. As analyzed in [51], random solutions do not provide a good starting point for improvement approaches. For that reason we initialize the solutions of NeuroLS and the meta-heuristics with the FDD/MWKR PDR [39] for scheduling and the savings construction heuristic [7] for the CVRP.

5.3 Results

JSSP. We evaluate all methods on the well-known benchmark dataset of Taillard [41]. It consists of 80 instances of size 15×15 up to 100×20. We compare our

Table 1. Results of state-of-the-art machine learning based construction methods and local search approaches (100 iterations) on the Taillard benchmark [41]. For instances of size 50×15, 50×20 and 100×20 we use the NeuroLS model trained on instances of size 30×15 and 30×20 respectively. Percentages are the average gap to the best known upper bound. Best gap is marked in **bold**.

Model	Instance size								Avg
	15×15	20×15	20×20	30×15	30×20	50×15	50×20	100×20	
ML-based									
L2D [53]	25.92%	30.03%	31.58%	32.88%	33.64%	22.35%	26.37%	13.64%	27.05%
L2S [37]	20.12%	24.83%	29.25%	24.59%	31.91%	15.89%	21.39%	9.26%	22.16%
SN [36]	15.32%	19.43%	17.23%	18.95%	23.75%	13.83%	13.56%	6.67%	16.09%
Meta-heuristic + Local Search									
SA	13.92%	17.01%	17.16%	17.53%	21.59%	12.50%	13.11%	6.61%	14.93%
$\text{SA}_{\text{restart}}$	13.77%	17.01%	17.57%	17.62%	21.78%	12.54%	13.22%	6.75%	15.03%
ILS	11.57%	13.57%	13.85%	16.07%	18.72%	12.65%	12.15%	6.72%	13.16%
ILS+SA	13.32%	16.05%	15.38%	16.93%	19.74%	13.07%	13.43%	7.08%	14.37%
VNS	9.96%	13.71%	14.51%	15.77%	18.69%	11.64%	11.92%	6.26%	12.81%
NeuroLS									
NLS_{A}	**9.76%**	13.33%	13.02%	15.29%	17.94%	11.81%	11.96%	6.33%	12.43%
NLS_{AN}	10.32%	**13.18%**	**12.95%**	**14.91%**	17.78%	11.87%	12.02%	6.22%	**12.41%**
NLS_{ANP}	10.49%	16.32%	15.24%	15.35%	**17.64%**	**11.62%**	**11.76%**	**6.09%**	13.06%

[1] https://github.com/jokofa/NeuroLS.

Table 2. Results of state-of-the-art machine learning based methods and local search approaches (200 iterations) on the Uchoa benchmark [45]. For all instances sizes we use the NeuroLS model trained on instances of size 100. Percentages are the average gap to the best known solution. Best gap is marked in **bold**.

Model	Instance group								Avg
	n100		n150		n200		n250		
	Cost	Time	Cost	Time	Cost	Time	Cost	Time	
ML-based									
POMO [25]	17.24%	0.3	12.81%	0.4	20.52%	0.4	15.14%	0.4	16.43%
POMO [25] (aug)	6.32%	**0.1**	9.41%	**0.1**	13.47%	**0.1**	9.21%	**0.1**	9.60%
DACT [29]	13.19%	15.1	20.26%	21.2	16.91%	27.2	24.93%	33.0	18.82%
DACT [29] (aug)	11.52%	16.6	17.75%	23.1	15.34%	29.8	21.86%	37.8	16.62%
Meta-heuristic + Local search									
ORT [38] GLS	6.91%	9.4	10.46%	10.9	**7.80%**	13.4	12.12%	5.0	9.32%
ORT [38] TS	6.78%	39.0	10.55%	109.5	7.85%	126.6	12.13%	7.0	9.33%
SA	6.92%	0.7	5.79%	1.3	16.63%	2.1	5.92%	3.3	8.81%
$SA_{restart}$	6.96%	0.7	5.79%	1.3	16.67%	2.0	5.99%	3.0	8.85%
ILS	6.56%	0.8	5.96%	1.5	16.73%	2.4	6.08%	3.7	8.83%
ILS+SA	6.94%	0.6	6.01%	1.2	16.72%	1.9	6.04%	2.8	8.93%
VNS	7.99%	0.4	6.55%	0.7	17.27%	0.9	6.36%	1.4	9.54%
NeuroLS									
NLS_A	5.43%	1.5	5.23%	2.4	15.97%	3.6	5.22%	5.3	7.96%
NLS_{AN}	**5.42%**	1.7	**4.90%**	2.7	15.85%	4.0	**5.08%**	5.9	**7.81%**
NLS_{ANP}	**5.42%**	1.7	**4.90%**	2.7	15.85%	4.0	**5.08%**	6.1	**7.81%**

model against common meta-heuristic baselines including SA, SA with restarts, ILS, ILS with SA acceptance and VNS. Moreover, we report the results of three recent state-of-the-art ML-based approaches: *Learning to dispatch* (L2D) [53], *Learning to schedule* (L2S) [37] and *ScheduleNet* (SN) [36]. We follow [53] in training different models for problem sizes 15×15 up to 30×20 and apply the 30×15 and 30×20 model to larger instances of size 50×15, 50×20 and 100×20 to evaluate its generalization capacity. All models are trained for 100 LS iterations but evaluated for 50–200. The aggregated results per group of same size are shown in Table 1 (for per instance results see appendix F). Results are reported as percentage gaps to the best known solution.

First of all, the results show that NeuroLS is able to outperform all other meta-heuristics on all sizes of instances. The different policies differ in how well they work for different problem sizes. While NLS_A works best for the smallest 15×15 instances, it is outperformed by NLS_{AN} on medium sized instances and NLS_{ANP} achieves the best results for large instances. This is to some extent expected, since the effect of specific LS operators and more precise perturbations is greater for larger instances while this does not seem to be necessary for rather small instances. VNS is the best of the meta-heuristic approaches which is able to beat NLS_{ANP} on the smaller instances and NLS_{A} and NLS_{AN} at least on some of the larger ones. In general, the iterative methods based on LS achieve much better results than the ML-based auto-regressive methods, which can be seen by the large improvements that can be achieved in just 100 iterations, reducing the gap by an average of 3.7% compared to the best ML-method SN [36].

CVRP. For the CVRP we use the recent benchmark dataset of Uchoa et al. [45] and select all instance up to 300 customer nodes. We define four groups of instances with 100–149 nodes as *n100*, 150–199 nodes as *n150*, 200–249 as *n200* and 250–299 as *n250*. We compare against the same meta-heuristics mentioned above and additionally to GLS and TS provided by the OR-Tools (ORT) library [38]. Furthermore, we compare to the recent state-of-the-art ML approaches POMO [25] and DACT [29] which outperformed all other ML methods mentioned in Sect. 2.3 in their experiments and provide open source code, which is why we consider them to be sufficient for a suitable comparison.

Since most ML-based methods (POMO, DACT, etc.) do not respect the maximum vehicle constraint for all instances of the Uchoa benchmark, we follow [29] in removing this constraint and treat the benchmark dataset as a highly diverse test set w.r.t. the distributions of customers and number of required vehicles. This is also consistent with the general goal of the ML-based methods, which is not to achieve the best known results but to find sufficiently good results in reasonable time. In this case we control the computational resources spent on the search by specifying a particular number of iterations for the search. Furthermore, we evaluate all models with a batch size of 1.

The results presented in Table 2 show that NeuroLS is able to outperform the considered meta-heuristic approaches on all instance groups. Moreover, our approach also outperforms the state-of-the-art ML-based methods on all groups but *n200*, where POMO and DACT with additional instance augmentations (aug) outperform NeuroLS by a small margin. The OR-Tools implementation of GLS and TS outperforms our method only on the *n200* instances, although they require prohibitively large runtimes (wall-clock time). In terms of runtimes we also outperform DACT by a magnitude, while the learned auto-regressive construction method POMO is the fastest overall. Finally, the experiment results also show that our method is able to generalize to problem sizes as well as numbers of iterations unseen in training. We show this on the JSSP instances of size 50×15, 50×20 and 100×20 and for all Uchoa instances for the CVRP, which are all larger than the 100 node instances used during training.

6 Conclusion

In this paper we identify three important intervention points in meta-heuristics for local search on COPs and incorporate them in a MDP. We then design a GNN-based controller which is trained with RL to parameterize three types of learned meta-heuristics. The resulting methods learn to control the LS by deciding about acceptance, neighborhood selection and perturbations. In comprehensive experiments on two common COPs in scheduling and vehicle routing,

NeuroLS outperforms several well-known meta-heuristics as well as state-of-the-art ML-based approaches, confirming the efficacy of our method.

For future work we consider more fine-grained interventions, e.g. to restrict the search neighborhood and to replace the problem graph with a graph representation of the corresponding LS graph, in which every node represents a feasible solution together with its respective cost.

Acknowledgements. This work was supported by the German Federal Ministry of Education and Research (BMBF), project "Learning to Optimize" (01IS20013A:L2O) and the German Federal Ministry for Economic Affairs and Climate Action (BMWK), within the IIP-Ecosphere project (01MK20006D).

References

1. Aarts, E., Aarts, E.H., Lenstra, J.K.: Local search in Combinatorial Optimization. Princeton University Press, Princeton (2003)
2. Ba, J.L., Kiros, J.R., Hinton, G.E.: Layer normalization. arXiv preprint arXiv:1607.06450 (2016)
3. Bello, I., Pham, H., Le, Q.V., Norouzi, M., Bengio, S.: Neural combinatorial optimization with reinforcement learning. arXiv preprint arXiv:1611.09940 (2016)
4. Bengio, Y., Lodi, A., Prouvost, A.: Machine learning for combinatorial optimization: a methodological tour d'horizon. Eur. J. Oper. Res. **290**(2), 405–421 (2021)
5. Blackstone, J.H., Phillips, D.T., Hogg, G.L.: A state-of-the-art survey of dispatching rules for manufacturing job shop operations. Int. J. Prod. Res. **20**(1), 27–45 (1982)
6. Chen, X., Tian, Y.: Learning to perform local rewriting for combinatorial optimization. In: Advances in Neural Information Processing Systems, vol. 32 (2019)
7. Clarke, G., Wright, J.W.: Scheduling of vehicles from a central depot to a number of delivery points. Oper. Res. **12**(4), 568–581 (1964)
8. Dabney, W., Ostrovski, G., Silver, D., Munos, R.: Implicit quantile networks for distributional reinforcement learning. In: International Conference on Machine Learning, pp. 1096–1105. PMLR (2018)
9. Falkner, J.K., Schmidt-Thieme, L.: Learning to solve vehicle routing problems with time windows through joint attention. arXiv preprint arXiv:2006.09100 (2020)
10. Gasse, M., Chételat, D., Ferroni, N., Charlin, L., Lodi, A.: Exact combinatorial optimization with graph convolutional neural networks. In: Advances in Neural Information Processing Systems, vol. 32 (2019)
11. Gendreau, M., Potvin, J.Y., et al.: Handbook of Metaheuristics, vol. 2. Springer, New York (2010). https://doi.org/10.1007/978-0-387-74759-0
12. Glover, F.: Future paths for integer programming and links to artificial intelligence. Comput. Oper. Res. **13**(5), 533–549 (1986)
13. Groër, C., Golden, B., Wasil, E.: A library of local search heuristics for the vehicle routing problem. Math. Program. Comput. **2**(2), 79–101 (2010)
14. Hendrycks, D., Gimpel, K.: Gaussian error linear units (GELUs). arXiv preprint arXiv:1606.08415 (2016)
15. Hottung, A., Kwon, Y.D., Tierney, K.: Efficient active search for combinatorial optimization problems. arXiv preprint arXiv:2106.05126 (2021)
16. Hottung, A., Tierney, K.: Neural large neighborhood search for the capacitated vehicle routing problem. arXiv preprint arXiv:1911.09539 (2019)

17. Hu, H., Zhang, X., Yan, X., Wang, L., Xu, Y.: Solving a new 3D bin packing problem with deep reinforcement learning method. arXiv preprint arXiv:1708.05930 (2017)
18. Hudson, B., Li, Q., Malencia, M., Prorok, A.: Graph neural network guided local search for the traveling salesperson problem. arXiv preprint arXiv:2110.05291 (2021)
19. Joshi, C.K., Laurent, T., Bresson, X.: An efficient graph convolutional network technique for the travelling salesman problem. arXiv preprint arXiv:1906.01227 (2019)
20. Karalias, N., Loukas, A.: Erdos goes neural: an unsupervised learning framework for combinatorial optimization on graphs. Adv. Neural. Inf. Process. Syst. **33**, 6659–6672 (2020)
21. Khalil, E., Dai, H., Zhang, Y., Dilkina, B., Song, L.: Learning combinatorial optimization algorithms over graphs. In: Advances in Neural Information Processing Systems, vol. 30 (2017)
22. Kirkpatrick, S., Gelatt, C.D., Jr., Vecchi, M.P.: Optimization by simulated annealing. Science **220**(4598), 671–680 (1983)
23. Kool, W., van Hoof, H., Gromicho, J., Welling, M.: Deep policy dynamic programming for vehicle routing problems. arXiv preprint arXiv:2102.11756 (2021)
24. Kool, W., Van Hoof, H., Welling, M.: Attention, learn to solve routing problems! arXiv preprint arXiv:1803.08475 (2018)
25. Kwon, Y.D., Choo, J., Kim, B., Yoon, I., Gwon, Y., Min, S.: Pomo: policy optimization with multiple optima for reinforcement learning. Adv. Neural. Inf. Process. Syst. **33**, 21188–21198 (2020)
26. Lawler, E.L., Wood, D.E.: Branch-and-bound methods: a survey. Oper. Res. **14**(4), 699–719 (1966)
27. Lourenço, H.R., Martin, O.C., Stützle, T.: Iterated local search: framework and applications. In: Gendreau, M., Potvin, J.-Y. (eds.) Handbook of Metaheuristics. ISORMS, vol. 272, pp. 129–168. Springer, Cham (2019). https://doi.org/10.1007/978-3-319-91086-4_5
28. Lu, H., Zhang, X., Yang, S.: A learning-based iterative method for solving vehicle routing problems. In: International Conference on Learning Representations (2019)
29. Ma, Y., Li, J., Cao, Z., Song, W., Zhang, L., Chen, Z., Tang, J.: Learning to iteratively solve routing problems with dual-aspect collaborative transformer. In: Advances in Neural Information Processing Systems, vol. 34 (2021)
30. Mazyavkina, N., Sviridov, S., Ivanov, S., Burnaev, E.: Reinforcement learning for combinatorial optimization: a survey. Comput. Oper. Res. **134**, 105400 (2021)
31. Mladenović, N., Hansen, P.: Variable neighborhood search. Comput. Oper. Res. **24**(11), 1097–1100 (1997)
32. Morris, C., et al.: Weisfeiler and leman go neural: higher-order graph neural networks. In: Proceedings of the AAAI Conference on Artificial Intelligence, vol. 33, pp. 4602–4609 (2019)
33. Nair, V., et al.: Solving mixed integer programs using neural networks. arXiv preprint arXiv:2012.13349 (2020)
34. Nazari, M., Oroojlooy, A., Snyder, L., Takác, M.: Reinforcement learning for solving the vehicle routing problem. In: Advances in Neural Information Processing Systems, vol. 31 (2018)
35. d O Costa, P.R., Rhuggenaath, J., Zhang, Y., Akcay, A.: Learning 2-opt heuristics for the traveling salesman problem via deep reinforcement learning. In: Asian Conference on Machine Learning, pp. 465–480. PMLR (2020)

36. Park, J., Bakhtiyar, S., Park, J.: Schedulenet: learn to solve multi-agent scheduling problems with reinforcement learning. arXiv preprint arXiv:2106.03051 (2021)
37. Park, J., Chun, J., Kim, S.H., Kim, Y., Park, J.: Learning to schedule job-shop problems: representation and policy learning using graph neural network and reinforcement learning. Int. J. Prod. Res. **59**(11), 3360–3377 (2021)
38. Perron, L., Furnon, V.: Or-tools. https://developers.google.com/optimization/
39. Sels, V., Gheysen, N., Vanhoucke, M.: A comparison of priority rules for the job shop scheduling problem under different flow time-and tardiness-related objective functions. Int. J. Prod. Res. **50**(15), 4255–4270 (2012)
40. Sutton, R.S., Barto, A.G.: Reinforcement Learning: An Introduction. MIT Press, Cambridge (2018)
41. Taillard, E.: Benchmarks for basic scheduling problems. Eur. J. Oper. Res. **64**(2), 278–285 (1993)
42. Thyssens, D., Falkner, J., Schmidt-Thieme, L.: Supervised permutation invariant networks for solving the CVRP with bounded fleet size. arXiv preprint arXiv:2201.01529 (2022)
43. Toth, P., Vigo, D.: The vehicle routing problem. SIAM (2002)
44. Tsang, E.: Foundations of constraint satisfaction: the classic text. BoD-Books on Demand (2014)
45. Uchoa, E., Pecin, D., Pessoa, A., Poggi, M., Vidal, T., Subramanian, A.: New benchmark instances for the capacitated vehicle routing problem. Eur. J. Oper. Res. **257**(3), 845–858 (2017)
46. Van Hasselt, H., Guez, A., Silver, D.: Deep reinforcement learning with double q-learning. In: Proceedings of the AAAI Conference on Artificial Intelligence, vol. 30 (2016)
47. Vaswani, A., et al.: Attention is all you need. In: Advances in Neural Information Processing Systems, vol. 30 (2017)
48. Veličković, P., Cucurull, G., Casanova, A., Romero, A., Lio, P., Bengio, Y.: Graph attention networks. arXiv preprint arXiv:1710.10903 (2017)
49. Vinyals, O., Fortunato, M., Jaitly, N.: Pointer networks. In: Advances in Neural Information Processing Systems, vol. 28 (2015)
50. Voudouris, C., Tsang, E.P., Alsheddy, A.: Guided local search. In: Gendreau, M., Potvin, J.Y. (eds.) Handbook of Metaheuristics, pp. 321–361. Springer, Boston (2010). https://doi.org/10.1007/978-1-4419-1665-5_11
51. Wu, Y., Song, W., Cao, Z., Zhang, J., Lim, A.: Learning improvement heuristics for solving routing problems. IEEE Trans. Neural Netw. Learn. Syst. **33**(9), 5057–5069 (2021)
52. Wu, Z., Pan, S., Chen, F., Long, G., Zhang, C., Philip, S.Y.: A comprehensive survey on graph neural networks. IEEE Trans. Neural Netw. Learn. Syst. **32**(1), 4–24 (2020)
53. Zhang, C., Song, W., Cao, Z., Zhang, J., Tan, P.S., Chi, X.: Learning to dispatch for job shop scheduling via deep reinforcement learning. Adv. Neural. Inf. Process. Syst. **33**, 1621–1632 (2020)

Branch Ranking for Efficient Mixed-Integer Programming via Offline Ranking-Based Policy Learning

Zeren Huang[1], Wenhao Chen[1], Weinan Zhang[1(✉)], Chuhan Shi[1], Furui Liu[2], Hui-Ling Zhen[2], Mingxuan Yuan[2], Jianye Hao[2], Yong Yu[1], and Jun Wang[2,3]

[1] Shanghai Jiao Tong University, Shanghai, China
{sjtu_hzr,wnzhang}@sjtu.edu.cn
[2] Huawei Noah's Ark Lab, Shenzhen, China
{liufurui2,zhenhuiling2}@huawei.com
[3] University College London, London, UK

Abstract. Deriving a good variable selection strategy in branch-and-bound is essential for the efficiency of modern mixed-integer programming (MIP) solvers. With MIP branching data collected during the previous solution process, learning to branch methods have recently become superior over heuristics. As branch-and-bound is naturally a sequential decision making task, one should learn to optimize the utility of the whole MIP solving process instead of being myopic on each step. In this work, we formulate learning to branch as an offline reinforcement learning (RL) problem, and propose a long-sighted hybrid search scheme to construct the offline MIP dataset, which values the long-term utilities of branching decisions. During the policy training phase, we deploy a ranking-based reward assignment scheme to distinguish the promising samples from the long-term or short-term view, and train the branching model named Branch Ranking via offline policy learning. Experiments on synthetic MIP benchmarks and real-world tasks demonstrate that Branch Ranking is more efficient and robust, and can better generalize to large scales of MIP instances compared to the widely used heuristics and state-of-the-art learning-based branching models.

Keywords: Combinatorial optimization · Reinforcement learning · Deep learning

1 Introduction

Mixed-integer programming (MIP) has a wide range of real-world applications such as scheduling, manufacturing and routing [13,18]. A universally applicable method for solving MIPs is branch-and-bound (B&B) [14], which performs decomposition of the solution set iteratively, and thus building a search tree with nodes corresponding to MIP problems.

Node selection and variable selection are two important sequential decisions to be made at each iteration of the B&B algorithm. The node selection strategy decides which node to process next, while the variable selection strategy decides

© The Author(s), under exclusive license to Springer Nature Switzerland AG 2023
M.-R. Amini et al. (Eds.): ECML PKDD 2022, LNAI 13717, pp. 377–392, 2023.
https://doi.org/10.1007/978-3-031-26419-1_23

which fractional variable to branch on at the current node. In this work, we focus on variable selection, which can dramatically influence the performance of the B&B algorithm, as discussed in [14].

The variable selection strategies in modern MIP solvers are generally based on manually designed heuristics, which are strongly dependent on the problem property. Therefore, when the structure or the scale of the problem changes, an adjustment of the heuristics is often required, which is time-consuming and labor-intensive [6]. To tackle the above issues, machine learning naturally becomes a candidate for constructing a more efficient and generalizable variable selection policy [3]. The similarity among many previous learning-based methods lies in the way to train the branching policy [1,7,10,12,15]. Specifically, the policy models are usually trained via imitation learning to mimic an effective but slow heuristic, i.e., *strong branching* (SB), which performs a one-step branching simulation for each candidate variable and greedily selects one with the highest one-step utility, the SB score, which is related to the dual bound improvements. As B&B is naturally a sequential decision making task, such a heuristic is considered to be myopic, that is, it only focuses on the short-term utility while neglecting the long-term impact on the bottom levels of the B&B tree. Under some circumstances, the optimal short-term branching decision can lead to poor long-term utility.

To address the above issues, in this paper, we formulate variable selection in B&B as an offline reinforcement learning (RL) problem, and propose a top-down hybrid search scheme to construct an offline dataset which involves branching information of both long-term and short-term decisions. Using the offline dataset, we deploy a ranking-based reward assignment scheme to distinguish the promising samples from other samples. Finally, the variable selection policy, which is named BRANCH RANKING (BR), is trained via ranking-based policy learning method, which is derived equivalent as maximizing the log-likelihood of samples with the corresponding rewards as the weights.

We conduct extensive experiments on four classes of NP-hard MIP benchmarks and deploy our proposed policy in the real-world supply and demand simulation tasks, and the results demonstrate that BRANCH RANKING is superior over widely used heuristics as well as offline state-of-the-art learning-based branching models, regarding the solution time and the number of nodes processed in the B&B tree. Furthermore, the evaluation results on different scales of MIPs show that BRANCH RANKING also has better generalization ability over larger scales of problems compared to other learning-based policies.

BRANCH RANKING has been deployed in real-world *demand and supply simution* tasks of Huawei, a global commercial technology enterprise, where the experiments demonstrate that BRANCH RANKING is applicable in decision problems encountered in practice.

In summary, the main technical contributions of this work are threefold:

- We formulate the variable selection in B&B as an offline RL task, and design a novel search scheme to construct the offline dataset which involves both long-term and short-term branching information.

- Based on the constructed offline dataset, we propose a ranking-based reward assignment scheme to distinguish the promising samples, and train the variable selection policy named Branch Ranking.
- Extensive experiments demonstrate that Branch Ranking can make better improvements to the optimization algorithm compared to other state-of-the-art heuristics and learning-based branching polices. Branch Ranking also shows better generalization ability to MIP problems with different scales.

2 Related Work

The branching decision is one of the most important decisions to be made in the MIP solvers, which can significantly influence the efficiency of the optimization algorithm. The core of branching for mixed-integer programming (MIP) is the evaluation of the branching scores and cost via tuning hyper-parameters. It is not surprising that the most of previous branching methods are devoted to imitate strong branching (SB), because with one step forward, SB can often effectively reduce the number of search tree nodes.

Khalil et al. [12] extract 72 branching features artificially including static and dynamic ones, and mimicked the strong branching strategy via a pair-wise variable ranking formulation. Similarly, our work also adopts a ranking-based scheme, however, we apply it in reward assignment which assigns higher implicit rewards to the top-ranking promising branching samples from the long-term or short-term view, and the assigned reward can be regarded the sample weight in the learning objective. More recently, Gasse et al. [7] train their policy model via imitation learning from the strong branching expert rules. The inspiring innovation is a novel graph convolutional network (GCN) integrated into Markov decision process (MDP) for variable selection which is beneficial to generalization and branching cost. Based on a similar policy architecture and learning method, Nair et al. [15] propose neural branching, which enables the expert policy to scale to large MIP instances through hardware acceleration. To attain a computationally inexpensive model compared to GCN [7], Gupta et al. [10] propose a hybrid architecture which combines the expressive GCN and computationally inexpensive multi-layer perceptrons (MLP) for efficient branching with limited computing power. With the aim at learning a branching policy that generalizes across heterogeneous MIPs, Zarpellon et al. [17] incorporate an explicit parameterization of the state of the search tree to modulate the branching decision. From another perspective, Balcan et al. [2] propose to learn a weighted version of several existing variable scoring rules to improve the branching decision. Different from the previous works, Sun et al. [16] employ reinforcement learning with a novelty based evolutionary strategy for learning a branching policy. Another reinforcement learning approach to tackle the branching decision problem comes from [5], which uses approximate Q-learning and the subtree size as value function. Note that the main algorithms of the above two reinforcement learning methods are on-policy, which evaluates and improves the same policy for making decisions, thus cannot leverage the abundant pre-collected branching data to train the model.

3 Background and Preliminaries

In this section, we first introduce the background of mixed-integer programming and the branch-and-bound algorithm. Then, we introduce several typical and effective variable selection heuristics for branching.

3.1 Mixed-Integer Programming Problem

The general combinatorial optimization task is usually formulated as a mixed-integer programming (MIP) problem, which can be written as the following form:

$$\arg\min_{\mathbf{x}} \left\{ \mathbf{z}^\top \mathbf{x} \mid \mathbf{A}\mathbf{x} \leq \mathbf{b}, \mathbf{x} \in \mathbb{Z}^p \times \mathbb{R}^{n-p} \right\}, \tag{1}$$

where $\mathbf{x}$ is the vector of n decision variables, $\mathbf{x} \in \mathbb{Z}^p \times \mathbb{R}^{n-p}$ means p out of n variables have integer constraints, $\mathbf{z} \in \mathbb{R}^n$ is the objective coefficient vector, $\mathbf{b} \in \mathbb{R}^m$ is the right-hand side vector, $\mathbf{A} \in \mathbb{R}^{m \times n}$ is the constraint matrix.

3.2 Branch-and-Bound

B&B is a classic approach for solving MIPs, which adopts a search tree consisting of nodes and branches to partition the total set of feasible solutions into smaller subsets.

The procedure of B&B can be described as follows: denote the optimal solution to the LP relaxation of Eq. (1) as x^*, if it happens to satisfy the integrality requirements, then it is also the solution to Eq. (1); else some component of x^* is not integer (while restricted to be integer), then one can select a fractional variable x_i and decompose the LP relaxation into two sub-problems by adding rounding bounds $x_i \geq \lceil x_i^* \rceil$ and $x_i \leq \lfloor x_i^* \rfloor$, respectively. By recursively performing the above binary decomposition, B&B naturally builds a search tree, within which each node corresponds to a MIP.

During the search process, the best feasible solution of the MIP provides an upper bound (or primal bound) for the optimal objective value; and solving the LP relaxation of the MIP provides the lower bound (or dual bound). The B&B algorithm terminates when the gap between the upper bound and the lower bound reduces to some tolerance threshold.

3.3 Variable Selection Heuristics for Branching

In the B&B framework for solving MIPs, for a given node, variable selection refers to the decision to select an integer variable to branch on. Designing a good variable selection strategy is essential for the efficiency of solving MIPs, which can lead to a much smaller search tree, that is, the number of processed nodes is significantly reduced and thus the whole solution process speeds up.

In modern MIP solvers, the variable selection module for branching is regarded as a core component of the optimization algorithm, and is generally based on manually designed heuristics. One of the classic heuristics is *strong branching*, which involves computing the dual bound improvements for each candidate variable by solving two resulting LP relaxations after temporarily adding bounds. Strong branching can often yield the smallest search trees while bringing much more computational costs, therefore, it is impractical to apply strong branching at each node.

Another more efficient heuristic, *pseudo cost branching*, is designed to imitate strong branching, which estimates the dual bound improvements of candidate variables based on historical information gathered in the tree. The combination of the above two heuristics is called *hybrid branching*, which employs strong branching at the beginning of the algorithm, and performs pseudo cost branching when more branching history information is available.

The current state-of-the-art variable selection heuristic is *reliability branching*, which is a refinement of pseudo cost branching, and is deployed as the default branching strategy by many modern MIP solvers (e.g., CPLEX [4], SCIP [9], etc.). During the solution process, reliability branching applies strong branching to those variables whose pseudo costs are uninitialized or unreliable. The pseudo costs are considered to be unreliable if there is not enough historical branching data aggregated.

4 Methodology

In this section, we first formulate learning variable selection strategy for branching as an offline RL problem, and introduce our method to derive a more efficient and robust variable selection policy named BRANCH RANKING.

4.1 Offline RL Formulation

The variable selection problem in branch-and-bound can be formulated as a sequential decision-making process with tuple (S, A, Π, g), where S is the state space, A is the action space, Π is the policy space and the function $g(s, a)$ is the dynamics function of the environment. At time step $t \geq 0$, a policy $\pi \in \Pi$ maps the environment state $s_t \in S$ to an action $a_t \in A$: $\pi(s_t) = a_t$, then the next state is $s_{t+1} = g(s_t, a_t)$. In the following, we clarify the state space S, the action space A, the transition function $g(s, a)$ and the roll-out trajectory τ in the context of branch-and-bound.

State Space S. The representation of S consists of the whole search tree with previous branching decisions, the LP solution of each node, the processing leaf node and also other statistics stored during the solution process. In practice, we encode the state using GCN as [7].

Action Space A. At the tth iteration, the action space A contains all the candidate variables $X^t_{cand} = \{x^t_1, x^t_2, \ldots, x^t_d\}$ to branch on at the currently processing node.

Transition Function $g(s, a)$. After a policy π selects an action $a_t = x \in X_{cand}^t$ in state s_t, the search tree is expanded and the LP relaxations of the two sub-nodes are solved. Then, the tree is pruned if possible, and eventually the next leaf node to process is selected. The environment then proceeds to the next state s_{t+1}.

Trajectory τ. A roll-out trajectory τ comprises a sequence of states and actions: $\{s_0, a_0, s_1, a_1, \ldots, s_T, a_T\}$. We define the return R_τ for the trajectory τ as the negative value of the number of visited (processed) nodes within it. Intuitively, each trajectory corresponds to a B&B tree. Note that there is no explicit reward defined for each state-action pair.

Based on the above MDP formulation, online RL algorithms usually suffer from low sample efficiency, which makes the training prohibitively long and leaves the pre-collected branching data of no use. To derive a more efficient scheme, we model the learning problem in the offline RL settings, where a batch of m roll-out trajectories $D = \{(\tau_i, R_{\tau_i}), i = 1, 2, \ldots, m\}$ is collected using some policy π_D. Using this dataset (without interaction with the environment), our objective is to construct a policy $\pi(a|s)$ which incurs the highest trajectory return when it is actually applied in the MDP.

4.2 Collecting Offline Data Through Top-Down Hybrid Search

To construct the offline dataset, a commonly-used roll-out policy is strong branching, which is time-consuming while often resulting in a small search tree. Under this case, the roll-out policy can be regarded as the expert policy, which directly provides the supervised data. As shown in the top dotted box in Fig. 1, the blue node represents the action selected by strong branching, which has the largest dual bound improvement after a one-step branching simulation. Strong branching can be viewed as a greedy method for variable selection, therefore, we consider such a strategy to be myopic, that is, it only focuses on the one-step return, which can lead to poor long-term outcome in some cases. As discussed in [8], a deeper lookahead can often be useful for making better branching decisions at the top levels of the search tree.

To derive a more farsighted and robust variable selection strategy for branching, we introduce a top-down hybrid search scheme to collect offline branching data which contains more information about the long-term impact of the branching decision. Figure 1 presents an overview of our proposed hybrid search scheme, which is a combination of short-term and long-term search. The basic procedure is as follows:

- At the tth iteration, for the currently processing node, the candidate variable set is $X_{cand}^t = \{x_1^t, x_2^t, \ldots, x_d^t\}$. First, we randomly sample the variables in X_{cand}^t for k times, which returns a subset of k variables $X_{exp}^t = \{x_{e_1}^t, x_{e_2}^t, \ldots, x_{e_k}^t\}$.
- For each variable $x \in X_{exp}^t$, we perform a one-step branching simulation, and obtain k simulation trees with different branches at the current node. Denote the set of k simulation trees as $T_{exp}^t = \{T_{e_1}^t, T_{e_2}^t, \ldots, T_{e_k}^t\}$.

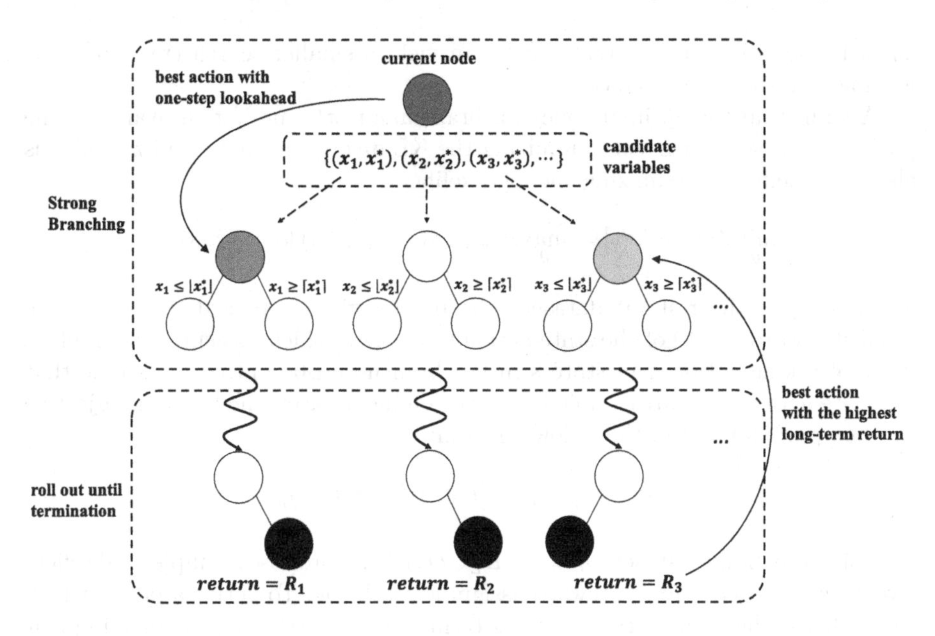

Fig. 1. Hybrid offline data collection

- As it shown in the lower dotted box in Fig. 1, for each simulation tree $T_{e_i}^t \in T_{exp}^t$, we continue the branching simulation, that is, to roll out the trajectory using some myopic policy (e.g. strong branching) until the termination node (black nodes), which returns a trajectory return $R_{e_i}^t$. Then, we obtain a set consisting of pairs of the top-level exploring action and the long-term trajectory return: $B^t = \{(x_{e_i}^t, R_{e_i}^t), i = 1, 2, \ldots, k\}$.
- Back up at the node processing at the tth iteration, we select the action $x \in X_{exp}^t$ with the highest long-term return in B^t (the green node in Fig. 1), and execute the corresponding branching in the real environment. The search tree then picks the next node to process using the node selection policy incorporated in the solver, and then the environment transits to the next state. The above process continues until the problem instance is solved.

Note that though collection branching data leads to multiple rounds of branching simulation, the whole process is conducted in an offline way, and thus the incurred cost is acceptable for practical use. Denote the collected offline dataset using the above hybrid search scheme as $D_{hyb} = \{\{(s_i^L, a_i^L, R_i^L)\}_{i=1}^M \cup \{(s_i^{SB}, a_i^{SB})\}_{i=1}^N\}$, in which $D_L = \{(s_i^L, a_i^L, R_i^L)\}_{i=1}^M$ is collected by the top-down long-term search as mentioned above, and $D_{SB} = \{(s_i^{SB}, a_i^{SB})\}_{i=1}^N$ is collected by strong branching during the simulation process.

4.3 Ranking-Based Policy Learning

Given the offline dataset $D_{hyb} = \{D_L \cup D_{SB}\}$ collected by our proposed hybrid search scheme, our goal is to derive a efficient and robust variable selection policy

π_θ (θ is the policy parameter) which can yield a smaller search tree, and thus leading to faster performance.

Assume that the optimal policy for branching is π^*, and if π^* is available, the policy π can be trained via minimizing the KL divergence of π^* and π, which is also equivalent to maximizing the log-likelihood:

$$\min_\theta KL\left[\pi^* \| \pi_\theta\right] = \max_\theta \mathbb{E}_{(s,a)\sim D_{exp}}\left[\pi^*(a|s)\log \pi_\theta(a|s)\right], \tag{2}$$

where D_{exp} is the roll-out dataset collected by π^*. We regard $\pi^*(a|s)$ as the implicit reward $\hat{r}(s,a)$ of the state-action pair (s,a), which measures the implicit value of taking action a in state s in the branch-and-bound tree. Assume that the state-action pairs are sampled from the offline dataset, the learning objective can be approximated in the following form:

$$\max_\theta \mathbb{E}_{(s,a)\sim D_{hyb}}\left[\hat{r}(s,a)\log \pi_\theta(a|s)\right]. \tag{3}$$

Using the offline dataset $D_{hyb} = \{D_L \cup D_{SB}\}$, we propose a simple while effective ranking-based implicit reward assignment scheme. To derive a policy which can balance the short-term and long-term benefits, we assign higher implicit rewards to the promising state-action pairs which lead to the top-ranked long-term or short-term return. We first give a ranking-based definition of long-term and short-term promising state-action pairs in Definition 1 and 2, respectively.

Definition 1 (Long-Term Promising). *For a state-action pair (s,a), its corresponding trajectory return $R^L(s,a)$ is the return of the trajectory starting from state s and selecting the action a. A state-action pair (s,a) is defined to be long-term promising if $R^L(s,a)$ ranks in the top $p\%$ trajectories starting from state s and with the highest trajectory returns. Note that p is a tunable hyper-parameter.*

Definition 2 (Short-Term Promising). *A state-action pair (s,a) is short-term promising if the one-step look-ahead return of the action a ranks the highest in state s in terms of strong branching score, which is related to the dual bound improvements after a one-step simulation.*

Based on the above ranking-based definitions, for a state-action pair (s,a), its implicit reward $\hat{r}(s,a)$ is defined as

$$\hat{r}(s,a) = \begin{cases} 1, & (s,a) \text{ is long-term or short-term promising} \\ 0, & \text{otherwise.} \end{cases} \tag{4}$$

After the reward assignment, the final constructed training dataset is a combination of long-term and short-term promising samples. We denote the proportion of short-term promising samples as h, which is a tunable hyper-parameter.

We then train our policy π_θ via Eq. (3), and evaluate π_θ on the new problem instances. The flowchart of our proposed method is shown in Fig. 2.

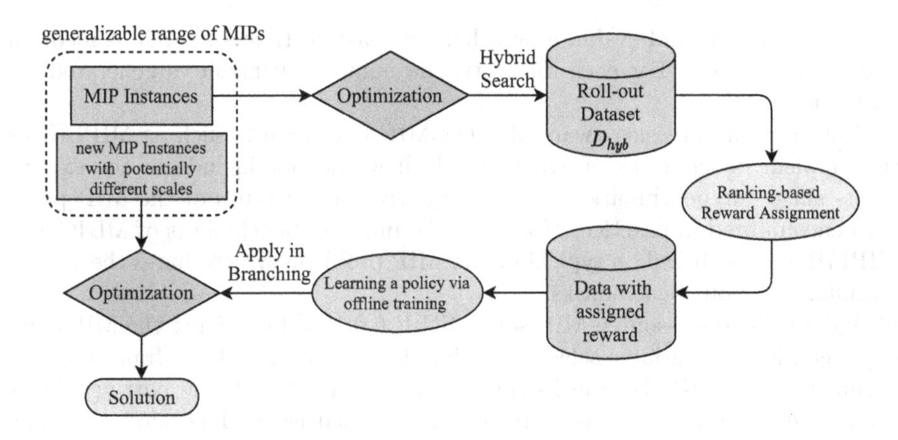

Fig. 2. The flowchart of BRANCH RANKING method.

5 Experiments

We conduct extensive experiments on the MIP benchmarks, and evaluate our policy compared to other state-of-the-art variable selection heuristics as well as learning-based branching models. By comparative analysis on the experimental results, we try to answer the following research questions:

RQ1: Is our policy more efficient and robust compared to typical heuristics, and learning-based models which imitate the strong branching strategy?

RQ2: Can our policy generalize to different scales of problem instances of the same MIP classes?

RQ3: How does different proportions of long-term and short-term promising samples influence the final performance of our trained polices?

RQ4: Can our policy be deployed in challenging real-world tasks and improve the solution quality given a solving time limit?

5.1 Experimental Setup

Benchmarks Used for Training and Evaluating. The synthetic MIP benchmarks consist of four classes of NP-hard MIP problems as Gasse et al. [7]: *set covering*, *combinatorial auction*, *capacitated facility location* and *maximum independent set*, which cover a wide range of real-world decision problems. For each class of MIPs, we randomly generate three scales of problem instances for training and evaluating. Since larger MIPs are usually more difficult to solve, the generated instances are referred to as easy, medium and hard instances according to the problem scales. For a more detailed description of the MIP problems, one can refer to [7]. For each MIP class, we train the learning-based policy only on easy instances, and evaluate on other new easy instances as well as medium and hard instances. For each difficulty, the number of instances generated for evaluation is 20.

Note that we are also aware of other MIP benchmarks such as MIPLIB or the problem benchmarks of ML4CO, which we do not include in the experiments since our benchmarks are representative of the types of the MIP problems encountered in practice. Moreover, the problem benchmarks of ML4CO or MIPLIB are also based on typical binary MIP problems which shares the similar formulation as our benchmarks.

We use the open-source MIP solver SCIP 6.0.1 [9] for solving the MIPs and implementing our variable selection policy. The main algorithmic framework for optimizing the MIPs is branch-and-bound, combined with the cutting planes enabled at the root node and some other MIP heuristics. The maximum time limit of solving a problem instance is set to 3600 s.

Metrics. We consider both the number of processed B&B nodes and the MIP solution time as the metrics to demonstrate the comprehensive performance of different variable selection policies. For each evaluated policy, we show the average value and the standard deviation of the evaluation results over a number of test instances. Note that the number of processed B&B nodes are reported only on solved instances (integrality gap is zero) within the time limit.

Baselines. We compare Branch Ranking with six baselines.

- **Branching heuristics**: we compare against four commonly-used and effective branching heuristics in MIP solvers: *inference* (Inference), *mostinf* (Mostinf), *pseudocost branching* (Pscost) and *reliability pseudocost branching* (Relpcost). One can find a detailed description of the heuristics in the documentation of [9].
- **Khalil** [12]: the state representation of MIP is based on available MIP statistics. A SVMrank [11] model is used to learn an approximation to strong branching.
- **GCN** [7]: the state representation of MIP is the bipartite graph, and the policy architecture is based on GCN which takes the MIP state as the inputs. The policy is trained to mimic the strong branching strategy via imitation learning.

Note that the method of [16] or [5] is not included as a learning-based baseline since the proposed algorithmic framework is on-policy, while in this work, we formulate learning to branch in the offline settings, and propose to derive a branching policy via offline learning, which we consider to be more suitable for the nature of the variable selection problem in B&B.

Hyper-Parameters. The architecture of our policy network is based on GCN as [7]. For other hyper-parameters, we set the exploring times k to 30, the ranking parameter p to 10. We fine-tune the data proportion parameter h for each MIP class, and set h to 0.7, 0.9, 0.95, 0.9, 0.9 for *set covering*, *combinatorial auction*, *capacitated facility location* and *maximum independent set*, respectively. For each learning-based branching model, the number of training and validation samples are 50000 and 5000, respectively.

5.2 Performance Comparison (RQ1 & RQ2)

As shown in Table 1, for medium and hard instances of each MIP class, our proposed Branch Ranking clearly outperforms all the baselines in terms of the solution time, and also leads to the smallest number of processed B&B nodes on three classes of the MIP problems, *set covering*, *combinatorial auction* and *maximum independent set*. For *capacitated facility location*, Branch Ranking produces a B&B tree with fewer nodes compared to other learning-based models and heuristics except Relpcost. However, Branch Ranking reduces the solution time notably.

As for the easy instances, considering the solution time, Branch Ranking is superior over other baselines on *set covering*, *capacitated facility location* and *maximum independent set*. For the *combinatorial auction* problems, Branch Ranking achieves the second minimum solution time. On account of the number of processed B&B nodes, Branch Ranking has also shown comparable performance on each MIP class. Note that though Relpcost leads to smaller B&B trees, it is not competitive regarding the solution time.

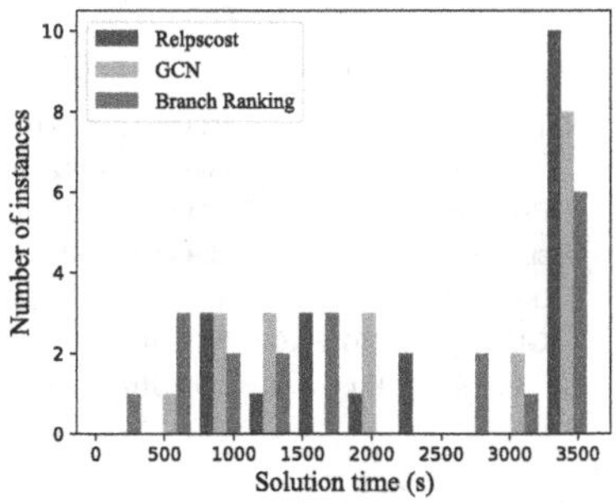

Fig. 3. Number of instances with different solution time on the hard set covering problems.

The comparative results demonstrate that our proposed policy Branch Ranking is more efficient compared to the state-of-the-art offline learning-based branching models, and also branching heuristics adopted in modern MIP solvers. Moreover, Branch Ranking is also more robust compared to the learning-based models which imitate the strong branching strategy. Specifically, Khalil and GCN suffer from performance fluctuations on the medium and hard *maximum independent set* instances, while Branch Ranking still achieves less solution time and fewer B&B nodes compared to other well-performed baselines.

Since the learning models are trained only on the easy instances, we are also able to evaluate the generalization ability of different polices. The evaluation results on medium and hard instances show that our proposed Branch Ranking performs remarkably well on larger instances with the same MIP class, and even gains higher performance improvements compared to other baselines when the problem size becomes larger. The results on different scales of instances indicate that Branch Ranking is capable of generalizing over problems with larger

Table 1. Evaluation results of different heuristics and learning-based policies in terms of the solution time and the number of processed B&B nodes. For each MIP class, the machine learning models are trained on easy instances only. The best evaluation result is highlighted. Among the compared methods, the learning-based ones are marked with $*$.

Set Covering

Method	Easy		Medium		Hard	
	Time	Nodes	Time	Nodes	Time	Nodes
INFERENCE	16.15±17.17	1366±1636	689.85±893.48	62956±91525	3600.00±0.00	N/A
MOSTINF	31.22±33.59	2397±2565	1216.57±1042.13	73955±91098	3600.00±0.00	N/A
PSCOST	6.96±2.38	209±134	139.28±178.43	13101±17965	2867.07±882.83	128027±53366
RELPSCOST	9.84±3.07	**105±125**	106.21±112.43	7223±9853	2530.76±1121.30	100952±65534
KHALIL *	8.18±3.76	118±88	146.58±185.44	6896±9330	2866.54±973.77	73780±33070
GCN *	6.99±1.80	168±138	86.76±98.56	5067±6512	2356.36±1189.37	71576±47806
BR (OURS) *	**6.93±1.67**	163±129	**80.76±87.97**	**4624±5912**	**2199.49±1170.82**	**67784±46808**

Combinatorial Auction

Method	Easy		Medium		Hard	
	Time	Nodes	Time	Nodes	Time	Nodes
INFERENCE	2.22±1.40	638±922	25.09±13.84	6043±4173	655.30±695.36	91328±99142
MOSTINF	3.20±3.36	938±1569	332.19±269.72	76349±62942	2786.27±73.02	282830±42980
PSCOST	**1.97±0.98**	374±432	22.91±17.12	3843±3593	386.61±351.55	39135±37257
RELPSCOST	3.18±1.58	**28±54**	19.53±7.15	987±843	213.66±229.91	14788±19115
KHALIL *	2.17±1.16	123±129	24.43±13.66	1154±782	612.89±784.94	19291±25459
GCN *	**1.97±0.60**	103±105	12.74±6.92	970±793	201.78±260.23	14529±20819
BR (OURS) *	1.98±0.62	106±102	**12.12±5.72**	**902±627**	**198.02±248.70**	**14380±20057**

Capacitated Facility Location

Method	Easy		Medium		Hard	
	Time	Nodes	Time	Nodes	Time	Nodes
INFERENCE	75.54±75.50	254±455	356.02±400.02	1040±1214	756.98±408.78	547±384
MOSTINF	43.88±33.26	128±194	280.38±281.19	673±740	662.62±375.54	393±273
PSCOST	40.80±31.99	107±161	271.28±241.35	673±637	644.65±400.78	433±360
RELPSCOST	42.29±29.15	**72±131**	320.14±286.33	**383±513**	703.75±425.64	**233±239**
KHALIL *	41.88±30.64	106±155	282.97±342.77	681±913	664.58±420.50	369±275
GCN *	34.03±33.05	160±214	255.35±259.19	584±647	638.07±410.58	479±444
BR (OURS) *	**33.92±33.90**	169±239	**249.02±222.21**	570±563	**633.59±406.13**	459±399

Maximum Independent Set

Method	Easy		Medium		Hard	
	Time	Nodes	Time	Nodes	Time	Nodes
INFERENCE	32.99±103.48	16105±58614	1807.13±750.10	116957±84167	3600.00±0.00	N/A
MOSTINF	30.34±91.58	10084±35418	1547.43±901.47	60128±52244	3600.00±0.00	N/A
PSCOST	10.11±13.88	1565±3935	1042.47±798.46	43256±38143	3181.14±455.73	66258±23654
RELPSCOST	8.83±5.11	**209±603**	136.78±113.04	4911±5774	2462.67±1391.50	53636±38140
KHALIL *	10.50±16.62	353±917	503.94±712.89	7983±10915	3023.02±798.52	60800±23837
GCN *	7.55±10.11	265±805	256.23±509.38	10948±23020	2485.27±1446.23	53804±37253
BR (OURS) *	**7.54±9.53**	260±742	**102.91±118.77**	**3908±5613**	**2320.20±1457.44**	**50460±37859**

scales, and has better generalization ability in comparison to other learning-based branching policies.

Furthermore, for a better visualization of the generalization results of the well-performed baselines and our proposed BRANCH RANKING, Fig. 3 shows the number of instances with different solution time for RELPCOST, GCN and BRANCH RANKING on the hard *set covering* problems. The visualization results show that more instances are solved with less solution time using BRANCH RANKING. Such results are consistent with our previous findings, which also proves that BRANCH RANKING can better improve the solution process for larger MIP problems.

5.3 Hyper-Parameter Study (RQ3)

To better understand the influence of the proportion of long-term or short-term promising samples on the final performance of BRANCH RANKING, we conduct a hyper-parameter study to evaluate BRANCH RANKING with different proportions of short-term promising samples on the *set covering* problems. The results are shown in Fig. 4, in which the dotted line represents the state-of-the-art offline learning-based branching model, GCN, which can be regarded as a policy trained using only the short-term promising samples.

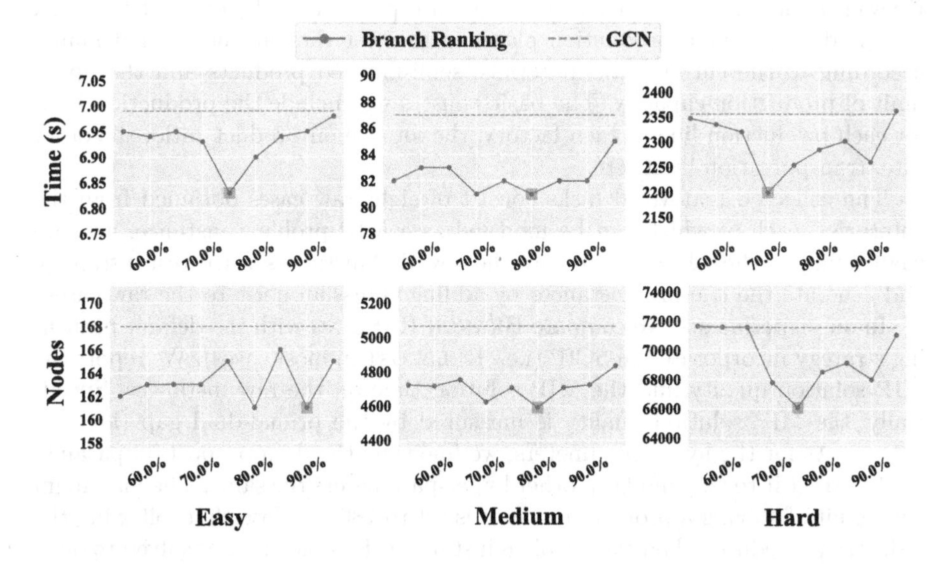

Fig. 4. Hyper-parameter study results of BRANCH RANKING with different proportions of short-term promising samples on the set covering problems.

As demonstrated in Fig. 4, for the easy instances, our proposed BRANCH RANKING slightly outperforms GCN in terms of solution time and number of processed B&B nodes under different proportions. Moreover, the policy with an intermediate proportion achieves the best performance considering both the two metrics.

For medium and hard instances, Branch Ranking gains remarkable performance improvements by combining long-term and short-term promising samples. Moreover, the empirical results show that too large or too small short-term sample proportion will worsen the performance of Branch Ranking. Precisely, for the Set Covering problems, we find that setting the short-term sample proportion between 70% and 80% often yields a better branching policy.

The results of our hyper-parameter study can be explained from the following aspects. First, as shown by Fig. 4, the long-term promising samples is beneficial for deriving a more efficient and robust branching policy which can make better decisions from a long-term view. Moreover, using excessive long-term promising samples will probably not produce the best policy, since long-term samples are also more difficult to learn for the policy models. To ease this situation, as introduced in this work, we propose to combine the long-term and short-term promising samples for learning a better variable selection policy for branching.

5.4 Deployment in Real-World Tasks (RQ4)

To further verify the effectiveness of our proposed branching policy, we deploy Branch Ranking in the real-world *supply and demand simulation* tasks of Huawei, a global commercial technology enterprise. The objective of this task is to find the optimal production planning to meet the current order demand, according to the current raw materials, semi-finished products and the upper limit of production capacity. The basic constraints include the production limit for each production line in each factory, the lot size on product ratio, the order rate, transportation limit, etc.

The collected real-world tasks consist of eight raw cases obtained from the enterprise, each of which can be modeled as a MIP problem instance, and are challenging for the solvers. We regard the raw MIP instances as the test instances, and generate the training instances by adding Gaussian noise to the raw cases.

In our experiments, we compare Branch Ranking with the default branching strategy incorporated in SCIP (i.e., Relpcost in most cases). We report the MIP solution quality and the MIP solution time on the raw instances. Specifically, the MIP solution quality is measured by the primal-dual gap (lower is better). As for the hyper-parameters, we fine-tune the data proportion parameter h and set it to 0.9, and keep other hyper-parameters the same. The maximum time limit of solving a problem instance is set to 1800 s. Note that following the industry procedure, when the problem instance is feasible and not solved to optimal within the time limit, the solver will return an approximate MIP solution.

As shown in Table 2, since the real-world supply and demand simulation problems are challenging for the solvers, most of the cases are not solved to optimal within the time limit. For case 3 and 4, the MIPs are found infeasible. For case 5, the optimal MIP solution is found for both the baseline and Branch Ranking, while Branch Ranking leads to less solution time. For other cases,

Table 2. Evaluation results of BRANCH RANKING and the default branching strategy in SCIP in terms of the solution time and the solution quality (primal-dual gap, the lower is better). Case 3 and 4 are found infeasible.

Method	Case 1		Case 2		Case 3		Case 4		Case 5		Case 6		Case 7		Case 8	
	Time	Gap	Time	Gap	Time	Gap	Time	Gap	Time	Gap	Time	Gap	Time	Gap	Time	Gap
DEFAULT	1800.0	98.59	1800.0	1.36	N/A	inf.	N/A	inf.	460.98	0	1800.0	1.14	1800.0	0.07	1800.0	0.21
BR (OURS)	1800.0	62.26	1800.0	0.87	N/A	inf.	N/A	inf.	451.30	0	1800.0	1.14	1800.0	0.06	1800.0	0.19

though the optimal solution is not found within the time limit, BRANCH RANKING improves the solution quality notably: the average primal-dual gap reduction ratio has reached 22.74% compared to the branching strategy incorporated in SCIP. The results further indicate the significance of making good branching decisions. Our proposed BRANCH RANKING has also shown to be more efficient on difficult real-world tasks.

6 Conclusion

In this paper, we present an offline RL formulation for variable selection in the branch-and-bound algorithm. To derive a more long-sighted branching policy under such a setting, we propose a top-down hybrid search scheme to collect the offline samples which involves more long-term information. During the policy learning phase, we deploy a ranking-based reward assignment scheme which assigns higher implicit rewards to the long-term or short-term promising samples, and learn a branching policy named BRANCH RANKING by maximizing the log-likelihood weighted by the assigned rewards. The experimental results on synthetic benchmarks and real-world tasks show that our derived policy BRANCH RANKING is more efficient and robust compared to the state-of-the-art heuristics and learning-based policies for branching. Furthermore, BRANCH RANKING can also better generalize to the same class of MIP problems with larger scales.

Acknowledgements. The SJTU team is supported by Shanghai Municipal Science and Technology Major Project (2021SHZDZX0102) and National Natural Science Foundation of China (62076161). The work is also sponsored by Huawei Innovation Research Program.

References

1. Alvarez, A.M., Louveaux, Q., Wehenkel, L.: A machine learning-based approximation of strong branching. INFORMS J. Comput. **29**(1), 185–195 (2017)
2. Balcan, M.F., Dick, T., Sandholm, T., Vitercik, E.: Learning to branch. In: International Conference on Machine Learning, pp. 344–353. PMLR (2018)
3. Bengio, Y., Lodi, A., Prouvost, A.: Machine learning for combinatorial optimization: a methodological tour d'horizon. Eur. J. Oper. Res. **290**(2), 405–421 (2021)
4. Cplex, I.I.: V12. 1: user's manual for CPLEX. Int. Bus. Mach. Corp. **46**(53), 157 (2009)

5. Etheve, M., Alès, Z., Bissuel, C., Juan, O., Kedad-Sidhoum, S.: Reinforcement learning for variable selection in a branch and bound algorithm. In: Hebrard, E., Musliu, N. (eds.) CPAIOR 2020. LNCS, vol. 12296, pp. 176–185. Springer, Cham (2020). https://doi.org/10.1007/978-3-030-58942-4_12
6. Fischetti, M., Lodi, A.: Heuristics in mixed integer programming. In: Wiley Encyclopedia of Operations Research and Management Science (2010)
7. Gasse, M., Chételat, D., Ferroni, N., Charlin, L., Lodi, A.: Exact combinatorial optimization with graph convolutional neural networks. In: Proceedings of the 33rd International Conference on Neural Information Processing Systems, vol. 2, pp. 15580–15592 (2019)
8. Glankwamdee, W., Linderoth, J.: Lookahead branching for mixed integer programming. Technical report, Citeseer (2006)
9. Gleixner, A., et al.: The SCIP Optimization Suite 6.0. ZIB-Report 18-26, Zuse Institute Berlin (2018). http://nbn-resolving.de/urn:nbn:de:0297-zib-69361
10. Gupta, P., Gasse, M., Khalil, E., Mudigonda, P., Lodi, A., Bengio, Y.: Hybrid models for learning to branch. Adv. Neural. Inf. Process. Syst. **33**, 18087–18097 (2020)
11. Joachims, T.: Optimizing search engines using clickthrough data. In: Proceedings of the eighth ACM SIGKDD International Conference on Knowledge Discovery and Data Mining, pp. 133–142 (2002)
12. Khalil, E.B., Bodic, P.L., Song, L., Nemhauser, G., Dilkina, B.: Learning to branch in mixed integer programming. In: Proceedings of the Thirtieth AAAI Conference on Artificial Intelligence, pp. 724–731 (2016)
13. Maraš, V., Lazić, J., Davidović, T., Mladenović, N.: Routing of barge container ships by mixed-integer programming heuristics. Appl. Soft Comput. **13**(8), 3515–3528 (2013)
14. Morrison, D.R., Jacobson, S.H., Sauppe, J.J., Sewell, E.C.: Branch-and-bound algorithms: a survey of recent advances in searching, branching, and pruning. Discret. Optim. **19**, 79–102 (2016)
15. Nair, V., et al.: Solving mixed integer programs using neural networks. arXiv preprint arXiv:2012.13349 (2020)
16. Sun, H., Chen, W., Li, H., Song, L.: Improving learning to branch via reinforcement learning. In: Learning Meets Combinatorial Algorithms at NeurIPS 2020 (2020)
17. Zarpellon, G., Jo, J., Lodi, A., Bengio, Y.: Parameterizing branch-and-bound search trees to learn branching policies. In: Proceedings of the AAAI Conference on Artificial Intelligence, vol. 35, pp. 3931–3939 (2021)
18. Zhu, Z., Heady, R.B.: Minimizing the sum of earliness/tardiness in multi-machine scheduling: a mixed integer programming approach. Comput. Ind. Eng. **38**(2), 297–305 (2000)

Learning Optimal Decision Trees Under Memory Constraints

Gaël Aglin(✉), Siegfried Nijssen, and Pierre Schaus

ICTEAM - UCLouvain, Louvain-la-Neuve, Belgium
{gael.aglin,siegfried.nijssen,pierre.schaus}@uclouvain.be

Abstract. Existing algorithms for learning optimal decision trees can be put into two categories: algorithms based on the use of Mixed Integer Programming (MIP) solvers and algorithms based on dynamic programming (DP) on itemsets. While the algorithms based on DP are the fastest, their main disadvantage compared to MIP-based approaches is that the amount of memory these algorithms may require to find an optimal solution is not bounded. Consequently, for some datasets these algorithms can only be executed on machines with large amounts of memory. In this paper, we propose the first DP-based algorithm for learning optimal decision trees that operates under memory constraints. Core contributions of this work include: (1) strategies for freeing memory when too much memory is used by the algorithm; (2) an effective approach for recovering the optimal decision tree when parts of the memory are freed. Our experiments demonstrate a favorable trade-off between memory constraints and the run times of our algorithm.

Keywords: Decision trees · Optimization · Memory management

1 Introduction

Decision trees (DTs) are among the most popular predictive machine learning models. They consist of tree structures in which internal nodes are labeled with tests and leaves are labeled with predictions. A key characteristic of DTs is their interpretability. A DT can be used to perform a prediction for an instance by performing a top-down traversal of the tree: the outcome of a test on an instance determines in which child node the traversal continues; the tests on the path towards a leaf explain why the model performs the prediction in that leaf.

Recent years have witnessed an increasing interest in explainable predictive models: in some crucial applications, such as medicine, bank loan repayment prediction or criminal recidivism prediction tools, it is important that the predictions of the models are understood in order to trust them.

One measure of the complexity of a DT model is its *depth*: the maximum number of tests performed to reach a leaf. Arguably, a DT model is easier to

This work was supported by bpost.

© The Author(s), under exclusive license to Springer Nature Switzerland AG 2023
M.-R. Amini et al. (Eds.): ECML PKDD 2022, LNAI 13717, pp. 393–409, 2023.
https://doi.org/10.1007/978-3-031-26419-1_24

interpret if its depth is limited. Traditionally, such trees would be learned using heuristic algorithms such as CART. However, in 2017, Bertsimas et al. [5] showed that depth-limited *optimal decision trees* (ODTs) generalize better to unseen data than heuristic DTs. In this context, ODTs are decision trees that obtain the best possible accuracy on training data under a depth constraint.

Finding decision trees that optimize accuracy is not a trivial problem. For instance, [8] showed that the problem of identifying a DT under a size constraint is NP-complete. This has led to a renewed interest in developing efficient algorithms for finding optimal decision trees.

Approaches for finding ODTs can be put into two categories: (1) the category of approaches based on the use of mixed integer programming (MIP) solvers [1,13]; (2) the category of specialized search algorithms that use some form of *dynamic programming* (DP) or *caching* [2,6,7,9–12]. Among these two categories, the specialized DP approaches have been demonstrated to be significantly faster [2,6]. Unfortunately, however, an important drawback of the DP approaches is their high memory consumption: in the worst case their memory consumption is exponential in the number of features. MIP approaches, at least theoretically, do not have this weakness.

The key aim of this paper is to address this situation. We propose a new DP-based approach that allows a user to determine the trade-off between time and space, by introducing a parameter that determines the amount of memory the algorithm is allowed to use.

The intuition for the good run time performance of DP approaches is that these approaches do not perform a search for *trees*, but a search for *paths*. As the number of paths is smaller than the number of trees, this reduces the size of search space in DP algorithms significantly. Their high memory consumption derives from the fact that the DP-based algorithms store information about paths in a *cache*, from which they construct ODTs.

The contributions of this work are the following:

C0 We study several caching strategies proposed in the literature, identifying which has the best characteristics when focusing on memory consumption.

C1 We propose a simple modification of existing DP-based approaches that amounts to removing, from time to time, elements from the cache if the cache becomes too large.

However, removing elements from the cache can have two undesirable consequences: (1) an ODT can no longer be recovered from the cache, and (2) search can become slower, as we can no longer use results in the cache. Our next contributions address these critical problems.

C2 We propose strategies for the order in which elements are deleted from the cache.

C3 We present a strategy to recover deleted elements if these are required to output the ODT.

In the resulting approach, a parameter determines the trade-off between run time and memory consumption. Our final contribution (**C4**) is an experimental evaluation of this trade-off, as well as other dimensions important for the memory consumption of DP-based algorithms.

This paper is structured as follows. Section 2 presents the state of the art of different caching architectures used for ODTs learning and some search techniques used to reduce memory consumption. Then, Sect. 3 presents the technical details of caching systems implemented as a trie and a hash table. We present in Sect. 4 our new size-limited cache. Finally, present experimental results before concluding.

2 Related Work

There are two classes of approaches for finding ODTs: (1) approaches that rely on solvers, such as MIP solvers [1,13], and (2) approaches that rely on dynamic programming [2,6,7,9–12]. The first class of methods relies on solvers of which the memory use is bounded; however, recent studies on DP-based approaches showed that the run time performance of these methods is significantly better. On the other hand, they suffer from a memory problem related to the use of a cache. Indeed, the size of the cache increases with the number of features and the depth of the ODT that needs to be found. For these algorithms it is hence important to minimize the amount of memory used. In the literature, two types of memory-based optimization have been studied. The first one focuses on the caching system itself and concerns the data structure used to implement the cache and the representation of the stored elements. The second optimization is related to the search and mainly consists in reducing the number of cache entries that are stored. Below we provide a high-level perspective on these optimizations; their details will be discussed in the next section.

2.1 Caching Optimizations

In principle to identify an ODT, some form of exhaustive search needs to be performed. A core idea underlying the DP-based approaches is to store intermediate results during this exhaustive search. Using a cache key, these intermediate results are then used later on to avoid repeating the same search twice.

A key distinguishing factor between various approaches is the key that is used to associate intermediate results in the cache to.

The first algorithm to take this approach was DL8 [10,11]. In DL8, a cache is built in which *itemsets* serve as the key to which information is associated. Here, every path in a decision tree corresponds to an itemset: an itemset is essentially the set of conditions on a path. For trees of limited depths, these keys are short. This technique is also used by the authors of an optimized version of DL8, DL8.5 [2], and works in the presence of depth constraints.

An optimisation was presented in DL8, in which *closed itemsets* were used. A closed itemset in this context is obtained by adding to an itemset I all other conditions that hold in the same instances in which the conditions of I are true. It was shown that this leads to more cache reuse; however, calculating this key takes more time and the key cannot be used in the presence of depth constraints, and hence was not used in DL8.5.

Keys based on instances were used in the GOSDT [9] and Murtree algorithms [6]. In particular, MurTree proposes to use as key a combination of instance identifiers and path length (of which more details will be provided in the next section). In particular for large datasets, this leads to much larger keys than a representation based on itemsets, but there are more opportunities for cache reuse.

Other differences between these algorithms concern the data structures used to store the cache and the keys within the cache. GOSDT and Murtree rely on the use of hash tables; GODST uses bit vectors to represent the sets of identifiers, while Murtree uses a list of IDs of instances.

DL8 and DL8.5, on the other hand, use a *trie*, aiming to exploit the overlap between itemsets that are stored in the cache.

None of these systems allow the user to impose a limit on the size of the cache.

2.2 Search Optimization

The higher the number of elements to be stored, the larger the cache. Hence it is important to minimize the number of elements that need to be stored as much as possible. Various improvements have been studied to reduce the size of the search space and hence the number of elements in the cache.

One core is to use a form of branch-and-bound search to limit which parts of the search need to be considered. Aglin et al. [2] proposed a hierarchical upper bound and an infeasibility lower bound to avoid exploring some nodes of a search tree over paths. [7] proposes a similarity lower bound. In particular, a lower bound is derived for a path on the basis of the error found for another path. This is computed based on common and distinct instances of both paths. This bound is inspired by [4] and is also used in [6].

Using better bounds can effectively reduce the number of elements that need to be stored in the cache, and hence can have an impact on the memory used by the algorithm; however, while better bounds can make the search more feasible for larger datasets, they do not resolve the problem that for some datasets, the search algorithms will run out of memory.

An additional strategy for reducing the size of the cache was taken in the Murtree algorithm [6], which proposes a specialized algorithm for finding depth-2 ODTs without creating any cache entries. This reduces the number of elements needed to be stored to find an ODT.

The memory reduction technique proposed in this paper is not a search algorithm optimization. Rather, we propose an improvement to the cache system that allows to set an upper bound on the maximum number of cache entries. Hence, our optimizations are orthogonal to possible bounds used during the search, and we skip over most of the details of how bounds are used in these algorithms, even though also in our algorithm we use state-of-the-art bounds.

3 Caching in DP-Based ODT Learning

In this section, we describe how the trie and hash table are used to cache subproblem solutions when learning ODTs in existing algorithms; while doing so, we also contribute a comparison that will motivate our choices in this paper.

3.1 Caching in DP-Based Algorithms

Let $\mathcal{D} = \{\boldsymbol{x}, y\}^n$ be a binary dataset of n instances and $\mathcal{F} = \{F_1, \ldots, F_m\}$ be the set of m features that describe $\mathcal{D}$. For each instance $\boldsymbol{x} = (x_1, \ldots, x_m)$ in $\mathcal{D}$, x_i takes value in $\{0, 1\}$. A decision tree recursively partitions a dataset into different groups following paths $p \in \mathcal{P}$. A decision tree can be seen as a collection $\mathcal{DT} \subseteq \mathcal{P}$ of paths, where each path starts at the root of the tree. While in decision tree, features are tested in a given order, it is the set of tests that determines which instances end up in a given node in the decision tree. For this reason, we will see a path as a *set* of tests on features. In the context of binary decision trees, a path p is a set of tests over binary features, $p \subseteq \bigcup_{F \in \mathcal{F}} tests(F)$, where $tests(F)$ returns the two possible tests for the feature F, $F = 1$ (abbreviated with f) and $F = 0$ (abbreviated with $\overline{f}$); we assume $|p \cap tests(F)| \leq 1$ for all F.

At a high level, dynamic programming-based approaches for solving the ODT problem are based on recursive equations. For approaches that use itemsets as keys, this is the recursive equation in its simplest form:

$$min\_error(p) = \begin{cases} \min_{F \in \mathcal{F}} \sum_{t \in tests(F)} min\_error(p \cup \{t\}) & \text{if } |p| < maxdepth; \\ leaf\_error(p) & \text{if } |p| = maxdepth, \end{cases}$$

where the recursion starts at $min\_error(\emptyset)$. In other words, to determine the error made by a decision tree of minimal error, we need to pick the feature in the root of the tree that minimizes error, when summing up the lowest possible errors for the left-hand and right-hand subtrees. Note that in a tree the tests are ordered; in a tree we can first test $A = 1$ followed by $B = 1$, or, alternatively, first test $B = 1$ and then $A = 1$. This order is not important when determining $min\_error(\{A = 1, B = 1\})$. DP approaches are based on the idea of storing information for p, such as $min\_error(p)$, such that we can reuse the information for all possible orders in which the tests can be put in a tree.

For the leafs of the tree, we assume a prediction is made, and a class is associated. In the case of our paper, the associated class is the majority class, while the associated error is the misclassification error defined by $leaf\_error(p) = |\mathcal{D}| - max_{c \in \mathcal{C}} |\mathcal{D} : c|$, where $\mathcal{D}$ is the set of instances falling in the leaf p.

In its more complex form, the recursive equation can take into account other constraints, and can be rephrased to return the optimal tree itself as well.

DP-based algorithms perform a depth-first search using the recursive equation, reusing information that is stored already, and using bounds to limit the cases for which the recursion is executed.

For approaches that use instances as keys, the recursive equation is slightly different:

$$min\_error(\mathcal{D}, d) = \begin{cases} \min_{F \in \mathcal{F}} \sum_{t \in tests(F)} min\_error(\sigma_t(\mathcal{D}), d+1) & \text{if } d < maxdepth; \\ leaf\_error(\mathcal{D}) & \text{if } d = maxdepth, \end{cases}$$

where $\mathcal{D}$ is a dataset and $\sigma_t(\mathcal{D})$ selects the instances of $\mathcal{D}$ which satisfy the condition in test t. The recursion starts for the full dataset and depth $d = 0$. In other words, to determine the error of an optimal tree for a dataset $\mathcal{D}$, we need to determine which test to put in the root of the tree, such that error for the datasets resulting from the split is minimal. Compared to using paths as keys, instances can allow for more reuse, as multiple paths may select the same set of instances.

In this work our aim is to strictly monitor the memory consumption of DP-based algorithms. Hence it is important to understand which of these two approaches leads to better memory use; however, earlier studies did not address this question. We study this in the next subsection.

3.2 Comparison of Caching Strategies

Figure 1 shows a comparison of the memory consumption for different cache implementations for some datasets, when implemented in a state-of-the-art DP-based algorithm. The red curves represent the cache implementation that uses the set of instances as key. The other curves denote the cache implementation that uses paths as key. The difference between the green and blue curves will be explained in the next subsection. Notice that the number of cache entries for the

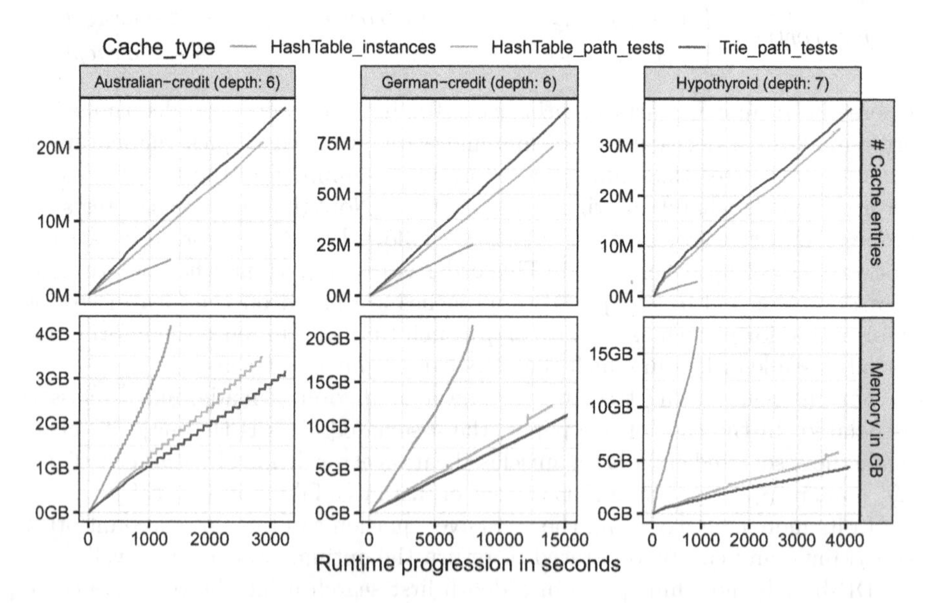

Fig. 1. Comparison of memory use given the cache type (Color figure online)

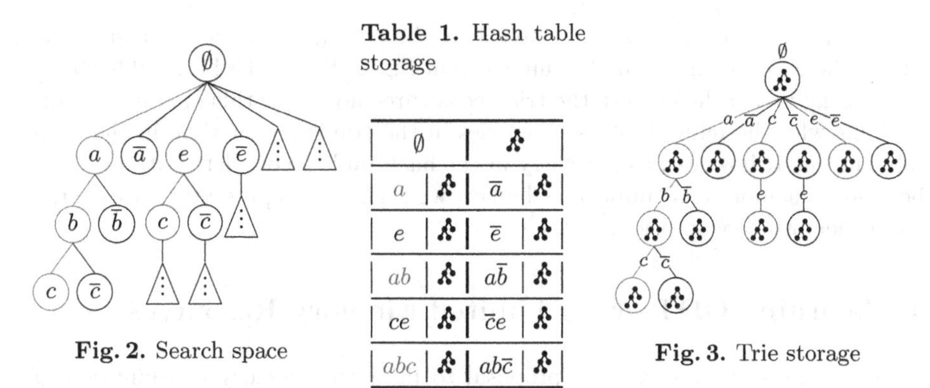

Fig. 2. Search space

Table 1. Hash table storage

$\emptyset$			
a		$\bar{a}$	
e		$\bar{e}$	
ab		$a\bar{b}$	
ce		$\bar{c}e$	
abc		$ab\bar{c}$	

Fig. 3. Trie storage

instance-based representation is indeed low compared to the other approach. At the same time, it consumes much more memory as a key that consists of instance identifiers is much longer than a key that consists of tests. The rest of the paper focuses on the test-based representation, as it requires less memory than the instance-based one.

3.3 Caching Data Structure

To store the cache, DP-based algorithms require a data structure. Two implementations of the cache have been used in the literature. Most algorithms use a hash table, while DL8 and DL8.5 use a *trie* (or prefix tree).

The difference between these two data structures is illustrated in the following example, where a path is used as key; here we sort the tests in the path to obtain an ordered representation for the path. Let us assume a dataset with at least four (4) features and an ODT learning algorithm for trees of depth 3. Figure 2 shows a part of the search space to explore to find the ODT. Table 1 and Fig. 3 show cache implementations based on a hash table and a trie, respectively. In the hash data structure, the path is passed through a hash function, but to avoid collisions, every path needs to be stored with its associated information. In the trie, nodes are created for every prefix of a path.

The size of these data structures can differ, as illustrated by the example. Consider the path abc (in red) during the search; this path is reached in the search by first considering two other paths: $\{a, ab\}$. To save these three paths $\{a, ab, abc\}$ in the cache, in the case of a hash table, an entry is created for each path, and all tests in each path are saved. This is represented by red entries in Table 1: Six (6) tests must be saved to store the paths leading to abc. In the case of trie, the tests in common for parent paths are shared in such a way to avoid saving duplicate tests. To store abc and its parent paths, only three (3) tests are necessary for a trie data structure, as the trie uses a compressed way to store paths. In practice, this is beneficial: the memory consumption in Fig. 1 of the trie data structure (in blue) is lower than that of the hash table (in green), and motivates our choice for a trie in our experiments. However, note that in terms of the number of paths stored in the cache, the trie can be larger. For instance, this is the case for the paths leading to *ec* in Fig. 2. While the hash table stores them using two cache entries, the trie needs three nodes for this operation. This explains why the number of cache entries in the trie is higher than in the hash table in Fig. 1; however, as per entry in the hash table a complete path needs to be stored, memory consumption is higher; we prefer the representation with the lower memory consumption.

4 Learning ODTs with Limited Memory Resources

In this section, we present our proposal to limit the memory consumption of DP-based learning algorithms.

Our core objective is to make sure the size of the cache is limited, while making sure that the performance of the search is not affected too much. Moreover, we wish to do so in a manner that can be integrated in DP-based algorithms with minimal modifications.

The core idea of our approach is simple: instead of keeping all elements in the cache, when necessary we will remove elements from the cache; when the search algorithm encounters these elements later, it will recalculate the results.

Hence, limiting the memory size will certainly impact the run time of the search algorithm, as its speed depends mainly on the reusability of the cache. The fewer entries in the cache, the less likely it is to find an existing solution. However, a good strategy specifying the order in which entries are deleted can limit the impact on the overall run time. In the next subsection, we will present our proposal for a bounded cache based on one deletion strategy. Subsequently, we will introduce additional deletion strategies and how to integrate them in the maintenance of the cache.

4.1 Implementation of a Bounded Cache

As the modifications of the DP-based systems we propose only concern the maintenance of the cache, we focus the description of our contribution to the maintenance of the cache. Pseudo-code for this can be found in Algorithm 1, for the trie data structure. Later, we explain how the pseudocode can be adapted to the simpler case of hash tables. Compared to the cache used by other systems, our new cache system requires two parameters. The first, `maxcachesize`, specifies the maximum number of entries that the cache can store. The second parameter; `wipeFactor` defines the percentage of the cache that will be cleared when the cache is full.

Any DP-based algorithm requires functionality that for a given key returns the data associated to the key, and if no such information is available, will add information to the cache; this is the function `insertOrGetEntry` in Algorithm 1. This function traverses the tests in a path, and either finds back the path in the trie, or adds the necessary nodes. Compared to the original trie-based system DL8.5, there are two differences in the insertion method. First, there is a condition (line 14) that checks that there is enough place in the cache to insert the

Algorithm 1: Class Bounded_Cache(maxcachesize, wipeFactor)

```
1  struct PathEntry{ useful: bool; curFeat: int; nDscPaths: int; solution:
     BestTree; childEntries: HashSet<int, PathEntry>}
2  struct BestTree {lb : float; feat : int; error : float }
   // list of pairs <pathEntry, parentPathEntry>
3  deletionQueue ← pair<PathEntry, PathEntry> [maxcachesize]
4  rootEntry, cachesize ← newEntry(), 0
5  Method newEntry()
6      return PathEntry(false,0, −1, (0, NO_FEAT, +∞), {})
7  Method insertOrGetEntry(p : array of int) // p: path
8      pEntry ← rootEntry
9      for t ∈ p do // foreach test in path
10         if t ∈ pEntry.childEntries then
11             pEntry ← pEntry.childEntries.get(t)
12         else // remainingTests() returns the number of tests in
13              // the path which are not in the cache
14             if cachesize + remainingTests() > maxcachesize then wipe()
15             childEntry ← newEntry()
16             pEntry.childEntries.push({t, childEntry})
17             deletionQueue.push({childEntry, pEntry})
18             cachesize, pEntry ← cachesize + 1, childEntry
19     return pEntry
20 Method wipe()
       // mark nodes that required by the current state of the search
21     setCurUsefulTag(rootEntry, {})
22     countDscpaths(rootEntry)
23     setOptiUsefulTag(rootEntry, {})
24     sortDeletionQueue() // desc order with useful nodes at beginning
25     nDel, counter ← (int)(maxcachesize * wipeFactor), 0
26     for path, parPath ∈ reverse(deletionQueue) do
27         if counter = nDel or path.useful is true then break
28         parPath.childEntries.remove(path); delete(path)
29         counter, cachesize ← counter + 1, cachesize − 1
30     unSetUsefulTag() // Remove tags
31 Method countDscpaths(pE : PathEntry)
32     pE.nDscPaths ← 0
33     for t ∈ pE.childEntries do
34         pE.nDscPaths ← pE.nDscPaths+countDscpaths(pE.childEntries.get(t))+1
35     return pE.nDscPaths
36 Method getChildPathEntries(p : array of int, feat : int)
37     t1, t2 ← tests(feat)
38     return {getEntry(sort(p∪t1)), p∪t1}, {getEntry(sort(p∪t2)), p∪t2}
39 Method setCurUsefulTag(pEntry : PathEntry, p : array of int)
40     for pE, p ∈ getChildPathEntries(p, pEntry.curFeat) do
41         pE.useful ← true; setCurUsefulTag(pE, p)
42 Method setOptiUsefulTag(pEntry : PathEntry, p : array of int)
43     pEntry.useful ← true
44     for pE, p ∈ getChildPathEntries(p, pEntry.solution.feat) do
45         setOptiUsefulTag(pE, p)
```

path being created, otherwise the `wipe` function is called; this `wipe` function is responsible for removing elements from the cache. Second, each path added to the trie is also added to a deletion queue (line 17). This queue is maintained to keep track of the nodes in the cache, which is used when wiping the cache.

Since a node in a trie has a parent that links to it, when a path is deleted, it is necessary to delete the edge from the parent in the trie. For this reason, its parent path entry in the trie is stored to perform the edge deletion.

A critical part of our contribution is the `wipe` function. At its core, this function sorts the nodes in the cache, and subsequently deletes the desired number of nodes from the cache according to this order. Of critical importance is here how the nodes are sorted for removal. To determine this order, our wipe function first calls a number of functions to compute necessary information for the entries in the cache. This information is subsequently used when sorting the nodes. The remaining steps of the `wipe` function are straightforward. The deletion queue is traversed from the end to the beginning, and each path entry is deleted (line 26) until the number of entries to be deleted is reached, or all possible paths have been deleted (line 27). Note that the number of entries to delete is calculated according to the percentage provided by the `wipeFactor` parameter (line 25). Before deleting a path, the edge that connects it to its parent path is deleted (line 28) to inform the parent path that its child path no longer exists.

The information that is collected for elements in the cache is the following.

(1) Information concerning which paths the search is currently considering. Please remember that DP-based approaches perform a recursive search over paths, by adding tests to paths. In the process of calculating the result for the path p, they assume that path p is present in the cache. The `setCurUsefulTag` method is used to mark the paths p that are currently under evaluation. These paths will be put first in the order, and will never be deleted. Please note that the number of such nodes is very small. The `setCurUsefulTag` method uses a `curFeat` field in the entries in the cache. This field is initialized by a small modification of the recursive search function. For DL8.5 this is illustrated in Algorithm 2, where in green the code is indicated that is added in the recursive search function of DL8.5; parts of the code of DL8.5 that are not modified are skipped.

(2) Statistics concerning nodes in the cache. One such statistic which can be used to order elements, illustrated in our pseudo-code, is the number of descendants a node has in the trie. In the trie case, when we delete a node in a trie, we also need to delete its children in the trie. This implies that each path must be deleted before its parent path. As a parent has more descendants than its children, by ordering nodes on the number of descendants, we can assure children are deleted before their parents. Moreover, an intuition is that paths with numerous descendants are more expensive to evaluate than those with fewer descendant paths, and hence should be removed less quickly. To count the number of descendant paths per path, a simple post-order traversal is performed through the trie using the `countDscpaths` method. The result is stored in the variable `nDscPaths` (line 1).

(3) Information concerning which paths are part of the current optimal solution; we will return to this issue later on.

4.2 Different Wipe Strategies

In the pseudo-code above, the statistic we used to order nodes was *the number of descendant path entries in the cache.* We also consider the following alternatives.

Number of Reuses of Solutions. The intuition behind the number of reuses of elements is that a path that has been reused many times is more likely to be needed again. This strategy removes the less frequently reused solutions before the more frequently used ones. To calculate the number of times a path solution has been reused, a variable initialized to 0 must be added to the path entry structure. Whenever in the method `insertOrGetEntry`, when an existing path is returned, the variable must be incremented for the path. Note that this deletion criterion does not satisfy the requirement of deleting the deepest nodes first in a trie. To enforce this behavior, there are two possibilities. The first is to set the method `sortDeletionQueue` so that the sort is performed according to two parameters. First, the length of the path and then the number of reuses. Another possibility is to increment the number of reuses for each ancestor of a path as well. In this work, we use the second option because it is based solely on the number of reuses, rather than using an additional criterion. Note that for this strategy, the statistic is not computed in the `wipe` method.

All Not Required Paths. This strategy, as the name implies, deletes all paths except from the *useful* ones, as defined by criterion (1) and (3) above. No variables need to be added to the path entry structure, and the deletion queue is not required. When the cache is full, after setting *useful* nodes, a post-order traversal is sufficient to delete all nodes without the *useful* tag.

4.3 Returning an ODT that Relies on Deleted Elements

Until now, our recursive equations focused on returning the error of the most accurate tree. However, in practice we also wish to return the tree that obtains this error. In DL8 an approach was proposed that allows to do so with minimal additional memory use: for every path p, only the feature F is stored in the cache that should be used to split optimally for p. The observation is that the optimal split for $p \cup \{f\}$ and $p \cup \{\overline{f}\}$ can also be found in the cache.

Unfortunately, if we wipe part of the cache, this strategy can no longer be used: if the optimal tree relies on a path that is no longer in the cache, we can no longer recover this tree from the cache.

One solution could be to associate a complete optimal tree to every path in the cache, but this would blow up the memory required for the cache, which we would like to avoid. Hence, a critical contribution of this work is an alternative solution that works well in practice: it avoids the use of large amounts of memory, while being fast at the same time. This solution consists of two components.

Algorithm 2: Bounded-DL8.5(maxdepth, maxcachesize, wipeFactor)

```
1  cache ← Bounded_Cache(maxcachesize, wipeFactor)
2  bestSolution ← DL8.5−Recurse({}, +∞)
3  while tree(bestSolution) is incomplete do
     reconstituteWipedNodes({})
4  return tree(bestSolution)
   // p is the path whose solution must be found
5  Procedure DL8.5−Recurse(p : array of int, ub : int)
6      pEntry ← cache.insertOrGetEntry(sort(p))
7      ... // if entry exists and has been solved, return its
           solution
8      for each feature F in a well-chosen order and split in tests f and f̄ do
9          pEntry.curFeat ← F
10         ... // compute error of best tree rooted by F
11     pEntry.curFeat ← −1
12     ... // return error and root of the tree with the lowest
           error
13 Procedure reconstituteWipedNodes(p : array of int, ub : int)
14     pSol ← cache.insertOrGetEntry(sort(p)).solution
15     if pSol.feat = NO_FEAT then return void
16     (pE1, p1), (pE2, p2) ← cache.getChildPathEntries(p, pSol.feat)
17     found_pE1 ← pE1 ≠ NULL and pE1.solution.feat ≠ NO_FEAT
18     found_pE2 ← pE2 ≠ NULL and pE2.solution.feat ≠ NO_FEAT
19     if found_pE1 is false or found_pE2 is false then
20         if found_pE1 is false and found_pE2 is true then
21             pE1.solution.lb ← pSol.error − pE2.solution.error
22             DL8.5−Recurse(p1, pE1.solution.lb + 1)
23             reconstituteWipedNodes(p2)
24         else if found_pE2 is false and found_pE1 is true then
25             pE2.solution.lb ← pSol.error − pE2.solution.error
26             DL8.5−Recurse(p2, pE2.solution.lb + 1)
27             reconstituteWipedNodes(p1)
28         else
29             pE1.solution.lb ← 0
30             DL8.5−Recurse(p1, pSol.error + 1)
31             pE2.solution.lb ← pSol.error − pE1.solution.error
32             DL8.5−Recurse(p2, pE2.solution.error + 1)
33     else
34         reconstituteWipedNodes(p1)
35         reconstituteWipedNodes(p2)
```

First, at the moment that we wipe the cache, using the `setOptiUsefulTag` function we determine which paths in the cache are part of the currently optimal solution; we do not remove these paths from the cache. In the best case, this optimal solution does not change any more and hence we can recover (most of) the solution from the cache.

Unfortunately it cannot be excluded that when the search continues, we find that a tree with a better quality exists that relies on paths that have been removed from the cache. In this case, we propose to *recalculate* the optimal tree for these paths. This algorithm is executed at the end of the original search to avoid making calculations for paths that later in the search may no longer be considered part of the final solution. This algorithm is described by the procedure `reconstituteWipedNodes` in the Algorithm 2. For each existing path, an attempt is made to obtain its two children paths from the cache (lines 16–18). If they exist (line 19), the DFS traversal continues (lines 34–35). Otherwise, the search is restarted. However, an important difference with the original search is that from the known errors (including the error of the optimal tree), much better bounds can be deduced than in the original search. In the case where a right child path exists but a search must be rerun for the left child path (lines 20–23), a simple subtraction between the errors of the parent path and the right child path provides the exact error of the left child path to be found. This error is used as a lower bound and a small value ϵ is added to it to define the upper bound. In the context of the misclassification rate, $\epsilon = 1$ is used. The same process is performed in the case of an existing left child path and a non-existing right child path (lines 24–27). When both child paths are nonexisting (lines 29–32), a search is performed for a first child path and the bounds for the second are derived from its solution. For the first path, the unknown lower bound is set to 0 while the upper bound is at most the error of the parent path added to ϵ. This procedure to reconstruct the wiped paths of the final ODT is very efficient in practice.

4.4 Adapting to Hash Tables

The implementation of the bounded trie-based cache can easily be adapted to a hash table and is even simpler in this case. In order to sort and delete paths in a specific order, the deletion queue no longer needs to store pointers towards parents. Moreover, non-useful cache entries can be deleted in any order without hierarchy constraints, leading to a greater freedom in the choice of deletion criteria. For example, for the *reuses number* strategy that we propose, in the case of a hash table, we can rely only on this number to define the deletion order of paths without having to increment the *reuses number* of ancestor paths.

5 Results

In this section we answer five main questions:

Q1 Which of the wipe strategies proposed has the lowest impact on the run time?

Q2 What is the impact of the memory usage when using a bounded cache?
Q3 What is the impact of a bounded cache on the run time?
Q4 How fast is the algorithm to recover removed nodes from the ODT?
Q5 How does the use of a bounded cache compare to a MIP approach?

The implementation of the learning algorithm without cache size restriction that we use in our experiments is `DL8.5` [2,3]. This means that the cache system used is a trie. However, we add some improvements to this `DL8.5` implementation to reduce the baseline memory consumption. These are the specialized algorithm proposed in MurTree to find depth-2 ODTs and the similarity lower bound introduced by OSDT. We call the final algorithm `DL8.5` in our experiments because the main search features originate from `DL8.5`. `Bounded-DL8.5` is the `DL8.5` version using our bounded cache. It is available at https://github.com/aia-uclouvain/pydl8.5. Experiments were run on a Linux Rocky 8.4 server with an Intel Xeon Platinum 8160 CPU @ 2.10 Ghz and 320 GB of memory.

In the first experiment, we use 20 binary datasets from CP4IM[1]. We run `DL8.5` on these datasets with a time limit of 5 min. To avoid comparing too easy instances, we learned ODTs of depth 5. Then we run `Bounded-DL8.5` with the same parameters until we find the ODT, or we reach the same point in the search as `DL8.5` when the timeout was raised. We show in Fig. 4 two ratios of `Bounded-DL8.5` over `DL8.5`: the run time and the total number of entries (including those created after they have been removed). Each wipe strategy is considered for `Bounded-DL8.5`. For the wipe parameters, we consider 75% and 50% of the memory as value for the `maxcachesize` parameter and a wipe factor of 40% and of 30%. Note that the results shown are representative for the results for other choices of the parameters. It is interesting to notice that the time spent by the algorithm is proportional to the number of cache entries created. It can also be seen that the number of cache entries and the run time increase as the bound of the cache is restricted. On the other hand, it is difficult to observe a concrete trend in these values when a change is applied to the percentage of memory to free at each cache wipe. Note, however, that it becomes easy to answer **Q1** thanks to the Fig. 4. As expected, the strategy of removing all non useful entries from the cache is the one increasing the most the run time of `Bounded-DL8.5`. Instead, the intuition of keeping as long as possible the entries often reused shows the best reduction of time impact. It performs better than removing the paths based on the number of descendants.

After this experiment, we select the best wipe strategy (*number of reuses*) to evaluate how `Bounded-DL8.5` can impact the memory usage on situations in which `DL8.5` requires a lot of memory to find the ODTs. To highlight these cases while ensuring reasonable run times, we experimentally select five specific datasets and depths. In the same way, we limited the cache size of `Bounded-DL8.5` to 30 million (30M) entries at most and set the percentage of entries to wipe to 40%. The impact of these values is already discussed above. The results are reported in Table 2. The memory usages are obtained by using

[1] https://dtai.cs.kuleuven.be/CP4IM/datasets/.

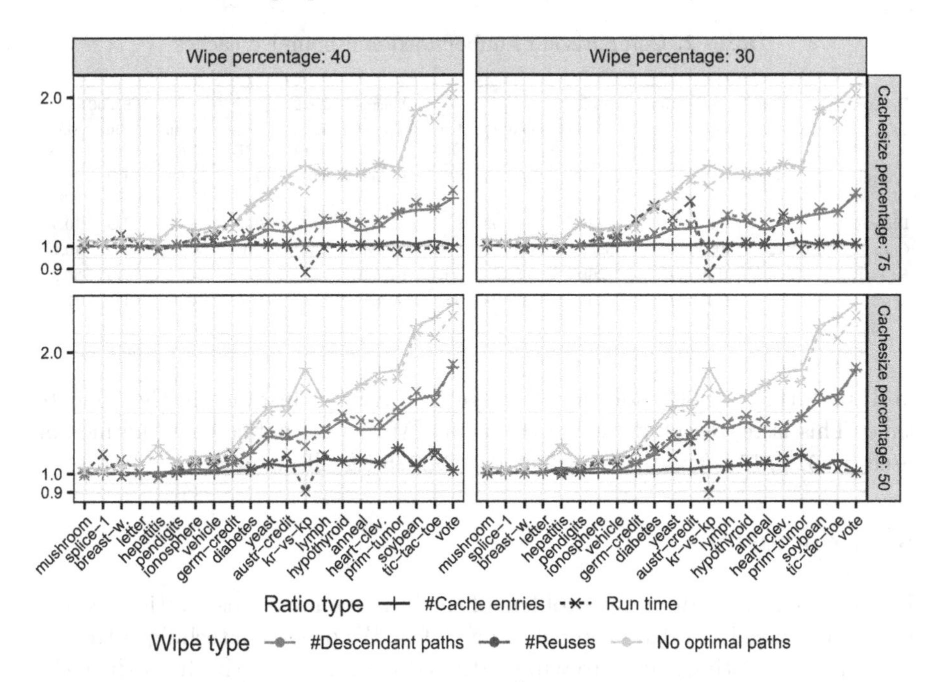

Fig. 4. Time factor of `Bounded-DL8.5` per wipe strategy

the program *top* available on Unix operating systems. Notice that the memory needed to solve the problems using `Bounded-DL8.5` never reaches 10 GB while up to 78 GB is required using `DL8.5`. Regarding the dataset *anneal*, more than 600 million of entries are created with `DL8.5` to find an ODT of depth 8. The advantage of using `Bounded-DL8.5` is that this number can be reduced to a desired quantity, here 30 million. As an answer to **Q2**, this reduces significantly the memory needed to find an ODT. Note, however, that this reduction depends on the limit on the number of cache entries. Moreover, it also has an impact on the run time. The time factor is recorded in Table 2. To answer **Q3**, notice that our strategy managed to achieve a time factor less than 1.5 on an instance that lasts over 3 h with `DL8.5`. In the worst case, it almost reaches a time factor of 3 on an instance that originally required 78 GB, which it reduces to 9 GB. Regarding our algorithm to recover the nodes of the ODT that have been removed during the search, the time used by our algorithm is reported in milliseconds in the column *rTime*. Notice that for all instances, our algorithm uses less than 5 milliseconds to recover the removed nodes from the ODT. This answers the question **Q4**.

To answer **Q5**, we finally compare `Bounded-DL8.5` to a MIP approach. For this, we use `BinOCT`[2] model. The model is solved using CPLEX[3] 22.1.0. As MIP approaches are time consuming, we set a time limit to 70000 s, which is greater than the maximum time used by `Bounded-DL8.5` to solve an instance. Notice

[2] https://github.com/SiccoVerwer/binoct.

[3] https://www.ibm.com/analytics/cplex-optimizer.

Table 2. Comparison of unbounded and bounded caches

Dataset	nFeats	nInsts	Depth	DL8.5			Bounded–DL8.5				BinOCT	
				|Cache|	Time (s)	Mem (GB)	Time (s)	Time factor	Mem (GB)	rTime (ms)	Time	Mem (GB)
Anneal	93	812	8	648M	21907.28	78.5	65453.05	2.99	**9**	2.63	TO	9.8
Diabetes	112	768	7	238M	18890.09	28	31534.13	1.67	**6.3**	0.21	TO	10.5
German-credit	112	1000	6	92M	15198.32	11	21898.92	1.44	**5.1**	0.10	TO	10.1
Kr-vs-Kp	73	3196	8	136M	3865.28	16.1	6365.04	1.65	**5.3**	4.68	TO	35.8
Yeast	89	1484	7	419M	34977.98	50.5	62483.77	1.79	**7.8**	4.70	TO	17.5

in Table 2 that BinOCT does not manage to solve any instance in the allocated time. This is represented by TO (timeout). Moreover, notice that the memory used by BinOCT is greater than Bounded-DL8.5 for all instances.

6 Conclusions

In this paper, we address the problem of the huge memory consumption required by caching based algorithms for learning ODTs. We propose a technique that can be added to existing caches to wipe a desired number of entries from the cache. This leads to a significant reduction in memory consumption, with a smaller impact on run time. We propose strategies to reduce the impact of the wipe on the overall run time. Finally, we show that our approach finds ODTs more quickly than MIP approaches, while also consuming less memory.

References

1. Aghaei, S., Gómez, A., Vayanos, P.: Strong optimal classification trees. arXiv Preprint arXiv:2103.15965 (2021)
2. Aglin, G., Nijssen, S., Schaus, P.: Learning optimal decision trees using caching branch-and-bound search. In: Proceedings of AAAI, vol. 34, pp. 3146–3153 (2020)
3. Aglin, G., Nijssen, S., Schaus, P.: PyDL8.5: a library for learning optimal decision trees. In: Proceedings of the 29th Conference on IJCAI, pp. 5222–5224 (2021)
4. Angelino, E., Larus-Stone, N., Alabi, D., Seltzer, M., Rudin, C.: Learning certifiably optimal rule lists for categorical data. arXiv Preprint arXiv:1704.01701 (2017)
5. Bertsimas, D., Dunn, J.: Optimal classification trees. Mach. Learn. **106**, 1039–1082 (2017). https://doi.org/10.1007/s10994-017-5633-9
6. Demirović, E., et al.: MurTree: optimal decision trees via dynamic programming and search. J. Mach. Learn. Res. **23**, 1–47 (2022)
7. Hu, X., Rudin, C., Seltzer, M.: Optimal sparse decision trees. In: Advances in Neural Information Processing Systems, vol. 32 (2019)
8. Laurent, H., Rivest, R.: Constructing optimal binary decision trees is NP-complete. Inf. Process. Lett. **5**, 15–17 (1976)
9. Lin, J., Zhong, C., Hu, D., Rudin, C., Seltzer, M.: Generalized and scalable optimal sparse decision trees. In: ICML, pp. 6150–6160 (2020)
10. Nijssen, S., Fromont, E.: Mining optimal decision trees from itemset lattices. In: KDD, pp. 530–539 (2007)

11. Nijssen, S., Fromont, E.: Optimal constraint-based decision tree induction from itemset lattices. Data Min. Knowl. Discov. **21**, 9–51 (2010). https://doi.org/10.1007/s10618-010-0174-x
12. Verhaeghe, H., Nijssen, S., Pesant, G., Quimper, C., Schaus, P.: Learning optimal decision trees using constraint programming. Constraints **25**, 226–250 (2020). https://doi.org/10.1007/s10601-020-09312-3
13. Verwer, S., Zhang, Y.: Learning optimal classification trees using a binary linear program formulation. In: Proceedings of AAAI, vol. 33, pp. 1625–1632 (2019)

SaDe: Learning Models that Provably Satisfy Domain Constraints

Kshitij Goyal[1(✉)], Sebastijan Dumancic[2], and Hendrik Blockeel[1]

[1] KU Leuven, Leuven, Belgium
{kshitij.goyal,hendrik.blockeel}@kuleuven.be
[2] TU Delft, Delft, The Netherlands
S.Dumancic@tudelft.nl

Abstract. In many real world applications of machine learning, models have to meet certain domain-based requirements that can be expressed as constraints (for example, safety-critical constraints in autonomous driving systems). Such constraints are often handled by including them in a regularization term, while learning a model. This approach, however, does not guarantee 100% satisfaction of the constraints: it only reduces violations of the constraints on the training set rather than ensuring that the predictions by the model will *always* adhere to them. In this paper, we present a framework for learning models that *provably* fulfill the constraints under *all* circumstances (i.e., also on unseen data). To achieve this, we cast learning as a maximum satisfiability problem, and solve it using a novel *SaDe* algorithm that combines constraint satisfaction with gradient descent. We compare our method against regularization based baselines on linear models and show that our method is capable of enforcing different types of domain constraints effectively on unseen data, without sacrificing predictive performance.

Keywords: Domain constraints · Constrained optimization · Satisfiability modulo theories

1 Introduction

There is increasing interest in using machine-learned models in contexts where strict requirements exist about the model's behavior. For instance, in a criminal sentencing context, a fairness constraint might express that all else being equal, two people of a different ethnicity should have an equal probability of ending up in jail [2]. In another example, when automating parts of an aircraft system, the model may be required to satisfy certain safety-critical requirements [20]. We call such requirements *domain constraints*, as they constrain the behavior of the learned model over its whole domain.

Machine learning methods often deal with such constraints by including them in the cost function they optimize (for example, in a regularization term) [4]. This

Supplementary Information The online version contains supplementary material available at https://doi.org/10.1007/978-3-031-26419-1_25.

© The Author(s), under exclusive license to Springer Nature Switzerland AG 2023
M.-R. Amini et al. (Eds.): ECML PKDD 2022, LNAI 13717, pp. 410–425, 2023.
https://doi.org/10.1007/978-3-031-26419-1_25

approach has the effect of *encouraging* the learner to learn a model that satisfies the imposed constraints on the *training* data, but it does not *guarantee* that the learned model satisfies the constraints over the whole *input space*. While this may be good enough when the constraints are intended to help the learner obtain better models from less data [10,31], it is insufficient for applications where constraint satisfaction is imperative under all circumstances (such as safety-critical systems).

For this reason, research has been conducted on approaches that can guarantee constraint satisfaction even on unseen data, as in [27] where a counter-example guided approach is used to enforce monotonicity constraints, and [16] where a multiplexer layer is used as the output layer in a neural network to enforce domain constraints. The existing literature, however, still lacks a general approach that can be used to enforce a variety of domain constraints on different learning problems with provable guarantees.

In this work, we present a framework for learning parametric models that are guaranteed to satisfy domain constraints. Rather than including the constraints in a cost function and using a standard learning approach, the machine learning problem is cast into a constraint satisfaction problem [25], more specifically a Maximum Satisfiability Modulo Theories (MaxSMT) problem [14]. Domain constraints are formulated as hard constraints, which must provably be satisfied, whereas the model's fit with the training data is evaluated using soft constraints, of which as many as possible should be satisfied. Thus, a model is found that optimally fits the data within the hard constraints imposed by the user.

Unfortunately, solving the obtained MaxSMT problem does not scale beyond a few dozen training instances. To resolve this, we propose *Satisfiability Descent* (*SaDe*), a variant of gradient descent [15] in which each step consists of solving a small MaxSMT problem to find a local optimum in the general direction of the negative gradient (rather than moving in the exact direction) that satisfies all domain constraints. We show experimentally that *SaDe* scales to realistically-sized datasets and that it finds models with similar performance as other learners while guaranteeing satisfaction of all domain constraints.

In Sects. 2–4, we consecutively introduce preliminaries, the MaxSMT-based approach, and *SaDe*. We position SaDe with respect to related work in Sect. 5 and present the empirical evaluation in Sect. 6. Section 7 concludes.

2 Preliminaries

We systematically use boldface for vectors and italics for their components, for example, $\mathbf{x} = (x_1, x_2, \ldots, x_n)$.

2.1 SAT, MaxSAT, SMT, MaxSMT, COP

Let $\mathbf{w}$ denote a vector of decision variables, and C_i a (hard or soft) constraint, i.e., a boolean function of $\mathbf{w}$. We say that $\mathbf{w}$ satisfies C_i if and only if $C_i(\mathbf{w})$ returns true. We call $\mathbf{w}$ *admissible* if it satisfies all hard constraints. We can then distinguish the following types of problems:

- SAT: given a set of constraints $C_i(\mathbf{w})$, $i = 1, \ldots, k$, determine whether an admissible $\mathbf{w}$ exists
- MaxSAT: given a set of hard constraints $\mathcal{H}$ and a set of soft constraints $\mathcal{S}$, find an admissible $\mathbf{w}$ that satisfies as many $C_i \in \mathcal{S}$ as possible
- SMT: Satisfiability modulo theories: this setting is identical to SAT, except that not only logical reasoning is used, but also a theory on the domain of $\mathbf{w}$. For instance, an SMT solver knows $x < y \wedge y < x$ is unsatisfiable; a SAT solver does not, because it does not know the meaning of $<$.
- MaxSMT: similar to MaxSAT, but satisfiability is determined modulo theories
- COP: constraint optimization: given a set of constraints and a function f, find the admissible $\mathbf{w}$ with smallest $f(\mathbf{w})$ (among all admissible $\mathbf{w}$).

MaxSAT reduces to SAT in the sense that any MaxSAT problem can be solved by iteratively solving SAT problems. The Fu-Malik algorithm [14] is an example of such an approach. Similarly, MaxSMT reduces to SMT. This implies that if we know how to solve the SMT problem for a particular type of theory, we can automatically solve the corresponding MaxSMT problem.

COP problems can be approximately solved by turning them into a MaxSMT problem, as follows: make all the original constraints hard constraints, and add soft constraints of the form $f(\mathbf{w}) < c_i$, $i = 1, \ldots, k$ where f is the function to be minimized and $c_i < c_{i+1}$. The solution is approximate in the sense that if $\hat{\mathbf{w}}$ is the returned solution and $\mathbf{w}^*$ the actual optimum, $f(\hat{\mathbf{w}}) - f(\mathbf{w}^*) \leqslant c_i - c_{i-1}$ for some i (i.e., closer thresholds guarantee a better solution).

Due to these properties, the solving power of SMT solvers can be lifted towards (approximate) constrained optimization. This is a key insight behind our approach.

2.2 Universally Quantified Constraints

Constraint solvers assume a finite set of constraints. Different solvers may use different languages in which these constraints can be expressed. Some solvers allow for the constraints to contain universal quantifiers, for instance (expressing monotonicity of f in some input variable x_i):

$$\forall \mathbf{x}, \mathbf{x}' \in \mathcal{X} : x_i \leqslant x_i' \wedge (\forall j \neq i : x_j = x_j') \implies f_{\mathbf{w}}(\mathbf{x}) \leqslant f_{\mathbf{w}}(\mathbf{x}') \tag{1}$$

When the universal quantification is over a variable with finite domain, such a constraint can always be handled by *grounding* it: making a separate copy for each value of the domain. For infinite domains, however, this is not possible. SMT solvers typically handle such cases by turning the quantified variable into a decision variable, and then, through reasoning, eliminating the quantifier. For example, the constraint $\forall x > 0 : f(x) > 0$ with $f(x) = ax + b$ cannot be turned into a finite set of constraints of the form $f(1) > 0, f(2) > 0, \ldots$ but an SMT system can deduce an equivalent constraint on the model parameters, namely $a > 0 \wedge b > 0$.

With this approach, the extent to which universally quantified constraints can be handled clearly depends on the strength of the mathematical reasoning engine. In this work we use Z3 [7], one of the more powerful systems in this respect. Z3 implements an SMT(NRA) solver: an SMT solver that can reason with non-linear equations (NRA = Nonlinear Real Arithmetic), and uses this to deal with universally quantified constraints. The ability to handle nonlinear functions is crucial for our approach, even when learning linear models. That is because turning the quantified variable x into a decision variable gives rise to formulas in which products of decision variables occur. For example, we typically think of $ax+b$ as linear because we think of a and b as constants, but to the solver, a, x, and b are all variables, and $ax+b$ is no more linear than $f(x, y, z) = xy+z$.

3 From Constrained Parametric Machine Learning to MaxSMT

In this section, we propose a framework for formulating supervised parametric machine learning as a MaxSMT problem.

3.1 The Learning Problem

We consider the following learning problem:

Definition 1. ***Learning problem.*** *Given a training set* $D \subseteq \mathcal{X} \times \mathcal{Y}$, *a set of constraints* $\mathcal{K}$, *a loss function* $\mathcal{L}$, *and a hypothesis space containing functions* $f_{\mathbf{w}} : \mathcal{X} \rightarrow \mathcal{Y}$; *find* $\mathbf{w}$ *such that* $f_{\mathbf{w}}$ *provably satisfies constraints* $\mathcal{K}$ *and* $\mathcal{L}(f_{\mathbf{w}}, D)$ *is minimal among all such* $f_{\mathbf{w}}$.

The language in which the constraints in $\mathcal{K}$ are expressed is essentially a subset of first-order logic. Formulas can contain universal quantification ($\forall$) over known sets, notably the training set D and the input space $\mathcal{X}$; arithmetic operators are defined, as well as operators that extract a component from a tuple; and the formula can refer to the function $f_{\mathbf{w}}$ for a given value of $\mathbf{w}$. The variable $\mathbf{w}$ is free: depending on its value, the function $f_{\mathbf{w}}$ either fulfills or violates the constraint $\mathcal{K}$. Examples of expressible constraints are monotonicity (see Eq. 1) and conditional bounds, for example (inspired by safety-critical applications [19]): $\forall \mathbf{x} \in \mathcal{X} : x_i > a \implies f_{\mathbf{w}}(\mathbf{x}) > 0$. For binary classification problems, we assume a real value prediction with $f_{\mathbf{w}}$, which is then translated into a binary decision using a sigmoid function.

3.2 Translation to MaxSMT

Though the above-defined learning problem looks quite standard, we could not find any constraint-based optimization approach that can handle it, among many we considered (which includes constraint programming and mixed integer linear programming). This finding is actually consistent with earlier work [11]. Solving the problem required a combination of the ability to handle universal quantification over continuous domains, non-linear real arithmetic, and optimization that no system offers. The easiest way out was to drop the "optimization" aspect and reduce the COP problem to a MaxSMT(NRA) and ultimately an SMT(NRA) approach. We have implemented such an approach on top of the Z3 solver.

Z3 contains algorithms for solving COP problems directly, but these cannot deal with universally quantified constraints. We therefore convert the COP problem to MaxSMT(NRA) in a way that is similar to the procedure explained in Sect. 2.1. We approximately encode the loss function $\mathcal{L}$ using soft constraints that we call *decision constraints*. Decision constraints impose a certain quality of fit on $f_{\mathbf{w}}$. They are typically of the form $C(f_{\mathbf{w}}(\mathbf{x}), y)$, with C some condition that is fulfilled when $f_{\mathbf{w}}(\mathbf{x})$ is "sufficiently consistent" with the observed y, for a given $(\mathbf{x}, y) \in D$. In this paper, we consider two different forms of C, depending on whether y is boolean (binary classification) or numerical (regression).

For regression, decision constraints are of the following form:

$$y - e \leqslant f_{\mathbf{w}}(\mathbf{x}) \leqslant y + e$$

with e some threshold. Multiple such constraints, each with a different threshold, can be introduced for each data point: the closer $f_{\mathbf{w}}(\mathbf{x})$ is to y, the more such constraints are satisfied for the data point $(\mathbf{x}, y)$. Depending on the context, the threshold e can be set relative to the value of y, for example, $e = 0.1 * y_{max}$ where $y_{max} = \max_D |y|$.

For binary classification, we assume that the sign of $f_{\mathbf{w}}(\mathbf{x})$ indicates the class, and its magnitude indicates the model's certainty about the prediction. Hence, we use decision constraints of the following form:

$$\begin{aligned} f_{\mathbf{w}}(\mathbf{x}) &> \tau \qquad \text{if}\, y = 1 \\ f_{\mathbf{w}}(\mathbf{x}) &< -\tau \qquad \text{if}\, y = -1 \end{aligned}$$

for some threshold τ. Again, multiple such constraints can be used, with varying thresholds.

3.3 Solving the MaxSMT Problem

Z3 natively supports a number of MaxSMT(NRA) algorithms, but this module of Z3 does not support universal quantifiers over real variables. It does support such quantifiers for SMT(NRA). We therefore made our own MaxSMT(NRA) solver by implementing the Fu-Malik algorithm [14] on top of the SMT(NRA) solver that is provided in Z3. The Fu-Malik algorithm solves MaxSAT problems iteratively: it consecutively identifies minimal sets of constraints that are jointly unsatisfiable and relaxes the problem by allowing exactly one of these to be violated; it keeps doing this until the relaxed problem is satisfiable.

One more change was needed to make this approach work. With our experiments, we realise that unbounded continuous domains make the learning very slow with Z3. To mitigate this issue, our approach assumes a bounded input space, where vectors $\mathbf{l}$ and $\mathbf{u}$ exist such that $l_i < x_i < u_i$ for all i, for all

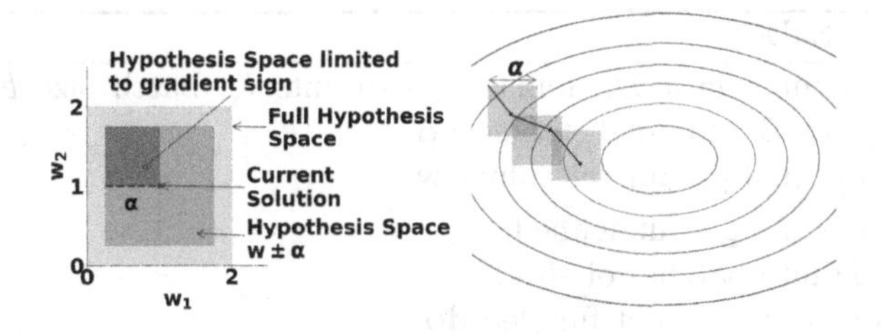

Fig. 1. An illustration of the SaDe algorithm for two parameters. **Left figure:** For the current solution $\{w_1 = 1, w_2 = 1\}$, assuming the gradients $\{\frac{\delta \mathcal{L}}{\delta w_1} > 0, \frac{\delta \mathcal{L}}{\delta w_2} < 0\}$, the hypothesis space for the next solution is the grey quadrant; **Right Figure:** Every grey quadrant represents the search space from one iteration to the next (maximal step size: α), decided by the gradients of the loss.

$\mathbf{x} \in \mathcal{X}$. These bounds can be provided by the user, or we can use as defaults $l_i = \min_D(x_i)$ and $u_i = \max_D(x_i)$. Enforcing such bounds is also practical: a continuous feature in a machine learning task always has a range of values it can realistically take. For example, the *age* of a person can only be in the range of [0, 150] and a value of, say 10000, is unrealistic. Hence, it is sensible to enforce the domain constraints in such realistic ranges, and the most straightforward way to get these ranges is the training data itself. In the remainder of this paper, when we have quantification over $\mathcal{X}$, it should be kept in mind that we actually assume a bounded $\mathcal{X}$.

4 SaDe: Satisfiability Descent

The approach explained above is straightforward and intuitive, but unfortunately not scalable. In our preliminary experiments, we observed that the above approach became prohibitively slow beyond a few dozen training instances. The reason for this is the combinatorial nature of MaxSMT: increasing the number of instances, and consequently the number of soft constraints, makes the problem exponentially more complex.

To overcome this limitation, we have devised an algorithm called *Satisfiability Descent* (*SaDe*). Essentially, *SaDe* just performs gradient descent, like other learning algorithms. However, it cannot simply "take a step in the direction of the negative gradient", as the point where it arrives may not be admissible. Instead, the MaxSMT procedure is used to find an admissible point near the point that gradient descent would lead to. More precisely: Let $\mathcal{L}$ be the loss measured on the whole training set, and $\mathcal{L}_B$ the number of violated constraints in a *batch*, a small subset of the training data (small enough that MaxSMT is feasible). The gradient descent principle makes *SaDe* move in the direction of a local optimum of $\mathcal{L}$, while the MaxSMT procedure makes sure the next point is admissible and minimizes $\mathcal{L}_B$ in a local region. A motivating assumption behind minimizing $\mathcal{L}_B$ is that the loss function correlates with the number of violations. This is true

Algorithm 1. SaDe

input: training data D, domain constraints $\mathcal{K}$, batch size b, number of epochs e, loss $\mathcal{L}$, maximal step size $\boldsymbol{\alpha}$

output: optimal parameter values $\mathbf{w}$

1: $W = \{\}$, $\mathcal{H} = \mathcal{K}$, $\mathbf{g}$ = undefined
2: partition D into batches of size b
3: **while** stop_criterion not fulfilled **do**
4:     **for** each batch B in D **do**
5:         $\mathcal{S}$ = DECISION_CONSTRAINTS(B)
6:         $sol = MaxSMT(\mathcal{S}, \mathcal{H})$
7:         **if** $sol.label$ is SAT **then**
8:             $\hat{\mathbf{w}} = sol.params$
9:             $W = W \cup \{\hat{\mathbf{w}}\}$
10:            $\mathbf{g} = \nabla\mathcal{L}(\hat{\mathbf{w}})$
11:        **else if** $\mathbf{g}$ is defined **then**
12:            $\mathbf{g} = -\mathbf{g}$
13:        **end if**
14:        **if** $\mathbf{g}$ is defined **then**
15:            $\mathcal{H} = \mathcal{K} \cup \{\mathbf{w} \in Box(\hat{\mathbf{w}}, \hat{\mathbf{w}} - \alpha \cdot \text{sgn}(\mathbf{g}))\}$
16:        **end if**
17:    **end for**
18: **end while**
19: **return** $\arg\min_{\hat{\mathbf{w}} \in W}(\mathcal{L}(D, \hat{\mathbf{w}}))$

for commonly used losses (for example, mean squared error, cross-entropy loss) and for the decision constraints introduced here.

Algorithm 1 shows pseudocode for *SaDe*. The algorithm runs for multiple epochs, each time processing all batches sequentially. It starts with finding a solution for the first batch; this solution must satisfy all domain constraints (stored in the set of hard constraints $\mathcal{H}$) and as many soft constraints ($\mathcal{S}$) as possible. The solution is stored in an object *sol* with a field *label* that equals SAT if the problem is satisfiable and a field *params* that in that case contains the solution. It adds this solution to a set W, computes the gradient of the loss function at this point, stores this gradient in variable $\mathbf{g}$, and extends the hard constraints with a "box" constraint, which states that the next solution must be inside the axis-parallel box defined by $\hat{\mathbf{w}}$ and $\hat{\mathbf{w}} - \alpha \cdot \text{sgn}(\mathbf{g})$, where the sign function is applied component-wise to a vector.[1] In other words, each w_i will be confined to the interval $[\hat{w}_i, \hat{w}_i + \alpha]$ or $[\hat{w}_i - \alpha, \hat{w}_i]$, depending on the sign of g_i. This forces the algorithm to move, not in the exact direction of the negative gradient, but in a direction that lies in the same orthant; see Fig. 1 for an illustration.

[1] We use a modified sign function where $\text{sgn}(0) = 1$, so that the box never reduces to a lower-dimensional box.

The box constraint may render the problem unsatisfiable. In that case, the algorithm takes a step back (lines 12, 15) and continues with the next batch from there. We call this a *restart*. Such a restart is also made when the solver does not find a solution within reasonable time (5 s, in our implementation).

Each intermediate solution is stored (line 9) and the one that minimizes the loss is returned as the final solution (line 19). *SaDe* runs until some stopping criterion is fulfilled. Our current implementation checks every 100th iteration (starting from the 400th) whether the loss is improved by at least 2%, compared to 200 iterations ago. If the improvement is less than 2%, or if a maximum number of iterations is reached, the process stops. This criterion is not crucial to the algorithm and can be replaced with another one.

5 Related Work

There is a substantial body of work on imposing constraints on machine-learned models. We distinguish *syntactic* constraints, which constrain the structure of the model (for example, maximal depth, for a decision tree), and *semantic* constraints, which constrain its behavior. The first type is easier to impose, and can serve as a proxy for the second. An example are the feature interaction constraints in XGBoost[2]: avoiding co-occurrence of two attributes in the same tree precludes interaction (in the statistical sense) between them. In neural networks, the architecture of the network can be chosen so that it enforces certain semantic constraints [4,16]. What semantic constraints can be imposed through syntactic constraints depends on the model format, but in general, the set is limited and ad hoc. For example, the multiplexnet [16] is relatively versatile, but still limited to quantifier-free formulas.

Multiple approaches have been proposed that enforce constraints through regularization (for example, [10,13,31]). These approaches typically allow for a much wider range of constraints to be expressed. However, they treat these constraints as soft constraints and cannot handle universal quantification over the domain.

Convex optimization based methods (for example, support vector machines) inherently include hard constraints in the optimization task. Given that they already deal with such constraints, one can just as well add more constraints to express domain knowledge. However, the type of constraints that can be expressed is again limited; for example, no quantifiers over continuous domains can be used.

Methods that rely on combinatorial optimization are closest to our work. Such methods have been proposed for decision trees (for example, [8,17,28,29]), but typically with syntactic constraints (for example, find an optimal decision tree of depth at most 5). There are some optimal decision tree methods that impose semantic constraints [1,30], but without guaranteed constraint satisfaction on unseen data. [27] proposes a counter-example guided approach to enforce

[2] https://xgboost.readthedocs.io/en/stable/tutorials/feature_interaction_constraint.html.

monotonicity constraints for all possible unseen instances, but lacks a general framework for other types of constraints. [22] proposes an approach to include logical constraints in neural network training using ProbLog, but is limited to classification problems and does not guarantee constraint satisfaction. MaxSAT has previously been used in various machine learning tasks, like Bayesian networks [3,6], interpretable classification rules [21] and optimal decision sets [32]. These approaches, however, learn in a discrete domain and do not support imposing domain constraints. To the best of our knowledge, ours is the first work that learns parametric models (with a continuous domain) in a MaxSMT framework.

Apart from all these approaches, our work also relates to work on verification of learned models, such as neural nets [18,20,26] and tree based models [5,9,12]. That work uses similar methods, but merely checks that a learned model meets certain requirements, rather than enforcing this through the learner.

6 Experimental Evaluation

The fact that SaDe *guarantees* compliance with domain constraints crucially distinguishes it from other systems, such as regularization-based methods. Even then, a number of questions can be raised:

Q1 Does it matter in practice? Perhaps other methods often learn admissible models anyway, even if they do not guarantee it.
Q2 Does this affect predictive performance?
Q3 What is the cost in terms of learning efficiency?

We address these questions empirically. We first describe the use-cases, then the evaluation methodology, and finally the results.

6.1 Use-Cases

SaDe supports any constraint and model expressible in the SMT-LIB language [24]; consequently, SaDe supports any machine learning task and setting that can be expressed in the same language. To demonstrate this ability, we design three use-cases with different tasks and settings: a binary classification problem of loan prediction, a multi-class classification problem of music genre prediction, and a multi-target regression problem of expense prediction. Despite this variety, the use-cases have the following in common: they include universally quantified domain constraints, and some of the input data violate these constraints. The later is motivated by the fact that learning robust models is more challenging, but also more useful, when training data may violate constraints (for example, data may contain undesirable bias that we explicitly do not want to model).

For readability, we use names rather than numerical indices for tuple components; for example, $artist(\mathbf{x})$ refers to the component of $\mathbf{x}$ that indicates the artist.

Our first use-case is a **music genre identification** problem. Data, consisting of 793 songs, comes from a music streaming company Tunify[3]. Each song is represented by 13 features and belongs to one of 5 classes: *rock, pop, classical, electronic, metal.* This is a *multi-class classification problem*, which we convert to several binary classification problems using a one-versus-all approach. The final prediction for a new instance is the class corresponding to the binary classifier with the highest confidence. We impose the domain constraint requiring that *a Beatles song can only be classified to either Pop or Rock*, encoded as:

$$\forall \mathbf{x} \in \mathcal{X} : artist(\mathbf{x}) = \textit{The Beatles} \implies ((rock(f_{\mathbf{w}}(\mathbf{x})) > 0 \vee pop(f_{\mathbf{w}}(\mathbf{x})) > 0) \wedge \\ classical(f_{\mathbf{w}}(\mathbf{x})) < 0 \wedge electronic(f_{\mathbf{w}}(\mathbf{x})) < 0 \wedge metal(f_{\mathbf{w}}(\mathbf{x})) < 0)$$

The dataset contains 60 violations. Our second use case is the **loan approval problem**[4]. The data consists of 614 instances with 6 categorical and 5 numerical features. This is a binary classification problem: predict whether the loan should be approved or not. We impose the domain constraint requiring that *everyone with no credit history (ch) and income lower than 5000$ should be denied a loan*, which is encoded as:

$$\forall \mathbf{x} \in \mathcal{X} : ch(\mathbf{x}) = 0 \wedge income(\mathbf{x}) < 5000 \implies f_{\mathbf{w}}(\mathbf{x}) < 0$$

The dataset contains 30 violations. Our final use-case is the **expense prediction problem**[5]. which consists of predicting multiple types of expenses for a household. The data consists of 1000 instances with 5 target expenses and 13 predictors. This is a *multi-target regression problem* which is converted into a collection of single target regression problems, one for each target (we use *exp* to represent a target in the expressions below). We enforce two domain constraints requiring that *the sum of all expenses must be smaller than the household income and going-out expense must not be more than 5% of the household income.* Domain constraints are encoded as

$$\forall \mathbf{x} \in \mathcal{X} : (\sum_{exp} exp(f_{\mathbf{w}}(\mathbf{x})) \leqslant income(\mathbf{x})) \bigwedge (going\_out(f_{\mathbf{w}}(\mathbf{x})) \leqslant 0.05 * income(\mathbf{x}))$$

The dataset contains 862 violations.

6.2 Evaluation Methodology

Evaluation Metrics: To answer question **Q1**, we need to measure "reliability": how certain are we that the model will not violate any constraints? To define a measure for this, we consider *counterexamples*: instances for which the model's prediction violates at least one domain constraint. We define the **adversity index (AdI)** as the percentage of training instances for which a counterexample

[3] https://www.tunify.com/en-gb/.
[4] https://www.kaggle.com/altruistdelhite04/loan-prediction-problem-dataset.
[5] https://www.kaggle.com/grosvenpaul/family-income-and-expenditure.

can be constructed in an l_∞ ball with radius δ centered around the instance. Note that the counterexample need not be part of the training set itself, but it must be similar to a training instance. This avoids the construction of "unrealistic" counterexamples that are totally different from anything ever seen and might not exist in practice. Counterexamples are constructed by simply using the SMT solver.

Measuring predictive performance (for question **Q2**) requires some care. We use accuracy (for classification) and mean squared error (MSE) (for regression) as performance metrics. But we should not simply compute these on the whole test set: some labels in the test set may violate the constraints and in such cases the model should explicitly *not* predict the same value. It is not known, however, what value should be predicted instead. For this reason, predictive performance is computed on the subset of the test data that satisfies the constraints.

Evaluation Procedure: We use nested 5-fold cross validation, in which the inner cross-validation is used to select the hyper-parameters. SaDe's hyper-parameters are the maximal step size α, which is selected from $\{0.5, 1, 2\}$ and the thresholds used to define the decision constraints. For classification, these thresholds are selected from $\{[0,1], [0,1,2], [1,2]\}$; for regression, they are $c * \max_D |y|$ with c selected from $\{[0.1], [0.1, 0.2], [0.1, 0.2, 0.3]\}$. The model class that SaDe uses for $f_{\mathbf{w}}$ is linear models, and the loss function is cross-entropy for classification, and sum of mean squared error (MSE) over all target variables for regression. The regularisation-based baselines that we compare SaDe to use the same model class and loss functions. They have one hyper-parameter, λ, which is the standard trade-off between the loss function and regularisation term ($loss + \lambda \cdot regularisation$). The value of λ that leads to minimum number of violations on a validation set is selected via cross-validation. All the features are scaled to $[0,1]$ using min-max normalization. The experiments are repeated 10 times, with each model being trained for 10 epochs and a batch size of 5. We use the SMT(NRA) solver z3 (version 4.8.10) for SaDe and an Intel(R) Xeon(R) Gold 6230R CPU @ 2.10 GHz machine with 256 GB RAM.

Baselines: For classification, we compare SaDe to Semantic-Based Regularisation (SBR) [10] and Semantic Loss (SL) [31] regularisation-based approaches. Note that these do not support universally quantified constraint over infinite domains: they simply ground such constraints over the training examples. For regression, we compare SaDe with a baseline model where we regularize the mean squared error loss with an additional penalty whenever the constraint is violated on the training data. For example, for use case 3, this regularized loss is:

$$\mathcal{L}_R = MSE + \lambda * \frac{1}{||D||} * \sum_{\mathbf{x} \in D} (\max(0, \sum_{exp} exp(f_{\mathbf{w}}(\mathbf{x})) - income(\mathbf{x})) \\ + \max(0, going\_out(f_{\mathbf{w}}(\mathbf{x})) - 0.05 * income(\mathbf{x})))$$

We will refer to this baseline as SBR in the remaining text.

Table 1. Adversity indices for all models. While it is not possible to construct a counter-example for SaDe models, regularisation-baselines are susceptible to them. As evident in the loan approval use-case, regularisation-based approach can occasionally result in reliable models that obey constraints, but that is not a rule

Use-case	Radius (δ)	SaDe	SBR	SL	PP
Music genre	0.01	0 ± 0	0.007 ± 0.004	0.007 ± 0.005	0 ± 0
	0.1	0 ± 0	0.089 ± 0.031	0.024 ± 0.014	0 ± 0
Loan approval	0.01	0 ± 0	0 ± 0	0 ± 0	0 ± 0
	0.1	0 ± 0	0 ± 0	0 ± 0	0 ± 0
Expense prediction	0.01	0 ± 0	0.056 ± 0.009	–	–
	0.1	0 ± 0	0.775 ± 0.029	–	–

For the classification task, we additionally consider a "post-processing" (PP) baseline: train the model without regard for any constraints; at prediction time, check whether the prediction violates a constraint, and if it does, change it. For classification, we assume that PP flips the prediction to the highest-scoring class that satisfies the domain constraint. Note that, while PP provides a trivial way to enforce domain constraints at prediction time, it is not a generally applicable method: it requires that we know how to "fix" the prediction, which is not always the case (as will be illustrated for the regression use case in the next section).

6.3 Results

We consecutively interpret the experimental results in the light of the three research questions listed before. The Post-Processing approach is discussed separately after that.

Q1. Do other methods return inadmissible models? Table 1 shows adversity indices for all models. The used values for δ are chosen to be small compared to the average ℓ_∞ distance between a pair of training instances (0.89 and 0.77 in the Music and Expense datasets, respectively), so that the constructed counter-examples can be said to be similar to some training instances. The results indicate that regularisation-based approaches are highly sensitive to counter-examples (while SaDe, by construction, is not). For SBR models and a radius of $\delta = 0.01$, it is possible to construct a counter-example in the neighbourhood of 0.7% and 5% of instances in the Music Genre and Expense Prediction use-cases, respectively. When the radius is increased to $\delta = 0.1$, it is possible to construct counter-examples in the neighbourhood of 9% and 77% of training instances in the Music Genre and Expense Prediction use-cases, respectively. SL seems slightly more robust, but counter-examples can still be found.

The loan approval use-case, on the other hand, shows that regularisation-based techniques can produce models that satisfy all constraints (no counter-examples

Table 2. Performance of all models. SaDe performs comparably to the baselines.

Use-case	SaDe	SBR	SL	PP
Music (accuracy)	80.76 ± 5.15	82.94 ± 2.47	82.96 ± 2.47	82.97 ± 2.49
Loan (accuracy)	78.03 ± 4.91	78.36 ± 4.32	78.32 ± 4.33	78.54 ± 5.08
Expense (MSE)	192.14 ± 102.96	243.49 ± 107.51	–	–

Table 3. SaDe requires more modelling time than the regularisation-based models.

Use-case	SaDe	SBR	SL	PP
Music genre	3146 ± 962	519 ± 38	513 ± 42	515 ± 39
Loan approval	134 ± 66	85 ± 3	86 ± 3	88 ± 3
Expense prediction	3297 ± 514	123 ± 53	–	–

could be constructed); they just do not guarantee it. It is not known under which conditions SBR and SL result in admissible models.

Overall, these results answer **Q1 positively**: learners that do not *guarantee* that the learned models are admissible often return models that indeed are not.

Q2. Does SaDe's restriction to admissible models affect predictive performance? Table 2 compares the predictive performance (accuracy/MSE on test data that do not violate constraints) of the learned models. For Loan Approval, SaDe performs comparably with the baselines. For Music Genre Identification, it performs slightly worse, while for Expense Prediction it performs better. The differences are not significant though.

These results show that SaDe has the potential to return admissible models without a substantial cost to predictive performance.

Q3. Is there a price to pay in terms of learning time? Table 3 shows the runtimes of SaDe and the baselines. For these use-cases, SaDe takes about 2, 6, or 30 times longer to learn a model, compared to the regularisation based approaches. This is not unexpected: SaDe solves the more complex task of not only finding models but also proving their admissibility.

For safety-critical applications, such an increase in learning time would often be considered acceptable, given the guarantees one gets in return. Where this is not the case, there is room for investigating variants of SaDe that are potentially faster. For instance, SaDe's stopping criterion was not optimized in this work; a more sophisticated criterion might make the approach considerably faster. Also, recent advances in developing SMT solvers capable of verifying neural networks [20] suggest that improvements in SMT solver technology may also positively affect SaDe's computational efficiency.

The Post Processing Baseline. We should devote some discussion to the post-processing baseline PP. For the classification uses cases, PP works well: the combination of model and post-processing step satisfies the domain constraints (Table 1) with a similar performance (Table 2) as the baselines. For these specific cases, SaDe does not have an advantage over PP.

However, it is important to realize that PP is not a generally applicable approach. It only works when there exists a trivial way to fix an individual prediction. For instance, when domain constraints enforce relationships between multiple targets, this kind of approach is not feasible. This is showcased in the expense prediction use case. The constraint "*sum of all expenses must be smaller than the household income*" does not translate to constraints on individual expenses, and there is more than one way in which individual predictions can be fixed in order to satisfy the domain constraint. Even if a fixed procedure were introduced (for example, reduce all of them proportionally), other constraints may interfere with this procedure, rendering it invalid; "*going-out expense must not be more than 5% of the household income*" is such a constraint.

6.4 Limitations of SaDe

Our current implementation of SaDe still has a number of limitations. Learning models with high degree of non-linearity (for example Neural Nets) has not been feasible up till now. The solver technology we are using was either too slow or its reasoning engine was simply too weak to be able to solve such problems. Future improvements in solver technology may make it possible to learn more complicated models using SaDe.

Additionally, our approach is not directly applicable to discrete models (for example Decision Trees) because SaDe relies on a differentiable loss function. A possible solution to this could be based on the ideas in Norouzi et al. [23]: they learn a decision tree as a parametric model by approximating the global non-differentiable loss with a differentiable one. Such an approach could be explored in conjunction with SaDe.

7 Conclusion

We proposed a new learning framework based on maximum satisfiability and a novel learning algorithm *SaDe* that can learn parametric models that provably satisfy user-provided domain constraints. The framework is general enough to handle a wide range of learning problems (classification, regression, ...) and constraints. To our knowledge, our approach is the first to guarantee admissibility of learned models for such a wide class of symbolically expressible constraints. While the approach is in principle generic and does not depend on the format of the model (as long as it has continuous parameters), there may be practical hurdles for complex model formats. We have empirically shown that the approach is feasible at least for linear models, that it guarantees admissibility where other approaches do not, and that this is often possible without a cost in predictive performance and acceptable cost in terms of training time. This makes the approach very relevant in application contexts that are safety-critical, governed by law or company policies, etc. Our approach is just a first step in a direction in which there is much opportunity for further work.

Acknowledgement. This research was jointly funded by the Flemish Government (AI Research Program) and the Research Foundation - Flanders under EOS No. 30992574.

References

1. Aghaei, S., Azizi, M.J., Vayanos, P.: Learning optimal and fair decision trees for non-discriminative decision-making. In: AAAI (2019)
2. Barocas, S., Hardt, M., Narayanan, A.: NIPS 2017 tutorial on fairness in machine learning, **1** (2017)
3. Berg, O.J., Hyttinen, A.J., Järvisalo, M.J., et al.: Applications of MaxSAT in data analysis. In: Pragmatics of SAT (2019)
4. Berner, J., Grohs, P., Kutyniok, G., Petersen, P.: The modern mathematics of deep learning. arXiv preprint arXiv:2105.04026 (2021)
5. Chen, H., Zhang, H., Si, S., Li, Y., Boning, D., Hsieh, C.J.: Robustness verification of tree-based models. In: NeurIPS (2019)
6. Cussens, J.: Bayesian network learning by compiling to weighted MAX-SAT. arXiv preprint arXiv:1206.3244 (2012)
7. de Moura, L., Bjørner, N.: Z3: an efficient SMT solver. In: Ramakrishnan, C.R., Rehof, J. (eds.) TACAS 2008. LNCS, vol. 4963, pp. 337–340. Springer, Heidelberg (2008). https://doi.org/10.1007/978-3-540-78800-3_24
8. Demirović, E., et al.: MurTree: optimal classification trees via dynamic programming and search. arXiv preprint arXiv:2007.12652 (2020)
9. Devos, L., Meert, W., Davis, J.: Versatile verification of tree ensembles. In: International Conference on Machine Learning, pp. 2654–2664. PMLR (2021)
10. Diligenti, M., Gori, M., Sacca, C.: Semantic-based regularization for learning and inference. Artif. Intell. **244**, 143–165 (2017)
11. Dumancic, S., Meert, W., Goethals, S., Stuyckens, T., Huygen, J., Denies, K.: Automated reasoning and learning for automated payroll management. In: Proceedings of the Thirty-Third Annual Conference on Innovative Applications of Artificial Intelligence (2020)
12. Einziger, G., Goldstein, M., Sa'ar, Y., Segall, I.: Verifying robustness of gradient boosted models. In: AAAI (2019)
13. Fischer, M., Balunovic, M., Drachsler-Cohen, D., Gehr, T., Zhang, C., Vechev, M.: DL2: training and querying neural networks with logic. In: ICML (2019)
14. Fu, Z., Malik, S.: On solving the partial MAX-SAT problem. In: Biere, A., Gomes, C.P. (eds.) SAT 2006. LNCS, vol. 4121, pp. 252–265. Springer, Heidelberg (2006). https://doi.org/10.1007/11814948_25
15. Gori, M.: Machine Learning: A Constraint-Based Approach (2017)
16. Hoernle, N., Karampatsis, R.M., Belle, V., Gal, K.: MultiplexNet: towards fully satisfied logical constraints in neural networks (2021)
17. Hu, H., Siala, M., Hébrard, E., Huguet, M.J.: Learning optimal decision trees with MaxSAT and its integration in AdaBoost. In: IJCAI (2020)
18. Huang, X., Kwiatkowska, M., Wang, S., Wu, M.: Safety verification of deep neural networks. In: Majumdar, R., Kunčak, V. (eds.) CAV 2017. LNCS, vol. 10426, pp. 3–29. Springer, Cham (2017). https://doi.org/10.1007/978-3-319-63387-9_1

19. Katz, G., Barrett, C., Dill, D.L., Julian, K., Kochenderfer, M.J.: Reluplex: an efficient SMT solver for verifying deep neural networks. In: Majumdar, R., Kunčak, V. (eds.) CAV 2017. LNCS, vol. 10426, pp. 97–117. Springer, Cham (2017). https://doi.org/10.1007/978-3-319-63387-9_5
20. Katz, G., et al.: The marabou framework for verification and analysis of deep neural networks. In: Dillig, I., Tasiran, S. (eds.) CAV 2019. LNCS, vol. 11561, pp. 443–452. Springer, Cham (2019). https://doi.org/10.1007/978-3-030-25540-4_26
21. Maliotov, D., Meel, K.S.: MLIC: a MaxSAT-based framework for learning interpretable classification rules. In: Hooker, J. (ed.) CP 2018. LNCS, vol. 11008, pp. 312–327. Springer, Cham (2018). https://doi.org/10.1007/978-3-319-98334-9_21
22. Manhaeve, R., Dumancic, S., Kimmig, A., Demeester, T., De Raedt, L.: DeepProbLog: neural probabilistic logic programming. In: NeurIPS (2018)
23. Norouzi, M., Collins, M., Johnson, M.A., Fleet, D.J., Kohli, P.: Efficient non-greedy optimization of decision trees. In: Advances in Neural Information Processing Systems, vol. 28 (2015)
24. Ranise, S., Tinelli, C.: The SMT-LIB standard: version 1.2. Technical report, Department of Computer Science, The University of Iowa (2006)
25. Rossi, F., Van Beek, P., Walsh, T.: Handbook of Constraint Programming (2006)
26. Singh, G., Gehr, T., Püschel, M., Vechev, M.: Boosting robustness certification of neural networks. In: ICLR (2018)
27. Sivaraman, A., Farnadi, G., Millstein, T., Van den Broeck, G.: Counterexample-guided learning of monotonic neural networks. arXiv preprint arXiv:2006.08852 (2020)
28. Verhaeghe, H., Nijssen, S., Pesant, G., Quimper, C.G., Schaus, P.: Learning optimal decision trees using constraint programming. In: BNAIC/BENELEARN (2019)
29. Verwer, S., Zhang, Y.: Learning optimal classification trees using a binary linear program formulation. In: Proceedings of the AAAI Conference on Artificial Intelligence, vol. 33, pp. 1625–1632 (2019)
30. Vos, D., Verwer, S.: Robust optimal classification trees against adversarial examples. arXiv preprint arXiv:2109.03857 (2021)
31. Xu, J., Zhang, Z., Friedman, T., Liang, Y., Van Den Broeck, G.: A semantic loss function for deep learning with symbolic knowledge. In: ICML (2018)
32. Yu, J., Ignatiev, A., Stuckey, P.J., Le Bodic, P.: Learning optimal decision sets and lists with SAT. J. Artif. Intell. Res. **72**, 1251–1279 (2021)

On the Generalization of Neural Combinatorial Optimization Heuristics

Sahil Manchanda[1,2](✉), Sofia Michel[1](✉), Darko Drakulic[1](✉), and Jean-Marc Andreoli[1](✉)

[1] NAVER LABS Europe, Meylan, France
sahil.manchanda@cse.iitd.ac.in,
{sofia.michel,darko.drakulic,jean-marc.andreoli}@naverlabs.com
[2] Indian Institute of Technology Delhi, Hauz Khas, India

Abstract. Neural Combinatorial Optimization approaches have recently leveraged the expressiveness and flexibility of deep neural networks to learn efficient heuristics for hard Combinatorial Optimization (CO) problems. However, most of the current methods lack generalization: for a given CO problem, heuristics which are trained on instances with certain characteristics underperform when tested on instances with different characteristics. While some previous works have focused on varying the training instances properties, we postulate that a one-size-fit-all model is out of reach. Instead, we formalize solving a CO problem over a given instance distribution as a separate learning task and investigate meta-learning techniques to learn a model on a variety of tasks, in order to optimize its capacity to adapt to new tasks. Through extensive experiments, on two CO problems, using both synthetic and realistic instances, we show that our proposed meta-learning approach significantly improves the generalization of two state-of-the-art models.

Keywords: Neural combinatorial optimization · Generalization · Heuristic learning · Traveling salesman problem · Capacitated vehicle routing problem

1 Introduction

Combinatorial optimization (CO) aims at finding optimal decisions within finite sets of possible decisions; the sets being typically so large that exhaustive search is not an option [5]. CO problems appear in a wide range of applications such as logistics, transportation, finance, energy, manufacturing, etc. CO heuristics are efficient algorithms that can compute high-quality solutions but without optimality guarantees. Heuristics are crucial to CO, not only for applications where optimality is not required, but also for exact solvers, which generally exploit

S. Manchanda—Work done while interning at NAVER LABS Europe.

Supplementary Information The online version contains supplementary material available at https://doi.org/10.1007/978-3-031-26419-1_26.

© The Author(s), under exclusive license to Springer Nature Switzerland AG 2023
M.-R. Amini et al. (Eds.): ECML PKDD 2022, LNAI 13717, pp. 426–442, 2023.
https://doi.org/10.1007/978-3-031-26419-1_26

numerous heuristics to guide and accelerate their search procedure [7]. However, the design of such heuristics heavily relies on problem-specific knowledge, or at least experience with similar problems, in order to adapt generic methods to the setting at hand. This design skill that human experts acquire with experience and that is difficult to capture formally, is a typical signal for which statistical methods may help. In effect, machine learning has been successfully applied to solve CO problems, as shown in the surveys [3,4]. In particular, *Neural Combinatorial Optimization* (NCO) has shown remarkable results by leveraging the full power and expressiveness of deep neural networks to model and automatically derive efficient CO heuristics.

Among the approaches to NCO, supervised learning [12,17,28] and reinforcement learning [2,14,20] are the main paradigms.

Despite the promising results of end-to-end heuristic learning, a major limitation of these approaches is their lack of generalization to out-of-training-distribution instances for a given CO problem [3,11]. For example, models are generally trained on graphs of a fixed size and perform well on unseen "similar" graphs of the same size. However, when tested on smaller or larger ones, performance tends to degrade drastically. Although size variation is the most reported case of poor generalization, in our study we will show that instances of the same size may still vary enough to cause generalization issues. This limitation might hinder the application of NCO to real-life scenarios where the precise target distribution is often not known in advance and can vary with time. A natural way to alleviate the generalization issue is to train on instances with diverse characteristics, such as various graph sizes [13,16,18]. Intuitively this amounts to augmenting the training distribution to make it more likely to correctly represent the target instances.

In this paper, we postulate that a one-size-fit-all model is out of reach.

Instead, we believe that one of the strengths of end-to-end heuristic learning is precisely its capacity to adapt to specific data and the exploitation of the underlying structure to obtain an effective specialized heuristic. Therefore we propose to use instance characteristics to define distributions and consider solving a CO problem over a given *instance distribution* as a separate *learning task*. We will assume a prior over the target task, by assuming it is part of a given *task distribution*, from which we will sample the training tasks. Note that this is a weaker assumption than most current NCO methods that (implicitly) assume knowing the target distribution at training. At the other extreme, without any assumption on the target distribution, the No Free Lunch Theorems of Machine Learning [30] tell us that we cannot expect to do better than a random policy. In this context, meta-learning [22,24] is a natural approach to obtain a model able to adapt to new unseen tasks. Given a distribution of tasks, the idea of meta-learning is to train a model using a sample of those tasks while optimizing its ability to adapt to each of them. Then at test time, when presented with an unseen task from the same distribution, the model only needs to be fine-tuned using a small amount of data from that task.

Contributions: We focus on two representative state-of-the-art NCO approaches: (i) the reinforcement learning-based method of [14] and the supervised learning approach of [12]. In terms of CO problems, we use the well-studied Traveling

Salesman Problem (TSP) and Capacitated Vehicle Routing Problem (CVRP). We first analyze the NCO models' generalization capacity along different instance parameters such as the graph size, the vehicle capacity and the spatial distribution of the nodes and highlight the significant drop in performance on out-of-distribution instances (Sect. 3). Then we introduce a model-agnostic meta-learning procedure for NCO, inspired by the first-order meta-learning framework of [21] and adapt it to both the reinforcement and supervised learning-based NCO approaches (Sect. 4).

Finally, we design an extensive set of experiments to evaluate the performance of the meta-trained models with different pairs of training and test distributions. Our contributions are summarized as follows:

- **Problem formalization:** We give the first formalization of the NCO out-of-distribution generalization problem and provide experimental evidence of its impact on two state-of-the-art NCO approaches.
- **Meta-learning framework:** We propose to apply a generic meta-training procedure to learn robust NCO heuristics, applicable to both reinforcement and supervised learning frameworks. To the best of our knowledge we are the first to propose meta-learning in this context and prove its effectiveness through extensive experiments.
- **Experimental evaluation:** We demonstrate experimentally that our proposed meta-learning approach does alleviate the generalization issue. The meta-trained models show a better zero-shot generalization performance than the commonly used multi-task training strategy. In addition, using a limited number of instances from a new distribution, the fine-tuned meta-NCO models are able to catch-up, and even frequently outperform, the reference NCO models, that were specifically trained on the target distribution. We provide results both on synthetic datasets and the well-established realistic Operations Research datasets *TSPlib* and *CVRPlib*.
- **Benchmarking datasets:** Finally, by extending commonly used datasets, we provide an extensive benchmark of labeled TSP and CVRP instances with a diverse set of distributions, that we hope will help better evaluate the generalization capability of NCO methods on these problems.

2 Related Work

Several papers have noted the lack of out-of-training-distribution generalization of current NCO heuristics, e.g. [3,4]. In particular, [11] explored the role of certain architecture choices and inductive biases of NCO models in their ability to generalize to large-scale TSP problems. In [18], the authors proposed a curriculum learning approach to train the attention model of [14], assuming good-quality solutions can be accessed during training and using the corresponding optimality gap to guide the scheduling of training instances of various sizes. The proposed curriculum learning in a semi-supervised setting helped improve the original model's generalization on size. Recently, [8] proposed a method able to generalize to large-scale TSP graphs by combining the predictions of a learned model on small subgraphs and using these predictions to guide a Monte Carlo Tree Search, successfully generalizing to instances with up to 10,000 nodes. Note that both [8,18] are specifically designed to deal with size variation.

One can note that hybrid approaches combining learned components and classical CO algorithms tend to generalize better than end-to-end ones. For example, the learning-augmented local search heuristic of [16] was able to train on relatively small CVRP instances and generalize to instances with up to 3000 nodes. Also recent learned heuristics within branch and bound solvers show a strong generalization ability [19,31]. Other approaches that generalize well are based on algorithmic learning. For instance, [9] learns to imitate the Ford-Fulkerson algorithm for maximum bipartite matching, by neural execution of a Graph Neural Network, similar to [27] for other graph algorithms. These methods achieve a strong generalization to larger graphs but at the expense of precisely imitating the steps of existing algorithms.

In this paper we focus on the generalization of end-to-end NCO heuristics. In contrast to previous approaches, we propose a general framework, applicable to both supervised and reinforcement (unsupervised) learning-based NCO methods, and that accounts for any kind of distribution shift, including but not restricted to graph size. To the best of our knowledge, we are the first to propose meta-learning as a generic approach to improve the generalization of any NCO model.

3 Generalization Properties

To analyze the generalization properties of different NCO approaches, we focus on two wide-spread CO problems: (i) the Euclidean Traveling Salesman Problem (TSP), where given a set of nodes in a Euclidean space (typically the plane), the goal is to find a tour of minimal length that visits each node exactly once; and (ii) the Capacitated Vehicle Routing Problem (CVRP), where given a depot node, a set of customer nodes with an associated demand and a vehicle capacity, the goal is to compute a set of routes of minimal total length, starting and ending at the depot, such that each customer node is visited and the sum of demands of customers in each route does not exceed the vehicle capacity. Note that the TSP can be viewed as a special case of the CVRP where the vehicle capacity is infinite.

3.1 Instance Distributions as Tasks

To explore the effect of variability in the training datasets, we consider a specific family $\mathcal{T}_{N,M,C,L}$ of instance distributions (tasks), indexed by the following parameters: the *graph size* N, the *number of modes* M, the *vehicle capacity* C and the *scale* L. Given these parameters, an instance is generated by the following process. When $M{\neq}0$: first, M points, called the modes, are independently sampled by an ad-hoc process which tends to spread them evenly in the unit square; then N points are independently sampled from a balanced mixture of M Gaussian components centered at the M modes, sharing the same diagonal covariance matrix, meant to keep the generated points within relatively small clusters around the modes; finally, the node coordinates are rescaled by a factor L. When M=0: the N points are instead directly sampled uniformly in the unit square then rescaled by L. Additionally, in the case of the CVRP problem, the depot is chosen randomly, the vehicle capacity is fixed to C and customer demands are generated as in [20]. Examples of spatial node distributions for various TSP tasks are displayed in Fig. 1.

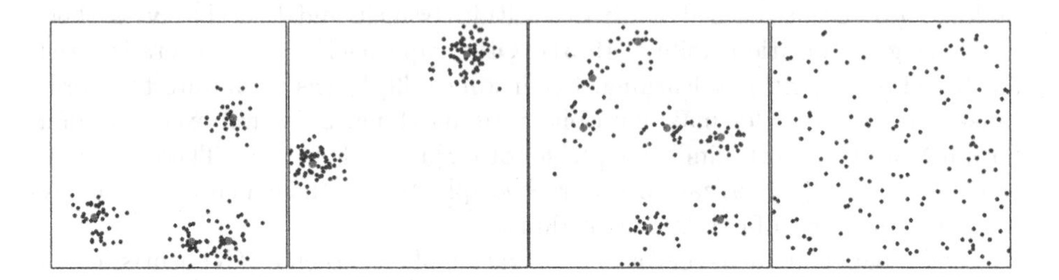

Fig. 1. A sample from each of 4 tasks $\mathcal{T}_{N=150,L=1,M}$ (blue points) with M=4, 2, 7, 0, respectively, from left to right. The red dots are the generated modes. (Color figure online)

3.2 Measuring the Impact of Generalization on Performance

To measure the performance of different algorithms on a given task, we sample a set of test instances from that task and apply each algorithm to each of these instances. Since the average length of the resulting tours is biased towards longer lengths, we measure instead the average "*gap*" with respect to reference tours. For the TSP, reference is provided by the Concorde solver [1], which is exact, so what we report is the true optimality gap; for the CVRP, we use the solutions computed by the state-of-the-art LKH heuristic solver [10], which returns high-quality solutions at the considered instance sizes (near optimality).

We measure the performance (gap) deterioration on generalization of the reinforcement learning based Attention Model of [14], subsequently abbreviated as AM, and the supervised Graph Convolutional Network model of [12], subsequently abbreviated as GCN. We consider several classes of tasks of the form $\mathcal{T}_{N,M,C,L}$ obtained by varying, in each class, only one of the parameters[1] N, M, C, L. For each class and each task in that class, we train each model on that task only and test it on each of the tasks in the same class, thus including the training one. The main results for the AM model are reported in Table 1.

As already observed in several papers, varying the number of nodes degrades the performance (columns (a) and (d)). Interestingly, varying the number of modes only also has a negative impact (columns (b) and (e)), and the same holds when varying the scaling of the node coordinates in the TSP (column (c)) or the vehicle capacity in the CVRP (column (f)). Similar results of performance degradation on generalization of the GCN model are given in Table 2 for TSP.

[1] Except with CVRP where, as in previous work [20], changes to C and N are coupled.

Table 1. Performance deterioration of AM(TSP and CVRP): Average gap of the AM model (in percentage, over 5000 test instances) when trained and tested on TSP instances with different (a) number of nodes N (b) number of modes M and (c) scale L; and CVRP instances with different (d) number of nodes N, (e) number of modes M and (f) vehicle capacities C.

$N\frac{\text{test}\rightarrow}{\text{train}\downarrow}$	N=20	N=50	N=100
N=20	**0.08**	1.78	22.61
N=50	0.35	**0.52**	2.95
N=100	3.78	2.33	**2.26**

(a) N (M=0, L=1)

$M\frac{\text{test}\rightarrow}{\text{train}\downarrow}$	M=0	M=3	M=6
M=0	**1.47**	32.17	2.74
M=3	26.38	**1.86**	7.32
M=6	6.91	6.01	**2.0**

(b) M (N=40, L=1)

$L\frac{\text{test}\rightarrow}{\text{train}\downarrow}$	L=1	L=5	L=10
L=1	**1.48**	282.55	292.39
L=5	32.84	**1.44**	13.83
L=10	98.62	7.12	**1.53**

(c) L (N=40, M=0)

$N\frac{\text{test}\rightarrow}{\text{train}\downarrow}$	N=20	N=50	N=100
N=20	**4.52**	12.61	20.23
N=50	7.99	**6.93**	8.47
N=100	12.90	9.75	**7.11**

(d) N (M=0, C=func(N))

$M\frac{\text{test}\rightarrow}{\text{train}\downarrow}$	M=1	M=3	M= 8
M=1	**4.39**	51.02	102.07
M=3	5.67	**6.32**	16.14
M=8	14.91	8.67	**7.85**

(e) M (N=50, C=40)

$C\frac{\text{test}\rightarrow}{\text{train}\downarrow}$	C=20	C=30	C=50
C=20	**5.83**	8.25	12.23
C=30	6.13	**7.37**	9.39
C=50	12.27	8.56	**7.99**

(f) C (N=func(C), M=0)

Table 2. Performance deterioration of GCN(TSP): Average gap of the GCN model, when varying (a) the number of nodes N (b) the number of modes M and (c) the scale L.

$N\frac{\text{test}\rightarrow}{\text{train}\downarrow}$	N=20	N=50	N=100
N=20	**1.83**	38.66	77.31
N=50	22.05	**5.10**	43.76
N=100	43.86	37.26	**14.79**

(a) N (M=0, L=1)

$M\frac{\text{test}\rightarrow}{\text{train}\downarrow}$	M=0	M=3	M=8
M=0	**5.05**	35.86	26.01
M=3	35.40	**6.96**	28.71
M=8	32.74	36.29	**5.48**

(b) M (N=50, L=1)

$L\frac{\text{test}\rightarrow}{\text{train}\downarrow}$	L=1	L=5	L=10
L=1	**5.10**	28.15	32.46
L=5	272.58	**5.23**	25.41
L=10	289.51	66.28	**5.46**

(c) L (N=50, M=0)

These results confirm the drastic lack of generalization of both models, even on seemingly closely related instance distributions. In the next section, we propose an approach to tackle this problem.

4 Meta-learning of NCO Heuristics

The goal of this paper is to introduce an NCO approach capable of out-of-distribution generalization for a given CO problem. Since NCO methods tend to perform well on fixed instance distributions, our strategy to promote out-of-distribution generalization is to modify the way the model is trained without changing its architecture.

Concretely, given a CO problem (e.g. the TSP), we assume that we have a prior over the relevant tasks (instance distributions), possibly based on historical data. For instance, we may know that the customers in our TSP are generally clustered around city centers, but without knowing how many clusters. Our underlying assumption is that it is easier and more realistic to obtain a prior distribution on target tasks, rather than the target task itself. We propose to first train a model to learn an efficient heuristic on a sample of tasks (e.g. TSP

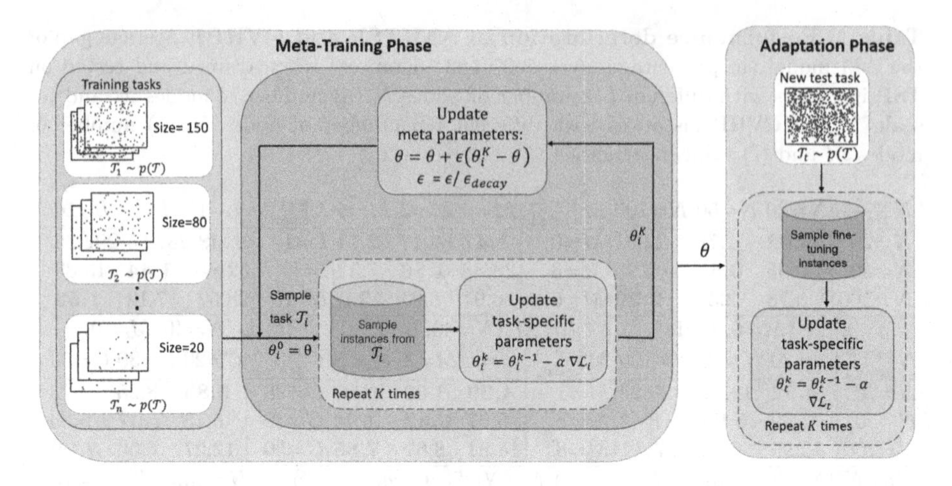

Fig. 2. Overview of our proposed method. Note that in the training phase, instead of size variation, one can have different types of distribution shifts.

instances with different numbers of modes). Then, considering a new unseen task (unseen number of modes), we would use a *limited number of samples* (few-shots) from that task to specialize the learned heuristic and maximize its performance on it. Figure 2 illustrates our proposed approach.

Formally, given an NCO model with parameter θ and a distribution of tasks $\mathcal{T}$, our goal is to compute a value of θ such that, given an unseen task $t \sim \mathcal{T}$ with associated loss $\mathcal{L}_t$, after K gradient updates, the fine-tuned parameter minimizes $\mathcal{L}_t$, i.e.

$$\min_\theta \mathbb{E}_{t \sim \mathcal{T}}[\mathcal{L}_t(\theta_t^K)], \tag{1}$$

where θ_t^K is the fine-tuned parameter after K gradient updates of θ using batches of instances from task t. Problem (1) can be viewed as a few-shot meta-learning optimization problem. We approach it in a model-agnostic fashion by leveraging the generic Reptile meta-learning algorithm [21]. Given a task distribution, Reptile is a surprisingly simple algorithm to learn a model that performs well on unseen tasks of that distribution. Compared to the seminal MAML framework [6], Reptile is a first-order method that does not differentiate through the fine-tuning process at train time, making it feasible to work with larger values of K. And we observed experimentally that in our context, to fine-tune a model to a new task, we need up to $K = 50$ steps, which is beyond MAML's practical limits. Furthermore, since Reptile uses only first-order gradients with a very simple form, it is more efficient, both in terms of computation and memory. Using Reptile, we meta-train each model on the given task distribution to obtain an effective initialization of the parameters, which can subsequently be adapted to a new target task using a limited number of fine-tuning samples from that task.

The first step to optimize Eq. 1 consists of K updates of task specific parameters for a task $\mathcal{T}_i \sim \mathcal{T}$ as follows:

$$\begin{cases} \theta_i^0 = \theta, \\ \theta_i^k = \theta_i^{k-1} - \alpha \nabla \mathcal{L}_i(\theta_i^{k-1}), \quad \forall k \in [1 \ldots K]. \end{cases} \tag{2}$$

In the above equation, the hyper-parameter α controls the learning rate. Then, using the updated parameters θ_i^K obtained at the end of the K steps, we update the meta-parameter θ as follows:

$$\theta = \theta + \epsilon \left(\theta_i^K - \theta \right). \tag{3}$$

This is essentially a weighted combination of the updated task parameters θ_i^K and previous model parameters θ. The parameter ε can be interpreted as a step-size in the direction of the Reptile "gradient" $\theta_i^K - \theta$. It controls the contribution of task specific parameters to the overall model parameters. We iterate over $\mathcal{T}_i \sim \mathcal{T}$ by computing Eq. 3 for different tasks and then using it for optimizing Eq. 1.

Scheduling ε: first specialize then generalize. As mentioned above, parameter ε controls the contribution of the task specific loss to the global meta parameters θ update in Eq. 3. A high value of ε leads to overfitting on the training task while a low value leads to underfitting, i.e. inefficient learning of the task itself. In order to tackle such scenario, in this work we utilize a simple decaying schedule for ε which starts close to 1 (i.e. we deliberately let the model forget a lot, but not all, after each new task) and tends to 0 as the training proceeds, thus stabilizing the meta-parameter that is more likely to work well for all tasks.

Fine-tuning for target adaptation: Once the model is meta-trained on a diverse set of tasks, given a new unseen task $\mathcal{T}_t$, we initialize the model parameter to the meta-trained value θ and do a number a fine-tuning steps to get the specialized parameter for that new task. Essentially,

$$\begin{cases} \theta_t^0 = \theta, \\ \theta_t^k = \theta_t^{k-1} - \alpha \nabla \mathcal{L}_t(\theta_t^{k-1}), \quad \forall k \in [1 \ldots K]. \end{cases} \tag{4}$$

We can now detail the meta-training procedure of NCO models for the TSP and CVRP problems over our two state-of-the-art reinforcement learning (AM) and supervised learning (GCN) approaches to NCO heuristic learning. For simplicity, we use as default problem the TSP in this section, while the adaptation of the algorithms for the CVRP is presented in Sect. A.2 in Supplementary Material.

4.1 Meta-learning of RL-based NCO Heuristics (AM Model)

The RL based model of [14] (AM) consists of learning a policy that takes as input a graph representing the TSP instance and outputs the solution as a sequence of graph nodes. The policy is parameterized by a neural network with attention

Algorithm 1. Meta-training of the Attention Model

Require: Task set $\mathcal{T}$, # updates K, threshold β, step-size initialization $\varepsilon_0 \approx 1$ and decay $\varepsilon_{decay} > 1$
1: Initialize meta-parameters θ randomly, baseline parameters $\theta_i^b = \theta$ for $\mathcal{T}_i \in \mathcal{T}$ and step-size $\varepsilon = \varepsilon_0$
2: **while** not done **do**
3: Sample a task $\mathcal{T}_i \in \mathcal{T}$
4: Initialize adapted parameters $\theta_i \leftarrow \theta$
5: **for** K times **do**
6: Sample batch of graphs g_k from task $\mathcal{T}_i$
7: $\sigma_k \leftarrow \text{SampleRollout}(g_k, \pi_{\theta_i}) \quad \forall k$
8: $\sigma_k^b \leftarrow \text{GreedyRollout}(g_k, \pi_{\theta_i^b}) \quad \forall k$
9: $\nabla_\theta \mathcal{L}_i \leftarrow \sum_k (c(\sigma_k) - c(\sigma_k^b)) \nabla_\theta \log \pi_{\theta_i}(\sigma_k)$
10: $\theta_i \leftarrow \text{Adam}(\theta_i, \nabla_\theta \mathcal{L}_i)$ {Update for task $\mathcal{T}_i$}
11: **end for**
12: **if** OneSidedPairedTTest$(\pi_{\theta_i}, \pi_{\theta_i^b}) < \beta$ **then**
13: Update baseline $\theta_i^b \leftarrow \theta_i$ {Update task specific baseline}
14: **end if**
15: Update $\theta \leftarrow (1 - \varepsilon)\theta + \varepsilon\theta_i, \quad \varepsilon \leftarrow \varepsilon/\varepsilon_{decay}$ {Update meta parameters, step size}
16: **end while**

based encoder and decoder [26] stages. The encoder computes nodes and graph embeddings; using these embeddings and a context vector, the decoder produces the sequence of input nodes in an auto-regressive manner. In effect, given a graph instance G with N nodes, the model produces a probability distribution $\pi_\theta(\sigma|G)$ from which one can sample to get a full solution in the form of a permutation $\sigma = (\sigma_1, \ldots, \sigma_N)$ of $\{1, \ldots, N\}$. The policy parameter θ is optimized to minimize the loss: $\mathcal{L}(\theta|G) = \mathbb{E}_{\pi_\theta(\sigma|G)}[c(\sigma)]$, where c is the cost (or length) of the tour σ. The REINFORCE [29] gradient estimator is used: $\nabla_\theta \mathcal{L}(\theta|G) = \mathbb{E}_{\pi_\theta(\sigma|G)}[(c(\sigma) - b(G))\nabla_\theta \log \pi_\theta(\sigma|G)]$. As in [14], we use as baseline b the cost of a greedy rollout of the best model policy, that is updated periodically during training.

Meta-training of AM: Algorithm 1 describes our approach for meta-training the AM model for the TSP problem. For simplicity, the distribution of tasks that we consider here is uniform over a finite fixed set of tasks. Otherwise, one just needs to define the task-specific baseline parameters θ_i^{BL} on the fly when a task is sampled for the first time. The training consists of repeatedly sampling a task (line 3), doing K update of the meta-parameters θ using samples from that task to get fine-tuned parameters θ_i, then updating the meta-parameters as a convex combination of their previous value and the fine-tuned value (line 15). Note that the baseline need not be updated at each step (line 12), but only periodically, to improve the stability of the gradients.

4.2 Meta-learning of Supervised NCO Heuristics (GCN Model)

The supervised model of [12] (GCN) consists of a Graph Convolution Network that takes as input a TSP instance as a graph G and outputs, for each edge ij in G, predicted probabilities $\hat{s}_{ij}$ of being part of the optimal solution. It is trained using a weighted binary cross-entropy loss between the predictions and the ground-truth solution s_{ij} provided by the exact solver Concorde [1]:

$$\mathcal{L}(\theta|G) = \sum_{ij \in G} w_0 s_{ij} \log(\hat{s}_{ij}) + w_1(1 - s_{ij}) \log(1 - \hat{s}_{ij}), \tag{5}$$

where w_0 and w_1 are class weights meant to compensate the inherent class imbalance, and B is the batch size. The predicted probabilities are then used either to greedily construct a tour, or as an input to a beam search procedure. For simplicity, and because we are interested in the learning component of the method, we only consider here the greedy version.

Meta-training of GCN: Algorithm 2 in Supplementary Material describes our approach for meta-training the GCN model. In contrast to Algorithm 1, we need here to fix the training tasks since the ground-truth optimal solutions must be precomputed in this supervised learning framework.

5 Experiments

The goal of our experiments is to demonstrate the effectiveness of meta-learning for achieving generalization in NCO. More precisely, given a prior distribution of tasks, we aim to answer the following questions: (i) How does the (fine-tuned) meta-trained NCO models perform on unseen tasks, in terms of *optimality gaps* and *sample efficiency*? (ii) How does the meta-trained models perform on unseen tasks that are *interpolated* or *extrapolated* from the training tasks? (iii) How effective is our proposed *decaying step-size* strategy in the Reptile meta-learning algorithm for our NCO tasks?

Experimental setup. Experiments were performed on a pool of machines running Intel(R) CPUs with 16 cores, 256GB RAM under CentOS Linux 7, having Nvidia Volta V100 GPUs with 32GB GPU memory. All the models were trained for 24 h on 1 GPU. The detailed hyperparameters are presented in Sect. A.4 of the Supp. Material. Our code and datasets are available at: https://github.com/ncometa/meta-NCO.

Task distributions. For the TSP (resp. CVRP) experiments, we consider four task distributions (Sect. 3.1) which are obtained from $\mathcal{T}_{N=40,M=0,L=1}$ (resp. $\mathcal{T}_{N=50,M=0,C=40,L=1}$) as follows: (i) a *var-size* distribution is obtained by varying N only, and for training tasks within this distribution we use $N \in \{10, 20, 30, 50\}$; (ii) *var-mode* distribution by varying M only, and for training $M \in \{1, 2, 5\}$; (iii) *mixed-var* distribution by varying both N and M and training with $(N, M) \in$

$\{20, 30, 50\} \times \{1, 2, 4\}$; and (iv) only for CVRP: *var-capacity* distribution by varying C only, for training $C \in \{10, 30, 40\}$. As test tasks, we use values that are both within the training tasks range to evaluate the *interpolation* performance (e.g. $M=3$ for *var-mode*) and outside to evaluate the *extrapolation* performance (e.g. $N=100$ for *var-size*). More details about the distributions are presented in Sect. A.3 of the Supp. Mat.

Datasets. We generate synthetic TSP and CVRP instances, according to the previously described task distributions. For AM training, samples are generated on demand while for the GCN model, we generate for each task a training set of 1M instances, a validation and test set of 5K instances each and use the Concorde solver [1] and LKH [10] to get the associated ground-truth solutions for TSP and CVRP respectively (as was done in the original work). In order to fine-tune the meta-trained models, we sample a set of instances from the new task, containing either 3K (AM) or 1K (GCN) samples; these numbers were chosen as approximately 0.01% and 0.1% of the number of samples used during the 24 h training of the AM and GCN models respectively (see details in Sect. A.5 in Supp. Mat.). In addition to synthetic datasets, we evaluate our models on the realistic datasets: TSPlib and CVRPlib. The precise settings and results are presented in Sect. 5.1.

Models. We use the AM-based heuristics of [14] for TSP and CVRP. For the GCN model, we use the model provided by [12] for the TSP and its adaptation by [15] for the CVRP. For a given task distribution (e.g. *var-size*) we consider the following models:

- `meta-AM` (resp. `meta-GCN`): the AM (resp. GCN) model meta-trained (following Algorithm 1 or 2 for TSP). E.g. for the *var-size* distribution, we denote this model `meta-AM-N` (resp. `meta-GCN-N`).
- `multi-AM` (resp. `multi-GCN`): the AM (resp. GCN) model trained with instances coming equiprobably from the training tasks. E.g. for the *var-mode* distribution, we denote this model `multi-AM-M` (resp. `multi-GCN-M`).
- `oracle-AM` (resp. `oracle-GCN`): original AM (resp. GCN) model trained on the *test* instance distribution, that is unseen during training of both the meta and multi models. Note that although the meta-models are not meant to improve over the oracles' performance, we will see that it happens sometimes.

To simplify the notations, we only explicitly differentiate between TSP and CVRP if it is not clear from the context. Since we are interested in the generalization of the neural models, regardless of the final decoding step (greedy, sampling, beam-search, etc), we use a simple greedy decoding for all the models. Besides, because our training is restricted to 24 h for all the models (which is sufficient to ensure convergence of the training, see Fig. 3 of Supp. Mat.), the results may not be as good are those reported in the original papers. To evaluate the impact of the meta-training on generalization when everything else fixed, we focus on the relative gap in performance between the different models.

Table 3. Average optimality gaps over 5,000 instances of the target tasks (e.g. N=100) coming from different prior task distributions (e.g. *var-size* distribution). `oracle-AM/GCN` denote the AM/GCN models trained on the target task. `multi-AM/GCN` and `meta-AM/GCN` are trained on a set of tasks from the prior distribution that does not contain the target tasks. K is the number of fine-tuning steps. In bold: for each model (AM or GCN) and each problem (TSP or CVRP), the best generalization result among the methods that were not trained on the target task.

	Tasks →	*var-size* distrib.		*var-mode* distrib.		*mixed-var* distrib.	
	Models ↓	N=100	N=150	M=3	M=8	(N,M)=(40,6)	(N,M)=(40,8)
TSP	`oracle-AM`	5.96%	12.08%	1.87 %	1.83%	2.00%	1.83%
TSP	Farthest Ins.[23]	7.48%	**8.55%**	2.08%	2.27%	16.32%	11.70%
TSP	`multi-AM` (K=0)	8.73%	14.40%	5.57%	6.20%	10.70%	15.18%
TSP	`multi-AM` (K=50)	7.25%	10.87%	5.26%	4.60%	7.59%	10.26%
TSP	`meta-AM` (K=0)	7.10%	12.25%	1.96%	2.16%	2.41%	3.50%
TSP	`meta-AM` (K=50)	**5.58%**	9.84%	**1.82%**	**1.70%**	**2.15%**	**2.93%**
	Tasks →	*var-size* distrib.		*var-mode* distrib.		*var-capacity* distrib.	
	Models ↓	N=100	N=150	M=3	M=8	C=20	C=50
CVRP	`oracle-AM`	8.71%	11.56%	6.32 %	7.85%	5.83%	8.01%
CVRP	`multi-AM` (K=0)	18.82%	18.76%	7.87%	12.65%	9.15%	14.28%
CVRP	`multi-AM` (K=50)	9.18%	11.41%	7.58%	10.20%	8.09%	10.16%
CVRP	`meta-AM` (K=0)	11.50%	16.42%	6.05%	9.38%	6.26%	8.94%
CVRP	`meta-AM` (K=50)	**7.71%**	**9.91%**	**5.96%**	**8.45%**	**6.05%**	**8.82%**

	Tasks →	*var-size* distrib.		*var-mode* distrib.		*mixed-var* distrib.	
	Models ↓	N=80	N=100	M=3	M=8	(N,M)=(40,6)	(N,M)=(40,8)
TSP	`oracle-GCN`	12.34%	14.72%	7.65%	6.21%	6.06%	3.22%
TSP	`multi-GCN` (K=0)	28.40%	34.29%	9.22%	7.89%	28.01%	5.05%
TSP	`multi-GCN` (K=50)	16.73%	30.80%	8.43%	6.59%	5.99%	4.42%
TSP	`meta-GCN` (K=0)	19.70%	32.01%	8.19%	7.32%	6.62%	3.72%
TSP	`meta-GCN` (K=50)	**13.73%**	**18.42%**	**7.72%**	**6.45%**	**5.67%**	**3.17%**

Generalization performance: To evaluate the generalization ability of the meta-trained models, we present in Table 3 the performance of the different models at 0-shot generalization (K=0) and after K=50 fine-tuning steps, for various pairs of prior task distributions and unseen test tasks. We observe that in all cases the fine-tuned `meta-AM` clearly outperforms the fine-tuned baseline `multi-AM` and even outperforms the `oracle-AM` model in 7 out of 12 tasks.

Similar observations hold for the `meta-GCN` model: it is better both at 0-shot generalization and after fine-tuning than the `multi-GCN` baseline, and it outperforms the oracle in 2 out of 6 tasks. These results show that `meta-AM` is able to achieve impressive quality while using a negligible amount of training data of the target task compared to the original model (`oracle-AM`). More results on different target tasks as well as plots of the evolution of the performance with the number of fine-tuning steps are presented in Sect. A.7 of the Supp. Mat.

Time and sample efficiency. For a complete evaluation of the proposed meta-training and then fine-tuning approach for NCO, we discuss here its cost in terms of the fine-tuning time and number of training samples from the target task required to reach the optimality gaps of Table 3. Regarding the fine-tuning time, the 50 fine-tuning steps took 2 to 6m for `meta-AM` and 43s to 2m for `meta-GCN`. Further, generating the 1k optimal solutions for fine-tuning the supervised `meta-GCN model` took up to 17m for TSP150 and 20h for CVRP150. These values should be compared to the generation time of the 1M solutions for training the `oracle-GCN` model on the target instance distribution. Besides, for example for TSP with $M{=}3$, we observed that `oracle-AM` needs around 23 h and more than 30 Million samples of the target task to reach the optimality gap of 1.82%. On the other hand, `meta-AM-M` only used 3000 samples from the target task and achieved a better performance after a few fine-tuning steps and less than 6 min. The baseline approach `multi-AM-M` was still far away at 5.2% optimality gap after fine-tuning. Similar observations hold for `meta-GCN` on TSP with $M{=}3$: `Oracle-GCN-M` needs around 22 h and 1 Million instances of labeled data (with optimal solutions) to reach an optimality gap of 7.72%, while `meta-GCN-M` reaches the same performance in just 16 s, using 500 solved instances. Hence, one model trained using our prescribed meta-learning approach can be used to adapt to different tasks efficiently within a short span of time and using few fine-tuning samples. More details on training time and number of samples used for different tasks can be found in the Table 7 of the Supp. Mat. Additionally, Fig. 3 in the Supp. Mat. presents the performance of different models w.r.t time on test tasks during their course of training.

5.1 Experiments on Real-World Datasets

To evaluate the performance of our approach beyond synthetic datasets, we ran experiments on two well-established OR datasets: TSPlib[2] and CVRPlib[3]. From TSPlib we took the 28 instances of size 50 to 200 nodes. Note that in this context, the RL approach which does not rely on labeled data for fine-tuning is more appropriate. Since these instances are heterogeneous (i.e. no clear underlying distribution), we directly fine-tune the models on each test instance.

This is an extreme case of our setting where the target task is reduced to 1 instance. We tested the models that were (meta-)trained on the *variable-size* distribution of synthetic instances for `meta-AM` and `multi-AM`. For AM we took the pretrained model on graphs of size $N{=}100$. Because of space limitation, we grouped the instances per size range and report in Table 4 the average optimality gap obtained after $K{=}100$ fine-tuning steps, taking 20s to 1m (detailed per-instance results in Sect. A.8 of Supp Mat). Note that in this case we also fine-tune the AM model since it was not trained on the target instances distribution.

From CVRPlib we used the 106 instances of size up to 200 nodes. Since instances are grouped by sets, we apply our few-shot learning setting: fine-tuning

[2] http://comopt.ifi.uni-heidelberg.de/software/TSPLIB95/.
[3] http://vrp.atd-lab.inf.puc-rio.br/index.php/en/.

for 50 steps on approximately 10% of the instances of a set and testing on the rest. In Table 4, we report the average optimality gap over 5 random fine-tuning/test splits for each set. The results are consistent with our previous observations and illustrate the superior performance of our proposed meta-learning strategy in this realistic setting. It also shows that even if the prior task distribution is not perfect (in the sense that it does not include the target task), the meta-training gives a strong parameter initialization which one can fine-tune effectively on the target task.

Table 4. Average optimality gaps on realistic instances

Dataset→ Model↓	TSPlib			CVRPlib				
	50–100	101–150	151–200	Set A	Set B	Set E	Set P	Set X
AM	8.52%	7.97%	17.35%	4.54%	5.69%	31.17%	5.45%	12.39%
multi-AM	11.95%	13.32%	26.04%	5.03%	5.73%	**13.00%**	6.13%	15.72%
meta-AM	**5.95%**	**5.91%**	**13.22%**	**3.56%**	**5.07%**	14.07%	**5.03%**	**11.87%**

5.2 Ablation Study

Fixed vs decaying step-size ε. In this section, we study the impact of our proposed decaying ε approach during meta-training. Specifically, Table 5 presents the results of using a standard fixed step-size ε versus a decaying ε. We see that the decaying ε version of `meta-AM` and `meta-GCN` outperforms the fixed ε one, both in terms of 0-shot generalization (i.e. $K = 0$) and after K=50 steps of fine-tuning. This supports our argument for performing task specialization in the beginning and generalization at the end of the meta-training procedure.

Table 5. **(Fixed vs decaying step-size ε)** Average optimality gap, on 5000 TSP instances sampled from a set of test tasks, using the meta-trained models `meta-AM` (resp. `meta-GCN`) when trained with a fixed step-size $\varepsilon = \varepsilon_0$ or a "decaying ε" where ε is close to 1 initially and tends to 0 at the end of the training.

	Test task	Fine-tuning	ε=0.1	ε=0.3	ε=0.5	ε=0.7	ε=0.9	*decaying* ε
`meta-AM`	N=100	before (K=0)	9.91%	8.33%	7.52%	6.94%	6.63%	**7.10%**
		after (K=50)	7.83%	6.50%	6.03%	5.95%	5.96%	**5.58%**
	M=8	before (K=0)	5.99%	3.07%	3.38%	2.35%	2.52%	**2.16%**
		after (K=50)	4.78%	2.27%	2.63%	1.87%	2.04%	**1.70%**
`meta-GCN`	M=6	before (K=0)	13.08%	11.90%	11.92%	12.90%	10.11%	**6.01%**
		after (K=50)	9.86%	8.27%	9.52%	10.80%	13.16%	**5.71%**
	M=8	before (K=0)	9.78%	8.81%	9.20%	11.05%	11.96%	**7.39%**
		after (K=50)	8.32%	7.37%	8.23%	9.76%	11.80%	**6.45%**

6 Conclusion

In this paper, we address the well-recognized generalization issue of end-to-end NCO methods. In contrast to previous works that aim at having one model perform well on various instance distributions, we propose to learn a model that can efficiently *adapt* to different distributions of instances. To implement this idea, we recast the problem in a meta-learning framework, and introduce a simple yet generic way to meta-train NCO models. We have shown experimentally that our proposed meta-learned RL-based and SL-based NCO heuristics are indeed robust to a variety of distribution shifts for two CO problems. Additionally, the meta-learned models also achieve superior performance on realistic datasets. We show that our approach can push the boundary of the underlying NCO models by solving instances with up to 200 nodes when the models are trained with only up to 50 nodes. While the known limitations of the underlying models (esp. the attention bottleneck, and fully-connected GCN) prevent tackling much larger problems, our approach could be applied to other models. Finally note that there are several possible levels of generalization in NCO. In this paper, we have mostly focused on improving the generalization to instance distributions for a fixed CO problem. To go further, one could investigate the generalization to other CO problems. For this more ambitious goal, domain adaptation approaches, which explicitly account for the domain shifts (e.g. using adversarial-based techniques [25]) could be an interesting direction to explore.

Acknowledgments. We wish to thank Pankaj Pansari and Anilkumar Swamy for preliminary experiments on the generalization of existing models. We are grateful to Julien Perez for helpful discussions and advice throughout the project. We also thank the anonymous reviewers for comments that helped improve the paper.

References

1. Applegate, D.L., Bixby, R.E., Chvátal, V., Cook, W.J.: The traveling salesman problem: a computational study. Princeton University Press (Sep 2011)
2. Bello, I., Pham, H., Le, Q.V., Norouzi, M., Bengio, S.: Neural combinatorial optimization with reinforcement learning. arXiv:1611.09940 (Jan 2017)
3. Bengio, Y., Lodi, A., Prouvost, A.: Machine learning for combinatorial optimization: A methodological tour d'horizon. Eur. J. Oper. Res. **290**(2), 405–421 (2021)
4. Cappart, Q., Chételat, D., Khalil, E., Lodi, A., Morris, C., Veličković, P.: Combinatorial optimization and reasoning with graph neural networks. arXiv:2102.09544 (Feb 2021)
5. Cook, W.J., Cunningham, W.H., Pulleyblank, W.R., Schrijver, A.: Combinatorial Optimization. Wiley-Interscience, New York, 1st edition edn. (Nov 1997)
6. Finn, C., Abbeel, P., Levine, S.: Model-agnostic meta-learning for fast adaptation of deep networks. In: Proceedings of the 34th International Conference on Machine Learning - Vol. 70. pp. 1126–1135. ICML 2017, JMLR.org, Sydney, NSW, Australia (Aug 2017)
7. Fischetti, M., Lodi, A.: Heuristics in mixed integer programming. In: Wiley Encyclopedia of Operations Research and Management Science. American Cancer Society (2011)

8. Fu, Z.H., Qiu, K.B., Zha, H.: Generalize a small pre-trained model to arbitrarily large TSP instances. arXiv:2012.10658 (Dec 2020)
9. Georgiev, D., Liò, P.: Neural bipartite matching. arXiv:2005.11304 (Jun 2020)
10. Helsgaun, K.: An extension of the Lin-Kernighan-Helsgaun TSP solver for constrained traveling salesman and vehicle routing problems, p. 60 (2017)
11. Joshi, C.K., Cappart, Q., Rousseau, L.M., Laurent, T., Bresson, X.: Learning TSP requires rethinking generalization. arXiv:2006.07054 (Jun 2020)
12. Joshi, C.K., Laurent, T., Bresson, X.: An efficient graph convolutional network technique for the travelling salesman problem. arXiv:1906.01227 (Jun 2019)
13. Joshi, C.K., Laurent, T., Bresson, X.: on learning paradigms for the travelling salesman problem. arXiv:1910.07210 (Oct 2019)
14. Kool, W., van Hoof, H., Welling, M.: attention, learn to solve routing problems! In: International Conference on Learning Representations (2019)
15. Kool, W., van Hoof, H., Gromicho, J., Welling, M.: Deep policy dynamic programming for vehicle routing problems. arXiv:2102.11756 (Feb 2021)
16. Li, S., Yan, Z., Wu, C.: Learning to delegate for large-scale vehicle routing. In: Advances in Neural Information Processing Systems 34 Pre-Proceedings (2021)
17. Li, Z., Chen, Q., Koltun, V.: Combinatorial optimization with graph convolutional networks and guided tree search. In: Bengio, S., Wallach, H., Larochelle, H., Grauman, K., Cesa-Bianchi, N., Garnett, R. (eds.) Advances in Neural Information Processing Systems 31, pp. 537–546. Curran Associates, Inc. (2018)
18. Lisicki, M., Afkanpour, A., Taylor, G.W.: Evaluating curriculum learning strategies in neural combinatorial optimization. arXiv:2011.06188 (Nov 2020)
19. Nair, V., Dvijotham, D., Dunning, I., Vinyals, O.: Learning fast optimizers for contextual stochastic integer programs. In: UAI 2018 (2018)
20. Nazari, M., Oroojlooy, A., Snyder, L., Takac, M.: Reinforcement learning for solving the vehicle routing problem. In: Advances in Neural Information Processing Systems, vol. 31. Curran Associates, Inc. (2018)
21. Nichol, A., Achiam, J., Schulman, J.: On first-order meta-learning algorithms. arXiv:1803.02999 (Oct 2018)
22. Ravi, S., Larochelle, H.: optimization as a model for few-shot learning, p. 11 (2017)
23. Rosenkrantz, D.J., Stearns, R.E., Lewis, P.M.: An analysis of several heuristics for the traveling salesman problem. In: Ravi, S.S., Shukla, S.K. (eds.) Fundamental Problems in Computing: Essays in Honor of Professor Daniel J. Rosenkrantz, pp. 45–69. Springer, Netherlands, Dordrecht (2009). https://doi.org/10.1007/978-1-4020-9688-4_3
24. Schmidhuber, J.: Evolutionary principles in self-referential learning, or on learning how to learn: The meta-meta-... hook (1987)
25. Tzeng, E., Hoffman, J., Saenko, K., Darrell, T.: Adversarial discriminative domain adaptation. In: Proceedings of the IEEE Conference on Computer Vision and Pattern Recognition, pp. 7167–7176 (2017)
26. Vaswani, A., et al.: Attention is all you need. arXiv:1706.03762 (Jun 2017)
27. Veličković, P., Ying, R., Padovano, M., Hadsell, R., Blundell, C.: Neural execution of graph algorithms. In: International Conference on Learning Representations (Sep 2019)
28. Vinyals, O., Fortunato, M., Jaitly, N.: Pointer networks. In: Cortes, C., Lawrence, N.D., Lee, D.D., Sugiyama, M., Garnett, R. (eds.) Advances in Neural Information Processing Systems 28, pp. 2692–2700. Curran Associates, Inc. (2015)
29. Williams, R.J.: Simple statistical gradient-following algorithms for connectionist reinforcement learning. Mach. Learn. **8**(3), 229–256 (1992)

30. Wolpert, D., Macready, W.: No free lunch theorems for optimization. IEEE Trans. Evol. Comput. **1**(1), 67–82 (1997)
31. Zarpellon, G., Jo, J., Lodi, A., Bengio, Y.: Parameterizing branch-and-bound search trees to learn branching policies. arXiv:2002.05120 (Jun 2021)

Time Constrained DL8.5 Using Limited Discrepancy Search

Harold Kiossou[1(✉)], Pierre Schaus[1(✉)], Siegfried Nijssen[1], and Vinasetan Ratheil Houndji[2]

[1] ICTEAM, Université Catholique de Louvain, Louvain-la-Neuve, Belgium
{harold.kiossou,pierre.schaus,siegfried.nijssen}@uclouvain.be
[2] LRSIA, Institut de Formation et de Recherche en Informatique, Abomey-Calavi, Benin
vratheilhoundji@gmail.com

Abstract. Decision trees that minimize the error on the training set with a depth limit have been found to be generally superior to those found by more standard greedy algorithms. However, when the search space to be explored is too large, the depth-first search used by exact algorithms can get trapped in left most branches. Consequently, when the user stops the algorithm, the best tree found so far may be unbalanced and poorly minimize the error. Our work aims to improve the anytime behavior by introducing the limited discrepancy search ingredient in these algorithms. This allows to explore the search space by waves increasingly deviating from standard heuristics such as information gain. Our experimental results show that the anytime behavior of the state-of-the-art exact method DL8.5 is greatly improved.

Keywords: Optimal decision trees · Limited discrepancy search · Knowledge discovery

1 Introduction

Decision trees are among the most popular models in machine learning. In particular, their simplest form, the boolean decision trees are considered in this paper since every dataset can be binarized. Each node represents an attribute or feature of the dataset, and each branch represents the selection made for the boolean attribute. The classification of a new instance is obtained by following the path from the root to the leaf node that gives the predicted class to the instance. Decision trees have become increasingly popular since their introduction [21]. Their simplicity, interpretability, and the number of algorithms to induce decision trees make them a preferred method for many applications. Learning a decision tree that minimizes the error on a training set is NP-hard. This is why, since their introduction, greedy algorithms have been used mainly to induce decision trees from a training set [7,20]. Despite the lack of optimality guarantees, these algorithms, choosing top-down recursively the feature to split based on a heuristic such as information gain [15], offer a good trade-off between accuracy and scalability.

© The Author(s), under exclusive license to Springer Nature Switzerland AG 2023
M.-R. Amini et al. (Eds.): ECML PKDD 2022, LNAI 13717, pp. 443–459, 2023.
https://doi.org/10.1007/978-3-031-26419-1_27

Recent advances in hardware and mathematical optimization libraries have made it possible to reconsider exact approaches to induce decision trees [5,19]. Beyond the theoretical and algorithmic aspects, this field is getting more interest, mainly motivated by the fact that minimizing the error of the tree on the training set also allows to reduce the error on unseen data [5]. Several approaches based on Mathematical Programming [2,5,8,19], Constraint Programming [18], and SAT solvers [16] have been proposed. Solver-based approaches are flexible and require less expertise to develop, but dedicated algorithms such as DL8.5 [3] and Murtree [9] based on branch-and-bound and dynamic programming have achieved the best results so far. They used a depth-first search to explore the search space of decision trees branching on the feature decision variables at each node. The performances for finding a provable optimal tree are generally good when the depth limit of the decision tree to discover is not too high (typically 3 or 4). However, for larger depths, when the training set is large and has many features, there is little hope to find and prove the optimal tree. In such cases, the search can get trapped in the left parts of the search tree exploration without having enough time to reconsider decisions close to the root. Stopping the search before its completion can therefore result in unbalanced decision trees (leaning to the left), with an error that is even larger than the ones the user would obtain with a greedy algorithm.

This work aims to improve the anytime aspect of the exact algorithm to induce decision trees by incorporating a limited discrepancy search [11], a well-known technique in combinatorial optimization, to improve depth-first search when an efficient heuristic is available. The article focuses on adapting the strategy to DL8.5 [3] but this idea can be applied to similar algorithms such as MurTree [9] or the AND-OR search Constraint Programming approach [18].

The adapted algorithm is called LDS-DL8.5. In this setting, the depth-first search also takes the decisions guided by a standard heuristic (information gain [15]) but does not allow to deviate too much from it according to a budget (called the discrepancy limit) per branch. With a discrepancy limit of zero, the algorithm discovers the same decision tree as C4.5. Then by gradually increasing the budget and restarting the search, the approach is able to deviate from the greedy tree and even possibly discover an optimal tree and prove its optimality when no pruning occurred because of the discrepancy limit. We show experimentally that the advantage of this approach is that the user can set an optimization time budget and still obtain a tree that is generally of better quality than the one obtained with a pure greedy algorithm such as C4.5 and CART. The trees discovered with LDS-DL8.5 are also in general better than those with DL8.5 for hard settings where it is not possible to find and prove optimality in a reasonable amount of time. Our implementation is publicly available at https://github.com/haroldks/lds-dl85.

This paper is organized as follows. In the next section, we present the related works, mainly works on optimal decision trees. We then explain the technical background by briefly discussing some notions of frequent itemset mining and the functioning of DL8.5 and the Limited Discrepancy search. Finally, we provide some experimental results that show the efficiency and interest of LDS-DL8.5 w.r.t. DL8.5 and the state-of-the-art greedy algorithms CART and C4.5.

2 Related Works

Decision trees are built in most cases using heuristic algorithms such as CART [7] and C4.5 [20]. While highly scalable, the constructed tree may not be the most accurate, in particular in the presence of constraints, such as on the depth of the trees. Optimal decision tree algorithms aim to address this issue by exhaustively exploring the search space at runtime cost. They have seen a resurgence in prominence in recent years as algorithms and technology have improved. Most popular methods use a mixed integer programming-based approach. Bertsimas and Dunn [5] in their work encoded the problem of finding optimal decision trees with respect to misclassification error by fixing a maximum depth in advance and creating a number of variables to represent the predicates for each node. Verwer and Zhang [19] later proposed BinOCT, which reduces the number of variables and constraints present in the model by taking advantage of the binarization of the data. Aghaei et al. [2] suggested a MIP-based approach based on maximum flow formulation and Benders decomposition to tighten the relaxation of binary decision trees. Boutilier et al. [8] introduced valid inequalities for learning optimal multivariate decision trees. Other approaches, such as the work of Verhaeghe et al. [18], induced optimal decision trees with constraint programming principles. They developed models with maximum depth and minimum support constraints while using a branch-and-bound strategy to prune the search space.

Another class of methods used SAT solvers to induce optimal decision trees. Narodytska et al., [16] modeled the decision tree as a propositional logic to construct the smallest tree in terms of the total number of nodes that perfectly describes the given dataset. At first, a tree is learned by using some heuristic method. The SAT-solver is then called multiple times to find each time a perfect tree with one less node.

Researchers also develop specialized algorithms for decision trees. In their work, Nijssen and Fromont [17] developed DL8, an algorithm inspired by ideas from the pattern mining literature that can support a wide range of constraints. Their approach allowed to evaluate the different branches of a node individually while saving the obtained subtrees using a new caching technique to reuse them later. In a later work, Aglin et al. [3] developed DL8.5, an improved version of DL8. The main contributions are the introduction of an upper bound that limits the tree error allowed for a child node as soon as an optimal subtree has been determined for one of its siblings and a lower-bound technique that allows the algorithm to store information on both optimal and pruned subtrees to provide a lower bound on the tree error. These improvements lead to a method that outperformed previous approaches by several margins when used with a depth constraint. Demirovic et al. [9] advanced the DL8.5 algorithm by adding support to limit the number of nodes in the tree, an efficient procedure to compute the tree of depth two, and a novel similarity-based lower bounding approach.

3 Technical Background

DL8.5 induces boolean decision trees by relying on itemset mining concepts. It starts with the transactional dataset, where each transaction is an itemset indicating the existence or absence of each feature. Formally, it is defined as a collection $\mathcal{D} = \{(t, I, c) \mid t \in \mathcal{T}, I \subseteq \mathcal{I}, c \in \mathcal{C}\}$, where $\mathcal{T}$ represents the transaction sets or row identifiers, $\mathcal{I}$ is the set of possible items, and $\mathcal{C}$ is the set of class labels; within $\mathcal{I}$ there are two items (one positive, the other negative) for each original Boolean feature, and each itemset I contains either a positive or a negative item for every feature. As an illustration, Table 1 shows the transactional database representation of the binary matrix of Table 1. The tids are the identifiers of the transactions, which can also be row numbers. For each itemset I:

- the cover of an itemset is the set of transactions that contain this itemset: $cover(I) = \{(t, X, c) \mid (t, X, c) \in \mathcal{D}\, and\, I \subseteq X\}$
- the class-based support of an itemset is the number of examples in its cover with the given class c: $Sup(I, c) = |\{(t, X, c') \in cover(I)\, and\, c = c'\}|$.

Table 1. Example of database formats.

(a) Binary matric

Features			class
A	B	C	
1	1	1	1
1	1	0	0
1	0	1	1
1	0	0	1

(b) Transactional database

tid	items	class
1	abc	1
2	$ab\neg c$	0
3	$a\neg bc$	1
4	$a\neg b\neg c$	1

DL8.5 Algorithm

For the sake of completeness, we first explain the DL8.5 algorithm and then introduce the changes related to the limited discrepancy search. Algorithm 1 shows the pseudocode of DL8.5. This algorithm, in the general case, performs a recursive depth-first branch and bound search at each node (itemset) to select the feature at that node that will extend the itemset. The left and right subtrees are collected, each obtained with a recursive call with the exclusion and inclusion of the considered item. The base cases ending the recursion occur:

- when the maximum depth constraint is reached, $\mid I \mid = maxdepth$ in line 7.
- when the node (itemset) error is already 0 ($leaf\_error(I) = 0$) in line 7. The leaf error here is the misclassification rate, defined as : $leaf\_error(I) = \mid cover(I) \mid - \max_{c \in \mathcal{C}} \{Sup(I, c)\}$.
- when the itemset support is below a user-defined threshold on line 16.

Algorithm 1: $DL8.5(maxdepth, minsup)$

```
1  struct BestTree {ub : float; tree : Tree; error : float}
2  cache ← Trie < Itemset, BestTree >
3  bestSolution ← DL8.5 − Recurse(∅, +∞)
4  return bestSolution.tree
5  Procedure DL8.5 − Recurse(I, ub)
6      leaf_error ← leaf_error(I)
7      if leaf_error = 0 or |I| = maxdepth or timeout is reached then
8          if leaf_error ≤ ub then
9              return BestTree(ub, make_leaf(I), leaf_error)
10         return BestTree(ub, NO_TREE, leaf_error)
11     solution ← cache.get(I)
12     if solution was found and ((solution.tree ≠ NO_TREE) or (ub ≤
         solution.ub)) then
13         return solution
14     (τ, b, base_ub) ← (NO_TREE, +∞, ub)
15     for all attributes i sorted by heuristic do
16         if |cover(I ∪ {i})| ≥ minsup and |cover(I ∪ {¬i})| ≥ minsup then
17             sol1 ← DL8.5 − Recurse(I ∪ {¬i}, base_ub)
18             if sol1.tree = NO_TREE then continue
19             sol2 ← DL8.5 − Recurse(I ∪ {i}, base_ub − sol1.error)
20             if sol2.tree = NO_TREE then continue
21             feature_error ← sol1.error + sol2.error
22             τ ← make_tree(i, sol1.tree, sol2.tree)
23             b ← feature_error
24             base_ub ← b − 1
25             if feature_error = 0 then break
26     solution ← BestTree(ub, τ, b)
27     cache.store(I, solution)
28     return solution
```

In addition, DL8.5 uses an upper bound specified as a parameter of the recursion procedure. This bound at the root node is initially set to $+\infty$, but is tightened each time the algorithm finds a better tree. The update is made at line 21 and the algorithm will continue the search with this new bound (lines 17 and 19). The upper bound ensures the pruning of the search space using the test in line 18. Here, exploring the second branch for the current attribute is useless if the quality of the first branch tree is worse than the authorized upper bound. The error of the left subtree is also used to tighten the maximum error allowed for the right subtree in line 19.

As several recursion paths can lead to the same itemset and thus the same cover, DL8.5 avoids useless recomputation by storing the itemset, its associated upper bound and tree, even if no solution was found. In doing so, DL8.5 will not continue the search for a stored itemset and bound if it encounters it again later, considering that for the given bound, a good enough tree could not be found.

Weakness of DL8.5

DL8.5 may require high execution times for large datasets with high depth. The user can specify a timeout to limit the computation time and still get the best tree found during this computation time. Unfortunately, this decision tree can be very unbalanced when the search gets stuck in the deepest branches, as illustrated in Fig. 1. In this example, the time limit occurs when exploring the subtree below $\neg a$. As a result, all the examples that fell in nodes a had no chance of further splitting, and the error in this node can be quite large. Although DL8.5 is a very powerful algorithm when the search can be terminated with a chance to optimally split all nodes at best, this is not the case when the search is interrupted before the end of the search. For each successor of a node, the search procedure is called on the left and right branches. The depth-first search can get stuck for a dataset with many features and high depth in the recursive calls of the line 17. When a timeout occurs at line 7, the search will return all remaining successors without any further exploration, even when line 19 is called for the right branch. If an efficient heuristic is available, an extension of the depth-first search with limited discrepancy allows it to avoid getting stuck in the depths by allowing a certain number of deviations to each node according to a budget.

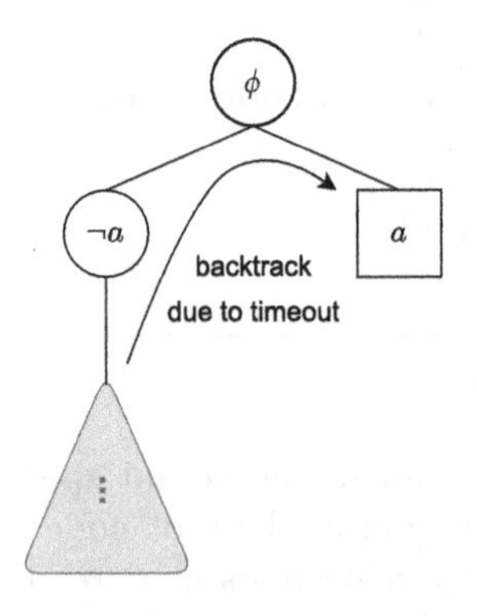

Fig. 1. Stuck DL8.5.

Limited Discrepancy Search

Many problem-solving approaches in AI use tree-depth-first search methods. It is common to employ a heuristic to guide the search towards the more promising search space regions first. For some problems, a good heuristic may directly lead to the optimal solution in the leftmost leaf node, but, in general, it has no guarantee of making no mistakes. This means that the search should have taken a few other rare decisions instead of always trusting the heuristic to discover the best solution.

By enumerating solutions in increasing order of the number of decisions that do not agree with the heuristic, the discrepancy search hopes to discover the best solution quickly. This strategy can be enforced using a depth-first search with a discrepancy budget along each branch, forcing the search to backtrack when the limit is reached. The completeness of the approach is ensured by gradually increasing the discrepancy limit along the iterations. The first iteration does not allow any discrepancy and thus is only able to discover the leftmost leaf node without any backtrack. The next iteration allows one discrepancy and will enumerate all the possibilities with one allowed discrepancy, and so on for the subsequent larger discrepancy. Note that we can augment the limit by more than one increment between consecutive iterations to speed up the process, since each iteration may visit in theory a super-set[1] of the nodes at the previous iteration.

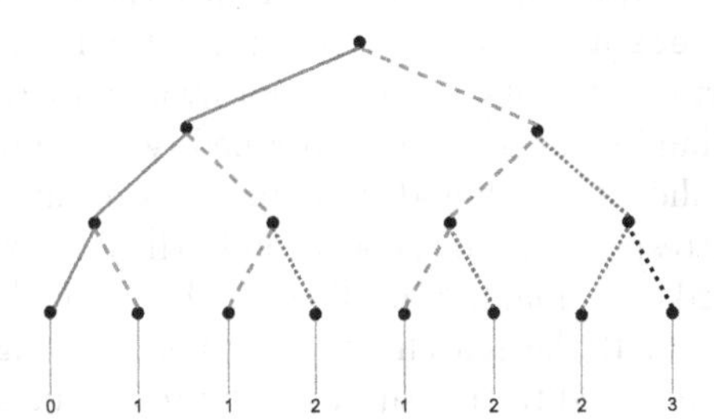

Fig. 2. Limited discrepancy search on a binary tree.

Figure 2 shows the result obtained with LDS searching for a binary tree of depth 3. The numbers at each leaf level give the total number of discrepancies needed to reach them. At the beginning of the search, the leftmost branch (blue) is traversed with a discrepancy of 0. The nodes of this branch correspond to the best results according to the heuristic used. The branches of the discrepancy 1 (green) are traversed when there is no solution at the discrepancy 0. At the root, it is possible to explore the right node with the discrepancy of 1 and then traverse the left branch for this node without a discrepancy budget as it corresponds to the best heuristic value.

Our main contribution is to include the LDS idea into the DL8.5 algorithm in the next section.

[1] The branch and bound may also prune the search space.

4 LDS-DL8.5

We propose LDS-DL8.5, a decision tree algorithm that improves the anytime behavior of DL8.5 by using limited discrepancy search. When there is not a time limit and the search is completed LDS-DL8.5 returns the optimal solution as the optimal approach like DL8.5.

Algorithm 1 describes LDS-DL8.5. The main loop of the algorithm corresponds to the lines 5 to 9. There, the `Search` procedure is called with a discrepancy budget k that increases from 0 to a maximum value K determined by:

$$K = \sum_{i=0}^{d} \mathbf{A} - i - 1, \tag{1}$$

where $\mathbf{A}$ is the number of attributes in the dataset, and d is the maximum depth of the search. The algorithm can iteratively or exponentially increase the discrepancy with the function `augment_discrepancy`. By doing so, we are able to limit the number of times the algorithm can restart, which can improve the runtime. The search loop is stopped when the allowed execution time is reached or when a null-error optimal solution is found.

Increasing the discrepancies means increasing the number of successors that the algorithm is allowed to visit at each node. For the classical LDS, *zero discrepancies* consists in exploring at each level only the first attribute of the list returned by `get_successors`. The function returns the node successors based on the minimum support constraints, sorted or not by a heuristic. If the time budget allows it and an optimal solution is not obtained, we progressively increase the discrepancy budget, allowing each node to explore more successors, each having a discrepancy value corresponding to its position in the list. When the discrepancy of an attribute exceeds the maximum allowed, the algorithm stops the search, as indicated on line 26. If the search can continue, the algorithm reduces the discrepancy budget allocated to this successor by removing its position from the current maximum. Moreover, contrary to the classical LDS, each branch of an attribute (for example, x and $\neg x$) has the same limit of discrepancy (lines 27 and 29) because an attribute is selected only if these two branches respect the imposed constraints.

The cost of a given iteration of LDS-DL8.5 is higher than the cost of the previous one due to the recomputing and re-exploration of previously visited nodes. To mitigate this cost, we use the cache by adding a parameter named *discrepancy* to the structure *NodeTree*. It corresponds to the discrepancy budget given to the node. This budget is re-evaluated on line 21. The actual number of successors along with the remaining depths ($maxdepth - |I|$) are used in the function `discrepancy_limit` to determine the current node discrepancy budget value using Eq. 1. If the computed budget is larger than the passed budget, there is a high chance that this node will not be fully explored. The opposite means that the node will be fully explored. The real budget of the node is set to the minimum between the computed value and the allowed discrepancy budget for this node (line 22). When a node error is 0, or the computed budget is the same as

Algorithm 2: LDS-DL8.5(*maxdepth*, *minsup*, *K*)

```
1  struct NodeTree {ub : float; tree : Tree; error : float, discrepancy : int}
2  cache ← Trie < Itemset, NodeTree >
3  result ← NodeTree{+∞, NO_TREE, 0, 0}
4  k ← 0
5  while k ≤ K do
6      result ← Search(root, result.error, k)
7      if result.error = 0 or timeout is reached then
8          return result.tree
9      k ← augment_discrepancy(k, K)
10 return result.tree
11 Procedure Search(I, ub, k)
12     leaf_error ← leaf_error(I)
13     if leaf_error = 0 or |I| = maxdepth or timeout is reached then
14         if leaf_error ≤ ub then return NodeTree(ub, make_leaf(I), leaf_error, k)
15         return NodeTree(ub, NO_TREE, leaf_error, k)
16     node ← cache.get(I)
17     if node was found then
           /* The node is full explored                                    */
18         if node.tree ≠ NO_TREE and node.discrepancy = K and ub ≤ node.ub then
19             return node
       /* List of the current node successors sorted by a heuristic        */
20     successors ← get_successors(I, minsup)
       /* Node real discrepancy budget                                     */
21     d ← discrepancy_limit(size(successors), maxdepth − |I|)
22     k ← min(d, k)
23     (τ, b, base_ub) ← (NO_TREE, +∞, ub)
24     for i in successors do
25         c ← successors.index(i)
26         if c > k then break // Discrepancy budget reached
27         first ← Search(I ∪ {¬i}, base_ub, k − c)
28         if first.tree = NO_TREE then continue
29         second ← Search(I ∪ {i}, base_ub − first.error, k − c)
30         if second.tree = NO_TREE then continue
31         feature_error ← first.error + second.error
32         τ ← make_tree(i, first.tree, second.tree)
33         b ← feature_error
34         base_ub ← feature_error − 1
35         if feature_error = 0 then break
       // Current discrepancy budget allows to reach the last successor or node is pure
36     if b = 0 or k = d then k ← K
37     result ← NodeTree(ub, τ, b, k)
38     cache.store(I, result)
39     return result
```

the budget limit for that node not further exploration is needed its discrepancy is set to the maximum possible K in the *NodeTree* structure in line 36. The saved value avoids exploring the same nodes in the future iterations of the search if the upper-bound is worse than the stored value (line 18). Furthermore, a cached solution cannot be used unless it has not been proven to be optimal without a discrepancy limit. Thus, as long as the discrepancy is not set to the maximum value the node is explored.

With this scheme, the cost of each search iteration is reduced, mainly the latter one, because it is more likely that the left part of the search space will be fully explored from some value of the discrepancy.

5 Results

This section presents the results of various experiments that we conducted. Experiments were carried out to answer the following questions:

- **Q1** How does the performance of LDS-DL8.5 compare with DL8.5 and greedy algorithms in a time-limited configuration?
- **Q2** What happens when the DL8.5 recursion budget is limited to match a number of LDS-DL8.5 discrepancies?
- **Q3** How does LDS-DL8.5 perform compared to DL8.5 in the search for the optimal solution?

All experiments were carried out on 23 CP4IM datasets[2]. For the comparison of DL8.5 and LDS-DL8.5, the information gain was used as a heuristic. The algorithms were run on a server with an Intel Xeon Platinum 8160 CPU, 320 GB of memory, running Rocky Linux version 8.4.

To answer **Q1**, all algorithms were run with a minimum support of 1 and a maximum depth of 9. Each method runs with different time intervals of 30, 60, and 90 seconds on the whole dataset to compare the optimal methods with the greedy ones. Table 2 compares the error obtained by each algorithm according to the allowed time limit (TL). The datasets are sorted by their number of attributes (Feat.) and we show for each of them the number of transactions (Trans.) and the errors for each method. For LDS-DL8.5, two discrepancy augmentation schemes were used:

- `inc` where the discrepancy increases iteratively by one at each restart;
- `exp` where the discrepancy doubles at each restart.

These tests show the efficiency of LDS-DL8.5 in a time-sensitive configuration. Regardless of the discrepancy augmentation scheme used, LDS-DL8.5 always has the lowest errors on the 23 instances. The greedy algorithms CART and C4.5 are fast enough to end in a few seconds. The errors will remain the same regardless of the time allocated to these algorithms. CART has the lowest performance among all algorithms, with a higher error in all instances except 2 where it was

[2] https://dtai.cs.kuleuven.be/CP4IM/datasets/.

able to find the optimal solution (with an error of 0) together with the other methods, thanks to the heuristics used. C4.5 performs better than CART, as it was able to find the optimal solution for 8 instances. Next, DL8.5 and LDS-DL8.5 find the optimal solution on 12 instances, but DL8.5 has higher errors on the remaining 11 instances. Moreover, C4.5 performs better than DL8.5 on 10 instances. This confirms that DL8.5 might get stuck in the deeper branches of the left part of the search space, as it has to go through all the successors to select the best. This search costs time and, when the time limit is reached, DL8.5 will return the current node as the leave. On the contrary, LDS-DL8.5 ensures a minimal quality of the trees obtained. When run with a discrepancy limit of 0, the algorithm discovers the same tree as C4.5, allowing an immediate restriction of the upper bound for the next discrepancies and thus a better pruning of the search space. LDS-DL8.5(`inc`) generally has better results than LDS-DL8.5(`exp`) in this configuration. The way the discrepancy increases with `exp` allows the search to explore more nodes, increasing the risk of the search being stuck.

Regarding the question **Q2**, we have run LDS-DL8.5 with a limited number of discrepancies from 1 to 4. For each limit, the number of recursive calls is evaluated and defined as an additional constraint for DL8.5. The tests were carried out with the support of 1 and a maximum depth of 3. Table 3 summarizes the results of the experiment for the ten largest datasets in terms of features. The RB column corresponds to the recursion budget obtained by LDS-DL8.5 with the discrepancy limits mentioned in the *Disc.* column. Within the recursion budget, DL8.5 has greater difficulty in reducing the error, which remains higher than the one obtained by LDS-DL8.5. LDS-DL8.5 updates the error faster than DL8.5 and is more reliable for critical problems with time limits. Moreover, the trees obtained by LDS-DL8.5 are more balanced than those of DL8.5, as illustrated in Fig. 4. This figure compares the trees obtained by LDS-DL8.5 for the discrepancy limits from 1 to 4 with those of DL8.5 with an equivalent recursion budget on the `mushroom` dataset. The trees obtained by DL8.5 are not balanced, unlike those of LDS-DL8.5. Furthermore, the trees do not change from discrepancy 1 to 3 with DL8.5 using the recursion budget, whereas LDS-DL8.5 will quickly improve the results. This is in line with the assumption that LDS-DL8.5 updates the upper bound and tree error more quickly.

To answer **Q3**, different algorithms were run to discover an optimal solution with a time limit of 10 min. Experiments were carried out with support of 1 and maximum depths of 3 and 4. For DL8.5, two versions were used: one with the information gain as heuristic and the second without any heuristic. The `inc` and `exp` versions of LDS-LD8.5 were also used. Figure 3 presents the performance profile [10] plots on the 23 instances with a maximum depth of 3 and 4 respectively. A performance profile is a cumulative distribution of the improved performance of an algorithm $s \in S$ compared to other algorithms of S over a set P of problems: $p_s(\tau) = \frac{1}{|P|} \times |\{p \in P : r_{p,s} \leq \tau\}|$ where the performance ratio is defined as $r_{p,s} = \frac{t_{p,s}}{\mathtt{min}\{t_{p,s} | s \in S\}}$ with $t_{p,s}$ the execution time of each algorithm.

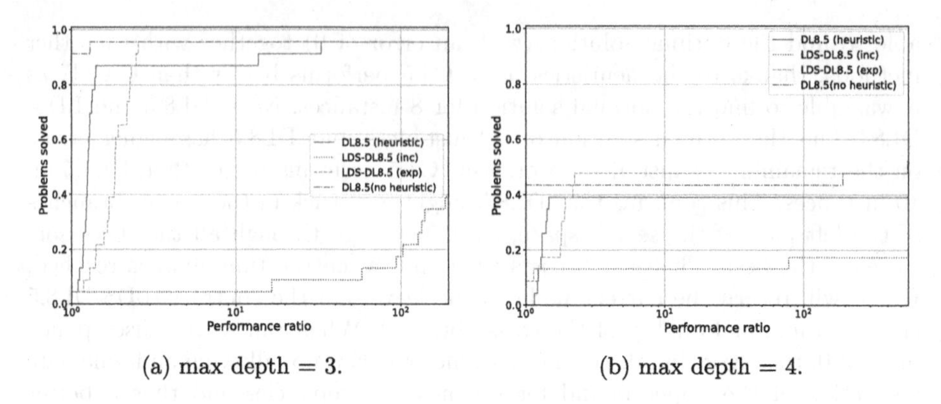

(a) max depth = 3. (b) max depth = 4.

Fig. 3. Runtime performance profile plots.

Table 2. Comparison of tree errors in the time-limited configuration for CART, C4.5, DL8.5 & LDS-DL8.5.

Datasets	Feat.	Trans.	TL(s)	Errors				
				CART	C4.5	DL8.5	LDS-DL8.5	
							inc	exp
ionosphere	445	351	30	26	**0**	**0**	**0**	**0**
			60	26	**0**	**0**	**0**	**0**
			90	26	**0**	**0**	**0**	**0**
splice-1	287	3190	30	258	21	68	**1**	**1**
			60	258	21	68	**1**	**1**
			90	258	21	68	**1**	**1**
vehicle	252	846	30	62	1	**0**	**0**	**0**
			60	62	1	**0**	**0**	**0**
			90	62	1	**0**	**0**	**0**
segment	235	2310	30	21	**0**	**0**	**0**	**0**
			60	21	**0**	**0**	**0**	**0**
			90	21	**0**	**0**	**0**	**0**
letter	224	20000	30	813	171	475	**37**	**37**
			60	813	171	475	**22**	37
			90	813	171	475	**22**	37
pendigits	216	7494	30	175	**0**	**0**	**0**	**0**
			60	175	**0**	**0**	**0**	**0**
			90	175	**0**	**0**	**0**	**0**
audiology	148	216	30	**0**	**0**	**0**	**0**	**0**
			60	**0**	**0**	**0**	**0**	**0**
			90	**0**	**0**	**0**	**0**	**0**
australian-credit	125	653	30	84	23	81	**3**	4
			60	84	23	81	2	**0**
			90	84	23	81	2	**0**
breast-wisconsin	120	683	30	24	1	**0**	**0**	**0**
			60	24	1	**0**	**0**	**0**
			90	24	1	**0**	**0**	**0**

(continued)

Table 2. (*continued*)

Datasets	Feat.	Trans.	TL(s)	Errors				
				CART	C4.5	DL8.5	LDS-DL8.5	
							inc	exp
mushroom	119	8124	30	544	**0**	**0**	**0**	**0**
			60	544	**0**	**0**	**0**	**0**
			90	544	**0**	**0**	**0**	**0**
german-credit	112	1000	30	265	120	174	29	**25**
			60	265	120	174	**22**	25
			90	265	120	174	**22**	25
diabetes	112	768	30	170	58	139	**18**	**18**
			60	170	58	49	**18**	**18**
			90	170	58	45	**18**	**18**
heart-cleveland	95	296	30	63	5	**0**	**0**	**0**
			60	63	5	**0**	**0**	**0**
			90	63	5	**0**	**0**	**0**
anneal	93	812	30	149	87	140	**68**	72
			60	149	87	140	**67**	72
			90	149	87	140	**67**	72
yeast	89	1484	30	436	251	432	184	**175**
			60	436	251	432	183	**173**
			90	436	251	432	175	**173**
hypothyroid	88	3247	30	54	34	63	**25**	**25**
			60	54	34	63	**24**	25
			90	54	34	63	**23**	25
kr-vs-kp	73	3196	30	189	18	54	**15**	**15**
			60	189	18	54	**15**	**15**
			90	189	18	54	**14**	15
lymph	68	148	30	18	**0**	**0**	**0**	**0**
			60	18	**0**	**0**	**0**	**0**
			90	18	**0**	**0**	**0**	**0**
hepatitis	68	137	30	15	**0**	**0**	**0**	**0**
			60	15	**0**	**0**	**0**	**0**
			90	15	**0**	**0**	**0**	**0**
soybean	50	630	30	50	12	31	**2**	**2**
			60	50	12	31	**2**	**2**
			90	50	12	31	**2**	**2**
vote	48	435	30	19	1	**0**	**0**	**0**
			60	19	1	**0**	**0**	**0**
			90	19	1	**0**	**0**	**0**
zoo-1	36	101	30	**0**	**0**	**0**	**0**	**0**
			60	**0**	**0**	**0**	**0**	**0**
			90	**0**	**0**	**0**	**0**	**0**
primary-tumor	31	336	30	40	24	59	**17**	**17**
			60	40	24	59	16	**15**
			90	40	24	59	16	**15**

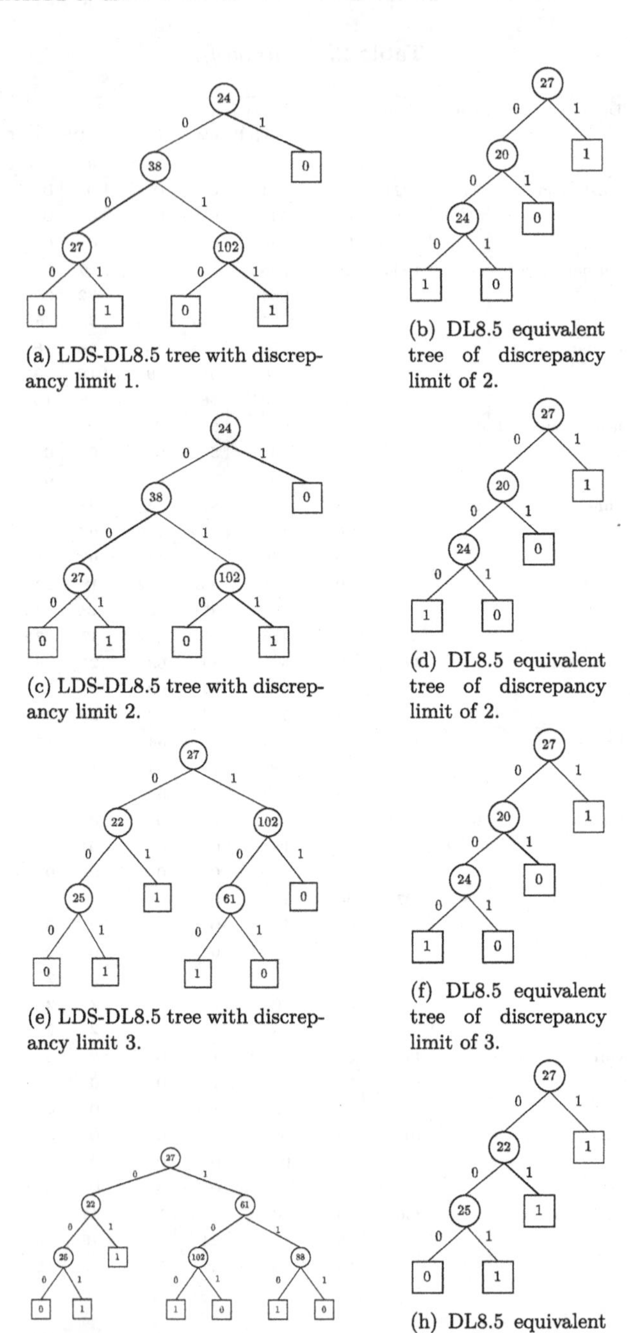

(a) LDS-DL8.5 tree with discrepancy limit 1.

(b) DL8.5 equivalent tree of discrepancy limit of 2.

(c) LDS-DL8.5 tree with discrepancy limit 2.

(d) DL8.5 equivalent tree of discrepancy limit of 2.

(e) LDS-DL8.5 tree with discrepancy limit 3.

(f) DL8.5 equivalent tree of discrepancy limit of 3.

(g) LDS-DL8.5 tree with discrepancy limit 4.

(h) DL8.5 equivalent tree of discrepancy limit of 4.

Fig. 4. Generated LDS trees with discrepancy limits of 1, 2, 3 and 4 with the DL8.5 equivalents on `mushroom` dataset.

Table 3. Experiments with recursion budget results on the 10 most large datasets in terms of features.

Datasets	Disc.	RB	Errors	
			DL8.5	LDS-DL8.5
ionosphere	1	50	**32**	**32**
	2	127	**32**	**32**
	3	265	**32**	**32**
	4	491	32	**30**
splice-1	1	63	574	**268**
	2	167	574	**268**
	3	358	574	**268**
	4	661	513	**267**
vehicle	1	55	216	**94**
	2	143	216	**76**
	3	261	216	**76**
	4	450	214	**63**
segment	1	24	**5**	**5**
	2	59	**5**	**5**
	3	121	**5**	**5**
	4	220	**5**	**5**
letter	1	53	**801**	813
	2	140	801	**686**
	3	236	801	**686**
	4	395	801	**686**
pendigits	1	57	88	**84**
	2	142	88	**60**
	3	268	88	**60**
	4	476	83	**51**
audiology	1	34	10	**6**
	2	87	10	**6**
	3	174	**6**	**6**
	4	319	6	**5**
australian-credit	1	51	**87**	**87**
	2	139	**87**	**87**
	3	304	**87**	**87**
	4	581	**87**	**87**
breast-wisconsin	1	61	50	**23**
	2	161	50	**16**
	3	297	30	**16**
	4	521	30	**16**
mushroom	1	37	376	**180**
	2	67	376	**180**
	3	128	376	**24**
	4	196	120	**8**

In Fig. 3 (depth = 3), D8.5 without heuristics has the best performance by solving the most problems in the least time. If a time factor of 2.5 is allowed, LDS-DL8.5(exp) solves the same amount of problem as the best solver. When the maximum depth is 4 DL8.5 without heuristics also has the best performance but in a time factor of 2.5 LDS-DL8.5(exp). LDS-DL8.5(inc) has the worst performance over time due to the higher number of restarts in this case. LDS-DL8.5(exp) is faster to prove optimality compared to LDS-DL8.5(inc) but will have more difficulties in the early time to update the error due to the large increase of the discrepancy budget at each iteration, leading to exploring more of the search space. This experiment shows that LDS-DL8.5 is able to find the optimal solution in a reasonable amount of time.

6 Conclusion

This paper investigated the interest of using the limited discrepancy search to improve the anytime aspect of DL8.5. The LDS-DL8.5 algorithm, introduced in this paper, allows one to set low time limits and get good and balanced decision trees. Moreover, it mitigates the cost iteration by taking advantage of the cache, allowing the method to be sufficiently reliable when looking for optimal trees. Experimentation with 23 different datasets clearly showed the efficiency of LDS-DL8.5 w.r.t. DL8.5 and the state-of-the-art greedy algorithms CART and C4.5. LDS-DL8.5 is a reliable approach for finding good decision trees in a limited amount of time. As a future work, it could be interesting to study other restarting schemes such as the Luby strategy to improve LDS-DL8.5.

References

1. Aghaei, S., Azizi, M., Vayanos, P.: Learning Optimal and Fair Decision Trees for Non-Discriminative Decision-Making. ArXiv:1903.10598 [cs, Stat]. (2019)
2. Aghaei, S., Gomez, A., Vayanos, P.: Learning optimal classification trees: Strong max-flow formulations. ArXiv Preprint ArXiv:2002.09142 (2020)
3. Aglin, G., Nijssen, S., Schaus, P.: Learning Optimal Decision Trees Using Caching Branch-and-Bound Search. In: Proceedings Of the AAAI Conference on Artificial Intelligence, pp. 3146–3153 (2020)
4. Agrawal, R., Mannila, H., Srikant, R., Toivonen, H., Verkamo, A.: Fast Discovery of Association Rules. In: Advances In Knowledge Discovery And Data Mining, pp. 307–328 (1996)
5. Bertsimas, D., Dunn, J.: Optimal classification trees. Mach. Learn. **106**, 1039–1082 (2017)
6. Bessiere, C., Hebrard, E., O'Sullivan, B.: Minimising decision tree size as combinatorial optimisation. Principles Pract. Constraint Programm.- CP **2009**, 173–187 (2009)
7. Breiman, L., Friedman, J., Stone, C., Olshen, R.: Classification and Regression Trees. Taylor and Francis (1984)
8. Boutilier, J., Michini, C., Zhou, Z.: Shattering Inequalities for Learning Optimal Decision Trees. In: International Conference on Integration of Constraint Programming, Artificial Intelligence, and Operations Research, pp. 74–90 (2022)

9. Demirovićc, E., et al.: MurTree: Optimal Decision Trees via Dynamic Programming and Search. J. Mach. Learn. Res. **23** 1–47 (2022)
10. Dolan, E., More, J.: Benchmarking optimization software with performance profiles. Math. Programm. **91** 201–213 (2002). https://doi.org/10.1007/s101070100263
11. Harvey, W., Ginsberg, M.: Limited Discrepancy Search. Proceedings of the 14th International Joint Conference on Artificial Intelligence, vol. 1,D pp. 607–613 (1995)
12. Hu, X., Rudin, C., Seltzer, M.: Optimal sparse decision trees. In: Advances in Neural Information Processing Systems (2019i)
13. Hu, H., Siala, M., Hebrard, E., Huguet, M.: Learning optimal decision trees with MaxSAT and its integration in AdaBoost. IJCAI-PRICAI 2020, In: 29th International Joint Conference on Artificial Intelligence and the 17th Pacific Rim International Conference on Artificial Intelligence (2020)
14. Langley, P.: Systematic and Nonsystematic Search Strategies. In: Artificial Intelligence Planning Systems, pp. 145–152 (1992)
15. Mitchell, T.: Machine Learning. McGraw-Hill Education (1997)
16. Narodytska, N., Ignatiev, A., Pereira, F., Marques-Silva, J.: Learning Optimal Decision Trees with SAT. In: Proceedings of the Twenty-Seventh International Joint Conference on Artificial Intelligence, IJCAI-18, pp. 1362–1368 (2018)
17. Nijssen, S., Fromont, E.: Optimal Constraint-Based Decision Tree Induction from Itemset Lattices. In: Data Mining And Knowledge Discovery, pp. 9–51 (2010). https://doi.org/10.1007/s10618-010-0174-x
18. Verhaeghe, H., Nijssen, S., Pesant, G., Quimper, C.-G., Schaus, P.: Learning optimal decision trees using constraint programming. Constraints **25**(3), 226–250 (2020). https://doi.org/10.1007/s10601-020-09312-3
19. Verwer, S., Zhang, Y.: Learning Optimal Classification Trees Using a Binary Linear Program Formulation. In: Proceedings of the AAAI Conference On Artificial Intelligence, vol.33, pp. 1625–1632 (2019)
20. Quinlan, J.: C4.5: Programs for Machine Learning, Morgan Kaufmann (1992)
21. Quinlan, J.: Induction of decision trees. Mach. Learn. **1** 81–106 (1986). https://doi.org/10.1007/BF00116251

Quantum, Hardware

Block-Level Surrogate Models for Inference Time Estimation in Hardware-Aware Neural Architecture Search

Kurt Stolle[1,2], Sebastian Vogel[2], Fons van der Sommen[1], and Willem Sanberg[2(✉)]

[1] Eindhoven University of Technology, Eindhoven, The Netherlands
{k.h.w.stolle,fvdsommen}@tue.nl
[2] NXP Semiconductors, Eindhoven, The Netherlands
{sebastian.vogel,willem.sanberg}@nxp.com

Abstract. Hardware-Aware Neural Architecture Search (HA-NAS) is an attractive approach for discovering network architectures that balance task accuracy and deployment efficiency. In an iterative search algorithm, inference time is typically determined at every step by directly profiling architectures on hardware. This imposes limitations on the scalability of search processes because access to specialized devices for profiling is required. As such, the ability to assess inference time without hardware access is an important aspect to enable deep learning on resource-constrained embedded devices. Previous work estimates inference time by summing individual contributions of the architecture's parts. In this work, we propose using block-level inference time estimators to find the network-level inference time. Individual estimators are trained on collected datasets of independently sampled and profiled architecture block instances. Our experiments on isolated blocks commonly found in classification architectures show that gradient boosted decision trees serve as an accurate surrogate for inference time. More specifically, their Spearman correlation coefficient exceeds 0.98 on all tested platforms. When such blocks are connected in sequence, the sum of all block estimations correlates with the measured network inference time, having Spearman correlation coefficients above 0.71 on evaluated CPUs and an accelerator platform. Furthermore, we demonstrate the applicability of our Surrogate Model (SM) methodology in its intended HA-NAS context. To this end, we evaluate and compare two HA-NAS processes: one that relies on profiling via hardware-in-the-loop and one that leverages block-level surrogate models. We find that both processes yield similar Pareto-optimal architectures. This shows that our method facilitates a similar task-performance outcome without relying on hardware access for profiling during architecture search.

Keywords: AutoML · Inference time estimation · Neural network design

© The Author(s), under exclusive license to Springer Nature Switzerland AG 2023
M.-R. Amini et al. (Eds.): ECML PKDD 2022, LNAI 13717, pp. 463–479, 2023.
https://doi.org/10.1007/978-3-031-26419-1_28

1 Introduction

Neural networks consistently achieve competitive results in a wide variety of machine learning contexts. It is thus not surprising that both academia and industry address challenging tasks in multiple domains with neural networks. Furthermore, with cloud services offering specialized neural network training infrastructure, a network can be trained and deployed with minimal operational investment. Since both environment perception and the interpretation of sensor data on-device benefit from the use of deep learning, neural networks are deployed outside data centers with increasing frequency. This shift to edge devices brings a new challenge: hardware cost of deployed neural networks, such as inference latency (i.e. execution time), memory usage, bandwidth utilization, etc. These hardware metrics must be reduced such that neural networks can effectively be executed on more computationally and power constrained devices [23].

Designing machine learning models targeting multiple objectives is a tedious task, which traditionally entails many hours of manual tuning by human experts. As a consequence, the adoption of hardware-optimal neural network design for practical innovations is often deferred due to domain knowledge scarcity. As a response to an increasing demand for task-specific solutions, automated machine learning (AutoML) has emerged to successfully address this limitation [30]. In the case of neural networks, the field of Neural Architecture Search (NAS) studies AutoML that yields optimal architectures. Latest state-of-the-art performance on a number of challenging computer vision benchmarks has been achieved with neural networks found via architecture search [31].

One of the toughest challenges for realizing NAS in practice is the amount of computational resources and power required per search. An important reason for this, is that many NAS methods estimate the accuracy of candidate networks by training them for several epochs [12,15]. Recent research has revealed that this can be alleviated by estimating the task performance directly from the architecture using a surrogate model. As a consequence, the amount of training required is drastically reduced [2,4].

As indicated previously, there is increasing demand for models that meet challenging hardware deployment cost requirements, such as inference time, memory requirements, or power consumption. This is addressed by hardware-aware neural architecture search (HA-NAS) which aims to optimize for both task-performance and hardware costs [3]. In this work, we consider the co-optimization of both classification accuracy and execution time on hardware. The hardware-aware search strategy determines architectures that perform optimally on both metrics. In the remainder of this paper, we use 'inference time', 'latency', and 'execution time' interchangeably.

Figure 1 shows a diagram of an iterative HA-NAS process. The latency of an architecture must be assessed at every step in the search process. Three categories of assessment methods are commonly found in the literature. Firstly, hardware-in-the-loop (HIL) methods profile a given architecture ad-hoc on the targeted hardware whenever a new architecture is emitted [7,24]. Secondly, lookup table (LUT) methods query latencies of parts of a neural networks based

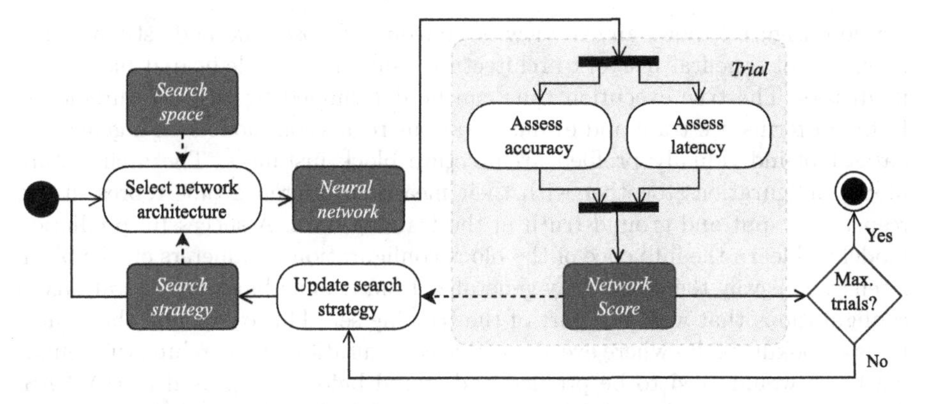

Fig. 1. UML activity diagram of hardware-aware neural architecture search. Neural networks are selected from the search space using a search strategy. The combined task and hardware performance scores are used to determine the next trial network. This work proposes a surrogate-model-based latency assessment.

on their building blocks [1,6,28,29,32,34]. The LUT contains an entry for each architecture block configuration in the search space. These part latencies are then combined, usually via summation, to yield the overall architecture latency. Thirdly, surrogate model (SM) methods define a prediction model to infer the approximate latency of a network [2,4,14,33].

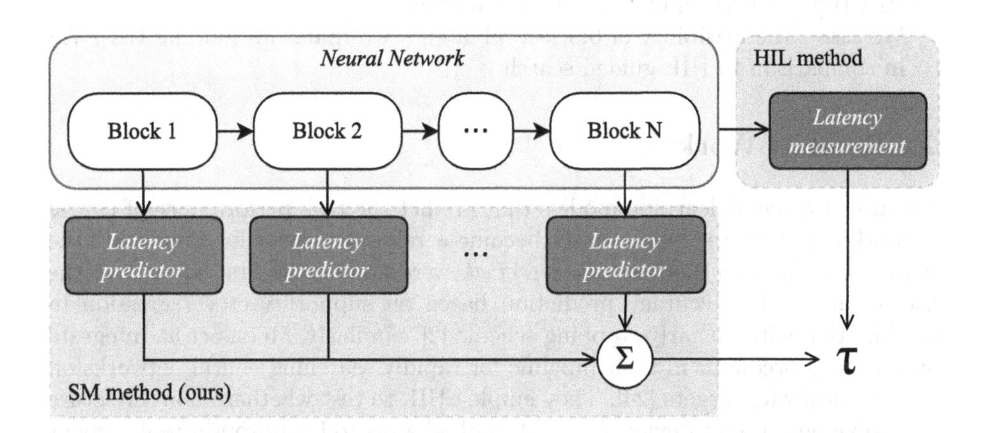

Fig. 2. Schematic showing the assessment methodolgy of full-network latency τ using HIL profiling and our proposed block-level based estimate.

We aim to combine various machine learning methodologies to engineer a latency estimation framework for block-based convolutional neural networks (CNN) in order to overcome the limitations of HIL and lookup table based latency assessment systems. As such, our approach can be categorized as an SM method. Figure 2 shows an overview of how our proposed method estimates the latency τ of a neural network architecture using a set of dedicated block-level predictors. The true execution time can be determined by HIL-measurements. In broad terms, we train and evaluate various regression models on a generated dataset of individually profiled architecture block instances. The architecture block configurations together with their measured inference time represent the respective input and ground truth in the training data. A successful prediction model will learn the influence of the block configuration parameters on the block latency. This way the model may generalize from trained block-configurations to configurations that were not part of the training set. This overcomes the limitations of lookup tables where every possible configuration of an architecture building block would need to be profiled and stored before being used in HA-NAS. As such, our proposed block-level surrogate model (BLSM) greatly increases the cardinality of a search space in a HA-NAS algorithm without requiring hardware access at search-time.

Our key contributions can be summarized as:

- We introduce a block-level surrogate-model methodology abbreviated BLSM that uses trained inference time estimators to overcome the need for hardware access during HA-NAS while preserving the flexibility benefits of HIL-methodologies.
- We propose a novel definition of blocks using a bijective relation between instantiation parameters and a computational sub-graph to make the BLSM pipeline generalize to any block type that may be uniquely described by a structured collection of constructor variables.
- We assess the feasibility of block-level latency estimates for guiding HA-NAS in comparison to HIL-guided search.

2 Related Work

The use of machine learning models that predict the task performance of neural network architectures has recently become a relevant aspect for improving the search time of NAS methods. Baker *et al.* were among the first to explore the use of an SM for accuracy prediction based on support vector regression in combination with an early stopping scheme [2]. Similarly, Moons *et al.* integrate an accuracy predictor in their pipeline for rapidly searching neural networks on various hardware targets [24]. They employ HIL to test whether an architecture is feasible on a target device. Our SM method aims to be a drop-in replacement for such a HIL approach. This would allow a fully prediction-based optimization scheme when optimizing for both accuracy and latency.

Different strategies have been employed to estimate hardware metrics. For example, Dong *et al.* define an analytical model to estimate DNN latency on an FPGA platform [10]. This analytical strategy requires detailed understanding of low-level execution of DNN blocks on the target hardware, which is not feasible for a wide variety of deployment scenarios, e.g., if specialized black-box inference engines are involved. Alternatively, Gupta *et al.* implement an

SM that uses hardware virtualization. Such cycle-accurate simulations can accurately predict the performance of a hardware target. However, cycle-accurate simulations are often prohibitive in HA-NAS for assessing execution time due to their significant computational effort. Additionally, profiling a neural network via simulation requires cycle-accurate emulation of the platform. Such software may not be available for the target hardware. Our method overcomes the limitations of simulation and analytical models by predicting the latency through a machine-learning-based surrogate model trained on a-priori profiled data of the target hardware platform.

Bouzidi *et al.* compare a variety of modeling algorithms to predict the latency of CNN-architectures on edge GPU platforms [4]. Whereas the method proposed in their work aims to make a single prediction on the network level, our approach uses block-level latency predictions to infer the network latency.

HELP and BRP-NAS are similar approaches for introducing predictor-based latency estimators in NAS [11,21]. Especially the focus of HELP lies on generalizing the latency estimations towards multiple HW-targets. However, the individual neural networks for latency estimation stem from one search space only. Our work focuses on block-level latency estimation and therefore enables latency estimation in arbitrary search spaces derived form a composition of blocks.

We evaluate our method as a drop-in module for the *AutoKeras* framework proposed in [19], which renders the NAS process into a hyperparameter optimization context. In their system, the authors programatically define a parameterized template network called the 'hypermodel', which we can represent as a mapping from a set of network instantiation parameters to a network architecture. By means of hyperparameter optimization, trials can be performed until an optimal configuration is found [19]. By contrast, the canonical term 'search space' is used for the remainder of this paper to refer to the set of all possible network instances that are in the range of a specific hypermodel. To clarify, the range of a hypermodel entails all architectures that can be defined on the cross product of the domain of each instantiation parameter.

HW-NAS-Bench is a recent dataset that includes latency measurements of networks for benchmarking *hardware-aware* NAS approaches [22]. Our work is based on block-level latency estimates. Unfortunately, HW-NAS-Bench only provides the latency of entire networks and not of their composing blocks. Therefore, our approach cannot be assessed on the HW-NAS-Bench dataset.

3 Method

3.1 Block Instances and Parameters

Before proceeding to formalize our models, let us first introduce the concept of block-based neural networks. This work has a focus on image classification networks that consist of sequentially connected block instances.

Definition 1. *A block instance is a sub-graph of a neural network that is the specific realization from the space of all block variants.*

Table 1. Evaluated block types

Block name	Reference implementation	Parameters cardinality(i)
`MBConv`	MobileNet convolutional block [18]	$3 + 5$
`Res V1`	ResNet original implementation [16]	$3 + 4$
`Res V2`	ResNet with identity mappings [17]	$3 + 4$

i Cardinality includes three input dimensions plus the number of searchable block configuration hyperparameters.

The networks investigated in this paper can be represented as a list of *parameters* that unambiguously define a series of block instances. Inversely, these parameters can be uniquely inferred from the network graph by inspecting the block instances it contains.

Definition 2. *Block parameters are a collection of mixed-domain values that uniquely map to a block instance.*

The bijective properties of the mapping from parameters to instances enables constructing a dataset of block parameters together with the measured inference latency of the respective instance for training a block-level latency predictor. Additionally, the block parameters of each block in a candidate neural network can be determined during search-time by parsing the computational graph of the full network. By design, this effectively isolates the training of latency predictors from the search process. The blocks defined in this work are parameterized versions of blocks commonly found in literature. Figure 1 summarizes the block types evaluated in this paper.

As is common practice in HA-NAS methods [3,6,26,27] and also experimentally confirmed [27], we assume that network latency can be sufficiently approximated by the sum of stacked block-instance latencies. As such, blocks can be profiled individually and mixed-and-matched to yield new networks that retain the summed-latency property. Thus, summing latency predictions for each block of a neural network yields an estimate of the entire architecture inference time. We hypothesize that there exist regression models that can predict the block-level latency from the block serialization. This requires a predictor model for each pair of block families and hardware targets, which can be trained on profiling data of single blocks.

3.2 Predictor Model

Having defined the relation between blocks and the full-network latency, the estimator model can now be formalized. Let $\tau_{\mathbf{c}} \in \mathbb{R}_+$ be the measured latency of a block B with parameters $\mathbf{c} \in \mathcal{C}_\mathrm{B}$, where $\mathcal{C}_\mathrm{B}$ represents the set of all possible block configurations. The objective of our block-level estimators is to learn the transformation

$$\hat{\tau} : \mathcal{C}_\mathrm{B} \rightarrow \mathbb{R}_+, \tag{1}$$

such that $\hat{\tau}(\mathbf{c})$ predicts the measured latency. This is identified as a regression problem with error function

$$\zeta = \hat{\tau}(\mathbf{c}) - \tau_{\mathbf{c}}. \tag{2}$$

We propose to use a separate estimator $\hat{\tau}(\cdot)$ for each block type and hardware platform. This enables to two degrees of freedom for designing inference time surrogate models. First, each block predictor can leverage the machine learning algorithm best suited for the domain of instantiation parameters that it defines. Second, it is not required to train predictors that generalize across all hardware-platforms. Instead, predictors may be fine-tuned to fit specific hardware characteristics.

This paper evaluates four estimator models, namely Linear, Random Forest, Boosted Trees, and Dense NAS. These different estimator models should address the mixed-domain characteristics of block parameters and are further explained in the following.

Linear Regression. This model is best suited for configuration spaces that consist of continuous independent variables. We make use of an exponentially activated linear regression model

$$\hat{\tau}(\mathbf{c}) = \exp(\mathbf{W}\mathbf{c} + b) \tag{3}$$

with weighting matrix $\mathbf{W}$ and bias b. We iteratively approximate the weighting matrix using gradient descent. Hyperparameters are the learning rate (0.01) and batch size (64), which were determined using 20 trials of grid search.

Decision Forest Regression. These models are expected to achieve high performance when the block configuration consists of mostly categorical parameters or when the input values are from mixed domains. In a decision forest model, the (regression) output is given by taking the mean output of a decision tree ensemble. We explore two flavors of decision-forest construction algorithms. First, the Random Forest predictor uses the method described in [5], which adds uncorrelated trees to the forest that minimize the prediction error. Second, the Boosted Treespredictor uses the gradient boosting algorithm [13], which iteratively reduces the error of the forest by adding trees that minimize the error of prediction as a product of the learning rate. For both models, hyperparameters were configured as suggested in the reference implementation [8].

Deep Neural Network Regression. The fourth estimator method under consideration is a deep neural network. We exploit a NAS algorithm to find an optimal fully connected deep neural network, tailored towards each block and hardware platform. Models of this class are best suited to inputs where each element is from the same domain. Additionally, inputs that have a high degree of interdependent relations are likely to benefit from deep neural networks. The network is trained using gradient descent, with the learning rate tuned by the NAS algorithm.

3.3 Experimental Set-Up

The methodology proposed in this paper is evaluated on four platforms. These platforms cover a wide range of hardware architectures to illustrate the extent to which our methodology is applicable. More specifically, we employ a high-performance CPU for high-end compute (CPU Cloud), a low-power CPU that is typically used in energy-constrained edge platforms (CPU Edge), a general-purpose GPU designed for parallel processing (GPU), and a specialized parallel compute platform (ASIC)[1]. For each platform, latency is measured in a profiling experiment consisting of 10 000 invocations on isolated block instances with random data and batch size 1. Note that in edge devices, achieving low latency is often more important than realizing high throughput. We therefore choose a batch size of 1. A number of external factors may affect the measured latency, such as processes executed in parallel on the targeted hardware and general profiling inaccuracies. To mitigate these effects in our approach, we rely only on the minimal measured latency of the block profiling dataset. This represents a feasible best-case execution time and is more robust than, e.g., a mean value, since we observed that inference measurements typically follow an exponential distribution.

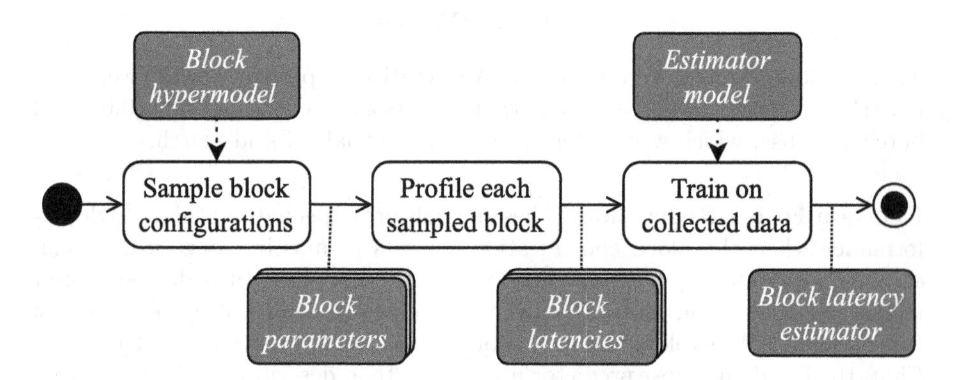

Fig. 3. UML activity diagram showing the end-to-end process of sampling configurations of a block, profiling these samples, and fitting a predictor model on this data. This yields a latency predictor that is specifically designed for a single block type and the hardware platform it was profiled on.

Figure 3 shows the process of sampling configurations of a particular block, profiling these block instances on hardware and fitting a regression model. This yields a block latency predictor as can be used in Fig. 2.

We evaluate out proposed block-level surrogate models (BLSM) for latency in a HA-NAS on a small-scale image classification task. For this, we use the

[1] Note that the experiments are meant to demonstrate our methodology. The experiments are not intended to benchmark specific hardware and deployment toolchains, hence those details are left out.

CIFAR-10 dataset [20]. This dataset covers 60 000 colored images, each having a width and height of 32 pixels. The objective is to classify each image into one of ten classes. The classification accuracy is defined as the percentage of correctly predicted classes over a test set of 10 000 samples.

3.4 Applicability of Block-Level Surrogate Models in HA-NAS

To assess the applicability of a block-level surrogate model for inference time, we compare the distribution of found neural networks of a HIL-based search with the distribution of architectures generated by a search that relies on our BLSM. If the found Pareto-optimal neural architectures are sampled in similar accuracy-latency regimes in both setups, we consider the use of a BLSM in HA-NAS is *feasible* for a block type and hardware platform. As a result, we can compare the optimal networks from a session guided by a predictor with the optimal networks found when HIL-measurements were used.

For the purpose of evaluating our proposed BLSM, we sample 200 architectures with random block configurations and assess the predicted latency with *true* latency measured in a HIL-setup. Note, random sampling of architectures reflects an unguided search, i.e. each trial is chosen independently of any previously evaluated accuracy or latency in a search process.

In contrast, a HA-NAS with a Bayesian Optimization search strategy samples architectures for training and profiling based on previous trials' accuracy and latency. In order to render the Bayesian Optimization as a hardware-aware multi-objective optimization, we use a scalarization paradigm [25]. To this end, we propose using the balanced sum of the task-related objective (accuracy) and the hardware performance metric (inference time). Specifically, we investigate the multi-objective-optimization of classification accuracy α on CIFAR-10 and inference latency τ on target hardware platforms. Thus, we define the following multi-objective scalarization to derive a *balanced accuracy-latency score*

$$\text{BALS} = \frac{\alpha}{\alpha_{\text{ref}}} + \frac{\log(\tau - \tau_{\text{ref}})}{\log \tau_{\text{ref}}} - 1, \tag{4}$$

where constants α_{ref} and τ_{ref} can be tuned according to the importance of each objective. For experiments, we employ an adapted version of the Bayesian Optimization search algorithm from the AutoKeras framework [19].

Tabel 2 summarizes the hyperparameters of the HA-NAS benchmark. We use 20 epochs of training as a proxy for the final validation accuracy in accordance with NAS-bench-201 [9].

Table 2. HA-NAS benchmark hyperparameters

Hyperparameter	Value
Primary objective	CIFAR-10 classification accuracy
Input image dimensions	$32 \times 32 \times 3$
Training epochs to estimate accuracy score	20
Batch size	64
Optimizer	Adam
Learning rate	0.01
Reference accuracy α_{ref}	1.0
Reference latency τ_{ref}	100 ms

4 Results and Discussion

Our latency predictors are trained on a generated dataset of block parameters and their corresponding measured latency on a specific target platform. To assess performance and compare results, each block-hardware-dataset is split into a training, validation, and test set. The training set is used to fit the model, while the hyperparameters of this model are tuned to maximize accuracy on the validation split. Finally, the test split is used to quantitatively assess the quality of predictors with respect to measured values and in relation to other predictors.

4.1 Block-Level Surrogate Model Performance

Figure 4 shows the predicted latency versus the measured latency. The amount of deviation from the $x = y$ line is a proxy for the quality of the predictor. From this visualization, it is evident that the Boosted Trees estimator appears to yield the least amount of predictive error compared to the other estimator models. Overall, the GPU hardware is most difficult to predict. This could be explained by inaccuracies in profiling at the small-scale range of GPU inference times. In order for a NAS algorithm to be guided towards optimal solutions using a predictor, the prediction and measured values ideally have a monotonic relation. To assess the strength of all trained predictors in this regard, Spearman's ρ is calculated, and results are summarized in Table 3. All predictor methods produce highly correlated results.

4.2 Comparing Recall of Optimal Networks Under Prediction

Previously, we computed the Spearman ρ to quantitatively assess the monotonic relation between block-level latency estimates and the measured latency of an isolated block. In this section, we analyze both the monotonic relation strength between network-level latency estimates using summed block-level SM estimators and mean measured latency. For our analysis, we use a set of 200 networks

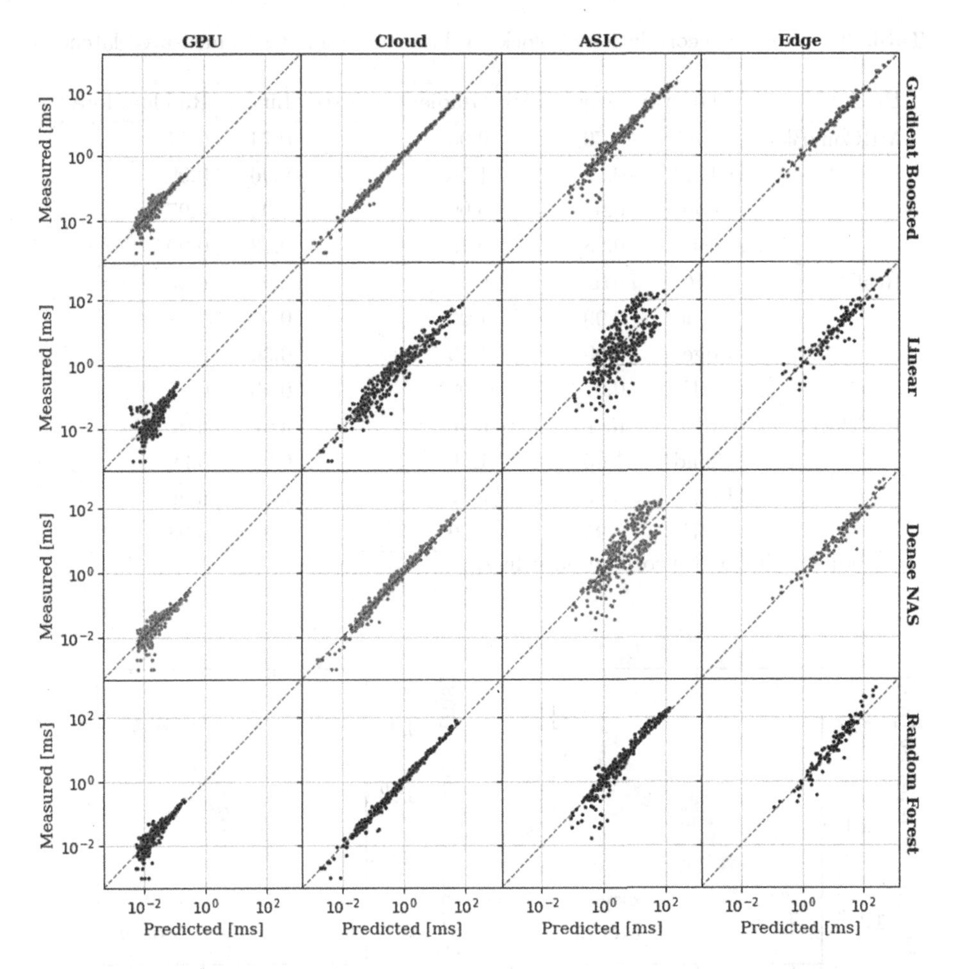

Fig. 4. Predicted latency versus measured latency for all MobileNet convolutional block estimators in isolation on different hardware platforms. Every point corresponds to a configuration of the ResBlock1 in the set of configuration samples not used in the training process.

randomly sampled from the search space, which consists of sequentially placed instances of each block type. While a block-level latency estimate may produce network-level latencies with a large degree of error, we expect a guided NAS algorithm to still find the set of Pareto-optimal networks when errors are proportionally incorrect. Figure 5 visually illustrates the relation between measured versus BLSM-predicted inference time on the network-level. Table 4 summarizes the resulting metrics for each predictor on the set of randomly sampled networks using the ASIC platform. On average, the Boosted Trees predictor produces the largest correlation coefficients, closely followed by the Random Forest predictor. This is expected, as the block-level Spearman's ρ correlation is similarly large when evaluated on isolated blocks (Table 3).

Table 3. Spearman's correlation of block-level latency estimates to measured latencies

Block	Platform	Dense NAS	Gradient boosted	Linear	Random forest
MBConvBlock	ASIC	0.79	0.98	0.74	0.97
	Cloud	0.99	1.00	0.96	0.99
	Edge	0.97	0.99	0.94	0.97
	GPU	0.88	0.89	0.78	0.90
ResBlock1	ASIC	n/a	n/a	n/a	n/a
	Cloud	0.99	1.00	0.97	0.99
	Edge	0.99	1.00	0.96	0.99
	GPU	0.91	0.98	0.93	0.97
ResBlock2	ASIC	n/a	n/a	n/a	n/a
	Cloud	1.00	1.00	0.97	0.99
	Edge	n/a	n/a	n/a	n/a
	GPU	0.95	0.99	0.94	0.98

$p < 0.001$ for all r and ρ correlation values.

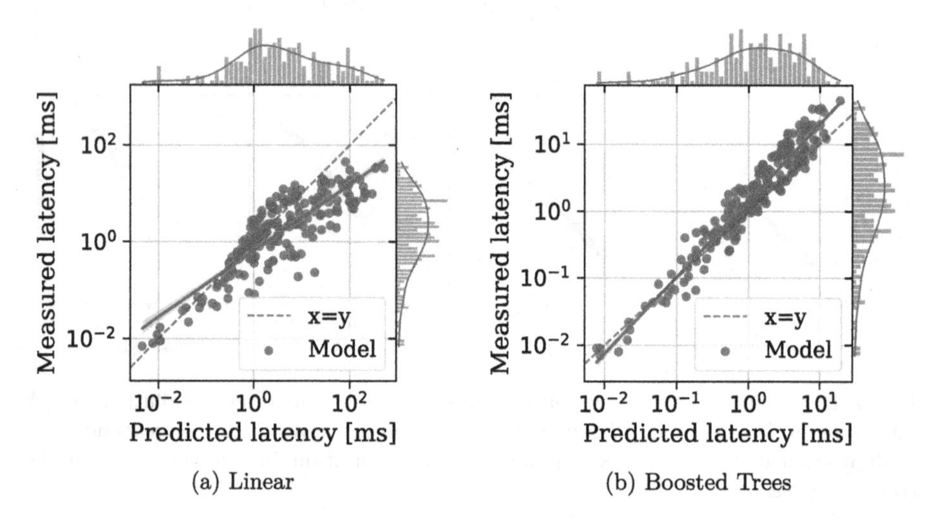

(a) Linear

(b) Boosted Trees

Fig. 5. Normal probability (center axis) and KDE (marginal axes) plots of measured versus predicted latency on full neural network architecture scale. Networks (blue dots) were discovered via random search consisting of stacked MBConvBlock($Color figure online$) profiled on the CPU Cloud platform. While distributions estimated on both axes appear similar, deviation from the dotted line shows the error of prediction accumulates exponentially.

4.3 Effects of Prediction on Guided Search Results

A quantitative comparison between search results becomes more involved when a strategy selects subsequent trials based on the previous trial's results, as is the case in guided search. Figure 6 shows networks from a search guided respectively

Table 4. Evaluation metrics of BLSM predictors for MBConvBlock based-networks.

Platform	Predictor	r	ρ
Cloud CPU	Linear	0.86	0.83
	Random forest	0.96	0.95
	Gradient boosted trees	**0.97**	**0.96**
	Dense NAS	0.96	0.95
Edge CPU	Linear	0.83	0.79
	Random forest	0.87	0.85
	Gradient boosted trees	**0.91**	**0.89**
	Dense NAS	0.87	0.8
GPU	Linear	0.57	0.54
	Random forest	0.45	0.41
	Gradient boosted trees	**0.58**	**0.57**
	Dense NAS	0.53	0.48
ASIC	Linear	0.80	0.60
	Random forest	**0.96**	**0.71**
	Gradient boosted trees	0.94	0.70
	Dense NAS	0.81	0.70

$p < 0.001$ for all r and ρ correlation values.

Table 5. Distance between SM and HIL guided sets of MBConvBlock networks.

Platform	Predictor	W_{overall}	W_{optimal}	HNQ
Cloud CPU	Linear	0.1052	0.3118	2.9
Cloud CPU	Boosted Trees	0.1278	0.0198	**0.2**
ASIC	Linear	0.2032	0.1574	0.8
ASIC	Boosted Trees	0.0923	0.0142	**0.2**

by HIL and SM assessments with the Bayesian Optimization strategy in the accuracy versus latency plane. To gain insight into the score of networks selected by each search, an estimate of the HA-score distribution of networks is shown in Fig. 7. We compute the Wasserstein distance W of HA-score distributions as a measure of similarity between the networks sampled in the HIL-setup and in the SM-setup. We report both the Wasserstein distance W_{overall} between all sampled networks and W_{optimal} between the found Pareto-optimal networks. A small value of W_{pareto} indicates that the BLSM-based HA-NAS process is *feasible*, because the SM set of optimal networks is similar to the HIL set of optimal networks. The range of the Wasserstein distance metric does not give an inherent qualitative insight, but requires comparison against a baseline distance to yield an interpretable result. We use the overall distance between SM and HIL results as a baseline for each qualitative assessment. The search process can

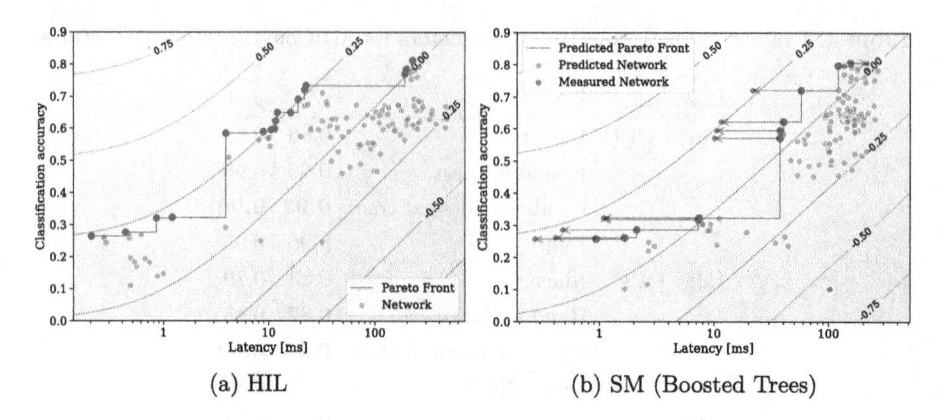

(a) HIL (b) SM (Boosted Trees)

Fig. 6. Guided HA-NAS results under different latency assessment sources (HIL or SM) visualized in an accuracy versus latency scatter plot.

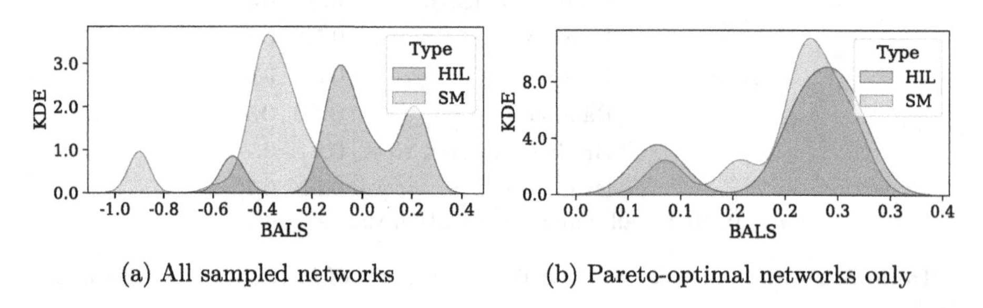

(a) All sampled networks (b) Pareto-optimal networks only

Fig. 7. KDE plot of HA-scores found via HIL and SM (Boosted Trees) Bayesian Optimization search. A stack of MBConvBlock models on the ASIC platform was used. Visually, the distances between both sets is much smaller, indicating the search strategy yields similar networks.

thus be asserted as *feasible* when the distance between optimal results is less than the overall distance of networks trialed during search. Therefore, we define the *HA-NAS quality ratio*

$$\mathrm{HNQ} = \frac{W_{\mathrm{optimal}}}{W_{\mathrm{overall}}} \in \mathbb{R}_{+}, \tag{5}$$

and the *necessary condition for feasibility* of the search process,

$$\mathrm{HNQ} < 1. \tag{6}$$

Table 5 summarizes the resulting metrics for each search experiment. Results indicate that the Boosted Trees BLSM is the most feasible on the CPU Cloud and ASIC platform with HNQ = 0.2. Additionally, the linear regression SM does not meet the necessary condition for feasibility Eq. (6) on the Cloud CPU platform and has a relatively large NHQ ratio on the ASIC platform.

While the measured latencies of networks found via SM-based guided search are close to that of networks found via HIL-based search, there is an offset between predicted and measured latencies in BLSM-based search that is likely due to the accumulation of predictive errors in the search strategy algorithm. An explanation for the ability of the SM-based search to find optimal networks despite a large degree of error between the predicted and measured latency follows from the random search experiment. Namely, the monotonic relation between predictions and measured latencies on network-level causes the search strategy to propose networks in the correct relative order, albeit with a large absolute error. These results demonstrate the utility of block-level latency predictors in guided search.

5 Conclusion

This paper presents a block-level surrogate model for inference latency prediction and shows its applicability by integrating it in a HA-NAS method. Our BLSM overcomes the need for hardware access and latency lookup tables during a neural architecture search process, thereby greatly improving both flexibility and scalability of HA-NAS applications. As a key design choice, our BLSM operates on block-parameters, which by definition are available for any block, current or new. As a result, our method generalizes over many architectures and hardware platforms using a simple training procedure.

Predictors are robust to varying block domains by choosing an appropriate regression model from a diverse set of model types. Experiments on a representative set of target platforms and block types have validated our methodology, with three key findings. First, results demonstrate that predictors based on decision trees, Random Forest and Boosted Trees, perform optimally for the evaluated block types. Second, when blocks are placed in sequence, the sum of block predictions correlates with the measured latency of the full network, achieving Spearman coefficients of 0.96, 0.89, 0.57 and 0.71 on the Cloud CPU, Edge CPU, GPU, and ASIC platforms respectively. Third, NAS algorithms guided by our proposed predictor were able to find Pareto-optimal neural networks with similar HA-score to those found via a HIL-based process.

Combined, the experiments confirm that our BLSM enables HA-NAS to scale out by removing the burden of hardware access during search. More specifically, it paves the way to deploy HA-NAS in a distributed fashion, by allowing the latencies of multiple block-based architectures to be assessed anywhere and in parallel. Ultimately, this facilitates large-scale automation of designing efficient and effective neural networks for a wide variety of applications in resource constrained edge devices.

References

1. Abdelfattah, M.S., Dudziak, Ł., Chau, T., Lee, R., Kim, H., Lane, N.D.: Best of both worlds: Automl codesign of a cnn and its hardware accelerator. In: ACM/IEEE DAC. IEEE (2020)

2. Baker, B., Gupta, O., Raskar, R., Naik, N.: Accelerating neural architecture search using performance prediction. ICLR Workshop (2018)
3. Benmeziane, H., Maghraoui, K.E., Ouarnoughi, H., Niar, S., Wistuba, M., Wang, N.: A comprehensive survey on hardware-aware neural architecture search. arXiv preprint arXiv:2101.09336 (2021)
4. Bouzidi, H., Ouarnoughi, H., Niar, S., Cadi, A.A.E.: Performance prediction for convolutional neural networks on edge gpus. In: ACM ICCF, p. 54–62 (2021)
5. Breiman, L.: Random forests. In: Machine Learning. vol. 45, pp. 5–32. Springer Science and Business Media LLC (2001). https://doi.org/10.1023/a:1010933404324
6. Cai, H., Gan, C., Wang, T., Zhang, Z., Han, S.: Once for all: Train one network and specialize it for efficient deployment. In: ICLR (2020)
7. Cai, H., Zhu, L., Han, S.: ProxylessNAS: Direct neural architecture search on target task and hardware. In: ICLR (2019)
8. Dillon, J.V., et al.: Tensorflow distributions. arXiv preprint arXiv:1711.10604 (2017)
9. Dong, X., Yang, Y.: NAS-Bench-201: Extending the scope of reproducible neural architecture search. In: ICLR (2019)
10. Dong, Z., Gao, Y., Huang, Q., Wawrzynek, J., So, H.K., Keutzer, K.: HAO: Hardware-aware neural architecture optimization for efficient inference. In: IEEE FCCM, pp. 50–59. IEEE (2021)
11. Dudziak, L., Chau, T., Abdelfattah, M., Lee, R., Kim, H., Lane, N.: BRP-NAS: prediction-based NAS using GCNs. NeurIPS **33**, 10480–10490 (2020)
12. Elsken, T., Metzen, J.H., Hutter, F.: Neural architecture search: a survey. J. Mach. Learn. Res. **20**(1), 1997–2017 (2019)
13. Friedman, J.H.: Greedy function approximation: a gradient boosting machine. Ann. Statist. **29**(5), 1189–1232 (2001)
14. Gupta, S., Akin, B.: Accelerator-aware neural network design using automl. arXiv preprint arXiv:2003.02838 (2020)
15. Guyon, I., et al.: Analysis of the automl challenge series, pp. 191–236 2015–2018
16. He, K., Zhang, X., Ren, S., Sun, J.: Deep residual learning for image recognition. In: IEEE/CVF CVPR, pp. 770–778 (2016)
17. He, K., Zhang, X., Ren, S., Sun, J.: Identity mappings in deep residual networks. In: Leibe, B., Matas, J., Sebe, N., Welling, M. (eds.) ECCV 2016. LNCS, vol. 9908, pp. 630–645. Springer, Cham (2016). https://doi.org/10.1007/978-3-319-46493-0_38
18. Howard, A., et al.: Searching for mobilenetv3. In: IEEE/CVF CVPR, pp. 1314–1324 (2019)
19. Jin, H., Song, Q., Hu, X.: Auto-keras: An efficient neural architecture search system. In: ACM SIGKDD International Conference on Knowledge Discovery & Data Mining, pp. 1946–1956. ACM (2019)
20. Krizhevsky, A.: Learning multiple layers of features from tiny images (2009)
21. Lee, H., Lee, S., Chong, S., Hwang, S.J.: HELP: Hardware-adaptive efficient latency prediction for NAS via meta-learning. In: Advances in Neural Information Processing Systems (NeurIPS) (2021)
22. Li, C., et al.: HW-NAS-Bench: Hardware-aware neural architecture search benchmark. In: ICLR (2021)
23. Li, W., Liewig, M.: A survey of ai accelerators for edge environment. In: Rocha, Á., Adeli, H., Reis, L.P., Costanzo, S., Orovic, I., Moreira, F. (eds.) WorldCIST 2020. AISC, vol. 1160, pp. 35–44. Springer, Cham (2020). https://doi.org/10.1007/978-3-030-45691-7_4
24. Moons, B., et al.: Distilling optimal neural networks: Rapid search in diverse spaces. In: IEEE/CVF CVPR, pp. 12229–12238 (2021)

25. Roijers, D.M., Zintgraf, L.M., Nowé, A.: Interactive Thompson sampling for multi-objective multi-armed bandits. In: Rothe, J. (ed.) ADT 2017. LNCS (LNAI), vol. 10576, pp. 18–34. Springer, Cham (2017). https://doi.org/10.1007/978-3-319-67504-6_2
26. Shaw, A., Hunter, D., Iandola, F., Sidhu, S.: Squeezenas: Fast neural architecture search for faster semantic segmentation. arXiv preprint arXiv:1908.01748 (2019)
27. Stamoulis, D., et al.: Single-path NAS: Device-aware efficient convnet design. In: Joint Workshop on On-Device Machine Learning & Compact Deep Neural Network Representations with Industrial Applications (ODML-CDNNRIA) at ICML (2019)
28. Tan, M., Chen, B., Pang, R., Vasudevan, V., Le, Q.V.: Mnasnet: Platform-aware neural architecture search for mobile. IEEE/CVF CVPR, pp. 2815–2823 (2019)
29. Tsai, H., Ooi, J., Ferng, C.S., Chung, H.W., Riesa, J.: Finding fast transformers: One-shot neural architecture search by component composition. arXiv preprint arXiv:2008.06808 (2020)
30. Vanschoren, J.: Meta-Learning, pp. 35–61. Springer International Publishing (2019)
31. Wistuba, M., Rawat, A., Pedapati, T.: A survey on neural architecture search. arXiv preprint arXiv:1905.01392 (2019)
32. Wu, B., et al.: FBNet: Hardware-aware efficient convnet design via differentiable neural architecture search. In: IEEE/CVF CVPR, pp. 10726–10734 (2019)
33. Wu, J., et al.: Weak NAS predictors are all you need. arXiv preprint arXiv:2102.10490 (2021)
34. Yang, T.-J., et al.: NetAdapt: platform-aware neural network adaptation for mobile applications. In: Ferrari, V., Hebert, M., Sminchisescu, C., Weiss, Y. (eds.) ECCV 2018. LNCS, vol. 11214, pp. 289–304. Springer, Cham (2018). https://doi.org/10.1007/978-3-030-01249-6_18

FASE: A Fast, Accurate and Seamless Emulator for Custom Numerical Formats

John Osorio[1,2](✉), Adriá Armejach[1,2], Eric Petit[3], Greg Henry[3], and Marc Casas[1,2]

[1] Universidad Politecnica de Catalunya, Barcelona, Spain
[2] Barcelona Supercomputing Center, Barcelona, Spain
john.osorio@bsc.es
[3] Intel Corporation, Oregon, USA

Abstract. Deep Neural Networks (DNNs) have become ubiquitous in a wide range of application domains. Despite their success, training DNNs is an expensive task which has motivated the use of reduced numerical precision formats to improve performance and reduce power consumption. Emulation techniques are a good fit to understand the properties of new numerical formats on a particular workload. However, current state-of-the-art techniques are not able to perform this tasks quickly and accurately on a wide variety of workloads.

We propose FASE, a Fast, Accurate and Seamless Emulator that leverages dynamic binary translation to enable emulation of arbitrary numerical formats. FASE is *fast*; allowing emulation of large unmodified workloads, *accurate*; emulating at instruction operand level, and *seamless*; as it does not require any code modifications and works on any application or DNN framework without any language, compiler or source code access restrictions. We evaluate FASE using a wide variety of DNN frameworks and large-scale workloads. Our evaluation demonstrates that FASE achieves better accuracy than coarser-grain state-of-the-art approaches, and shows that it is able to evaluate the fidelity of multiple numerical formats and extract conclusions on their applicability.

Keywords: Numerical emulation · Reduced precision · DNN training

1 Introduction

Current trends on DNNs indicate that training costs will continue to grow as state-of-the-art DNNs feature increasingly large parameter counts [3]. There are already approaches on reducing the training computation costs via mechanisms that incur accuracy degradations [16,29,32]. Additionally, there are approaches able to reduce training costs without reducing DNNs accuracy. These approaches

Supplementary Information The online version contains supplementary material available at https://doi.org/10.1007/978-3-031-26419-1_29.

© The Author(s), under exclusive license to Springer Nature Switzerland AG 2023
M.-R. Amini et al. (Eds.): ECML PKDD 2022, LNAI 13717, pp. 480–497, 2023.
https://doi.org/10.1007/978-3-031-26419-1_29

rely on reduced computer number formats [10,11,33,34]. To decide among all potential format designs which ones display the best opportunities for efficient and accurate DNN training, it is critical to empirically evaluate them with as much fidelity as possible and on as many real neural net topologies and real input datasets as possible. The emulation of these reduced precision approaches becomes one of the most important and costly phases to evaluate the reliability of new numerical data types. The emulation helps to avoid cost overrun, by avoiding costly hardware implementations.

TensorQuant [26], proposes two source level approaches, intrinsic and extrinsic, to emulate low precision using Tensorflow. The extrinsic approach is an approximation where the rounding process is done just on high level operators like convolutions. This is the mode implemented in QPyTorch [39] to address the PyTorch framework. The intrinsic approach rounds each individual floating point operation and displays a latency of 50× with respect to native executions. It is a source level approach that can be used to evaluate all implementation of neural network based on Tensorflow. All of these approaches are designed targeting specific DNN frameworks and require changes on the framework and model source code. Other tools like Verificarlo [4] work at the compiler level, and can be applied to any Python framework; but, they do require complex recompilation.

To overcome these issues we propose FASE: a fast, accurate and seamless tool that enables the emulation of custom numerical formats on any application. FASE relies on dynamic binary instrumentation using PIN [27] to perform fine-grain instruction-level instrumentation. In addition, FASE seamlessly works on any application or DNN framework without any language, compiler or source access restrictions. Since no code modification or recompilation steps are necessary, FASE guarantees that the instrumented binary matches the original one. Therefore, FASE works on all DNN frameworks, such as: Caffe [17], Tensorflow [1] and PyTorch [30]. While fine-grain instrumentation can inject large latencies, we propose a set of optimizations that enable FASE to emulate unmodified applications on large input sets with latencies that range from 17× to 39×, which are comparable to other fine-grain state-of-the-art techniques. As a result, FASE enables hardware architects to understand numerical behaviour before committing to costly hardware implementations. This paper makes the following contributions:

- We propose FASE[1] an emulation tool for custom numerical formats that enables accurate emulation of large workloads without requiring any source code modifications or access to third-party dynamically linked libraries.
- We design performance optimizations that enable accurate emulation with low overhead to support large-scale experimentation.
- An exhaustive evaluation campaign that demonstrates that FASE achieves better accuracy with respect to other state-of-the-art coarser-grain approaches, as well as large-scale experiments using multiple numerical formats that demonstrate that FASE is able to evaluate the fidelity of numerical formats.

[1] Source code is publicily available at https://gitlab.bsc.es/josorio/fase.

Table 1. Comparison of state-of-the-art proposals.

Features	Emulators				
	RPE [7]	QPyTorch [39]	TensorQuant [26]	Verificarlo [4]	FASE
Fast	×	✓✓	✓	✓✓	✓
Accurate	✓	×	✓	✓	✓
Seamless	×	×	×	× (recompilation)	✓
Dynamic Libraries	×	×	×	× (Lib. recompilation)	✓
Independent	×	×	✓	× (compiler dep.)	✓

2 Background and Motivation

The increasing demand for computing power in machine learning training motivates the use of reduced numerical precision formats. It has lead to a myriad of proposals for custom reduced precision numerical formats, both floating-point and integer, to improve the large computational and energy costs of training DNN. These workloads can tolerate well low-precision formats in certain computations, with proposals that go as low as 4-bit numerical representations [10,11,34].

Machine learning and DNN workloads in particular heavily rely on linear algebra kernels that can greatly benefit from low-precision formats in order to reduce memory bandwidth and storage usage, as well as improve compute throughput by leveraging vectorisation or accelerators that can fit more elements per instruction. An example is the adoption of the new Brain Float 16-bit (BF16) numerical format, extensively used in DNN workloads, by most hardware vendors [2,28,31]; which may be used to substitute the IEEE 754 32-bit floating-point typically employed.

In order to evaluate new numerical formats without available hardware support, several tools and methodologies to emulate low-precision numerical formats have been proposed. Table 1 qualitatively compares multiple state-of-the-art proposals on a number of key features. We consider a proposal is *fast* if it is feasible to emulate unmodified applications on large input sets, i.e., if the workload does not need to be scaled down to have feasible emulation times. *Accurate* means that the emulation is done at a fine-grain granularity (e.g., per instruction), rather than at coarse-grain granularity (e.g., per function) which may lead to results that are more accurate than actual computations at low precision. *Seamless* means that the emulated code does not need to be modified while *dynamic libraries* means that the tool is able to emulate code from dynamically linked libraries which may not always be open source. Finally, *Independent* is for tools that can work on any programming language and are also compiler independent.

The Reduced Precision Emulator (`rpe`) [7] is an emulator which supports reduced precision that can be computed on the available hardware format and rounding. The tool operates in a fine-grain manner. They report overheads from 10–70× on small emulated workloads. However, like all other source level approaches, code modification interfere with compiler optimizations impacting numerical accuracy [8]. Furthermore, it is currently restricted to Fortran applications. `Verificarlo-Vprec` [4,8] propose an LLVM compiler pass, at the end of the optimization passes, replacing all floating-point operations by user defined ones. Vprec enables emulating reduced precision formats like BF16. It allows accurate per operation rounding, with a latency from 3 to 17× according to their experiments [4]. It handles all programming languages supported by LLVM. However it does require the recompilation of all the application and its static and dynamic dependencies. Which is a tedious process, and not always possible for closed source libraries. However, they support Python environment by proposing a prebuild linux docker image.

There are two main tools that focus specifically on DNN workloads, TensorQuant [26] and QPyTorch [39]. TensorQuant is a quantization toolbox for the Tensorflow framework that provides multiple methods to apply reduced precision formats. They propose a coarse-grain method that applies the rounding processes at the end of each DNN layer, all intermediate computations inside a layer are not altered. TensorQuant also has a fine-grain operation-by-operation method that enables accurate emulation with a reported latency increase of around 20×. Using the fine-grain method is complex, as the user needs to re-implement each composite operation using C++ calls. It only works on Tensorflow, requires code modifications on each workload and does not instrument dynamically linked libraries. Some low-precision DNN training schemes like [10,11] use QPyTorch [39] as reduced precision framework. QPyTorch is a fast reduced-precision emulation framework for PyTorch. QPyTorch first represents the low precision numbers as their corresponding floating point number, then operates using single-precision floating point computation and then removes the extra precision through a final quantization step. While it is a fast methodology, the reduced-precision transformations are done at coarse-grain level; it may not capture the real effects of using reduced precision; it requires code modifications on each workload; it does not instrument dynamically linked libraries.

In contrast, FASE seamlessly works on any ML framework and is able to emulate code in dynamically linked libraries. This is crucial in DNN training workloads as most low level compute kernels are implemented in highly optimized external libraries. In addition, we make FASE accurate by operating at fine-grain. To reduce the latency of having accurate emulation we implement multiple optimizations that enable FASE to emulate large workloads with overheads that are competitive with other state-of-the-art proposals. We detail our design choices in Sect. 3, the implementation and performance optimization in Sect. 4, the strategy we apply to evaluate the tool on machine learning frameworks in Sect. 5 and evaluate accuracy and performance in Sect. 6.

3 FASE Design

Our goal is to design FASE with simplicity in mind by enabling fast, accurate and seamless emulation of reduced precision formats. In addition, we want our

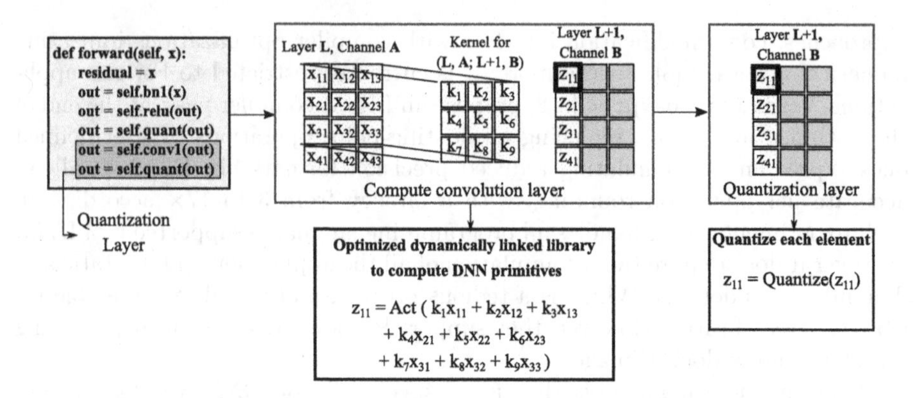

Fig. 1. Steps for coarse-grain emulation on a convolutional layer.

tool to be able to emulate code of external dynamically linked libraries, as many applications rely on such libraries which contain key optimized routines.

Figure 1 shows a forward pass example to demonstrate the operations of extrinsic coarse-grain reduced precision emulation. In this example, a convolution layer performs a dot product using a 3×3 filter to compute each element of the output layer *L+1*. These low level compute kernel implementations are typically found in optimized external libraries such as Intel oneDNN [19]. On the left side, the application needs to be modified to indicate where the conversion (quantization) takes place prior to the kernel. After the output layer *L+1* computation, a quantization and rounding step is performed over each element to obtain the desired reduced precision representation. This is a simple and fast methodology that allows to use well-known optimized libraries to compute the convolution. However, this method is not accurate as all operations within the layer employ the original single-precision format, leading to optimistic results not as accurate than using a fine-grain approach or real hardware.

FASE aims to provide an accurate and seamless method. To achieve this fine grain emulation, we propose to leave the target application unmodified and operate at binary level intercepting the executed machine instruction. By identifying key floating-point instructions, for which we can modify the input and output operands, FASE can seamlessly work on any application and DNN framework including dynamically linked external libraries.

4 FASE Implementation

4.1 Overview

In order to provide a fast, accurate and seamless experience; FASE relies on Dynamic Binary Translation (DBT). DBT enables modifications in the dynamic instruction flow of any application binary, as well as on any dynamically linked

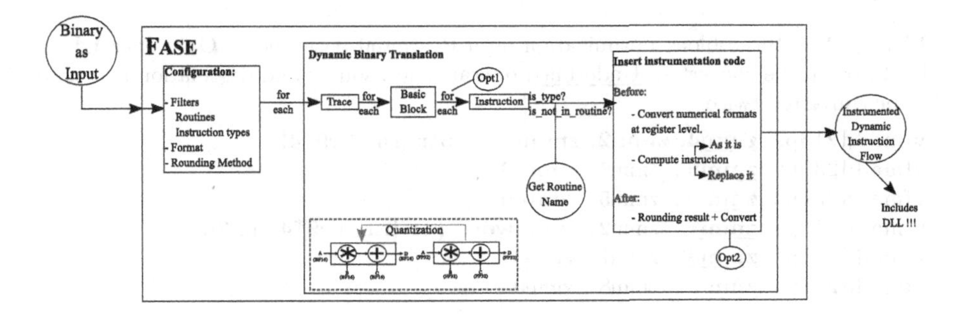

Fig. 2. FASE implementation overview

libraries the binary invokes. These modifications are done during the *instrumentation* step, which is executed only once.

Figure 2 shows an overview of the DBT instrumentation step on FASE. FASE can be attached to any binary, and is configured through a simple configuration file that specifies the desired instrumentation parameters in terms of routines and instructions to be instrumented as well as the emulated reduced precision format and rounding method. The DBT step which performs the instrumentation goes through each statically defined basic block once, and for each instruction it can insert instrumentation code. In our context, for each instruction of interest, we want to perform up to three code insertions:

1. **Before:** Insert code that converts the source registers of the instruction to the desired reduced precision format and applies the desired rounding.
2. **Instruction:** In most cases the instruction can be executed as is with the modified source registers. In some cases, when the numerical format will not execute as expected on the existing instruction or available hardware, the instruction needs to be replaced by equivalent code that emulates the intended behaviour. For example, when employing compound data types or custom formats that cannot be represented with the original numeric representation.
3. **After:** Insert code that converts the output to the desired reduced precision format and applies the rounding mechanism.

Once the code has been instrumented at basic block level, the next step is *analysis*. During the analysis step the instrumented dynamic instruction flow, which includes any external libraries, is executed. Analysis is the most compute expensive step as the modified instruction flow with code insertions is executed.

4.2 Features and Configuration Options

FASE has a number of built-in features and configuration options that simplify the use of the tool and enable fine tuning of the emulation process.

Listing 1.1. Basic block optimization on a ResNet50 basic block. Only operands in bold need to be converted. Underlined operands are source and destination and need to be converted twice.

```
vfmadd213ps zmm4, zmm2, zmmword ptr [rax+r9*4]
vfmadd231ps zmm5, zmm4, zmm3
vfmadd231ps zmm6, zmm5, zmm0
vfmadd213ps zmm7, zmm2, zmmword ptr [rax+r9*4+0x20]
vfmadd231ps zmm8, zmm7, zmm3
vfmadd231ps zmm9, zmm8, zmm0
```

Filters: There are two main types of filters: routine names and instruction types. Users can specify routines that should not be instrumented, i.e., routines that require high precision or that are not of interest for the target application. In terms of instruction types, FASE provides easy tags to identify most types, for example, all floating-point instructions or just specific instruction types like FMAs. Different instruction types can be defined to use different reduced precision numerical formats or rounding methods.

Dynamically Changing Precision During Analysis Step: FASE supports an inter-process communication (IPC) method that enables signaling FASE from the emulated application to dynamically change emulation behaviour. This does require modifications to the emulated application, in the form of simple function calls, to signal FASE to change its operation mode.

Numerical Formats and Rounding Methods: FASE can support any custom low precision numerical format and rounding method. If the format is compatible with the original instruction binary size of exponent and mantissa, then the inserted code in the instrumentation phase is simpler, as it just has to convert the source and destination registers. If the format cannot be operated by the original instruction, it is replaced by code that can perform the operation.

4.3 Optimizations

In this section we explain the different optimizations we apply to FASE to match state-of-the-art proposals while achieving high emulation accuracy. For all optimizations, FASE is performing as much work as possible in the instrumentation step to lower analysis overheads, as instrumentation is performed only once statically per basic block. Therefore, we apply all the filters during the instrumentation step and only insert the necessary code for the selected instructions and routines, which will run in the analysis step.

We started from a straight forward implementation where each FP instruction is instrumented and the computation in the analysis phase in FASE is not optimized. This `unopt` version will be our upper bound for performance against

Listing 1.2. Vectorized BF16 with RNE conversion

```
1 inline __m512 ToBFloatRNEVec (__m512* input)
2 {
3     __m512i MSB_mask = _mm512_set1_epi32(0x80000000);
4     __m512i LSB_mask = _mm512_set1_epi32(1);
5     __m512i mask = _mm512_set1_epi32(0xFFFF0000);
6     __m512i qnan_mask = _mm512_set1_epi32(0x7FC00000);
7     __m512i rounding_mask = _mm512_set1_epi32(0x7FFF);
8
9     __m512i tmp = _mm512_srli_epi32(*(__m512i*)input, 16);
10    tmp = _mm512_and_si512(tmp, LSB_mask);
11    __m512i rounding_bias = _mm512_add_epi32(tmp, rounding_mask);
12
13    __m512i MSB_set = _mm512_and_si512(*(__m512i*)input, MSB_mask);
14
15    tmp = _mm512_xor_si512(*(__m512i*)input, MSB_set);
16    tmp = _mm512_add_epi32(tmp, rounding_bias);
17    tmp = _mm512_or_si512(tmp, MSB_set);
18
19    __mmask16 not_nan_mask = _mm512_cmp_ps_mask(*input, *input, _CMP_EQ_OQ);
20
21    tmp = _mm512_mask_and_epi32(qnan_mask, not_nan_mask, tmp, mask);
22
23    *input = *(__m512*)&tmp;
24
25    return *input;
26 }
```

which the following optimization will be evaluated in Sect. 6. On the other end, the fully optimized version will be referred as `full opt`.

Basic-Block Level Optimization: During the instrumentation step we perform a basic-block level optimization that enables a substantial reduction of inserted code. We keep track of all source and destination register names that will be converted and rounded, if one of these registers is used as source in a subsequent instruction within the basic-block, it is safe to skip the conversion and rounding of that register as it is already in the desired target numerical format. Since it is quite common for destination and source registers to be reused in subsequent instructions, this optimization is very effective at reducing the overheads during the analysis step, as no work needs to be done for many source operands. Listing 1.1 shows an example of the traces generated by DNN frameworks. Only the highlighted operands need to be converted (underlined need to be converted twice as they are source and destination registers), saving 29.2% of the time in this particular basic block. In Fig. 2 and Sect. 6.2 we refer to this optimization as `Opt1`.

Vectorization: When instrumenting vectorized code, which is common in HPC and DNN low-level optimized kernels, FASE has support to do the numerical conversions and rounding methods also in a vectorized manner. This optimization greatly reduces the latency of instrumented vector instructions.

Listing 1.2 presents the vectorized optimization FASE implements to boost the performance, reducing the emulation latency as Sect. 6.2 shows. In this example, we implement the rounding process using AVX512 Intel Intrinsics, but 256bit, 128bit and scalar implementations are also available. This allows us to round the elements in the AVX vector register in a data parallel manner. Lines 3–7 define the whole set of masks we need to do the rounding. Lines 9–11 compute the rounding bias. Then, we need to do an unsigned integer addition between the rounding bias and the input, however AVX512 does not support it. Due to this issue FASE uses a few additional instructions to achieve it: we save the MSB bits of each element of the AVX512 vector (line 13), then we set all MSB of the input to zero in line 15, then compute a signed integer addition (line 16) and finally reset the MSB bits to its original value in line 17. Finally, FASE just needs to check for NaN values and return the AVX512 vector. In Fig. 2 and 6.2 section we refer to this optimization as `Opt2`.

5 Applying FASE to DNN Training Workloads

In this paper, the main use case for FASE is its applicability to DNN training workloads. These workloads have high computational cost while tolerating reduced precision formats that FASE can emulate accurately. Multiple proposals to employ reduced precision training methodologies for DNN workloads exist. Some are based on emulation [21], while others target existing hardware [28].

5.1 Reduced Precision Formats

The need for reduced precision formats for DNN training has lead to numerous proposals. The most prominent to date, which is being adopted by most hardware vendors, BF16 format. BF16 retains the same dynamic range as FP32 as it has the same number of exponent bits (8), but has a shorter mantissa of just 7 bits. The use of a 16-bit format can alleviate memory storage and bandwidth requirements as well as increase computational throughput.

With FASE we can emulate multiple numerical formats to understand the behaviour of DNN training. For example:

- **Floating-point and integer formats:** FASE can easily support emulation of BF16, FP8, or other FP layouts by converting the necessary source and destination registers of floating-point instructions to these formats. Similarly, integer formats such as INT8 can also be emulated.
- **Compound numerical formats:** Compound datatypes based on the BF16 format have been proposed recently [14]. These formats link several (two or three) BF16 literals to increase precision while just operating using BF16 arithmetic. With FASE we can also emulate the use of these compound datatypes, as it is possible to change the semantics of the instrumented

instruction to perform the necessary computation required. However the final result cannot always be stored in memory with the compound datatype and must be converted to the original type. This could be alleviated by using a shadow memory mechanism in future works [6].

5.2 DNN Training Strategies

FASE enables the implementation of popular DNN training approaches as well as experimenting with new methodologies.

Static Strategies: For example, one can test the accuracy of a DNN model training when using BF16, FP8 or any other FP representation on the entire workload. Or emulate the already proposed mixed precision [21,28] training technique, which is similar to using BF16 but uses the FP32 representation to do the accumulation step on FMA instructions.

Using Routine Filters: Certain functions (or DNN layers) require higher accuracy than others. For this reason FASE enables applying different numerical format conversions or avoid emulation altogether of certain routines. In DNN training, the *weight updates* and *batch normalization* layers are known to require FP32 precision to ensure network convergence. FASE enables this behaviour via simple configuration options.

Dynamic Precision Schemes and Compound Datatypes: FASE also enables to use of dynamic precision schemes that dynamically adapt to workload state at runtime. For example, it enables to adapt the numerical precision of the emulated format depending on how training convergence progresses in order to achieve the desired result.

6 FASE Evaluation

Our experimental methodology considers the evaluation of FASE on several DNN frameworks. We consider: Caffe, Tensorflow, PyTorch and an additional test using the C programming language. Our experiments are performed on an Intel Xeon Platinum 8160 processors. We compile each framework from source enabling AVX512 Intel optimizations on all of them.

6.1 Emulation Accuracy

Methodology: To evaluate emulation accuracy we use a common kernel present both in DNN training as well as a single precision matrix multiply (SGEMM). We implement this benchmark that multiplies two matrices using the Intel Math Kernel Library (oneMKL) [20]. We compiled the source code using GCC 8.1 with

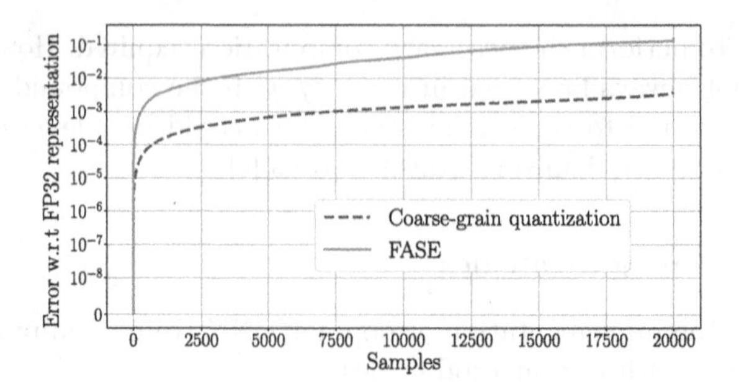

Fig. 3. Relative error of fine-grain and coarse-grain emulation methodologies (for BF16 and RNE rounding) with respect to a native FP32 execution.

all the AVX512 optimizations active on the platform we use. We use as input two matrices: $A = 20000 \times 2000$ and $B = 2000 \times 10000$.

We execute this benchmark with regular FP32 precision to get the reference output. We then emulate the use of BF16 with RNE rounding using two approaches. Firstly, we apply quantization for each element of the output matrix to represent the numbers using BF16 and RNE rounding (*coarse-grain quantization* label) over the reference result. This is akin to the coarse-grain methods used by QPyTorch [39] and TensorQuant [26]. Secondly, we attach FASE on top of the benchmark binary, which instruments the code from the dynamically linked Intel MKL library. This enables to execute the workload using FASE fine-grain emulation at instruction level, representing both the input and output numbers in BF16 with RNE rounding (FASE label). Finally, we compare the relative error with respect to the FP32 reference of the two emulation strategies.

The following results illustrate that the fine-grain approach is a much more accurate approach to emulate reduced precision numerical formats.

Results: Figure 3 compares the relative error when employing fine-grain and coarse-grain emulation on the Intel MKL SGEMM kernel. The *x-axis* represents 20000 samples (elements) of the result matrix, sorted in terms of the absolute numerical error for the fine-grain and coarse-grain techniques. The *y-axis* displays the magnitude of the relative error with respect to the reference FP32 result. As can be seen in the figure, the relative error with FASE, which is close to what would be observed on a real hardware implementation, is consistently one order of magnitude higher than with the coarse-grain approach. Therefore, using the coarse-grain approach may lead to wrong assumptions about a particular reduced precision numerical format, as it delivers results that are more accurate than they should. A coarse-grain method cannot capture the errors that accumulate per instruction; however, FASE is able to track these errors and deliver a result that is much closer to reality.

In Sect. 6.3 we demonstrate FASE on full DNN training workloads and show that using BF16 exclusively fails to deliver state-of-the-art training accuracy for certain neural networks, demonstrating the importance of fine-grain accurate emulation of reduced precision formats.

6.2 FASE Emulation Overhead Measurement

Methodology: To evaluate FASE latency overheads, and the impact of our optimizations, we propose an incremental evaluation process using the different FASE versions described in Sect. 4.3. For all benchmarks, we compare each version against a reference native FP32 execution without instrumentation. Additionally, we report FASE's instrumentation overhead, which just increments a counter per instruction of interest, i.e., without computing any of the conversions or rounding processes. This instrumentation overhead allows us to get a lower bound of the tool overhead and estimate the cost of the conversion and rounding process in the fully optimized `full opt` version.

We evaluate each FASE version on several benchmarks. First we evaluate the SGEMM computation, as described in Sect. 6.1. Then we evaluate FASE on the following machine learning workloads:

- ResNet50 for one batch of size 64, with Intel-Caffe [17] 1.1.6a. We use Intel MKL-DNN [18] 0.18.0, and Intel MKL [20] 2019.0.3 to run the numerical kernels. To define and run the experiments we use the *pyCaffe* Python interface. The learning rate, gamma hyperparameter, momentum value, and weight decay are set to 0.05, 0.1, 0.9, and 0.0001, respectively.
- CERN 3DGAN [22] with Tensorflow [1] 1.15 and Keras [5] 2. We use the same MKL and MKLDNN libraries as in the ResNet50 case. The 3DGAN network is trained for one batch using the Adam optimizer and a batch size of 128. The training dataset consists of 180,000 $25 \times 25\text{x}25$ three-dimensional images generated using HPC simulation for high-energy particles [22].

Finally, we consider two natural language processing models, for which we use a source-compiled version of PyTorch [30] version 1.8.0, Intel MKL-DNN [18] version 1.22.0, and Intel MKL library [20] version 2019.4.

- LSTMx2 model [38] on the PTB dataset. Following Zaremba et al. [38] we train one batch of the medium-sized model using the associated source code in [9] with a batch size of 20, an initial learning rate of 1, 2 LSTM layers, a hidden size of 650, a sequence length of 35, and a dropout value of 0.5.
- A transformer-based model [37] applied to the IWSLT16 dataset to translate between Dutch and English. We train for a batch size of 12000 using the Adam Optimizer with $\beta_1 = 0.9, \beta_2 = 0.98$, and $\epsilon = 10^{-9}$. We use the available code [12] and follow the author's additional instructions [12,37].

Results: Table 2 shows the emulation latencies introduced by FASE when converting in a fine-grain manner the input and output operands to BF16 with RNE rounding. We show the latency introduced by the *instrumentation* step of

Table 2. FASE instrumentation latency and latencies for FASE unoptimized, after applying each optimization and fully optimized.

Workload (framework)	FASE Instr.	Latency			
		`Unopt`	`Opt1` Basic block	`Opt2` Vectorization	**`Full Opt`**
SGEMM (MKL)	15×	1809×	880×	82×	39×
ResNet50 (Caffe)	11×	1131×	553×	76×	30×
3DGan (Tensorflow)	7×	714×	340×	66×	28×
LSTM (PyTorch)	18×	1096×	551×	70×	29×
Transformer (PyTorch)	8×	818×	423×	36×	17×

FASE in the "FASE Instrum." column, which on average is of 12×. This is the latency introduced just by counting the number of instructions of interest, and is therefore a lower bound of the overhead imposed by Intel Pin dynamic binary translation in FASE.

Regarding the latencies that include the emulation of the reduced precision format, we first show the latencies for an unoptimized version of FASE (Unopt.). This approach leads to latencies of up-to 1809×, which may deem the execution of large workloads unfeasible in a reasonable amount of time.

The *Basic-block* optimization, which refers to the *Opt1* version that avoids redundant rounding of registers, reduces FASE overhead by around half ranging from 340× to 880×. The observed latency reductions are inline with the amount of operands that need to be modified, as this optimization reduces by 50.89% the number of operands that FASE needs to convert in ResNet50.

The *Opt2* version in the *Vectorization* column is measuring the improvement we propose with custom AVX512 conversion and rounding process at analysis level using Intel Intrinsics. It results in a substantial speed up reducing FASE overhead latencies to the 36× to 82× range, emphasizing the importance of vectorizing the code on modern wide vector architectures.

Finally, we apply both the basic block *Opt1* and vectorization *Opt2* optimizations to our final *Full Opt* FASE version. It further reduces the final overhead down to 17× to 39×. It makes our fine-grain approaches very competitive to the state-of-the-art without any language, compiler or source access restrictions and the guarantee that the instrumented binary matches the original one.

6.3 Large-scale Experiments

Methodology: To show FASE supports real workloads we perform a set of large-scale experiments. These tests consider the use of several DNN models, datasets and numerical datatypes. We report the validation accuracy after training, BLEU Score, or perplexity depending on the workload type. We compare

Table 3. Large-scale experiments using FASE

Model	Dataset	Accuracy			
		FP32	BF16	MP	BF16 × 2
ResNet18	CIFAR100	71.91%	71.46%	71.89%	71.95%
ResNet34	CIFAR100	73.21%	72.83%	73.86%	72.66%
ResNet50	CIFAR100	74.78%	69.24%	74.25%	72.57%
ResNet101	CIFAR100	75.93%	67.10%	75.65%	76.00%
MobileNetV2	CIFAR100	75.04%	73.92%	75.16%	74.82%
AlexNet	ImageNet	60.79%	57.80%	60.18%	N/A
Inception	ImageNet	74.01%	72.03%	73.73%	N/A
LSTMx2 (Perplexity)	PTB	86.86	137.69	87.09	86.90
Transformers (BLEU)	IWSLT16	34.53	34.86	34.66	34.65

the obtained accuracies against the reference implementation using FP32. We use FASE to emulate three different numerical formats in order to demonstrate the versatility of our tool:

- **BF16** with RNE rounding used until now.
- The mixed-precision (**MP**) [21,28] approach that employs FP32 precision in batch normalization and weight update layers. And performs FMA instructions using BF16 source inputs for the multiplication and an FP32 input for the accumulator, returning an FP32 value as output.
- A compound datatype that represents FP32 values using a tuple of BF16 values (**BF16** × **2**) [14]. Note that this format requires changing the original instruction with ad-hoc code that performs the operation using BF16 × 2.

We consider the following object classification models: ResNet18, ResNet34, ResNet50, ResNet101 [13], and MobilenetV2 [32] on CIFAR100 datasets [25]. FASE attaches to Caffe framework to train AlexNet [24] and InceptionV2 [35] models, we use the same versions of tools as with the ResNet50 test on Sect. 6.2. Finally, we consider a full test on the same two natural language processing models as in Sect. 6.2. The whole set of hyper-parameters to train all of the models are detailed in the Supplementary Material.

Results: Table 3 shows the results of using FASE for several full DNN training workloads. We compare the accuracy of each network using our tool emulating different numerical formats (BF16, MP and BF16 × 2), and FP32.

With FASE we can determine if a reduced precision format is able to achieve the desired level of accuracy. When training object classification models on CIFAR100 with the BF16 numerical datatype, we observe significant drops in accuracy because the reduced number of mantissa bits in the BF16 numerical format fails to capture important information, especially on accumulations between distant numbers [15]. These drops are even higher on deeper models, for example, in ResNet101 there is an accuracy loss of 8.82% with respect to FP32. However, when FASE emulates MP using BF16 inputs and FP32 accumulators, these drops disappear, keeping the same levels of accuracy as FP32. The column BF16 $\times$ 2 shows results for a new compound datatype proposed by Henry et al. [14] that we emulate using FASE, it enables computing using BF16 arithmetic exclusively. In this case, we also observe good accuracy, on par with FP32.

Additionally, we emulate AlexNet and Inception training processes, FASE's results again show that using the BF16 numerical datatype is not enough to achieve comparable accuracy with respect to FP32. For AlexNet we measure an accuracy drop of 2.99%, while Inception model loses 1.98%. When emulating MP using FASE, we measure a boost on the accuracy reaching similar levels as FP32 for AlexNet and Inception, having drops of just 0.61% and 0.2% respectively.

Finally, FASE emulates the training of two natural language processing models. For the Transformer model we measure the BLEU score, higher is better. We observe that all the emulated numerical formats lead to accurate BLEU scores when compared to FP32. Transformer-based models are known to display robust numerical properties and are resilient to numerical noise [36]; therefore, we can obtain state-of-the-art results using BF16. For the LSTMx2 model we measure *perplexity* (lower is better); the BF16 approach stops converging after 13 epochs giving NaN as result, we register the last perplexity value of 137.69, this confirms that LSTM models are not good candidates to use BF16 exclusively.

However, when we emulate approaches such as MP or BF16 $\times$ 2, we again obtain results comparable with FP32. These set of results on large-scale workloads illustrates the potential of FASE to emulate different numerical formats and to extract conclusions on their applicability. FASE can also be employed to study scenarios where numerical precision is changed at runtime depending on application progress, and to study other custom floating-point representations; making it a compelling fast, accurate and seamless tool.

7 Conclusions

The use of reduced precision numerical formats to lower computational costs and increase compute throughput has shown good results in the context of HPC workloads. More recently, the same principle is leading to a myriad of proposals for custom reduced precision numerical formats, both floating-point and integer, to improve the large computational and energy costs of training DNN.

Prior tools and methodologies to emulate reduced precision formats cannot deliver a fast, accurate and seamless experience when training DNN workloads. In this paper we propose FASE, an emulation tool for custom numerical formats. FASE is: (i) *accurate* by leveraging DBT techniques to emulate formats at instruction operand level; (ii) *fast* as it enables emulation of unmodified applications on large input sets thanks to a set of optimizations that lower its overheads significantly; and (iii) *seamless* as it works on any application or DNN framework without any language, compiler or source access restrictions and the guarantee that the instrumented binary matches the original one.

Our evaluation demonstrates that FASE is more accurate than other state-of-the-art proposals that employ coarse-grain emulation, uncovering relative errors that appear only in fine-grain emulation. We demonstrate that by applying both the *basic block* and *vectorization* optimizations, FASE latency overheads are manageable, ranging between 17× to 39× for a wide variety of workloads. These latencies enable the evaluation of large-scale unmodified workloads, which illustrate the potential of FASE to emulate different numerical formats and to extract conclusions on their applicability.

Acknowledgements. Marc Casas has been partially supported by the Grant RYC-2017-23269 funded by MCIN/AEI/10.13039/501100011033 and by ESF Investing in your future. Adriá Armejach is a Serra Hunter Fellow and has been partially supported by the Grant IJCI-2017-33945 funded by MCIN/AEI/ 10.13039/501100011033. John Osorio has been partially supported by the Grant PRE2019-090406 funded by MCIN/AEI/ 10.13039/501100011033 and by ESF Investing in your future. This work has been partially supported by Intel under the BSC-Intel collaboration and European Union Horizon 2020 research and innovation programme under grant agreement No 955606 - DEEP-SEA EU project.

References

1. Abadi, M., Agarwal, A., Barham, E.A.: TensorFlow: large-scale machine learning on heterogeneous distributed systems. arXiv preprint arXiv:1603.04467 (2016)
2. ARM: SVE Instructions (sep 2021). https://bit.ly/3xC2B8E
3. Brown, T.B., Mann, B., Ryder, N., et al.: Language models are few-shot learners. Adv. Neural Inf. Process. Syst. **33**, 1877–1901 (2020)
4. Chatelain, Y., Petit, E., de Oliveira Castro, P., Lartigue, G., Defour, D.: Automatic exploration of reduced floating-point representations in iterative methods. In: Yahyapour, R. (ed.) Euro-Par 2019. LNCS, vol. 11725, pp. 481–494. Springer, Cham (2019). https://doi.org/10.1007/978-3-030-29400-7_34
5. Chollet, F., et al.: Keras. https://github.com/fchollet/keras (2015)
6. Courbet, C.: Nsan: A floating-point numerical sanitizer. In: Proceedings of the 30th ACM SIGPLAN (2021)
7. Dawson, A., Düben, P.D.: rpe v5: an Emulator for Reduced Floating-Point Precision in Large Numerical Simulations. Geoscientific Model Dev. **10**(6), 2221–2230 (2017)
8. Denis, C., de Oliveira Castro, P., Petit, E.: Verificarlo: checking floating point accuracy through monte carlo arithmetic. In: 2016 IEEE 23nd Symposium on Computer Arithmetic (ARITH) (2016)
9. Durmus, A.U.: Replication of Recurrent Neural Network Regularization by Zaremba (Sep 2019). https://github.com/ahmetumutdurmus/zaremba/
10. Fu, Y., Guo, H., Li, M., et al.: CPT: efficient DNN training via cyclic precision. In: International Conference on Learning Representations (ICLR) (2021)
11. Fu, Y., You, H., Zhao, Y., et al.: FracTrain: fractionally Squeezing Bit Savings Both Temporally and Spatially for Efficient DNN Training. In: Conference on Neural Information Processing Systems (NeurIPS) (2020)

12. Gordic, A.: Original PyTorch Transformer Model (Oct 2020). https://github.com/gordicaleksa/pytorch-original-transformer
13. He, K., Zhang, X., Ren, S., Sun, J.: Deep Residual Learning for Image Recognition. CoRR (2015)
14. Henry, G., Tang, P.T.P., Heinecke, A.: Leveraging the BFLOAT16 Artificial Intelligence Datatype for Higher-Precision Computations (2019)
15. Higham, N.J.: The accuracy of floating point summation. SIAM J. Sci. Comput. **14**, 783–799 (1993)
16. Huang, G., Liu, Z., van der Maaten, L., Weinberger, K.Q.: Densely Connected Convolutional Networks (2018)
17. Intel: Intel Caffe Framework Optimization. https://github.com/intel/caffe
18. Intel: Intel Deep Neural Network Library. https://github.com/intel/mkl-dnn
19. Intel: OneAPI DNN Library. https://oneapi-src.github.io/oneDNN/v1.4/index.html
20. Intel: Intel Math Kernel Library (2020). https://software.intel.com/en-us/mkl
21. Kalamkar, D., Mudigere, D., Mellempudi, E.A.: A Study of BFLOAT16 for Deep Learning Training (2019)
22. Khattak, G.R., Vallecorsa, S., Carminati, F., Khan, G.M.: Particle Detector simulation using generative adversarial networks with domain related constraints. In: 2019 18th IEEE International Conference On Machine Learning And Applications (ICMLA) (2019)
23. Kiar, G., et al.: (December 2021). https://github.com/verificarlo/fuzzy
24. Krizhevsky, A., Sutskever, I., Hinton, G.E.: ImageNet Classification with Deep Convolutional Neural Networks. In: Advances in Neural Information Processing Systems (NIPS) **25** (2012)
25. Krizhevsky, A.: Learning Multiple Layers of Features from Tiny Images. (2009)
26. Loroch, D.M., Wehn, N., Pfreundt, F.J., Keuper, J.: Tensorquant - A Simulation Toolbox for Deep Neural Network Quantization (2017)
27. Luk, C.K., et al.: Pin: Building customized program analysis tools with dynamic instrumentation. ACM SIGPLAN Notices (2005)
28. Micikevicius, P., Narang, S., Alben, R.A.: Mixed Precision Training. In: International Conference on Learning Representations (ICLR) (2018)
29. Molchanov, P., Tyree, S., Karras, T., Aila, T., Kautz, J.: Pruning Convolutional Neural Networks for Resource Efficient Inference (2017)
30. Paszke, A., Gross, S., Chintala, S., Chanan, G.: PyTorch (2020). https://github.com/pytorch/pytorch
31. Rodriguez, A., Ziv, B., Fomenko, E., Meiri, E., Shen, H.: Lower Numerical Precision Deep Learning Inference and Training (Oct 2018). https://intel.ly/32G5WrT
32. Sandler, M., Howard, A., Zhu, M., Zhmoginov, A., Chen, L.C.: MobileNetv2: Inverted Residuals and Linear Bottlenecks (2019)
33. Sun, X., et al.: Hybrid 8-bit floating point (HFP8) training and inference for deep neural networks. In: Advances in Neural Information Processing Systems (NeurIPS) (2019)
34. Sun, X., et al.: Ultra-low precision 4-bit training of deep neural networks. In: Advances in Neural Information Processing Systems (NeurIPS) (2020)
35. Szegedy, C., et al.: Going Deeper with Convolutions. In: 2015 IEEE Conference on Computer Vision and Pattern Recognition (CVPR) (2015)
36. Tambe, T., et al.: AdaptivFloat: A Floating-Point Based Data Type for Resilient Deep Learning Inference (2019)
37. Vaswani, A., et al.: Attention is All You Need (2017)

38. Zaremba, W., Sutskever, I., Vinyals, O.: RNN Regularization. arXiv preprint arXiv:1511.08400 (2015)
39. Zhang, T., Lin, Z., Yang, G., Sa, C.D.: QPyTorch: A Low-Precision Arithmetic Simulation Framework (2019)

GNNSampler: Bridging the Gap Between Sampling Algorithms of GNN and Hardware

Xin Liu[1,2], Mingyu Yan[1(✉)], Shuhan Song[1,2], Zhengyang Lv[1,2], Wenming Li[1,2], Guangyu Sun[3], Xiaochun Ye[1,2], and Dongrui Fan[1,2]

[1] SKLP, Institute of Computing Technology, CAS, Beijing, China
{liuxin19g,yanmingyu,songshuhan19s,lvzhengyang19b, liwenming,yexiaochun,fandr}@ict.ac.cn
[2] University of Chinese Academy of Sciences, Beijing, China
[3] School of Integrated Circuits, Peking University, Beijing, China
gsun@pku.edu.cn

Abstract. Sampling is a critical operation in Graph Neural Network (GNN) training that helps reduce the cost. Previous literature has explored improving sampling algorithms via mathematical and statistical methods. However, there is a gap between sampling algorithms and hardware. Without consideration of hardware, algorithm designers merely optimize sampling at the algorithm level, missing the great potential of promoting the efficiency of existing sampling algorithms by leveraging hardware features. In this paper, we pioneer to propose a unified programming model for mainstream sampling algorithms, termed GNNSampler, covering the critical processes of sampling algorithms in various categories. Second, to leverage the hardware feature, we choose the data locality as a case study, and explore the data locality among nodes and their neighbors in a graph to alleviate irregular memory access in sampling. Third, we implement locality-aware optimizations in GNNSampler for various sampling algorithms to optimize the general sampling process. Finally, we emphatically conduct experiments on large graph datasets to analyze the relevance among training time, accuracy, and hardware-level metrics. Extensive experiments show that our method is universal to mainstream sampling algorithms and helps significantly reduce the training time, especially in large-scale graphs.

Keywords: Graph neural network · Sampling algorithms · Acceleration · Hardware feature · Data locality

1 Introduction

Motivated by conventional deep learning methods, graph neural networks (GNNs) [1] are proposed and have shown remarkable performance in graph learning, bringing about significant improvements in tackling graph-based tasks [2–4].

© The Author(s), under exclusive license to Springer Nature Switzerland AG 2023
M.-R. Amini et al. (Eds.): ECML PKDD 2022, LNAI 13717, pp. 498–514, 2023.
https://doi.org/10.1007/978-3-031-26419-1_30

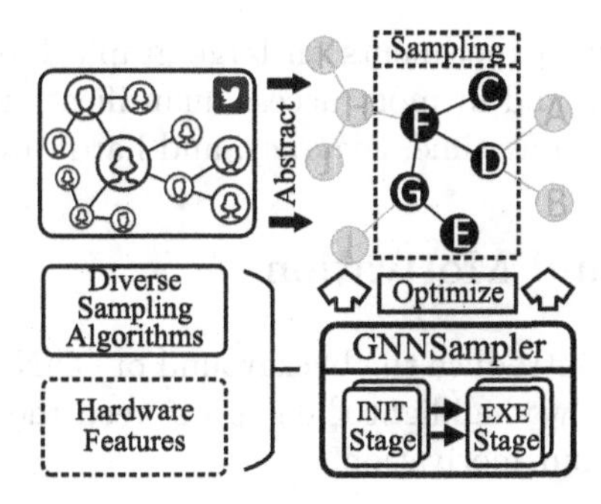

Fig. 1. Overview of efficient sampling in GNN training based on GNNSampler, where GNNSampler consists of multiple steps in the INIT and EXECUTE stage, making it possible that some substeps of sampling can be optimized in a fine-grained manner by leveraging hardware features.

Whereas, a crucial issue is that real-world graphs are extremely large. Learning large-scale graphs generally requires massive computation and storage resources in practice, leading to high cost in training GNN [5]. To this end, sampling algorithms are proposed for efficient GNN training, by conditionally selecting nodes to reduce the computation and storage costs in GNN training.

However, abundant irregular memory accesses to neighbors of each node are generally required in all sampling algorithms, introducing significant overhead due to the irregular connection pattern in graph [6]. Previous sampling-based models [4,7–13] leverage mathematical and statistical approaches for improvement, but they do not alleviate the high cost caused by irregular memory access. Thereby, algorithm designers merely optimize sampling at the algorithm level without considering hardware features. The efficient execution of sampling, and even the efficiency of GNN training, are limited by the gap between algorithm and hardware.

To this end, we target to bridge the gap between sampling algorithms of GNN and hardware. In this paper, as illustrated in Fig. 1, we pioneer to build GNNSampler, a unified programming model for sampling, by abstracting diverse sampling algorithms. Our contributions can be summarized as follows:

- We propose a unified programming model, termed GNNSampler, for mainstream sampling algorithms, which covers key procedures in the general sampling process.
- We choose data locality in graph datasets as a case study to leverage hardware features. Moreover, we explore the data locality among nodes and their neighbors in a graph to alleviate irregular memory access in sampling.
- We implement locality-aware optimizations in GNNSampler to improve the general sampling process, helping reduce considerable cost in terms of time. Notably, the optimization is adjustable and is performed once and for all, providing vast space for trading off the training time and accuracy.

- We conduct extensive experiments on large graph datasets, including time-accuracy comparison and memory access quantification, to analyze the relevance among the training time, accuracy, and hardware-level metrics.

2 Background and Motivation

In this section, we first introduce the background of GNN and mainstream sampling algorithms. Then, we highlight the gap between the algorithms and hardware, and put forward our motivation.

2.1 Background of GNN

GNN [1] was first proposed to apply neural networks to graph learning. It learns a state embedding $\mathbf{h}_v$ to represent neighborhood information of each node v in a graph. Generally, the $\mathbf{h}_v$ can be represented in the following form:

$$\mathbf{h}_v = f_{\mathbf{w}}(\mathbf{x}_v, \mathbf{x}_{e[v]}, \mathbf{h}_{n[v]}, \mathbf{x}_{n[v]}), \tag{1}$$

$$\mathbf{o}_v = g_{\mathbf{w}}(\mathbf{h}_v, \mathbf{x}_v), \tag{2}$$

where $\mathbf{x}_v$, $\mathbf{x}_{e[v]}$, and $\mathbf{x}_{n[v]}$ denote the features of v, v's edges, and v's neighbors, respectively. $f_{\mathbf{w}}(\cdot)$ and $g_{\mathbf{w}}(\cdot)$ are the functions defined for local transition and local output. And $\mathbf{o}_v$ is the output generated by the embedding and feature of node v. In this way, the hidden information of a graph is extracted by the following approach:

$$\mathbf{h}^{l+1} = f_{\mathbf{w}}(\mathbf{h}^l, \mathbf{x}), \tag{3}$$

where l denotes the l-th iteration of the embedding computation. Many variants take the idea from the original GNN and add some particular mechanisms to modify the models for handling various graph-based tasks. Herein, we highlight the form of graph convolutional networks (GCNs) [2] since most sampling algorithms are applied to GCNs for efficient model training. Generally, GCNs use a layer-wise propagation rule to calculate the embedding in the following form:

$$\mathbf{H}^{l+1} = \sigma\left(\widetilde{\mathbf{D}}^{-1/2}\widetilde{\mathbf{A}}\widetilde{\mathbf{D}}^{-1/2}\mathbf{H}^l\mathbf{W}^l\right), \tag{4}$$

where $\mathbf{H}^l$, $\mathbf{W}^l$ are the hidden representation matrix and trainable weight matrix in the l-th layer of the model. And $\sigma(\cdot)$ is the nonlinear activation function, such as ReLU and Softmax. GCNs represent neighborhood information with a renormalized adjacency matrix and extract the hidden information in such an iterative manner.

2.2 Sampling Algorithms in Training

Training GNNs, especially GCNs, generally requires full graph Laplacian and all intermediate embeddings, which brings about extensive storage cost and makes it hard to scale the training on large-scale graphs. Moreover, the conventional

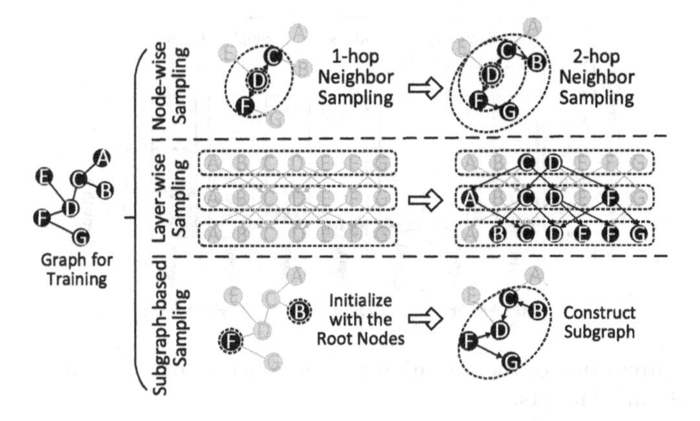

Fig. 2. Illustration on typical processes of various sampling algorithms.

training approach uses a full-batch scheme to update the model, leading to a slow convergence rate.

To overcome these drawbacks, sampling algorithms are proposed to modify the conventional training through a mini-batch scheme and conditionally select partial neighbors, reducing the cost in terms of storage and computation. Specifically, **mainstream sampling algorithms** can be divided into multiple categories according to the granularity of the sampling operation in one sampling batch [14]. As illustrated in Fig. 2, we respectively show typical sampling processes of multiple categories, that is, **node-wise**, **layer-wise**, and **subgraph-based** sampling algorithms.

As typical of **node-wise** sampling algorithms, GraphSAGE [4] randomly samples the neighbors of each node in multiple hops recursively; VR-GCN [7] improves the strategy of random neighbor sampling by restricting sampling size to an arbitrarily small value, which guarantees a fast training convergence. **Layer-wise** sampling algorithms, e.g., FastGCN [8] and AS-GCN [9], generally conduct sampling on a multi-layer model in a top-down manner. FastGCN presets the number of nodes to be sampled per layer without paying attention to a single node's neighbors. AS-GCN executes the sampling process conditionally based on the parent nodes sampled in the upper layer, where the layer sampling is probability-based and dependent among layers. As for **subgraph-based** sampling algorithms, multiple subgraphs, which are generated by partitioning the entire graph or inducing nodes (edges), are sampled for each mini-batch for training. Cluster-GCN [10] partitions the original graph with a clustering algorithm and then randomly selects multiple clusters to construct subgraphs. GraphSAINT [11] induces subgraphs from probabilistic sampled nodes (edges) by leveraging multiple samplers.

Unfortunately, the cost of sampling is gradually becoming non-negligible in the training of some sampling-based models, especially on large datasets. As proof, we conduct experiments on datasets with a growing graph scale (amount of nodes & edges) using GraphSAGE [4], FastGCN [8] and GraphSAINT (node

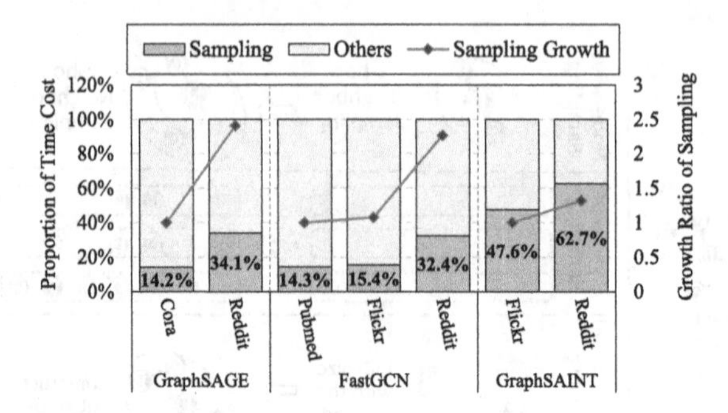

Fig. 3. The proportion of the execution time in different parts of training across different models and datasets.

sampler) [11], and quantify the proportion of sampling time to the training time, **based on the official setting of all parameters in their works**. The sampling part includes selecting nodes, constructing an adjacency matrix, and some subsequent processes. The other part denotes the rest of processes in training, e.g., feature aggregation and update. For each sampling algorithms, we consider the proportion of sampling time in the smallest dataset (e.g., Cora [15]) as the baseline and plot the growth trend of sampling time on each dataset. Detailed information about datasets is shown in Table 1. Distinctly, for all these sampling-based models, as the number of nodes and edges in a graph dataset grows, the cost of sampling becomes increasingly large and intolerable. As illustrated in Fig. 3, the proportion of sampling time becomes larger as the graph scale of a dataset increases, even up to 62.7%.

2.3 The Gap Between Algorithm and Hardware

As a critical process in training, sampling is becoming non-trivial, and its cost mainly derives from the gap between sampling algorithms and hardware. Algorithm designers do not consider hardware features and merely improve sampling at the algorithm level. On the other hand, hardware designers have not improved sampling algorithms since they do not know specific implementations of sampling algorithms. Therefore, mining of the improvement space for sampling algorithms is restricted by the gap. Thereby, recent literature proposes to leverage the hardware feature for improving the efficiency regarding training and inference of GNNs. For example, NeuGraph [16] makes the best of the hardware features of platforms (CPU and GPU), achieving excellent improvements in GNN training. For another example, HyGCN [17] tailors its hardware features to the execution semantic of GCNs based on the execution characterization of GCNs [18], greatly improving the performance of GCN inference.

We argue that sampling is also a process of algorithm and hardware coordination. We observe that existing sampling algorithms vary in their mechanisms. And an efficient improvement should be universal to most sampling algorithms of different mechanisms, urging the demand to put forward a general sampling

Algorithm 1: Abstraction for node-wise sampling

```
1  INPUT Graph G(V, E, X)
2  OUTPUT adjacency_matrix
3  get_minibatch(G)
4  S ← get_sample_standard()
5  for each batch B^k do
6      inputs ← get_batch_input(B^k)
7      n ← get_sample_num();
8      for each node v in batch inputs do
9          nei_idx ← get_neighbor(v)
10         sampled_idx ← sample_neighbor(nei_idx, n, S)
11         update_adjacency_matrix(sampled_idx)
12     end
13 end
```

(a) Abstraction for node-wise sampling

Algorithm 2: Abstraction for layer-wise sampling

```
1  INPUT Graph G(V, E, X)
2  OUTPUT adjacency_matrix
3  get_minibatch(G)
4  S ← get_sample_standard()
5  for each batch B^k do
6      inputs ← get_batch_input(B^k)
7      for each layer l do
8          n^l ← get_sample_num(l)
9          P ← get_prob_distribution()
10         sampled_idx ← sample_node(inputs, n^l, S, P)
11         update_adjacency_matrix(sampled_idx)
12         inputs ← update_input(sampled_idx)
13     end
14 end
```

(b) Abstraction for layer-wise sampling

Fig. 4. (a) Abstraction for node-wise sampling algorithms. (b) Abstraction for layer-wise sampling algorithms. To be as unified as possible, we use the same name for steps (functions) in the abstractions, despite some slight distinctions of required parameters.

programming model. To support our argument, we propose GNNSampler, a unified programming model for mainstream sampling algorithms. Moreover, we choose data locality as a case study and implement locality-aware optimizations to improve the general sampling process. By improving the process of sampling, we eventually reduce the time consumption of GNN training.

3 Unified Programming Model

In this section, we abstract sampling algorithms in different categories and propose the unified programming model.

3.1 Abstractions for Sampling Algorithms

Abstraction for Node-wise Sampling Algorithms. In Fig. 4(a), lines 3–4 of **Algorithm 1** correspond to the initialization phase of sampling. In this phase, nodes needed for training are first divided into multiple mini-batches. Lines 5–13 correspond to the execution phase of sampling. First, nodes in a mini-batch are first obtained in the form of an indices list. Next up, the sampling size for each node is obtained before sampling. Then, sampling is executed on each node in a batch to get indices of neighbors by leveraging some pre-calculated parameters. Finally, the adjacency matrix is constructed based on the sampled nodes and updated periodically per batch.

Abstraction for Layer-Wise Sampling Algorithms. In Fig. 4(b), layer-wise samplings do not need to focus on a single node since they sample a fixed number of nodes together in each layer based on the pre-calculated probability distribution. In **Algorithm 2**, line 9 corresponds to the calculation of probability distribution. Line 10 represents that n^l nodes are sampled together in the l-th layer

Algorithm 3: Abstraction for subgraph sampling

```
1  INPUT Graph G(V, E, X)
2  OUTPUT adjacency_matrix
3  get_minibatch(G)
4  S ← get_sample_standard()
5  for each batch B^k do
6      inputs ← get_batch_input(B^k)
7      P ← get_prob_distribution()
8      while node_num ≤ subgraph_size do
9          sampled_idx ← sample_node(inputs, S, P)
10         node_pool ← append_node(sampled_idx)
11     end
12     adjacency_matrix ← processing(node_pool)
13     induce_subgraph(adjacency_matrix)
14 end
```

(a) Abstraction for subgraph-based sampling

Algorithm 4: GNNSampler

```
1  INPUT Graph G(V, E, X)
2  OUTPUT adjacency_matrix
                                   ◁ INIT ▷
3  training_batch ← soft_partition(G)
4  L ← construct_locality(G)
                                   ◁ EXECUTE ▷
5  for each batch B^k in training_batch do
6      inputs ← get_batch_input(B^k)
7      info^k ← get_model_info()
8      M ← get_sample_metrics(L, info^k)
9      while not terminate do
10         sampled_idx ← sample_node(inputs, info^k, M)
11         update_sampled_node(sampled_idx)
12     end
13     update_batch_adj()
14 end
```

(b) Pseudocode of GNNSampler

Fig. 5. (a) Abstraction for subgraph-based sampling algorithms. (b) Pseudocode of GNNSampler. Please note that, in subplot (b), we add a particular step, termed "construct_locality", to the INIT stage to exploit the data locality among nodes and their neighbors in a graph. In the EXECUTE stage, sampling can be executed with less irregular memory access since the sampling weight is computed according to the data locality among nodes.

by leveraging some pre-calculated parameters. Generally, the indices of sampled nodes are used to update the adjacency matrix. For some layer-wise sampling algorithms (e.g., AS-GCN [9]) in which nodes are sampled according to parent nodes (already sampled nodes) in the upper layer, the indices are further used to update the input of sampling for the next layer.

Abstraction for Subgraph-Based Sampling Algorithms. In Fig. 5(a), mini-batches obtained in line 3 of **Algorithm 3** correspond to the training nodes partitioned artificially or with a clustering algorithm. Lines 8–11 correspond to the main loop of sampling, and it does not stop until the number of sampled nodes equals the number needed to form a subgraph. The sampled nodes are temporarily stored in a "node_pool". The function named "processing" in line 12 denotes a further process after sampling, e.g., sorting nodes' indices and removing the duplicate nodes. Finally, a subgraph is induced from the adjacency matrix generated by sampled nodes. Multiple subgraphs are generated by repeating the above process.

3.2 GNNSampler and Workflow

Design: Based on the above abstractions, we propose the unified programming model, i.e. GNNSampler, in Fig. 5(b). We divide the sampling process into two stages, namely INIT and EXECUTE. The target of the INIT stage is to obtain necessary data, e.g., batched nodes, in advance for the EXECUTE stage. In the EXECUTE stage, line 7 denotes that information of model structure is obtained to help configure the sampling. In line 8, the obtained metrics denote critical

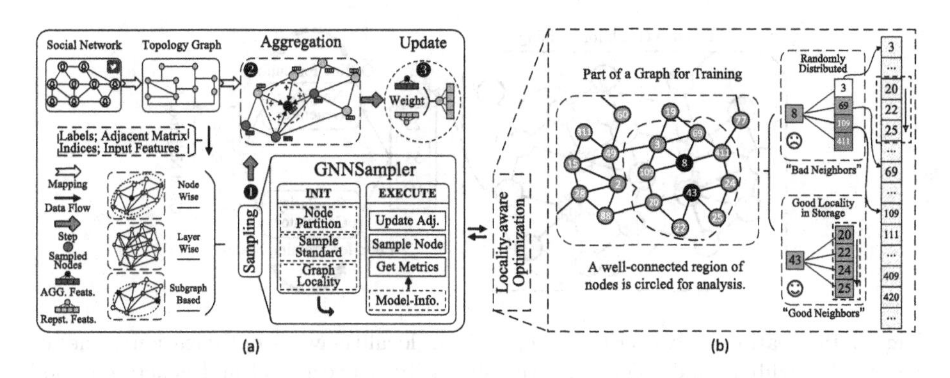

Fig. 6. (a) Workflow of learning large-scale graphs with GNN embedded with GNNSampler (some general processes, e.g., loss compute and model update, are not specially shown). (b) An exemplar for sampling nodes leveraging hardware feature.

factors for sampling, e.g., sampling size and probability. We also add the influence of data locality in the calculation of the sampling probability (i.e., L that computed by function construct_locality). Lines 9–12 denote an iterative sampling process requiring significant computation and storage resources. Finally, the batched adjacency matrix is updated after the batched sampling. And subsequent steps, e.g., subgraphs induction and model training, can directly use the generated adjacency matrix.

Workflow: To embed GNNSamlper to GNN training, we first introduce the steps of pre-processing, sampling, aggregation, and update in GNN training, where sampling is a tight connection between other steps. Figure 6 (a) illustrates the workflow of learning large-scale graphs with GNN, where GNNSampler is embedded for optimizing sampling. To begin with, the graph data, such as social network, is transformed into a topology graph and is further converted into elements directly used during sampling, in the pre-processing step. ❶ In the sampling step, GNNSampler decomposes sampling into INIT and EXECUTE stages. The processed elements are fed into the INIT stage to compute critical metrics for sampling. In the EXECUTE stage, sampling is performed according to the critical metrics and acquires an adjacency matrix. ❷ In the aggregation step, the aggregation feature (abbreviated as AGG. Feats.) of one node is aggregated from features of the sampled nodes, after which a concatenation operation is applied to the AGG. Feats. and the representation feature (abbreviated as Repst. Feats.) of the node in the upper layer. ❸ In the update step, the Repst. Feats. of one node is updated by transforming the weighted concatenate feature with a nonlinear function [4]. The most critical one, i.e., the sampling process with GNNSampler embedded, is designed to be universal for all categories of sampling algorithms. Based on this universal design, we can propose a generic and highly compatible optimization to benefit all categories of sampling algorithms.

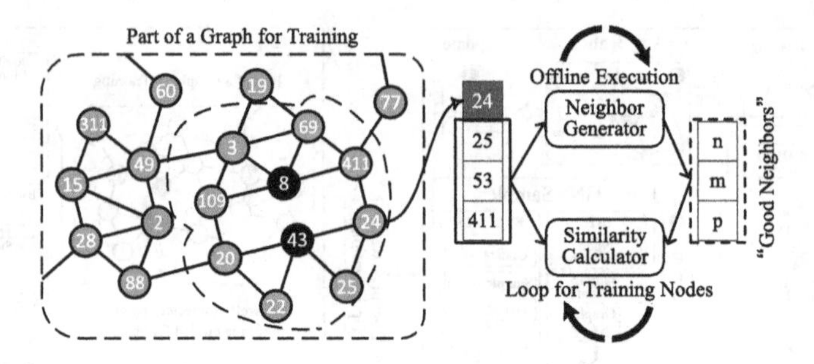

Fig. 7. Illustration of two critical steps of the locality-aware optimization: generating good neighbors and calculating the similarity between real and generated good neighbors. Please note that, this process is offline performed for each training node.

4 Case Study: Leveraging Locality

In this section, to leverage the hardware feature, we choose the data locality as a case study and implement locality-aware optimizations in GNNSampler to improve sampling. Please note that we refer to data locality as locality in brief in the rest of the paper.

4.1 Exploring Locality in Graph

Locality, i.e., principle of locality [19], is a particular tendency that processors have more opportunity to access data in the same set of memory locations repeatedly during a shorter interval of time. Specifically, locality can be embodied in two basic types: temporal locality for denoting the repeated access to the same memory location within a short time, and spatial locality for denoting the access to particular data within close memory locations.

Inspired by the success of locality exploitation in graph processing [20], we exploit locality to alleviate the irregular memory access in the sampling which helps reduce the execution time. Figure 6(b) gives an exemplar for analyzing neighbors of two nodes (i.e., "No.8" and "No.43") in a well-connected region. Neighbors of node "No.8" are randomly distributed, whilst neighbors of node "No.43" are almost contiguous. Notably, neighbors of "No.43" have a greater probability of being accessed within a short time according to the feature of locality since they are stored in adjacent location in memory. We would sample node "No.43" rather than "No.8".

Moreover, a new graph is constructed before sampling ends based on the sampled nodes in some cases. In this process, all nodes for training are required to search whether their neighbors are stored in the pool of the sampled nodes. For a node like "No.43", it is immediate to search the adjacent locations in the pool to verify their connections and add connections in storage (e.g., CSR format). When the graph is large, locality-aware optimizations can reduce considerable time cost in the above process since irregular memory accesses in searching neighbors are avoided as possible.

Algorithm 5: Locality-aware Optimizations

```
1  INPUT Adjacency matrix for training: adj_train; Similarity threshold: s;
     Minimum number of neighbors: n
2  OUTPUT Locality-based sampling weight: L
3  Function good_neighbor_generation(neighbors of v: nei_v)
4  avg_v ← get_average_neighbor_idx(nei_v)
5  num_nei_v ← get_neighbor_number(nei_v)
6  good_nei_v ← [···, avg_v − k × step, avg_v, avg_v + k × step, ···]
     (where k = 1, 2, ···, num_nei_v/2, and step can be specified)
7  Return good_nei_v
8  Function construct_locality(adj_train, s, n)
9  L ← Initialize(adj_train)
10 for each node v in training node set do
11   | nei_v ← get_neighbor(adj_train, v)
12   | if number of nei_v < n then
13   |   | L ← update_weight(0, v)
14   | end
15   | else
16   |   | good_nei_v ← good_neighbor_generation(nei_v)
17   |   | similarity_v ← Σ_{i=0}^{n−1} (nei_v[i] × good_nei_v[i]) / Σ_{i=0}^{n−1} nei_v[i]^2
18   |   | if similarity_v > s then
19   |   |   | L ← update_weight(1, v)
20   |   | end
21   |   | else
22   |   |   | L ← update_weight(0, v)
23   |   | end
24   | end
25 end
26 Return L
```

Fig. 8. Pseudocode of locality-aware optimizations.

4.2 Implementations of Locality-Aware Optimization

Design flow: The distribution of "No.43" and its neighbors is an exactly ideal case in a graph. In most situations, nodes are randomly distributed in general. Therefore, to estimate which node is suitable for sampling, we design two modules, i.e., a generator to yield (virtual) good neighbors and a calculator to compute the similarity between real neighbors and the generated good neighbors. As illustrated in Fig. 7, for one node v used in training, based on our prescript of good neighbors, i.e., neighbors are contiguous and adjacently located in memory, we use a generator to yield good neighbors for the given v. However, there may exist a gap between real neighbors and the generated good neighbors. We then utilize a calculator to quantify the gap. Specifically, we calculate the similarity between two neighbor sequences via a dot product ratio scheme

$$Similarity_v = \frac{\sum_{i=0}^{n-1} n_v[i] \cdot gn_v[i]}{\sum_{i=0}^{n-1} n_v^2[i]} \tag{5}$$

where n_v and gn_v denote sequences of real and the generated good neighbors of v. Through evaluation, we discover that the calculated similarity is generally larger as the resemblance of sequences' distributions increases. By setting a suitable threshold for the similarity, nodes whose real neighbors meet our standard (i.e., exceed the threshold) are chosen for sampling. The sampling process is uniformly performed (i.e., the sampling probability is uniform) on these nodes.

Implementation: We implement locality-aware optimizations for all categories of sampling algorithms. For node-wise and subgraph-based sampling algorithms, we use the function "construct_locality" given in Fig. 8 to construct locality among nodes and generate sampling weight. The input are the adjacency matrix, the minimum number of neighbors per node (abbreviated as `n`), and the similarity threshold (abbreviated as `s`). `n` is used to filter nodes with sparse connections, and `s` is used to measure the quality of locality among neighbors of each node by comparing the similarity between real neighbors and the generated good neighbors. The function "good_neighbor_generation" is used to generate neighbor. The output of the function "construct_locality" is the locality-based sampling weight `L` filled with "0" or "1" value which is used to represent whether a node v is suitable to be sampled. The subsequent sampling is performed based on `L`. By this means, nodes whose neighbors are stored closely in memory (i.e., with good locality) are more likely to be sampled, helping alleviate irregular data access. For layer-wise sampling algorithms, number of nodes to be sampled in each layer is relatively small and fixed for all datasets, which tends to form sparse connections between layers, especially in large datasets. We thereby explore leveraging the number of neighbors and previously sampled nodes to construct locality in two continuous layers. We first initialize a weight vector `L` for training nodes and set l_v as the weight in the corresponding position in `L`, where l_v is directly proportional to the number of v's neighbors. Moreover, nodes sampled in the upper layer are partly added to the candidate set to be sampled in the current layer to increase the sampling probability of the frequently accessed nodes.

Once-for-all: The proposed optimization is flexible to be embedded in the preprocessing step of mainstream sampling-based models. Moreover, two parameters, i.e., `n` and `s`, can be adaptively adjusted to achieve the desired trade-off. Notably, the computation of `L` merely requires the connections among training nodes and their neighbors (i.e., an adjacency matrix for training: adj_train) and **can be performed offline**. The pre-computed `L` can be reused in each batch of sampling, making the computation of `L` a **once-for-all process for each dataset**. Please refer to our code [1] for more details.

5 Experiment

To analyze the advance of our method, we conduct experiments on all categories of sampling algorithms to compare **vanilla** training methods (i.e., original models) and **our** improved approaches with locality-aware optimizations.

[1] https://github.com/TeMp-gimlab/GNNSampler.

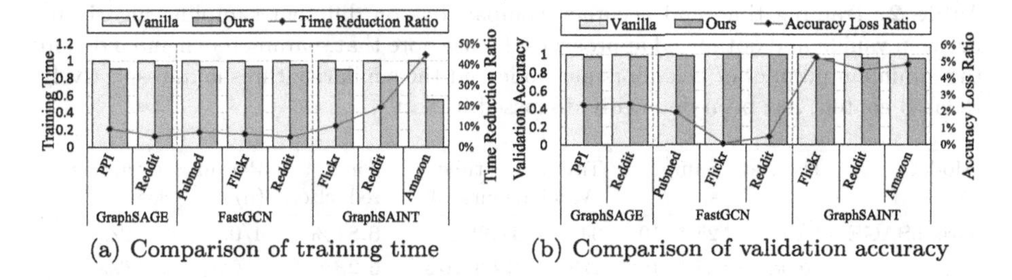

(a) Comparison of training time (b) Comparison of validation accuracy

Fig. 9. Comparisons between vanilla and our optimized approaches (normalized to vanilla) among three models using various datasets.

Table 1. Comprehensive analysis on the datasets.

Dataset	#Node	#Edge	ANN	MNN	NRR
Pubmed [15]	19717	44338	4	12	77.76%
PPI [21]	14755	225270	15	39	93.54%
Flickr [11]	89250	899756	5	9	80.16%
Reddit [4]	232965	11606919	50	113	97.99%
Amazon [11]	1598960	132169734	85	101	98.82%

5.1 Experimental Setup

Since the GNNSamper is general and compatible with mainstream sampling algorithms, we choose sampling algorithms in all categories as representatives, including GraphSAGE [4] (node-wise sampling), FastGCN [8] (layer-wise sampling), and GraphSAINT [11] (subgraph-based sampling). For all sampling-based models, **we use their official configurations in both sampling and training,** especially batch size, sampling size, and learning rate, to guarantee similar performance compared to their reported values. The basic GCN used in all cases is a two-layer model. For GraphSAINT, we choose the serial node sampler implemented via Python to randomly sample nodes for inducing subgraphs. These sampling-based models are regarded as comparison baselines, and we apply locality-aware optimizations to these models for improvement. By referring to the practice of previous works, we mainly focus on five benchmark datasets distinguishing in graph size and connection density as shown in Table 1. All experiments are conducted on a Linux server equipped with dual 14-core Intel Xeon E5-2683 v3 CPUs and an NVIDIA Tesla V100 GPU (16 GB memory).

5.2 Experimental Result and Analysis

Preliminary: We first analyze the datasets in multiple aspects. As given in Table 1, we make statistics on the (round-off) average number of neighbors per node (**ANN**) and the maximum number of neighbors of 90% nodes (**MNN**)

Table 2. Training time and accuracy comparisons on different sampling algorithms between vanilla and optimized approaches. Please note that, parameters **n** and **s** denote the minimum number of neighbors per node and the similarity threshold, respectively. Related content has been detailedly discussed in Sect. 4.2.

Model	Dataset	Sampling size	Training time (Vanilla/Ours)	Time reduction	Param. (n/s)	Accuracy loss
GraphSAGE	PPI	25 & 10	34.97 s/**31.89 s**	**8.81%**	4/0.95	2.41%
	Reddit	25 & 10	299.21 s/**283.40 s**	**5.28%**	4/0.87	2.48%
FastGCN	Pubmed	100	35.84 s/**33.30 s**	**7.09%**	-	1.93%
	Flickr	100	110.29 s/**103.34 s**	**6.30%**	-	0.06%
	Reddit	100	671.60 s/**639.16 s**	**4.83%**	-	0.48%
GraphSAINT	Flickr	8000	16.04 s/**14.39 s**	**10.29%**	3/0.865	5.25%
	Reddit	4000	229.47 s/**185.67 s**	**19.09%**	3/0.77	4.51%
	Amazon	4500	2263.92 s/**1253.33 s**	**44.64%**	4/0.775	4.83%

in datasets to reflect density of connections among nodes. We count neighbor reusing rate (**NRR**) by calculating the number of reused neighbors as a proportion of the total number of neighbors of all nodes. The collected statistics will help establish a relationship among attributes (e.g., size, density of connection) of graph datasets and experimental results.

Result: As illustrated in Fig. 9(a), we compare the converged training time and validation accuracy on diverse sampling-based models and datasets. The overall average time reduction is 13.29% with a 2.74% average accuracy loss. Specifically, the average time reduction on GraphSAGE and FastGCN is 7.06% and 7.14%. Notability, the optimization achieves an average 24.67% time reduction in GraphSAINT, while the peak of time reduction is 44.62% on Amazon dataset. We also observe a trivial decline in accuracy in Fig. 9(b), which varies by model. Considering characteristics of locality, nodes whose neighbors are adjacently distributed have a higher probability of being sampled, which yields a non-uniform sampling distribution. Therefore, biased sampling can result in a sacrifice in accuracy despite the considerable time reduction. Detailed performance and parameters (n and s) are given in Table 2.

Analysis: Based on the result, we analyze the relevance among the training time, accuracy, and hardware-level metrics, and summarize our findings as follows:

Time Reduction Introduced by the Optimization Varies by Dataset Because of the Distinct Sampling Size and Graph Scale. Distinctly, the percentage of time reduction in GraphSAINT is larger than in GraphSAGE and FastGCN, since the sampling size used in GraphSAINT is quite large. Moreover, for GraphSAINT, the percentage of time reduction on Amazon dataset is larger than on Flickr and Reddit since Amazon has an enormous graph scale (amount of nodes & edges). We also argue that higher **NRR** is another reason for achieving significant time reduction on Amazon. As shown in Table 1, **NRR** is a metric

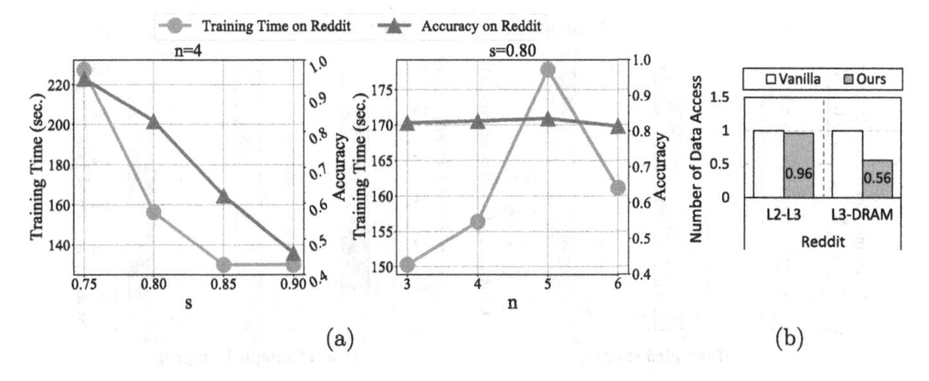

Fig. 10. (a) Exploration of the trade-off between training time and accuracy in GraphSAINT on Reddit by adjusting n & s. (b) Comparison of the number of data access.

to reflect neighbor reusing rate. Generally, a dataset with higher **NRR** includes node regions in which multiple nodes have many common neighbors. If neighbors of one node are frequently accessed, it is likely that for other nodes in such regions, their neighbors are also frequently accessed since they share many common neighbors. Thus, locality among nodes is more easier to explore in this case. Consequently, models using large sampling sizes and large-scale graphs are more likely to benefit from locality-aware optimizations.

A Good Trade-Off Between Training Time and Accuracy can be Achieved by Adjusting Parameters n and s. Since we merely retain the "good nodes" for sampling, our target is to find a trade-off point with considerable time reduction and tolerable accuracy loss. As illustrated in the left subplot of Fig. 10(a), as s increases (with n fixed), we have a tighter restriction on the quality of locality among neighbors per node, causing the nodes to be sampled are a minor part of total. This leads to accuracy loss since reducing nodes implies losing connections in a graph. Moreover, we sample nodes with good locality to reduce irregular memory access, indirectly saving training time. In the right subplot of Fig. 10(a), as n increases (with s fixed), the training time is generally increasing before reaching a peak. When n is set to 5, we can obtain a competitive accuracy with an undesirable training time, implying a compromise of choosing a smaller n is acceptable under variations in the accuracy are trivial. Thus, there is a correlation between training time and accuracy. By adjusting the parameters, one can derive a comparable accuracy with an acceptable training time.

Alleviating Irregular Memory Accesses to Neighbors Helps Reduce the Training Time. As shown in Fig. 10(b), we quantify the number of data access from L2 cache to L3 cache (L2-L3) and L3 cache to DRAM (L3-DRAM) with Intel PCM Tools [22] to analyze the sampling process. Definitely, locality-aware optimizations can significantly reduce the number of data access in L3-DRAM under the condition that L2-L3 is almost similar. By reducing data

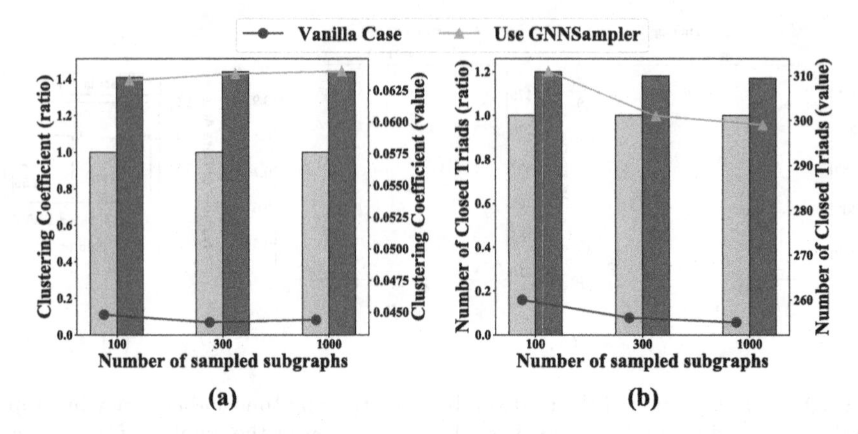

Fig. 11. Comparisons between vanilla and our (use GNNSampler) cases on Reddit among various metrics: (a) Comparisons on clustering coefficient; (b) Comparisons on the number of closed triads. Curves in subplots denote value variations of metrics.

access to DRAM, time of sampling is saved, which eventually accelerates the training.

Locality can be Empirically Reflected by the Topology of Sampled Subgraphs. With locality-Aware optimizations, we argue that subgraphs sampled are more concentrated in the local structure, avoiding irregular or highly stochastic pattern in the graph topology. To reflect such properties, we conduct analysis via Stanford Network Analysis Platform (SNAP) [23] tools. Specifically, by quantifying two metrics, i.e., the clustering coefficient (CC) and the number of closed triads (NCT), we analyze sampled subgraphs on Reddit using GraphSAINT model. CC is a local property that measures the cliquishness of a typical neighbourhood in a graph [24]. In Fig. 11(a), as the number of sampled subgraphs increases, the average CC of Ours is 1.42X of Vanilla, implying our sampled subgraphs are more concentrated and have a higher tendency of clustering. NCT is a typical structure among three nodes with any two of them connected, which is used to reflect to a balanced group pattern in social networks [25]. In Fig. 11(b), NCT of Ours is 1.18X of Vanilla, indicating highly interconnected structures are sampled under locality-aware optimizations.

6 Conclusion

In this paper, we propose a unified programming model for mainstream GNN sampling algorithms, termed GNNSampler, to bridge the gap between sampling algorithms and hardware. Then, to leverage hardware features, we choose locality as a case study and implement locality-aware optimizations in algorithm level to alleviate irregular memory access in sampling. Our work target to open up a new view for optimizing sampling in the future works, where hardware should be well considered for algorithm improvement.

Acknowledgment. This work was partly supported by the Strategic Priority Research Program of Chinese Academy of Sciences (Grant No. XDA18000000), National Natural Science Foundation of China (Grant No.61732018 and 61872335), Austrian-Chinese Cooperative R&D Project (FFG and CAS) (Grant No. 171111KYSB20200002), CAS Project for Young Scientists in Basic Research (Grant No. YSBR-029), and CAS Project for Youth Innovation Promotion Association.

References

1. Scarselli, F., Gori, M., Tsoi, A.C., Hagenbuchner, M., Monfardini, G.: The graph neural network model. IEEE Trans. Neural Netw. **20**(1), 61–80 (2008)
2. Kipf, T.N., Welling, M.: Semi-supervised classification with graph convolutional networks. In: ICLR (2017)
3. Schlichtkrull, M., Kipf, T.N., Bloem, P., van den Berg, R., Titov, I., Welling, M.: Modeling relational data with graph convolutional networks. In: Gangemi, A., et al. (eds.) ESWC 2018. LNCS, vol. 10843, pp. 593–607. Springer, Cham (2018). https://doi.org/10.1007/978-3-319-93417-4_38
4. Hamilton, W., Ying Z., Leskovec, J.: Inductive representation learning on large graphs. In: Advances in Neural Information Processing Systems, vol. 30 (2017)
5. Liu, X., Yan, M., Deng, L., Li, G., Ye, X., et al.: Survey on graph neural network acceleration: an algorithmic perspective. arXiv preprint arXiv:2202.04822 (2022)
6. Yan, M., Hu, X., Li, S., et al.: Alleviating irregularity in graph analytics acceleration: a hardware/software co-design approach. In: MICRO (2019)
7. Chen, J., Zhu, J., Song, L.: Stochastic training of graph convolutional networks with variance reduction. In: ICML (2018)
8. Chen, J., Ma, T., Xiao, C.: FastGCN: fast learning with graph convolutional networks via importance sampling. In: ICLR (2018)
9. Huang, W., Zhang, T., Rong, Y., Huang, J.: Adaptive sampling towards fast graph representation learning. In: Advances in Neural Information Processing System, vol. 31, pp. 4563–4572 (2018)
10. Chiang, W.L., Liu, X., Si, S., et al.: Cluster-GCN: an efficient algorithm for training deep and large graph convolutional networks. In: SIGKDD (2019)
11. Zeng, H., Zhou, H., Srivastava, A., Kannan, R., Prasanna, V.: GraphSAINT: graph sampling based inductive learning method. In: ICLR (2020)
12. Zeng, H., Zhou, H., Srivastava, A., Kannan, R., Prasanna, V.: Accurate, efficient and scalable graph embedding. In: IPDPS (2019)
13. Zeng, H., Zhang, M., et al.: Decoupling the depth and scope of graph neural networks. In: Advances in Neural Information Processing Systems, vol. 34, pp. 19665–19679 (2021)
14. Liu, X., Yan, M., Deng, L., Li, G., Ye, X., Fan, D.: Sampling methods for efficient training of graph convolutional networks: a survey. IEEE/CAA J. Automatica Sin. **9**(2), 205–234 (2021)
15. Sen, P., Namata, G., Bilgic, M., Getoor, L., Galligher, B., Eliassi-Rad, T.: Collective classification in network data. AI Mag. **29**(3), 93–93 (2008)
16. Ma, L., Yang, Z., Miao, Y., et al.: NeuGraph: parallel deep neural network computation on large graphs. In: USENIX ATC (2019)
17. Yan, M., et al.: HyGCN: a GCN accelerator with hybrid architecture. In: HPCA (2020)
18. Yan, M., Chen, Z., Deng, L., et al.: Characterizing and understanding GCNs on GPU. IEEE Comput. Archit. Lett. **19**(1), 22–25 (2020)

19. Denning, P.J.: The locality principle. In: Communication Networks And Computer Systems: A Tribute to Professor Erol Gelenbe (2006)
20. Mukkara, A., Beckmann, N., Abeydeera, M., et al.: Exploiting locality in graph analytics through hardware-accelerated traversal scheduling. In: MICRO (2018)
21. Zitnik, M., Leskovec, J.: Predicting multicellular function through multi-layer tissue networks. Bioinformatics **33**(14), i190–i198 (2017)
22. Thomas, W., Roman, D.: Intel performance counter monitor - a better way to measure CPU utilization (2018). https://github.com/opcm/pcm
23. Leskovec, J., Sosič, R.: SNAP: a general-purpose network analysis and graph-mining library. ACM Trans. Intell. Syst. Technol. (TIST) **8**(1), 1–20 (2016)
24. Watts, D.J., Strogatz, S.H.: Collective dynamics of 'small-world' networks. Nature **393**(6684), 440–442 (1998)
25. Easley, D., Kleinberg, J.: Networks, Crowds, and Markets: Reasoning about a Highly Connected World. Cambridge University Press, Cambridge (2010)

Training Parameterized Quantum Circuits with Triplet Loss

Christof Wendenius, Eileen Kuehn(✉), and Achim Streit

Karlsruhe Institute of Technology, Hermann-von-Helmholtz Platz 1, 76344 Eggenstein-Leopoldshafen, Germany
{christof.wendenius,eileen.kuehn,achim.streit}@kit.edu

Abstract. Training parameterized quantum circuits (PQCs) is a growing research area that has received a boost from the emergence of new hybrid quantum classical algorithms and Quantum Machine Learning (QML) to leverage the power of today's quantum computers. However, a universal pipeline that guarantees good learning behavior has not yet been found, due to several challenges. These include in particular the low number of qubits and their susceptibility to noise but also the vanishing of gradients during training. In this work, we apply and evaluate Triplet Loss in a QML training pipeline utilizing a PQC for the first time. We perform extensive experiments for the Triplet Loss based setup and training on two common datasets, the MNIST and moon dataset. Without significant fine-tuning of training parameters and circuit layout, our proposed approach achieves competitive results to a regular training. Additionally, the variance and the absolute values of gradients are significantly better compared to training a PQC without Triplet Loss. The usage of metric learning proves to be suitable for QML and its high dimensional space as it is not as restrictive as learning on hard labels. Our results indicate that metric learning provides benefits to mitigate the so-called barren plateaus.

Keywords: Metric learning · Quantum machine learning · Parameterized quantum circuits

1 Introduction

In recent years, Quantum Machine Learning (QML) has become a highly active and promising field of research starting to leverage the enormous potential of Quantum Computers (QC) [7]. Current devices of the noisy intermediate-scale quantum (NISQ) era are still error-prone and unable to process large-scale datasets due to the coupling between algorithm complexity and noise [22]. However, through tremendous efforts, multiple advances have been made in both theory and practice to use QCs in and for machine learning [26,28].

Parameterized quantum circuits (PQCs), also referred to as Quantum Neural Networks (QNNs), are one of the most studied aspects of QML as their structure and trainable parameters are reminiscent of neural networks [18]. Hoping

© The Author(s), under exclusive license to Springer Nature Switzerland AG 2023
M.-R. Amini et al. (Eds.): ECML PKDD 2022, LNAI 13717, pp. 515–530, 2023.
https://doi.org/10.1007/978-3-031-26419-1_31

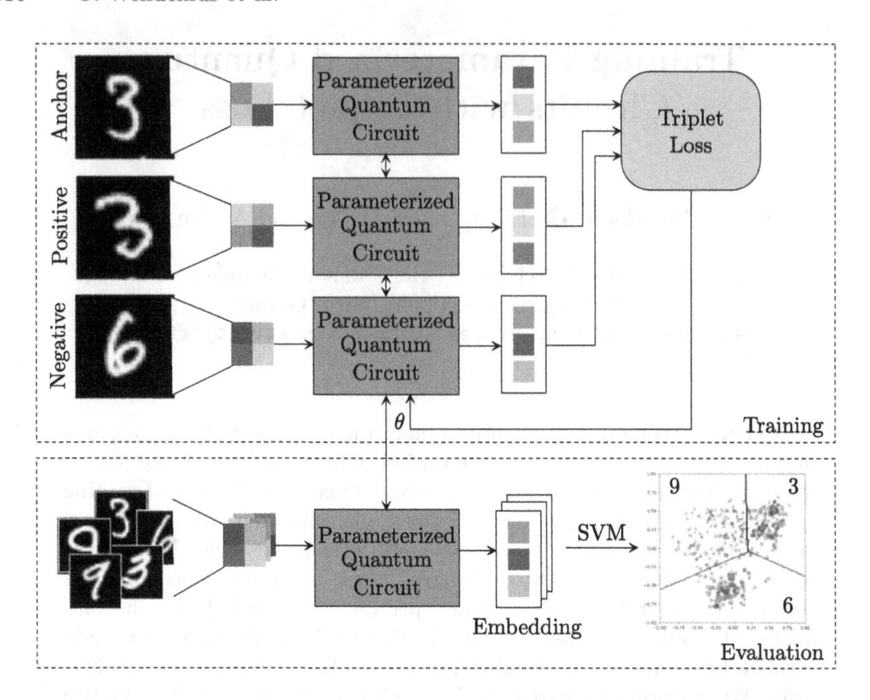

Fig. 1. Overview of our proposed Triplet Loss approach and the two pipelines for training and evaluation. A parameterized quantum circuit is used to create an embedding of the inputs for anchor, positive, and negative that are required for Triplet Loss. During training, these embeddings are taken to minimize the Triplet Loss. In the evaluation, the embeddings of the training set are used for training a linear SVM classifier. The accuracy of the trained quantum model is evaluated based on the embeddings of a dedicated test set.

to achieve similar leaps in machine learning tasks as their classical counterparts, particular attention has been focused on training PQCs. So-called barren plateaus, where the gradients vanish towards zero with an increasing number of qubits, often hinder successful training [17]. Although advances have been made towards mitigating these effects [5,9,26], there is still no universal answer to guarantee good learning behavior. Despite many proposed methods, effectively using classical data on QCs remains a challenge as well. Since PQCs are part of a hybrid quantum-classical training loop, the embedding of features into the quantum model and the interpretation of measurements for the loss function have to be studied further.

In this work, we propose to approach training PQCs from a new angle by applying a ranking loss function from the task of metric learning. These loss functions do not aim to get a model output that corresponds to a label for classification but optimize the relative distances between different inputs in the embedding space of the model [13]. Intuitively, we expect this to be a viable application of quantum computers for two reasons: On the one hand, entanglement of qubits allows to represent a huge embedding space internally, quadratic in the number of qubits. On the other hand, the measurement used to transition from quantum to classical state corresponds to a scalar product between quantum states, suitable as a native metric computation.

In particular, we propose to apply the so-called *Triplet Loss* which has been used to great effect for the training of classical neural networks [23]. This choice is somewhat arbitrary with respect to similar loss functions from metric learning such as contrastive loss. While we expect triplet loss to better populate the embedding space, a thorough evaluation of different metric loss functions is out of scope for this paper.

For triplet loss, as for any ranking loss function, similar inputs, i.e. inputs with the same class, should generate model outputs in proximity to each other. Analogously, non-matching inputs should be further apart in the embedding space. In this way, the model does not need to create a specific output for each input but can make use of the entire embedding space. By transferring this kind of loss function and training scheme onto quantum devices, training of PQCs is more unrestricted and can make use of the full range of the underlying computational space, the so-called expressibility.

To demonstrate the applicability of Triplet Loss for training of PQCs, we run several experiments on the moon [1] and MNIST [15] datasets. For this purpose, we implement two pipelines for training and evaluating our approach, which can be seen in Fig. 1. During training, the Triplet Loss is used to optimize the embeddings calculated by using a PQC by maximizing distances between dissimilar samples and minimizing distances for similar samples. With our work, we aim to show that Triplet Loss can achieve comparable results to training as it is currently common in the literature. In addition, higher gradient values indicate that the approach could achieve better learning behavior with larger circuits on NISQ devices. While metric learning is not new in the area of quantum computing [16,27], it has neither been used in the regime of PQCs nor has it been evaluated with respect to variances of gradients during training.

The paper is organized as follows. In Sect. 2 we give a brief overview on QNN and QML as well as the training of PQCs and discuss related work. This is followed by the basics on metric learning and Triplet Loss in Sect. 3, which are necessary to understand our approach in Sect. 4. Our experiments, datasets, and results are presented in detail in Sect. 5 We conclude in Sect. 6.

2 Quantum Neural Networks and Quantum Machine Learning

Quantum Neural Networks (QNNs) can be thought of as a generalization of Deep Neural Networks (DNNs). While in both cases a classical optimizer updates the models parameters θ to minimize a predefined loss function $\mathcal{L}$, the main difference lies in the model to be trained, as illustrated in Fig. 2. In the case of QML, the model is based on the principles of quantum mechanics such as superposition, entanglement and interference. In practice, there are numerous realizations of QNN models but most architectures share the same structure:

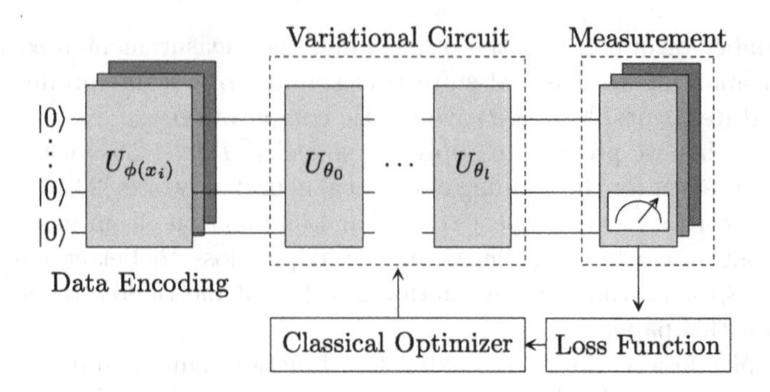

Fig. 2. Common quantum machine learning pipeline to train a parameterized quantum circuit. The parameterized quantum circuit is composed of three parts: the data encoding $U_{\phi(x)}$, variational circuit U_θ and measurement. The variational circuit has up to l layers. The training runs on n qubits. Measurement results are fed into a classical loss function that is used by a classical optimizer for minimization and update of θ. The QML pipeline is denoted as hybrid as a quantum and classical part are jointly used to harness the power of quantum computers.

Parameterized Quantum Circuits [2,4]. The hope is that with QNNs the power of quantum computers can be harnessed to outperform DNNs for specific use cases or quantum data.

A circuit contains any number of gates that are assigned to one or more qubits. Gates can transform the quantum state of each qubit they are assigned to corresponding to computational steps on a classical computer. In the visual representation of a circuit, the individual qubits are shown as a horizontal line. The structure of the circuit and thus the arrangement and composition of the gates is often referred to as an ansatz.

A PQC constitutes three parts: an encoder circuit $U_{\phi(x)}$ parameterized by the vector $\phi(x)$, a variational circuit U_θ parameterized by a vector θ, and the measurement part M. In this case $\phi(\cdot)$ relates to a preprocessing scheme that maps the input data to a vector that can be applied on the circuit. For each of these parts different gates can be chosen to describe transformations of the quantum state. While the selection, composition, and evaluation of a proper ansatz is not part of this paper, this is still a highly relevant field of research [3,4] impacting the trainability and practical usefulness of PQCs.

Purpose of the encoding part is loading classical data by encoding it into the quantum state of the qubits. In literature, there are many data encoding methods [14,19] and research is still ongoing as the data encoding itself is relevant to enable high expressibility of PQCs [24]. Some well-known approaches are for example (i) basis encoding, (ii) angle encoding, or (iii) amplitude encoding. For further details regarding data encoding as well as theoretical aspects of quantum computing we refer the interested reader to the work by Nielsen and Chuang [19].

Once the classical data sample is prepared, the variational circuit U_θ is applied to this state. This trainable part U_θ consists of trainable single-qubit quantum gates and fixed two-qubit quantum gates. Here, the weights known from classical NNs are in the form of rotational parameters. As with the encoding part, the variational circuit can be composed in different ways where using more gates enables higher expressibility but also challenged trainability [11]. One important concept that has established in literature is the hardware-efficient ansatz that introduces a layered structure where the placement of gates is equal in each layer l but parameterized by an own set of trainable parameters θ_l. The variational circuit U_θ is the sum of all its layers, that is $U_\theta = \prod_{l=1}^{L} U_{\theta_l}$ where L is the number of layers.

This processing step is followed by the quantum measurement to extract quantum information into its classical form. During measurement of a single qubit, its quantum state collapses to one classical single value, 0 or 1. Therefore, most QML approaches require a number of shots for measurements to estimate the expectation value with high probability, e.g. 1,000. The extracted information can either be used directly as a predicted label e.g. in case of quantum classifiers [14,27] or a hidden feature depending on the appropriate QML pipeline.

The whole QNN pipeline is commonly trained and executed in a hybrid fashion to accommodate for current NISQ devices while still being able to explore the potential advantages such as speed-ups in training and processing. While the PQC is purely based on principles of quantum mechanics, other parts including optimization are primarily done in a classical fashion. Much development in combining quantum and classical models has already been done by the community, but systematic research on the trade-offs and potential advantages is still ongoing. Furthermore, especially the generalization error of QML is of interest that directly depends on the training error. Thus, our approach complements these studies by evaluating the use of Triplet Loss in the QNN training pipeline.

2.1 Related Work on Training Parameterized Quantum Circuits

The trainability of PQCs is paramount to advance the field of QML. However, as mentioned before the expressibility of a quantum circuit challenges its trainability and comes along with the effect of exponentially vanishing gradients in the number of qubits, as a function of the number of layers, the so-called barren plateaus [17].

Different strategies have been proposed to mitigate barren plateaus, e.g. by initializing the parameters of variational circuits in form of identity blocks [9], or by pre-training a variational circuit based on layerwise learning strategies [26] as has been shown to provide good results in classical machine learning [10]. However, most of these strategies do not strictly guarantee to prevent barren plateaus during training.

The authors in [8,21] show that there are specific hierarchical architectures of hybrid quantum-classical models that are immune to barren plateaus. However, this massively restricts the available ansätze in a way that has not been explored yet.

But not only barren plateaus have an influence on the trainability of PQCs but also the composition of ansatz and handling of local minima. However, we do not want to focus on those criteria but want to evaluate the loss function. Here, we focus on a well-functioning method from metric learning, the Triplet Loss. Further we also care about noise robustness, that is the functioning with lower number of layers and small number of qubits as deep circuits are not useful in the NISQ era.

3 Metric Learning and Triplet Loss in Classical Machine Learning

The task of metric learning is closely related to ranking loss functions. In contrast to most loss functions where the output of a model is compared to a corresponding label or specific value, they predict relative distances between inputs, $x \in X$. The model is a function $f_\theta(x) : \mathbb{R}^D \mapsto \mathbb{R}^C$ that maps the D-dimensional input into a C-dimensional embedding space. During training, its parameters θ are updated according to the current loss value. The objective of the loss function is to ensure that semantically close inputs are also metrically close in the embedding space [13]. To do so, a similarity score between the points of the dataset has to be known. This can be realized for example based on class labels. The class labels can be interpreted as a binary score, i.e. the points belong to the same class or a different one, which is sufficient for many ranking loss functions [12].

This binary score is also utilized for Triplet Loss which was introduced by Schroff et al. [23]. In their work, the authors introduce Triplet Loss as a metric learning approach in the context of face recognition with a Deep Neural Network. In each training step i, a triplet of face images is drawn from the training set X. The triplets consist of a so-called *anchor* x_i^{a}, a *positive* x_i^{p}, and a *negative* x_i^{n}. Aim of the approach is that the image of the anchor x_i^a representing a specific person is closer to all images x^p of the same person than to images of another person x^n. Following the idea of metric learning, the loss function is:

$$\mathcal{L}_{\mathrm{TL}}(x_i^{\mathrm{a}}, x_i^{\mathrm{p}}, x_i^{\mathrm{n}}) = \max(\|f_\theta(x_i^{\mathrm{a}}) - f_\theta(x_i^{\mathrm{p}})\|^2 - \|f_\theta(x_i^{\mathrm{a}}) - f_\theta(x_i^{\mathrm{n}})\|^2 + \alpha, 0). \quad (1)$$

The relative distance between the embedding of the anchor and the positive should be close and the distance between the anchor and the negative should be large. The parameter α enforces a margin between the two pairs. When the distance between the two pairs is greater than α, the loss becomes zero in this step.

Schroff et al. [23] cut the error rate in comparison to the best published result by 30% on two datasets achieving new state-of-the-art results. The comparison is especially impressive since their approach has a much better representational efficiency using an embedding dimension of only 128.

Deep Metric Learning has become a highly active field of research achieving state-of-the-art results on many datasets [12].

4 Training a Parameterized Quantum Circuit with Triplet Loss

Our approach aims to combine the idea of Triplet Loss with the training of a parameterized quantum circuit. In regular training with a Cross-Entropy Loss, PQCs operate in high-dimensional space that is reduced to a single value by the measurement and then compared to the corresponding label. By applying Triplet Loss, our circuit is less restricted and more easily able to leverage the high-dimensionality of Quantum Computing.

As shown in Fig. 1, we split our approach into two parts: i) the training of a model based on a PQC with Triplet Loss; and the evaluation of the model with a classical linear support vector machine (SVM) [6]. In each training step, we run the PQC with the same parameters θ and encode three different inputs given a specific labeled input dataset: an anchor, a positive of the same class, and a negative of a different class. The PQC is used to create an embedding of each of the inputs. The dimensionality C of these embeddings is defined by the number of measurement operators, i.e. the number of measured qubits, of the circuit. The calculation of the Triplet Loss based on the three embeddings is unchanged in comparison to the classical approach, see Eq. 1. The computation of gradients and update of circuit parameters are done with the parameter-shift rule [25]. Roughly speaking, this means the gradient of parameters $\nabla_\theta f_\theta(x)$ can be calculated from a finite parameter shift s as $\|f_{\theta-s}(x) - f_{\theta+s}(x)\|$. In effect, this allows us to derive the gradients of parameters of a parameterized quantum circuit using the same circuit.

To evaluate the model and perform classification on unknown data, we train a linear SVM with the embedded training data which we obtain by running each data point through the trained PQC. For evaluation of the test data, we use the same circuit pipeline to obtain the respective embeddings which are subsequently classified by the trained SVM.

Since current NISQ devices only have a limited amount of qubits, training on high-dimensional data is not possible. Therefore, one either has to use datasets with low dimensionality or perform a dimensionality reduction for the circuit's input. For the latter, we include a preprocessing step into our training and evaluation pipelines. This preprocessing step reduces the inputs to D-dimensional features to not exceed the number of available qubits and thus enables encoding them onto the circuit.

The model represented by the PQC can be expressed as follows:

$$f_\theta(q^D) = \langle 0^{\otimes n} | U^\dagger(q^D, \theta) M^{\otimes C} U(q^D, \theta) | 0^{\otimes n} \rangle \,, \tag{2}$$

Here, $\langle 0^{\otimes n}|$ and $|0^{\otimes n}\rangle$ are the initial state of n qubits in the quantum computer. The operations to transform this initial state, according to our input vector q^D and the parameters θ to optimize, are expressed by the unitary operator $U(q^D, \theta)$ equivalent to the gates of our circuit. Finally, $M^{\otimes C}$ is a measurement M applied to C output qubits of interest. Notably, $C \leq n$ since some qubits are used for computations only.

5 Experiments

In the following, we present the setup of our experiments and the corresponding results. For a better understanding and reproducibility of our work, we first discuss the two datasets and the chosen constraints. We then analyze the accuracy of the trained models in light of our proposed approach. Finally, we examine the behavior of the gradients in the circuit. All our used datasets as well as the source code is publicly available for better reproducibility [29].

5.1 Datasets

For our work, we perform experiments on the moon dataset [1] provided by scikit-learn [20] and the well-known MNIST dataset [15] from classical ML. The moon dataset is a toy dataset of two interleaving half-circles on a two-dimensional plane. Each of the half-circles represents a different class and consists of 1250 data points in our experiments. Depending on a settable variable, noise can be added when the dataset is created to make the learning task more difficult. As the number of features of this dataset is two, no dimensionality reduction is required to encode the data onto the circuit. The only preprocessing step required is scaling the values to the range $[0, \pi]$.

The MNIST dataset consists of grayscaled images of handwritten digits with respective labels zero to nine. Each image has a size of 28×28 pixels and therefore is too large to be directly encoded on current QCs. Still, we chose to perform our experiments with this dataset as it is commonly used for various quantum classifiers and QML experiments. To reduce the number dimensionality of the dataset, we train a classical autoencoder. We prefer this preprocessing pipeline over using a principal component analysis (PCA) for dimensionality reduction as commonly done [26] as an autoencoder evenly distributes the sample information in the features. Our tests show that data generated by PCA is unbalanced leading to an easy classification task as the model can concentrate on a few or even one feature for good classification. After reducing the images, we scale the feature values to the range $[0, 2\pi]$ allowing better encoding.

Both datasets are divided into a training set and a test set. The training set consists of 2500 and the test set consists of 500 samples for both datasets. In both cases, the class labels are evenly distributed.

5.2 Parameter and Ansatz Selection

During the training phase (see Fig. 1), the triplets are chosen randomly from the training set. To make our experiments comparable, we run each experiment with $n = 12$ qubits. This is sufficient for up to 8 output qubits plus auxiliary qubits for complex computations. We chose to limit the number of features of the MNIST dataset to eight for every ansatz. We keep the Triplet Loss specific parameter α (see Eq. 1) at 1.0 since our experiments showed that different values appear to have no positive effect on our selected setup.

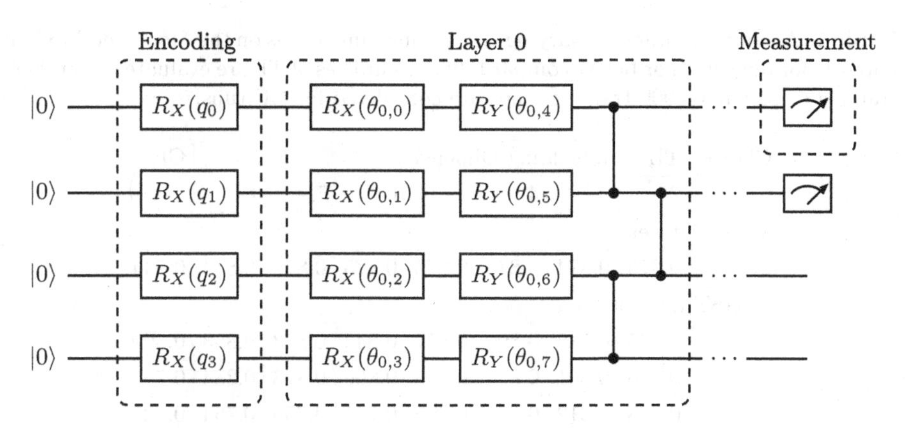

Fig. 3. Our chosen ansatz exemplarily visualized for four qubits and an embedding dimension of two. All qubits are initialized to the zero-state. The ansatz consists of the encoding part and one variational circuit layer with parameterized R_X and R_Y gates as well as entangling CZ gates. The variational layer can be repeated several times before measurements take place. Whithout any changes, this circuit can equally be used with Cross-Entropy Loss as well as Triplet Loss for a dataset with two classes.

As shown in Fig. 3, our ansatz is composed of a data encoding, a variational circuit as well as a measurement. For data encoding, we use angle encoding with R_X gates to encode the classical data. Following the data encoding, we employ a variational circuit with parameterized gates R_X and R_Y as well as non-parameterized entangling CZ gates. We use ten layers of the variational circuit during experiments and the analytical mode to get the exact expectation values of the measurements. Although our experiments show a better results for ZZ observables for measurements, we use a simple Z measurement as this allows for direct comparison with the associated ansatz for Cross-Entropy Loss. Furthermore, a simple Z measurement requires only one qubit while a ZZ measurement needs two qubits. This allows to perform extensive experiments with varying numbers of embedding dimensions. We conduct experiments for embedding dimensions from two to eight for up to three classes.

To evaluate the proposed Triplet Loss (TL) approach, we set up a second pipeline training a circuit commonly done in the literature and described in Sect. 2. So for each class, we measure one corresponding qubit and calculate a regular Cross-Entropy (CE) Loss in each training step:

$$\mathcal{L}_{\text{CE}}(x, y) = -\sum y * log(f_\theta(x)), \tag{3}$$

where x is the input, y the corresponding label and f_θ the model represented by the PQC. Therefore, this pipeline does not need a subsequent SVM to perform classification and evaluate the model as the highest expectation value is interpreted as the model's prediction. In contrast to our Triplet Loss approach, the output dimension of the circuit is fixed by the number of classes. For the moon dataset, the number of classes is 2. For the MNIST dataset we conduct exper-

Table 1. Average accuracy for varying embedding dimensions on the Moon and MNIST datasets for ten runs. For better comparibility, accuracies of TL are evaluated after 2000 training steps while 6000 training steps are considered for CE runs.

Classes	TL - Embedding Dimension							CE
	2	3	4	5	6	7	8	
Moon dataset								
2	0.873	0.874	0.875	0.875	0.875	0.875	0.874	0.844
MNIST dataset								
2	0.775	0.788	0.803	0.817	0.843	0.843	0.853	0.800
3	0.786	0.801	0.834	0.853	0.866	0.873	0.884	0.767
4	0.578	0.612	0.651	0.695	0.735	0.755	0.754	0.620

iments for 2, 3, and 4 classes. To ensure comparability, the number of qubits, layers, and the ansatz is identical for TL and CE runs. Although the number of gradient updates is the same with 2000 training steps respectively, it is in the essence of Triplet Loss that the circuit has to be called three times as much. Therefore, we ensure that each CE run uses three times as many steps to make the results better comparable.

5.3 Results

Table 1 shows the accuracy for up to eight embedding dimensions averaged over ten runs for each experiment for the Moon and MNIST dataset. The Triplet Loss approach can always outperform the regular Cross-Entropy training given an appropriate number of embedding dimensions. In all cases, even for small embedding dimensions of no more than four, the Triplet Loss rivals the Cross-Entropy approach within a few percentage points. Additional embedding dimensions improve the performance of Triplet Loss consistently beyond Cross-Entropy.

For both datasets, a higher number of embedding dimensions for the Triplet Loss leads to better separation and thus better accuracy. However, this effect seems to stagnate in case of limited expressibility of the model. This can be seen for the MNIST dataset with four classes at seven or more embedding dimensions and the Moon dataset. The accuracy seems to converge, suggesting no further improvements with increasing number of embedding dimensions. However, a much larger number of experiments would be necessary to identify dependencies and regularities. But since a higher number of measurement operators would also lead to a higher number of qubits and thus larger circuits, trainings on current hardware and simulators would be computationally expensive and hard to compare.

A drop in accuracies can be noted for the MNIST dataset when comparing the different classes grouped by their respective embedding dimensions. While the accuracies slightly improve from two and three classes, there is a decrease of the accuracy for four classes. This suggests that the expressibility of the model is

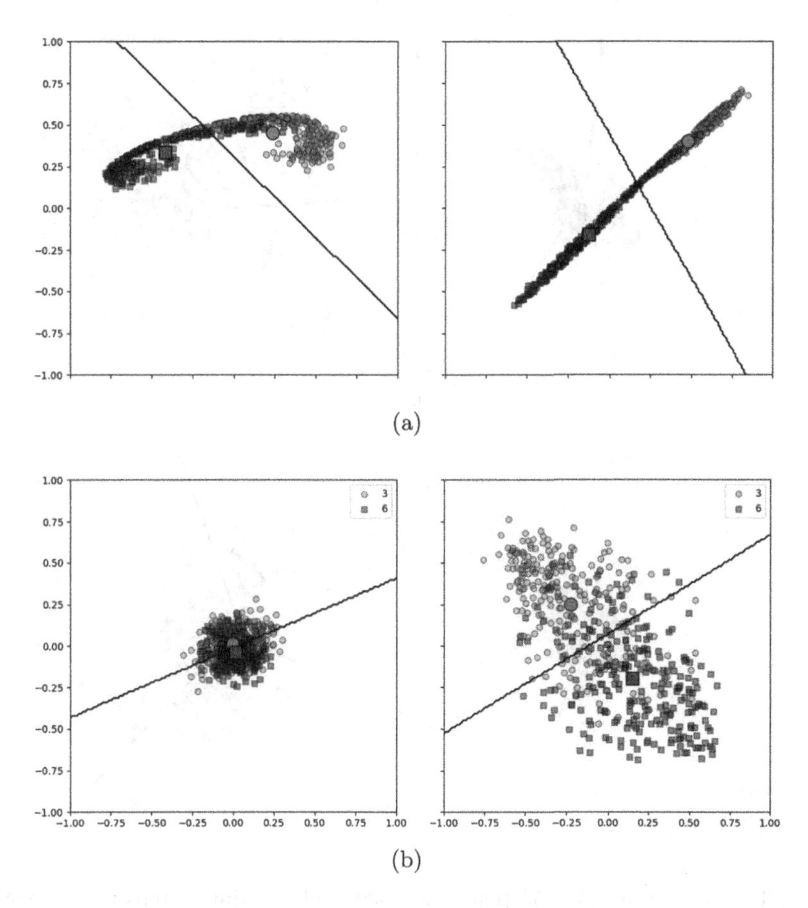

Fig. 4. Test set output after 50 (left) and 2000 (right) training steps for the moon dataset (a), and digits 3 and 6 of the MNIST dataset (b). The top and bottom as well as left and right figures share the same x- and y-axes. The bigger symbols represent the respective center of the clusters. The black line is the linear decision boundary determined by the SVM.

not sufficient to learn an efficient embedding. Future experiments should consider increasing the number of layers.

As shown in Fig. 4 for two classes on both datasets, the learning behavior corresponds to the intended one: In the course of the training, the two centers of the clusters are pushed further and further apart and the separation of the two classes becomes clearer. The two different runs also underline that Triplet Loss does not specify the location of the clusters and the mapping. While for the moon dataset shown in Fig. 4a the whole test dataset is mapped on a straight line, the two digits of the MNIST dataset shown in Fig. 4b form two separate round clusters. This behavior can also vary for comparable runs of the same dataset with a different parameter initialization.

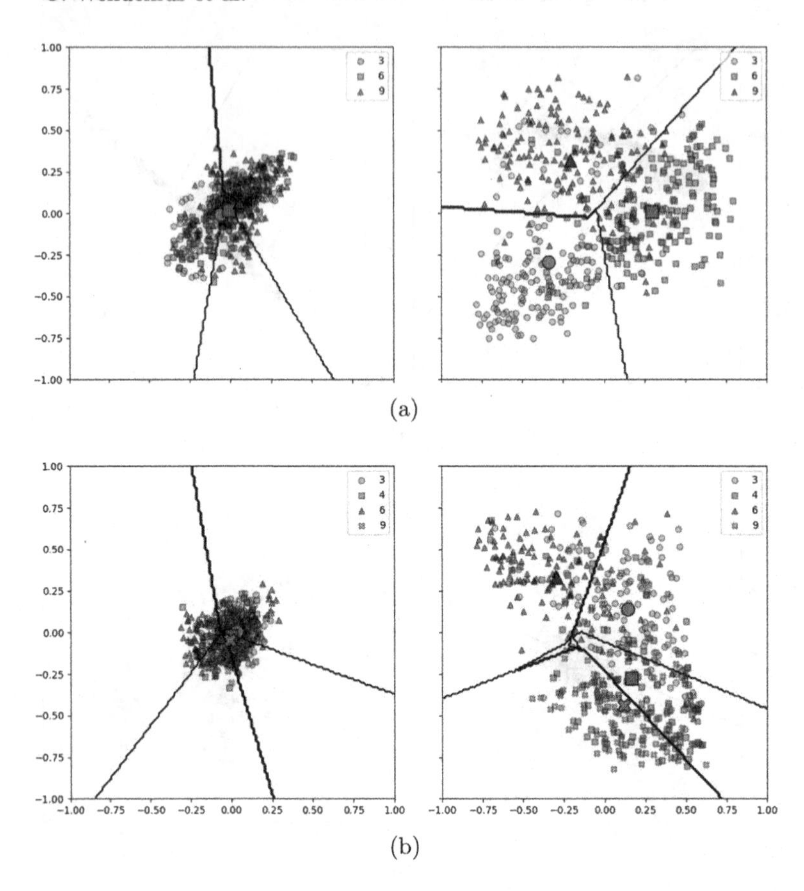

Fig. 5. Test set output after 50 (left) and 2000 (right) training steps for the MNIST dataset for the three digits 3, 6, and 9 (a) and the four digits 3, 4, 6, and 9 (b). The top and bottom as well as left and right figures share the same x- and y-axes. The bigger symbols represent the respective center of the clusters. The black line is the linear decision boundary determined by the SVM.

Figure 5 shows the SVM output after 50 and 2000 training steps. Figure 5a and 5b confirm the numerical results for three and four classes respectively. While the separation of three classes is fairly distinct with good numerical accuracy, Triplet Loss cannot clearly separate four digits of the MNIST dataset. For four classes, three classes are separated well but the fourth one is spread across all regions. A modification of the original Triplet Loss equation to encourage a broader mapping might be a possible solution.

During the training of classical models, the triplet selection has a direct influence on the performance and progression of the training and is one of the most studied questions in the field of metric learning [30,31]. Although this was not the focus of our work, we notice that a random selection seems to work best. Selecting hard triplets, where the negative is close to the anchor and the positive

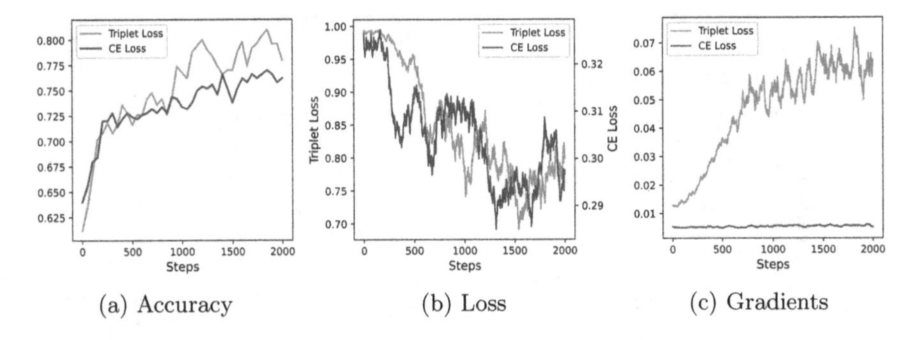

(a) Accuracy (b) Loss (c) Gradients

Fig. 6. Comparison of Triplet Loss (red) with a Cross-Entropy (blue) training regarding the accuracy, the loss curve and the average absolute gradient during 2000 steps. (Color figure online)

is further away, does not lead to any learning. But to further improve scores and derive conclusions, more experiments are needed.

Figure 6 shows a comparison of various metrics during a training with Triplet Loss and with a Cross-Entropy Loss. In both experiments, the accuracy (see Fig. 6a) quickly reaches a very high value, which can hardly be improved in the course of the training. This similar behavior also becomes apparent in the two loss curves in 6b although both curves differ significantly in their absolute values which is due to the different calculation of the two losses. However, it can be seen that the Triplet Loss continues to show large fluctuations during training, whereas the Cross-Entropy Loss remains much more steady. An explanation for this is provided by looking at the average absolute gradients in the entire circuit, shown in Fig. 6c. Gradients for Triplet Loss are much higher and allow further adjustments of the weights during the entire training.

5.4 Gradients

To further investigate the behavior of the gradients, we undertake similar experiments to McClean et al. [17] in their work on barren plateaus. We therefore analyze the variance of the first parameter in the first layer across different runs for the two experiment setups. The resulting variances of gradients are shown in Fig. 7. The blue lines show the behavior of the Cross-Entropy based training while the red lines show the behavior for the Triplet Loss based training. As can be clearly seen from the blue lines, we can reproduce the behavior of exponentially vanishing gradients towards zero as a function of the number of qubits as reported by McClean et al. [17]. This implies that random initializations can cause the gradient-based optimization to fail.

For the Triplet Loss based training, a different behavior can be observed: While the gradient continues to decrease with an increasing number of qubits, it increases again with a higher number of layers. First, for qubit counts up to eight, the Triplet Loss based approach shows significantly higher variances of gradients. For ten qubits, the variances are comparable but improve significantly

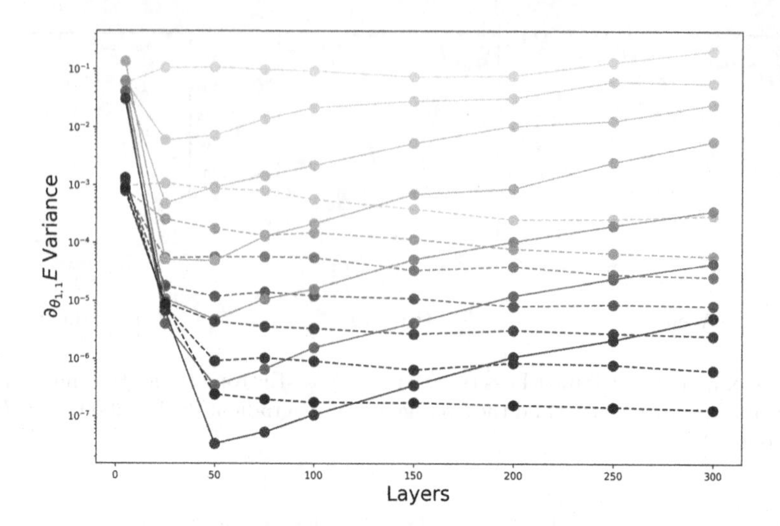

Fig. 7. Variance of the gradient for the first parameter over 100 runs. For the Triplet Loss approach (red) and the regular Cross-Entropy approach (blue). The different lines correspond to all even numbers of qubits between 4 and 16, starting from 4 quits at the top of the respective color panel. (Color figure online)

for layer counts bigger than 50. This can also be observed for bigger qubit counts. Thus, training with a higher number of layers may work with Triplet Loss, while a regular pipeline may not be able to learn because of effects of a barren plateau. This is also true for training circuits of arbitrary size on NISQ devices, where a small gradient is hardly distinguishable from noise. To investigate this further and to get insights on the scalability, a large number of computationally intensive simulations and extensive training on NISQ devices are required.

6 Conclusion

In this work, we have demonstrated the applicability of Triplet Loss for training a parameterized quantum circuit in a Quantum Machine Learning pipeline. For this purpose, we compared our pipeline in extensive experiments with a Cross-Entropy approach that is currently prominently used in literature. On two datasets, we were able to demonstrate competitive accuracies for both loss functions. As intended, using metric learning our models were able to consistently separate the given classes. As our evaluation used a fixed circuit layout and thus computational power, for higher numbers of classes modifications of the Triplet Loss function and ansatz are conceivable to further improve the scores.

The observed high gradient values during training are particularly promising: These are significantly higher than those observed in conventional training on a QC. Still, owing to the current limitations of hardware and simulations our experiments were limited to a small number of layers. It remains a task for future work to explore scalability and generalizability on larger circuits. In the best case, Triplet Loss reduces the impact of barren plateaus and thus improves training in the area of QML. Especially in the NISQ era, a stronger gradient would improve the robustness against noise and thus increase the applicability of QML.

Acknowledgements. The authors acknowledge support by the state of Baden-Württemberg through bwHPC.

References

1. Moon dataset. https://scikit-learn.org/stable/modules/generated/sklearn.datasets.make_moons.html. Accessed 04 Apr 2022
2. Benedetti, M., Lloyd, E., Sack, S., Fiorentini, M.: Parameterized quantum circuits as machine learning models. Quant. Sci. Technol. **4**(4), 043001 (2019). https://doi.org/10.1088/2058-9565/ab4eb5
3. Bilkis, M., Cerezo, M., Verdon, G., Coles, P.J., Cincio, L.: A semi-agnostic ansatz with variable structure for quantum machine learning (2021). https://arxiv.org/abs/2103.06712
4. Cerezo, M., et al.: Variational quantum algorithms. nature reviews. Physics **3**(9), 625–644 (2021). https://doi.org/10.1038/s42254-021-00348-9
5. Cerezo, M., Sone, A., Volkoff, T., Cincio, L., Coles, P.J.: Cost function dependent barren plateaus in shallow parametrized quantum circuits. Nature Commun. **12**(1), 1791 (2021). https://doi.org/10.1038/s41467-021-21728-w
6. Cortes, C., Vapnik, V.: Support-vector networks. Mach. Learn. **20**(3), 273–297 (1995). https://doi.org/10.1007/BF00994018
7. Dunjko, V., Wittek, P.: A non-review of quantum machine learning: trends and explorations. Quantum Views **4**, 32 (2020)
8. Grant, E., et al.: Hierarchical quantum classifiers. NPJ Quantum Inf. **4**(1), 65 (2018). https://doi.org/10.1038/s41534-018-0116-9
9. Grant, E., Wossnig, L., Ostaszewski, M., Benedetti, M.: An initialization strategy for addressing barren plateaus in parametrized quantum circuits. Quantum **3**, 214 (2019). https://doi.org/10.22331/q-2019-12-09-214
10. Hettinger, C., Christensen, T., Ehlert, B., Humpherys, J., Jarvis, T., Wade, S.: Forward thinking: building and training neural networks one layer at a time (2017)
11. Holmes, Z., Sharma, K., Cerezo, M., Coles, P.J.: Connecting ansatz expressibility to gradient magnitudes and barren plateaus. PRX Quantum **3**, 010313, Published 24 January 2022 (2021). https://doi.org/10.1103/PRXQuantum.3.010313, https://arxiv.org/abs/2101.02138
12. Kaya, M., Bilge, H.Ş.: Deep metric learning: a survey. Symmetry **11**(9), 1066 (2019)
13. Kulis, B., et al.: Metric learning: a survey. Found. Trends® Mach. Learn. **5**(4), 287–364 (2013)
14. LaRose, R., Coyle, B.: Robust data encodings for quantum classifiers. Phys. Rev. A **102**, 032420 (2020). https://doi.org/10.1103/PhysRevA.102.032420, https://arxiv.org/abs/2003.01695
15. LeCun, Y., Cortes, C.: MNIST handwritten digit database (2010). https://yann.lecun.com/exdb/mnist/
16. Lloyd, S., Schuld, M., Ijaz, A., Izaac, J., Killoran, N.: Quantum embeddings for machine learning. arXiv preprint arXiv:2001.03622 (2020)

17. McClean, J.R., Boixo, S., Smelyanskiy, V.N., Babbush, R., Neven, H.: Barren plateaus in quantum neural network training landscapes. Nat. Commun. **9**(1), 1–6 (2018)
18. McClean, J.R., Romero, J., Babbush, R., Aspuru-Guzik, A.: The theory of variational hybrid quantum-classical algorithms. New J. Phys. **18**(2), 023023 (2016)
19. Nielsen, M.A., Chuang, I.L.: Quantum Computation and Quantum Information. Cambridge University Press, Cambridge (2000)
20. Pedregosa, F., et al.: Scikit-learn: machine learning in python. J. Mach. Learn. Res. **12**, 2825–2830 (2011)
21. Pesah, A., Cerezo, M., Wang, S., Volkoff, T., Sornborger, A.T., Coles, P.J.: Absence of barren plateaus in quantum convolutional neural networks. Phys. Rev. X **11**, 041011 (2021). https://doi.org/10.1103/PhysRevX.11.041011
22. Preskill, J.: Quantum computing in the NISQ era and beyond. Quantum **2**, 79 (2018)
23. Schroff, F., Kalenichenko, D., Philbin, J.: FaceNet: a unified embedding for face recognition and clustering. In: Proceedings of the IEEE Conference on Computer Vision and Pattern Recognition, pp. 815–823 (2015)
24. Schuld, M.: Effect of data encoding on the expressive power of variational quantum-machine-learning models. Phys. Rev. A **103**(3), 032430 (2021). https://doi.org/10.1103/PhysRevA.103.032430
25. Schuld, M., Bergholm, V., Gogolin, C., Izaac, J., Killoran, N.: Evaluating analytic gradients on quantum hardware. Phys. Rev. A **99**(3), 032331 (2019)
26. Skolik, A., McClean, J.R., Mohseni, M., van der Smagt, P., Leib, M.: Layerwise learning for quantum neural networks. Quantum Mach. Intell. **3**(1), 1–11 (2021). https://doi.org/10.1007/s42484-020-00036-4
27. Thumwanit, N., Lortaraprasert, C., Yano, H., Raymond, R.: Trainable discrete feature embeddings for variational quantum classifier (2021). https://arxiv.org/abs/2106.09415
28. Wecker, D., Hastings, M.B., Troyer, M.: Progress towards practical quantum variational algorithms. Phys. Rev. A **92**(4), 042303 (2015)
29. Wendenius, C., Kuehn, E.: Quantum-triplet-loss, July 2022. https://doi.org/10.5281/zenodo.6786443
30. Xuan, H., Stylianou, A., Liu, X., Pless, R.: Hard negative examples are hard, but useful. In: Vedaldi, A., Bischof, H., Brox, T., Frahm, J.-M. (eds.) ECCV 2020. LNCS, vol. 12359, pp. 126–142. Springer, Cham (2020). https://doi.org/10.1007/978-3-030-58568-6_8
31. Yu, B., Liu, T., Gong, M., Ding, C., Tao, D.: Correcting the triplet selection bias for triplet loss. In: Ferrari, V., Hebert, M., Sminchisescu, C., Weiss, Y. (eds.) ECCV 2018. LNCS, vol. 11210, pp. 71–86. Springer, Cham (2018). https://doi.org/10.1007/978-3-030-01231-1_5

Immediate Split Trees: Immediate Encoding of Floating Point Split Values in Random Forests

Christian Hakert[1(✉)], Kuan-Hsun Chen[2], and Jian-Jia Chen[1]

[1] TU Dortmund University, Dortmund, Germany
christian.hakert@tu-dortmund.de, jian-jia.chen@cs.tu-dortmund.de
[2] University of Twente, Enschede, Netherlands
k.h.chen@utwente.nl

Abstract. Random forests and decision trees are increasingly interesting candidates for resource-constrained machine learning models. In order to make the execution of these models efficient under resource limitations, various optimized implementations have been proposed in the literature, usually implementing either *native trees* or *if-else trees*. While a certain motivation for the optimization of if-else trees is to benefit the behavior of dedicated instruction caches, in this work we highlight that if-else trees might also strongly depend on data caches. We identify one crucial issue of if-else tree implementations and propose an optimized implementation, which keeps the logic tree structure untouched and thus does not influence the accuracy, but eliminates the need to load comparison values from the data caches. Experimental evaluation of this implementation shows that we can greatly reduce the amount of data cache misses by up to $\approx 99\%$, while not increasing the amount of instruction cache misses in comparison to the state-of-the-art. We additionally highlight various scenarios, where the reduction of data cache misses draws important benefit on the allover execution time.

1 Introduction

Increasing focus on resource-constrained machine learning throughout the last years brings up random forests and decision trees in various shapes and flavours as premier candidates for resource-limited classification or regression models. Although decision trees can be tuned toward resource limitation on the tree structure already, by for instance limiting the amount of nodes and thus the total memory consumption, considering the underlying computing architecture turns out to allow even further optimization, e.g., [3,4,7,11,12]. As one particular aspect of the computing architecture, memory hierarchies and caches turn out to allow a certain interplay with decision trees. Caches are organized in a way, that they store a copy of intensively used memory contents in a small and fast memory, in order to reduce memory access latencies. The decision, which memory contents are stored in the cache, is made by the hardware, following certain prefetching

© The Author(s), under exclusive license to Springer Nature Switzerland AG 2023
M.-R. Amini et al. (Eds.): ECML PKDD 2022, LNAI 13717, pp. 531–546, 2023.
https://doi.org/10.1007/978-3-031-26419-1_32

and preemption strategies. Since the probability of accessing certain nodes in a decision tree can be profiled at training time, a certain prediction of memory accesses and memory access sequences can be made. This can be consequently used to shape the trees in a way that the hardware caches automatically prefetch the data objects, which are likely used during further execution.

Towards this, implementation of decision trees yield two common realizations: 1) *native trees* and 2) *if-else trees* [1]. The former ones aim to encode tree nodes in a large data array and build up the tree structure by following pointers to the right or left child node at every node. The latter ones aim to translate every node of a tree to an if-else construct in the programming language and build up the tree structure by extensively nesting these if-else constructs. Generally, although most computers implement the von Neumann architecture and contain a unified data and instruction memory, the CPU implements different methods for data and instruction memory accesses. Within the CPU pipeline, data and instruction memory accesses can be performed at different stages and therefore utilize different hardware units. Furthermore, systems with cache hierarchies usually partition caches (mostly the first level caches) into instruction and data cache, leading to distinct access behavior and access latencies. Due to these aspects, considering data and instruction memory accesses of a program separately and focus on their optimization can lead to important performance improvements.

Since native trees intend to intensively utilize data memory and data caches, optimization of native trees targets to modify the node array structure in order to comfort data cache prefetching [4]. Contrarily, if-else trees intend to intensively use instruction memory and therefore optimization of if-else tree tries to reshape the structure of nested if-else blocks in order to comfort the prefetching of instruction caches [4]. When random forest models exceed the capacity of a cache, which likely happens for first level caches, compulsory cache misses in the instruction cache and data cache happen during the execution and cause increased latencies since caches are usually faster by orders of magnitude than main memory. Optimizing the prefetching of caches, as described before, can help to lower the introduced additional latency. When caches are partitioned into instruction and data caches, prefetching as well is separated for both caches, which highlights the need for separate optimization of instruction and data memory accesses.

Problem: In this work, we take a closer look to the memory behavior of if-else tree implementations and highlight that, counter intuitively to the design principle, if-else trees depend on intensive use of data memory and data caches as well, causing data cache misses, which is not considered and targeted by existing optimizations [3,4].

Solution: Consequently, we propose an optimized implementation of if-else trees with floating point split values, where we eliminate large parts of the use of data caches. Thus, we can apply existing optimizations for if-else trees subsequently and improve their execution.

Experimental evaluation on X86 and ARMv8 based server and embedded systems highlights that our proposed optimization can reduce the amount of data cache misses by up to 99%, while not increasing the amount of instruction cache misses. In addition to the great reduction of data cache misses, we show that the reduction of data cache misses directly contributes to a reduction of the allover execution time in various scenarios. End-to-end timing measurements demonstrate that we can reduce the allover execution time of decision trees with our proposed optimization, especially on server systems.

Our Novel Contributions:

- Analysis of the state-of-the-art optimization for if-else trees regarding the usage of the data cache.
- An optimized implementation, reducing large portions of data cache misses in if-else trees
- Experimental evaluation of the proposed implementation and comparison to the state of the art

2 Related Work

Several techniques have been proposed in the literature to speed up the execution of inference for tree-based ensembles. For binary search trees, Kim et al. in [7] presents an optimized realization by using vectorization units on X86, considering the register sizes, cache sizes, and page sizes specifically. However, such a technique requires a specific support from the underlying architectures. The concept of vectoring the tree structures is also applied to the context of ranking models in [9], which enhances the QuickScorer algorithm for gradient boosted trees [5,8]. Ye et al. in [13] further improve the scalability of such vectorization methods by encoding the node representation to compact the memory footprint. These techniques decompose the tree-ensembles into different data structures based on the feature values, which is especially effective for large ensembles of smaller trees. Without traversing trees one by one, however, the target applications are mainly limited to batch-processing.

Architecture-aware implementations for decision trees have started from [12], which optimizes the implementations of decision trees on different architectures, i.e., CPUs, FPGAs, and GPUs. By fixing the tree-depth, Prenger et al. in [11] further show an effective pipelining approach over these computing units, based on the CATE algorithm during training. However, the impact of cache misses was not taken into account. The two common implementations for decision trees, i.e., native trees and if-else trees, are first distinguished in [1], which provides the first attempt to increase data locality for native trees. By leveraging the probability model of accessing nodes during tree traversal in [2], Buschjäger et al. in [3] propose several optimizations for memory layout over different tree implementations to improve the memory locality and show the potential speed-ups can be up to 2–4× over different architectures. Chen et al. enhance this method further by compiler based binary size estimation [4]. Hence, we consider the proposed optimization for if-else trees from [4] as the state-of-the-art approach.

3 Problem Analysis

After training of a random forest model (e.g. with scikit-learn [10]), the model is derived in a logic representation (e.g. encoded in JSON). Executing this model without special operating system or library support requires a realization in a programming language and compilation to machine code. The realization of decision trees and random forests as if-else trees, as introduced by [1] intensively utilizes instruction caches during the execution, as the entire tree structure is encoded in instructions itself. Only loading of the data point for inference is mandatory from data memory and therefore uses data caches. Listing 1.1 depicts an example of the implementation of a single tree node as an if-else tree in C++.

```
1 if(pX[3] <= (float) 1.500000){
2     return 1;
3 } else { ...
```

Listing 1.1. C++ node example

It can be seen that the loading of the data point (stored in pX) is an array access and therefore a data memory access. The split value, which is used to decide in combination with the data point if the left or right subtree should be further traversed, is immediately encoded in the source code, also the prediction value is immediately encoded with the return statement. To illustrate the conversion to assembly code, we investigate the assembly code for an X86 machine, produced by the gcc compiler in version 11.1.0 in the following. Later in this paper we consider both X86 and ARMv8 architectures.

```
1 movss    0xe50(%rip),%xmm9
2 comiss   0xc(%rdi),%xmm9
3 jmp      2fa0
```

Listing 1.2. Assembly node example

Listing 1.2 illustrates the relevant assembly code for the implementation of the node from Listing 1.1. Line 1 is responsible for loading the split value, Line 2 loads the feature value from the data point and performs the comparison to the split value. Line 3 then performs the according jump. Counter intuitively to the C++ implementation, the split value is not encoded as an immediate value, but rather leads to a data memory load within the `comiss` (compare scalar ordered single-precision floating point) instruction. Since the X86 instruction set does not offer immediate values for floating point instructions, the compiler decides to place the split values at a central position in data memory and translate the accesses to regular data loads[1]. Please note that the `movss` instruction (which loads a floating point number to a register) uses the immediate encoding for the offset within main memory, but not for the floating point constant itself. In consequence, two out of the three relevant instructions for an if-else tree perform

[1] This observation is not necessarily bounded to the X86 architecture, the ARMv8 architecture neither does offer such a feature.

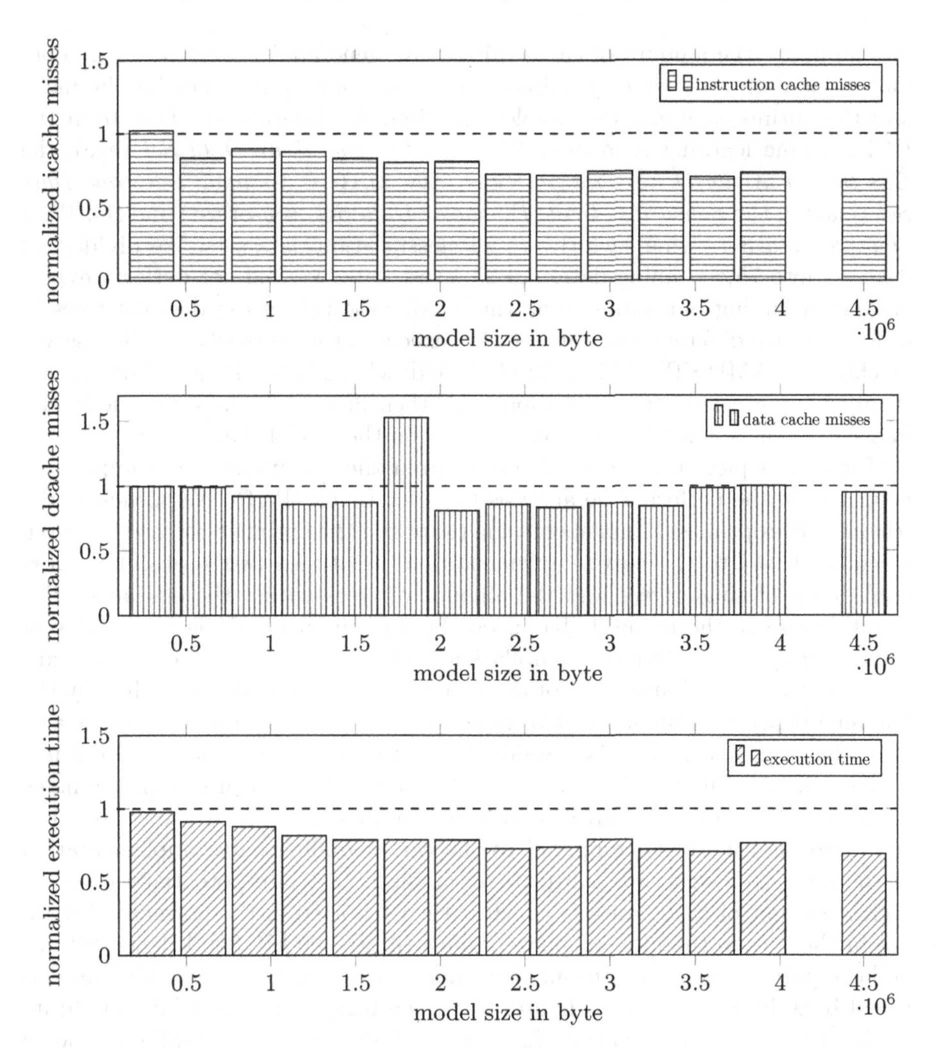

Fig. 1. Execution time, icache misses and dcache misses for if-else tree optimization

data accesses and utilize the data cache. The motivational concept of intensively utilizing instruction memory and caches for if-else trees does not hold all along.

To further illustrate the impact of this condition, we investigated the state-of-the-art implementation of if-else trees from Chen et al. [4]. Specifically, we studied two possible implementations of the same logic model in the following: 1) a naive implementation of if-else trees, where every node in the logic tree structure becomes an if-else block and the left and right subtree is placed within the corresponding if or else block. 2) the generated trees with the state-of-the-art optimization [4], where the tree is reordered with regards to the branch probability within every node. This reordering aims to optimize cache prefetching and minimize the amount of cache misses. We generated a large set of random forests for data sets, which resulted in floating point split values for the naive and the optimized if-else tree implementation. As datasets we chose from the UCI machine learning repository [6]: the *EEG Eye State Data Set* (eye), the *Gas Sensor Array Drift Data Set* (gas), the *MAGIC Gamma Telescope Data Set* (magic), the *Sensorless Drive Diagnosis Data Set* (sensorless) and the *Wine Quality Data Set* (wine), which are all classification data sets. We divided all datasets into 75% training data and 25% test data. We did not perform hyper-parameter tuning but rather tune the maximal depth of the decision trees in order to derive different sized models. We executed these models on a X86 server machine (2x AMD EPYC 7742, 32 kB L1 i/dcache, 256 GB RAM) and compare them with respect to their execution time, their amount of misses in the level 1 instruction cache and the amount of misses on the level 1 data cache.

Figure 1 depicts the recorded results from the execution of the implementations with the performance analysis tools for Linux (Perf). We normalize the results of the optimized implementation (applying the optimization method from Chen et al. [4]) to the naive implementation. We tuned two knobs: The maximal depth of single trees and the amount of trees within the ensemble. The resulting size of the model is based on the measurement of the binary size of the implementation after the compilation, which is illustrated along the x axis. Please not that the binary size of the model is only indirectly controlled by the maximal depth and the amount of trees, hence not for every size on the x axis also a model is generated. We consider the optimized implementation, even if it may result in a different binary size, to the original binary size from the naive implementation. Hence, even if optimizing the model increases the binary size, the performance still is compared to the corresponding naive implementation of the same tree structure. An increase in the binary size potentially causes a higher amount of cache misses, which is then reported in the normalized data. We further group the models in size groups (0 kB–300 kB, 300 kB–600 kB, ...) and compute the geometric mean of the normalized improvements. This value is the ultimately depicted in the figure. The green bars with diagonal lines indicate the reduction in total execution time, the red bars with horizontal lines indicate the reduction in L1 icache misses, and the blue bars with vertical lines indicate the reduction in L1 dcache misses.

It can be observed that, though the optimization reduces the total execution time and amount of icache misses for large models[2], the amount of L1 dcache misses is not reduced similarly. This stems from the fact that the if-else tree optimization proposed by Chen et al. [4] only modifies the sequence of the source code in order to reduce the amount of L1 icache misses. The placement and loading of the split values is not considered and thus not handled in the optimization. When the dcache misses can be reduced as well, a further reduction of

[2] The optimization targets to optimize the memory layout, such that cache misses are reduced. Thus, effects likely only can be observed when the model size exceeds the cache capacity, which is only for larger models the case.

the execution time can be possible. Furthermore, load can be released from the instruction memory, which may comfort other applications within the system.

Observing this shortcoming in the existing optimization motivates us to develop a new optimization technique, which specifically focuses on the optimization of dcache misses by handling the loading of the split values in a dedicated manner. One trivial method is to round the floating point split values to integer values and subsequently encode them in the immediate field of the instructions itself, such that they do not need to be loaded from data memory at all. This, however, potentially induces a loss in accuracy due to the rounding of the split values. In this paper, we alternatively present an implementation, where the full floating point split value can be encoded in the immediate field of instructions and therefore also omits the need to load the split values from data memory.

4 Immediate Encoding in If-Else Trees

As mentioned before, when it comes to the optimization of the cache behavior of if-else implementations of decision trees, both cache types, i.e. the instruction cache and the data cache, need to be handled. In general, optimization methods profile the execution of the decision tree on the training dataset and determine empirical branch probabilities. These probabilities are used subsequently to shape the tree implementation in an optimized manner. When the total model size exceeds the capacity of a cache, which likely happens for kilobyte sized level 1 caches, cache misses cannot be avoided during execution of the tree.

Hence, the optimization target is to reduce the amount of cache misses in order to improve the total execution time of the decision tree. Such optimizations usually can exploit two aspects: 1) the tree is shaped in a way that frequently accessed parts of the decision tree are less likely evicted from the cache as in a naive implementation and therefore do not cause cache misses on access, and 2) the tree is shaped in a way that automatic prefetching of (spatial) local memory contents is utilized to load parts of the tree into caches before they are accessed and thus omit cache misses at the access time itself. To shape the tree itself, data memory and instruction memory needs to be distinguished. Data memory is usually used to store variables and arrays. If a tree implementation uses large arrays, changing the layout of the array allows to shape the tree. Instruction memory is used to store the instruction sequence of the tree itself. If the tree implementation uses many instructions, changing the sequence of instructions allows to shape the tree regarding the behavior of instruction caches.

As motivated before, the naive implementation of an if-else tree in C++ uses data memory to load both feature values and split values. Access to the feature values cannot be omitted and hardly be optimized, since the input tuple is not created by the tree implementation itself. Thus, data memory accesses to the feature values are compulsory. In consequence, optimization of the data memory accesses for the split values is challenging, since these accesses are necessarily interleaved with the accesses to the feature values. Therefore, the implementation we propose in this paper alters the loading of the split value from data memory to

```
1  __rtitt_lab_27_0:
2      movss   12(%1),          %%xmm1
3      //0x3fc00000=1.5
4      mov     $0x3fc00000,     %%eax
5      movd    %%eax,           %%xmm2
6      comiss  %%xmm1,          %%xmm2
7      jnb     __rtitt_lab_29_0
```

Listing 1.3. Optimized assembly implementation (X86)

```
1  "__rtitt_lab_27_0:"
2      ldr     s1,              [%1, 12]
3      //0x3fc00000=1.5
4      movz    w2,              #0x0000
5      movk    w2,              #0x3fc0,      lsl 16
6      fmov    s2,              w2
7      fcmp    s1,              s2
8      b.le    __rtitt_lab_29_0
```

Listing 1.4. Optimized assembly implementation (ARMv8)

instruction memory. Subsequently, the tree is shaped by ordering the instruction sequence with respect to the behavior in the instruction cache.

Based on the arch-forest framework[3], used in [4], we implement a new code generator module for the generation of our optimized if-else tree. The code generator does not generate C or C++ code, but rather directly generates X86 or ARMv8 assembly code, which is embedded by inline assembly to the rest of the framework. In order to explain the assembly implementation, we focus on the example node from Listing 1.1.

Listing 1.3 illustrates the output of our code generator for the example node. In line 2, similarly as in the compiler generated code, the feature value is loaded from data memory, which cannot be omitted. Afterwards, the split value (1.5) is converted to IEEE-754 32 bit representation in line 4 and loaded as a bitmask to a general purpose register[4]. The `movd` instruction subsequently copies the register content without conversion to a floating point register and in line 6 and 7 the according comparison and jumps are executed.

Listing 1.4 similarly depicts an example of the ARMv8 code, which is generated by our code generator. The key difference is that ARMv8 does not offer pseudo instructions to load 32 or 64 bit immediate values, thus we decompose these into a set of `movz` (move and zero contents before) and `movk` (move and keep contents) instructions with according bit shifts. The `fmov` instruction in

[3] https://github.com/tudo-ls8/arch-forest.

[4] Our generator also supports double precision floating points; the code is generated accordingly on demand.

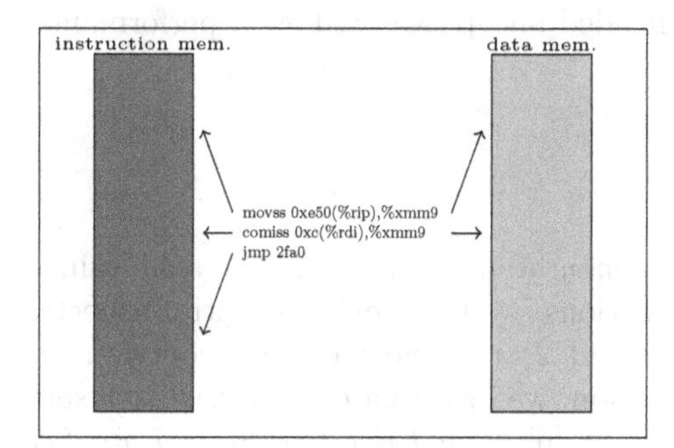

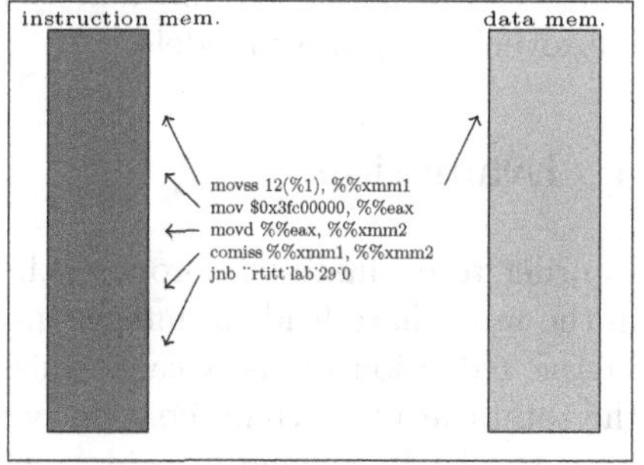

Fig. 2. Optimized loading of constants from memory

ARMv8 is the respective instruction to move contents from a general purpose register to a floating point register without conversion.

Figure 2 illustrates the difference in memory accesses between the compiler generated code and our generated code for X86. All instructions, by default, access instruction memory, since the instruction has to be loaded from instruction memory. In the compiler generated version, two out of three instructions in addition access data memory, in our optimized version only one out of five instructions additionally accesses data memory. Despite moving the split value entirely to the instruction memory, we also inherit the code sequence optimization from [4] in our code generator. For every node, the relative probability to visit the left or right child is compared and the more probable child is placed as the subsequent instructions. The less probable child hence is labeled and targeted by the jump/branch instruction. Implementation wise, this requires a swap of the branch condition, since the branch must be taken either on the $\leq$ or on the $>$ condition. We achieve this by either generating a `jnb` (jump if not below)/`b.le` (branch if less or equal) or a `jb` (jump if below)/ `b.gt` (branch if greater than) instruction.

In order to integrate our code generation in a generally applicable shape, we implement all possible combinations for datatypes within if-else trees in the code generator. This includes various combinations of datatypes for the feature and for the split values, since the comparison has to be realized accordingly. Our code generator allows to generate if-else trees for 32 and 64 bit floating point split values, including the optimization from [4], in assembly code and eliminates a large portion of data memory loads, at the cost of few additional instructions, which are used to encode the data directly in the immediate field. Thus, the data cache misses are likely reduced when employing this implementation.

Beyond our implementation, this method is applicable to other models and structures as well. Floating point constants are required for a large set of machine learning models, e.g. neural networks or simple regression models. Such models are usually trained by adjusting a set of constants (weights, parameters, etc.), which are then incorporated for computation during inference. Since the computation is implemented as code execution in a CPU based variant, constants can be similarly immediately encoded and possibly allow a performance improvement of other models.

5 Evaluation

In order to evaluate our proposed implementation of encoding the split values in the immediate fields of integer instructions, we focus on two central aspects: 1) the reduction of data cache misses and 2) the effect of the reduction on the total execution time. For the evaluation, we investigate again the data sets from the UCI machine learning repository [6]: The *EEG Eye State Data Set* (eye), the *Gas Sensor Array Drift Data Set* (gas), the *MAGIC Gamma Telescope Data Set* (magic), the *Sensorless Drive Diagnosis Data Set* (sensorless) and the *Wine Quality Data Set* (wine). These data sets are all classification data sets. We used the arch-forest framework together with our custom code generator to generate ensembles of different amount of trees and tree sizes for all data sets. We generated subsequently three implementations for every tree: 1) a naive if-else tree implementation without any optimization, 2) the optimized if-else tree implementation from Chen et al. [4] as the state of the art and 3) our assembly-based implementation, as presented in Sect. 4. As test platforms, we chose four different systems, two server systems with X86 and ARMv8 architectures and two embedded systems with X86 and ARMv8 architectures. The system details can be found in Table 1.

Table 1. Test system details

	CPU	L1 icache	L1 dcache	Memory
X86 Server	2x AMD EPYC 7742	32 kB	32 kB	256 GB DDR4
X86 Embedded	Intel Atom x5-Z8350	32 kB	24 kB	2 GB DDR3
ARMv8 Server	2x Cavium Thunder X2	32 kB	32 kB	256 GB DDR4
ARMv8 Embedded	Amlogic S9052	32 kB	32 kB	2 GB DDR3

We executed all generated ensembles on all of the systems and used the performance analysis tools for Linux (Perf) to record instruction cache misses, data cache misses and the total execution time for every configuration. We again determined the model size in bytes after compilation to compare the different configurations regarding their final size. Similarly, we consider the optimized implementations for the binary size of the naive implementation and build size groups, which we use to compute the geometric mean and present the results. Thus, even if the model size is increased by the optimization, the normalized ratio still is depicted for the same logic model structure.

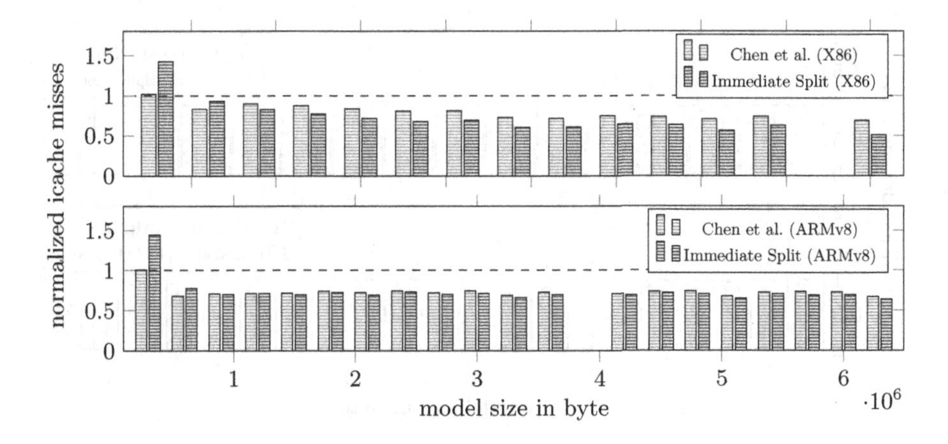

Fig. 3. Instruction cache misses of immediate encoded split values - server

Figure 3 depicts the icache misses for the server systems, Fig. 4 depicts the dcache misses, and Fig. 5 depicts the execution time, respectively. Figure 6 depicts the icache misses for the embedded systems, Fig. 7 the dcache misses and Fig. 8 the execution time, respectively. We again compute the normalized ratio between the optimized and the naive implementation. Thus, a number larger than 1 indicates worse performance in comparison to the naive implementation. Each figure includes results for the X86 architecture and for the ARMv8 architecture. Comparing the reduction for the optimization from the state of the art and our optimization leads to another, relative improvement, which is illustrated in Table 2. The #IMPROVED and #IMPROVED($> 900k$) values describe in how many of the tested models our optimization performs better regarding instruction cache misses, data cache misses or execution time than the optimization from Chen et al. The latter value only considers models, which lead to a binary size of more than 900 kB. We further compute the improvement ratio by $1 - \frac{\text{Immediate Split}}{\text{Chen et al.}}$. Hence, a number of +100% for the cache misses would mean that our optimization eliminates all cache misses, which are left after the optimization from the state of the art. We compute this improvement for all data sets and report the geometric mean for models larger than 900 kB and the peak value in the table.

5.1 Discussion

Generally, it can be observed that for rather small ensembles (up to $\approx$ 900 kB) a diminished performance can be observed for most configurations. If a small model anyway can be held entirely in the level 1 cache, there is no requirement for any optimization. The optimization, however, induces certain overheads by introducing more instructions, which leads to an ultimate performance decrease. In consequence, this draws the conclusion that the optimization should only be applied in meaningful scenarios, where the ensemble size exceed the level 1

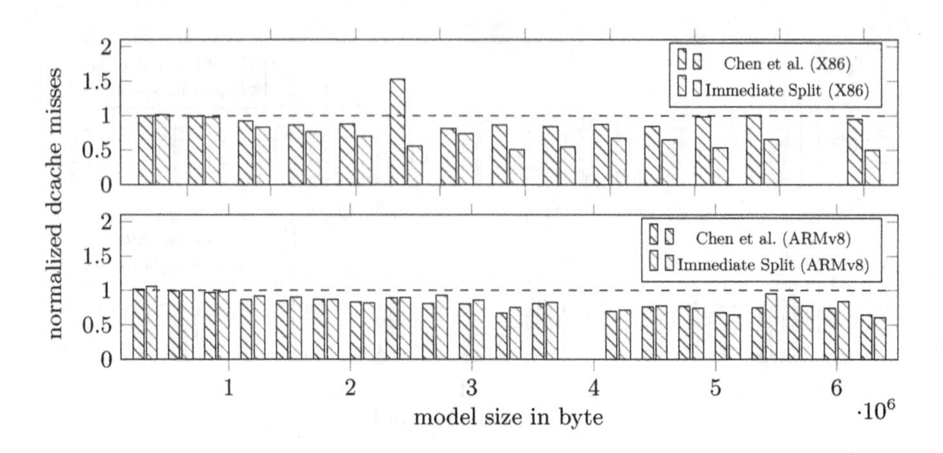

Fig. 4. Data cache misses of immediate encoded split values - server

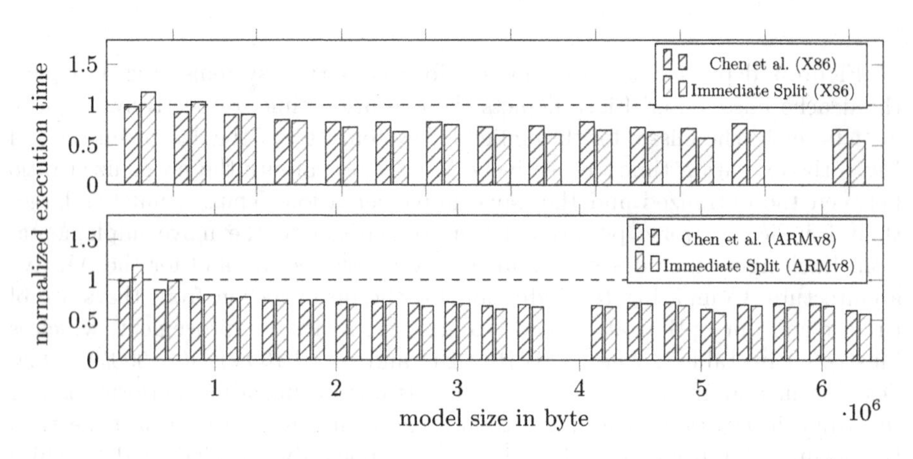

Fig. 5. Execution time of immediate encoded split values - server

cache size and necessarily produces cache misses. Therefore, we focus on these meaningful scenarios only.

Focusing on the instruction cache misses only, it can be seen that for most configurations with large model sizes the amount of icache misses is further decreased by our proposed optimization, compared to the state of the art (on the X86 server system in 95% of the relevant cases in geomean by 14.7%). Considering the data cache misses, considerable reductions can be observed for larger ensembles in comparison to the state of the art as well. For the X86 server, the amount of data cache misses for large ensembles is even reduced in 84% of the relevant cases by up to $\approx$ 92% in peak. In case of the ARMv8 server, a slighter reduction of dcache misses can be observed, up to $\approx$ 65% in peak and even an increase of $\approx$ 4% in geomean for large ensembles. Focusing on the embedded sys-

Table 2. Average and peak improvements compared to [4] (The geomean values in this table are computed for models only, which are larger than 900 kB) for server and embedded systems

	X86		ARMv8	
	Server	Embd.	Server	Embd.
Time				
#IMPROVED	33.7%	09.6%	34.6%	06.1%
#IMPROVED(>900k)	80.9%	10.7%	68.3%	06.5%
GEOMEAN(>900k)	+08.4%	−16.4%	+01.5%	−17.8%
Peak	+39.5%	+36.7%	+29.7%	+38.7%
ICache				
#IMPROVED	60.0%	33.7%	55.5%	–
#IMPROVED(>900k)	95.2%	60.7%	76.9%	–
GEOMEAN(>900k)	+14.7%	+00.3%	+02.6%	–
Peak	+90.4%	+26.8%	+61.8%	–
DCache				
#IMPROVED	64.1%	76.8%	41.6%	37.5%
#IMPROVED(>900k)	84.5%	100.0%	39.4%	27.8%
GEOMEAN(>900k)	+26.1%	+96.1%	−4.7%	+4.6%
Peak	+92.3%	+99.8%	+65.5%	+87.7%

tems, similar behavior can be observed for the data cache, the behavior for the icache misses contrarily differs[5]. Data cache misses are reduced by up to $\approx 99\%$ in peak and $\approx 96\%$ in average for X86. Instruction cache misses, however, are not significantly reduced on the X86 embedded system. For the ARMv8 embedded system, the improvement of dcache misses as well is comparably lower to the X86 embedded system.

Despite reducing icache and dcache misses, the allover execution time of the optimized implementation matters. In general, it can be observed that a high reduction in dcache misses does not necessarily result in a high reduction in execution time. For large ensemble sizes on the server machines, a consistent reduction of execution time can be however observed for our proposed optimization. The majority of relevant cases (more than 65%) yields an improvement in execution time on the X86 and ARMv8 servers. The improvement is up to $\approx 40\%$ in peak for X86 and ARMv8, compared to the state of the art. For small ensemble sizes, it can be observed that the execution time is increased beyond the naive implementation with our optimized implementation. In theses cases, the additional overheads due to the immediate encoding cannot be leveraged by the improvement. Investigating the embedded systems, the execution time can only be improved for few cases ($\approx 10\%$ on the X86 and $\approx 6\%$ on the ARMv8 sys-

[5] The ARMv8 system we use does not allow tracking of icache misses with perf.

tem). In geomean, the execution time is enlarged for the relevant cases, although the amount of dcache misses is drastically reduced for X86. This suggests that dcache misses are not the limiting factor for the execution in this scenario. Furthermore, this also implies that the CPU architecture is an important factor to the intended reduction of dacache misses with our optimization.

Although the results reveal that our proposed optimization cannot improve performance unconditionally, especially for small model sizes and embedded systems, we can report scenarios with a massive reduction of dcache misses and also a reduction of icache misses. Such a reduction can be useful to comfort parallel running applications. In several cases, the reduction of cache misses further directly relates to reduction of total execution time. When generating implementations, various versions can be profiled on the training data set so the best implementation can be chosen. Thus, for the cases where we achieve a worse result, the implementation of Chen et al. can still be chosen. Similarly for small models, where the optimized implementation induces a high overhead, the native implementation can be chosen.

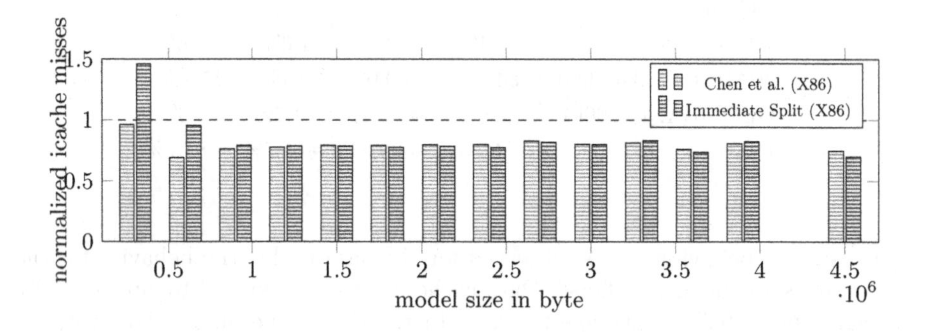

Fig. 6. Instruction cache misses of immediate encoded split values - embedded

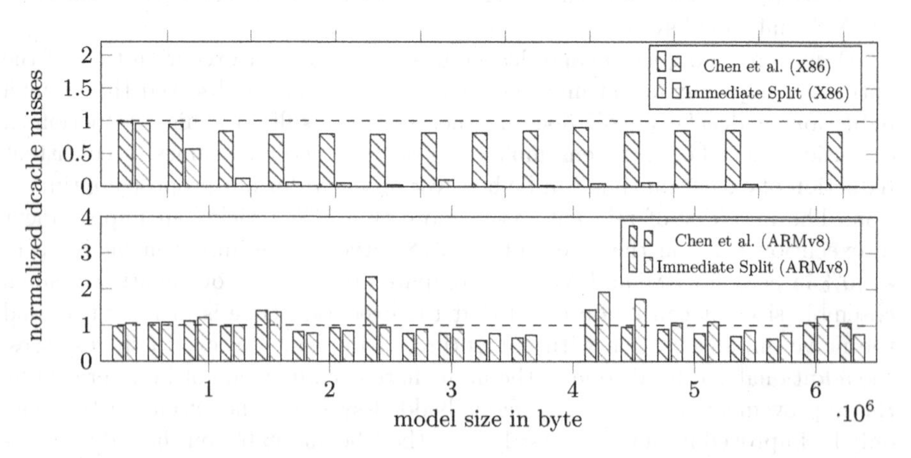

Fig. 7. Data cache misses of immediate encoded split values - embdedded

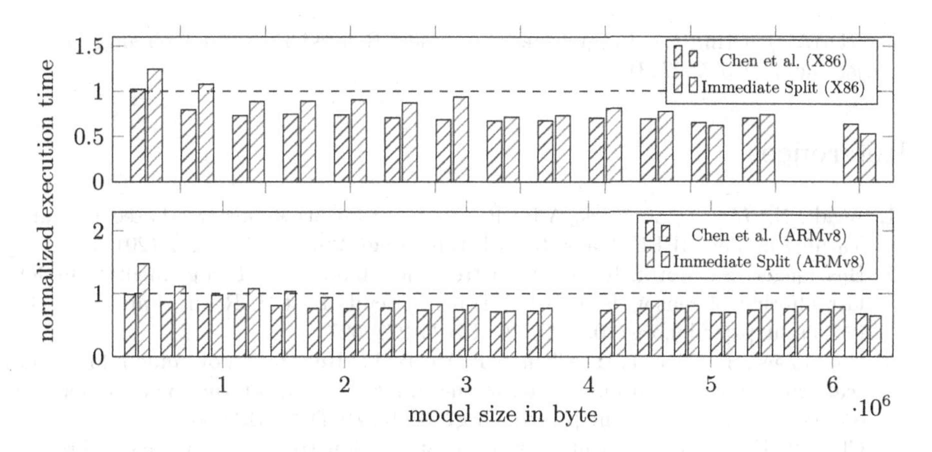

Fig. 8. Execution time of immediate encoded split values - embedded

6 Conclusion and Outlook

In this paper, we investigate the realization of random forests as if-else trees, which is one popular way of implementing random forests. We show that, counter intuitively to the design principle, data memory accesses and therefore potentially data cache misses play a big role in if-else tree implementations. The state-of-the-art optimization for if-else trees does not target data memory accesses specifically, therefore we propose an optimized implementation in this paper, where we eliminate a huge portion of data memory accesses. Experimental evaluation shows that our optimization can reduce the amount of data cache misses by up to 99% upon the state-of-the-art and can even lower the allover execution time by up to 40% on server systems. We further conclude that the overheads, which are introduced by our optimization, can only be leveraged for model sizes, which exceed the size of the level 1 caches. Thus, the optimization should be only applied in these cases. On embedded systems, the execution time is overall not significantly lowered, although the amount of cache misses can be drastically reduced. Hence, different aspects should also be explored. The implementation of our code generation fully supports X86 and ARMv8 architectures with different width integer and floating point data types. The source code is available at https://github.com/tu-dortmund-ls12-rt/arch-forest/tree/immediatesplittrees.

For future work, we plan to include model-based optimizations to our implementation, where we try to estimate the cache behavior during compile time with precise models and layout the trees accordingly. We also plan to investigate the relation between cache misses and execution time more intensively.

Acknowledgement. This work has been supported by Deutsche Forschungsgemeinschaft (DFG) within the project OneMemory (project number 405422836), the SFB876 A1 (project number 124020371), and Deutscher Akademischer Austauschdienst (DAAD) within the Programme for Project-Related Personal Exchange (PPP) (project number 57559723).

References

1. Asadi, N., Lin, J., de Vries, A.P.: Runtime optimizations for tree-based machine learning models. IEEE Trans. Knowl. Data Engi. **26**(9), 2281–2292 (2014)
2. Buschjäger, S., Morik, K.: Decision tree and random forest implementations for fast filtering of sensor data. IEEE Trans. Circuits Syst. I: Regul. Pap. **65**, 1–14 (2017). https://doi.org/10.1109/TCSI.2017.2710627
3. Buschjäger, S., Chen, K.H., Chen, J.J., Morik, K.: Realization of random forest for real-time evaluation through tree framing. In: 2018 IEEE International Conference on Data Mining (2018). https://doi.org/10.1109/ICDM.2018.00017
4. Chen, K.H., et al.: Efficient realization of decision trees for real-time inference. ACM Trans. Embed. Comput. Syst. (TECS) **21**, 1–26 (2022)
5. Dato, D., et al.: Fast ranking with additive ensembles of oblivious and non-oblivious regression trees. ACM Trans. Inf. Syst. **35**, 1–31 (2016)
6. Dua, D., Graff, C.: UCI machine learning repository (2017). https://archive.ics.uci.edu/ml/index.php
7. Kim, C., et al.: FAST: Fast architecture sensitive tree search on modern CPUs and GPUs. In: Proceedings of the 2010 ACM SIGMOD International Conference on Management of data. ACM (2010)
8. Lucchese, C., Nardini, F., Orlando, S., Perego, R., Tonellotto, N., Venturini, R.: Quickscorer: a fast algorithm to rank documents with additive ensembles of regression trees. In: Proceedings of the 38th International ACM SIGIR Conference on Research and Development in Information Retrieval, pp. 73–82. ACM (2015)
9. Lucchese, C., Perego, R., Nardini, F.M., Tonellotto, N., Orlando, S., Venturini, R.: Exploiting CPU SIMD extensions to speed-up document scoring with tree ensembles. In: SIGIR 2016 - Proceedings of the 39th International ACM SIGIR Conference on Research and Development in Information Retrieval (2016). https://doi.org/10.1145/2911451.2914758
10. Pedregosa, F., et al.: Scikit-learn: machine learning in python. J. Mach. Learn. Res. **12**(85), 2825–2830 (2011)
11. Prenger, R., Chen, B., Marlatt, T., Merl, D.: Fast map search for compact additive tree ensembles (cate). Technical report, Lawrence Livermore National Laboratory (LLNL), Livermore, CA (2013)
12. Van Essen, B., Macaraeg, C., Gokhale, M., Prenger, R.: Accelerating a random forest classifier: multi-core, GP-GPU, or FPGA? In: 2012 IEEE 20th Annual International Symposium on Field-Programmable Custom Computing Machines (FCCM), pp. 232–239. IEEE (2012)
13. Ye, T., Zhou, H., Zou, W.Y., Gao, B., Zhang, R.: RapidScorer: Fast tree ensemble evaluation by maximizing compactness in data level parallelization. In: Proceedings of the ACM International Conference on Knowledge Discovery and Data Mining (2018). https://doi.org/10.1145/3219819.3219857

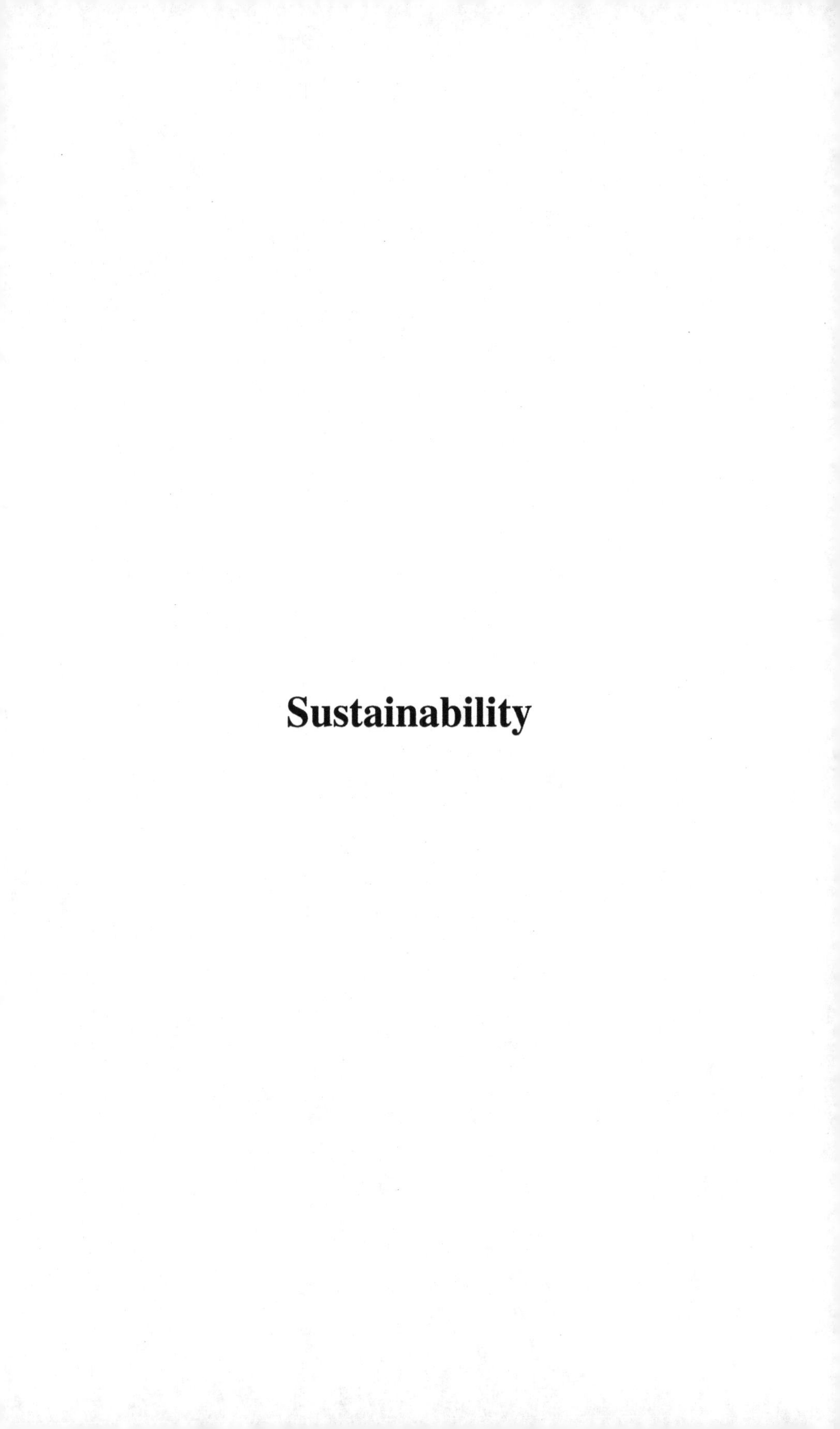

Sustainability

CGPM: Poverty Mapping Framework Based on Multi-Modal Geographic Knowledge Integration and Macroscopic Social Network Mining

Zhao Geng[1], Gao Ziqing[2], Tsai Chihsu[1], and Lu Jiamin[1](✉)

[1] Department of Mathematics, Department of International Economics and Trade, College of Artifical Intelligence, Jinan University, Guangzhou, China
zg1063316621@outlook.com, tttjlu@jnu.edu.cn
[2] Department of Chinese Language and Literature, Xi'an Jiaotong University, Xi'an, China

Abstract. Having high-precision and high-resolution poverty map is a prerequisite for monitoring the United Nations Sustainable Development Goals(SDGs) and for designing development strategies with effective poverty reduction policies. Recent deep-learning-related studies have demonstrated the effectiveness of the geographically-fine-grained data composed with satellite images, geolocated article texts and Open-Street-Map in poverty mapping. Unfortunately, there is no presented method which considers the multimodality of data composition or the underlying macroscopic social network among the investigated clusters in socio-geographic space. To alleviate these problems, we propose CGPM, a novelty end-to-end socioeconomic indicator mapping framework featured with the cross-modality knowledge integration of multi-modal features, and the generation of macroscopic social network. Furthermore, considering the deficiency of labeled clusters for model training, we proposed a weak-supervised specialized framework CGPM-WS to overcome this challenge. Extensive experiments on the public multimodality socio-geographic data demonstrate that CGPM and CGPM-WS significantly outperforms the baselines in semi-supervised and weak-supervised tasks respectively of poverty mapping.

Keywords: Sustainability · Poverty mapping · Multi-modality · Social networks mining

1 Introduction

Recently, the application of data mining in the field of sustainable development and global human rights protection has attracted a lot of attention [12,18,22]. One of the important applications is intelligent poverty mapping [1,19]. The main content of poverty mapping is to obtain high-precision key socioeconomic indicators [4] that measure the wealth level or poverty level of geographically

© The Author(s), under exclusive license to Springer Nature Switzerland AG 2023
M.-R. Amini et al. (Eds.): ECML PKDD 2022, LNAI 13717, pp. 549–564, 2023.
https://doi.org/10.1007/978-3-031-26419-1_33

distributed clusters. Having high-precision poverty maps is a crucial prerequisite for monitoring the UN Sustainable Development Goals(SDGs) and designing development strategies [10,22] for effective poverty reduction policies.

Driven by this, with the continuous progress of cutting-edge research in data mining and the increasing availability of geospatial data, many recent studies have proposed frameworks that combine deep learning algorithms with geospatial information as a highly-accurate, low-cost, and scalable technical system [2] to conduct poverty mapping.

Existing work on mapping poverty generally collects and uses multimodal geospatial data. Among the many data sources, the image data from Google's satellite map [9,26], Geolocated Article Texts data [3,15] and Open Street Map data [10,14] are currently used by the mainstream. Open-sourced Demographic and Health Survey(DHS) data is often used as professional socioeconomic indicators.

However, we must emphasize that most of the existing A.I. poverty mapping algorithms simply aggregate the training results of various specialized models. Such algorithms cannot overcome the two challenges faced by the current poverty mapping: 1) The ideal poverty mapping algorithm should realize the integration and complementary enhancement of multimodal data, to obtain the optimal cluster-level features representations. However, the proposed integrated or end-to-end splicing frameworks do not have this function. 2) The various clusters on the poverty map do not exist in isolation, and they constitute a potential social network with multiple semantics. We hope to be able to intelligently mine and generate the potential macro-social network structure at the cluster level, and synergistically apply it to the optimization of node (cluster) level feature representation, which will help to improve the accuracy of the poverty mapping framework.

Therefore, to alleviate the above problems, we propose CGPM, a novelty Poverty Mapping Framework simultaneously considers the Integration of Multi-Modal Geographic data and the mining/generating of the underlying Macroscopic Social Network. The framework mainly includes two core components, Cross-modal Feature Integration Module and Feature-based Macroscopic Social Network Generating Module. In the Cross-modal Feature Integration Module, we construct a cross-modality feature transformer based on the cross-modality attention mechanism, and use it to conduct cross-modal feature transformation and integration. In the Macroscopic Social Network Generating module, we generate multiple feature-based candidate macroscopic social network structure graphs. Furthermore, we jointly train the integrated representations, the generated social network structure (parameterized), and the networks' semantic embedding (parameterized) under the task, and eventually conduct the high-precision poverty mapping. Meanwhile, to alleviate limitedness of the DHS labels that we often face in actual surveying and mapping, we improved the CGPM and proposed a specialized architecture for weakly supervised scenarios (CGPM-WS). CGPM-WS performs refinement operations based on the pseudo-labeling technique to obtain a better feature representation, and effectively overcome the above challenges. In conclusion, our contributions are as follows:

(a) CGPM is the first method which realizes cross-modal fusion and the underlying macroscopic social networks generating simultaneously, thus to augment the representations of the multimodal data in this field. Whether it is in the field of poverty mapping or the related frontiers of data mining, CGPM is significantly ahead of the baselines in terms of novelty and design, with relatively strong academic significance.

(b) We designed a specialized architecture for weakly supervised scenarios-CGPM-WS, as a migration variant of CGPM, to alleviate the problem of the low coverage of the cluster areas marked by DHS.

(c) Extensive experiments in six typically developing countries demonstrate that CGPM has a significant accuracy advantage over current baselines. CGPM-WS has a significant accuracy advantage while maintaining granularity. It is validated that CGPM has considerable application prospects in intelligent poverty mapping research.

2 Related Work

2.1 Existing Poverty Mapping Framework

In recent years, some progress has been made in the research on integrated/migratory poverty mapping based on artificial intelligence & data mining algorithms. Dohyung Kim [19] et al. proposed a migration algorithm based on satellite image data to achieve high-precision poverty prediction. Evan Sheehan [15] et al. established an integrated poverty-mapping-oriented deep neural network architecture based on Wiki geographic text comment data and high-definition satellite data. Chiara Ledesma [9] et al. introduced textual data and statistics from social media to optimize the accuracy of poverty prediction from the perspective of interpretable learning. Masoomali Fatehkia et al. [3] transferred the interpretability scheme to a wider range of socioeconomic indicator predictions and assessed the generalization of existing methods compared to baselines. Kumar Ayush [1] designed a dynamic mapping framework for poverty maps based on Reinforcement Learning, which achieved high-precision poverty detection with extremely high computational overhead. Lee. K [10] et al. proposed a technical approach to anchor clusters to sample alignment under multimodal data, while designing a simple architecture suitable for weakly supervised environments. However, existing research has significant shortcomings in multimodal data fusion and representation optimization, as well as the isolation assumption of individual clusters. Therefore, to alleviate the above two problems has become the motivation of our research.

2.2 Discussion: Underlying Macroscopic Social Network Mining and Generating—Why and How

In our research, we note that the existing A.I. poverty mapping frameworks treat each cluster as an isolated sample point. However, we know that in a spatial geographic area, due to the existence of social and economic ties, the clusters in the area are not completely independent from each other, and their subsets will form multiple macro-social networks based on these linkages and ties. We interpret the implications of the underlying macro-social network structure in poverty mapping as mobility and homogeneity. The mobility structure refers to the existence of important linkages or ties between two clusters in terms of economic and social activities. For example, the satellite city structures and core industrial chains in the metropolitan areas. Due to the existence of the social network formed economic circles, economic complexes and other entities among the clusters [24], the social functions of each cluster often show different characteristics. There exists uneven developments of the transformation, such as satellite cities tend to undertake more housing and basic medical care, but lack of other positive socio-geographical characteristics. This economic phenomenon is not clearly revealed in either the OSM features, the satellite images, or the lighting data. It is difficult for the existing frameworks to compare the low-poverty clusters with high-poverty clusters while recognizing their with uneven values in such indicators with precise distinctions. The realization of feature sharing and dissemination among nodes in such a social network structure can effectively alleviate this problem. Homogeneity structure means that, for a aggregated feature representation measured by weight, if the values of two clusters show a sufficiently high similarity, we can consider the two as homogeneous clusters. Therefore, we can smooth the node feature vector of each cluster in such a homogeneous social network to a certain extent [8], so as to alleviate the observation error. caused by the operation in data collection and cluster anchoring.

Meanwhile, in sociological investigations, researchers often determine the semantics of social network structures by means of subjective definitions [27]. However, in the A.I. application scenarios, this would not be an optimal graph-structured representation. Therefore, we hope that the framework adaptively learns a better latent social network structure representation from the node (cluster) features, and cooperates them with the spatial node-level message passing layer (with graph structure and node features as trainable inputs) training to optimize the feature representations of each cluster. Such practices can be realized with the Graph Structure Learning [13,17,21,28] and Spatial Encoding [25] techniques.

3 Data Acquisition and Preprocessing

Open Street Map Data and Preprocessing

OpenStreetMap (OSM) contains open-source geospatial and infrastructure data open to the world [18]. A recent study shows that user-generated road maps in OSM are about 86% complete by 2020, and more than 40% of countries have a complete OSM street network [7]. We can obtain OpenStreetMap (OSM) data for the target area from Geofabrik, an online repository for OSM data [6]. From this, we extract extensive information about the number of roads, buildings and points of interest in a specific area, which will be presented as tabular features. We further discretize it to obtain Categorical Features.

In our feature engineering, OSM feature extraction ranges from rural areas with a 5 km radius and urban areas with a 2 km radius, each centered on the cluster location. We identified five road types in the dataset: arterial, arterial, paved, unpaved, and intersection. In terms of engineering road features, the preprocessing techniques we employ are as follows: for each type of road, we calculate the distance from the current cluster to the nearest road, the total number of roads, and the total road length for each cluster.

Satellite Imagery Data and Preprocessing

High-definition satellite image data is the most used geospatial data with the most stable acquisition channels in Poverty Mapping researches. We use the Google Static Maps API to obtain satellite images of clusters in rural areas within a radius of 0.5–5 km, and satellite images of clusters in urban areas within a radius of 0.5–3 km. To support this study, we prepared a total of 77960 images for download with a zoom level of 17, a scale of 1, and a pixel resolution of about 1.25 m. And after matching it with the area covered by a single data point of nighttime light data, typically each image can cover a land area of 0.25 km.

The Night Light (NTL) data we use is from the 2019 Visible Infrared Imaging Radiometer Suite Day/Night Band Dataset(VIIRS DNB). The VIIRS DNB dataset has a nighttime luminance resolution of 15 arcseconds, and contains geophotometric data on the ground at continuous photometric levels from 0 to 122. It can be used with Satellite or by calculating statistics (for example, nighttime light intensity within 1×1 square kilometers around the area) as tabular features. In this study, we choose the latter.

To more fairly demonstrate the efficiency of our proposed framework, align with most Poverty Mapping methods, we deploy the trained open-source VGG16 model accepting 400 × 400 pixel images. Further, we augment the data with random horizontal mirroring and use 30% dropout on the convolutional layers instead of the fully connected layers. Finally, we start to fine-tune the entire network using the Adam optimizer, and get a preliminary 3200-dimensional imagery embeddings.

Wiki Text Data and Preprocessing

In terms of text encoding, we completely follow the preprocessing scheme of MMPM [15]. We use the pre-trained Doc2vec model open sourced by Genism to encode text data from about 1.2K articles. In terms of parameters, we set the Windows Size as 8, and obtain a 400-dimensional text embeddings.

Labeling: Demographic and Health Survey Indicator

In this poverty mapping study, we use the Resident Wealth Index (DHS-WI) published by the International Population and Health Organization as a dependent variable indicator to measure the poverty level of the clusters. Through past researches [16], scholars from various countries have widely recognized that DHS-Program data can be used as the basic fact for constructing indicators to measure social economic activities.

DHS-WI is a comprehensive wealth measurement index constructed by DHS Program officials based on its surveys. In the existing work [23], the researchers proved that the DHS-WI indicator has a strong correlation with the international wealth index [10] (IWI, a common set of asset weighting calculations which is widely accepted as a measure of wealth index or poverty level, and is difficult to be calculated).

Eventually, the representations of our pre-trained multi-modal features (tabular-categorical features, image embeddings, text embeddings) can be written as $F \in \mathbb{R}^{n \times d_f}$, $I \in \mathbb{R}^{n \times d_I}$, $T \in \mathbb{R}^{n \times d_t}$, which is the inputs of CGPM.

4 Methodology

Framework: Figure 1 demonstrate the architecture of CGPM, which is the core algorithm of our proposed A.I. poverty mapping module based on cross-modality integration and graph structure learning (social network generating). We firstly construct a cross-modality feature transformer based on the cross-modality attention mechanism in order to implement the cross-modal feature transformation and integration. In the social network generating module, we generate multiple feature-based candidate macroscopic social network structure graphs. Furthermore, we jointly train the integrated representations, the generated social network structure (parameterized), and the semantic embedding (relation embedding, parameterized) of each graph structure under the task, eventually achieve/conduct? the high-precision poverty mapping.

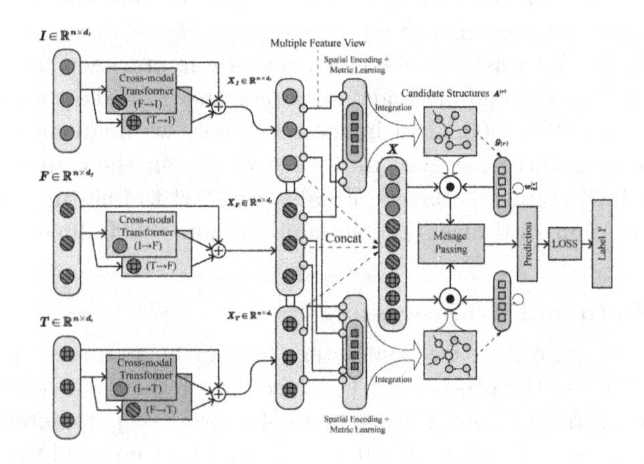

Fig. 1. Framework of CGPM

4.1 Cross-modality Feature Integration

The motivation of the Cross-modality Feature Integration module is to utilize the features from other source modalities to achieve the augmentation and supplementation of the target modality [11]. Therefore, we deploy a multi-head cross-modality transformer to achieve the integration of multi-modal information of clusters to be mapped (Fig. 2).

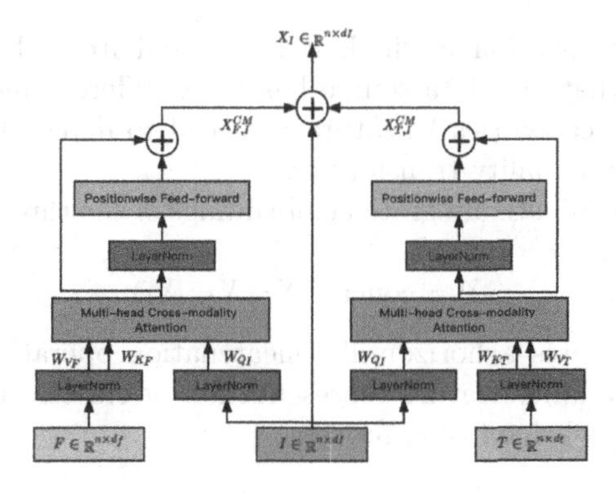

Fig. 2. Structure of cross-modal transformer

Note that we use Image as the target modality in our discussion. While in the model architecture, all source modalities will be treated as target modality separately. In the computation of each channel of multi-head cross-modality attention, we construct an attention matrix, then obtain the transferred representations of features from source modalities, that can be written as:

$$X_{F,I}^{AttT} = MulH\left(\sigma\left(\frac{(IW_{Q_I})^T (FW_{K_F})}{\sqrt{d_k}}\right)(FW_{V_F})^T\right) \tag{1}$$

where $W_{Q_I} \in \mathbb{R}^{d_I \times d_k}$, $W_{K_F} \in \mathbb{R}^{d_f \times d_k}$ and $W_{V_F} \in \mathbb{R}^{d_f \times d_k}$ denotes the weighted parameter matrix of cross-modality attention. $\sigma(\cdot)$ is defaulted to a softmax activation function. $MulH(\cdot)$ is a multi-head function. $X_{F,I}^{Att} \in \mathbb{R}^{n \times d_k}$ is the output of multi-head cross modality attention. We added residual connections into the calculation of the mapping and deployed position-wise feed-forward to form a complete cross-modality transformer, which is written as:

$$X_{F,I}^{CM} = \text{conv}\,1d\left(X_{F,I}^{Att}\right) + \text{relu}\left(X_{F,I}^{Att} W_1^{F,I} + b_1^{F,I}\right) W_2^{F,I} + b_2^{F,I} \tag{2}$$

where $\text{conv}\,1d(\cdot)$ is a 1-dimension convolutional layer deployed to adjust the residual dimension, and others are the formulation description of the positionwise feed-forward networks. $X_{F,I}^{CM}$ denotes the output of the cross-modal transformer about the transition from the source modality of discretized tabular features to the target modality of the image embedding. Eventually we set an attenuation coefficient to fuse the cross-modality information with the original information of the target modality, which is:

$$X_I = \alpha I + (1 - \alpha)\left(\alpha^{ad} X_{F,I}^{CM} + \left(1 - \alpha^{ad}\right) X_{T,I}^{CM}\right) \tag{3}$$

α_I and α_I^{ad} are settable attenuation coefficients. $X_{T,I}^{CM}$ denotes the output of the cross-modality module on the transition from text-modality features to the image modality. In the calculation process of the entire cross-modality transformer, the BatchNorm operation is deployed in each specific layer. Since it is a conventional optimization method, it is omitted from the formula description. Considering that our data source has three different modalities: Imagery, Text, and (tabular/categorical) Feature, we need to deploy 6 parallel channels of multi-head cross-modality transformers.

Finally, we concatenate integrated embeddings of the three target modalities, which is written as:

$$X = \text{concat}(X_I, X_F, X_t) \tag{4}$$

where concat(·) denotes a horizontal concatenation operation, X denotes the integrated cross-modality feature representation of clusters that we expect for the multimodal knowledge fusion process.

4.2 Feature-Based Macroscopic Social Network Mining and Generating Module

The core function of this module is to generate the underlying macroscopic social network structure and optimize the feature representation of clusters(referred as nodes) using the learned graph structure. For nodes with high-dimensional features, inspired by Graphformer [25], we deploy the spatial encoding of the transformer layer to extract the global information of latent interactions, then generate candidate graphs to embed each cluster into high-dimensional space of underlying Macroscopic social networks, which can be written as:

$$G_{i,j}^r = \sigma\left(\frac{\left(x_i W_r^Q\right)\left(x_j W_r^K\right)^T}{\sqrt{d}}\right) \tag{5}$$

where $W_r^Q, W_r^K \in \mathbb{R}^{d_k \times d_g}$ denotes weighted parameter matrices of the spatial coding, $G^r \in \{G^r\}_{r=1}^R$ denotes the r-th generated graph of the underlying social networks. (In order to modeling multiple semantics of interactions in social networks, we project to obtain $|R|$ candidate graph in total) Meanwhile, considering the assumptions of social networks that linkages tend to exist between pair-wise nodes with significant homogeneity, we implement a H-head multi-channel metric-based approach to compute the similarity of features/embeddings of pair-wise nodes using a cosine kernel function as a weight for candidate edges, which is:

$$S_{i,j}^r = \frac{1}{|H|}\sum_h^{H^r} \cos\left(w_s^{r,h} \odot x_i^T, w_s^{r,h} \odot X_j^T\right) \tag{6}$$

While elements of $w_s^{r,h} \in \mathbb{R}^{d_k}$ represent the importance of features in the measurement of similarity.

Eventually, we integrate the spatial encoding graph and node similarity graph by a structure propagation operation, and to augment the representation of underlying macroscopic social networks among clusters, which is

$$\tilde{A}^r = \text{spar}\left(\sigma\left(\left(\eta_g G^r + (1-\eta_g) I\right)\left(\eta_s S^r + (1-\eta_s) I\right)\right), \varepsilon\right) \tag{7}$$

While η_g, η_s denote restriction coefficients of the propagations. spar$(\cdot)$ is a sparsification function that enhances the sparsity of learned graph structure through dropping elements smaller than ε. $\left\{\tilde{A}^1, \ldots, \tilde{A}^R\right\}$ describes the generated underlying macroscopic social network obtained by CGPM. Furthermore, we initially distribute a semantic embedding g_r for each candidate graph/social network (defined as basis vectors or obtained by random walk). We compute node-level message passing [5,20] under the precondition of considering the heterogeneous semantics of the underlying social networks, which is:

$$x_i^{(l)} = \sigma\left(\sum_{(j,r)\in\mathcal{N}(i)} A_{i,j}^r W_{mp}^{r,(l)T}\left(x_j^{(l-1)} \odot g_r^{(l-1)}\right) + h\left(x_i^{(l-1)}\right)\right) \quad (8)$$

where l represents the depth of the message passing layer (deployed as a spatial graph convolutional layer), $\odot$ denotes a composition multiply operation, $\mathcal{N}(i)$ represents the set of clusters that have any social network structural connection with the cluster i under various semantics. $h(\cdot)$ and $g_r^{(l)} = W_{rel}^{(l)} g_r^{(l-1)}$ both denote a dimension alignment operation.

Eventually, the loss function of CGPM can be written as:

$$\mathcal{L}_{task} = \mathcal{L}_{rmse}\left(X^{\text{logit}}, Y\right) + \lambda\mathcal{L}_{re.g.}\left(X^{(l)}, \{A_r\}_{r=1}^R\right) \quad (9)$$

While $\mathcal{L}_{rmse}(\cdot)$ denotes a standard RMSE loss function and $\mathcal{L}_{reg}(\cdot)$ denotes a constraint function that prevents over-smoothing the vector representation of nodes and the learned macroscopic social network from being excessively dense.

4.3 CGPM-WS: For Weak Supervised Learning

Pseudo Labeling technology is the current solution to alleviate the challenge of excessive weak supervision in poverty mapping. Unlike existing general solutions which pseudo labels generated by other sub-models provide pseudo labeling refinement for CNNs (not mentioned previously), our proposed framework CGPM-WS is discussed as follows (Fig. 3):

Cold-Start Stage: We train 50 iterations of CGPM to obtain cross-modality integrated representations, as well as the parameters of each layer. Subsequently, CGPM-WS (Weak Supervision System) starts. The Cross-modality Transformer module (Paragraph 4.1) will be supplemented with the MLP(not mentioned previously layer and downstream tasks and moved to the **C area (Cross-modality Embedding Area)**, and the Macroscopic Social Network Mining module (Paragraph 4.2) will be moved to the **G area (Social Networks Generating Area)**. Meanwhile, CGPM-WS includes an **F area (Feature-based Model Area)**.

F Area: Take OSM data, and the features obtained from statistical description and artificial feature engineering of nightlight image and text data as inputs, construct LightGBM model (Adaboost's ensemble mode) for only DHS-WI labeled

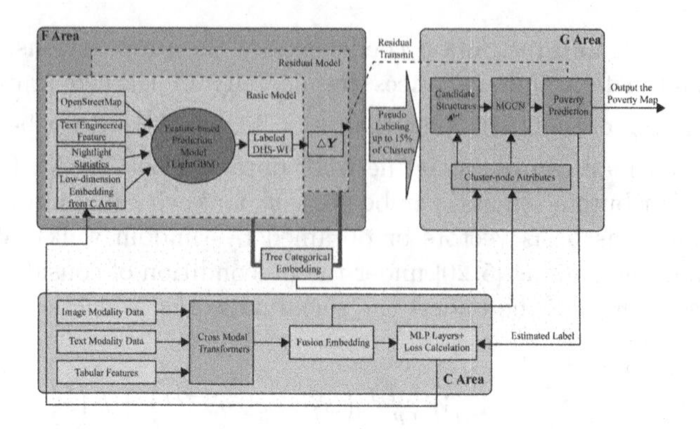

Fig. 3. Framework of CGPM-WS

clusters. Furthermore, only when the first generation of the framework is executed, LightGBM will make predictions for all clusters, and select clusters whose predictions are the closest to the output of the pre-training, then utilize the predictions as pseudo labels, so that the proportion of labeled clusters can be supplemented to 15%. In all iterations of the system, 15% (ground-truth+ pseudo) will be input into the Macroscopic Social Network Generating module as estimated labels. In addition to the cross-modality representation output by the Cross-modality Integration Module, we concatenate the Tree Categorical Embedding (Pred_leaf parameter) output by LightGBM with it to complement the discretized tabular features.

G Area: We train the Social Networks Generating Module with received labels and features. The output will be submitted to the C area as the Refinement Label. After completing all training epochs, the output of the G Area is applied to draw the poverty map.

C Area: We directly connect MLP layers, Refinement Labels and standard loss functions to the downstream of the Cross-modality Transformer Module to obtain the updated features embedding of the sampled clusters. We weight and fuse the embedding vector on the Cross-modality Transformer side with the corresponding part of the node representation in the G area at a decay rate of 0.1, and submitted this embedding vector output from one layer of the MLP layers? (the second layer was selected in the experiment) to the F area. , concatenate with the features of the F area to provide a more informative representation for the LightGBM model.

Residual Transmittal: In early iterations of training, we make a residual between the output of the current round and the output of the previous round of the ground-truth labeled clusters and input the residual into the F area. The F area utilize the residual as labels, and take all the feature vectors received in this area as the input to append an additional LightGBM sub-model. When the F area provides estimated labels to the G area, the output is the summation of the prediction of all LightGBM sub-models.

Meanwhile, we emphasize/conclude? that CGPM-WS is a specialized model suitable for typical weak supervised environment.

5 Experiment

5.1 Implementation Details:

We select six representative developing countries in southern Asia and Africa as the experimental subjects: the Philippines(PHL), India(IND), Bangladesh(BAN), Tanzania(TZA?), Uganda(UGA), and Nigeria(NGA). The statistics of the sample points we collected and used in the experiments are shown in Table 1. For baselines, we choose TMPM [19] (transferable model modeled with satellite images and OSM data), WI-MMPM [15] (the most recognized end-to-end multimodal model, short for MMPM or WIPM) and HEPM [10] (specialized poverty estimation framework for weakly supervised learning), the above three methods are the most representative open source baselines.

For parameters, we set α_I and α_I^{ad} as 0.8, 0.5. The depth of Cross-modal Transformers and Message Passing are both set to 2. $\varepsilon = 0.1$, $H = 4$, $|R| = 5$ to ensure fair evaluation, the linear layer depth of all models is set to 2 (Fig. 4).

Table 1. Clusters in our experiment.

	PHL	BAN	IND	NGA	UGA	TZ
Total	9970	8705	33076	23649	10700	8935
Labelled	1213	600	2058	1681	1011	1044
Precisely geolocated	6488	5729	8914	11057	5163	5597

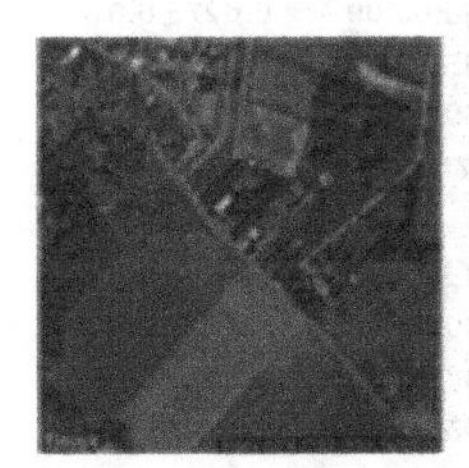

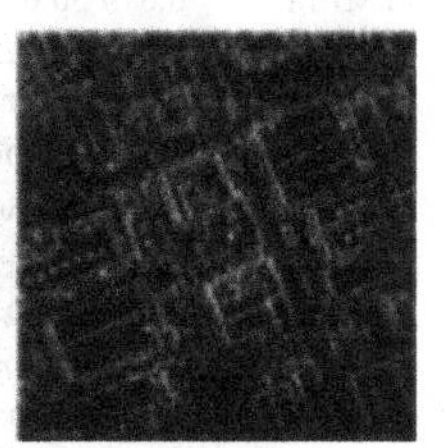

Fig. 4. Satellite data

5.2 Evaluation

Following previous research, we measured the coefficient of determination (R-squared) between the estimated wealth index of the model and the observed wealth index in the recent 5-year DHS surveys. The R-squared can be interpreted as the proportion of the variance for the observed wealth index that is explained by the estimated wealth index. Although the R-squared does not represent the accuracy of prediction precisely, it conveys a degree of performance in an intuitive way. Compared to RMSE, R-squared can be derived from econometrics showing whether our estimated index can be used to replace the DHS-WI index for relevant empirical research.

In the experimental setting, we set up three groups:

1) In the Semi-Supervised Group, 50% of the clusters are labeled (containing all labeled data points we have). We use 80% labeled clusters as a training set, 10% labeled clusters as a validation set, and 10% labeled and other unlabeled clusters as de facto validation set.

2) In the Weakly-Supervised Group, 8%–18% of the clusters are labeled (contains all the labeled data points we have, since the number of unlabeled data points varies, this ratio fluctuates between 8%–18%). We use 80% labeled clusters as training set, 10% labeled clusters as validation set, and 10% labeled and other unlabeled clusters as de facto validation set.

3) Cross-National Group: the experimental group which is to verify the generalization performance of our proposed method. In this group, we use the Philippines as the training set, transfer the trained model to other countries, and verify the performance. In CGPM and CGPM-WS, since the Macroscopic Social Network cannot be migrated, during the training process, we merge the target country and the Philippine clusters as the input, while using the labeled Philippine clusters as the training set and the validation set, the target country as the test set, thus to fairly evaluate the generalization of the model.

Table 2. Comprehensive evaluation.

Performance Evaluation: Metric: Mean Pearson's R-square + Values Deviation						
	PHL	BAN	NGA	UGA	TZ	IND
Semi-Supervised Group: Labeled Ratio: 50%						
TMPM	0.679±0.004	0.713±0.017	0.641±0.004	0.732±0.026	0.68±0.009	0.627±0.01
MMPM	0.715±0.011	0.74±0.006	0.675±0.002	0.717±0.009	0.765±0.005	0.695±0.006
HEPM	0.73±0.007	0.726±0.003	0.626±0.001	0.768±0.011	0.669±0.013	0.633±0.006
CGPM	**0.824±0.005**	**0.78±0.002**	**0.737±0.004**	**0.793±0.003**	**0.775±0.007**	**0.764±0.001**
Weakly-Supervised Group: Labeled Ratio: 10–20%						
TMPM	0.656±0.019	0.593±0.017	0.597±0.016	0.674±0.021	0.64±0.011	0.639±0.033
MMPM	0.732±0.024	0.575±0.022	0.625±0.012	0.665±0.037	0.669±0.009	0.652±0.025
HEPM	0.783±0.009	0.718±0.006	0.682±0.008	0.721±0.009	0.683±0.006	0.707±0.011
CGPM	0.819±0.014	0.705±0.015	0.696±0.008	0.726±0.014	0.705±0.005	0.716±0.02
CGPM-WS	**0.845±0.005**	**0.759±0.003**	**0.731±0.005**	**0.773±0.012**	**0.724±0.002**	**0.748±0.017**
Semi-Supervised Cross National Experiment on Cross-Modality Feature Integration Training (on PHL)						
TMPM	/	0.691±0.016	0.476±0.031	0.575±0.024	0.557±0.016	0.598±0.006
MMPM	/	0.713±0.028	0.569±0.029	0.668±0.035	0.541±0.02	0.622±0.014
HEPM	/	0.699±0.009	0.413±0.025	0.585±0.018	0.462±0.013	0.617±0.022
CGPM	/	**0.747±0.013**	**0.63±0.01**	**0.692±0.004**	**0.575±0.007**	**0.709±0.008**

The results of the evaluation are reported in Table 2, from which we have the following observations: (a): In the Semi-Supervised Group, the multimodal end-to-end approach significantly outperforms the multimodal model ensemble method, while the performance improvement ratio is as high as eight percentage points. At the same time, CGPM significantly outperforms all other baseline methods in all country experiments, with an average advantage of about 6% points. And considering the setting in this experiment, each module of our CGPM uses a lightweight deployment scheme, implying that there is still considerable room for improvements in the performance of CGPM. (b): In the Weakly-Supervised Group, the weakly supervised learning methods are significantly more suitable for this experimental setting. Both TTMPM and MMPM suffer a large performance loss, while HEPM and CGPM-WS based on CGPM improvement expands the performance advantage by about 5% points, compared to the previous set of experiments. Among them, CGPM-WS outperforms the baseline schemes in all countries and has higher stability. (c) Our proposed CGPM exhibits good generalization performance and stability beyond the baselines in all multinational experiments. According to the experimental results, the effect of transnational migration experiments is relatively good between countries in the same geographic region or countries(effect is good?) with similar economic patterns and social development levels. Considering the performance and the stability in the three sets of experiments, CGPM is more suitable for poverty mapping deployed in countries with highly complex socioeconomic environments (Philippines, Bangladesh, and India).

In order to visually demonstrate the accuracy and stability of CGPM, we calculate the average of the model's prediction results for the residential areas by province in the Philippines and displayed it in Fig. 5. From this, we can conclude that the accuracy of TMPM is slightly insufficient, MMPM has an overall shift in the predicted value, and the prediction result of CGPM is the closest to the Ground Truth(not mentioned previously or you used another term in the previous text) and has the highest accuracy.

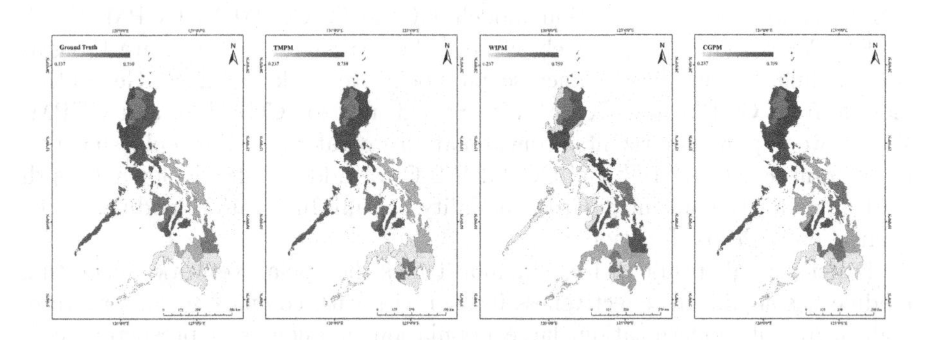

Fig. 5. Case visualization, Philippines

Meanwhile, to demonstrate the advantages of CGPM-WS over CGPM in the weakly supervised learning domain, we run the CGPM and CGPM-WS models in the Weakly-Supervised Group respectively, and present them in the form of scatter plots in Fig. 6. The visualization results show that CGPM-WS has higher

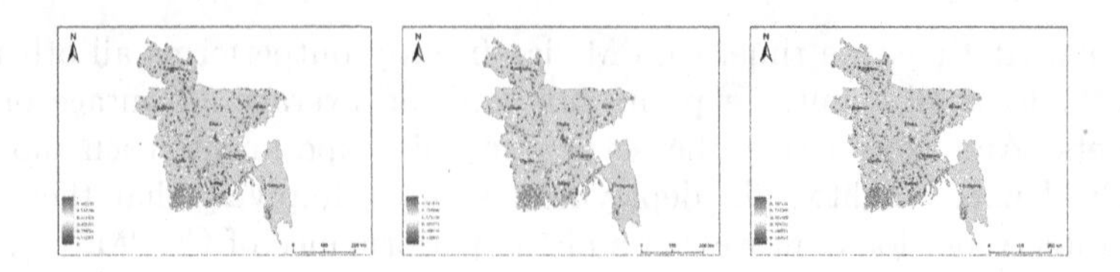

Fig. 6. Case visualization for weakly supervised learning, Bangladesh (Ground truth, CGPM, CGPM-WS)

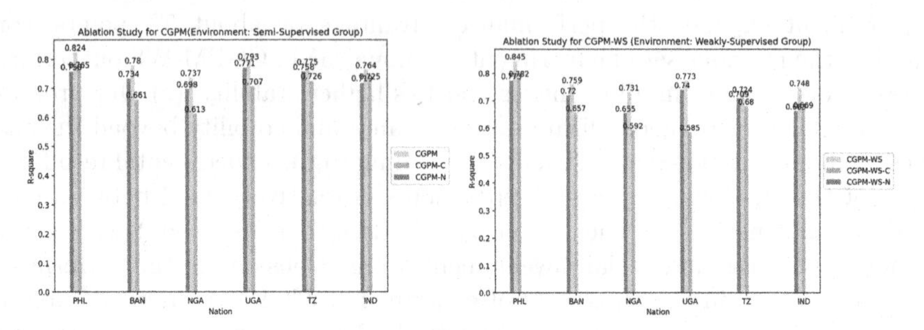

Fig. 7. Ablation study

accuracy in general, and the number of sample points with excessive prediction bias is significantly less than that of CGPM.

5.3 Ablation Study

In this section, we will verify the validity of each module of CGPM and CGPM-WS. Therefore we set reduction models CGPM-C, CGPM-N, CGPM-WS-C, CGPM-WS-N, which respectively remove the Cross-modality Feature Integration module, Feature-based Macroscopic Social Network Mining & Generating module from CGPM and CGPM-WS. Specifically, for CGPM-C, and CGPM-WS-C, we deploy a horizontal concatenation operation as the replacement of the margin. For CGPM-C, and CGPM-WS-C, we additionally deploy a 2-depth MLP layer for the output of Cross-modality Feature Integration module, as the connection with loss function.

Experimental Result in Fig. 7 demonstrates the Social Network Generating module has significant effectiveness in countries with complex social structure, high degree of modernization, large population, frequent and prosperous economic activities. One possible reason is that there are numerous and important underlying macro-social network structures among the clusters in such countries, and they have interrelated socioeconomic effects that cannot be neglected. The Cross-modality Feature Integration module can stably improve the model performance in all countries, which means that modeling with knowledge-fused multimodal data will hopefully become a beacon for future poverty mapping research.

6 Conclusion

In this paper, we propose an end-to-end Poverty Mapping framework, CGPM, which is ahead of the academic frontier. CGPM innovatively realizes the cross-modal transformation and integration of multimodal data in this field, as well as the mining of underlying macroscopic social network structure, thus to optimize the representations of clusters. Meanwhile, we propose a variant of CGPM, CGPM-WS, to specialize to overcome the weakly supervised learning challenges commonly found in Poverty Mapping. Extensive experiments demonstrate that CGPM and CGPM-WS significantly outperform the current baselines, and show more promising research prospects.

References

1. Ayush, K., Uzkent, B., Tanmay, K., Burke, M., Lobell, D., Ermon, S.: Efficient poverty mapping from high resolution remote sensing images. In: Proceedings of AAAI Conference on Artificial Intelligence, vol. 35, pp. 12–20 (2021)
2. Belhadj, B., Kaabi, F.: New membership function for poverty measure. Metroeconomica **71**(4), 676–688 (2020)
3. Fatehkia, M., et al.: Mapping socioeconomic indicators using social media advertising data. EPJ Data Sci. **9**(1), 1–15 (2020). https://doi.org/10.1140/epjds/s13688-020-00235-w
4. Flechtner, S.: Poverty research and its discontents: review and discussion of issues raised in dimensions of poverty. measurement, epistemic injustices and social activism. Rev. Income Wealth **67**(2), 530–544 (2021). (beck, v., h. hahn, and r. lepenies eds., springer, cham, 2020)
5. Hamilton, W., Ying, Z., Leskovec, J.: Inductive representation learning on large graphs. In: Advances in Neural Information Processing Systems, vol. 30 (2017)
6. Htet, N.L., Kongprawechnon, W., Thajchayapong, S., Isshiki, T.: Machine learning approach with multiple open-source data for mapping and prediction of poverty in myanmar. In: 2021 18th International Conference on Electrical Engineering/Electronics, Computer, Telecommunications and Information Technology (ECTI-CON), pp. 1041–1045. IEEE (2021)
7. Hu, S., Ge, Y., Liu, M., Ren, Z., Zhang, X.: Village-level poverty identification using machine learning, high-resolution images, and geospatial data. Int. J. Appl. Earth Obs. Geoinf. **107**, 102694 (2022)
8. Jin, W., Ma, Y., Liu, X., Tang, X., Wang, S., Tang, J.: Graph structure learning for robust graph neural networks. In: Proceedings of the 26th ACM SIGKDD International Conference on Knowledge Discovery & Data Mining, pp. 66–74 (2020)
9. Ledesma, C., Garonita, O.L., Flores, L.J., Tingzon, I., Dalisay, D.: Interpretable poverty mapping using social media data, satellite images, and geospatial information. arXiv preprint arXiv:2011.13563 (2020)
10. Lee, K., Braithwaite, J.: High-resolution poverty maps in sub-saharan africa. arXiv preprint arXiv:2009.00544 (2020)
11. Liu, F., Wu, X., Ge, S., Fan, W., Zou, Y.: Federated learning for vision-and-language grounding problems. In: Proceedings of the AAAI Conference on Artificial Intelligence. vol. 34, pp. 11572–11579 (2020)

12. Martínez, S., Rueda, M., Illescas, M.: The optimization problem of quantile and poverty measures estimation based on calibration. J. Comput. Appl. Math. **405**, 113054 (2020)
13. Pilco, D.S., Rivera, A.R.: Graph learning network: a structure learning algorithm. arXiv preprint arXiv:1905.12665 (2019)
14. Roghani, H., Bouyer, A., Nourani, E.: PLDLS: a novel parallel label diffusion and label selection-based community detection algorithm based on spark in social networks. Expert Syst. Appl. **183**, 115377 (2021)
15. Sheehan, E., et al.: Predicting economic development using geolocated Wikipedia articles. In: Proceedings of the 25th ACM SIGKDD International Conference on Knowledge Discovery & Data Mining, pp. 2698–2706 (2019)
16. Steele, J.E., et al.: Mapping poverty using mobile phone and satellite data. J. Roy. Soc. Interface **14**(127), 20160690 (2017)
17. Tang, J., Qian, T., Liu, S., Du, S., Hu, J., Li, T.: Spatio-temporal latent graph structure learning for traffic forecasting. arXiv preprint arXiv:2202.12586 (2022)
18. Thornton, P., et al.: Mapping poverty and livestock in the developing world, vol. 1. ILRI (aka ILCA and ILRAD) (2002)
19. Tingzon, I., et al.: Mapping poverty in the Philippines using machine learning, satellite imagery, and crowd-sourced geospatial information. In: AI for Social Good ICML 2019 Workshop (2019)
20. Vashishth, S., Sanyal, S., Nitin, V., Talukdar, P.: Composition-based multi-relational graph convolutional networks. arXiv preprint arXiv:1911.03082 (2019)
21. Wang, L., Chan, R., Zeng, T.: Probabilistic semi-supervised learning via sparse graph structure learning. IEEE Transa. Neural Netw. Learn. Syst. **32**(2), 853–867 (2020)
22. Watson, D., Whelan, C.T., Maître, B., Williams, J.: Non-monetary indicators and multiple dimensions: the ESRI approach to poverty measurement. Econ. Soc. Rev. **48**(4), 369–392 (2017)
23. Xie, M., Jean, N., Burke, M., Lobell, D., Ermon, S.: Transfer learning from deep features for remote sensing and poverty mapping. In: Thirtieth AAAI Conference on Artificial Intelligence (2016)
24. Xu, K., Hu, W., Leskovec, J., Jegelka, S.: How powerful are graph neural networks? arXiv preprint arXiv:1810.00826 (2018)
25. Ying, C., et al.: Do transformers really perform badly for graph representation? In: Advances in Neural Information Processing Systems, vol. 34 (2021)
26. Zhang, H., Xu, Z., Wu, K., Zhou, D., Wei, G.: Multi-dimensional poverty measurement for photovoltaic poverty alleviation areas: evidence from pilot counties in china. J. Cleaner Prod. **241**, 118382 (2019)
27. Zhu, Y., et al.: A survey on graph structure learning: progress and opportunities (2021)
28. Zhu, Y., Xu, W., Zhang, J., Liu, Q., Wu, S., Wang, L.: Deep graph structure learning for robust representations: A survey. arXiv preprint arXiv:2103.03036 (2021)

Bayesian Multi-head Convolutional Neural Networks with Bahdanau Attention for Forecasting Daily Precipitation in Climate Change Monitoring

Firas Gerges[1](✉), Michel C. Boufadel[2], Elie Bou-Zeid[3], Ankit Darekar[1], Hani Nassif[4], and Jason T. L. Wang[1]

[1] Department of Computer Science, New Jersey Institute of Technology, University Heights, Newark, NJ 07102, USA
{fg92,apd8,wangj}@njit.edu

[2] Center for Natural Resources, Department of Civil and Environmental Engineering, New Jersey Institute of Technology, University Heights, Newark, NJ 07102, USA
boufadel@njit.edu

[3] Department of Civil and Environmental Engineering, Princeton University, Princeton, NJ 08544, USA
ebouzeid@princeton.edu

[4] Department of Civil and Environmental Engineering, Rutgers University - New Brunswick, Piscataway, NJ 08854, USA
nassif@soe.rutgers.edu

Abstract. General Circulation Models (GCMs) are established numerical models for simulating multiple climate variables, decades into the future. GCMs produce such simulations at coarse resolution (100 to 600 km), making them inappropriate to monitor climate change at the local regional level. Downscaling approaches are usually adopted to infer the statistical relationship between the coarse simulations of GCMs and local observations and use the relationship to evaluate the simulations at a finer scale. In this paper, we propose a novel deep learning framework for forecasting daily precipitation values via downscaling. Our framework, named Precipitation CNN or PCNN, employs multi-head convolutional neural networks (CNNs) followed by Bahdanau attention blocks and an uncertainty quantification component with Bayesian inference. We apply PCNN to downscale the daily precipitation above the New Jersey portion of the Hackensack-Passaic watershed. Experiments show that PCNN is suitable for this task, reproducing the daily variability of precipitation. Moreover, we produce local-scale precipitation projections for multiple periods into the future (up to year 2100).

Keywords: Machine learning · Convolutional neural networks · Statistical downscaling · Climate change

© The Author(s), under exclusive license to Springer Nature Switzerland AG 2023
M.-R. Amini et al. (Eds.): ECML PKDD 2022, LNAI 13717, pp. 565–580, 2023.
https://doi.org/10.1007/978-3-031-26419-1_34

1 Introduction

Climate change refers to the phenomena where regional and global climate patterns change over time. Following the industrial revolution, greenhouse gas emissions from human activities are the primary driver of climate change [4]. This anthropogenic climate change is severely impacting communities' infrastructure and ecosystems, causing sea level rise, species extinction, in addition to disturbance in the operation of key infrastructures, such as bridges and power supply. Although national and international policies and research centers are mainly focused on the changes in global climate patterns, climate change involves regional/local climate variability as well. Subsequently monitoring climate change at the local regional level is of utmost importance. General Circulation Models (GCMs) are numerical models for simulating the physical processes taking place on land and ocean surfaces, as well as in the atmosphere. The very large spatial resolutions of GCMs make them too coarse to monitor climate change at a smaller scale. As such, to analyze and project regional climatic changes, one would need to perform spatial downscaling of GCM outputs to desired finer scales. Spatial downscaling, in the context of GCMs, refers to enhancing the coarse spatial resolution of the GCM simulations, where the simulations are downscaled either to a local weather station level, to a finer grid resolution, or to a local region level (say a watershed). Multiple attempts exist in the literature that leverage machine learning for downscaling climate variables, such as temperature [5,25], wind [6,12], and precipitation [20,25]. Machine learning methods used in the previous studies include classical techniques such as support vector regression (SVR) [9], random forests (RF) [24], decision trees (DT) [27], and multi-layer perceptron (MLP) [1], as well as deep learning techniques such as convolutional neural networks (CNNs) [18,23]. With the GCM simulations, we present a novel deep learning method, named PCNN, for the downscaling of daily precipitation. PCNN employs a multi-head CNN framework, with embedded self-attention and stacked Bahdanau attention layers, that aims to implicitly capture the spatial relationship between multiple GCM grid points. Moreover, we adopt the Monte-Carlo dropout sampling technique [13,21,22] to quantify aleatoric and epistemic uncertainties. The CNNs used in the previous studies differ from ours in that they attempt to downscale a GCM grid to multiple grid locations at the same time where the output is also a grid. Furthermore, the previous CNNs adopt one type of simulations. In contrast, our PCNN downscales a GCM grid to a local region/area while employing multiple types of simulations (e.g., temperature, humidity, wind, and so on). As a case study, we apply PCNN to downscale the daily precipitation over the Hackensack-Passaic watershed in New Jersey. The main contributions of our work are summarized below.

- We develop a multi-head CNN framework (PCNN) with embedded self-attention and stacked Bahdanau attention blocks to downscale/forecast the daily precipitation in a local region (i.e., the area above the New Jersey portion of the Hackensack-Passaic watershed).
- We incorporate the Monte-Carlo dropout sampling technique into our framework to quantify the aleatoric and epistemic uncertainties.

- Experimental results obtained from forecasting/downscaling the daily precipitation in the local region show that our framework is suitable for the downscaling task.
- We apply our trained deep learning model to future climate simulations to produce local-scale projections for multiple periods into the future (up to year 2100).

2 Problem Formulation and Data Collection

2.1 Problem Formulation

To perform statistical downscaling, one would need to relate large-scale GCM simulations of weather patterns to local observations. In this study, we aim to downscale the daily precipitation values using coarse-resolution simulations of multiple climate variables (temperature, heat flux, humidity, wind, and so on). To capture the different interactions of weather patterns, we extract the GCM simulations from multiple grid points, surrounding the area of interest. Figure 1 shows the New Jersey portion of the Hackensack-Passaic watershed (highlighted in orange color), for which we attempt to downscale precipitation. The GCM grid points we use in our downscaling are represented by black circles. We select a total of $7 \times 7 = 49$ points, covering around 1.5 million km2 surrounding the watershed area. We formulate the downscaling task as a regression problem, and we aim to use GCM-simulated climate variables from the 49 grid points, treated as input features, to forecast the daily precipitation, treated as the label, above the watershed.

2.2 Local Observations

Our study area, highlighted in orange color in Fig. 1, consists of the New Jersey portion of the Hackensack-Passaic watershed. We aim to downscale the average daily precipitation over the whole area instead of a single station. We extracted the daily precipitation values from the meteorological data set (NCEI Accession 0129374) provided by the National Centers for Environmental Information (NCEI), of the National Oceanic and Atmospheric Administration (NOAA) [19]. These data are provided as a grid, with a 1/16°C resolution. We selected the grid points contained within the watershed and computed their average for each day from January 1st, 1950, to December 31st, 2005 (daily data for 56 years). These local observations are used as labels (ground truth) in our study.

2.3 GCM Simulations

There are multiple GCMs included in CMIP5. We opted-in to select the CM3 model of the Geophysical Fluid Dynamics Laboratory (GFDL) of NOAA [11]. GFDL CM3 simulations are at grid with a 2 deg × 2.5 deg resolution (220 km × 270 km). For each of the 49 grid points shown in Fig. 1, we extract, from

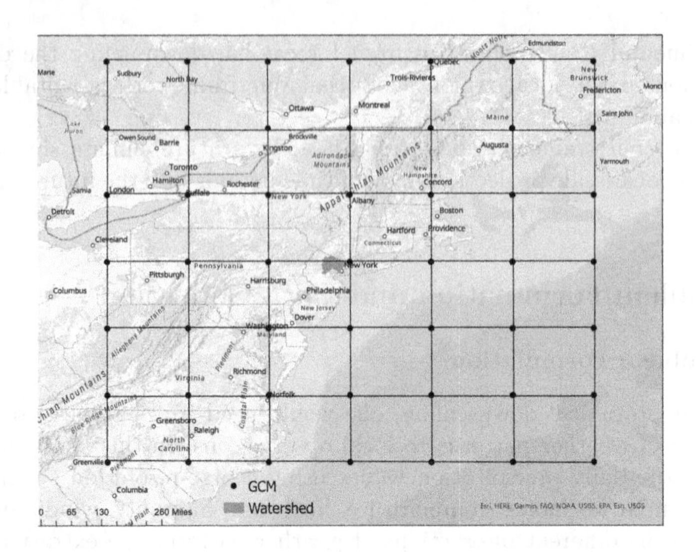

Fig. 1. GCM grid points (black circles) selected for downscaling. The area highlighted in orange color depicts the New Jersey portion of the Hackensack-Passaic watershed, for which we attempt to downscale/forecast daily precipitation. (Color figure online)

the CM3 model, 26 climate variables, which are listed in Table 1. As such, we have a total of $49 \times 26 = 1{,}274$ input features. We retrieved the daily data from January 1st, 1950, to December 31st, 2005, obtaining a total of 20,418 data records. Moreover, we retrieved the simulations for the periods between 2030–2040, 2060–2070, and 2090–2100, which are to be used for the long-term local-scale projections. We selected these future periods because we have the corresponding simulations available.

3 The PCNN Framework

We present a novel deep learning framework, named PCNN. This framework aims to apply convolution neural networks (CNNs) to each climate variable independently, where each climate variable is represented by a matrix containing values for all the 49 grid points, followed by a sequence of Bahdanau attention blocks. We apply dropout steps across the PCNN framework to perform uncertainty quantification using the Monte-Carlo dropout sampling technique. The main architecture of our deep learning framework is illustrated in Fig. 2.

3.1 Multi-head Convolutional Neural Networks

Convolution neural networks (CNNs) [7,8] have gained popularity for their performance in computer vision and image analysis. CNNs are based on a grid-like topology [10], and consist of a set of convolution and pooling layers. Convolution is a specialized matrix operation, which is the core of CNNs, used to extract local

Table 1. Large-scale climate variables extracted from GCM simulations.

Climate variable	Description	Unit
clt	Total cloud fraction	%
$hfls$	Surface upward latent heat flux	W/m2
$hfss$	Surface upward sensible heat flux	W/m2
hus_{250}	Specific humidity at 250 hPa	–
hus_{500}	Specific humidity at 500 hPa	–
hus_{850}	Specific humidity at 850 hPa	–
$huss$	Specific humidity at near-surface	–
pr	Precipitation	Kg/m2/s
psl	Sea level pressure	Pa
rhs	Relative humidity at near-surface	%
$sfcWind$	Daily mean wind speed at near-surface	m/s
ta_{250}	Air temperature at 250 hPa	K
ta_{500}	Air temperature at 500 hPa	K
ta_{850}	Air temperature at 850 hPa	K
tas	Air temperature at near-surface	K
ua_{250}	Eastward wind at 250 hPa	m/s
ua_{500}	Eastward wind at 500 hPa	m/s
ua_{850}	Eastward wind at 850 hPa	m/s
uas	Eastward wind at near-surface	m/s
va_{250}	Northward wind at 250 hPa	m/s
va_{500}	Northward wind at 500 hPa	m/s
va_{850}	Northward wind at 850 hPa	m/s
vas	Northward wind at near-surface	m/s
zg_{250}	Geopotential height at 250 hPa	m
zg_{500}	Geopotential height at 500 hPa	m
zg_{850}	Geopotential height at 850 hPa	m

patterns (learnable features) of the corresponding feature group. The convolution layer applies filters to the input features using a set of kernels. A pooling layer often follows each convolution layer and aims to extract the patterns by focusing on the maximum, minimum, or average-based statistical summary of the neighborhood. In our architecture, we utilize a self-attention layer after each pooling layer, which allows our network to learn to choose a subset of the pooling output by giving selective attention to the features. On each input matrix representing a climate variable, we apply a sequence of convolution-pooling-attention two times, where the first pooling layer applies maximum pooling, and the second is average pooling. These sequences are applied independently to each climate

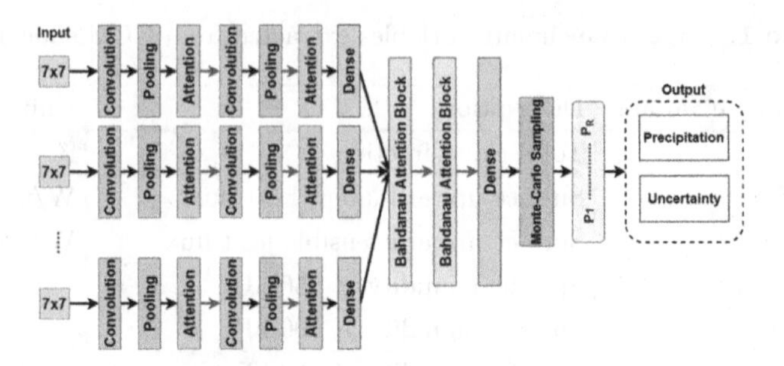

Fig. 2. Architecture of the proposed PCNN model. The input consists of the 26 7×7 matrices corresponding to the 26 climate variables considered in the study. Each 7×7 matrix contains the values of the corresponding climate variable taken from the 49 grid points shown in Fig. 1. Each matrix is independently fed to a sequence of convolution, pooling, and self-attention layers, followed by a dense layer. Red arrows represent dropout steps. The outputs of the 26 independent sequences are concatenated and fed to a sequence of Bahdanau attention layers. By utilizing the Monte-Carlo sampling technique, the proposed model outputs R prediction samples (P_o for $1 \leq o \leq R$). The mean of the R prediction samples is the predicted daily precipitation. Our model also outputs the aleatoric and epistemic uncertainty values associated with the input data and the model.

variable where the sequences are followed by a dense layer. The outputs of the dense layer are concatenated and fed to the next component. We refer to our architecture as "multi-head," which is not related to the multi-head attention mechanism but rather the multiple input heads of our model (reminiscent of the "multi-convolutional heads" [14,16]). Specifically, we have 26 climate variables, so there are totally 26 input heads. Each climate variable is represented by a 7 × 7 matrix corresponding to the 49 grid points in Fig. 1. As shown in Fig. 2, each input matrix is fed to a different sub-model, each starting with a convolution layer.

3.2 Bahdanau Attention Blocks

The Bahdanau attention methodology [2] was originally proposed to improve the performance of conventional encoder-decoder models. The main difference between the Bahdanau attention and the conventional attention is that the former aims to use a variable-length vector instead of the fixed-length one, to enhance the translation performance of the models. Similar to other attention approaches, the components of the Bahdanau attention include hidden decoder states s, context vectors c, weights a, attention scores e, as well as an annotation vector h. In the Bahdanau attention, the encoder uses the input sequence to generate the annotation sets h_i, which are combined with the hidden decoder state of the previous step s_{t-1}, and fed to an alignment model to evaluate the attention scores $e_{t,i}$. These attention scores are normalized into wights $a_{t,i}$, which are used to evaluate the context vector c_t. Similar to other attention methodologies, the context vector is used with the previous target hidden state to produce the final output. In our PCNN framework, we adopt two Bahdanau attention blocks followed by a dense layer as shown in Fig. 2 to improve the learning capability of the proposed model.

3.3 Uncertainty Quantification

In many real-world applications, uncertainty quantification is important [17,26]. Uncertainty occurs in different components (or steps) within a deep neural network. We are particularly interested in quantifying two uncertainty types: aleatoric, which portrays the intrinsic randomness in the input data, and epistemic, which portrays the uncertainty of the deep learning model itself. We incorporate the Monte-Carlo dropout sampling technique to quantify the two uncertainties, following the methodology in [13,15]. This methodology employs the Bayes' theorem $P(W \mid D) = (P(D \mid W)P(W))/P(D)$ to calculate the probability $P(W)$ over the network weights W. However, evaluating the posterior probability was shown to be a difficult task, and we could opt-in to leverage the parameterized variational distribution $q_\theta(W)$ over the weights (variational inference [3]). This could be achieved by using dropout mechanisms across the network during training. Still, one would need to minimize the relative entropy of $q_\theta(W)$, which could be done by using the Adam optimizer, with the cross-entropy loss function. It is standard to use dropout mechanisms within deep learning models to tackle the problem of over-fitting. The methodology behind dropout is to randomly (following a certain rate) drop neurons in selected layers. Using dropout during training would enable better generalization, and subsequently better performance on unseen data. To quantify the uncertainties in our case, we apply the dropout mechanisms during testing and perform the prediction R times (where R is set to 50 in our study) to produce R Monte-Carlo samples for each test case. We calculate the mean and variance over the R samples and leverage them to quantify the aleatoric and epistemic uncertainties [13,15].

4 Experiments and Results

4.1 Experimental Setup

We followed the 80/20 split procedure, in which, 80% of the data, from January 1950 to August 1994 with 16,334 data records, are used to train and calibrate our model, and 20% of the data, from September 1994 to October 2005 with 4,084 data records, are used for testing. We adopt multiple performance metrics that are often employed in the precipitation downscaling literature. These are the root mean squared error (RMSE) reported in millimeter (mm), the same unit as precipitation, mean absolute error (MAE) reported in the same unit as precipitation (mm), and error in daily mean (Mean Bias) which denotes the absolute difference (reported in %) between the mean of the predicted and observed precipitation values (over the evaluation period), viz: Mean Bias = | observed mean

Table 2. Hyperparameters used by PCNN.

Hyperparameter	Value	Description
Epochs	2000	Number of epochs
Batch size	64	Size of each batch
Activation (Conv)	ReLU	Activation function in the convolution layers
Filters	32	Number of filters in convolution layers
Optimizer	Adam	Optimization algorithm used
Loss Function	MSE	Loss function used

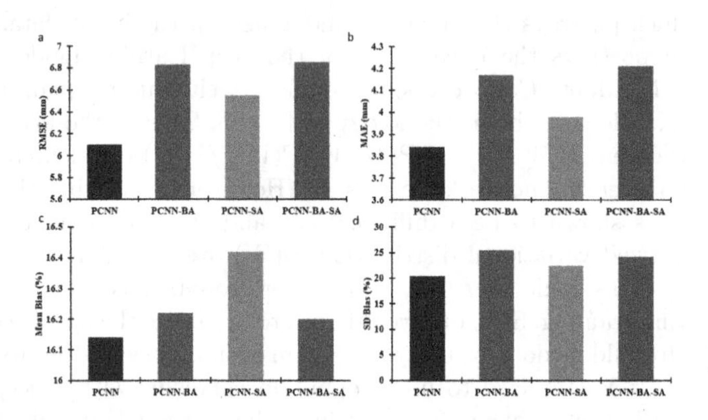

Fig. 3. Results of the ablation studies, in terms of a) RMSE reported in mm, b) MAE reported in mm, c) Mean Bias reported in %, and d) Standard Deviation (SD) Bias reported in %.

- predicted mean| / observed mean x 100%. We also compute the error in daily standard deviation (SD Bias), which depicts the absolute difference (reported in %) between the standard deviation of the predicted and observed precipitation values (over the evaluation period), viz: SD Bias = | observed SD - predicted SD|/observed SD $\times$ 100%. We used 20% of the training data for hyperparameter tuning. Table 2 summarizes the hyperparameters and their corresponding values that gave the best results, and that are used to configure PCNN.

4.2 Ablation Studies

To assess the contribution of each component of our model, we turned off the Bayesian inference in our model and compared the performance of four variants: (i) PCNN which refers to the original model; (ii) PCNN-BA which refers to the model without the Bahdanau attention layers; (iii) PCNN-SA which refers to the model without the self-attention layers; and (iv) PCNN-BA-SA which refers to the model without the Bahdanau attention and self-attention layers. We report in Fig. 3 the RMSE, MAE, Mean Bias, and Mean SD for the four variants. PCNN has better performance than the other three variants, indicating the importance of both the Bahdanau attention and self-attention components.

4.3 Comparative Studies

Next, we compared the performance of PCNN with that of widely used machine learning (ML) methods in the downscaling literature: support vector regression (SVR) [9], random forests (RF) [24], decision trees (DT) [27], multilayer perceptron (MLP) [1], and convolutional neural networks (CNN) [25]. Since the related ML methods cannot quantify uncertainties, we again turned off the Bayesian inference in our model when comparing with the related ML methods. Notice that the CNN in the downscaling literature is tailored for multiple input types but without the attention mechanisms, and therefore the CNN in the downscaling literature is equivalent to the PCNN-BA-SA used in our ablation studies. Figure 4 presents the results of the comparative studies. It can be seen from the figure that PCNN outperforms the other ML methods in terms of RMSE, MAE and Mean Bias. We note that decision trees (DT) produced the closest daily standard deviation (SD) to the observed data.

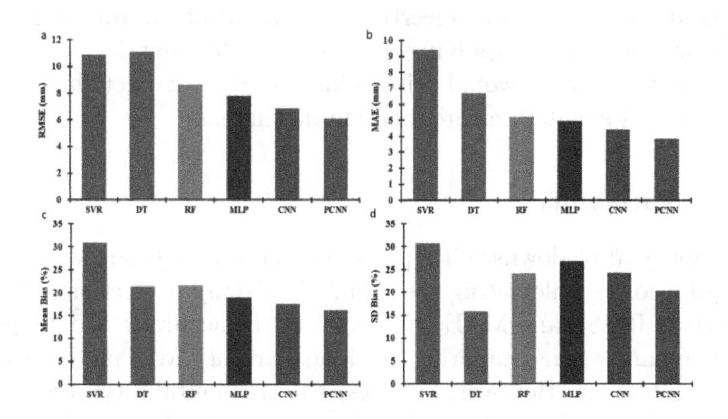

Fig. 4. Performance comparison between PCNN and the related ML methods in terms of a) RMSE reported in mm, b) MAE reported in mm, c) Mean Bias reported in %, and d) Standard Deviation (SD) Bias reported in %.

4.4 Uncertainty Quantification Results

The use of the Monte-Carlo dropout sampling technique allowed us to quantify the aleatoric and epistemic uncertainty when making predictions. Figure 5 presents predicted mean daily precipitation within each month for the evaluation period, as well as the obtained epistemic and aleatoric uncertainties. More uncertainties come from the data than from our model. For reference, the average aleatoric (data) and epistemic (model) uncertainties are 1.41 and 0.86 respectively. These values depict that the model and data uncertainties are relatively low over the evaluation period. One could further reduce these uncertainties by better tuning the hyperparameters of the model (for model uncertainty) and

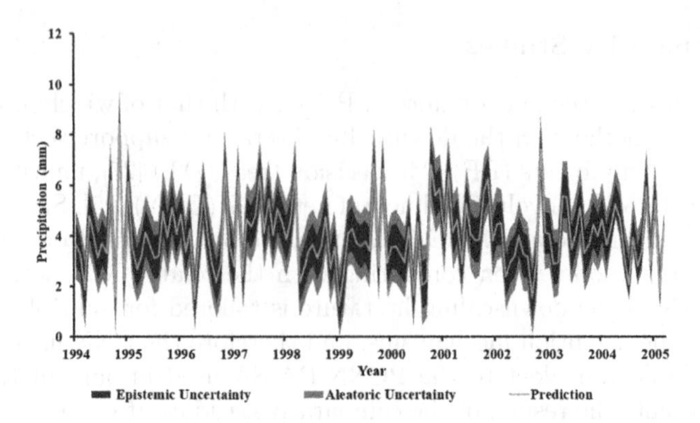

Fig. 5. Predicted monthly averages of daily precipitation as well as the epistemic and aleatoric uncertainties for the 1994–2005 period.

refining the dataset (for data uncertainty). The aleatoric uncertainty is data-dependent, as such, when applying PCNN on a different location, a different value for such uncertainty would arise. This is due to the fact that the bias and noise in the GCM simulations are location-dependent.

4.5 Daily Variability

An important goal of downscaling precipitation is to reproduce the daily variability in addition to calculating the accuracy of daily predictions. Performance metrics such as RMSE and MAE are related to the accuracy of daily predictions and are important when comparing machine learning (ML) methods to identify the most suitable one. However, to assess the true suitability of that identified ML method, one would need to analyze its performance in reproducing the daily variability and calculating certain precipitation measures, such as mean wet and dry spell lengths. Mean wet spell length is computed as the average number of consecutive days with precipitation (i.e., precipitation ≥ 1 mm). Similarly, mean dry spell length is the average number of consecutive days without precipitation (i.e., precipitation < 1 mm). In climatology, a total precipitation of 1 mm is often the cutoff to classify days as wet or dry. Figure 6 presents the mean wet and dry spell lengths (and standard deviations as error bars), as extracted from PCNN prediction results, as well as from NOAA observations. These results demonstrate the ability of PCNN to reproduce wet and dry spells, with relatively low bias errors (0.68% and 7.3% respectively). Here, Wet Bias error = |observed mean wet spell length - predicted mean wet spell length|/observed mean wet spell length $\times$ 100% and Dry Bias error = |observed mean dry spell length - predicted mean dry spell length|/observed mean dry spell length $\times$ 100%.

Moreover, we investigate the ability of our PCNN framework to reproduce the probability distribution of precipitation. Figure 7 shows the cumulative distribution functions (CDF) for daily precipitation as obtained from the PCNN prediction results and the NOAA observations respectively. It is apparent that

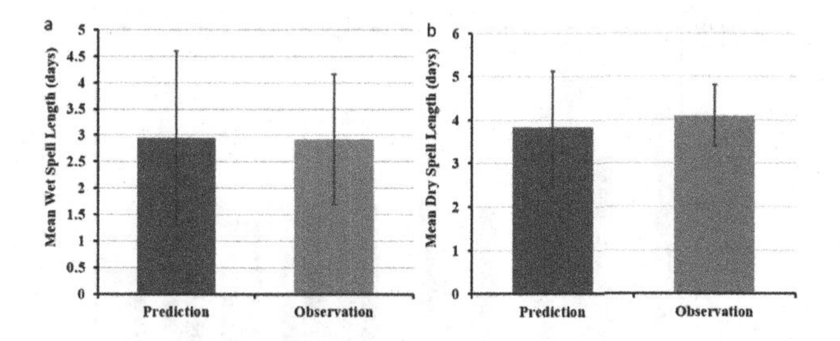

Fig. 6. Mean a) wet and b) dry spell length as observed (orange) and predicted using PCNN (blue) where error bars represent standard deviations. Here the mean wet (dry, respectively) spell length is the average number of consecutive days with (without, respectively) precipitation over the 1994–2005 period. (Color figure online)

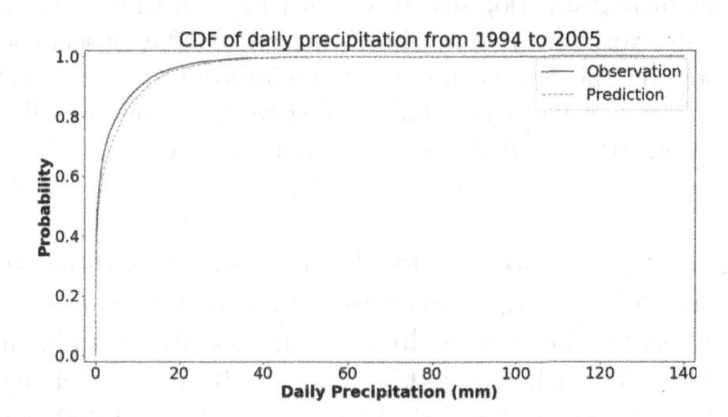

Fig. 7. Cumulative distribution function (CDF) for the daily precipitation using PCNN (dashed red line). The blue line represents the observed cumulative distribution function. (Color figure online)

the probability distribution of the prediction results from PCNN is tightly close to that of the NOAA observations. To further assess the ability of PCNN to reproduce variability, we cluster precipitation values into ranges and compute the number of days (frequency) predicted within each range. For example, PCNN predicted a precipitation value between 0 and 2 mm for 2520 days, and as such, the range 0–2 will have a frequency of 2520 days. Figure 8 compares the frequency (i.e., the number of days) of each precipitation range between those predicted by PCNN (gray bars) and the NOAA observations (yellow bars) where the histograms are log scaled for visibility. It can be seen from Fig. 8 that our PCNN framework reproduces the frequency distribution well, which supports our claim that PCNN is suitable for the downscaling task.

4.6 Long-Term Projections

Heavy precipitation has multiple impacts on the environment, leading to crop damage, increased flooding rate, as well as soil erosion. Furthermore, runoff from precipitation can wash pollutants into water bodies, affecting the water quality.

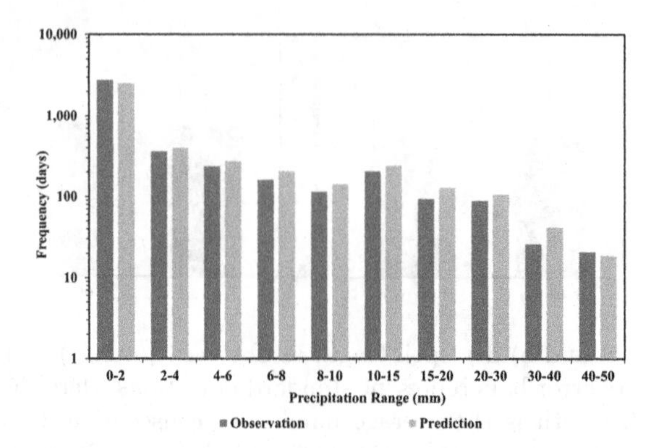

Fig. 8. Frequency histograms (log scaled) of daily precipitation (mm) values predicted by the PCNN framework (gray) and obtained from NOAA observations (yellow) for the 1994–2005 period. The X-axis shows the precipitation ranges used to compute the frequency. For instance, a histogram for the 0–2 range represents all those days with precipitation between 0 mm and 2 mm.(Color figure online)

As such, long-term projections of local-scale daily precipitation is crucial for water quality and risk management. Our experimental results demonstrate the suitability of PCNN for the downscaling of daily precipitation in the Hackensack-Passaic watershed. As such, we re-trained our PCNN model using the 20,418 data records obtained from the period between 1950 and 2005 and applied the trained model to produce future projections. In particular, we used the GCM simulations for the periods between 2031–2040, 2061–2070, and 2091–2100 where we have the data available for these periods. We downscaled the precipitation for each period independently where the RCP8.5 was the radiative heat scenario used in the study. Figure 9 shows the a) mean daily precipitation, b) mean wet spell length, c) mean dry spell length, and d) average of the number of wet days (with precipitation *geq* 1 mm) annually for each period where red error bars represent standard deviations. We note the large standard deviations of the mean daily precipitation values, which are expected given that the observed daily precipitation values from NOAA in the period between 1950 and 2005 have a standard deviation of 6.9 mm. The reported mean and standard deviation are for all the days within a study period (10 years each). As such, the large standard deviation depicts that the PCNN model is not predicting a near-mean value for each day, but with a variability that resembles that of the training period. A low standard deviation would imply that most days have precipitation values close to the mean daily precipitation over the 10-year period, which would be wrong, given the large number of dry days, as well as days with high rainfall rate, in the period between 1950 and 2005. Figure 10 shows the mean daily precipitation within each month for the selected future periods, as well as the epistemic and aleatoric uncertainties. We note that the increasing trend of the mean daily precipitation from 2031 to 2100 agree with existing climate change studies which argue that more rain and snow are expected in the future. This

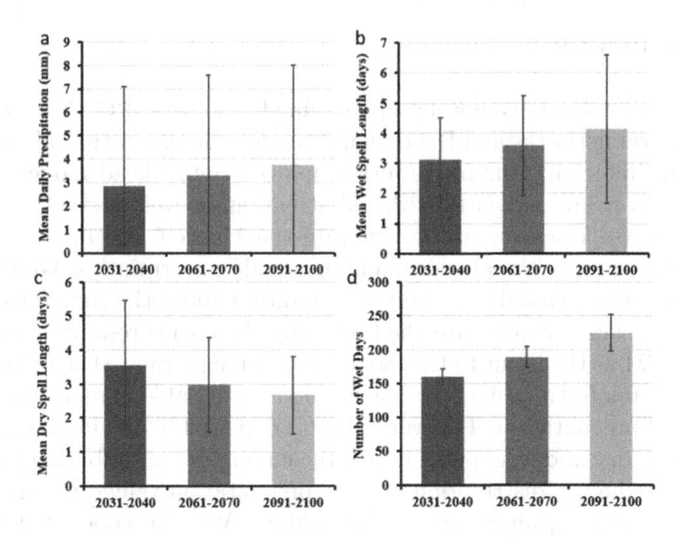

Fig. 9. Projected a) mean daily precipitation, b) mean wet spell length, c) mean dry spell length, and d) average of the number of wet days annually for three future periods where red error bars represent standard deviations. The standard deviations of the projected daily precipitation values are large, which means that we don't have a near-mean value for all the days within each period. This is expected given that the observed standard deviation on the training data is 6.9 mm. (Color figure online)

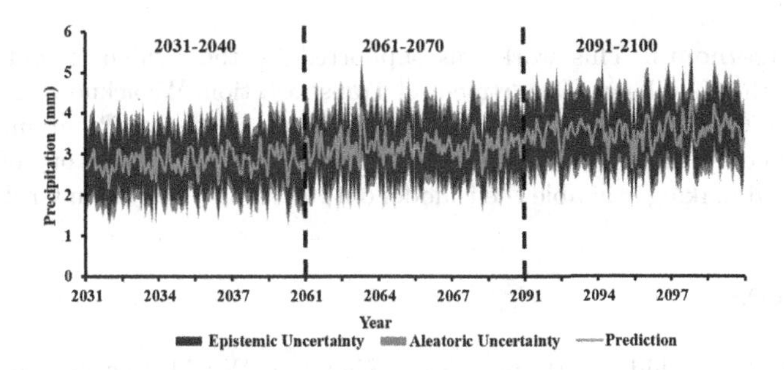

Fig. 10. Projections of precipitation as produced by PCNN with the aleatoric and epistemic uncertainties for the periods 2031–2040, 2061–2070, and 2091–2100 respectively. The periods are separated by dashed lines.

fact is further supported by Fig. 9d which shows an increasing trend in the average number of wet days annually. We note that an increasing trend in mean daily precipitation might not be visible when examining each 10-years period independently (see Fig. 5 and each period in Fig. 10). However, such a trend will become apparent when analyzing a relatively longer period (e.g., 2031–2100) as shown in Fig. 10.

5 Conclusions

In this paper, we develop a novel deep learning framework (PCNN) based on convolution neural networks and Bahdanau attention for the statistical downscaling of daily precipitation. Our framework employs a multi-head model, consisting of convolution neural networks with self-attention, as well as stacked Bahdanau attention layers. Moreover, we incorporate the Monte-Carlo dropout sampling technique to quantify the aleatoric and epistemic uncertainties. We trained the PCNN model to downscale the daily precipitation above the New Jersey portion of the Hackensack-Passaic watershed by using the coarse-resolution simulations from the GCM as the input to the model. Experiments show that PCNN outperforms closely related machine learning methods and is able to reproduce the daily variability of precipitation. This variability is depicted by the mean daily precipitation, mean dry and wet spell lengths, the cumulative distribution function, as well as the frequency distribution of daily precipitation values. Our results were obtained by utilizing an 80/20 split procedure. We also conducted additional experiments with five-fold cross validation and obtained similar results. We then trained the PCNN model by data from 1950 to 2005 and applied the trained model to project the long-term daily precipitation using future GCM simulations for the periods between 2031–2040, 2061–2070, and 2091–2100 respectively. We reported the projected statistics for each future period (mean, wet and dry spell lengths, number of wet days).

Acknowledgements. This work was supported by the Bridge Resource Program (BRP) from the New Jersey Department of Transportation. We acknowledge the Working Group on Coupled Modelling of the World Climate Research Program, responsible for CMIP, and we thank the Geophysical Fluid Dynamics Laboratory of NOAA for producing and making available their model output via Earth System Grid Federation.

References

1. Ahmed, K., Shahid, S., Haroon, S.B., Xiao-Jun, W.: Multilayer perceptron neural network for downscaling rainfall in arid region: a case study of Baluchistan, Pakistan. J. Earth Syst. Sci. **124**(6), 1325–1341 (2015). https://doi.org/10.1007/s12040-015-0602-9
2. Bahdanau, D., Cho, K., Bengio, Y.: Neural machine translation by jointly learning to align and translate. In: 3rd International Conference on Learning Representations (2015)
3. Blei, D.M., Kucukelbir, A., McAuliffe, J.D.: Variational inference: a review for statisticians. J. Am. Stat. Assoc. **112**(518), 859–877 (2017)
4. Fang, J., Zhu, J., Wang, S., Yue, C., Shen, H.: Global warming, human-induced carbon emissions, and their uncertainties. Sci. China Earth Sci. **54**(10), 1458–1468 (2011). https://doi.org/10.1007/s11430-011-4292-0
5. Gerges, F., Boufadel, M.C., Bou-Zeid, E., Nassif, H., Wang, J.T.L.: A novel deep learning approach to the statistical downscaling of temperatures for monitoring climate change. In: The 6th International Conference on Machine Learning and Soft Computing. ACM (2022). https://doi.org/10.1145/3523150.3523151

6. Gerges, F., Boufadel, M.C., Bou-Zeid, E., Nassif, H., Wang, J.T.: A novel Bayesian deep learning approach to the downscaling of wind speed with uncertainty quantification. In: Gama, J., Li, T., Yu, Y., Chen, E., Zheng, Y., Teng, F. (eds.) Advances in Knowledge Discovery and Data Mining, PAKDD 2022. LNCS, vol. 13282, pp. 55–66. Springer, Cham (2022). https://doi.org/10.1007/978-3-031-05981-0_5
7. Gerges, F., Shih, F., Azar, D.: Automated diagnosis of acne and rosacea using convolution neural networks. In: 2021 4th International Conference on Artificial Intelligence and Pattern Recognition, pp. 607–613 (2021)
8. Gerges, F., Shih, F.Y.: A convolutional deep neural network approach for skin cancer detection using skin lesion images. Int. J. Electr. Comput. Eng. **15**(8), 475–478 (2021)
9. Ghosh, S.: SVM-PGSL coupled approach for statistical downscaling to predict rainfall from GCM output. J, Geophys. Res. Atmos. **115**(D22) (2010)
10. Goodfellow, I., Bengio, Y., Courville, A., Bengio, Y.: Deep Learning, vol. 1. MIT press, Cambridge (2016)
11. Griffies, S.M., et al.: The GFDL CM3 coupled climate model: characteristics of the ocean and sea ice simulations. J. Clim. **24**(13), 3520–3544 (2011)
12. Hu, W., Scholz, Y., Yeligeti, M., von Bremen, L., Schroedter-Homscheidt, M.: Statistical downscaling of wind speed time series data based on topographic variables. In: EGU General Assembly Conference Abstracts, pp. EGU21-12734 (2021)
13. Jiang, H., et al.: Tracing h alpha fibrils through Bayesian deep learning. Astrophys. J. Suppl. Ser. **256**(1), 20 (2021)
14. Khan, Z.N., Ahmad, J.: Attention induced multi-head convolutional neural network for human activity recognition. Appl. Soft Comput. **110**, 107671 (2021)
15. Kwon, Y., Won, J.H., Kim, B.J., Paik, M.C.: Uncertainty quantification using Bayesian neural networks in classification: application to biomedical image segmentation. Comput. Stat. Data Anal. **142**, 106816 (2020)
16. Linmans, J., van der Laak, J., Litjens, G.: Efficient out-of-distribution detection in digital pathology using multi-head convolutional neural networks. In: MIDL, pp. 465–478 (2020)
17. Liu, J.: Variable selection with rigorous uncertainty quantification using deep Bayesian neural networks: posterior concentration and Bernstein-von mises phenomenon. In: International Conference on Artificial Intelligence and Statistics, pp. 3124–3132. PMLR (2021)
18. Liu, Z., Wan, M., Guo, S., Achan, K., Yu, P.S.: Basconv: aggregating heterogeneous interactions for basket recommendation with graph convolutional neural network. In: Proceedings of the 2020 SIAM International Conference on Data Mining, pp. 64–72. SIAM (2020)
19. Livneh, B., et al.: A spatially comprehensive, hydrometeorological data set for Mexico, the us, and southern Canada 1950–2013. Sci. Data **2**(1), 1–12 (2015)
20. Misra, S., Sarkar, S., Mitra, P.: Statistical downscaling of precipitation using long short-term memory recurrent neural networks. Theor. Appl. Climatol. **134**(3), 1179–1196 (2018)
21. Myojin, T., Hashimoto, S., Ishihama, N.: Detecting uncertain BNN outputs on FPGA using monte Carlo dropout sampling. In: Farkaš, I., Masulli, P., Wermter, S. (eds.) ICANN 2020. LNCS, vol. 12397, pp. 27–38. Springer, Cham (2020). https://doi.org/10.1007/978-3-030-61616-8_3
22. Myojin, T., Hashimoto, S., Mori, K., Sugawara, K., Ishihama, N.: Improving reliability of object detection for lunar craters using monte Carlo dropout. In: Tetko, I.V., Kůrková, V., Karpov, P., Theis, F. (eds.) ICANN 2019. LNCS, vol. 11729, pp. 68–80. Springer, Cham (2019). https://doi.org/10.1007/978-3-030-30508-6_6

23. Pan, X., Shi, J., Luo, P., Wang, X., Tang, X.: Spatial as deep: spatial CNN for traffic scene understanding. In: 32nd AAAI Conference on Artificial Intelligence (2018)
24. Pang, B., Yue, J., Zhao, G., Xu, Z.: Statistical downscaling of temperature with the random forest model. Adv. Meteorol. **2017** (2017)
25. Sun, L., Lan, Y.: Statistical downscaling of daily temperature and precipitation over china using deep learning neural models: Localization and comparison with other methods. Int.l J. Climatol. **41**(2), 1128–1147 (2021)
26. Wang, Y., Rocková, V.: Uncertainty quantification for sparse deep learning. In: International Conference on Artificial Intelligence and Statistics, pp. 298–308. PMLR (2020)
27. Xu, R., Chen, N., Chen, Y., Chen, Z.: Downscaling and projection of multi-cmip5 precipitation using machine learning methods in the upper HAN river Basin. Adv. Meteorol. **2020**, 1–17 (2020)

Cubism: Co-balanced Mixup for Unsupervised Volcano-Seismic Knowledge Transfer

Mahsa Keramati[1(✉)], Mohammad A. Tayebi[1], Zahra Zohrevand[1], Uwe Glässer[1], Juan Anzieta[2], and Glyn Williams-Jones[2]

[1] School of Computing Science, Burnaby, Canada
{mahsa_keramati,tayebi,zahra_zohrevand,glaesser}@sfu.ca

[2] Department of Earth Sciences, Simon Fraser University, Burnaby, Canada
{janzieta,glynwj}@sfu.ca

Abstract. Volcanic eruptions are severe global threats. Forecasting these unrests via monitoring precursory earthquakes is vital for managing the consequent economic and social risks. Due to various contextual factors, volcano-seismic patterns are not spatiotemporal invariant. Training a robust model for any novel volcano-seismic situation relies on a costly, time-consuming and subjective process of manually labeling data; using a model trained on data from another volcano-seismic setting is typically not a viable option. Unsupervised domain adaptation (UDA) techniques address this issue by transferring knowledge extracted from a labeled domain to an unlabeled one. A challenging problem is the inherent imbalance in volcano-seismic data that degrades the efficiency of an adopted UDA technique. Here, we propose a co-balanced UDA approach, called Cubism, to bypass the manual annotation process for any newly monitored volcano by utilizing the patterns recognized in a different volcano-seismic dataset with labels. Employing an invertible latent space, Cubism alternates between a co-balanced generation of semantically meaningful inter-volcano samples and UDA. Inter-volcano samples are generated via the mixup data augmentation technique. Due to the sensitivity of mixup to data imbalance, Cubism introduces a novel co-balanced ratio that regulates the generation of inter-volcano samples considering the conditional distributions of both volcanoes. To the best of our knowledge, Cubism is the first UDA-based approach that transfers volcano-seismic knowledge without any supervision of an unseen volcano-seismic situation. Our extensive experiments show that Cubism significantly outperforms baseline methods and effectively provides a robust cross-volcano classifier.

Keywords: Volcano-seismic event classification · Unsupervised domain adaptation · Imbalanced data · Mixup · Flow-based generative models

© The Author(s), under exclusive license to Springer Nature Switzerland AG 2023
M.-R. Amini et al. (Eds.): ECML PKDD 2022, LNAI 13717, pp. 581–597, 2023.
https://doi.org/10.1007/978-3-031-26419-1_35

1 Introduction

Monitoring volcanic unrest is a topic of significant interest, with volcanic hazards threatening the lives of more than 800 million people who live in the vicinity of active volcanoes [22]. Volcanic activity originates from physical processes related to fluid and energy transportation and ranges from gas or non-explosive lava emissions to extremely violent explosive bursts that may last many hours. These activities generally lead to breaks or cracks in rocks that surge seismicity beneath a volcano before an eruption [22]. Seismological observations are of vital importance for volcanic monitoring as they provide real-time internal data about volcano-seismic events with a high temporal resolution [22].

Rapidly increasing volumes of recorded seismic data call for a paradigm shift in volcano monitoring and forecasting the associated risks. While conventional manual solutions are not practical anymore due to their time-consuming and resource-intensive nature, advanced machine learning techniques and automated predictive analytics can assist risk mitigation and management by annotating seismic events in supervised [4,19] and unsupervised [3,7] manners. Unsupervised approaches are prone to low performance and applications of supervised methods are typically limited [2]. The reason is that supervised techniques suffer from low generalization power due to the challenges of unifying the data characteristics from different sources. The unification process is not trivial because of the contextual factors such as soil characteristics and source geometry in addition to potential noise introduced during signal recording [2]. Several methods [2,3] are proposed to generate a unified feature space for seismic events pursuing a general purpose solution; however, their unification process is highly subjective to signal standardization since these methods alleviate noise impact by relying on a manual selection of intrinsic mode function (IMF) components. This process disregards important aspects of domain shift. Note that the performance of these methods still depends on availability of several large labeled volcano-seismic datasets.

In this paper, we propose Cubism, an algorithmic method for robust contextual unification and effective knowledge transfer in the volcano-seismic domain. Cubism is an unsupervised domain adaptation (UDA) approach that effectively alleviates the negative impact of inherent class imbalance in volcano-seismic domains by introducing a novel co-balanced inter-volcano modeling. UDA imposes domain-invariance by mitigating the shift between the data distributions of a labeled and an unlabeled domain. Despite being a well-studied area of research, mitigating significant domain gaps is still challenging [20]. Recently, UDA approaches [20,26] were proposed to alleviate this gap by continuously modeling inter-domain latent space using an emerging vicinal risk minimization technique known as mixup training [28]. These approaches produce inter-domain samples through convex combinations of data and labels/pseudo-labels on input or latent manifold. Yet, these methods don't guarantee semantically meaningful inter-domain modeling [27] and are also subject to bias in case of skewed class distributions. Cubism addresses these two issues using a novel co-balanced mixup in the latent space of the flow-based Gaussian mixture model (FlowGMM) [10].

As a result, our proposed solution is capable of developing a robust UDA for volcano-seismic knowledge transfer.

Cubism employs FlowGMM because it encourages semantically meaningful inter-domain modeling through a sequence of invertible transformations as a characteristic of flow-based generative models [27]. In addition, FlowGmm assists with conditional mixup by providing a disentangled latent representation. The conditional generative model learned by FlowGMM accommodates inter-domain mixup concerning class imbalance in the labeled domain. Re-balancing that only considers the labeled domain disregards the conditional distribution of the unlabeled domain [25]. To address this issue, Cubism proposes co-balanced mixup to address bias and reverse-bias utilizing pseudo-labels of the unlabeled domain; it imposes a robust discriminative cross-volcano feature space with an interplay between a co-balanced mixup and an adversarial UDA.

Co-balanced mixup, first generates samples from a disentangled representation learned by FlowGMM. Then, it models inter-domain space by linear interpolation of these generated samples and samples from the unlabeled domain, considering the conditional skewness of both domains. Finally, for the adversarial adaptation, the training data and inter-domain samples are fed to a soft domain discriminator and a soft classifier so that the minimax game between the discriminator and flow model mitigates the domain gap smoothly [26].

To the best of our knowledge, Cubism is the first work that proposes unsupervised volcano-seismic knowledge transfer by employing unsupervised cross-domain classification. To evaluate Cubism, we use two real-world volcano-seismic datasets (see Sect. 3.2) from the Llaima and Deception Island volcanoes located in Chile and Antarctica. We define two unsupervised knowledge transfer tasks from each dataset to the other one and assess the cross-volcano classification performance of Cubism and several baselines. Our experimental results confirm the effectiveness of Cubism through a comparative study where Cubism significantly outperforms the strongest baseline by 9.4% in terms of F1-score.

2 Related Work

Volcano-seismic Data Analysis. Many works in the literature addressed the classification task of volcano-seismic events utilizing anomaly detection or other machine learning approaches [23,24]. Despite their promising performance for a specific volcano-seismic situation, they fail to generalize well to a different temporal or spatial volcano-seismic domain [3,19]. Limited works tackled the problem of designing a spatio-temporal invariant volcano-seismic recognition system. For example, [1] proposes a Bayesian-based approach to learn the mixture of events from two different volcanoes, or [3] trains a model on the standardized data from several volcanoes. The mentioned efforts not only expect large labeled data from multiple volcanoes but also do not aggregate the collected contexts into a unified contextual representation. These problems motivate others to apply unsupervised learning on unlabeled volcano-seismic datasets at the expense of accuracy [3,7]. In this work, we aim for a more challenging yet rewarding task, unsupervised knowledge transfer from a volcano with annotated data to a different volcano with unlabeled records. One of the only related studies [19] employs active learning to limit the number of required labeled data from a new volcanic setting for training. However, this work does not mitigate the volcano-seismic domain shift; in addition, it requires labeled data from the studied volcanoes and is subject to class imbalance. Several existing approaches mitigate the scarcity and imbalance in volcano-seismic datasets by applying traditional data augmentation techniques [4] or generative models [9]. However, these approaches are subject to underrepresented samples from low-density regions [15], and their effectiveness are highly dependent on large labeled datasets. Our proposed solution, using an imbalance-aware UDA technique, is the first work to effectively address the above-mentioned issues, to the best of our knowledge.

Unsupervised Domain Adaptation (UDA) is a subcategory of transductive transfer learning that generalizes a model from a labeled source to an unlabelled target under dataset shift. UDA is extensively-studied especially in image processing and computer vision, and existing methods can be categorized into distance-based [12,20] or adversarial approaches [8,13,17]. Distance-based methods train two separate classifiers with some shared layers for each domain. Domain-adversarial neural network (DANN) [8] proposes adversarial UDA as a group of methods that impose domain confusion by a minimax game between a feature extractor and a domain discriminator. Several UDA methods [20,26] effectively mitigate significant domain gaps by modeling locally Lipschitz in inter-domain space using an efficient regularization technique called mixup [28]. DM-ADA [26] is a mixup-based UDA method that learns the latent representation of both the two domains using variational auto-encoder (VAE) and jointly mixup input and latent space for robust cross-domain classification. Despite promising performances, these state-of-the-art methods are prone to data imbalance issue. Limited UDA approaches address adaptation under imbalanced settings [11]; however, their focus is on label shift that is not aligned with the imbalance problem in volcano-seismic datasets. In this work, we propose Cubism as an unsupervised knowledge transfer solution for the volcano-seismic domain considering the data imbalance issue using mixup-based adversarial UDA.

3 Basic Concepts and Problem Definition

This section elaborates on the volcano-seismic domain discrepancy problem and its empirical justification based on the characteristics of the datasets employed in our study. Subsequently, we formally define the imbalanced cross-volcano classification problem for volcano-seismic knowledge transfer.

3.1 Volcano-seismic Domain Discrepancy

Each Volcanic hazard has its specific seismic signature. Analyzing the categorical frequency of volcano-seismic activities is a principal step in forecasting volcanic eruptions [22]. Technical quantification of volcanic earthquakes, known as seismic

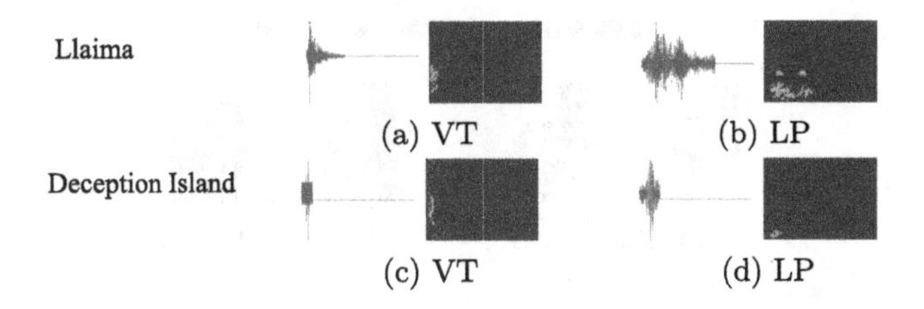

Fig. 1. Volcano-seismic events and their spectrograms

catalog, strengthens volcanic hazards monitoring. Manual detection and labeling of volcanic events by domain experts is not a feasible solution in the era of big data when dealing with massive and rapidly growing volumes of seismic data.

Existing unsupervised methods for seismic event annotation are subject to low performance, and efficient supervised approaches rely on large and high-quality manually labeled data. Still, the manual data annotation process is prone to human subjectivity and lacking unified contextual factors [2,19,22]. In addition, characteristics of observed seismicities depend highly on the geophysical properties of volcanoes and placement of sensors [18]. Despite providing high-quality catalogs for signals from a specific station, supervised volcano-seismic recognition (VSR) systems often fail to robustly generalize to a new situation regarding the volcanic state, quality of sensors, environmental noise, etc. [22].

To address the above issues, we propose here to leverage the robustness of the supervised models along with the self-dependency of unsupervised methods. This way, we deliver a model that provides promising catalogs without labeled data for an unseen setting. Utilizing semi-supervised approaches does not effectively fulfill our purpose since these approaches assume the same distribution for labeled and unlabeled data. Therefore, we propose to exploit unsupervised domain adaptation techniques to generalize seismic knowledge from a cataloged volcano-seismic setting to a non-annotated set of events from a different one. Noteworthy, VSR systems suffer from class imbalance issue considering the nature of volcanic activities. Even moderate data imbalance degrades the performance of UDA techniques more than intra-domain learning. Lacking conditional knowledge in unlabeled datasets, the learned cross-domain representation can be biased to the majority classes. None of the explicit data augmentation solutions (undersampling, oversampling, domain-specific approaches, generative models, etc.) efficiently compensate for underrepresented data distributions [28]. Furthermore, although there is an extensive body of research on UDA, just a few recent UDA methods did address the data imbalance. Thus, we propose a novel flow-based method, Cubism, that implicitly regularizes adversarial UDA to address imbalanced cross-volcano classification through vicinal risk minimization.

3.2 Imbalanced Cross-Volcano Classification

Exploratory Observations. In this work, we use datasets from two well-studied active volcanoes located in Chile and Antarctica: Llaima [4] and

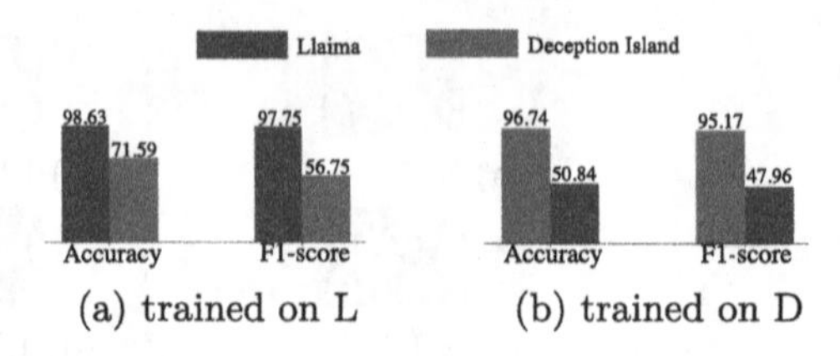

Fig. 2. Cross-volcano generalization power(L: Llaima, D: Deception Island)

Deception Island [3]. These datasets comprise labeled records of the two most recurrent volcano-seismic events, long period (LP) and volcano-tectonic (VT). Due to their geophysical origins, signals falling into the same categories share certain characteristics. VT events show an impulsive start and exponential decay, whereas LP signals are non-impulsive and decay slowly. LP events typically are more frequent than VT ones. Despite the common characteristics of all events in the same category, seismicity patterns belonging to different volcanoes do typically not match. Figure 1 illustrates that signals from the Deception Island volcano are nonidentical to signals from the Llaima volcano.

To empirically verify volcano-seismic domain discrepancy, we first train a classifier on each dataset and then test it on the other one. As Fig. 2 shows, a model trained on one volcano-seismic dataset can successfully be generalized to a test set extracted from the same dataset, while applying the model to another volcano-seismic situation results in poor performance in terms of accuracy and F1-score. This means that leveraging discriminative knowledge from the labeled dataset for annotating the unlabeled one is not feasible without unifying contextual factors.

Problem Definition. UDA techniques alleviate the cross-domain discrepancy between two different but related domains. We refer to volcano-seismic scenarios with labeled samples as the source domain D_s and the ones with unlabelled samples as the target domain D_t, assuming a class imbalance in both D_s and D_t. Suppose a given labelled samples $S = \{(x_i^s, y_i^s)\}_{i=1}^{N^s}$ from D_s along with unlabeled samples $T = \{(x_i^t)\}_{i=1}^{N^t}$ from D_t. Our intended task is binary classification: D_s and D_t share the same class set ($C_s = C_t = \{0, 1\}$) with label frequencies $W^s = \{w_0^s, w_1^s\}$ and $W^t = \{w_0^t, w_1^t\}$. W^t is unknown during training and $w_0^i \neq w_1^i$, for $i \in \{s, t\}$. The objective is to learn the Cubism function that mitigates the domain shift, and given a sample x^t from D_t, it accurately predicts label y_i^t:

$$\mathsf{Cubism}(S, T) \rightarrow \{(y_i^t)\}_{i=1}^{N^t}. \tag{1}$$

4 Preliminaries

This section gives an overview of two fundamental concepts our proposed method builds on: mixup regularization and Flow Gaussian Mixture Model.

4.1 Mixup

Mixup [28] is a simple yet effective method to regularize the training process by modelling both intra-class and inter-class vicinity relations. Mixup provides synthetic samples via linear interpolations of pairs of samples and their corresponding labels:

$$x^m = \lambda x_i + (1 - \lambda)x_j \ ; \ y^m = \lambda y_i + (1 - \lambda)y_j, \tag{2}$$

where (x_i, y_i) and (x_j, y_j) are randomly sampled pairs of $(\mathsf{instance}, \mathsf{one\text{-}hot\ label})$ and $\lambda \in [0, 1]$.

In the UDA problem, we can mixup $(\mathsf{sample}, \mathsf{label})$ pairs from the source domain with $(\mathsf{sample}, \mathsf{pseudo\text{-}label})$ pairs of the target domain and incorporate these intermediate samples into the adaptation process. This approach improves the efficiency of UDA techniques by modelling inter-domain vicinity relations [26].

4.2 Flow Gaussian Mixture Model (FlowGMM)

A flow-based generative model [5] is an unsupervised model that provides exact inference and density evaluation via seeking an invertible transformation from data space X to the latent space Z. This exact mapping from data probability distribution P_X to a tractable latent probability distribution P_Z is obtained via the change of variable formula. For the sake of computational simplicity, the latent distribution P_Z is usually a standard Gaussian.

FlowGMM [10] replaces the Standard Gaussian distribution in the latent space of flow-based models with a Gaussian mixture where each component $\mathcal{N}(\mu_k, \sigma_k)$ corresponds to class k in data space X. This model (F_θ) provides an exact joint likelihood $P_X(x, y)$ via modelling the exact conditional likelihood $P(x|y)$ using change of variable formula:

$$L_{GMM} = P_X(x|y = k) = \mathcal{N}(P_Z(\mathcal{F}_\theta(x)|\mu_k, \sigma_k)) \cdot |\det(\frac{\partial \mathcal{F}_\theta(x)}{\partial x})|, \tag{3}$$

and then p(y|x) is inferred through Bayes' rule as follows:

$$P_X(y|x) = \frac{\mathcal{N}(\mathcal{F}_\theta(x)|\mu_y, \sigma_y)}{\sum_{i=1}^{C} \mathcal{N}(\mathcal{F}(x)|\mu_i, \sigma_i)}. \tag{4}$$

5 Methodology

We propose Cubism as a robust framework for addressing the imbalanced cross-domain binary classification. Cubism addresses bias and reverse bias in domain alignment through an interplay between *co-balanced inter-domain mixup* and *adversarial UDA* with *conditional mapping*. Figure 3 illustrates the Cubism framework. Each training iteration comprises the following steps:

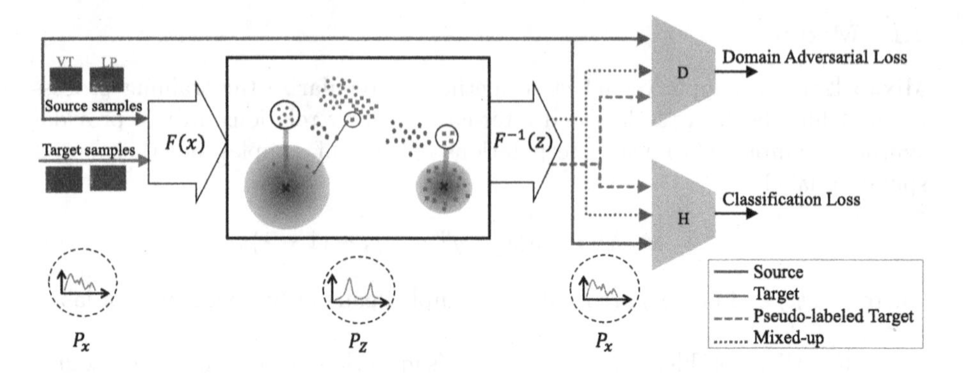

Fig. 3. The architecture of Cubism.

(a) Source samples are fed into a FlowGMM F that maps the complex data space to a latent Gaussian mixture model.
(b) Target samples are fed into F, and then corresponding pseudo-labels are assigned to target embeddings based on the learned Gaussians. Note that Cubism estimates the class imbalance ratios in the target domain regarding the conditional information of pseudo-labeled target embeddings and later combines the ratios from both domains producing a set of co-balanced ratios.
(c) Source-like samples are generated using learned Gaussians applying the co-balanced ratios determined in the previous step.
(d) Random Gaussian noise is injected to target samples in order to impose local Lipschitz and robustness.
(e) The generated source samples are linearly interpolated in the latent space along with noisy target embeddings to produce semantically-meaningful inter-domain samples. Invertibility of flow-based models provides a semantic preserving cross-domain data augmentation.
(f) Source, target and inter-domain samples are fed into an adversarial UDA framework which is slightly modified with a soft domain discriminator and soft source classifier. In addition, for a plausible conditional domain alignment, Cubism maps target samples that are further from the decision boundaries of the source classifier to their corresponding source Gaussians.

Algorithm 1 summarizes Cubism. We elaborate on different aspects of this approach in the following sections.

5.1 Co-balanced Inter-Domain Mixup

Co-balanced inter-domain mixup regularizes the inter-domain space in an unbiased manner. Cubism synthesizes inter-domain samples by co-balanced sampling from the source Gaussians and then mixing up the generated samples with the target samples.

Algorithm 1: Co-balanced domain alignment

Input: $S = \{(x_i^s, y_i^s)\}_{i=1}^{N^s s}$, $T = \{(x_i^t)\}_{i=1}^{N^t}$
Output: $M_{ST}(\{\Psi_F, \Psi_H, \Psi_D\})$: Final model

```
1  F = (Ψ_F) ← preTrain(S);
2  W^s ← classFreq(S);
3  for e ∈ EPOCH do
4      for b = {S_b, T_b} ∈ Batchs do
5          Z_b^t ← F(T_b);
6          T̂_b ← softPseudoLabel(Z_b^t);
7          T̄_b ← NoiseInjection(T̂_b);
8          Ŵ_b^t ← classFreq(T̂_b);
9          if e>0 then
10             Ŵ^t ← ξŴ^t + (1 − ξ)Ŵ_b^t;
11         else
12             Ŵ^t ← Ŵ_b^t;
13         W̄ ← coBalancedFreq(W^s, Ŵ^t);
14         p ← [1 − w̄_0, 1 − w̄_1];
15         Z_b^G ~^p G[(μ_0, σ_0), (μ_1, σ_1)];
16         λ ~ Beta(α, β);
17         M_b ← Mixup(Z_b^G, T̄_b, λ);
18         L_total ← computeLoss(T̄_b, S_b, M_b, λ);
19         M_ST.backpropagate(L_total);
20         M_ST.update();
21 return M_ST
```

Co-balanced Sampling. We train a FlowGMM model F on source data since this conditional generative model computes the exact likelihood and maps the source data to a Gaussian mixture model in latent space Z through optimizing the likelihood explained in Eq. 3: $Z^s = \{(z^s, Y^s)\}, \mid z^s = F(x^s)$, where Y^s is the one-hot label encoding.

Now, to address the class imbalance issue we want to generate samples from the learned Gaussians $G = [\mathcal{N}(\mu_0, \sigma_0), \mathcal{N}(\mu_1, \sigma_1)]$. Our empirical observations show that ignoring class imbalance in the target domain misleads the conditional alignment. The reason is that by incorporating imbalanced information from the source, the model gradually will be biased toward the minority class even if the first few training steps follow a normal process. Thus, to dynamically maintain an unbiased mixup, we sample from the set of Gaussians G considering the class imbalance in both domains.

Class frequency $W^s = [w_0^s, w_1^s]$ in the source domain is directly estimated via the labels $Y^s = \{y_i\}_{i=1}^{N^s}$ where $w^s{}_i = N^s{}_i / N^s$, and $N^s{}_i$ is the number of source samples from class i. However, due to the lack of labels in the target domain, assessing class-imbalance ratios is not as straightforward as it is in the source domain. We estimate these ratios through utilizing obtained pseudo-labels of the target data. For this, we first map target samples T to the latent space Z through network F, and then, we assign soft pseudo-labels to the embedded target samples representing their classification probability by employing the set of learned Gaussians G. This process results in a set of softly pseudo-labeled target samples $\hat{T} = \{(z^t, \hat{Y} = \{\hat{y}_j\}_{j\in\{0,1\}})\}$, where $y_j = P_X(y = j|x^t)$ and $P_X(y = j|x^t)$ is realized through Eq. 4 and $z^t = F(x^t)$.

Now, we estimate class ratios in the target domain $\hat{W}^t = \{\hat{w}_0^t, \hat{w}_1^t\}$ using hard pseudo-labels $\tilde{Y} = \{\tilde{y}_i\}_{i=1}^{N^t}$ where $\tilde{y}_i = \underset{j\in\{0,1\}}{argmax}\ \hat{Y}_i$. For a global estimation of $\hat{W}^t$, these weights are adapted during training by accumulating all previous local target ratios: $\hat{W}^t \leftarrow \xi\hat{W}^t + (1-\xi)\hat{W}_b^t$, where b is the batch number.

Cross-domain imbalance ratio $\bar{W} = \{\bar{w}_i\}_{j\in\{0,1\}}$ is obtained as follows:

$$\bar{w}_i = \frac{{w^s}_i + \hat{w}_i^t}{2}. \tag{5}$$

For a co-balanced sampling, we assign sampling probability $p = \{1-\bar{w}_0, 1-\bar{w}_1\}$ to the set of learned Gaussians G. Then, we use reparameterization trick [14] to generate new source-like samples Z^G in the latent space Z regarding the probability set p: $Z^G = \{(z^G, y)\},\ |\ z^G \overset{p}{\sim} \mathcal{N}(\mu_y, \sigma_y),\ y \in \{0,1\}$.

Manifold Mixup. To bridge the inter-domain gap, we generate intermediary instances via interpolating Z^G and $\hat{T}$ and incorporate them to the adaptation process along with source and target data.

Penalizing drastic changes of a classifier prediction affected by input perturbations (a.k.a locally Lipschitz) imposes the cluster assumption [21]. Before mixing up source and target samples, similar to NFM [16] we model locally-Lipschitz by injecting additive and multiplicative noises to the embedded target samples: $\bar{z^t}_i = (1 + \sigma_m\zeta_i^m)\cdot {z^t}_i + \sigma_a\zeta_i^a$, where ζ_i^m and ζ_i^a are random variables modeling the desired noise and $\bar{T} = \{\bar{z^t}_i, \hat{Y}_i\}_{i=1}^{N^t}$ is the set of noisy target embeddings.

Now, we randomly mix up generated source samples Z^G and noisy target samples $\bar{T}$: $z_i^m = \lambda z_i^G + (1-\lambda)\bar{z}_i^t,\ \ Y_i^m = \lambda Y_i^G + (1-\lambda)\hat{Y}_i^t$, where $\lambda \sim Beta(\alpha, \beta)$.

Due to the exact coding and decoding of F, these linear interpolations in expressive latent space Z model perceptual mixup on the complex data manifold. As depicted in Fig. 3, by using $M = \{(F^{-1}(z_i^m), Y_i^m)\}_{i=1}^{N^m}$ in addition to S and $\check{T} = \{(F^{-1}(z_i^t), \hat{Y}_i^t)\}_{i=1}^{N^t}$ we prepare a robust and enriched data as input of adversarial and conditional UDA that can smoothly direct the domain alignment.

5.2 Adversarial UDA and Conditional Mapping

The adaptation phase has two main components: a holistic alignment using adversarial UDA and a conditional alignment of target samples with high classification confidence. Holistic adaptation globally aligns the distribution of two domains and conditional alignment enforces discriminative domain transfer.

Holistic Alignment. Inspired by mixup-based adversarial UDA [26], we modified the DANN [8] architecture to incorporate inter-domain synthetic samples in the adaptation process. DANN is a minimax game between a domain discriminator and a feature extractor, alongside training a classifier for labeled data.

To minimize a cross-entropy loss (CE), source samples and their corresponding labels are fed into a two-way classifier H:

$$L_H = \mathbb{E}_{(x,y)\sim S}[CE(H(G(x)), y)]. \tag{6}$$

Optimizing the classifier H with respect to the inter-domain samples M in addition to the source samples encourages locally-Lipschitzness in the inter-domain space. Therefore, inter-domain samples and their soft pseudo-labels are fed to the classifier H to optimize the following objective:

$$L_H^m = \mathbb{E}_{(x,y)\sim M}[CE(H(G(x)), y)]. \tag{7}$$

For a holistic distribution alignment, a domain label l_D is assigned to each sample, where l_D is 1 for $x \sim S$, 0 for $x \sim \check{T}$, and λ for $x \sim M$.

Now, source and target samples along with their domain labels are fed to a domain discriminator D with a classification objective as follows:

$$L_d = \mathbb{E}_{x\sim S}[\log D(G(x))] + \mathbb{E}_{x\sim T}[\log(1 - D(G(x)))]. \tag{8}$$

In addition, we feed inter-domain samples and their domain labels to the domain discriminator to model the inter-domain space:

$$L_d^m = \mathbb{E}_{x\sim M}[l_D \log D(G(x)) + (1 - l_D)\log(1 - D(G(x)))]. \tag{9}$$

Aiming for an adversarial UDA, we maximize domain discriminator losses by optimizing the parameters of the flow model F, while parameters of D are trained via minimizing the objective functions of D. This minimax game imposes global domain confusion; therefore, a domain invariant representation in space Z is learned through optimizing L_H and L_H^m alongside a minimax game for the adversarial loss functions L_d and L_d^m:

$$L_{Adv} = \underset{F,H}{min}\ \underset{D}{max}\ \rho L_d + \gamma L_d^m + \eta L_H^m + L_H, \tag{10}$$

where ρ, γ and η are hyper-parameters to regulate the interplay of the modules over the course of the adaptation process.

Conditional Alignment. To impose a plausible conditional alignment, analogous to class-aware UDA [13], we encourage more confident target samples to be aligned with their corresponding class. First, we select a set of easy target samples with a classification confidence higher than a threshold τ as follows:

$$T_e = \{(x_e^t, y_e)\} \mid CP(x_e^t, y_e) > \tau, \tag{11}$$

where $CP(x_e^t, y_e)$ is the probability of sample x_e belonging to the class y. Next, we encourage conditional alignment by mapping easy target samples T_e to their corresponding Gaussians in space Z by optimizing Eq. 4 as follows:

$$L_c = P_X(x_e^t | y_e = k). \tag{12}$$

Finally, total objective function is:

$$L_{tot} = L_{adv} + \delta L_c = \underset{F,H}{min}\ \underset{D}{max}\ \rho L_d + \gamma L_d^m + \eta L_H^m + L_H + \delta L_c, \tag{13}$$

where δ is the regulating factor for conditional alignment. Throughout the training, alternating between the co-balanced mixup and domain alignment for optimizing L_{tot} effectively align the distribution of the source and target domains. Eventually, this step-by-step process enables the classifier H to correctly classify samples from the target domain.

6 Experiments

We assess here the effectiveness of Cubism on unsupervised volcano-seismic knowledge transfer. We first elaborate on the characteristics of the datasets and the feature extraction process. Then, after briefing implementation setup, we substantiate the efficacy of Cubism via a comprehensive analysis of our experiments.

6.1 Data Characteristics

As discussed in Sect. 3.2, we are using event records from Llaima Volcano and Deception Island Volcano as two non-identical volcano-seismic situations. These datasets have pairs of (records segment, event type), where each record segment is a raw stream of seismic signals. The seismic records of Llaima Volcano were captured between 2010 and 2016, while data from the Deception Island Volcano incorporates seismic records belonging to two different periods: 1994-1995 and 2009-2010. Table 1 summarizes the characteristics of these datasets. Note that both datasets suffer from data imbalance issues. To make the signals compatible, we first standardize each record segment with respect to its maximum value and then interpolate the data from Deception Island to match the sampling rate of the data from the Llaima volcano. This process helps to preserve the signals' temporal characteristics. Afterward, by zero-padding the signals, we maintain the same dimension for all the signals. After padding, all the records are set to 6,000 samples (an interval of 60 s). Finally, following [4], we utilize short-time fast Fourier transform (FFT) of 512 points to convert the raw datasets to sets of spectrograms. The resulting images are used as input for training Cubism.

Table 1. Characteristics of the studied datasets

Attribute	Volcano	
	Llaima	Deception Island
Number of LP events	1310	262
Number of VT events	304	77
Sampling Frequency (Hz)	100	50

6.2 Implementation Details

For exact disentangled coding, following [10], we use a RealNVP normalizing flow with two coupling layers, one hidden state and 128 hidden units. Both source classifier and domain discriminator comprise a backbone, dropout and two fully connected layers with the Relu activation function. A pre-trained Resnet-18 on ImageNet is adopted as the backbone while training the downstream layers from scratch. For training the network, we employ the Adam optimizer with a momentum of 0.9 and a decaying learning rate initiated by 0.01. We set α and β in Eq. 10 to 8 and 2 respectively, τ to 0.9, ξ to 0.9 and the batch size to 15. Towards an effective interplay of components of Cubism, we arrange the values of ρ, γ, η and δ (see Eq. 13) to gradually increase with an exponential schedule equal to $(\frac{2}{1+e^{(-\iota.b)}} - 1)$ [8], where $\iota = 10$ and b is increased linearly from 0 to 1.

6.3 Empirical Analysis

We define two volcano-seismic knowledge transfer tasks from Llaima Volcano to Deception Island Volcano and from Deception Island Volcano to Llaima Volcano, denoted as $L \to D$ and $D \to L$, respectively. To empirically evaluate the efficacy of Cubism, we compare the performance of Cubism to several baselines on these two tasks. Following is the list of comparison partners in our experiments:

- Source-Only: a pre-trained Resnet-18 plus two layers classifier for the tasks $L \to D$ and $D \to L$ on data from Llaima volcano and Deception Island Volcano, respectively.
- IMF-STD: an approach analogous to [2] that standardize volcano-seismic records by using a set of six intrinsic mode function components.
- DANN [8]: adversarially aligns distributions disregarding class imbalance.
- DM-ADA [26]: a mixup-based adversarial UDA that produces inter-domain samples in input and latent spaces of a VAE without tackling imbalanced data issue.
- DM-ADA-Flow: a version of DM-ADA that uses flow-based generative model instead of VAE.
- BMix-DA: a version of Cubism without addressing the data imbalance issue in the target domain.

Table 2. Methods performance (%) for the transfer tasks of $L \to D$ and $D \to L$

Method	Accuracy		Precision		Recall		F1-score	
	$L \to D$	$D \to L$	$L \to D$	$D \to L$	$L \to D$	$D \to L$	$L \to D$	$D \to L$
Source-Only	71.59	50.84	57.46	55.70	56.43	59.10	56.75	47.95
IMF-STD	81.06	71.29	84.46	69.27	59.35	81.17	60.45	67.36
DANN [8]	89.94	91.69	93.25	87.61	78.37	84.27	82.99	85.80
DM-ADA [26]	87.86	88.03	92.01	90.31	73.83	69.39	78.51	74.25
DM-ADA-Flow	88.46	90.51	91.36	92.35	75.59	75.97	80.11	81.04
BMix-DA	93.79	95.23	94.92	96.00	87.28	88.22	90.39	91.50
Cubism	94.98	96.15	95.70	95.75	89.87	91.44	92.38	93.41

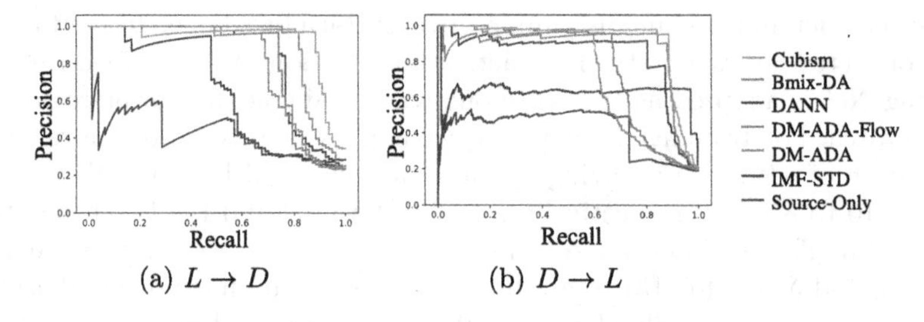

(a) $L \to D$ (b) $D \to L$

Fig. 4. Precision-Recall curve

Cross Domain Classification Performance. Table 2 presents the performance of Cubism and the baselines in terms of classification accuracy, precision, recall and F1-score. The results are reported as the average performances over five runs. Cubism significantly outperforms all the baselines since it robustly mitigates the domain gap via co-balanced inter-domain modeling. Cubism delivers a more performant model than BMix-DA, confirming the significant contribution of co-balanced mixup in addressing forward bias and reverse bias in the course of training. Besides, BMix-DA outperforms DANN and DM-ADA substantiating the significance of addressing the data imbalance issue using UDA. DM-ADA-Flow replaces the VAE in DM-ADA with a flow-based generative model to assess the effectiveness of an invertible generative model utilization. As shown in Table 2, DM-ADA-flow outperforms DM-ADA due to delivering semantic preserving mixup. Although using Mixup substantially improves the performance of UDA approaches, in the case of data imbalance, these solutions are biased toward the majority class. Thus, as shown in Table 2, DM-ADA has a lower performance compared to DANN. IMF-std is outperformed by all the UDA-based methods confirming the crucial role of UDA techniques in aligning the data distribution of the two volcanoes. Finally, the poor performance of source-only emphasizes the necessity of addressing the inter-volcano gap.

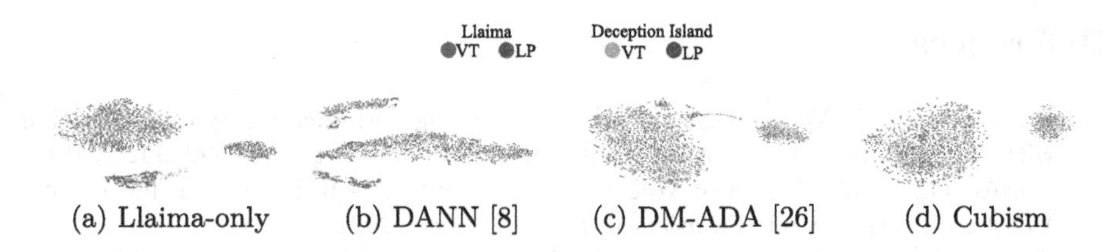

Fig. 5. t-SNE visualization of network activations generated by DANN, DM-ADA and Cubism for the transfer task $L \rightarrow D$

Furthermore, Fig. 4 compares the predictive power of the classifier learned by Cubism with the baselines using the Precision-Recall curve. Cubism considerably outperforms all the baselines confirming the performance analysis above.

Feature Visualization. t-SNE [6] is a widely-used approach to reduce high dimensional data to 2D. For a visualized comparison, we mapped the deep features learned by Source-Only, DANN, DM-ADA and Cubism to 2D space for the transfer task $L \rightarrow D$ employing t-SNE. Figure 5 demonstrates the t-SNE projections. There is a significant inter-volcano distribution gap for Source-Only as depicted in Fig. 5-a. In contrary to DANN and DM-ADA, Cubism is not subject to class imbalance, as one can see in Figures 5b-d. In other words, Cubism imposes the best conditional alignment between the two volcanic domains compared to all baselines.

7 Conclusions

Cubism is a novel framework for unsupervised cross-volcano classification that robustly models inter-volcano manifold in an invertible latent space. Cubism goes one step beyond the limited assumption of conditional balance in unsupervised domain adaptation methods by dynamic co-alleviation of bias and inevitable reverse bias. Cubism proposes co-balanced inter-volcano modeling and delivers well-rounded mitigation of the inter-volcano gap. This approach opens a new perspective to significantly less resource-intensive volcano-seismic knowledge transfer with a promising performance. We evaluate Cubism in an extensive comparative study on the knowledge transfer task for two well-studied volcanoes, showing that it outperforms all the baselines by a large margin, thus establishing the efficacy of this new approach. Cubism is a game changer in forecasting volcanic hazards by substantiating a fundamental step toward low-cost mining of volcano-seismic data. Our future work aims at extending Cubism to the more complicated task of unsupervised discriminative knowledge transfer given volcano-seismic stream data where there is no prior knowledge about the similarity of class sets in volcano-seismic situations.

Acknowledgements. This research has been funded by the Canadian Mountain Network as part of the Networks of Centres of Excellence program and the Natural Sciences and Engineering Research Council of Canada.

References

1. Bueno, A., et al.: Volcano-seismic transfer learning and uncertainty quantification with Bayesian neural networks. Trans. Geosci. Remote Sens. **58**, 892–902 (2019)
2. Cortés, G., et al.: Practical volcano-independent recognition of seismic events: VULCAN. ears project. Frontiers in Earth Science (2021)
3. Cortés, G., et al.: Standardization of noisy volcanoseismic waveforms as a key step toward station-independent, robust automatic recognition. SLR (2019)
4. Curilem, M., et al.: Using CNN to classify spectrograms of seismic events from Llaima volcano (Chile). In: IJCNN (2018)
5. Dinh, L., et al.: Nice: Non-linear independent components estimation. arXiv (2014)
6. Donahue, J., et al.: Decaf: A deep convolutional activation feature for generic visual recognition. In: ICML (2014)
7. Duque, A., et al.: Exploring the unsupervised classification of seismic events of cotopaxi volcano. J. Volcanol. Geothermal Res. **403**, 107009 (2020)
8. Ganin, Y., et al.: Unsupervised domain adaptation by backpropagation. In: ICML (2015)
9. Grijalva, F., et al.: Eseismic-GAN: a generative model for seismic events from cotopaxi volcano. J. Select. Top. Appl. Earth Observ. Remote Sens. **14**, 7111–7120 (2021)
10. Izmailov, P., et al.: Semi-supervised learning with normalizing flows. In: ICML (2020)
11. Jiang, X., et al.: Implicit class-conditioned domain alignment for unsupervised domain adaptation. In: ICML (2020)
12. Kang, G., et al.: Contrastive adaptation network for unsupervised domain adaptation. In: CVPR, pp. 4893–4902 (2019)
13. Keramati, M., et al.: Norma: a hybrid feature alignment for class-aware unsupervised domain adaptation. In: CIKM (2021)
14. Kingma, D.P., et al.: Auto-encoding variational Bayes. In: ICLR (2014)
15. Lee, J., et al.: Self-diagnosing GAN: diagnosing underrepresented samples in generative adversarial networks. In: NeurIPS (2021)
16. Lim, S.H., et al.: Noisy feature Mixup. arXiv (2021)
17. Madadi, Y., et al.: Deep visual unsupervised domain adaptation for classification tasks: a survey. IET Image Processing (2020)
18. Malfante, M., et al.: Machine learning for volcano-seismic signals: challenges and perspectives. Signal Process. Mag. **35**, 20–30 (2018)
19. Manley, G., et al.: A deep active learning approach to the automatic classification of volcano-seismic events. Front. Earth Sci. **10**, 807926 (2022)
20. Na, J., et al.: FixBi: bridging domain spaces for unsupervised domain adaptation. In: CVPR (2021)
21. Shu, R., et al.: A DIRT-T approach to unsupervised domain adaptation. In: ICLR (2018)
22. Thelen, W.A., et al.: Trends in volcano seismology: 2010 to 2020 and beyond. Bull. Volcanol. **84**, 26 (2022). https://doi.org/10.1007/s00445-022-01530-2
23. Titos, M., et al.: A deep neural networks approach to automatic recognition systems for volcano-seismic events. J. Select. Top. Appl. Earth Observ. Remote Sensing **11**, 1533–1544 (2018)
24. Venegas, P., et al.: Combining filter-based feature selection methods and gaussian mixture model for the classification of seismic events from Cotopaxi volcano. J. Select. Top. Appl. Earth Observ. Remote Sens. **12**, 1991– 2003 (2019)

25. Wei, C., et al.: Crest: A class-rebalancing self-training framework for imbalanced semi-supervised learning. In: CVPR (2021)
26. Xu, M., et al.: Adversarial domain adaptation with domain Mixup. In: AAAI (2020)
27. Yüksel, O.K., et al.: Semantic perturbations with normalizing flows for improved generalization. In: ICCV (2021)
28. Zhang, H., et al.: Mixup: beyond empirical risk minimization. In: ICLR (2018)

Go Green: A Decision-Tree Framework to Select Optimal Box-Sizes for Product Shipments

Karthik S. Gurumoorthy(✉) and Abhiraj Hinge

India Machine Learning, Amazon, Bangalore, India
karthik.gurumoorthy@gmail.com

Abstract. In package-handling facilities, boxes of varying sizes are used to ship products. Improperly sized boxes with box dimensions much larger than the product dimensions create wastage and unduly increase the shipping costs. Since it is infeasible to make unique, tailor-made boxes for each of the N products, the fundamental question that confronts e-commerce companies is: *"How many $K << N$ cuboidal boxes need to manufactured and what should be their dimensions?"* In this paper, we propose a solution for the *single-count* shipment containing one product per box in two steps: (i) reduce it to a clustering problem in the 3 dimensional space of length, width and height where each cluster corresponds to the group of products that will be shipped in a particular size variant, and (ii) present an efficient forward-backward decision tree based clustering method with low computational complexity on N and K to obtain these K clusters and corresponding box dimensions. Our algorithm has multiple constituent parts, each specifically designed to achieve a high-quality clustering solution. As our method generates clusters in an incremental fashion without discarding the present solution, adding or deleting a size variant is as simple as stopping the backward pass early or executing it for one more iteration. We tested the efficacy of our approach by simulating actual single-count shipments that were transported during a month by Amazon using the proposed box dimensions. By just modifying the existing box dimensions and not even adding a new size variant, we achieved a reduction of 4.4% in the shipment volume, contributing to the decrease in non-utilized, air volume space by 2.2%. The reduction in shipment volume and air volume improved significantly to 10.3% and 6.1% when we introduced 4 additional boxes.

Keywords: Box size selection · Clustering · Decision-trees · Product shipments

1 Introduction

E-commerce companies often deliver their product in brown corrugated boxes. Though there is a constant strive towards package free shipping due to environmental concerns, many product characteristics like its fragility, hazardous

© The Author(s), under exclusive license to Springer Nature Switzerland AG 2023
M.-R. Amini et al. (Eds.): ECML PKDD 2022, LNAI 13717, pp. 598–613, 2023.
https://doi.org/10.1007/978-3-031-26419-1_36

nature, sensitivity to public disclosure (e.g. adult diapers) precludes them from being shipped without any packaging to avoid degraded customer delivery experience. In these circumstances, the best approach is to keep packaging wastage to a minimum. One of the principal contributors to such packaging wastage is the size of the packaging material, a.k.a, the box dimensions in which the products are shipped. For instance, if the box dimensions are much bigger than the product dimensions, the non-utilized empty space is often stuffed with filler material like dunnages to keep the product in position, creating added waste. The image in Fig. 1 drives this point home.

Further, such empty spaces negatively impact the number of products that can be simultaneously transported, as the size of the individual boxes determine the quantity of shipments that can be loaded onto a container. Hence, the shipment cost per product is directly proportional to the volume of the box in which it is sent, which may be huge compared to the product volume. The ideal solution is to manufacture N boxes, one for each of the N products, with dimensions exactly matching the corresponding product dimension. However, this is practically infeasible due to the high fixed cost associated with making new box sizes alongside the operational difficulty involved in scaling the packaging process for a large number of box sizes, as they need to be placed in separate shelves, all in the vicinity of each other. Hence the problem of reducing the empty spaces within the box naturally breaks down into the following two sub-problems:

1. How many $K << N$ boxes need to be manufactured, bearing in mind the fixed cost and operational scalability?
2. Given that K boxes are manufactured, what should be their dimensions so that the overall shipment volume is minimized?

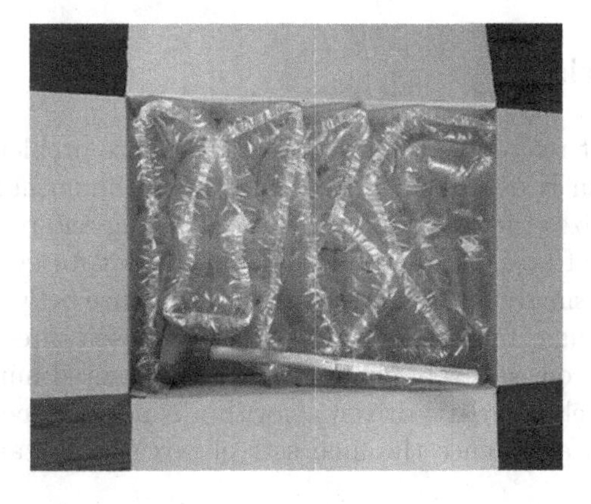

Fig. 1. Product shipped in a huge box causing excessive wastage.

1.1 Contributions

In this work, we propose an efficient algorithm to solve the problem of deciding the box dimensions of K boxes. Once the sizes are determined and the boxes made accordingly, it makes practical sense to ship every product in a box that fits it snugly with minimum air volume, as it reduces both the wastage and the shipping cost. We show that given K, the problem reduces to a clustering problem of grouping products into K clusters where each cluster specifies the set of products that will be shipped in the same sized box. Akin to $K-$means [1], determining the globally optimal solution for the clusters in computationally intractable as the problem is $NP-$complete. To this end, we propose a novel, forward-backward decision tree based method to determine the K clusters, which alongside simple heuristics like product-cluster reassignment and iterative dimension refinement as explained in Sect. 4 is able to arrive at a very good local minimum of our objective namely, *minimize the overall shipment volume across all product shipments.*

The theoretical time complexity of our algorithm is analyzed and derived to be $O(KN \log N + NK^2)$ as discussed in Sect. 5. The sub-quadratic growth rate of $O(N \log N)$ w.r.t. the number of products N is critically important from the scalability perspective, as the number of products sent in boxes could potentially be in hundred-thousands in established e-commerce companies like Amazon. This is the foremost advantage of our method when compared to techniques based on genetic algorithms [7,20]. As the total shipment volume will steadily decrease with increasing K from the fact that more size options are available to ship the product, the best K is that value where the benefit from the decreased shipment volume is maximum compared to the cost of increased fixed cost and operational hindrance. As it may not be feasible to bring these benefits and the costs into a comparable scale, we propose to set K as the *elbow-point* where the decrease in shipment volume plateaus with increasing K, as traditionally followed in $K-$means clustering [9].

2 Prior Work

One of the earliest references that studies the box-size problem in detail is [19], where the problem is described as selecting the optimum number and sizes of boxes that minimizes the total shipment, warehousing and related costs. A possibly large of set of boxes are initially created so that for every product, there exists at least one snugly fitting box where the difference between the box dimensions and the product dimensions are less than a chosen threshold τ. These size variants are then consecutively eliminated till the desired number of boxes are reached. This largely heuristic-driven algorithm is not designed to optimize any objective function and hence the final sets of box sizes obtained are generally sub-optimal.

In the recent past, genetic algorithms have been used to address the box-sizing problem. Specifically, Wong et. al in [20] introduce the use of multi-objective genetic algorithms (MOGA) to choose optimal box sizes for combined orders and demonstrate an application of their method to an actual industrial problem in [10]. These methods are designed to choose box dimensions where multiple items can be packed into a single box. We henceforth refer to them as *multis* where the orientation, the order, and the number of allowable items that can be packed into the same box influences the box dimensions. As majority of shipments in e-commerce conglomerates are *singles* where each box holds only one product, in our present work we deal only with single-count shipments. Hence the approach developed in [10] is less useful in our setting.

The MOGA technique with a problem definition very similar to ours is explored in [7] under the ambit of genetic algorithms which we pit against our clustering based method in Sect. 6.2. The work in [7] also present an optimal dynamic programming solution for one-dimensional variant of the box-sizing problem, which as explained in Sect. 6.1 is used as the baseline. Generally, these evolutionary methods are very time-consuming and not scalable as many different sets of candidate solutions (box dimensions) must be evaluated individually to choose the most optimal box dimensions among them. The subsequent generations of possible dimensions are not instantiated from the view point of minimizing the overall shipment volume. Rather, they are created as minor modifications of the parent solution (crossover and mutation) and are explicitly evaluated which is computationally very expensive.

The problem for fixing the box dimensions is studied in different fields with different names. In the apparel industry, it is called the standardization problem and is framed as finding standard sizes for a given population, while minimizing the adaption loss due to mismatch in dimension. The work in this field [4,14,16] focuses on solving the problem mainly for one-dimension using the distribution of the population and the interval bisection method [15,17] with different loss functions to find the optimal sizes.

The box sizing can also be treated as a special case of the assortment or catalogue problem, where the goal is to optimally choose a subset from a large discrete set of possible sizes to stock, taking into consideration the space and inventory costs along with the demand for a particular size. The survey work in [12] presents a detailed review of the methodologies designed for the assortment problem in the last 50 years. In particular, it identifies the sizing problem as a special case and discusses the techniques proposed in [3] in this regard. The author in [12] notes that while [3] does present an extension for solving the sizing problem in two-dimensions, its success is highly dependent on the dimensions of the products being correlated, which does not necessarily hold in the e-commerce industry. The work in [11] analyzes the uncapacitated and capacitated versions of the logit product assortment problem of a retail operation and proposes a computationally feasible branch-and-cut algorithm. Other recently developed approaches for determining the box sizes include [13] and [21].

3 Reduction to Clustering

In order to reduce the shipment volume, it is logical that frequently sold products be shipped in size variants which are very close to their product dimension. Hence the selection of the best size variants depends on two factors: (a) the dimensions of the N product that are shipped in the boxes, (b) the expected number of shipments per product, a.k.a the sales velocity. Recall that our goal is to determine the box dimensions $\{l^k, w^k, h^k\}$, $k \in \{1, 2, \ldots, K\}$ for the K different size variants will be introduced. Denote $\{l_j, w_j, h_j\}$ as the product dimensions of the product j for $j \in \{1, 2, \ldots, N\}$ and let s_j be its sales velocity. In most cases, the past shipment data can be leveraged to closely approximate s_j. Identifying the optimal size dimensions is tantamount to determining the set of products that will be shipped in each of the box size variant. Given any such partition of the products into K clusters, let $\mathcal{C}_k$ denote cluster k containing N_k products. The cluster $\mathcal{C}_k$ represents the group of products that will be shipped in the same size variant k. Then, it is easy to see that the optimal dimensions for the box k, namely $\{l^k, w^k, h^k\}$ having the least shipment volume will equal the largest length, width and height of the products in $\mathcal{C}_k$, i.e. $l^k = \max_{j \in \mathcal{C}_k} l_j$, $w^k = \max_{j \in \mathcal{C}_k} w_j$, and $h^k = \max_{j \in \mathcal{C}_k} h_j$. Identifying the best size variants reduces to a clustering problem, where the goal is to cluster N products into K clusters with the primary objective of reducing the total volume shipped. Let x_{jk} be the binary membership variable determining whether product j is shipped in the box k. The overall shipment volume can be mathematically expressed as:

$$V(X) = \sum_{j=1}^{N} s_j \sum_{k=1}^{K} x_{jk} \left(\max_{j} x_{jk} l_j \times \max_{j} x_{jk} w_j \times \max_{j} x_{jk} h_j \right) \tag{3.1}$$

where the j, k^{th} entry of the binary matrix X is x_{jk}. Our aim is to minimize $V(X)$ subject to the binary constraints:

$$\sum_{k} x_{jk} = 1 \; \forall j, \quad x_{jk} \in \{0, 1\} \; \forall j, k.$$

As $V(X)$ strictly decreases with increasing K, we solve for different K and select K as the *elbow-point* where we see diminishing returns with increasing K [9].

4 Solution Methodology

Obtaining the global optimal solution for the clustering formulation in Eq. (3.1) in the 3 dimensional space of length, width and height is computationally intractable because of the binary constraints on the membership variables x_{jk}. Instead, we propose the following decision tree based forward-backward algorithm to obtain a good local minimum. In each iteration, our method obtains C clusters (size variants) in an *incremental* fashion by using the $C-1$ clusters from the previous iteration as the starting point. Among the $C-1$ size variants, we then split one of the size variant into 2 to obtain C different boxes. As the current $C-1$ size variants are akin to the leaf nodes in a binary tree out of which one is chosen to be split further, our algorithm closely resembles the decision tree based methods [5]. However as explained below, our method is composed of multiple constituents, each of them meticulously designed to minimize the specific objective in Eq. (3.1). In the experimental section we highlight the utility of each of these parts. We obtain K clusters when the algorithm completes.

Our algorithm primarily consists of 4 operations that are performed in a specific order to reach the final optimal box-dimensions. They are: (1) Cluster splitting, (2) Product reassignment, (3) Iterative refinement and (4) Cluster combination. Below we discuss these operations in detail.

4.1 Cluster Splitting

The process of selecting and segmenting a cluster k, denoted by $\mathcal{C}_k$, of N_k products into two clusters of left and right child nodes, with the intention of minimizing the total volume shipped is called cluster splitting.

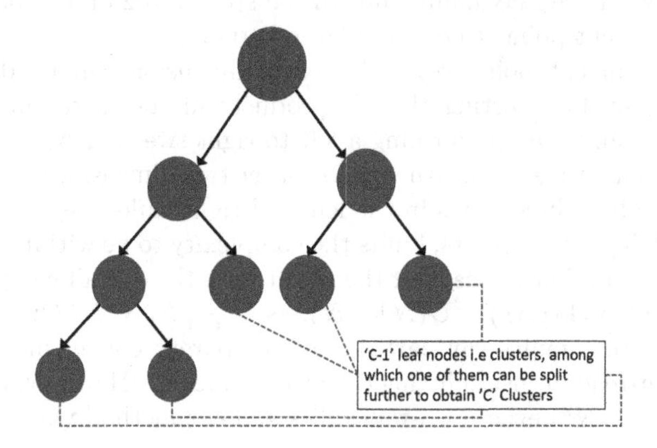

Fig. 2. Tree created by cluster splitting

The volume for $\mathcal{C}_k$ is the smallest box size that can fit every product in the cluster multiplied with the number of shipments in the cluster i.e.,

$$V(k) = \left(\max_{j \in \mathcal{C}_k} l_j\right) \left(\max_{j \in \mathcal{C}_k} w_j\right) \left(\max_{j \in \mathcal{C}_k} h_j\right) \left(\sum_{j \in \mathcal{C}_k} s_j\right) \tag{4.1}$$

and the total volume across all clusters is $V = \sum_{k=1}^{C} V(k)$. The aim is divide *one* of the current $C-1$ clusters, say $\mathcal{C}_k$, into two child clusters $\mathcal{L}_k$ and $\mathcal{R}_k$, assign a subset of the N_k products to $\mathcal{L}_k$ and the remainder to $\mathcal{R}_k$ so that the total volume shipped is minimized. Finding the optimal solution would require examining every possible segmentation of the products, resulting in a computationally intractable time complexity of $O(2^{N_k})$. Instead, we propose a greedy search method that finds a good local minimum.

Each cluster can be split across 3 possible dimensions of length, width and height to obtain new left and right child nodes. For instance, if we decide to split the size variant k containing N_k products on the length (l) dimension, then the objective is to decide the cut point $\tau_{k,l}$ so that the subset of products with length less than or equal to $\tau_{k,l}$ will be assigned to left node $\mathcal{L}_{k,l}$ and products with length greater than $\tau_{k,l}$ to $\mathcal{R}_{k,l}$. For each of the $C-1$ clusters and the 3 dimensions, we will individually determine the cut points $\tau_{k,dim}$ that maximally decreases the net shipment volume. In other words, choose the cut point $\tau_{k,dim}$ that results in maximum gain, where the gain is given by:

$$Gain(k, dim) = V(k) - [V(L_{k,dim}) + V(R_{k,dim})]$$

for $dim \in \{l, w, h\}$. The best dimension to split $\mathcal{C}_k$ is the one whose corresponding optimum cut point results in the maximum gain compared to splitting with the other two dimensions. The gain corresponding to the best dimension is the gain for splitting the cluster $\mathcal{C}_k$ and is given by: $Gain(k) = \text{argmax}_{dim}\ Gain(k, dim)$. That cluster with the maximum gain will be split into 2 on the best dimension at the optimum cut point, to obtain C size variants.

The optimum cut point $\tau_{k,dim}$ for each dimension can be determined in $O(N_k \log N_k)$, by first sorting the N_k products in the increasing value along that dimension and then performing a left to right sweep of $N_k - 1$ possible cut point values. The usage of max-heaps for other two dimensions to keep track of the largest product dimensions in the left and right child clusters, produced for each choice of $N_k - 1$ cut points, limits the complexity to be within $O(N_k \log N_k)$. Since we will be evaluating each of the C clusters, the overall complexity of this step is $\sum_{k=1}^{C} O(N_k \log N_k) = O(N \log N)$, as $\sum_{k=1}^{C} N_k = N$. Our approach can be generalized to a multi-way partition of the parent cluster into more than 2 child nodes, instead of just splitting $\mathcal{C}_k$ into $\mathcal{L}_k$ and $\mathcal{R}_k$. However, the time complexity will be $O(N^2)$ even for 3-way splits and the method will not be scalable for large N.

4.2 Product Reassignment

As mentioned earlier, once we have partitioned the products into C clusters, the box dimensions for the cluster $\mathcal{C}_k$ will equal the largest length, width and height among the products in $\mathcal{C}_k$. However, it is possible that products are not assigned to the most optimum box that snugly fits them and minimizes the shipment volume. So for each product, we will *reassign* it to that $\mathcal{C}_k$ whose box dimensions individually are at least as large as the product dimension and the box volume is closest to the product volume. The reassignment step is composed of iterating over each product and selecting the best cluster $\mathcal{C}_k$ in terms of lowest shipment volume and involves a linear time complexity of $O(NC)$.

4.3 Iterative Refinement

Given a set of box dimensions, it may be possible to tweak some of them by a small amount to arrive at a new set of box dimensions that lead to more efficient packing. Iterative refinement is a process that tests out this possibility by refining the box dimensions in a greedy manner. Each iteration of the algorithm works as follows. Assume that at iteration t, we have a clustering solution with the expected overall shipment volume V_t computed as per Eq. (3.1). Our objective is to obtain an improved clustering solution with volume cost V_{t+1} at iteration $t+1$ *by moving exactly one product between two clusters* such that the difference in volume between successive iterations namely, $V_t - V_{t+1}$ is maximized. To this end, note that the dimension of any box k can be changed only by moving the product with the largest length, width or height in $\mathcal{C}_k$ to a different cluster. So we have a maximum of three product choices per $\mathcal{C}_k$ and the chosen product can be moved to other $C-1$ clusters. In total, we have $3(C-1)C$ options to move one product between two clusters. We will evaluate all these $O(C^2)$ options, compute the gain in volume reduction for each of them, and greedily select the one that gives the minimal overall shipment volume V_{t+1} at iteration $t+1$. If a product is moved from cluster $\mathcal{C}_{k_1} \rightarrow \mathcal{C}_{k_2}$, then the reduction in volume equals: $V_t - V_{t+1} = V_t(k_1) + V_t(k_2) - [V_{t+1}(k_1) + V_{t+1}(k_2)]$, where $V_t(k)$ is the kth cluster volume (Eq. (4.1)) at iteration t. Note that, we need to evaluate all the $O(C^2)$ combinations only for the very first iteration. For subsequent iterations, the volume reduction gains need to be computed only among $C \setminus \{k_1, k_2\} \times \{k_1, k_2\}$ and between k_1 and k_2 which are only $2C-3$ new evaluations. Each evaluation is $O(1)$, equal to the time to compute Eq. (4.1) with the decreased (increased) sum of sales velocity on $\mathcal{C}_{k_1}$ ($\mathcal{C}_{k_2}$) as a product j with sales velocity s_j is moved from $\mathcal{C}_{k_1} \rightarrow \mathcal{C}_{k_2}$, and with either the present or the second largest product dimensions in $\mathcal{C}_{k_1}$ depending on which dimension(s) change and perhaps new largest product dimensions in $\mathcal{C}_{k_2}$. The algorithm stops at iteration T when all the possible moving options only increases the current shipment volume V_t.

The computational complexity of this step can be computed as follows. At the beginning, we construct a max-heap for each $\mathcal{C}_k$ containing N_k products in $O(N_k)$, one for each of the 3 dimension, to track the products with largest dimensions which could potentially be moved to other clusters. As $\sum_{k=1}^{C} N_k = N$, the total pre-processing time involved is $O(N)$. Once a product is moved from $\mathcal{C}_{k_1} \rightarrow \mathcal{C}_{k_2}$, decreasing for instance the length l^{k_1} of $\mathcal{C}_{k_1}$, we respectively delete the product from the 3 max-heaps for $\mathcal{C}_{k_1}$ and push these products to the max-heaps maintained for $\mathcal{C}_{k_2}$, so that the products with largest dimensions in the modified clusters $\mathcal{C}_{k_1}$ and $\mathcal{C}_{k_2}$ are updated. While the delete operation in $\mathcal{C}_{k_1}$ for the max-heap corresponding to the length will be $O(1)$, as only the root needs to be popped out, the delete operation for other two heaps could potentially be $O(N_{k_1})$. However, the push operations into the 3 max-heaps for $\mathcal{C}_{k_2}$ will all be $O(\log N_{k_2})$. Hence the total computation complexity is $O(N + C^2 + (C+N)*T)$ where we discount $O(\log N) << N$.

4.4 Cluster Combination

Cluster combination is the process of moving from C packaging boxes to $C-1$ packaging boxes by combining the pair of clusters that produce the least additional increase in total volume shipped. We iterate over all the ${}^{C}C_2$ possible combination and select that pair $\{\mathcal{C}_{k_1}, \mathcal{C}_{k_2}\}$ that gives the least total volume shipped when merged. The utility of this process may not be immediately apparent and will become clear in the next section. As we search over all possible pairs each in $O(1)$, the time complexity for evaluation is $O(C^2)$ and the final merging operation is $O(1)$.

4.5 Final Algorithm

Having explained the constituent parts of our solution, we proceed to put these parts together and describe the actual algorithm. Recall that our objective is to find K clusters that minimize Eq. (3.1). Our algorithm has two high-level phases, the forward pass and the backward pass. The forward pass is similar to the divisive clustering method [18], incrementally building up the tree using clustering splitting to generate $\tilde{K} \geq K$ clusters. This process is visualized in Fig. 2. The backward pass, following a process akin to agglomerative clustering [8], sequentially combines these $\tilde{K}$ clusters into the required K groups.

Creating more than the required number of clusters and then combining them in a bottom-up fashion tends to explore the solution space better leading to an improved clustering solution. For instance, let $\mathbb{C}_C$ denote the set of C clusters obtained in the forward step. Say a cluster $\mathcal{C}_k \in \mathbb{C}_C$ is further split into $\mathcal{L}_k$ and $\mathcal{R}_k$ to get $C+1$ clusters. It could be possible to combine $\mathcal{L}_k$ or $\mathcal{R}_k$ with another cluster $\mathcal{C}_{\hat{k}} \in \mathbb{C}_C$ to produce a new clustering solution $\mathbb{C}_C^{new}$ of C clusters which may be superior to the original solution $\mathbb{C}_C$. We test this hypothesis in Sect. 6, by comparing results with and without the backward pass and notice an improvement in performance in its presence. The beginning point $\tilde{K}$ for the backward pass is a hyper-parameter, chosen following the process described in Sect. 4.6.

As the iterative refinement tries to greedily refine and improve the current clustering solution without changing the number of clusters, it is invoked following both cluster splitting and cluster combination subroutines. Whenever the box dimensions change either because of the split or merge operations, or are refined by moving products between clusters, the product reassignment step ensures that product are placed in the best-fitting box. Hence it is invoked after each cluster split, cluster combination, and iterative refinement steps.

The forward pass starts off with one cluster, setting $C = 1$ containing all the products. The dimensions of this box will equal the corresponding largest dimension among all the products. The best possible split for every cluster is evaluated using the cluster splitting method and the cluster that leads to maximum reduction in shipment volume is broken into 2. At this point we have moved from C to $C+1$ clusters. After reassigning the products to better-fitting boxes, we iterative refine and fine-tune the dimensions of the $C+1$ boxes, followed by the product reassignment step as the box dimensions may have changed. This entire procedure is repeated till we reach $\tilde{K} \geq K$ clusters.

The backward pass begins at $\tilde{K}$ clusters proceeding in a bottom-up fashion. After reducing the number of clusters by 1 through merging the best two pairs using the cluster combination method, the products are reassigned, the clusters are refined by moving one product between two clusters in successive iterations to further optimize the box dimensions, followed by one more reassignment step. This agglomerative procedure is repeated till we reach K clusters. The maximum value of $\{l, w, h\}$ in each of the final K clusters will be the dimensions of the corresponding size variants.

4.6 Hyper-parameter Selection

The only hyper-parameter in our algorithm is the beginning point $\tilde{K}$ for the backward pass. Each $\tilde{K}$ may produce different size variants once once we reach K clusters from below. To choose the best $\tilde{K}$, we pursued the following validation process. We considered the actual single-count shipment data containing one product per box that occurred in a different time period, referred to as the validation set, and simulated these shipments by sending products in snugly-fit boxes whose dimensions are obtained by starting the backward pass on the training shipment set at a position K'. On the simulated shipments, we then determined the percentage of air in the box ξ as per Eq. (6.1) defined below. We set $\tilde{K}$ to that value of K' for which the K clusters and the corresponding K box sizes lead to minimum ξ in the validation data set. It is important to note that the box dimensions are determined from the training set and their performance is evaluated on a different, unseen validation data set.

5 Time Complexity Analysis

Denote $\tilde{K} = \alpha K$ for some α independent of N and K and let the iterative refinement step be executed for a maximum of T_{max} iterations. The time complexity for the forward pass equals:

$$O\left(\sum_{C=1}^{\alpha K} N\log N + NC + N + C^2 + (C+N) * T_{max}\right)$$
$$= O\left(KN\log N + NK^2 + K^3 + (K^2 + KN) * T_{max}\right).$$

Similarly, for the backward pass it will be:

$$O\left(\sum_{C=K+1}^{\alpha K} C^2 + NC + N + C^2 + (C+N) * T_{max}\right)$$
$$= O\left(K^3 + NK^2 + (K^2 + KN) * T_{max}\right).$$

As $K << N$ and T_{max} is a constant independent of N and K, the overall time complexity can be succinctly stated as $O(KN\log N + NK^2)$. It is worth emphasizing that the computational complexity of only $O(N\log N)$ on the number of products N, makes our algorithm scalable to even millions of products.

6 Experiments

Recall that our primary goal is to decide on the number and the sizes of the boxes, so that they snugly fit the products, minimizing the non-utilized space in each shipment and thereby the overall shipment volume. In order to determine the extent of empty space—the air in the box—across all shipments, we use the metric ξ described as follows. Let S_{TE} denote the number of shipments that occurred in test time period TE, equal to the sum of sales velocity of the products during that interval. This interval TE could be any non-overlapping period in the future, different from both the time TR of the training shipments which are used to learn the box dimensions and the validation period. Given the K box sizes, we first associated each product shipment i with the most snugly-fitting box and computed the product and box volumes, pv_i and bv_i respectively. Defining $P = \sum_{i=1}^{S_{TE}} pv_i$ and $V = \sum_{i=1}^{S_{TE}} bv_i$ to be the total product and shipment volumes, we determine the % air-in-box, denoted by ξ, by:

$$\xi \equiv 100 \times \left(1 - \frac{P}{V}\right). \tag{6.1}$$

As $V \geq P$, $\xi \in [0, 100)$, where a value close to 0 is indicative of the best possible box-dimensions across all products and a value near 100 is the worst case scenario.

As P is a constant, it is clear that ξ and V are commensurable and minimizing $V(X)$ in Eq. (3.1) is tantamount is achieving smallest value for ξ in Eq. (6.1). The business sensitive nature of the shipment volumes precludes us from disclosing their actual values. Hence we report the %air-in-box metric in all our experiments results. Since ξ and V are directly related, the inferences made using ξ are straight away applicable to V and vice versa.

The principal aim of our experiments is to answer the following question: *"For different methods/variants, how does ξ vary with K?"* To this end, we study the following variants of our clustering method to underscore the role played by each of different subroutines and compare it with two competing approaches.

1. Our algorithm in its entirety that includes all the 4 constituent parts namely, cluster splitting, product reassignment, iterative refinement and the cluster combination involved in the backward pass.
2. An alternative that comprises of the only forward pass to highlight the value addition from the backward phase.
3. Exclusion of the iterative refinement step both in the forward and the backward passes.
4. Another alternative that does not involve the product reassignment in both the phases.
5. The Genetic Algorithm (GA) based algorithm proposed in [7] tailored to our setting.
6. As a baseline, we also implemented the $1D$ clustering method on the product volumes as described below.

6.1 Baseline Method

Instead of clustering in the 3 dimensional space of length, width and height, we project the products into the single dimensional volumes $v_j = l_j w_j h_j$ and then cluster these N volumes $\{v_1, v_2, \ldots, v_N\}$ into K clusters such that the following alternative objective function is minimized:

$$\tilde{V}(X) = \sum_{k=1}^{K} \left(\max_{j} x_{jk} v_j \right) \sum_{j=1}^{N} x_{jk} s_j, \tag{6.2}$$

subject to the binary constraints on x_{jk}. The one dimensional clustering formulation can be solved in $O(N^2K)$ using Dynamic Programming method [6,7]. As before, the clustering output determines those set of products that will be shipped in a particular box variant k, whose dimensions will equal the largest length, width and height among the products in that cluster $\mathcal{C}_k$.

6.2 Set-up and Results

We considered about 2 million shipments, each containing one product, that occurred during June 2019, for training. Our training data set $\mathcal{D}_{TR} = \{l_j, w_j, h_j, s_j\}_{j=1}^{N}$ is the set of 4-tuples for about $N = 75,000$ products, containing its length l_j, width w_j, height h_j and number of shipments s_j known as the sales velocity. These products are currently shipped in $K = 14$ boxes of different dimensions. We set July 2019 as our validation period VD to determine the starting point $\tilde{K}$ as described in Sect. 4.6.

We evaluated the performance of each of the 4 different variants, the GA based approach [7] and the baseline method using the %air-in-box metric ξ on the test set shipments $\mathcal{D}_{TE}$, that took place in August 2019, for values of $K \in \{12, 13, \ldots, 19, 20\}$. The size of $\mathcal{D}_{TE}$ was about 2 million shipments. For every K we determined the value of $\tilde{K}$ following the process described in Sect. 4.6. The plot in Fig. 3 shows the ξ values computed on the validation data set $\mathcal{D}_{VD}$ for different starting points $\hat{K}$, and for different K box dimensions determined from the training shipments $\mathcal{D}_{TR}$. We would like to emphasize that generating the graph in Fig. 3 is computationally not expensive. When performing a backward pass starting from a point $\hat{K}$, the box dimensions (as a function of $\hat{K}$) for all the values of $K \leq \hat{K}$ can be obtained *along the way* after merging the chosen two clusters in the group $\mathcal{C}_{K+1}$ of $K+1$ clusters to get the set $\mathcal{C}_K$ of K clusters. It is *not necessary* to repeat this step once for each value of K, corresponding to the starting point $\hat{K}$. From Fig. 3 we note $\tilde{K} = 44$ as a good point to being the backward phase for most values of K.

Setting $\tilde{K} = 44$, we gauged the performance of the 6 methods. In Fig. 4 we show the %air-in-box values computed on the test shipments $\mathcal{D}_{TE}$ for each of these methods, across different K values. The horizontal line in red is the value of the metric $\xi_{curr} = 60.5\%$ for the test shipments when products are shipped in the currently used 14 box sizes. It is important to bear in mind that after manually analyzing the shipping data over several months, the current dimensions for 14 boxes are carefully handpicked to minimize %air-in-box. *So any improvement over ξ_{curr} is of high significance.* As expected, our complete method plotted in black, containing all the 4 sub-parts has the least %air-in-box among all variants, for all values of K barring $K = 15$. The minor deviation at $K = 15$ is because the

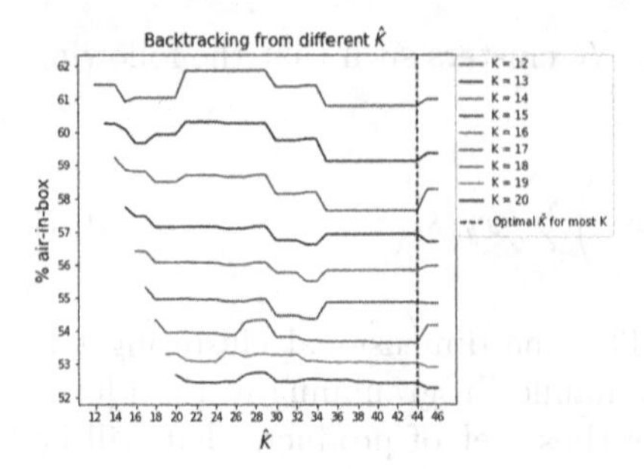

Fig. 3. The % air-in-box for different backtracking starting points.

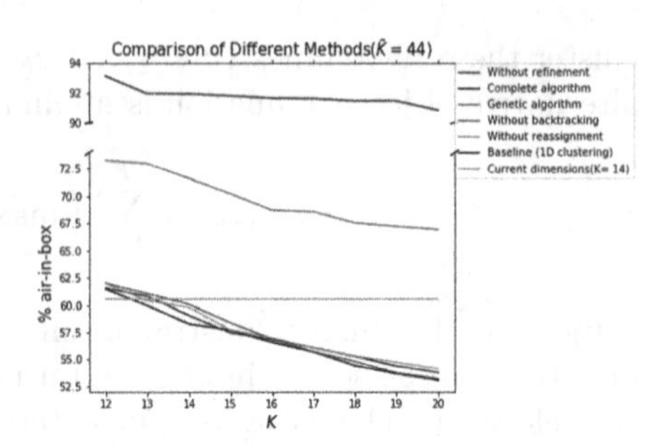

Fig. 4. Evaluation of the 4 variants, the genetic algorithm and the baseline methods.

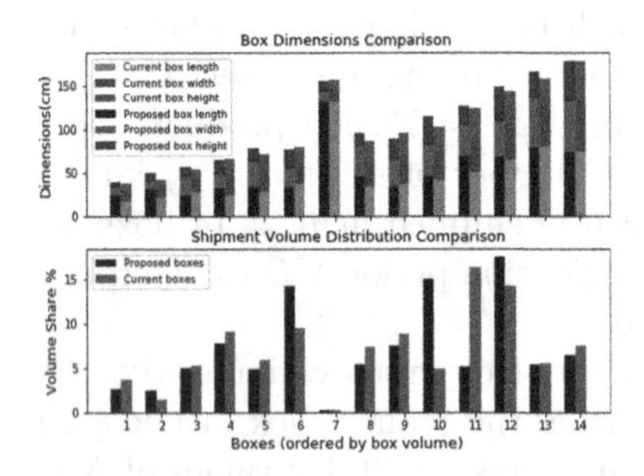

Fig. 5. (i) Top: Comparison of box dimensions, (ii) Bottom: Volume share distribution

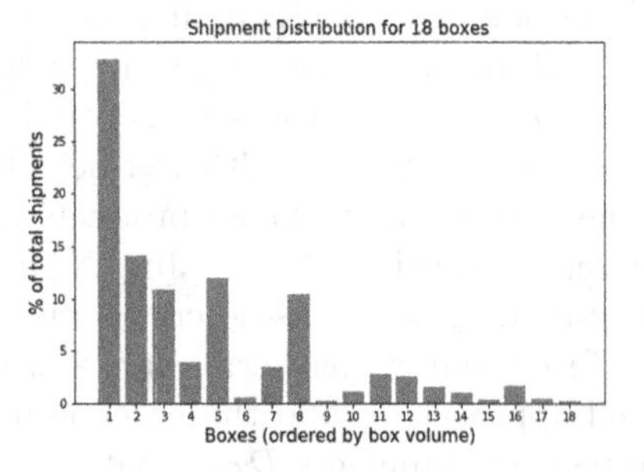

Fig. 6. Box usage distribution on the test shipments $\mathcal{D}_{TE}$

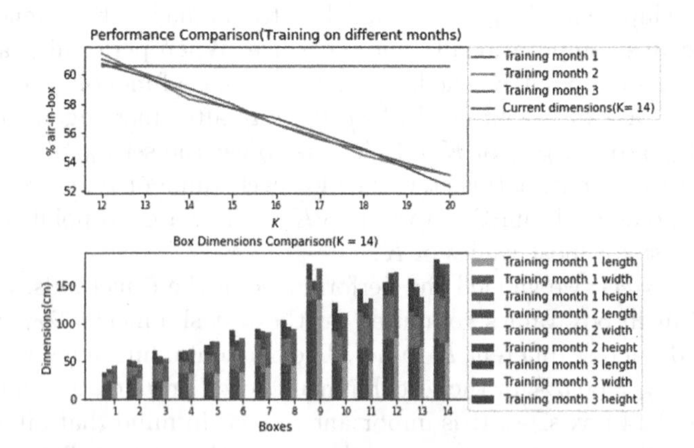

Fig. 7. Sensitivity of box dimensions to change in training data $\mathcal{D}_{TR}$

variant without the iterative refinement step, circumstantially had a marginally better local minimum value of 57.5% compared to our complete algorithm whose $\xi = 57.7\%$. While each constituent part contributes to decreasing the shipment volume, the product reassignment step is the most valuable, as ξ increases by more than 1% in its absence. Even with 13 boxes, 1 *less than* the current usage of 14, our method has a lower ξ value of 59.9% compared to ξ_{curr} and decreases further to $\xi = 58.3\%$ at $K = 14$.

The baseline method, where we perform clustering in the one-dimensional projected space of product volumes, invariably performs poorly with a very high ξ value of 90.8% even at $K = 20$ and is a poor alternative for the actual objective function in Eq. (3.1). Though the GA based approach performs better with respect to the baseline, it consistently yields higher %air-in-box values compared to our clustering based technique even after *multi-starting* the method from 5 different initial population size of 200, where each population is a set of K box sizes, and choosing the best out of the 5 solutions based on cross-validation using the shipments in $\mathcal{D}_{VD}$. The primary reason why GA based methods may result in poor local minimum is that subsequent generations of possible dimensions are not necessarily produced from the perspective of minimizing the overall shipment volume, but are instantiated by *crossing-over* the parent dimensions which could be sub-optimal.

In top half of Fig. 5 we ordered the boxes by their volume, and compared the dimensions of the currently used 14 boxes against the sizes variants suggested by our algorithm. In the bottom part we show the *volume share* of these boxes, where we plot %volume of shipments sent in each of the size variant. By slightly increasing the box dimensions of a box C, our method shifts a large amount of product volume from box $C + 1$, leading to smaller total shipment volume and lesser wastage of non-utilized space in the box. A prominent case of this observation is $C = 10$, where by increasing the dimensions of box 10, a huge share is taken out of the larger volume box 11.

On simulating the actual shipments in $\mathcal{D}_{TE}$ using the 14 box sizes produced by our method instead of the presently used size variants, we observed the overall shipment volume V to decrease by 4.4%, translating to shipment cost savings of tens of millions of dollars even in emerging marketplaces. As the elbow point occurs at $K = 18$ we recommend the usage of 18 size variants, where we estimated the shipment volume to reduce substantially by 10.3% and %air-in-box by 6.1% compared to the currently used 14 box sizes. Looking into the *shipment share* distribution plot in Fig. 6, where the boxes are numbered in increasing order of their volume and we plot the % shipments sent in each of them, we notice a skewed distribution in the sense that 88.1% shipments are sent in smaller boxes (number ≤ 8) and the usage of large boxes (≥ 9) are reserved only for 11.9% shipments.

6.3 Low Sensitivity to Training Data

Further, we analyzed the sensitivity of our algorithm to the choice of training shipments $\mathcal{D}_{TR}$, to study whether changing those leads to drastically different box dimensions. We independently executed our algorithm using 3 non-overlapping months of shipment data as $\mathcal{D}_{TR}$ for the same hyper-parameter value of $\tilde{K} = 44$, and obtained 3 sets of K box-dimensions for different values of K. As before, we simulated the test shipments $\mathcal{D}_{TE}$ on these 3 sets of K boxes and computed the %air-in-box shown in the top-half of Fig. 7. We observe that the ξ values, across different values of K, vary very little over different training sets. In the bottom-half of Fig. 7, we compare the box dimensions of the corresponding 14 boxes obtained from each training set and again do not see any significant variations. These results strongly point to the fact that our method favorably has low sensitivity, equivalent to a *low model variance* [2], w.r.t. changing the training shipments.

7 Conclusion

We proposed an approach for determining the box sizes used to ship products. After reducing it to an equivalent clustering problem, we presented a decision-tree based algorithm containing forward and backward phases, coupled with steps like product reassignment, iterative refinement etc. to arrive at the best dimensions for K boxes. In addition to minimizing the overall shipment volume leading to substantial savings in shipment cost, our algorithm also contributes to a greener environment by keeping the wastage as low as possible. Further, addition or deletion of a size variant is as straightforward as stopping the backward pass early or continuing it for one more iteration, *as our method creates clusters in an incremental fashion without discarding the present solution.*

Extending our approach to handle *multis* containing more than one product in the same shipment is a challenging task as they depend on: (i) type, the number of products and their dimensions that are shipped together (ii) the order and the orientation in which products are packed in the box. Deeper understanding of customer purchase patterns is required to identify such product groups that are bought and shipped collectively. Sparsity in the data further compounds this problem, as the number of shipments of large product groups are highly likely to be few in number. These are fruitful avenues that require further investigation.

References

1. Aloise, D., Deshpande, A., Hansen, P., Popat, P.: NP-Hardness of Euclidean sum-of-squares clustering. Mach. Learn. **75**(2), 245–248 (2009)
2. Bishop, C.M.: Pattern Recognition and Machine Learning (Information Science and Statistics). Springer-Verlag, Berlin (2006)
3. Bongers, C.: Standardization: Mathematical Methods in Assortment Determination. Springer, Netherlands (1980)
4. Bongers, C.: Optimal size selection in standardization: a case study. J. Oper. Res. Soc. **33**(9), 793–799 (1982)
5. Castin, L., Frénay, B.: Clustering with decision trees: divisive and agglomerative approach. In: European Symposium on Artificial Neural Networks, Computational Intelligence and Machine Learning (2018)

6. Cormen, T.H., Leiserson, C.E., Rivest, R.L., Stein, C.: Introduction to Algorithms, 3rd edn. The MIT Press, Cambridge (2009)
7. Jia, L.S.: A study on crate sizing, inventory and packing problem. Ph.D. thesis, National University of Singapore (2014)
8. Kaufman, L., Rousseeuw, P.: Finding Groups in Data: An Introduction to Cluster Analysis. Wiley Series in Probability and Statistics, Wiley (2009)
9. Ketchen, D., Shook, C.: The application of cluster analysis in strategic management research: an analysis and critique. Strateg. Manage. J. **17**(6), 441–458 (1996)
10. Leung, S., Wong, W., Mok, P.: Multiple-objective genetic optimization of the spatial design for packing and distribution carton boxes. Comput. Ind. Eng. **54**(4), 889–902 (2008)
11. Médez-Díaz, I., Miranda-Bront, J.J., Vulcano, G.J., Zabala, P.: A branch-and-cut algorithm for the latent class logit assortment problem. Discrete Appl. Math. **36**, 383–390 (2010)
12. Pentico, D.W.: The assortment problem: a survey. Eur. J. Oper. Res. **190**(2), 295–309 (2008)
13. Singh, M., Ardjmand, E.: Carton set optimization in E-commerce warehouses: a case study. J. Bus. Logistics **41**(3), 222–235 (2020)
14. Tryfos, P.: An integer programming approach to the apparel sizing problem. J. Oper. Res. Soc. **37**(10), 1001–1006 (1986)
15. Vidal, R.V.V.: Optimal Partition of an Interval - The Discrete Version, pp. 291–312. Springer, Berlin (1993)
16. Vidal, R.V.V.: On the optimal sizing problem. J. Oper. Res. Soc. **45**(6), 714–719 (1994)
17. Vidal, R.V.V., Ferreira, J.S.: Optimal partitioning of an interval: Some case studies. In: 2nd. APORS, Beijing, China, pp. 277–282. Peking University Press (1992)
18. Ward, J.H.: Hierarchical grouping to optimize an objective function. J. Am. Stat. Assoc. **58**(301), 236–244 (1963)
19. Wilson, R.C.: A packaging problem. Manage. Sci. **12**(4), B-135-B-145 (1965)
20. Wong, W.K., Leung, S.Y.S.: Carton box optimization problem of VMI-based apparel supply chain. In: 2006 IEEE International Conference on Management of Innovation and Technology, vol. 2, pp. 911–915 (2006)
21. Yang, G., Cun, M.M.: A machine learning approach to shipping box design. In: SAI Intelligent Systems Conference (2019)

An Improved Yaw Control Algorithm for Wind Turbines via Reinforcement Learning

Alban Puech[1,2(✉)] and Jesse Read[1]

[1] LIX, Ecole Polytechnique, Institut Polytechnique de Paris, Palaiseau, France
alban.puch@polytechnique.edu

[2] DEIF Wind Power Technology Austria GmbH, Klagenfurt, Austria

Abstract. Yaw misalignment, measured as the difference between the wind direction and the nacelle position of a wind turbine, has consequences on the power output, the safety and the lifetime of the turbine and its wind park as a whole. We use reinforcement learning to develop a yaw control agent to minimise yaw misalignment and optimally reallocate yaw resources, prioritising high-speed segments, while keeping yaw usage low. To achieve this, we carefully crafted and tested the reward metric to trade-off yaw usage versus yaw alignment (as proportional to power production), and created a novel simulator (environment) based on real-world wind logs obtained from a REpower MM82 2 MW turbine. The resulting algorithm decreased the yaw misalignment by 5.5% and 11.2% on two simulations of 2.7 h each, compared to the conventional active yaw control algorithm. The average net energy gain obtained was 0.31% and 0.33% respectively, compared to the traditional yaw control algorithm. On a single 2 MW turbine, this amounts to a 1.5 k–2.5 k euros annual gain, which sums up to very significant profits over an entire wind park.

Keywords: Wind turbine control · Multi-objective reinforcement learning · Yaw control

1 Introduction

As the world tries to move away from fossil fuels, wind energy appears to be one of the most promising renewable energy sources. Wind turbines convert kinetic energy from the wind into electricity. Energy output mainly depends on the wind speed, the turbine blade diameter and the generator size, but it is now known that the layout, as well as the operation of turbines, can have a significant impact on the power performance of wind parks [1,24]. In particular, the orientation of the wind turbine rotor against the wind, achieved by the yaw mechanism, is of key importance to ensure that the maximal amount of wind energy is extracted. To do so, one needs to minimise the yaw misalignment γ, or yaw error, measured as the difference between the wind direction ϕ and the nacelle position θ:

© The Author(s), under exclusive license to Springer Nature Switzerland AG 2023
M.-R. Amini et al. (Eds.): ECML PKDD 2022, LNAI 13717, pp. 614–630, 2023.
https://doi.org/10.1007/978-3-031-26419-1_37

$$\gamma = \phi - \theta \tag{1}$$

Based on a collection of 300 turbines, [17] showed that more than 50 % of wind turbines were misaligned to the wind direction, while in [5], the loss yield by a misalignment of only 7° on a 2 MW wind turbine was estimated to be around 6000 euros/year. Yaw misalignment also has an impact on the wind park as a whole since it increases the wake effect left by the rotor on the downstream turbines, as shown by [7]. Finally, yaw misalignment reduces the lifespan of the turbine, putting them under high constraint loads [22].

The yaw control algorithm that we use in the present paper as a baseline is a conventional yaw control algorithm (CYCA) deployed in most modern fixed and variable speed horizontal axis wind turbines. In this paper, the results of CYCA are obtained in two different ways. When needed, we make the distinction between the version of CYCA running on the wind turbine and which control sequence (sequence of control action) is constructed from the turbine logs (CYCA-L), and its simulated version CYCA-S, which control sequence is obtained using a yaw control simulator from DEIF. It is important to note that both CYCA versions correspond to the same algorithm, but their results differ due to external factors (e.g. maintenance, cable untwisting, emergency stop) that interfere with the correct execution of CYCA while running on the turbine. CYCA takes the wind direction measured over a small duration preceding the yaw control as an estimate for the nacelle position to reach during the next control cycle. It typically suffers from irreducible operation delays stemming from lags in communications and its fixed (relatively slow) yaw speed. Beyond slow reaction time to wind direction change, the yaw resources are intentionally limited by the controller because a too-frequent yaw usage risks damage to key components of the mechanism such as the bearings.

To limit unnecessary yaw actuation and load on the mechanism, CYCA actuates the yaw mechanism only after the cumulative error of yaw misalignment exceeds some threshold. As a result, the same importance is given to yaw alignment in low-speed segments, where the potential energy gain by reducing yaw misalignment is low, and to higher speed segments, where most of the energy gain can be achieved. We summarise the working principle of CYCA in Fig. 1. An optimal yaw system control algorithm would allocate yaw resources in order to minimise the yaw misalignment, with a priority given to the regions where most of the energy gain can be obtained.

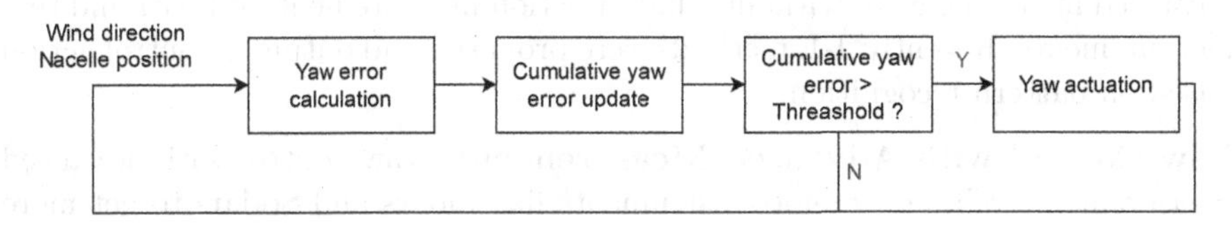

Fig. 1. Simplified control loop of CYCA. The cumulative yaw error is updated with the new yaw error computed from the wind direction measurement. The value of the threshold defines the sensibility of CYCA to small or short yaw misalignment. The length of the control cycle varies but is usually around 1 s.

Little research has been done on new yaw control strategies compared to other turbine operations, such as pitch control. Some yaw control strategies rely on direct or indirect measurement of the wind direction, while others rely on maximum power point tracking. Most of them depend on wind measurements that are sensitive to external factors. Moreover, yaw alignment is often achieved at the expense of yaw usage. They also do not consider the power output for the allocation of the yaw resources. To address those issues, we propose in this paper a new data-driven yaw control strategy that bases the control decision on prediction and pattern recognition rather than past measurements. This new strategy, relying on a Reinforcement-Learning (RL)-trained agent, increases the power output by reducing the overall yaw misalignment and by intelligently re-allocating yaw resources. The contributions of this paper are:

- A novel RL-based yaw control algorithm for wind turbines that optimally allocates yaw resources and better tracks wind direction change, keeping yaw usage low.
- A new realistic simulation yaw environment that allows training and benchmarking of our algorithm against several baselines (simulated and real-world conventional yaw control algorithm) using past wind conditions obtained from turbine logs.
- A new multi-objective reward function to handle short wind direction variations without the use of deadband functions and new evaluation metrics that take into account the yaw drive consumption changes.

We organise the remainder of the paper as follows: We will first introduce the novel Reinforcement-Learning control algorithm (RLYCA) and discuss its key components. We will then focus on the RL environment used for both training and assessment of the novel strategy. Finally, we will discuss the results of the proposed strategy and the possible improvement opportunities.

2 Related Work and Existing Yaw Control Strategies

There are mainly 4 different types of active yaw control strategies. Contrary to passive yaw control strategies, which are based on wind force to orient the rotor, active yaw control strategies use a servomechanism to change the rotor position. Among the active yaw strategies, three of them are deterministic: the ones with advanced measurements, without wind direction measurement and with indirect measurements. Recently, RL methods were proposed and output a control action based on pattern recognition.

Yaw Control with Advanced Measurements. Yaw control with advanced measurements relies on remote instruments like Lidars and Sodars to get more accurate wind direction and speed values, as in [5]. As the wind direction measure is made upstream of the turbine, it is not affected by the wake effect and gives a more accurate measure of the yaw misalignment, which is then used by an active yaw control algorithm. Alternatively, a nacelle-mounted Lidar can be used to learn a correction function that is then applied to the wind direction measurement obtained from the wind vane [12]. Both methods were shown to help at correcting the yaw angle, but they require expensive measurement tools.

Yaw Control without Wind Direction Measurement. Other methods, without wind direction measurement, have been proposed recently [11,15]. They rely on maximum power point tracking. The difference between the actual power output and the theoretical power output computed using the power curve gives information on the power loss coming from the yaw misalignment, and, in turn, gives an estimate of the yaw misalignment. A benefit of those methods is that they no longer depend on the wind direction measurements that are often, as discussed previously, disturbed by the operation of the wind turbine and by the wake effect. However, they are sensitive to wind speed variation since the theoretical output power is computed out of it and to any other external factor such as snow or dirt that could reduce the power output of the turbine.

Yaw Control with Indirect Measurement. The methods that gained the most attention recently are the ones based on indirect measurements. They rely on short term prediction of the wind direction to feed a model predictive control (MPC) algorithm that anticipates wind direction change [8–10]. Although the algorithm is no longer directly using the wind direction measure, its performance is still highly dependent on the accuracy of the wind direction prediction. In high wind variability conditions, the accuracy tends to be low (Dzulfikri, Nuryanti and Erdani [10] showed a mean absolute error of 9.9° for a 1-min ahead prediction of the wind direction on the testing set), which restricts the use of such methods to a relatively small space of wind conditions. It is important to note that, to this day, none of the proposed methods relying on wind direction prediction significantly outperforms the traditional active yaw control strategy at reducing the yaw misalignment.

RL-Based Yaw Control Strategies. Finally, the RL methods are an alternative to the other data-driven methods presented previously as they do not use the wind direction and speed measurements in a predefined, deterministic way. They rather determine the best control action based on the current wind and position state, learning from different simulation scenarios or online, while running on the turbine controller. Those techniques are the most promising as they always output the optimal control action to the best of the model knowledge acquired from the training. They work in a black box manner and are not impacted by the measurement inaccuracy since no control strategy needs to be designed in advance, and since no decision relies directly on the value of the wind direction or speed. The performance of the models then depends on the abstraction and generalisation potential of the algorithm and the quantity and relevance of the training data for the given task, which can be a limitation in situations where we do not have enough data. Those strategies make easier the task of reallocating yaw resources by prioritising segments of high potential gain since the algorithm can learn to identify those segments.

To this day, most of the research proposing an RL-trained agent focuses on long term control. Saenz-Aguirre [19] proposed a model that actuates the nacelle position every 24 h only. This present paper shows a method that is suitable for most wind turbines that usually spend 3 to 10 % of the time yawing.

In addition to that, the wind conditions used in the simulators often make the control decision easier to determine. The wind direction used in the simulations ran to test the control algorithms against CYCA often has the same mean over any segment of the dataset and shows periodic sudden change with some added noises of small magnitude. The wind speed is often taken to be almost stable with very small variations around its mean [19,20]. Instead, we base our simulations on real-world data obtained from turbine logs. Most of the papers proposing new yaw control techniques do not show their performances for different levels of yaw usage and the bench-markings against current yaw control strategies are often limited. In the present paper, we offer a clear comparison with both CYCA-S and CYCA-L and we take into account the power consumption change of the yaw drive in the computation of the energy gains.

The originality of this paper is that it offers a yaw control algorithm that can directly and realistically replace CYCA, increasing the power extracted while keeping the yaw usage reasonable.

3 Proposed Yaw Control Strategy

3.1 Goals and Constraints of the Strategy

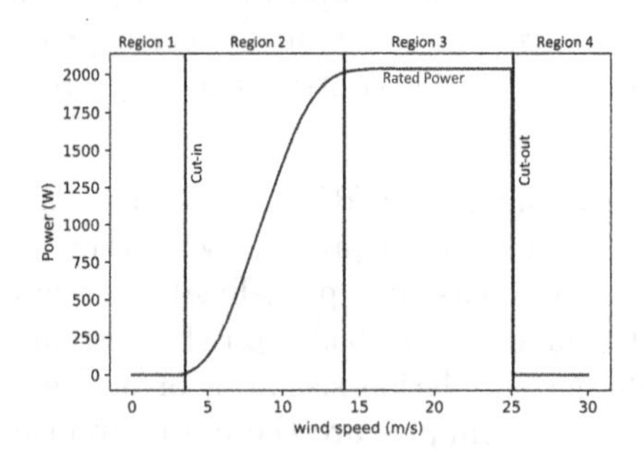

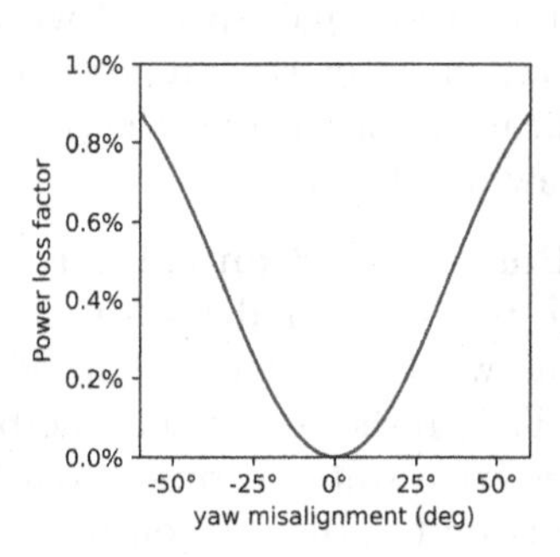

Fig. 2. Left: Power curve of 2MW turbine and its different regions. The turbine operates only after cut-in and before cut-out. The rated power corresponds to the maximum power the turbine can produce. Right: Power loss factor as a function of the yaw misalignment with $\alpha = 3$.

One of the main motivations for our new yaw control algorithm is to maximise the wind power extracted by the turbine. Power loss due to yaw misalignment occurs in region 2 of the power curve, shown in Fig. 2. Indeed, in region 1, that is, before cut-in, the wind is too slow and the turbine is not generating electricity. In region 4, after cut-out, the turbine is shut down and the rotor is braked to avoid over-speeding. Moreover, yaw misalignment does not have a significant impact on the power production in region 3, since the turbine is already outputting the maximum amount of power that it can output. However, the turbine will require a larger wind speed to reach rated power in case of yaw misalignment. In region 2, the power loss P_γ caused by yaw misalignment γ is expressed in degrees as:

$$P_\gamma = P_* \cos^\alpha \gamma \tag{2}$$

where P_* is the power extracted assuming perfect yaw alignment. Moreover, P_* is directly related to the wind speed v, the circular swept area of the turbine A, the air density ρ and the Betz factor [23] c:

$$P_* = \frac{1}{2}\rho A v^3 c \tag{3}$$

Many values have been proposed for the cosine exponent α. Its value significantly impacts the calculated loss caused by yaw misalignment and it is often assumed that the value of α is three, as supported by [3] and by analysis conducted by [14]. Most recent studies like [22] showed that the wake effect caused by upstream turbines can have an impact on the value of α. Dahlberg and Montgomerie [4] showed that the value of the cosine exponent could range from 1.7 to 5.1. Thus, we will stick $\alpha = 3$, corresponding to the theoretical analysis proposed in [14]. The power loss factor associated with a yaw misalignment in region 2 is shown in Fig. 2. The other benefit of reducing the yaw misalignment is to prevent the turbine from undergoing large mechanical and fatigue loads. Damiani [6] showed that misaligned turbines are more subject to damage in case of a storm but also see their lifespan reduced. Fleming [13] showed an increase in the blade, yaw bearings and tower loading in a situation of yaw misalignment.

Yaw misalignment minimisation should not be achieved at the expense of yaw usage. Indeed, yawing too often and for too long can harm the yaw mechanism. It is thus crucial to keep the time spent yawing under some limit mainly defined by the turbine safety guidelines, in our case, less than 10 % of the time should be spent yawing. Hence, the goal is not to achieve perfect yaw alignment, but rather to find the optimal trade-off between yaw correction and yaw actuation. The main constraint is to avoid actuating the yaw systems too often and for too long. As the loss induced by yaw misalignment is cubic, reducing the yaw misalignment when it is large already allows us to decrease most of the loss caused by the misalignment. Moreover, and as discussed earlier, most of the yaw usage should be done on high-speed segments, where yaw misalignment has a higher cost on the power output.

3.2 A Proposed Yaw Control Strategy via Reinforcement Learning

Reinforcement learning (RL) involves teaching an agent progressing in an environment to take actions that maximise expected gains defined by a reward function. The agent has limited information concerning its current state and learns its policy based on experience from repeated episodes.

We define a control cycle as the episode of fixed duration (e.g. 3 s) during which information is received by the agent which then outputs a control action.

In the case of RLYCA-A, during a control cycle, the agent takes an action A_t by applying its internal policy, upon observing its current state S_t. After the action is taken, a reward R_{t+1} and a new state S_{t+1} is computed. The corresponding RL task can be described by Fig. 3.

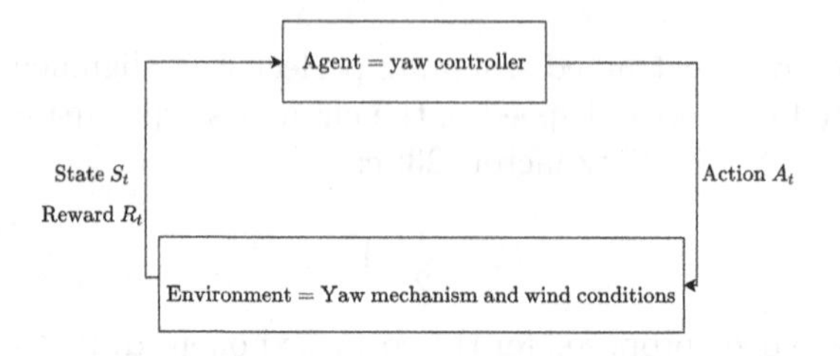

Fig. 3. Simplified block Diagram of the RL task. The new state S_{t+1} and the reward R_{t+1} is computed only after the agent took its action A_t.

We train the agent to learn the optimal control policy for the yaw system. Optimal control theory and reinforcement learning are closely related as they both refer to techniques used to find a control for a system that evolves in time and interacts with its environment, such that a certain objective function is maximised. However, an optimal control algorithm takes an action that maximises the predefined objective function in a deterministic way. In reinforcement learning, the agent does not know the reward function in advance and has to learn it through experiences to define its policy.

In this paper, the proposed yaw control strategy is based on a Proximal Policy Optimisation Algorithm (PPO) [21]. Unlike Deep Q-learning, used in [19,20], PPO can not learn from offline stored data and requires online training.

Designing the Action Space. At each control cycle, the agent outputs one of the following actions:

- 0: Clockwise rotation for the full duration of the control cycle
- 1: No rotation
- 2: Counterclockwise rotation for the full duration of the control cycle

Defining the Reward Function. The goal of the agent is to learn the optimal policy, which is the one that maximises the expected return, computed as the sum of rewards over an episode. We define a reward function in two parts, each accounting for one of our two goals: maximise power production, and limit the yaw mechanism usage.

The first term $R^{(1)}$ of the reward function accounts for the power production maximisation aspect. To maximise the power extracted during the control cycle t, we need to minimise the yaw misalignment γ_t as expressed in Eq. (1). However, more importance should be given to the control cycles where the wind speed v_t is high (but still lower than the rated speed), as Eq. (3) shows that the power extracted assuming perfect alignment P_0 increases with the cube of the wind speed v. Hence, we define $R^{(1)}_{t+1}$, the first negative component of the reward for the action taken during the control cycle t:

$$R^{(1)}_{t+1} = -\gamma^2_{t+1}\tilde{v}^3_{t+1} \quad (4)$$

Weighting the yaw misalignment γ_{t+1} squared by the standardised wind speed $\tilde{v}_{t+1}$ cubed allows us to penalise more yaw misalignment when the power output is high, and thus encourages a better yaw resource allocation

The second term of the reward function, denoted by $R^{(2)}$, accounts for the second objective of RLYCA-A: yaw usage limitation. To keep the yaw count low and reduce the time spent yawing, we should prevent the agent from reacting to short wind direction variations (in time) or to small yaw misalignments (in magnitude) that are not worth being corrected compared to larger yaw misalignment. Indeed, since the loss increases with the cube of the yaw misalignment, a correction of one degree will yield more gain when the yaw misalignment is large than when it is small. We thus want our agent to react in priority to large yaw misalignment, and to spare the yaw mechanism the rest of the time. Hence, we reward the agent for observing a period of at least k control cycles without taking any moving actions. We define $R^{(2)}_{t+1}$, the second component of the reward for the action a_t taken during control cycle t as

$$R^{(2)}_{t+1} = w \prod_{i=t-k+1}^{t} \mathbf{1}\{a_i = 1\} \quad (5)$$

where the weight $w \in \mathbb{R}^+$ encodes the importance given to the preservation of the yaw mechanism.To illustrate the principle of this loss function, let us assume that $w = 40$, that the controller has not taken a moving action in the last $k - 1$ control cycles, and that the wind conditions remain similar during the control cycle. Assume further that the maximum angle covered by the yaw mechanism in one control cycle is 3°C. If the agent takes the correct moving action (yaws in the correct direction), the yaw error can be corrected by a maximum of 3°C. The second component of the reward function will be 0, and the reward will be given by $R_{t+1}(moving) = -(\gamma_t - 3)^2\tilde{v}^3_t$.

If the agent chooses the stationary action, the reward will be given by the sum of the two reward components defined in Eqs. (4) and (5), where the yaw misalignment remains the same since no correction was made and since we assumed stable wind conditions. The reward will thus be: $R_{t+1}(stationary) = -(\gamma_t)^2\tilde{v}^3_t + 40$.

Figure 4 (left) shows the reward if a stationary action is taken (blue), and if a moving action is taken (orange). For the moving action to give a larger reward than the stationary action, the yaw error needs to be larger than 8.1°C, where the two curves intersect. This can be interpreted as the fact that the agent will decide to move the yaw mechanism only if the yaw misalignment is larger than this value, and will not actuate the yaw mechanism otherwise. Figure 4 (right) shows, for different standardised wind speed values, the minimum yaw misalignment for the moving action to give a larger reward than the stationary action. We designed the reward function such that, when the wind speed is high, the agent is more willing to correct small yaw misalignment. Note that in region 3 of the power curve, the reward function penalises yaw misalignment in the same manner as described before, but not for the same reason. Indeed, although there is no power loss due to yaw misalignment in this region, the constraint loads are higher, which justifies yaw misalignment to be equally penalised.

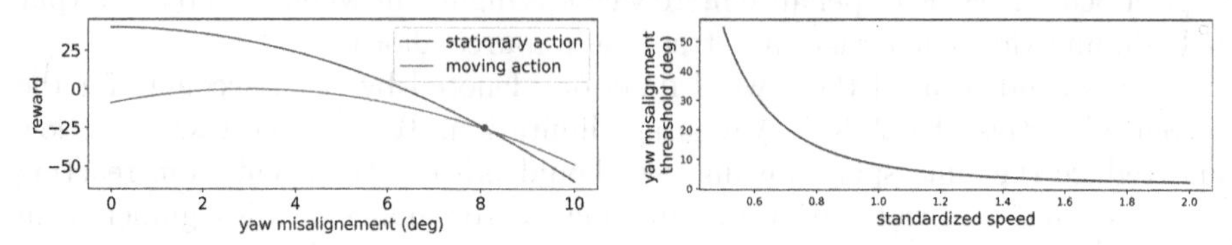

Fig. 4. Left: Reward for the stationary action and the moving action as a function of the yaw misalignment under fixed $\tilde{v} = 1$ and $w = 40$. Right: Minimum yaw misalignment (vertical axis) starting from which the moving action has a higher reward than the stationary action (for $w = 40$); as the wind speed increases, we encourage more and more yaw correction.

Choosing the State Space. The reward being dependent on the yaw misalignment, the wind speed and the past actions, we aim at including that information into the state S_t given to the agent to take its action A_t during the control cycle t. We define j as the number of lagged values or ancestor values given for each feature of the state. S_t is thus expressed as in Eq. (6), where γ_t, ϕ_t and $\tilde{v}_t$ denote the yaw misalignment, the wind direction and the standardised wind speed during the control cycle t, respectively.

$$S_t = \begin{bmatrix} a_{t-1} & \gamma_t & \phi_t & \tilde{v}_t \\ a_{t-2} & \gamma_{t-1} & \phi_{t-1} & \tilde{v}_{t-1} \\ a_{t-3} & \gamma_{t-2} & \phi_{t-2} & \tilde{v}_{t-2} \\ \vdots & \vdots & \vdots & \vdots \\ a_{t-j} & \gamma_{t-j+1} & \phi_{t-j+1} & \tilde{v}_{t-j+1} \end{bmatrix} \tag{6}$$

It is important to note that, in the control cycle t, state S_t contains the wind direction ϕ_t and wind speed v_t, but not v_{t+1} nor ϕ_{t+1} (the agent cannot know in advance the future wind direction and speed).

4 Yaw System Simulation Environment

We develop a yaw system environment that we use with both `RLYCA-A` and `CYCA-S`. In the case of `RLYCA-A`, we use the environment for both training and testing. In the case of `CYCA-S`, we use the yaw system environment for testing only as the latter algorithm does not require to be trained. Finally, `CYCA-L` testing does not require any simulator since the control sequence, as well as all the measurements needed for the bench-marking, are directly extracted from the turbine logs.

4.1 Datasets

The yaw system simulation environment simulates the wind conditions recorded at a real-life wind park. Namely, we collected two time-series data sets each consisting of 21000 turbine log points, recorded in September and February of 2021. Each t-th data point ($t = 1, \ldots, 21000$) contains the mean wind direction and the mean wind speed over 1 s, stacked into state S_t as illustrated in 6. Dataset 1 exhibits a relatively steady wind direction; of $34.1° \pm 9.7$ with an interquartile range of $13.7°$. Dataset 2 has similar statistical features (wind direction of $41.4° \pm 11.6$ with an interquartile range of $14.2°$) but shows fast wind direction changes of large magnitude between 10000 s and 12500 s and between 15000 s and 20000 s.

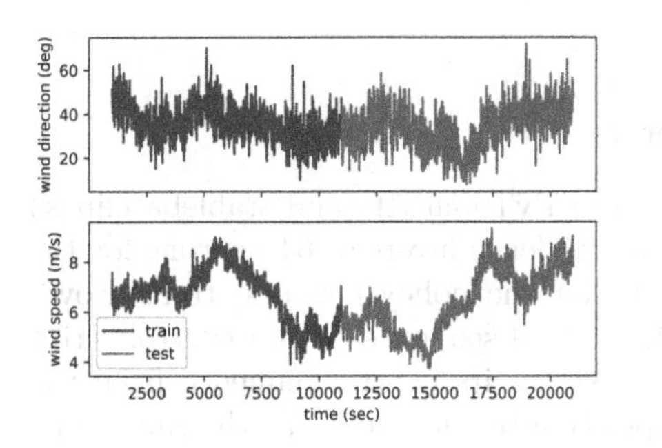

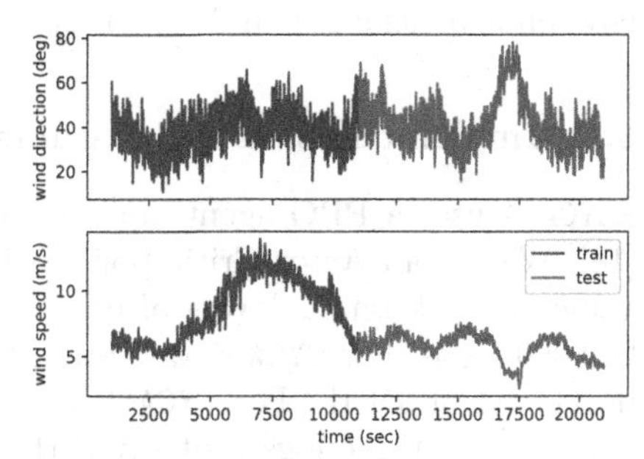

Fig. 5. The first dataset (left) shows a steady wind direction while the second one (right) shows 2 large wind direction changes. Both are used for our experiments. For each dataset, we train `RLYCA-A` on its first half of a dataset and test it on the other.

4.2 Working Principle of the Yaw System Simulation Environment

The working principle of the yaw system environment used for the RL agent is described in Fig. 6. We update the new nacelle position θ_{t+1} as in 7, where r denotes the yaw rate and p the duration of the control cycle.

$$\theta_{t+1} = \theta_t + p(action - 1)r \tag{7}$$

The new yaw misalignment is then computed using the new nacelle position and wind direction. We compute the power assuming perfect yaw alignment using the power curve shown in Fig. 2 and the power under the effect of yaw misalignment is obtained from Eq. (2), provided that P_* belongs to region 2 of the power curve. Otherwise, we assume no power loss ($P_\gamma = P_*$). We simulate a communication delay of a duration equal to the period of the control cycle. We can assume that the yaw mechanism receives the control signal a_t when the next control cycle ($t+1$) is executed. In practice, this means that we iterate in the wind time series before we compute the new yaw misalignment. This greatly contributes to the complexity of the task: the wind conditions can change drastically between the real-time where a given action is decided and its application.

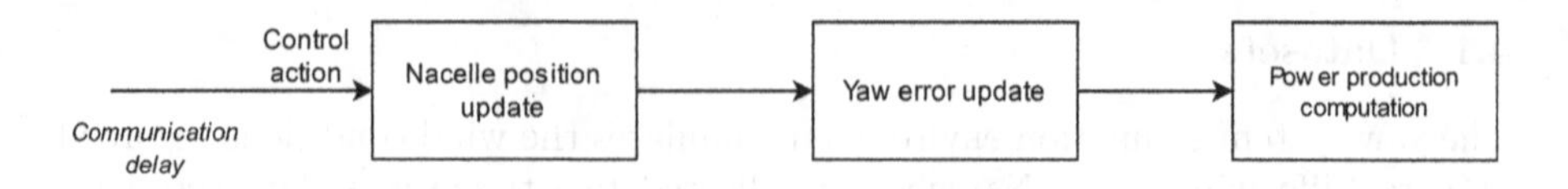

Fig. 6. Working principle of the yaw system environment

5 Experiments and Discussions

In this section, we compare our novel algorithm `RLYCA-A` with benchmarks `CYCA-L` and `CYCA-S` on the two datasets which we develop into a simulated environment using open AI gym [2].

5.1 Implementation and Parameters

`RLYCA-A` uses a PPO agent, implemented in PyTorch [16] and stablebaselines3 [18]. We use an Actor-Critic policy with a 2 dense layers of 64 neurons for the value network and 2 layers of 64 neurons for the policy network; trained over 200,000 steps. For `CYCA-S` we obtain the control sequence from a control script developed at DEIF. For `CYCA-L` we reconstruct its control sequence from the real-world turbine logs that saved the position of the nacelle at the same time as the wind condition used for the experiment was saved.

The power curves used for the yaw system simulated environment is shown in Fig. 2. Table 1 gives the parameters used for the yaw system simulated environment.

The yaw alignment performances of the algorithms are compared using the average yaw misalignment computed as stated in Eq. (1). We assess the power extraction performance using the estimate of the power produced given by the yaw system environment and stated in the previous subsection. Since the three experiments are conducted in the same wind conditions given by the testing dataset, the only variable that we use in the computation of the power output and that differs between the three approaches is the nacelle position, which is the direct result of the applications of the three algorithms. We compare the yaw usage of all methods using three metrics: The proportion of the time spent yawing over the simulated duration of the experiment, the angle covered by the yaw mechanism, and the number of yaw actuations. Additionally, we compute the difference in yaw mechanism power consumption $\Delta(t)$ as shown in Eq. (8), where $\delta_{\phi_{RL}}(t)$ and $\delta_{\phi_{simu}}(t)$ denote the change in nacelle position angle during the control cycle t obtained with `RLYCA-A` and with `CYCA-S`, respectively. Moreover, r denotes the yaw rate and $P_{yawdrive}$ denotes the power consumption of the yaw mechanism.

$$\Delta(t) = \frac{\delta_{\phi_{RL}}(t) - \delta_{\phi_{simu}}(t)}{r} P_{yawdrive} \tag{8}$$

Table 1. Parameters used for the experiment on both datasets. They are identical to those of the turbine from which the controller logs were obtained (REpower MM82).

Parameter name	Value
Environment parameters	
k number of consecutive stationary action to get reward	2
j number of lagged values in state	12
w weight of $R^{(2)}$	40
v_r rated speed	14 m/s
p cycle period	10 s
$comm$ communication delay	10 s
$P_{yawdrive}$	18 kWh
RL agent parameters	
Learning rate	0.003
Number of steps before update	2048
Batch size	64
Number of epochs	10
Discount factor	0.99
Generalised Advantage Estimator factor	0.95

5.2 Experiment in Steady Wind Direction Conditions (dataset/simulation 1)

We first compare the algorithms in steady wind direction conditions. Figure 7 shows the nacelle position obtained for the three algorithms. Most of the yaw misalignment decrease is achieved between 5000 and 7000 s (first bar plot from top). This results in a large power output increase in the same time interval with a cumulative gain of 1.5 kw between 6000 and 6500 s. This power production improvement requires more yaw actuations, which translates into a yaw consumption increase that is mainly concentrated in the segment where the wind direction is decreasing. The yaw usage is also increased in the first half of the dataset but it does not improve the power output. This phenomenon is something that should be corrected in a future version of `RLYCA-A`. The power output change, net of any increase in the yaw consumption, is negative (power loss) in the first half of the dataset because the increase in yaw usage does not lead to a power production increase. Indeed, the power output is increased, but the yaw consumption change is larger than this gain, which results in a negative net power change. The region where the net power output is positive corresponds to a high speed (and thus high power) region so that the relative energy gain is translated into large energy gains. On the opposite, the region where the net power change is negative is a low-speed region, so that the absolute power loss does not affect too much the overall performance of the algorithm. Table 2 shows the results obtained with the three algorithms, while Table 3 summarises the results achieved by RLYCA-A. All in all, the algorithm succeeded in allocating more resources to the high-speed segment, resulting in a large power output increase. The relative energy gain achieved by RLYCA-A compared to CYCA-S is 0.4 %, and the net energy gain amounts to 0.31 %. This gain is due to a better allocation of the yaw resources.

5.3 Experiment in Variable Wind Direction Conditions (dataset/simulation 2)

We now compare the algorithms in the simulation derived on the variable wind direction dataset. Figure 8 shows the nacelle position obtained for the three algorithms. Most of the yaw misalignment decrease is achieved in the second half of the dataset. RLYCA-A reacts quicker to the change in wind direction than CYCA-S.

The power output is increased in almost every region of the dataset. The yaw consumption is also more important but the increase in power output compensates for such additional power consumption everywhere except after 9000 s. Indeed, the additional yaw actuations after 9000 s do not translate into a better yaw alignment. All in all, the relative energy gain achieved by RLYCA-A is 0.75 %, but the net energy gain is 0.33, as the yaw usage is doubled. The wind speed is stable on the testing set, as shown in Fig. 5, so that, this time, the energy gain can not be explained by a better allocation of the yaw resources. In this example, it is the overall yaw misalignment decrease achieved by RLYCA-A (18.5 % compared to CYCA-S and 24.4 % compared to CYCA-L) that explains such a high energy gain.

Table 2. Result summary of the three algorithms. Only the power outputs for CYCA-S and RLYCA-A are shown as the power output for CYCA-L is impacted by external factors that are not considered in the simulation (e.g. dirt or snow).

	Steady wind			*Variable wind*		
	RLYCA-A	CYCA-S	CYCA-L	RLYCA-A	CYCA-S	CYCA-L
Average yaw error (deg)	**6.18**	6.52	6.91	**5.71**	7.01	7.56
Power output (kWh)	**1173.1**	1168.5		**741.5**	736.0	
Angle covered (deg)	111.0	53.5	**27.3**	261	112.1	**91.8**
Yaw count	34	15	**5**	73	27	**14**
Time spent yawing (%)	3.7	2.0	**0.4**	8.7	4.2	**1.6**

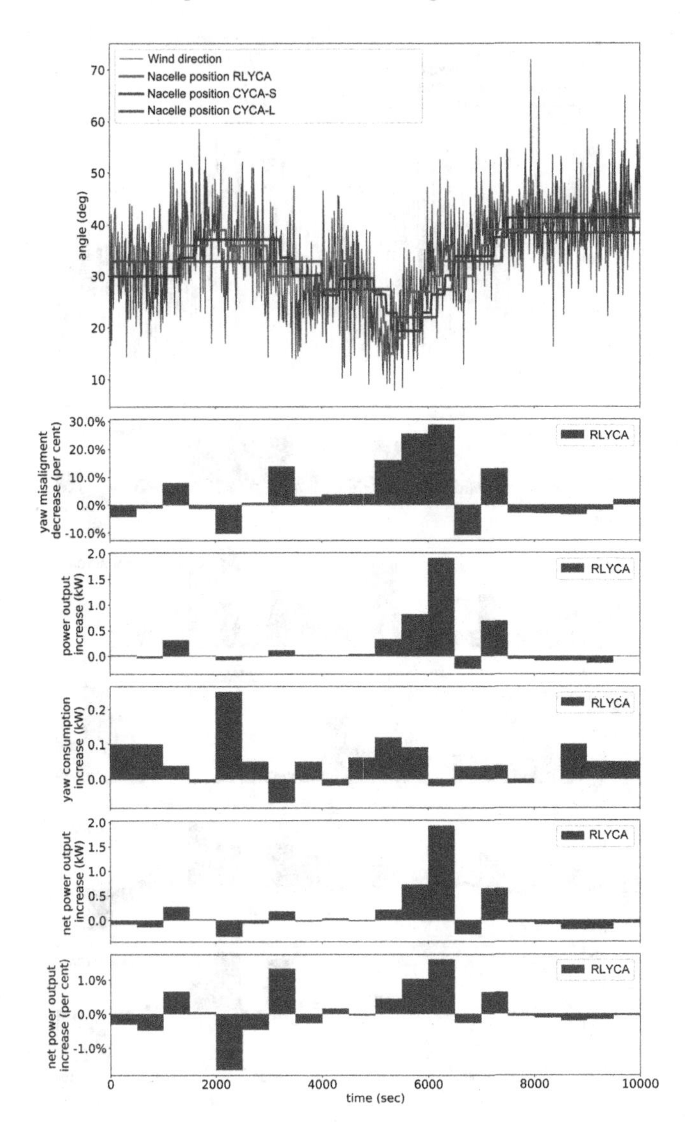

Fig. 7. Comparison of control signals of the three algorithms on the steady wind direction dataset. The power and yaw error comparisons shown on the bar plots are made against CYCA-S.

Table 3. Performance of RLYCA-A compared to CYCA-S. The values indicated between parenthesis indicate the comparison with CYCA-L.

	Steady wind	Variable wind
Average yaw error decrease (%)	5.5 (11.2)	18.5 (24.4)
Energy gain (%)	0.40	0.75
Net energy gain (%)	0.31	0.33

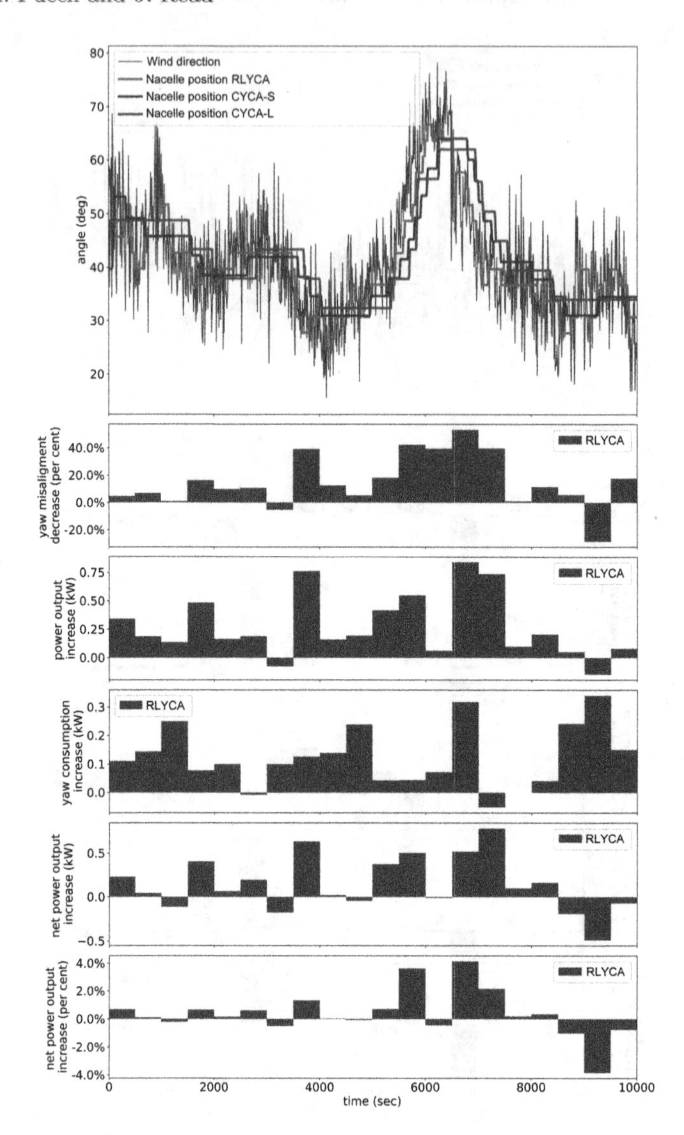

Fig. 8. Comparison of control signals of the three algorithms on the variable wind direction dataset. The power and yaw error comparisons are made against CYCA-S.

6 Conclusion and Future Work

In this paper, we proposed a reinforcement learning yaw control strategy to optimally allocate yaw resources and track wind direction change, with the goal to maximise the power extracted, while keeping yaw usage under a 10 % bound. We compared our algorithm to a simulation of the current algorithm running on a REpower MM82 and to the control signals gathered from the turbine logs. The comparison was done on two 10000 s-long test datasets obtained from past wind measurements logged by the turbine controller. The RL algorithm achieved better wind direction change tracking, which led to a decrease in the yaw misalignment. It also better allocates yaw resources, prioritising high power (and thus high potential gain) segments. Compared to the simulated baseline yaw control algorithm, our new algorithm improves by 5.5 and 18.5 % the yaw alignment in the steady wind and variable wind conditions, respectively. When comparing its performance to the control sequence obtained from the turbine logs, the last metrics become 11.2 and 24.4 %. Taking into account the increase in yaw drive consumption, the net energy gain observed on the two datasets was 0.31 and 0.33 %. For a single 2 MW wind class 2 turbine, this amounts to an annual 1500–2500 euros gain[1], which, at the scale of a wind park, represents substantial gains. Throughout our work, we not only showed the relevance of data-driven methods for wind turbines control, but we more generally demonstrated the potential of reinforcement learning to deal with the uncertain nature of renewable energy resources and renewable energy structure behaviour. In the future, we plan to use wind turbine digital twins to have an even more accurate simulation of the yaw mechanism, a model of the constraint loads, as well as a more accurate estimation of the power output. The power output could then be directly used in the reward function. We also plan to include data from upstream turbines to the state space, which could help in anticipating the wind changes. Finally, the next step would be to move to the wind farm level, as it has been shown that local yaw alignment can be detrimental at the farm level and because new intended yaw misalignment strategies are being discussed.

Reproducible Paper. The code and the data sets used for this paper are available at https://github.com/albanpuech/RL-yaw-control-algorithm-for-wind-turbines

Acknowledgements. We would like to thank Mohamed Alami Chehbourne from LIX as well as Walter Telsnig, Anatoliy Zabrovskiy and Martin Göldner from DEIF for the helpful discussions and insights on the topic during the preparation of this paper.

References

1. Al-Addous, M., Jaradat, M., Albatayneh, A., Wellmann, J., Al Hmidan, S.: The significance of wind turbines layout optimization on the predicted farm energy yield. Atmosphere **11**(1), 117 (2020)
2. Brockman, G., et al.: OpenAI gym (2016)
3. Burton, T., Sharpe, D., Jenkins, N., Bossanyi, E.: Wind Energy Handbook. John Wiley (2001). https://books.google.at/books?id=4UYm893y-34C
4. Dahlberg, J., Montgomerie, B.: Technical report, final report part 2, wake effects and other loads. Research program of the utgrunden demonstration offshore wind farm (2005)
5. Dalmas J., P.A.: Etude comparative de trois appareils pour la mesure d'alignement nacelle. In: 5th International Renewable Energy Congress (IREC) (2014)

[1] 82 m rotor diameter, 8–11 GWh annual power production, 80 % of the time spent in region 2, cosine-cube power loss law, 0.09 euros per kWh.

6. Damiani, R., et al.: Assessment of wind turbine component loads under yaw-offset conditions. Wind Energy Sci. **3**, 173–189 (2018)
7. Dijk, M., Wingerden, J.W., Ashuri, T., Li, Y., Rotea, M.: Yaw-misalignment and its impact on wind turbine loads and wind farm power output. J. Phys.: Conf. Ser. **753**, 062013 (2016)
8. Dongran, R., et al.: Wind direction prediction for yaw control of wind turbines. Int. J. Control Autom. Syst. **15**, 1720–1728 (2017). https://doi.org/10.1007/s12555-017-0289-6
9. Dongran, S., Li, L., Yang, J., Joo, Y.H.: A model predictive control for the yaw control system of horizontal-axis wind turbines. Energy Procedia **158**, 237–242 (2019)
10. Dzulfikri, Z., Nuryanti, N., Erdani, Y.: Design and implementation of artificial neural networks to predict wind directions on controlling yaw of wind turbine prototype. J. Robot. Control (JRC) **1**(1), 20–26 (2019)
11. Farret, F., Pfitscher, L., Bernardon, D.: Sensorless active yaw control for wind turbines. In: IECON 2001. 27th Annual Conference of the IEEE Industrial Electronics Society, vol. 2, pp. 1370–1375 (2001)
12. Fleming, P.A., et al.: Field-test results using a nacelle-mounted lidar for improving wind turbine power capture by reducing yaw misalignment. In: Journal of Physics: Conference Series, vol. 524 (2014)
13. Fleming, P., et al.: Simulation comparison of wake mitigation control strategies for a two-turbine case. Wind Energy **18**(12), 2135–2143 (2015)
14. Kragh, K., Hansen, M.: Potential of power gain with improved yaw alignment. Wind Energy **18**(6), 979–989 (2015)
15. Mademlis, C., Mesemanolis, A., Karakasis, N., Nalmpantis, T.: Active yaw control in a horizontal axis wind system without requiring wind direction measurement. IET Renew. Power Gener. **10**(9), 1441–1449 (2016)
16. Paszke, A., et al.: Pytorch: an imperative style, high-performance deep learning library. In: Advances in Neural Information Processing Systems, vol. 32, pp. 8024–8035 (2019)
17. Pedersen, M.: Yaw misalignment and power curve analysis. EWEA Analysis of operating wind farms (2016)
18. Raffin, A., Hill, A., Gleave, A., Kanervisto, A., Ernestus, M., Dormann, N.: Stable-baselines3: reliable reinforcement learning implementations. J. Mach. Learn. Res. **22**(268), 1–8 (2021). https://jmlr.org/papers/v22/20-1364.html
19. Saenz-Aguirre, A., et al.: Artificial neural network based reinforcement learning for wind turbine yaw control. Energies **12**, 436 (2019)
20. Saenz-Aguirre, A., Zulueta, E., Fernandez-Gamiz, U., Ulazia, A., Teso-Fz-Betono, D.: Performance enhancement of the artificial neural network-based reinforcement learning for wind turbine yaw control. Wind Energy **23**(3), 676–690 (2020)
21. Schulman, J., Wolski, F., Dhariwal, P., Radford, A., Klimov, O.: Proximal policy optimization algorithms. arXiv preprint arXiv:1707.06347 (2017)
22. Urbán, A.M., Liew, J., Dellwik, E., Larsen, G.C.: The effect of wake position and yaw misalignment on power loss in wind turbines. J. Phys.: Conf. Ser. **1222**(1) (2019)
23. Venkata, S., Krishnamurthy, S.: Wind energy explained: theory, design, and application. Power Energy Mag. IEEE **1**, 50–51 (2003)
24. Wan, S., Cheng, L., Sheng, X.: Effects of yaw error on wind turbine running characteristics based on the equivalent wind speed model. Energies **8**(7), 6286–6301 (2015)

Author Index

© The Editor(s) (if applicable) and The Author(s), under exclusive license
to Springer Nature Switzerland AG 2023

M.-R. Amini et al. (Eds.): ECML PKDD 2022, LNAI 13717, pp. 631–633, 2023.
https://doi.org/10.1007/978-3-031-26419-1

Printed in the United States
by Baker & Taylor Publisher Services